High Temperature Superconductivity

High Temperature Superconductivity

Proceedings of LT-19 Satellite Conference, Cambridge, 13–15 August 1990

EDITED BY

J Evetts

IRC in Superconductivity, Cambridge University

Adam Hilger
Bristol, Philadelphia and New York

British Library Catologuing in Publication Data

CIP catalogue record for this book is available from the British Library

ISBN 0-7503-0108-2

Library of Congress Catologing-in-Publication Data are available

Published under the Adam Hilger imprint by IOP Publishing Ltd
Techno House, Redcliffe Way, Bristol BS1 6NX, England
335 East 45th Street, New York, NY 10017-3483, USA
US Editorial Office: 1411 Walnut Street, Philadelphia, PA 19102

Printed and bound in Great Britain by Page Bros (Norwich) Ltd

List of Delegates

Abrikosov A A, Institute for High Pressure Physics, Troitsk, USSR
Acrivos J V, San José State University, USA
Adkins C J, Cambridge University, UK
Adrian G, Daimler-Benz AG, Frankfurt, Germany
Adrian H, Technical University Darmstadt, Germany
Allenspach P M, ETH-Zurich, Switzerland
Amelinckx S, RUCA University of Antwerp, Belgium
Aomine T, Kyushu University, Japan
Arabi H, Manchester University, UK

Baker A P, BDH Ltd (Advanced Materials), Poole, UK
Blunt F J, Cambridge University, UK
Browning N D, Cambridge University, UK
Buluggiu E, University of Parma, Italy
Bungre S S, Imperial College, London, UK

Camerlingo C, Instituto di Cibernetica, Arcofelice, Italy
Campari E G, University of Bologna, Italy
Campbell A M, Cambridge University, UK
Caplin A D, Imperial College, London, UK
Chaffron L, CEN Saclay, France
Chandler S J, Cambridge University, UK
Chang F, University of Bath, UK
Cheah H M, Cambridge University, UK
Cherenkov V, Institute of Metrological Science, Moscow, USSR
Claeson T, Chalmers University of Technology, Göteborg, Sweden
Clarke J, University of California, Berkeley, USA
Cohen D, Technion-Israel Institute of Technology, Haifa, Israel
Coldea M, University of Cluj, Romania
Collocott S J, CSIRO Applied Physics, Lindfield, Australia
Cooper J R, Cambridge University, UK
Crisan M, Uppsala University, Sweden
Cui G J, KFA Jülich, Germany
Cumberbatch T J, Cambridge University, UK

Dai U, Tel Aviv University, Israel
Dam B, Phillips Research Laboratories, Eindhoven, The Netherlands
Denhoff M, National Research Council, Ottawa, Canada
Diederichs J, University of München, Germany
Ding S, Nanjing University, China
Donaldson G, University of Strathclyde, UK
Doyle R A, University of the Witwatersrand, South Africa
Dubovskii L B, I V Kurchatov Institute of Atomic Energy, Moscow, USSR
de Groot P A J, Southampton University, UK

Ekino T, Hiroshima University, Japan
Elgy C, Morgan Materials Technology Ltd, Stourport on Severn, UK
Escudero R, University of Mexico, Mexico
Evetts J E, Cambridge University, UK

Fiorani D, ITSE-CNR-CP10, Monterotondo Stazione (Roma), Italy
Fisher R A, University of California, Berkeley, USA
Fossheim K, Norwegian Institute of Technology, Trondheim, Norway
Fradin F, Argonne National Laboratory, USA
Franck J P, University of Alberta, Edmonton, Canada
Franks P, ARE(MOD), Slough, UK
Freeman P A, Cambridge University, UK
Fukami T, Kyushu University, Japan
Fukase T, Tohoku University, Japan

Gallop J C, National Physical Laboratory, Teddington, UK
Gameson I, Cambridge University, UK
Gasperic J, Jozef Stefan Institute, Yugoslavia
Gauss S, Hoechst AG, Frankfurt, Germany
Gehring K, GEC Hirst Research Centre, Wembley, UK
Gerber A, CNRS Grenoble, France
Gijs A M, Phillips Research Laboratories, Eindhoven, The Netherlands
Ginzburg V L, P N Lebedev Physics Institute of the Academy of Sciences, Moscow, USSR
Glowacki B A, Cambridge University, UK
Goodrich R G, Louisiana State University, Baton Rouge, USA
Goodyear S W, RSRE, Malvern, UK
Gordon J E, University of California, Berkeley, USA
Gotoh S, ISTEC, Tokyo, Japan
Grenet T, CNRS, Grenoble, France
Gross R, Tubingen University, Germany

Grovenor C R M, Oxford University, UK
Gurevich A, Kernforschungszentrum Karlsruhe, Germany
Güttler B, Physikalisch-Technische Bundesanstalt, Braunschweig, Germany

Halbritter J, Kernforschungszentrum Karlsruhe, Germany
Hands B, Oxford University, UK
Harrison M R, GEC Hirst Research Centre, Wembley, UK
Hasegawa T, University of Tokyo, Japan
Hayashi Y, University of Osaka Prefecture, Japan
He Y, Tsinghua University, China
Herzog R, Atominstitut, Wien, Austria
Hibbs A, Quantum Magnetics Inc, San Diego, USA
Hiller W, Institut für Strahlen- und Kernphysik, Bonn, Germany
Hilscher G, Technical University of Vienna, Austria
Hiroi Z, Kyoto University, Japan
Hodby J, Oxford University, UK
Horng H-E, National Taiwan Normal University, Taipei, Taiwan
Horovitz B, Ben-Gurion University, Beer-Sheva, Israel
Hosseini S M, University of Bonn, Germany
Hudner J, Royal Institute of Technology, Kista, Sweden
Huebener R P, Tubingen University, Germany
Humphreys R G, RSRE, Malvern, UK

Ito T, University of Tokyo, Japan
Ivanov Z, Chalmers University of Technology, Göteborg, Sweden
Izuyama T, University of Tokyo at Komaba, Japan

James J H, IMO, Lausanne, Switzerland
Johnson J D, Cambridge University, UK
Jones P, ARE(MOD), Slough, UK
Jorgensen J D, Argonne National Laboratory, USA
Jung J, University of Alberta, Edmonson, Canada
Junod A, University of Geneva, Switzerland

Kadin A M, University of Rochester, New York, USA
Kadowaki K, University of Amsterdam, The Netherlands
Kamimura H, University of Tokyo, Japan
Kawai T, Osaka University, Japan
Keck K, University of Wuerzburg, Germany
Keene M N, University of Birmingham, UK
Kinder H, University of München, Germany
Kitazawa K, University of Tokyo, Japan
Kobayashi N, Tohoku University, Japan
Kober F, Tubingen University, Germany
Koch H, Physikalisch-Technische Bundesanstalt, Berlin, Germany
Koch R H, IBM Research Division, Yorktown Heights, USA
Koelle D, Tubingen University, Germany
Koike Y, Tohoku University, Japan
Konczykowski M, Ecole Polytechnique, Palaiseau, France
Kopitzki K, University of Bonn, Germany
Kritscha W, Atominstitut, Wien, Austria
Kubo Y, NEC Corporation, Kawasaki, Japan
Kumakura H, National Research Institute for Metals, Tsukuba, Japan
Kuroda K, Mitshibushi Electric Corp, Amagasaki, Japan
Kwok W K, Argonne National Laboratory, USA

Lai H-C, Oxford University, UK
Lancaster M, Birmingham University, UK
Laverty J R, Imperial College, London, UK
Lean H W, Cambridge University, UK
Lees J S, University of Nottingham, UK
Liang W Y, Cambridge University, UK
Lin C, Cambridge University, UK
Lin S-Y, Chinese Academy of Sciences, Beijing, China
Lindenfeld P, Rutgers University, Piscataway, USA
Litvinenko Y G, Institute for Single Crystals, Kharkov, USSR
Liu R-S, Cambridge University, UK
Lo W, Cambridge University, UK
Loegel B, LPSE-FST, Mulhouse, France
Lopez A J, University de Santiago, Spain
Loram J W, Cambridge University, UK

Mackie A G, UMIST, UK
Maeda A, University of Tokyo, Japan
Markiewicz R S, Northeastern University, Boston, USA
Masuda H, Superconductivity Research Laboratory, Nagoya, Japan
Matijasevic V, Stanford University, USA
Meilikhov E Z, I V Kurchatov Institute of Atomic Energy, Moscow, USSR
Mertler G, University of Bonn, Germany
Mitchell W, SERC, Swindon, UK
Mizuno F, Superconductivity Research Laboratory, Nagoya, Japan
Mizushima K, Toshiba Corporation, Kawasaki, Japan
Mocaër P, ESPCI, Paris, France
Mohamed M A-K, University of Alberta, Edmonton, Canada
Molokac S, Slovak Academy of Sciences, Solovjevova, Czechoslovakia
Moore J, Oxford University, UK
Mori N, University of Tokyo, Japan
Morita M, Nippon Steel Corporation, Kawasaki, Japan
Mott N, Cambridge University, UK
Murakami M, International Superconductivity Technology, Tokyo, Japan

Muto Y, Tohoku University, Japan
Mydosh J A, de Rijksuniversiteit, Leiden, The Netherlands
Mészaros S, Hungarian Academy of Sciences, Debrecen, Hungary

Nakamura T, Nihon University, Japan
Naqvi S M M R, University of Karachi, Pakistan
Nes O-M, Sintef, Trondheim, Norway
Neumueller H-W, Siemens AG, Erlangen, Germany
Niskanen K, Uppsala University, Sweden
Norling P, Uppsala University, Sweden

Ohlsen H, Swedish Institute of Microelectronics, Kista, Sweden
Okajima Y, Hokksido University, Japan
Onari S, University of Tsukuba, Japan
Oota A, Toyohashi University of Technology, Japan
Ozkan R, Cambridge University, UK

Pan V M, Institute of Metal Physics, Kiev, USSR
Park J-C, Korea Standards Research Institute, Seoul, South Korea
Parvin E M, University of Birmingham, UK
Pavuna D, Ecole Polytechnique, Lausanne, Switzerland
Peng N, Cambridge University, UK
Petrusenko Y T, Academy of Sciences of the USSR, Kharkov, USSR
Phillips N E, Lawrence Berkeley Laboratory, USA
Phillips W A, Cambridge University, UK
Pindoria G, Cambridge University, UK
Pohl J, Nottingham University, UK
Polturak E, Technion-Israel Institute of Technology, Haifa, Israel
Porch A, University of Birmingham, UK
Przelozny Z, Monash University, Australia

Raffy H, Université Paris-Sud, France
Ralston R W, MIT Lincoln Laboratory, Lexington, USA
Regnier P, CEN Saclay, France
Reissner M, Technische Universität Wien, Austria
Ricketts J, University of Birmingham, UK
Ries G, Siemens AG, Erlangen, Germany
Rietschel H, KFK Karlsruhe, Germany
Rogalla H, University of Twente, The Netherlands
Ross J W, University of Manchester, UK
Rowell J M, Conductus Inc, Sunnyvale, USA
Rush J D, British Telecom Research Laboratory, Martlesham Heath, UK
Ruvalds J, University of Virginia, Charlottesville, USA

Sagdahl L T, Norwegian Institute of Technology, Trondheim, Norway
Sager R, Quantum Magnetics Inc, San Diego, USA
Salje E K H, Cambridge University, UK
Sander M, KFK Karlsruhe, Germany
Sarnelli E, Istituto Cibernetica, Arcofelice, Italy
Satchell J S, RSRE, Malvern, UK
Sato M, NTT Transmission Systems Laboratory, Ibaraki-Ken, Japan
Schalk R M, Osterreich Universität, Wien, Austria
Schieber M, Xsirius Superconductivity Inc, Jerusalem, Israel
Selinder T I, Linköping University, Sweden
Shaltiel D, The Hebrew University of Jerusalem, Israel
Shen Z X, Imperial College, London, UK
Shimizu K, Toray Industries Inc, Shiga, Japan
Shrivastava K N, University of Houston, USA
Slaughter R, Consultant, Chislehurst, UK
Somekh R E, Cambridge University, UK
Sonin E B, A F Ioffe Physical Technical Institute, Leningrad, USSR
Staines M P, DSIR, Lower Hutt, New Zealand
Staub U, ETH-Zurich, Switzerland
Steel D, Admiralty Research Establishment, Poole, UK
Steiner W, Technische Universität Wien, Austria
Sundqvist B, Umea University, Sweden
Swanson J, Central Electricity Research Laboratories, Leatherhead, UK

Tamegai T, University of Tokyo, Japan
Taylor M E, Cambridge University, UK
Terasaki I, University of Tokyo, Japan
Thomas G A, A T & T Bell Labs, Murray Hill, USA
Tinchev S, Hans Kolbe & Co, Bad Salzdetfurth, Germany
Tinkham M, Harvard University, USA
Tonkin B, Portsmouth Polytechnic, UK

Uchinokura K, University of Tokyo, Japan
Umemura T, Mitshibushi Electric Corp, Kanagawa, Japan
Ustinov A, Tubingen University, Germany

Valladares A, Instituto de Investigaciones en Materiales-UNAM, Mexico
Vera A, Parma University, Italy

Waldram J R, Cambridge University, UK
Wang H-S, Max-Planck Institut Für Festkörperforschung, Stuttgart, Germany
Weber H W, Osterreich Universität, Wien, Austria
Weber M, ETH-Zurich, Switzerland
Webster M W J, University of St Andrews, UK
Welp U, Argonne National Laboratory, USA
Wnuk J J, University Nijmegen, The Netherlands
Woodall P, University of Birmingham, UK
Wu T, Norwegian Institute of Technology, Trondheim, Norway
Wu Z, University of Birmingham, UK

Yamamoto H, Nihon University, Japan
Yamamoto K, Stanford University, USA
Yan S S, Peking University, China

Yang H-C, National Taiwan University, Taipei, Taiwan
Yang T-J, National Chiao Tung University, Hsinchu, Taiwan
Yoshizaki R, University of Tsukuba, Japan
Yu M, University of Alberta, Edmonton, Canada
Yuan J, Cambridge University, UK

Zakharov A A, Institute of Atomic Energy, Moscow, USSR
Zammatio S, GEC Hirst Research Centre, Wembley, UK
Zayats V A, Academy of Sciences of the USSR, Moscow, USSR
Zhang L-Y, Peking University, China
Zhang Q, Zhejiang University, China
Zhang Y, University of Science and Technology of China, Hefei, China
Zheng D-N, Cambridge University, UK
Zhou W, Cambridge University, UK
Zvyagin A I, FTINT AN USSR, Kharkov, USSR

Contents

POSTER PRESENTATIONS

Supercond. Sci. Technol. **4** (1991). Printed in the UK

Foreword

This volume constitutes the Proceedings of the LT-19 Satellite Conference on High Temperature Superconductivity, organized by the Interdisciplinary Research Centre in Superconductivity at the University of Cambridge. The conference took place on 13–15 August 1990 at Queens College, Cambridge and immediately preceded the 19th International Conference on Low Temperature Physics (LT-19) held in Brighton, UK.

The purpose of the Satellite Conference was to provide an opportunity for a focused residential meeting on high temperature superconductivity with a remit that reached beyond low temperature physics to embrace the broader spread of disciplines that have been essential for research in this fast moving subject. The size of the conference was restricted to 180 participants and there were no parallel sessions; this was in order to enhance the interdisciplinary impact of the meeting. Furthermore, all delegates were resident in Queens College and oral and poster sessions were in close proximity, giving the meeting something of the atmosphere of a workshop.

Some 200 papers were received in the general areas of physics, chemistry, materials synthesis and processing, and thin film technology. A Programme Committee drawn from research workers in industry and academia together with the Local Committee undertook the selection and arrangements of papers and found excellent speakers to cover important areas of current research. Invited papers are distinguished by their length, six–eight pages, from the contributed papers, which are three pages.

The great majority of the manuscripts were refereed during the meeting and I am grateful to the many members of the conference who contributed their time; their help was essential because only a very short time was available for the refereeing process.

I am also indebted to all those who assisted in the very busy editorial office during the meeting. The very considerable task of assembling and organizing the manuscripts, contacting and 'encouraging' the referees and dealing with authors in connection with corrected or modified manuscripts at the meeting and in the subsequent weeks was carried through with characteristic efficiency by Paul Aslin. The smoothness of the arrangements during the meeting and the impeccable overall logistics of the conference organization were due to Ken Diffey. He was ably assisted by Margaret Hilton and Sarita Haggart, who contributed many hours labour to produce the abstract booklets and on the myriad of other tasks that are essential to the smooth running of a conference.

Thanks are also due to the staff at Institute of Physics Publishing Ltd for their support throughout in the rapid production of these proceedings to a very tight schedule.

Jan Evetts
Honorary Editor

Supercond. Sci. Technol. **4** (1991) S1–10. Printed in the UK

THERMOELECTRIC EFFECTS IN SUPERCONDUCTORS*

V.L.Ginzburg

P.N.Lebedev Physical Institute, USSR Academy of Sciences;

Leninsky Prospect 53, 117924, Moscow, USSR

A general review of thermoelectric effects in superconductors is given. Specific features of thermoelectric effects in high–temperature superconductors are also discussed.

Thermoelectric phenomena in superconductors have been studied for more than 60 years. It has been established, with the then accuracy of measurements, that in a superconducting state the thermoelectric effects are completely absent. Such statements can be found in books written 30–35 years ago (see e.g. [1,2]). It is therefore natural that thermoelectric effects in superconductors have been and are now studied almost exclusively in normal state, i.e., in the absence of a magnetic field, above the critical temperature T_c. Meanwhile, as far back as 1944 it was mentioned [3] that in a superconducting state the thermoelectric effects do not disappear either, although it is not generally easy to observe them. Since that time, thermoelectric effects in superconducting state have been actually discovered, and the reviews [4,5] are devoted to this problem. Nevertheless, the

---

* Prepared for LT–19 Satellite Conference of High Temperature Superconductivity, Cambridge, 13–15, August 1990.

0953-2048/91/01S001 + 10 $03.50

picture as a whole is not yet quite clear, and, as far as I know from publications, the works in this direction are either not frequent or is not accomplished. In such a situation, and proceeding from the conviction that the study of thermoelectric phenomena in superconductors is of great interest, I shall begin here from the very beginning. I think this approach is convenient especially for those who is unacquainted with the problem (the text of the paper [6] is partially used here).

1. In a normal conductor (for simplicity, let us assume that it is isotropic) the current density is defined by

$$\mathbf{j} = \sigma(\mathbf{E} - \nabla\mu) + b\nabla T \tag{1}$$

where $\mathbf{E}$ is the electric field strength, μ is the chemical potential appropriately normalized, and ∇T is the temperature gradient.

If the conducting circuit is not closed, $\mathbf{j} = 0$ and $\mathbf{E} = (b/\sigma)\nabla T = S\nabla T$ in equilibrium; here and below we disregard signs and neglect the role of μ (for a closed circuit this is justified). The Seeback coefficient $S = b(T)/\sigma(T)$ defines the thermo–emf in the circuit consisting of the conductors I and II (Fig.1). From (1) it is clear that the thermo–emf in such a circuit is equal to

$$\mathcal{E} = \int_{T_1}^{T_2} \left\{ \left[\frac{b}{\sigma}\right]_{\mathrm{II}} - \left[\frac{b}{\sigma}\right]_{\mathrm{I}} \right\} \mathrm{d}T \tag{2}$$

For a completely superconducting macroscopic circuit (where both conductors I and II are in the superconducting state), within the usual measurement accuracy $\mathcal{E} = 0$ in accordance with the above statement [1]. However, as we shall see below, in a superconducting circuit some current $I_s \neq 0$ is in fact flowing, but this current is $(\delta/d)^2$ times weaker compared with the current I_n flowing in an analogous normal circuit (δ is the magnetic field penetration depth in the superconductor and d is the characteristic thickness of the sample, i.e., the wire diameter; here we assume that $d>>\delta$). For $\delta \sim 10^{-5}$ cm and $d \sim 0.1$ cm, obviously $I_s \sim 10^{-8} I_n$; it is impossible to observe such a current in the course of usual thermoelectric measurements. However, the current and magnetic field produced by such a current can be measured using present techniques.

Before discussing superconducting circuits let us consider the simplest situation: a nonclosed superconducting specimen, for instance a rod (Fig.2). In the normal state the electric field $\mathbf{E} = (b/\sigma)\nabla T$ in this specimen, and this prevents current flow [see (1)]. In the superconducting state the field $\mathbf{E}$ (or, more precisely, the quantity $\mathbf{E} - \nabla\mu$) is equal to zero at equilibrium. Thus, if $\mathbf{E} - \nabla\mu \neq 0$, the "superconducting electrons" would be accelerated (this is evident from the well–known equation $\partial\Lambda\mathbf{j}_s/\partial t = \mathbf{E} - \nabla\mu$, where $\mathbf{j}_s$ is the density of the superconducting current). At the same time, as in the case of a normal conductor, the temperature gradient must give rise to a current of "normal electrons" with density $\mathbf{j}_n = b_n\nabla T$. Nothing prevents the flow of this current because at the boundaries of the specimen (at temperatures T_2 and T_1; see Fig.2) the normal current is transformed into the superconducting current, and inversely. As a result, in a superconductor $\mathbf{j}_s = -\mathbf{j}_n$, the total current density $\mathbf{j} = \mathbf{j}_s + \mathbf{j}_n = 0$, and a magnetic field does not appear. However, the existence of the currentsleads to an additional (circulatory) flow of heat, hence the analogy with HeII (this is just what stimulated the appearance of paper [3]).

The density of heat flow $\mathbf{q} = æ\nabla T$, $æ = æ_{ph} + æ_{el} + æ_c$, where $æ_{ph}$ is the thermal conductivity due to the lattice (phonons), $æ_{el}$ is the thermal conductivity due to normal electrons, and $æ_c$ is the thermal conductivity connected with the circulation. According to the Wiedemann–Franz law $æ_{el} = (\pi^2 k_B{}^2/3e^2)T\sigma_n$, where $\sigma_n = j_n/E$ is the electric conductivity of normal electrons. The transformation of normal current into superconducting current is connected with the formation of pairs with charge $2e$ and energy 2Δ per pair. Evidently the circulatory heat flow $q_c \sim æ_c \nabla T \sim j_n(\Delta/e)$ and $æ_c \sim b_n(T)\Delta(T)/e$. From this and the Wiedemann–Franz law we obtain

$$\frac{æ_c}{æ_{el}} \sim \frac{3eS_n\Delta}{\pi^2 k_B{}^2 T} \sim \frac{\Delta(T)}{E_F} \sim \frac{k_B T_c}{E_F} \qquad (3)$$

where, in the expression before the last one, the value

$$S_n = (b_n/\sigma_n) = \pi^2 k_B{}^2 T(3eE_F)^{-1}$$

correct for free electrons is used (E_F is the Fermi energy; of course $\Delta(T) \sim k_B T$ only for $T \sim T_c$).

In conventional superconductors $T_c \sim 1 - 10$ **K** and $E_F \sim 3 - 10$ eV, so that $æ_c/æ_{el} \lesssim 3 \cdot 10^{-4}$. This is why circulatory thermal conductivity is negligibly small (see, e.g., [4]). However, for high–temperature superconductors (HTS) with $T_c \sim 100$ **K** and $E \sim 0.1$ eV, we obtain $æ_c/æ_{el} \sim 0.1$. We note in addition that the thermoelectric coefficient b (or S) depends on the derivative of the electronic distribution function with respect to momentum or energy. Therefore, it is quite possible that, for HTS with s–pairing, $æ_c/æ_{el} \sim 1$, that is, the effect can be large enough [6,7]. Further, according to [8,9], for p– or d–pairing, the coefficient b_n increases considerably in the superconducting state (it is multiplied by a factor of the order $E_F/k_B T_c$). Hence, for p– or d–pairing, generally speaking, $æ_c/æ_{el} \sim 1$, and measurements of $æ_c$ can serve as a method to determine the type of pairing in HTS; the same is valid for heavy fermions. As is well known, this question is real and important.

2. For HTS an increase in æ at $T<T_c$ is sometimes observed (see, e.g. [10], where Fig.3 was taken). Such a dependence $æ(T)$ could be related to contributions from $æ_c(T)$. However, the observed nonmonotonic dependence $æ(T)$ could be related also to an increase in $æ_{ph}(T)$ due to a reduction in scattering of phonons by "dying" normal electrons. The problem obviously requires special analysis. Evidently, it is necessary, combining experiment and theory, to single out the contributions $æ_{ph}$, $æ_{el}$ and $æ_c$. This has not yet been achieved (in this connection, see [11–15]). So, the situation remains quite unclear. Especially promising, I think, is an account of anisotropy [13], but the investigation of the dependence on the duration of neutron irradiation [15] and, probably, on the strength of a superimposed magnetic field and on deformations is also of interest.

3. The total compensation of currents $\mathbf{j}_n$ and $\mathbf{j}_s$, that is, the absence of the total current $\mathbf{j}$, takes place, generally speaking, for homogeneous and isotropic superconductors only (Fig.2). If the specimen is homogeneous (we always assume homogeneity for $\nabla T = 0$) but anisotropic and ∇T does not coincide with the crystal symmetry axis, a certain total current $\mathbf{j} \neq 0$ appears; this current flows on the surface of the specimen in the layer with thickness of the order δ. We will not discuss this and other thermoelectric effects [3–5,16,17] here but will return to an inhomogeneous circuit of the type shown in Fig.1. In this case, as already mentioned, even for a completely superconducting circuit there is a certain current with density $\mathbf{j} \neq 0$, but only near the surface of the metal. It is easy to calculate the total magnetic flux $\mathbf{\Phi}$ through contour C (dotted line in Fig.1). The result is

$$\Phi = n\Phi_0 + \frac{4\pi}{c}\int_{T_1}^{T} \{(b_n\delta^2)_{II} - (b_n\delta^2)_I\} dT, \qquad \Phi_0 = \frac{\hbar c}{2e} \qquad (4)$$

In fact, in the bulk of the specimen (we assume that $d >> \delta$), on the contour C, we have $\mathbf{j} = \mathbf{j_s} + \mathbf{j_n}$, $\mathbf{j_n} = b_n \nabla T$, and $\mathbf{j_s} = (e\hbar n_s/2m)\,[\nabla\varphi - (2e/\hbar c)\mathbf{A}]$, where n_s is the "concentration" of the superconducting electrons ($\delta^2 = mc^2/4\pi e^2 n_s$), φ is some phase (scalar), and curl $\mathbf{A} = \mathbf{H}$. Integrating the equality $\mathbf{j_n} = -\mathbf{j_s}$, along the contour C, we obtain (4) because $\oint \mathbf{A}\, d\mathbf{S} = \int \mathbf{H}\, d\mathbf{S}$ and $\oint \nabla\varphi\, d\mathbf{S} = 2\pi n$.

Let us assume that $(b_n\delta^2)_{II} >> (b_n\delta^2)_I$ and $\delta^2_{II} = \delta^2_{0II}(1 - T/T_c)^{-1}$. Then near T_c the temperature–dependent part of this flux is

$$\Phi_T = \Phi - n\Phi_0 \approx \frac{4\pi}{c}\, b_{n_{II}}(T_c)\,\delta^2_{0II}\, T_c \ln\frac{T_c - T_1}{T_c - T_2} \qquad (5)$$

For tin, $\delta_{0II} \sim 2.5\cdot 10^{-6}$cm, $b_n(T_c) \sim 10^{11}$cgse/K, and $\Phi_T \sim 10^{-2}\Phi_0 \ln(T_c - T_1/T_c - T_2)$. Such a flux is easy to measure [4,5]. For a circuit such as a bimetallic loop, a formula similar to (4) was confirmed. However, for a circuit such as a bimetallic toroid, the measured flux was considerably larger [5] than that obtained using (4). The phenomenon remains unexplained. One can assume [18,19] that the trapped magnetic flux $n\Phi_0$ depends on the temperature and geometry of the circuit, and this explains the divergence of experimental results. To clarify the situation, the easiest and most straightforward way is to use the bimetallic plate without any hole (in this case, $n = 0$ always). We note that such a bimetallic plate, and only it, was considered at the end of the paper [3].

4. According to some measurements, for HTS the coefficient S $\sim 10^{-5}$V/K, but for tin S $\sim 10^{-7}$V/K. However, the conductivity σ for tin, for instance, is much larger than for the known HTS specimens. That is why the coefficient $b(T_c) = S(T_c)\sigma(T_c)$ for HTS is also considerably smaller than that for conventional superconductors. However, the product $b_n\delta^2$ [see (4)] differs not as much. Moreover, if the coefficient b_n increases at $T<T_c$ (see [8,9]), the product $b_n\delta^2$ can be even larger compared with that in conventional superconductors. It is clear that measurements of the thermoelectric coefficient b_n for HTS are very interesting for an estimate of $æ_c$ and determination of the type of pairing. Further, in HTS the coherence length is considerably smaller than in conventional superconductors. As a result, fluctuation effects near T_c in HTS are considerably increased (see [20]). In accordance with this, such a fluctuation increase of S near T_c (for $T>T_c$) has probably already been observed [21]. Properly speaking, the anomaly in [21] is definitely observed, and the only doubt can be connected with its interpretation as a fluctuation effect. It is noteworthy that in the literature there also exist a number of other indications [22–24] of some non–monotonous behaviour of S at $T>T_c$ near T_c. It would be very interesting to investigate in detail fluctuation effects near T_c at $T>T_c$, and also analogous effects at $T<T_c$. In the latter case, however, one should evidently study thermoelectric effects which can be observed below T_c (i.e., in superconducting state). I would also like to mention the circulatory thermoelectric effect in He II [25].

I am convinced that investigation of thermoelectric effects in superconductors deserves full attention.

References

1. D.Shoenberg. Superconductivity. Cambridge, Cambridge Univ. Press (1965)
2. A.C.Rose–Innes and E.H.Rhoderick. Introduction to Superconductivity. Oxford, Pergamon Press (1969)
3. V.L.Ginzburg. J. Phys. USSR 8, 148 (1944), Zh. Eksp. Teor. Fiz. 14, 177 (1944)
4. V.L.Ginzburg and G.F.Zharkov. Sov. Phys. – Uspekhi 21, 381 (1978)
5. D.J.Van Harlingen. Physica, 109–110B, 1710 (1982)
6. V.L.Ginzburg. Journal of Superconductivity 2, 323 (1989)
7. V.L.Ginzburg. Sov. Phys. JETP Lett. 49, 58 (1989)
8. B.Arfi, M.Bahlouli, C.J.Pethick and D.Pines. Phys. Rev. Lett. 60, 2206 (1988). Phys. Rev. B39, 8959 (1989)
9. P.J.Hirschfeld. Phys. Rev. B37, 9331 (1988).
10. A.Jezowski, A.J.Zaleski, M.Ciszek et al. Helvetica Physica Acta, 61, 438 (1988)
11. A.Jezowski. Solid State Commun. 71, 419 (1989)

11a. I.K.Kamilov, A.B.Batdalov and B.K.Chakalski (in press).

12. M.A.Izbizki, M.N.Regueiro, P.Esquinazi and C.Fainstein. Phys. Rev. B38, 9220 (1988)
13. S.J.Hagen, Z.Z.Wang and N.P.Ong. Phys. Rev. B40, 9389 (1989)
14. D.T.Morelli, G.L.Doll and J.Heremans. Phys. Rev. B41, 2520 (1990)
15. C.Uher and W.N.Huang. Phys. Rev. B40, 2694 (1989)
16. B.A.Mattoo and Y.Singh. Progress of Theoret. Physics 70, 51 (1983)
17. A.V.Ustinov, M.Hartman and R.P.Huebener. Europhysics Letters (in press). See also: Phys. Rev. B (in press)
18. V.L.Ginzburg, G.F.Zharkov and A.A.Sobyanin. J. Low Temp. Phys. 56, 195

(1984)

19. G.F.Zharkov, in "Superconductivity, Superdiamagnetism, Superfluidity". Ed. V.L.Ginzburg. Mir Press, Moscow (1987)

20. V.L.Ginzburg. Physica C153–155, 1617 (1988). See also: Physica C 152, 378 (1988). Sov. Phys. – JETP 68, 1499 (1988)

21. M.A.Howson, M.B.Salamon, T.A.Friedmann, J.P.Rice and D.Dinsberg. Phys. Rev. B41, 300 (1990)

22. Shousheng Yan et al. Modern Physics Letters B2, 1005 (1988)

23. T.K.Dey et al. Solid State Commun. 72, 525 (1989)

24. A.K.Bhatnagar et al. Solid State Commun. 73, 53 (1990)

25. V.L.Ginzburg and A.A.Sobyanin. Sov. Phys. JETP 58, 934 (1983)

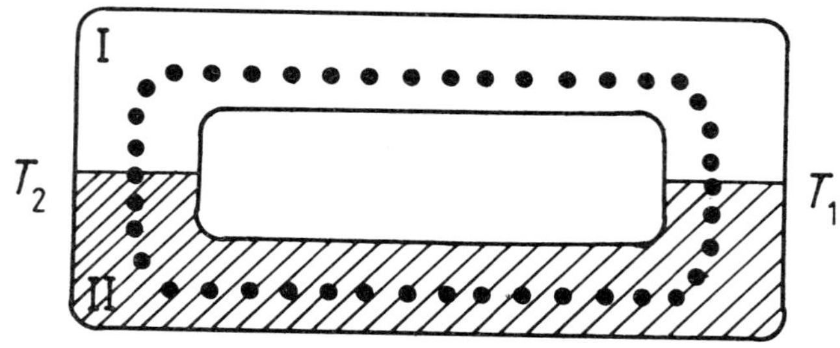

Fig. 1. Thermoelectric circuit; the dotted line is the contour C.

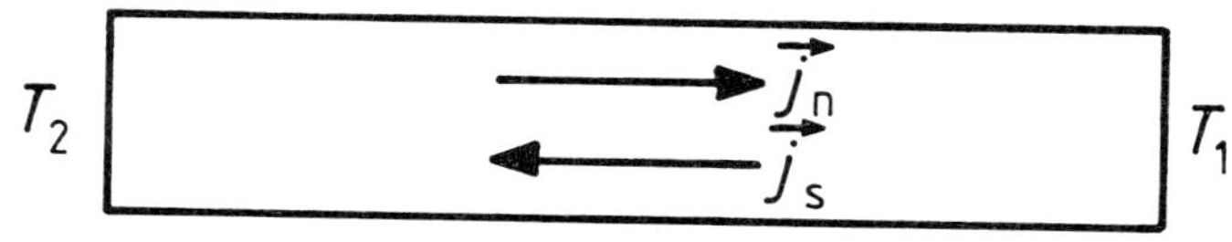

Fig. 2. The currents $\mathbf{j}_n$ and $\mathbf{j}_s$.

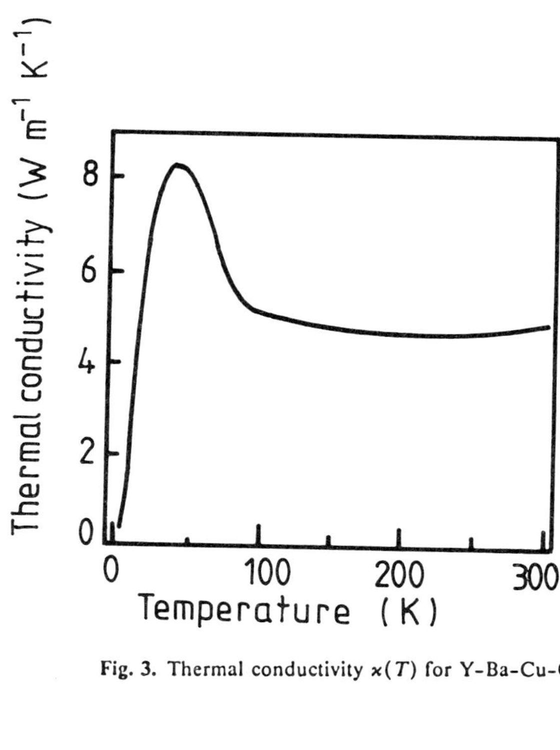

Fig. 3. Thermal conductivity $\varkappa(T)$ for Y-Ba-Cu-O.

Supercond. Sci. Technol. 4 (1991) S11–18. Printed in the UK

Structural Properties of the Oxide Superconductors: The Critical Role of Defects

J. D. Jorgensen*, P. Lightfoot**, and Shiyou Pei**

*Materials Science Division and **Science and Technology Center for Superconductivity, Argonne National Laboratory, Argonne, IL 60439

Abstract. Structural data have contributed greatly to our understanding of the oxide superconductors. However, it is increasingly clear that a knowledge of the average structure is not sufficient. Structural defects, which are present in essentially all of these systems, have a profound effect on the superconducting properties.

1. Introduction

Structural studies of the oxide superconductors have revealed a fascinating variety of novel features and have contributed substantially to our understanding of these new materials. For example, in the $YBa_2Cu_3O_{7-x}$ system, structural data played a major role in developing the concept of charge transfer between the one-dimensional CuO_{1-x} "chains", which act essentially as reservoirs of charge, and the two-dimensional CuO_2 "planes", which are responsible for the superconductivity (Cava et al 1990, Jorgensen et al 1990b). In some cases, structural studies can be used to quantitatively probe such behavior, since individual Cu-O bond lengths are sensitive to local charge distributions (Brown 1989, Tallon et al 1990). During this work, however, it has become increasingly clear that the superconducting behavior cannot be fully understood in terms of the average crystal structures. The structures of defects, which are present in most or all of the oxide superconductors, must also be understood (Jorgensen and Hinks 1990).

In general, the oxide superconductors can support a wide variety of structural defects, including metal-site substitution, metal anti-site defects, metal-site vacancies, oxygen vacancies, interstitial oxygen, and complex defects arising from commensurate or incommensurate structural modulations (Jorgensen and Hinks 1990, Schuller and Jorgensen 1989). A full discussion of all of these cases is beyond the scope of this paper. Attention will be focussed on the simpler systems where important new information has recently been obtained.

2. Defect Ordering in $YBa_2Cu_3O_{7-x}$

0953-2048/91/01S011 + 08 $03.50

The $YBa_2Cu_3O_{7-x}$ system provides us with one of the clearest examples of the importance of defects. In this structure, two of the five oxygen sites, both located in the CuO_x chain layer, can be partially occupied. Complete ordering, at the composition $YBa_2Cu_3O_7$, leads to the well-known one-dimensional CuO chains. At oxygen concentrations less than 7, vacancies form at the chain oxygen sites, and oxygen atoms can occupy the originally empty sites between the chains.

Recent experiments show that hopping of oxygen atoms between these chain and interchain sites can occur at room temperature and that the superconducting properties depend on the local configuration of the occupied sites (Jorgensen et al 1990a). Samples of $YBa_2Cu_3O_{6.41}$ quenched from 500° C. into liquid nitrogen and then annealed at room temperature exhibit time-dependent superconducting and structural properties due to this oxygen motion. T_c increases asymptotically as the oxygen atoms diffuse to form local chain fragments. This experiment teaches us that both the defect concentration and ordering must be known in order to correlate the superconducting and structural properties. Furthermore, ordering phenomena on a length scale not probed by conventional diffraction methods can have dramatic effects on the superconducting behavior, since the superconducting coherence length is unusually short (e.g., 10-100Å) in these materials.

3. Interstitial Oxygen Defects in the T Structure -- $La_2CuO_{4+\delta}$

$La_2CuO_{4+\delta}$ is the first oxide superconductor where it was conclusively shown that interstitial oxygen defects provide the doping required for metallic behavior (Jorgensen et al 1988). Because the solubility for excess oxygen is low, bulk superconducting samples can only be made using high oxygen pressures in the range 0.01 to 2.5 GPa (Jorgensen et al 1988, Beille et al 1987, Schirber et al 1988, Dabrowski et al 1989a&b, Zhou et al 1989).

One particularly novel feature of the interstitial oxygen defect in $La_2CuO_{4+\delta}$ is that it is highly mobile. This defect mobility leads to an unusual phase separation that occurs near room temperature (Jorgensen et al 1988). The oxygen defect segregates into an oxygen-rich phase of approximate composition $La_2CuO_{4.08}$ that is superconducting, leaving behind a nearly stoichiometric phase (i.e., $\delta \approx 0$) that is non-superconducting. Since the two phases have nearly identical structures, this phase separation was not seen in earlier diffraction studies because of insufficient resolution. Because of the phase separation, typical samples of $La_2CuO_{4+\delta}$ are mixtures of superconducting and nonsuperconducting domains. Such samples exhibit an essentially constant T_c of 30-40K (because the composition of the superconducting phase is constant) and a varying superconducting fraction that depends on the processing conditions. Significant diamagnetic shielding is not observed unless the size of the superconducting domains exceeds the penetration depth. Thus, the superconducting fraction as measured by diamagnetic shielding does not vary linearly with the volume fraction of oxygen-rich phase as measured by diffraction. Rather, the behavior shown in Fig. 1, where no significant diamagnetic shielding occurs until the volume fraction exceeds about 30%, is observed (Dabrowski et al 1989a).

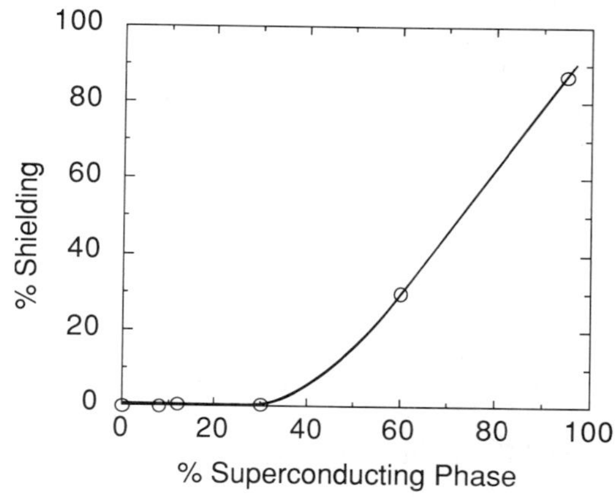

Fig. 1. Superconducting fraction determined from diamagnetic shielding versus the volume fraction of superconducting phase determined by neutron diffraction for powder samples of $La_2CuO_{4+\delta}$.

The interstitial oxygen defect forms in the La_2O_2 layer away from the two-dimensional CuO_2 planes. Hole carriers are presumably created in the CuO_2 planes by charge transfer, with each O^{2-} defect contributing two holes. Thus, the doping level of $\delta \approx 0.08$ is essentially equivalent to $x \approx 0.15$ in the analogous compound $La_{2-x}Sr_xCuO_4$, and it is not surprising that they exhibit similar T_c's.

Although there is general agreement about the structure of the interstitial oxygen defect, there is some controversy concerning the details of the local configuration. In the original neutron powder diffraction studies of phase-separated $La_2CuO_{4+\delta}$ (Jorgensen et al 1988) it was not possible to uniquely determine the defect structure because of the complications of refining structures for the two coexisting phases. However, the structure of the interstitial oxygen defect in the isostructural compound $La_2NiO_{4+\delta}$ was readily determined from neutron powder diffraction data for single-phase samples with $\delta = 0.18$ (Jorgensen et al 1989). The defect structure is shown in Fig. 2.

The similarities in the overall phase diagrams of the Cu and Ni compounds imply that their oxygen defect structures are the same. However, Chaillout et al (1989, 1990) proposed a slightly different defect structure based on neutron diffraction studies of $La_2CuO_{4+\delta}$ single crystals. They found the interstitial oxygen atom at essentially the same site, coordinated to four La atoms, but concluded that the interstitial oxygen atom had one close oxygen atom neighbor, suggesting the formation of a peroxide or superoxide ion.

The subtle difference in these two structural models for the oxygen defect centers around the conclusions concerning the four near-neighbor oxygen atoms. Jorgensen et al (1989) concluded that all four are displaced away from the interstitial oxygen atom, while Chaillout et al (1989, 1990) concluded that three are displaced away from and one toward the interstitial oxygen. In neither case does the conclusion derive uniquely from the diffraction data, since the analysis is based on a model that assumes translational (and point) symmetry and can only

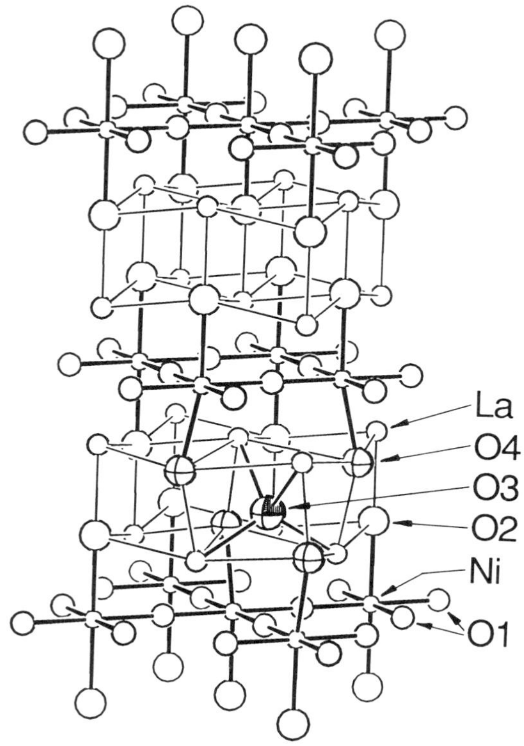

Fig. 2. Structure of the interstitial oxygen defect in $La_2NiO_{4+\delta}$ as determined from neutron powder diffraction data. The interstitial oxygen atom, O3, is tetrahedrally coordinated to four near-neighbor La atoms. Four near-neighbor O2 atoms are displaced from their normal sites into new sites, O4, to accommodate the defect.

give a spatial average of the true structure. One of the reasons for the conclusion of Chaillout et al stems from the fact that they find the space group of the oxygen-rich phase to be orthorhombic Bmab, rather that the closely related orthorhombic Fmmm proposed by Jorgensen et al. It should be realized, however, that the local structure of the defect is not required to obey the symmetry of the host lattice. Thus, the question of which space group is correct is irrelevant to understanding the local defect structure. In both studies, the structural refinements give a symmetry-equivalent set of displaced oxygen sites allowing choices displaced in various directions. In the absence of <u>local</u> structural information, conclusions about the local defect structure can only be based on chemical intuition. It is important to realize, however, that the interstitial O^{2-} defect proposed by Jorgensen et al is consistent with the doping level required for superconductivity, while models based on peroxide or superoxide defects are not.

4. Defects in the T' Structure -- $Nd_{2-x}Ce_xCuO_4$

The interstitial oxygen defect site in the T structure of $La_2CuO_{4+\delta}$ is equivalent to the normal oxygen site in the Nd_2O_2 layer of the T' Nd_2CuO_4 structure. It is, thus, logical to ask whether the Nd_2CuO_4 structure can incorporate an interstitial oxygen defect equivalent to the apical oxygen atom in La_2CuO_4. Refinements of the $Nd_{2-x}Ce_xCuO_4$ structure show no such defect (Izumi et al 1989a, Kwei et al 1989, Lightfoot et al 1990c). However, an unusual phase separation phenomena has been observed (Lightfoot et al 1990c), suggesting that defects may also play an important role in this system. In order for phase separation to occur, the two phases must differ in some way. Since the host structures are the same in both phases, it is logical to assume that the difference involves a defect structure or subtle structural ordering not yet elucidated by diffraction experiments. Such a hypothesis is consistent with the novel superconducting properties of $Nd_{2-x}Ce_xCuO_4$. Superconductivity is observed only for a very narrow range of Ce concentrations -- too narrow to be readily explained on the basis of a simple solid solution of Ce in $Nd_{2-x}Ce_xCuO_4$. Additionally, for the same Ce concentration ($x \approx 0.16$), superconductivity is observed only for samples annealed in a particular atmosphere. Superconducting and nonsuperconducting samples differ in oxygen composition by only a tiny amount -- again too small to explain the behavior in terms of variations in carrier concentration. Subtle differences in the defect structure may provide an explanation.

5. Interstitial Defects in the T* Structure -- $La_{1.25}Dy_{0.75}CuO_{3.75}F_{0.5}$

The T* structure is essentially a combination of the T and T' structures (Sawa et al 1989, Izumi et al 1989b). It consists of alternating T- and T'-like layers. Equivalently, it can be viewed as consisting of alternating rocksalt and fluorite Ln_2O_2 layers along the c axis so as to leave each Cu of the intervening CuO_2 planes fivefold coordinated to oxygen. Superconductivity for this structure type was first reported for a compound synthesized by a standard ceramic method (Akimitsu 1988). However, it was later reported that a large number of insulating compounds with the T* structure can be made superconducting by annealing in high-pressure oxygen (Hundley et al 1989, Fisk et al 1989). The implication of this work is that the insulating material is being doped by the incorporation of excess oxygen, analogous to the interstitial oxygen in $La_2CuO_{4+\delta}$.

Attempts to find the excess oxygen in these small high-pressure samples using neutron diffraction have thus far been unsuccessful. However, an interstitial defect has recently been identified in the T* structure of $La_{1.25}Dy_{0.75}CuO_{3.75}F_{0.5}$ (Lightfoot 1990a). Through the use of fluorine, the concentration of the excess anion defect is raised to a high level, making the structural solution of the defect straightfor-

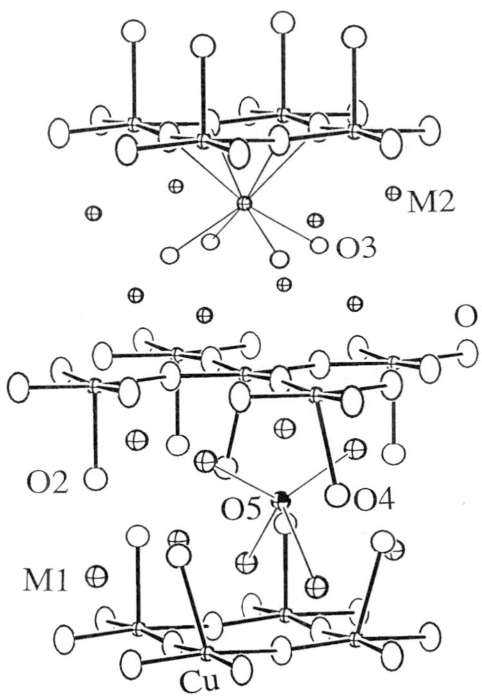

Fig. 3. Structure of the interstitial anion defect in the T* phase $La_{1.25}Dy_{0.75}CuO_{3.75}F_{0.5}$ (Lightfoot et al 1990b). As in $La_2NiO_{4+\delta}$, the defect anion, O5, is coordinated to four La/Dy atoms, M1, and four oxygen atoms, O4, that have been displaced from their normal sites, O2, to accommodate the defect.

ward by neutron powder diffraction techniques. The defect structure is shown in Fig. 3. The excess anions are accommodated in an interstitial site in the T-like Ln_2O_2 layer analogous to the defect in $La_2CuO_{4+\delta}$. The interstitial anion is coordinated to four La/Dy atoms and four near-neighbor O/F atoms are displaced into new sites. As in the case of $Nd_{2-x}Ce_xCuO_4$, no interstitial defects are found in the T'-like Ln_2O_2 layer (i.e., at the empty apical position). Although this defect has been identified in a nonsuperconducting member of the T* system, it seems likely the the same (or a similar) defect will be shown to be the doping mechanism in T* phases that are made superconducting by processing in high-pressure oxygen.

6. Interlayer Oxygen in Nonsuperconducting $La_{1.85}Sr_{1.15}Cu_2O_{6.25}$

The previous examples have been cases where excess oxygen defects that form outside the two-dimensional CuO_2 layer lead to superconductivity by creating hole carriers in the layer. One recent study suggests that superconductivity can be destroyed by excess oxygen defects that form at sites in or near the superconducting CuO_2 layer(s).

In recent work, Cava et al (1990) reported the discovery of superconductivity in the simplest double-layer cuprate $La_{2-x}Sr_xCaCu_2O_6$. Until their report, this family of layered cuprates had been an enigma in the field of high-T_c superconductivity because of their failure to exhibit superconductivity even though the average copper oxidation state appeared to be in the desired range (Torrance et al 1988).

A possible explanation for this seemingly inconsistent behavior has been proposed by Lightfoot et al (1990b). In a structural refinement for $La_{1.85}Sr_{1.15}Cu_2O_{6.25}$, based on neutron powder diffraction data, they show that the excess oxygen is incorporated at a site between the double CuO_2 layers, as shown in Fig. 4. This defect is equivalent to placing oxygen in the Y layer in the well-known $YBa_2Cu_3O_{7-x}$ structure. Lightfoot et al speculate that an oxygen defect in this site destroys supercon-

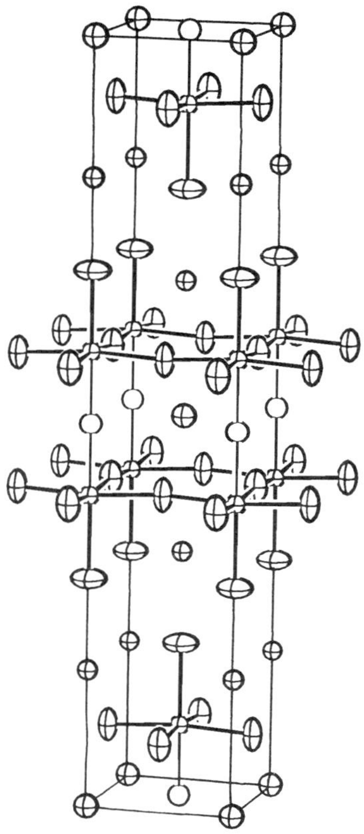

Fig. 4. Oxygen defect structure in $La_{1.85}Sr_{1.15}Cu_2O_{6.25}$. Excess oxygen is located in the partially occupied O3 site between the CuO_2 layers.

ducting behavior and that such defects account for the previous failure to find superconductivity in this family of compounds. They propose that in the superconducting compound reported by Cava et al, $La_{2-x}Sr_xCaCu_2O_6$, occupation of this site by oxygen is blocked by placing the small Ca atoms, rather that larger La or Sr atoms, at the interlayer position. This hypothesis has yet to be tested by structural studies of additional superconducting and nonsuperconducting compounds.

7. Conclusions

Clearly, the determination of defect structures in the oxide superconductors is still in its infancy. However, there is no question, that the defects must be understood before structural data can lead to an understanding of the superconducting behavior. As has been shown by the examples presented, the determination of these defect structures will be a tedious process, especially in light of the fact that ordering phenomena on a length scale not probed by conventional diffraction can affect the superconducting behavior.

Acknowledgements

This work is supported by the U. S. Dept. of Energy, Basic Energy Sciences - Materials Sciences, under contract No. W-31-109-ENG-38 (JDJ) and the National Science Foundation, Science and Technology Center for Superconductivity, under Grant No. DMR-88-09854 (PL, SP).

References

Akimitsu J., Suzuki S., Watanabe M., and Sawa H. 1988 Jpn. J. Appl. Phys. **27** L1859.

Beille J., Chevalier B., Demazeau G., Deslandes F., Etourneau J., Laborde O., Michel C., Lejay P., Provost J., Raveau B., Sulpice A., Tholence J. L., and Tournier R. 1987 Physica B **146** 307.

Brown I. D. 1989 J. Solid State Chem. **82** 122.

Cava R. J., Hewat A. W., Hewat E. A., Batlogg B., Marezio M., Rabe K. M., Krajewski J. J., Peck W. F. Jr., and Rupp L. W. Jr. 1990a Physica C **165** 419.

Cava R. J., Batlogg B., van Dover R. B., Krajewski J. J., Waszczak J. V., Flemming R. M., Peck W. F. Jr., Rupp L. W. Jr., Marsh P., James A. C. W. P. and Schneemeyer L. F. 1990b Nature **345** 602.

Chaillout C., Cheong S. W., Fisk Z., Lehmann M. S., Marezio M., Morosin B., Schirber J. E. 1989 Physica C 158 183.

Chaillout C., Chenavas J., Cheong S. W., Fisk Z., Marezio M., Morosin B., and Schirber J. E. 1990 Physica C (in press).

Dabrowski B., Hinks D. G., Jorgensen J. D., and Richards D. R. 1989a Mat. Res. Soc. Symp. Proc. Vol. 156, p. 69.

Dabrowski B., Jorgensen J. D., Hinks D. G., Pei S., Richards D. R., Vanfleet H. B., and Decker D. L. 1989b Physica C **162-164** 99.

Fisk Z., Cheong S.-W., Thompson J. D., Hundley M. F., Schwartz R. B., Kwei G. H., and Schirber J. E. 1989 Physica C **162-164** 1681.

Hundley M. F., Thompson J. D., Cheong S.-W., Fisk Z., Schwartz R. B., and Schirber J. E. 1989 Phys. Rev. B **40** 5251.

Izumi F., Matsui Y., Takagi H., Uchida S., Tokura Y., and Asano H. 1989a Physica C **158** 433.

Izumi F., Takayama-Muromachi E., Fujimori A., Kamiyama T., Asano H., Akimitsu J., and Sawa H. 1989b Physica C **158** 440.

Jorgensen J. D., Dabrowski B., Pei S., Hinks D. G., Soderholm L., Morosin B., Schirber J. E., Venturini E. L., and Ginley D. S. 1988 Phys. Rev. B **38** 11337.

Jorgensen J. D., Dabrowski B., Pei S., Richards D. R., and Hinks D. G. 1989 Phys. Rev. B **40** 2187.

Jorgensen J. D. and Hinks D. G. 1990 Neutron News Vol. 1, No. 2, p. 24.

Jorgensen J. D., Pei S., Lightfoot P., Shi H., Paulikas A. P., and Veal B. W. 1990a Physica C **167** 571.

Jorgensen J. D., Veal B. W., Paulikas A. P., Nowicki L. J., Crabtree G. W., Claus H., and Kwok W. K. 1990b Phys. Rev. B **41** 1863.

Kwei G. H., Cheong S.-W., Fisk Z., Garzon F. H., Goldstone J. A., and Thompson J. D. 1989 Phys. Rev. B **40** 9370.

Lightfoot P., Pei S., Jorgensen J. D., Tang X.-X., Manthiram A., and Goodenough J. B. 1990a Physica C **169** 15.

Lightfoot P., Pei S., Jorgensen J. D., Tang X.-X., Manthiram A., and Goodenough J. B. 1990b Physica C (in press).

Lightfoot P., Richards D. R., Dabrowski B., Hinks D. G., Pei S., Marx D. T., Mitchell A. W., Zheng Y., and Jorgensen J. D. 1990c Physica C **168** 627.

Sawa H., Suzuki S., Watanabe M., Akimitsu J., Matsubara H., Watabe H., Uchida S., Kokusho K., Asano H., Izumi F., and Takayama-Muromachi E. 1989 Nature **337** 347.

Schirber J. E., Morosin B., Merrill R. M., Hlava P. F., Venturini E. L., Kwak J. F., Nigrey P. J., Baughman R. J., and Ginley D. S. 1988 Physica C **152** 121.

Schuller I. K. and Jorgensen J. D., 1989 MRS Bulletin, January 1989, p. 27.

Tallon J. L. 1990 Physica C **168** 85.

Torrance J. B., Tokura Y., Nazzal A., and Parkin S. S. P. 1988 Phys. Rev. Lett. **60** 542.

Zhou J., Sinha S., and Goodenough J. B. 1989 Phys. Rev. B **39** 12331.

Supercond. Sci. Technol. 4 (1991) S19–34. Printed in the UK

THE STUDY OF HIGH T_C-SUPERCONDUCTING MATERIALS BY ELECTRON MICROSCOPY AND ELECTRON DIFFRACTION

S. Amelinckx, G. Van Tendeloo and J. Van Landuyt

University of Antwerp, RUCA, Groenenborgerlaan 171, B-2020 Antwerp,Belgium

SUMMARY

A survey is given of the application of different electron microscopic techniques to the study of structural features of high T_C-superconducting materials. Emphasis is laid in this contribution on those structural aspects for the study of which electron microscopy has been essential or has contributed to a significant extent.

1. INTRODUCTION

Electron diffraction and electron microscopy have become increasingly important as complementary tools to X-ray and neutron diffraction in the study of structures and microstructures of complicated compounds. This has been particularly striking in the case of the high T_C superconductors where these techniques have recently produced significant results. It is the purpose of this paper to review some of these contributions from the viewpoint of the potentialities of the technique, illustrating which type of information one can expect from such studies. In a general way one can state that electron microscopy is superior to most other techniques when "local" rather than "averaged" information is required.

2. IMAGING OF THE CRYSTAL STRUCTURES

The high T_C superconducting materials are very suitable objects for high resolution imaging since in these structures the average separation of heavy atom columns along the most relevant crystal zones is larger than the resolving power of present day microscopes (~1.7 Å). Heavy atom columns can thus be imaged as well separated dots, dark or bright, depending on the exact imaging conditions. In the thinnest part of the specimen and close to Scherzer focus the atomic columns are usually marked by dark dots. Oxygen atom columns are mostly not imaged under these conditions, although they clearly influence the image.

As an example the crystal structure of $Tl_2Ba_2Ca_3Cu_4O_{12}$ (n=3) [1] is imaged along the [100] zone in fig. 1; the inset shows the corresponding diffraction pattern. Superposed on the image is a model of the crystal structure using symbols for the atoms defined in fig. 2 The identification of the atom columns, imaged as dark dots, was made using simulated images of the average structure determined by X-ray diffraction.

The image of fig. 1 also shows a planar defect, marked by an arrow : only a single TlO layer is present whereas the other TlO lamellae all contain two layers. Such defects are discussed further below (§ 4.3).

0953-2048/91/01S019 + 16 $03.50 © 1991 IOP Publishing Ltd

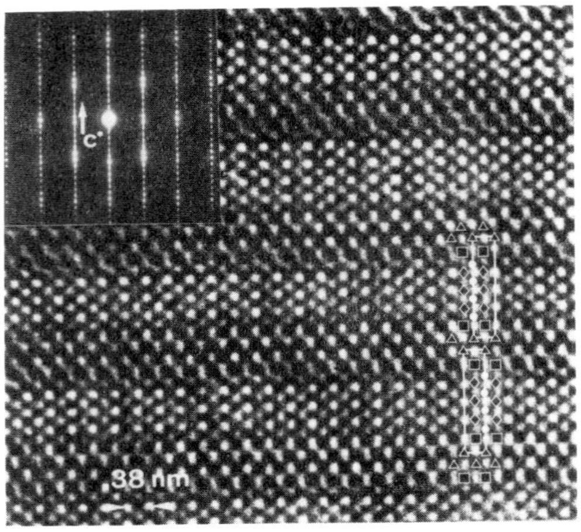

Figure 1 Structure image of $Tl_2Ba_2Ca_3Cu_4O_{12}$ (n=3) along the [100] zone. The corresponding diffraction pattern is shown as an inset. A structure model is superposed on the image. A planar defect consisting of a single TlO layer is indicated by an arrow.

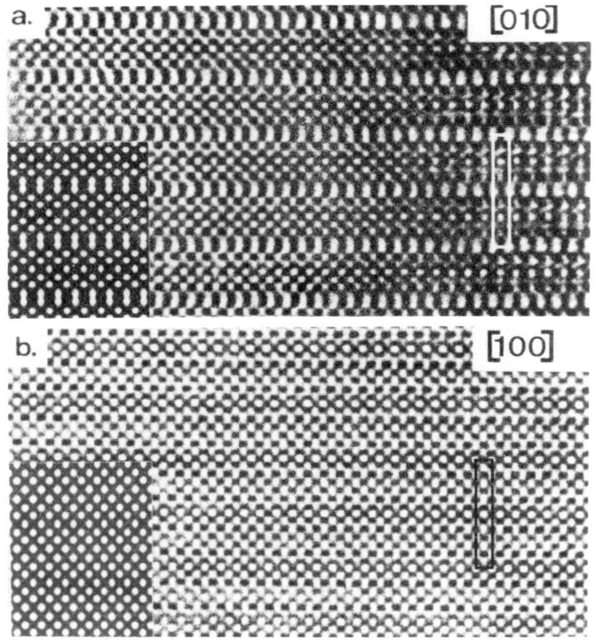

Figure 3 Structure image of the 1-2-4 structure
(a) along the [010] zone
(b) along the [100] zone
The insets show simulated images.

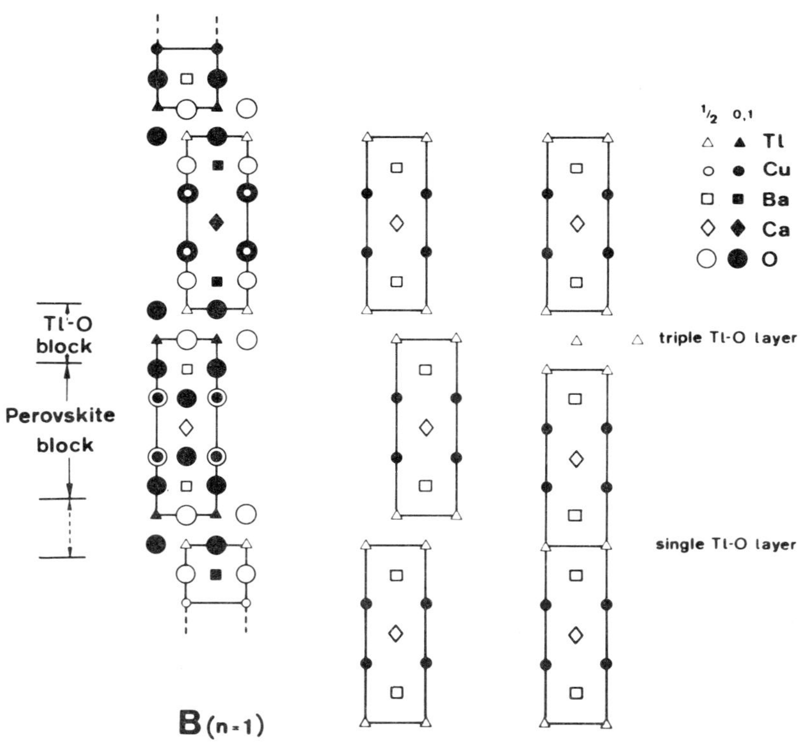

Figure 2 Schematic representation of the structure of $Tl_2Ba_2CaCu_2O_8$ (n=1). Successive perovskite blocks show an offset at double TlO layers; there is no offset at a single TlO layer.

The image of fig. 3 represents the 1-2-4 structure of $YBa_2Cu_4O_8$, a more recently discovered variant [2] of the well known 1-2-3 structure; in (a) along the [010] zone i.e. along the Cu-O-Cu-O chains, and in (b) along the [100] zone.
This 1-2-4 structure results if in the 1-2-3 structure each of the single CuO layers is replaced by a double layer $(CuO)_2$ as represented schematically in fig. 4. The insets in fig. 3 are simulated images obtained from the models represented in fig. 4(b) and (c). Note the prominent rows of elongated bright dots which mark the $(CuO)_2$ double layers.

The three most relevant zones of the structure of $Pb_2Sr_2(Ca,Y)Cu_3O_{8+\delta}$ [3] are reproduced in fig. 5 and compared with simulated images, based on the model of fig. 6, given as insets. These images can easily be interpreted in terms of the structure represented in fig. 6 The perovskite block $SrO-CuO_2-(Ca,Y)-CuO_2-SrO$ can clearly be distinguished from the PbO-Cu-PbO triplets, which are visible as dark bands with a bright dot row in the centre in fig. 5(c).

3. SMALL DEVIATIONS FROM THE AVERAGE STRUCTURE

3.1. General considerations

The structures as determined by X-ray or neutron diffraction are in a number of cases only "average" structures since the intensities of weak reflections, for instance due to displacive modulations with small amplitudes or due to ordering of light atoms, remain below the detection threshold or have intentionally been ignored in the past.
Since the interaction of electrons with matter is stronger than that of X-rays or neutrons, reflections which have very small structure factors as compared to those of the basic Bragg reflections can still be recorded. Moreover the diffraction is dynamical rather than kinematical which unfortunately restricts its application in the field of structure determination. However the atomic resolution images may reveal such small deviations from the average structure and suggest concrete geometrical models.

The size of the coherently scattering areas required to produce a single crystal diffraction pattern is much smaller in electron diffraction than in the other diffraction techniques. Electron diffraction and electron microscopy can thus provide "local" information as opposed to "averaged" information. Diffraction contrast images made in single reflections allow to determine the crystal area responsible for a particular reflection.

E.M. and E.D. are complementary to the other diffraction techniques; on the one hand they need the basic structural information provided by the other techniques for the interpretation of high resolution images but on the other hand with this knowledge they can go far beyond and detect subtle and localized deviations from the average structure. Electron diffraction is in the field of structure determination still in the age of "trial and error" as opposed to the direct methods used in X-ray diffraction but it is far superior for the study of long period structural variants and of defects. We shall now illustrate these considerations by means of examples taken from the study of high T_C superconductors.

3.2. Displacive modulation in the layered superconductors

The displacively modulated structures in the series of compounds $Bi_2Sr_2Ca_nCu_{n+1}O_{2n+6}$ [4] were first detected by means of E.M. and E.D. [5][6]. The relevant section of reciprocal space is shown in fig. 7(a) for the 2201 compound and in fig. 7(b) for the 2212 compound. The Bragg spots

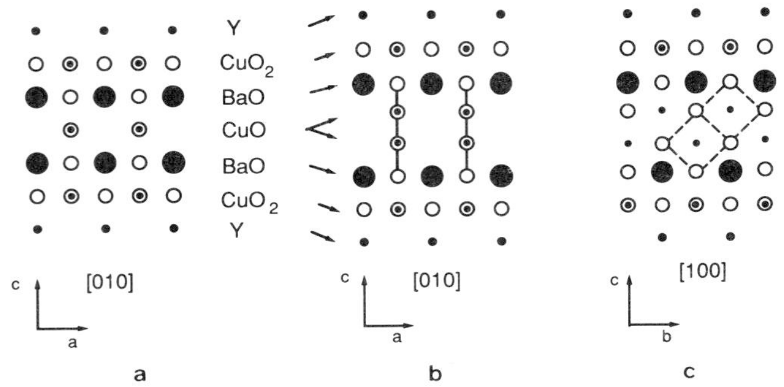

Figure 4 Schematic representation of the formation of the 1-2-4 structure from the 1-2-3 structure.
(a) 1-2-3 structure along [010]
(b) insertion of an additional CuO layer viewed along [010]
(c) same as (b) viewed along [100].

Figure 5
Structure images of $Pb_2Sr_2(Ca,Y)Cu_3O_{8+\delta}$ viewed along three different zones, each compared with a simulated image based on the model of fig. 6 : from left to right [001], [110], [100]-zones. Specimen thickness 3 nm; Δf 72, 30, 30 nm.

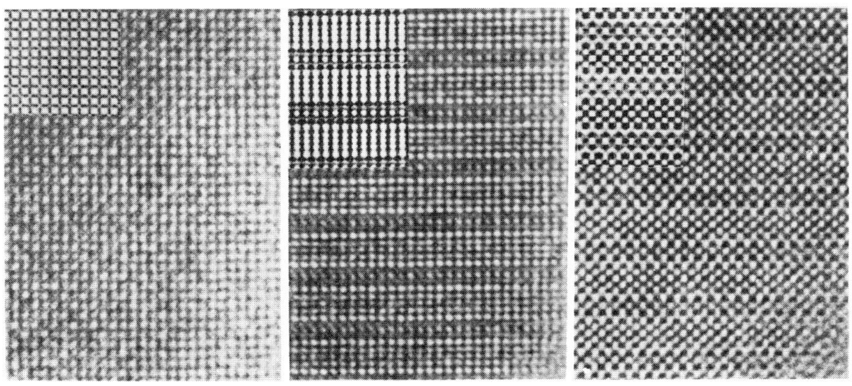

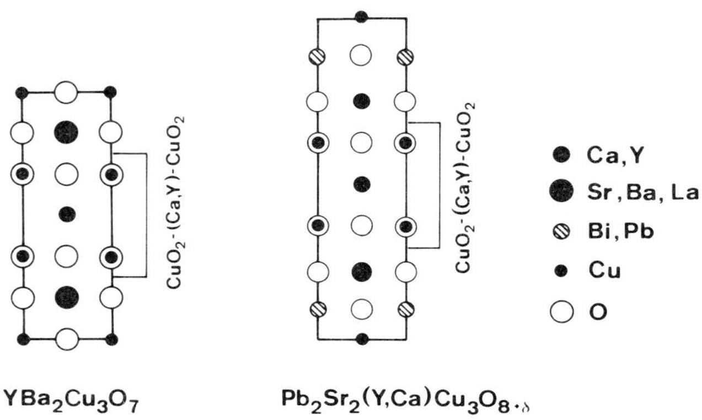

Figure 6
Model of the $Pb_2Sr_2(Y,Ca)Cu_3O_{8+\delta}$ structure in comparison with $YBa_2Cu_3O_{7-\delta}$.

associated with the average structure and which form closely spaced rows parallel with c* have acquired linear sequences of satellite spots enclosing an angle of around 60° with c*. Where the satellite sequences belonging to adjacent rows meet they do not match; the superstructure is in fact incommensurate with the basic structure. The diffraction pattern exhibits the characteristic features of a displacively modulated structure. The modulation in 2201 reduces the tetragonal symmetry of the average structure to a monoclinic one. The modulation pattern can be visualized directly in the image corresponding with this diffraction pattern (see e.g. fig. 6 in ref. [7]); it then becomes evident that the modulation amplitude is largest for the $(BiO)_2$ sandwich, but all lattice planes are more or less "wavy".

In the compounds 2 2 n n+1 with $n \geq 1$ the modulation pattern has orthorhombic symmetry. Fig. 7(b) shows the satellite pattern along the same zone as fig. 7(a) for the compound 2212; the (001)* plane is now a mirror plane for the pattern, which was not the case in the pattern of fig. 7(a). The image taken along this zone is reproduced in fig. 8; the inset is a simulated image using a model described in ref. [7].
The "wavy" nature of the lattice planes is essentially attributed to the mismatch between the sodiumchloride like $(BiO)_2$ lamellae and the perovskite lamellae occuring in these mixed layer compounds [7]. The stresses due to this mismatch can be relieved by the formation of regions of excess oxygen in the $(BiO)_2$ lamellae, which can formally be described as continuous distributions of dislocations in the contact plane between the two types of lamellae. The modulation pattern then results from the equilibrium configuration of these dislocation distributions.

Also the compounds in the $Tl_2Ba_2Ca_nCu_{n+1}O_{2n+6}$ family have displacively modulated structures [8]. Whereas in the Bi-family the modulation could be described as a 1-q state, the modulation in the Tl-compounds is a 2-q state i.e. the modulation takes place along two mutually perpendicular directions in the layer planes. The satellite sequences disappear after a few minutes of irradiation with the image-forming electrons (200 kV) suggesting that the modulation is related to the oxygen sublattice since only oxygen atoms in the structure can be displaced by collisions with 200 kV electrons. The calcium free Tl-compound shows a more pronounced and also more complicated modulation, reducing the symmetry to triclinic [9].

In a general way the strength of the modulation, as measured by the relative intensity of the satellite spots, seems to decrease with increasing n in both the Tl- and the Bi-compounds.
The compound $Pb_2Sr_{2-x}La_xCu_2O_{6+\delta}$ acquires a displacively modulated structure when loaded with oxygen ($\delta \geq 1$) by annealing at 400°C in an oxygen atmosphere [10]. The configuration of satellites is similar to that observed in the Bi-compounds with $n \geq 1$ fig. 7(b). From the HREM image follows that the modulation along b is mainly localised in the $PbO-CuO_\delta-PbO$ triplet lamellae. Annealing in argon allows to remove the modulation. Presumably the exchangeable oxygen is accomodated in the CuO_δ layers (fig. 6).

3.3. Commensurate superstructures

3.3.1. Vacancy ordered superstructures in $YBa_2Cu_3O_{7-\delta}$

The presence of vacancy ordered superstructures in oxygen deficient $YBa_2Cu_3O_{7-\delta}$ was first discovered by means of ED and HREM [12]. After mild "in-situ" heat treatment of the specimen by means of the electron beam in the microscope vacuum, the [001] zone diffraction pattern exhibits weak

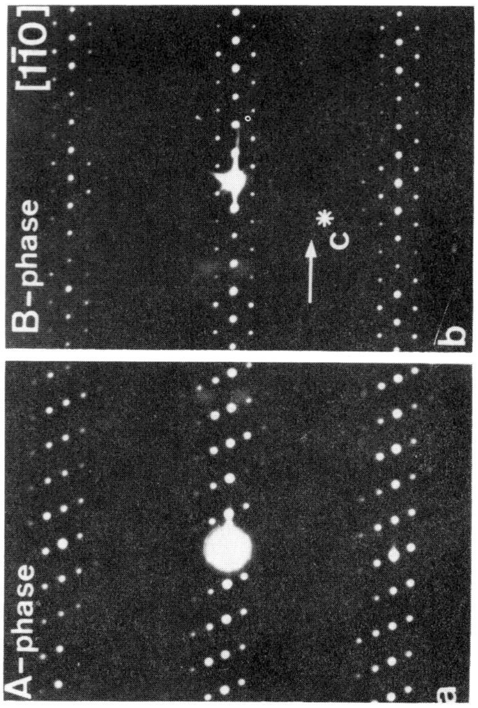

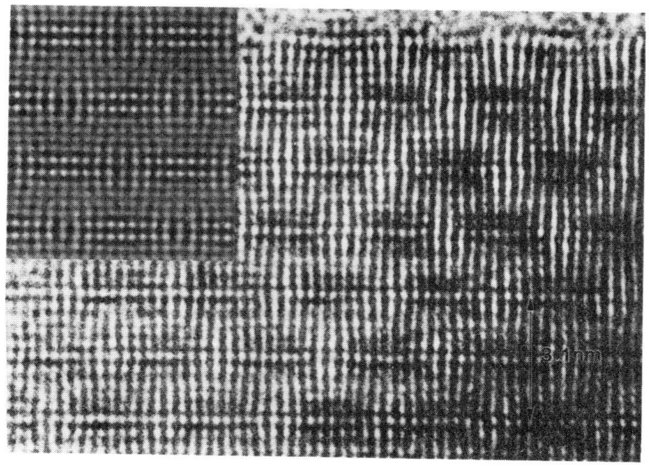

Fig. 8 Image of the modulated structure in the Bi-compound with n=1.The inset is a simulate image; after ref.[7].

Figure 7 Section of reciprocal space containing the satellite sequences due to the displacive modulations in the compounds $Bi_2Sr_2Ca_nCu_{n+2}O_{2n+6}$.
(a) in 2201; (b) in 2212
In 2201 the tetragonal symmetry is reduced to monoclinic; in 2212 it becomes orthorhombic.

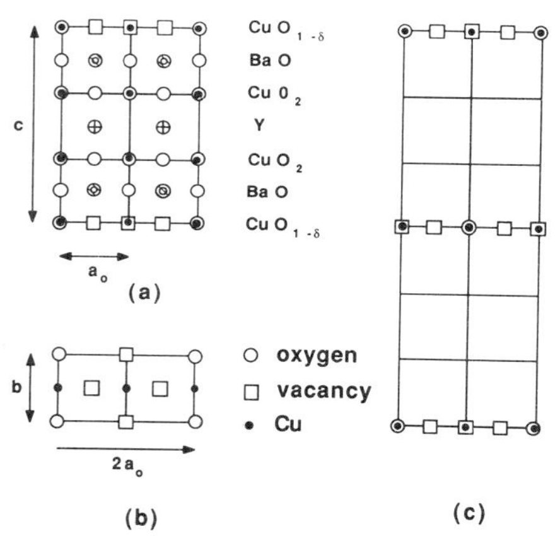

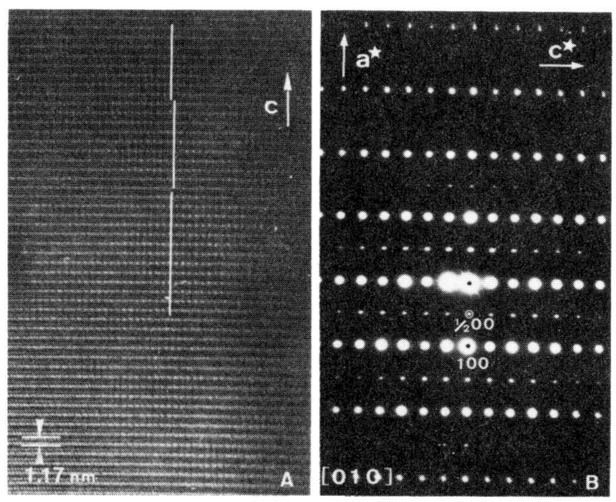

Figure 9 The $2a_0$ (orthorhombic II) superstructure in $YBa_2Cu_3O_{6.5}$.
(a) Arrangement in space : vertical stacking of $CuO_{1-\delta}$ layers.
(b) Arrangement of filled and empty Cu-O-Cu-O chains in the (001) plane.
(c) Alternative stacking of $CuO_{1-\delta}$ layers.

Figure 10 (a) High resolution image of the $2a_0$ structure in $YBa_2Cu_3O_{6.5}$ along the[010] zone. Note the presence of 1/2[100] antiphase boundaries.
(b) Corresponding diffraction pattern along [010] zone.

and often diffuse extra spots at positions 1/2 0 0 and 0 1/2 0, suggesting a doubling of the a_0 and/or b_0 lattice parameters in one or two directions. The high resolution image along [001] shows that only the a_0 parameter is doubled, the unit mesh in the (001)* plane being $2a_0 x\ b_0$; it proves that the DP is usually the superposition of the DP's due to two orientation variants of this superstructure. Since the described heat treatment is known to lead to oxygen loss it is clear that the period doubling is associated with the oxygen sublattice. Neutron diffraction [13] has shown that the oxygen positions in the CuO layers are only partly occupied and exhibit a large thermal vibration ellipsoid; they are the most loosely bound oxygen atoms. The superstructure must therefore be localized in the CuO layers and fig. 9(b) shows the arrangement of the oxygen vacancies in this layer. It can be described with reference to the $YBa_2Cu_3O_{7-\delta}$ structure noting that one out of two Cu-O-Cu-O chains is vacant the ideal composition then corresponding with $\delta = 1/2$.
The diffraction patterns along the [100] or [010] zone reveal continuous streaks along the rows n+1/2, o, ℓ or o, n+1/2, ℓ showing the lack of correlation between the arrangements in successive CuO layers, at least in specimens obtained "in-situ" which implies rapid cooling of the specimen.

The structure could be confirmed on specimens made under much better defined conditions, using the "constant-stoichiometry cooling" method [14]. The material is now slowly cooled from the equilibration temperature keeping the composition (i.e. the weight) constant by continuously adapting the oxygen partial pressure on the specimen. In such material the oxygen superstructure is much better ordered and single variants with a well defined spatial arrangement could now be obtained [15]. The resulting model is shown in fig. 9(a) whereas fig. 10 shows the high resolution image along the corresponding zone [14]. A part from the $2a_0$ structure (also called orthorhombic II) a number of other superstructures na_0 (n=integer) based on the same building principle (i.e. an alternation of parallel vacant and occupied chains have since been found by several authors in specimens prepared under slow cooling conditions [14][15]. The most frequently observed one is the $3a_0$ structure of which the diffraction pattern is shown in fig. 11(b). The two possible structures consistent with this pattern and corresponding with δ=1/3 and δ=2/3 apparently occur both [14].

More complicated "uniform" sequences of filled and empty chains occur for intermediate compositions. The theory of de Fontaine et al., [16] based on an ANNNI-model accounts for the occurence of this variety of one-dimensional superstructures.

3.3.2. Commensurate deformation modulated superstructure in $YBa_2Cu_3O_{6+\varepsilon}$ [17][18]

In specimens of Y-Ba-Cu-O heat treated in vacuum (partial pressure of oxygen 10^{-6} Tor) at progressively higher temperatures one always obtains as the last phase (~450°C) before the specimen is irreversibly decomposed, a tetragonal phase characterized by the [001] zone diffraction pattern of fig. 12. Under the formation conditions used, the composition must be close to O_6 i.e. the CuO layers must be nearly oxygen free and the material is no longer superconducting. The diffraction pattern along zones normal to the c-direction reveal that continuous streaks with reinforcements at positions ±3,5 c_0* cause the observed sharp diffraction spots in the [001] zone. Streaks are absent for $ho\ell$ (h = odd) and $ok\ell$ (for k = odd). The corresponding spots are nevertheless present in the [001] zone pattern as a result of double diffraction. The systematic extinctions allow to determine the possible space groups as $P42_12$, P4bm and $P\bar{4}b2$. The

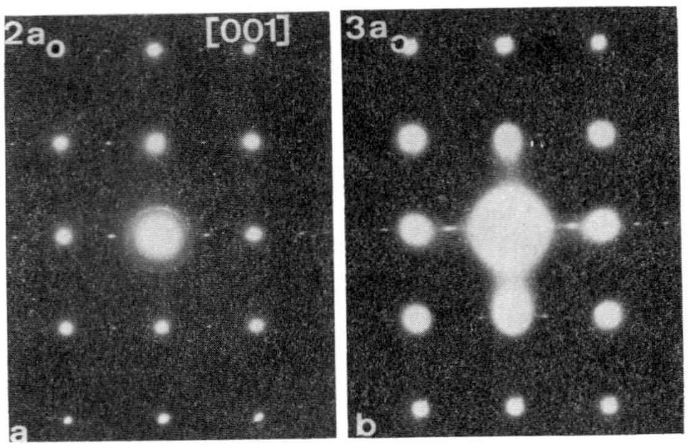

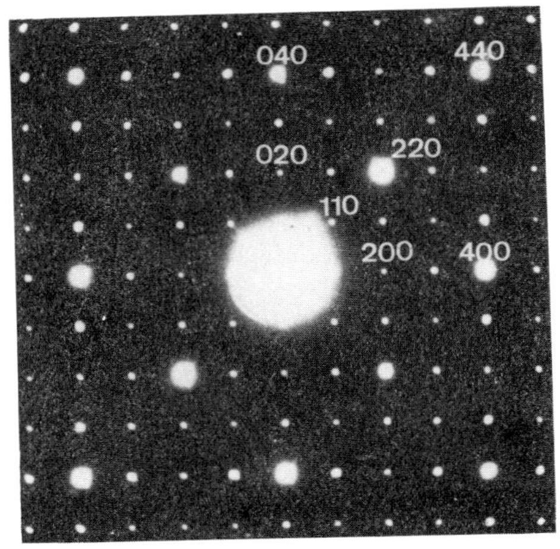

Figure 11 Diffraction patterns of two vacancy ordered superstructures in $YBa_2Cu_3O_{7-\delta}$.
(a) Two variants of the $2a_0$ structure
(b) One variant of the $3a_0$ structure

Figure 12 Diffraction pattern of a deformation modulated superstructure $YBa_2Cu_3O_{6+\varepsilon}$.

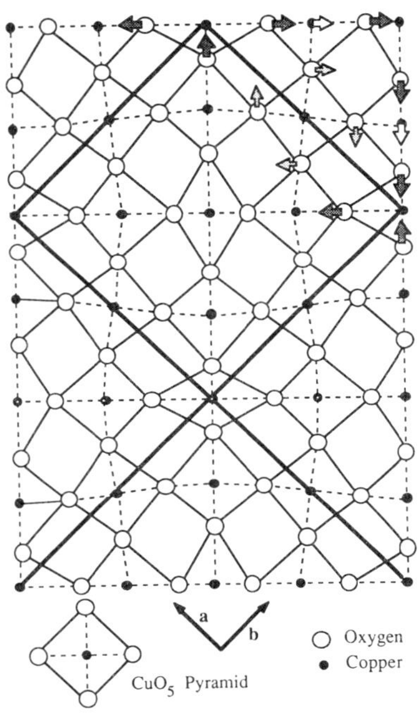

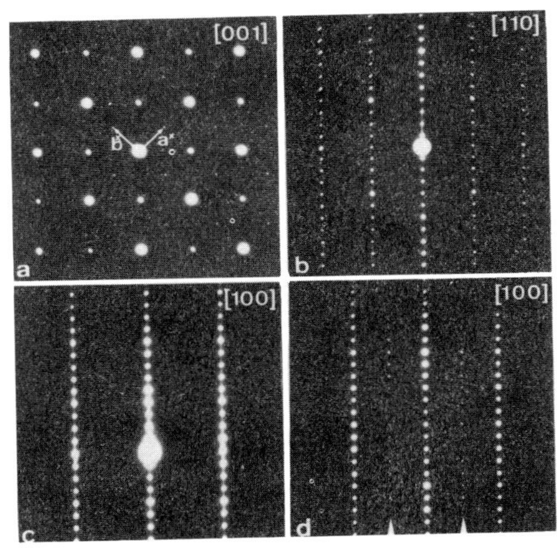

Figure 13 Proposed model for the deformation modulated superstructure of $YBa_2Cu_3O_{6+\varepsilon}$ of which fig. 12 is the diffraction pattern. The deformation is strongly exagerated.

Figure 14 Different sections of reciprocal space of $Pb_2Sr_2Ca_{0.5}Y_{0.5}Cu_3O_{8+\delta}$ exhibiting weak extra reflections violating the extinction criteria of the spacegroup Cmmm ((a) and (d)).
(a) [001], (b) [110], (c) [100] without extra reflections, (d) [100] with extra reflections.

reinforcements in the streaks suggest that the correlation among successive layers causing the superstructure is weak but that nevertheless a non-vanishing correlation exists between layers separated by (1/3.5) c_0 in direct space. This spacing corresponds with the separation between the two CuO_2 layers on either side of the Y-layer within the same unit cell. The described formation treatment further suggests that the CuO layers must be nearly oxygen free. It was shown in ref. [17] that the observed spacegroup and the assumed composition were incompatible with a vacancy ordered superstructure situated in the CuO layers.

The general features of the diffraction pattern and of the HREM image further suggested a deformation modulated structure as more likely. The observed, albeit weak correlation lead us to assume that the CuO_2 layers should be responsible for the superstructure spots. For different models, based on these ideas the theoretical diffraction pattern was compared with the observed one. From this it was concluded that a model in which the CuO layers are oxygen free and the CuO_2 layers are deformed in the way shown in fig. 13 leads to the best fit. The displacements in this figure have been strongly exagerated.

Such a deformed structure might for instance be due to the cooperative Jahn-Teller effect associated with copper, resulting in a deformation of the network of corner-sharing CuO_5 pyramids. It is well established that copper ions may cause pronounced Jahn-Teller effects.

3.3.3. Superstructure in $Pb_2Sr_2(Ca,Y)Cu_3O_{8+\delta}$

As an example of how electron diffraction can detect subtle deviations from an average structure we discuss the compound $Pb_2Sr_2(Ca,Y)Cu_3O_{8+\delta}$ which was first reported to be superconducting by Cava et al. [19]. These authors referred the structure to a C-centered orthorhombic, pseudo-tetragonal unit cell (a=0,5375; b=0.5407, c=1.5713) based on the systematic extinctions found by X-ray diffraction (spacegroup Cmmm). However electron diffraction patterns almost invariably exhibit weak reflections which violate the extinction criteria for C-centering [20]. Along the (001)*-section reflections are observed at the 100 and 010 positions; their intensity is variable from specimen to specimen and depends on the preparation method (fig. 14a). Along the [100] zone rows of weak $0k\ell$ spots with k=odd or $h0\ell$ spots with h=odd are observed. The image made along a [100] (or [010]) zone exhibiting "forbidden" reflection is shown in fig. 15. Rows of extra bright dots with double spacing of 0,54 nm (indicated by arrows) are visible along the PbO layers whereas all other layers show the 0,27nm dot spacing. Moreover the sequences of extra bright dots are interleaved in the two PbO layers of the PbO-Cu-PbO triplet, suggesting the presence of a (001) glide mirror plane at the level of the Cu-layer.

These observations suggest that the weak reflections should be attributed to a superperiod in the PbO layers. In ref. [20] it was suggested that due to the presence of a lone pair on the lead atoms the square symmetry is perturbed in the PbO layers. Image simulations support this interpretation.

4. STACKING DEFECTS

4.1. General considerations

EM and in particular HREM is especially powerful in revealing singular layers or lamellae in an otherwise periodic structure. The occurrence of

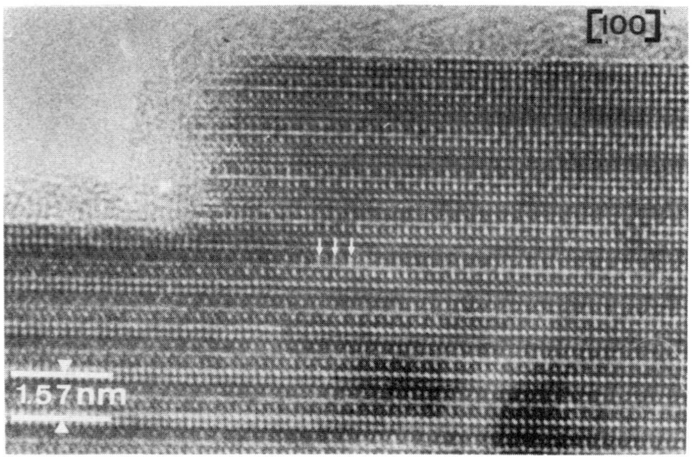

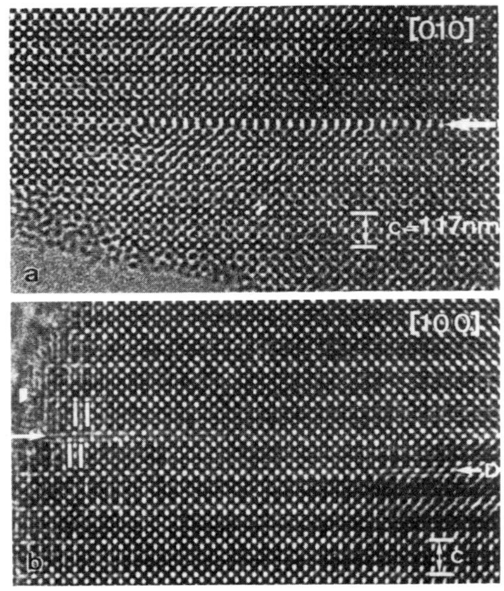

Figure 15 High resolution image along [100] of a specimen of $Pb_2Sr_2Ca_{0.5}Y_{0.5}Cu_3O_{8+\delta}$ exhibiting weak additional reflections such as in fig. 14d. Note the spacing doubling (0,54nm) along the PbO-Cu-PbO triplet.

Figure 16 Planar defect along (001) in 1-2-3 consisting of a double CuO layer. (a) [010], (b) [100].
This image can be compared with fig. 4(b) and (c)

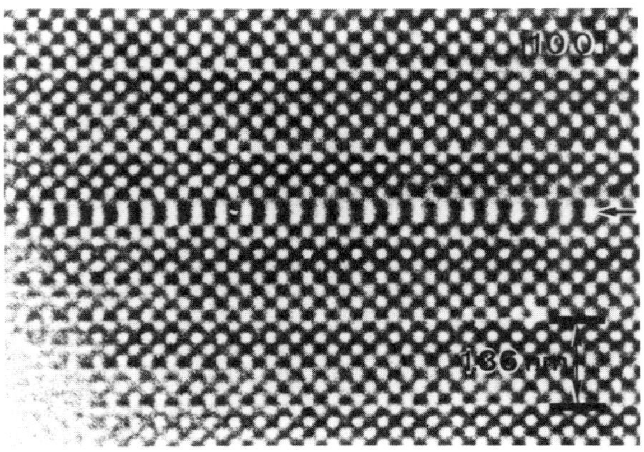

Figure 17 Single layer twin in the 1-2-4 compound. In one layer of unit cells the chains are observed end on; the rest of the crystal is viewed along [100].

Figure 20 90° Rotation twin in the 1-2-4 compound; the contact plane is (001) and the rotation axis [001].

such defects may suggest new structures in which such defects occur in a periodic fashion. The structure and composition of these hypothetical compounds can be guessed with reasonable confidence from the HREM image, possibly be confirmed by image simulation and eventually prepared in pure form. This course of events has for instance been followed in the preparation of the compound $YBa_2Cu_4O_8$ derived from $YBa_2Cu_3O_7$ and it also lead to the preparation of the higher members of the Bi- and Tl-families of compounds.

4.2. The Y-Ba-Cu-O system

The simplest example is the occurence of a double CuO layer instead of a single one in $YBa_2Cu_3O_{7-\delta}$ [21]. Such a defect is shown in fig. 16 as viewed along the b_0 zone, i.e. along the chain direction (fig. 16(a)) and along the a_0 direction (fig. 16(b)); these images can be compared with the models of fig. 4. The elongated bright dots correspond with the vacancy channels along b between ribbons of square planar CuO_4 arrangements. This could be concluded from image simulations.

It turned out to be possible to prepare crystals in which systematically all CuO layers are double, leading to an ideal composition $YBa_2Cu_4O_8$ [22]. The structure images due to this structure have been discussed above.

It also has been possible to produce crystals in which single and double CuO layers alternate regularly corresponding with an ideal composition $Y_2Ba_4Cu_7O_{15}$ [23] or $1\text{-}2\text{-}3^{1/2}$.

In crystals of $YBa_2Cu_4O_8$ planar defects occur in which a single layer of unit cells differs 90° in orientation with that in the rest of the crystal [24]. In the defect layer the rows of CuO-Cu-O chains are thus perpendicular to the chains in the adjacent crystal parts. Such a defect can easily be recognized in a high resolution image (fig. 17). This type of defect could in principle also occur in 1-2-3 material, but it has not yet been identified perhaps due to the much greater difficulty of revealing it. However the reason why it has more frequently been found in the 1-2-4 compound than in the 1-2-3 compound may also be related to the fact that the orthorhombicity i.e. the deviation from tetragonal symmetry is larger in the 1-2-3 compound than in the 1-2-4 compounds.

The 1-2-3, $1\text{-}2\text{-}3^{1/2}$ and 1-2-4 structures can be considered as the members of a family of "shear structures" derived from the basic 1-2-3 structure by the insertion of CuO layers along (001) crystallographic shear planes located at the CuO layers of the 1-2-3 structure and with a displacement vector $R = 1/6[031]$ expressed in terms of the 1-2-3 lattice. If crystallographic shear takes place at every CuO layer the 1-2-4 structure results; if shear takes place every other CuO layer one obtains the $1\text{-}2\text{-}3^{1/2}$ structure. The intensity distribution in the diffraction pattern of these structures can satisfactory be accounted for in terms of this model. It is to be expected that more complicated sequences of single and double CuO layers can occur for intermediate compositions. Indicating the numbers of 1-2-3 unit cells in the successive blocks the stacking symbol for the 1-2-4 structure would read : $1\ \bar{1}$ ...; it would be $2\ \bar{2}$ for the $1\text{-}2\text{-}3^{1/2}$ structure. The minus sign indicates that the corresponding block of 1-2-3 cells is displaced along the a direction over $^{1/2}a$ with respect to the reference block. Sequences such as $2\ \bar{1}\ 2\ \bar{1}$ ... or $2\ \bar{2}\ 2\ \bar{1}$ ... are plausible, but would require carefully controlled preparations.

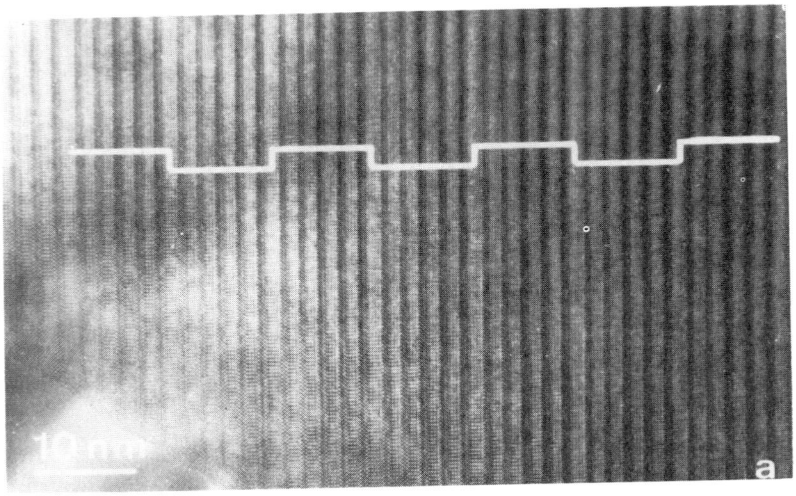

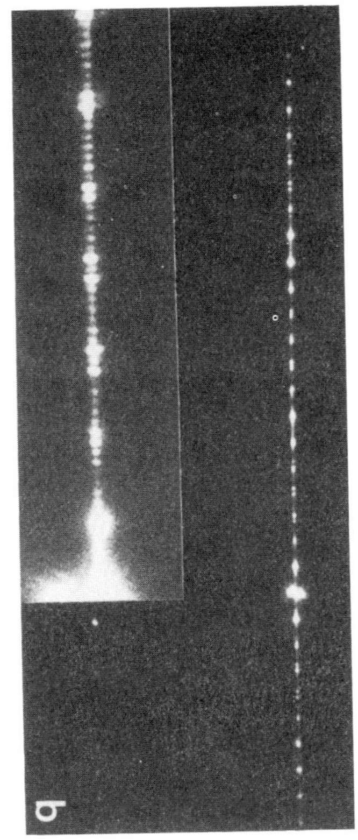

Figure 18 (a) Low magnification, high resolution image of $Tl_2Ba_2Ca_nCu_{n+1}O_{2n+6}$ exhibiting a stacking period of 21 nm.
(b) Diffraction pattern corresponding with this stacking. Note the sequence of well resolved sharp spots.

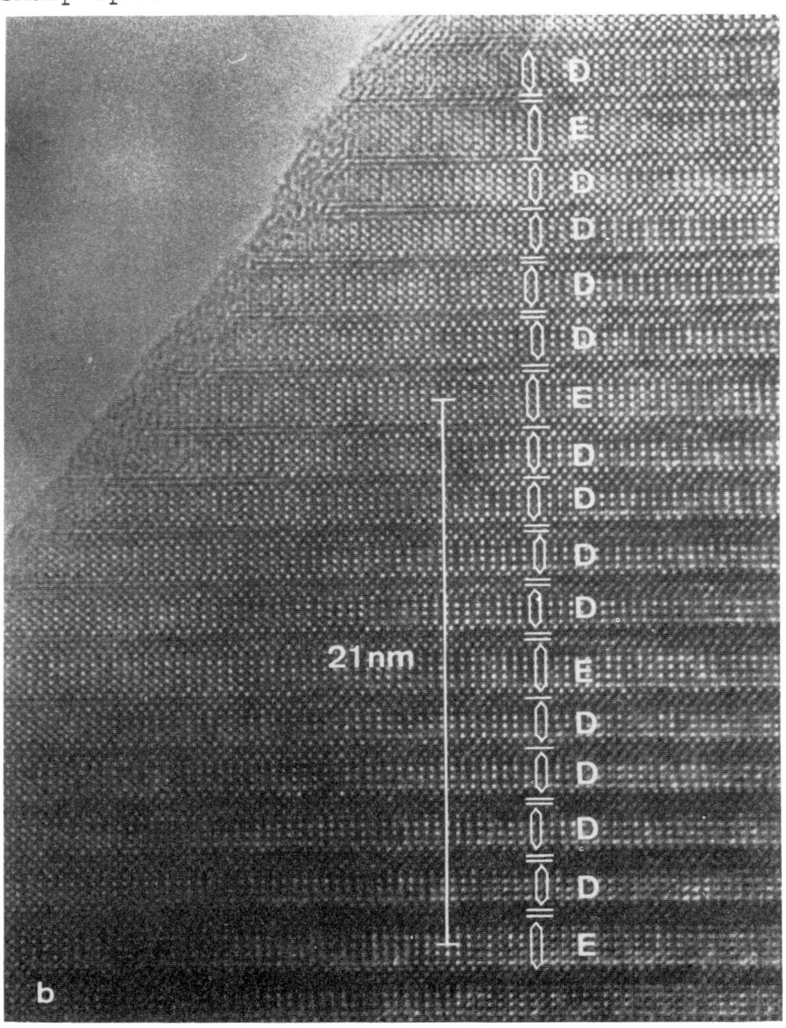

Figure 19
High resolution high magnification image corresponding with the low magnification image of fig. 18. The sequence can be deduced directly from the image. Two different block sizes occur D=2234; E=2245. Also single as well as double TlO layers occur; they are indicated by single and double horizontal bars.

4.3. Sequential defects in the layered superconductors

It was first shown by means of E.M. that in the bismuth as well as in the thallium family of compounds layers of unit cells containing different numbers of Cu-layers in the unit cell often occur intimitely intergrown either periodically or at random [25][5]. The fit along the layer planes between successive lamellae is perfect, only the c-parameter is different since it evidently increases with the number of Cu-layers. In the Bi-family the sequence of layers in one unit cell is :

$$BiO - SrO - CuO_2 - (Ca\text{-}CuO_2)_n - SrO - BiO$$

The larger the number of Cu-layers per unit cell i.e. the larger n the more difficult it turns out to prepare the pure phase. Fortunately lead doping leads to the stabilization of the 2223 phase in the Bi-family [26], modifying at the same time the geometry of the modulated structure as compared to that of the lead free compound without significantly affecting T_c however.

The Tl-compounds have average structures which are rather similar to those of the Bi-compounds the layer sequence along the c-direction being :

$$...TlO - BaO - CuO_2\text{-}(Ca\text{-}CuO_2)_n\text{-}BaO\text{-}TlO...$$

However whereas in the Bi-compounds the sodium chloride-like BiO layers always occur as double layers, the homologeous TlO layers in the Tl-compounds may either be double or single [27]. This provides one more degree of freedom for conceivable different stackings.

Particularly complicated periodic sequences consisting of lamellae with different n-values as well as containing single and double TlO layers have been studied by means of HREM [28]. An example is reproduced in fig. 18 at low magnification, showing the magnitude of the repeat distance ~ 210 Å) and the number of "giant" unit cells. The detailed stacking sequence could be deduced from high magnification high resolution images such as fig. 19. The number of Cu-layers can be obtained by direct counting of the atomic layers; the presence of a double or a single TlO layer can be deduced either in the same way or from the presence or the absence of a lateral offset between the images of successive perovskite blocks viewed along the [100] zone. This point is illustrated in fig. 2. The stacking sequence imaged in fig. 19 consists of a regular alternation of four n = 3 lamellae and one n=4 lamellae. Between the n=4 lamellae and the adjacent two n=3 lamella there is no lateral offset i.e. there is a single Tl-O layers whereas the other TlO lamellae are double. This is indicated in fig. 19 by single and double bars parallel with the layer planes.

5. TWINS

5.1. Growth twins

5.1.1. The Y-Ba-Cu-O system

The possibility for the Cu-O-Cu-O chains along b_0 to adopt two mutually perpendicular orientations within the same perovskite matrix leads in the 1-2-4 compound to the formation of growth twins with a (001) contact plane the two components of the twin being related by a rotation over 90° about the [001] axis. Such a twin can easily be recognized in the high resolution image of $YBa_2Cu_4O_8$ in fig. 20 [24].

In $YBa_2Cu_3O_{7-\delta}$ domains having mutually perpendicular c-axis have been observed occasionally. They may possibly be the result of the ordering of the cations Ba and Y, which determine the direction of the c-axis, along different cube directions in different parts of a common matrix of copper oxide with cubic symmetry. The interfaces are usually cube planes of the underlying perovskite lattice [12].

5.1.2. Modulation twins in $Bi_2Sr_2Ca_nCu_{n+1}O_{2n+6}$ [7][8][29][30]

The average structure of these compounds has tetragonal symmetry but the displacive modulation of the structure reduces this symmetry to monoclinic in the compound with n=o and to orthorhombic in the n=1,2,... compounds.

The modulated structure of the n=o compound can thus be formed in four different modulation variants, based on the same average structure, they are related by symmetry elements present in the average structure but absent in the modulated structure. Such symmetry elements are the mirror plane parallel with (001) and the rotation over 90° about the c-axis. Two essentially different twin interfaces, corresponding with these symmetry operations are predicted and have indeed been observed; for both the contact planes is (001) [29].

In the n=1,2 compounds the (001) plane is a mirror plane for the pointgroup of the modulated structure, and hence only one type of twin is possible : the 90° rotation twin with [001] as a rotation axis and (001) as a contact plane [7][30].

The twins discussed here form during the growth process as a result of a stacking mistake at the level of the $(BiO)_2$ layers, as was deduced from the localization of the contact plane of the twin in high resolution images. Unlike the transformation twins in $YBa_2Cu_3O_{7-\delta}$ their formation can be avoided be carefully growing the crystals.

5.2. Transformation twins in 1-2-3 compounds

"Single" crystals of 1-2-3 compounds are almost invariably fragmented by nearly coherent reflection twins on (110) and $(1\bar{1}0)$ the chain directions b_0 being roughly mutually perpendicular in the two components of the twin. The geometry of these twins is very well documented as a result of the extensive literature on the subject, mainly based on electron microscopic observations. In the very first papers [11][12] on 1-2-3 it was already pointed out that the twins resulted from a disorder-order transformation on the oxygen sublattice in the CuO layers of the structure. As a result of the oxygen ordering in chains along b_0 the tetragonal pointgroup of the high temperature disordered structure (4/m 2/m 2/m) is reduced to the orthorhombic pointgroup (mmm) of the low temperature, ordered 1-2-3 structure, which is a subgroup of the former. The resulting structural variants are then related by the symmetry elements lost in the transition i.e. either by the (110) and $(1\bar{1}0$ symmetry planes or by the 90° rotation of the tetragonal pointgroup. The strain free interfaces between variants in this paraelastic-ferro-elastic type of transition are entirely determined by symmetry; they are also the (110) and $(1\bar{1}0)$ planes as observed.

"In-situ" studies in the electron microscope have confirmed that in the initial stage the twin formation is due to ordering of the oxygen atoms but that the final coarsening of the twin domains occurs by the propagation of twinning dislocations along the interfaces, accompanied by local reshuffling of the oxygen atoms in the core of the twinning dislocation (so-called "synchro-shear") [31].
The application of an uniaxial compressive stress along a_0 (or b_0) produces the required shear stress on the twinning dislocations causing their propagation along twin interfaces in such a sense that the domains oriented with their a_0-axis along the stress axis increase in width at the expense

of the b_0-oriented domains. This mechanism accounts for the possibility of "detwinning" crystals subjected to uniaxial compression at high temperature under an oxygen partial pressure sufficient to maintain the composition constant [32].

Thanks are due to a number of colleagues for the use of photographs from joint papers : M. Verwerft, J. Reyes-Gasga, T. Krekels, H.W. Zandbergen, Y.K. Huang, M.J.V. Menken, J.N. Li, K. Kadowaki, A.A. Menovsky, P. Groen, L.T. De Jongh, W.T. Fu, J.M. Van Uitenbeek, W.H.M. Bruggink, H. Verweij, T. Okabe.

REFERENCES

In view of space limitation an exhaustive list of references is not attempted. A more detailed survey as well as a more extensive list of references is given in S. Amelinckx et al. "Physics and Materials of Science of high temperature superconductors" ed. by R. Kossowsky, S. Methfessel and D. Wohlleben, p. 189, Nato Asi Series E, vol. 181, Kluwer Academic Publishers, 1990 Dordrecht, Boston, London.

1. Cava R.J., Science 247, 656 (1990).
   Hewat A.W. et al. : Physica C 152, 438 (1988).
   Beyers R. et al. : Appl. Phys. Lett. 53, 432 (1988).
   Subramanian M.A. et al. : Nature 332, 420 (1988); Science 240, 631 (1988); Phys. Rev. B 38, 225 (1988).
   Bourgault D. et al., Physica C 158, 555 (1989).
2. Marsh P. et al., Nature 334, 142 (1988).
   Yamaguchi K. et al., Jpn. Journ. of Appl. Physics 28, (11) L1942 (1989).
3. Cava R.J. et al, Nature 336, 211 (1988).
   Zandbergen H.W. et al. Physica C 158, 155 (1989); Physica C, 159, 81 (1989).
4. Matsui Y. et al., Jpn. J. Appl. Phys. 27, 3, L372 (1988).
   Maeda H. et al., Physica C, 153-155, 602 (1988).
   Von Schnering H.G. et al., Angew. Chem. Int. 27, 574 (1988).
   Michel C. et al., Z. Phys. B 68, 421 (1987).
   Tarascon J.M. et al., Phys. Rev. B 37, 9382 (1988).
   Torrance J.B. et al., Sol. St. Comm. 66, 703 (1988).
5. Zandbergen H.W. et al., Sol. St. Comm. 66, 4, 397 (1988).
6. Van Tendeloo G. et al., Sol. St. Comm. 66, 9, 927 (1988).
7. Zandbergen H.W. et al., Physica C 156, 325 (1988).
8. Zandbergen H.W. et al., Appl. Phys. A 46, 233 (1988).
9. Hewat E.A. et al., Physica C 156, 369 (1988).
10. Zandbergen H.W. et al., Sol. St. Comm. 72, 445 (1989).
11. Van Tendeloo G. and Amelinckx S., Phys. stat. sol. (a) 103, K1 (1987).
    Van Tendeloo G. et al., Solid St. Comm. 63, 7, 603 (1987).
12. Zandbergen H.W. et al., Phys. stat. sol. (a) 103, 45 (1987).
13. Beno M.A. et al., Appl. Phys. Lett. 51, 57 (1987).
    Jorgensen J.D. et al., Physica C, 153, 578 (1988).
    Schuller I.K. et al., Sol. St. Comm. 63, 385 (1987).
14. Reyes-Gasga J. et al., Physica C 159, 831 (1989); Sol. St. Comm. 70, 269 (1989).
15. Werder D.J. et al., Phys. Rev. B37, 2317 (1988); Phys. Rev. B38, 5130 (1988).
16. de Fontaine D. et al., Phys. Rev. B36, 5709 (1987); Phys. Rev. Letters 60, 1065 (1988); Physica C 153-155, 598 (1988).
17. Krekels T. et al., Physica C 167, 677 (1990).

18. Alario-Franco M. et al., Physica C 156, 455 (1988).
19. Cava R.J. et al., Nature 336 211 (1988).
20. Zandbergen H.W. et al. Physica C 158, 155 (1989).
21. Zandbergen H.W. et al. Nature 331, 596 (1988); Hervieu M. et al., Europhys. Lett. 4(2) 205 (1987).
22. Fischer P. et al., Sol. St. Comm. 69, 531 (1989).
Kaldis E. et al., Physica C 159, 668 (1989).
23. Hewat A.W. et al., Physica C 167, 579 (1990).
Karpinski J. et al., Physica C 161, 618 (1989).
Karpinski J. et al., Physica C 160, 449 (1989).
24. Krekels T. et al., to be published in Physica C.
25. Zandbergen H.W. et al., Nature 332, 620 (1988).
Ramesh et al., Appl. Phys. Lett. 53(7), 615 (1988).
Matsui Y. et al., Jpn. J. Appl. Phys. 27(7), L1241 (1988).
Ganapalki L. et al., Sol. St. Comm. 67, 967 (1988).
Hetherington C.T.D. et al., Appl. Phys. Lett. 53(11), 1016 (1988).
26. Statt B.W. et al., Physica C., 156, 251 (1988).
Yuseng He et al., Phys. Rev. B40, 7384 (1989).
Liu H.K. et al., Physica C 157, 93 (1988).
Chen C. et al., Sol. St. Comm. 68(8), 749 (1988).
27. Morosin et al., Physica C 156, 587 (1988).
Sugise R. et al., Physica C 157, 131 (1988).
28. Iijima et al., Jpn. J. Appl. Phys. 27(6), L1054 (1988).
Verwerft M. et al., Physica C 156, 607 (1988).
Maignan A. et al., Mod. Phys. Lett. B2(5), 681 (1988).
Schilling A. et al., Physica C 157, 144 (1989).
29. Van Tendeloo G. et al., Applied Physics A 46, 153 (1988).
30. Eibl O., Physica C, 168, 215 (1990).
31. Van Tendeloo G. et al., Physica C 167, 627 (1990).
32. Welp U. et al., Physica C 161, 1, (1989).

Supercond. Sci. Technol. **4** (1991) S35–42. Printed in the UK

Anomalous Magnetic Behaviours of the Cuprate Superconductors --------Meissner Effect, H_{c1} and H_{c2}

K. Kitazawa, Y. Tomioka, T. Hasegawa, K. Kishio, M. Naito*, T. Matsushita**, I. Tanaka*** and H. Kojima***

*Department of Industrial Chemistry, University of Tokyo, Hongo, Tokyo 113, *Basic Research Laboratory, Nippon Telegraph and Telephone Corporation, Musashino, Tokyo 180, **Department of Computer Science and Electronics, Kyushu Institute of Technology, Kawazu Iizuka 820, ***Institute of Inorganic Synthesis, Yamanashi University, Kofu 400, Japan*

Abstract. In order to elucidate whether the superconductivity of the cuprate HTSC is "normal" or "abnormal", systematic studies have been performed on the Meissner effect, lower and higher critical fields. The Meissner fraction is strongly dependent on the size of sample and on the field, which is explained by a novel model considering the sharp increase in the vortex pinning force while the flux expulsion process proceeds as T is lowered. Another new method has been proposed to determine H_{c1} from the magnetization curve above H_{c1}, based on which the abrupt upturn of the H_{c1}-T curve frequently reported in the low T region has been attributed to the error induced by the strong pinning. The ambiguity associated with the H_{c2}-T curve is discussed in terms of the "giant fluctuation" of superconductivity.

1. Introduction

There have been reports of many anomalies in the normal state behaviours of the cuprate superconductors. On the other hand, it has generally been regarded that their superconducting properties are rather normal. However, there are some properties reported which are strange in the light of our understandings of the conventional metallic superconductors. Some of the examples are the great differences in the Meissner fraction from a study to another on specimens with apparently similar qualities, from which it has been concluded that some of the specimens are superior to the others, i.e., of better homogeneity (van Dover *et al* 1987, Bloom *et al* 1988). Also, some have claimed the temperature-dependence of H_{c1} to be abnormal, finding a sharp upturn of the H_{c1}-T curve in the low temperature (Strobel *et al* 1988) and theories have been proposed to interpret the anomaly (Hirashima *et al* 1990) while the other have reported it to be normal with respect to the mean field approach (Krusin-Elbaum *et al* 1989). Furthermore, we have been facing a serious ambiguity in the determination of H_{c2} because of the broadened resistive transition of superconductivity under the magnetic field (Malozemoff *et al* 1989, Oh *et al* 1988).

This report will summarize our recent studies on these apparent anomalies, claiming that intrinsic magnetic properties of the superconducting states can only be correctly obtained by quantitatively considering the irreversible vortex motion. Also an extraordinary strong superconducting fluctuation has been advocated in the strong magnetic field.

2. Meissner Effect

In order to systematically explain the experimental observations, we have modeled the Meissner flux expulsion process as follows. (Kitazawa *et al* 1989a, 1990, Matsushita *et al* submitted).

1) The flux expulsion takes place reversibly down to a certain temperature T* (irreversibility temperature) below T_c, and hence the process should follow the equilibrium GL

curve with a homogeneous flux distribution as has been pointed out by Krusin-Elbaum and Malozemoff (1988). This, however, can only explain a part of the whole expulsion process.

2) Below T* pinning of flux lines sets in and is gradually intensified. The flux expulsion further proceeds from the surface where the flux density is in equilibrium, creating a gradient of flux density towards the bulk. The gradient is equal to the pinning force in accord with the Bean's critical state model. Therefore, a typical final ($T<T(H_{c1}=H)$) flux density distribution in a singlecrystalline platelet would be as illustrated in Fig. 1 (Kitazawa *et al* 1989a, 1990).

3) The Meissner fraction ($f_M=-4\pi M/H$), the ratio of the shaded area to the total area in Fig.1, therefore, depends on the thickness D as $f_M= a + b/D$ as far as D is large enough. The constants a and b are dependent on T and the field H. This size effect has experimentally been proved as shown in Fig. 2 (Kitazawa *et al* 1990).

4) Quantitative prediction of the Meissner curve can hence be done by knowing the equilibrium flux density and the pinning force (or J_C) as a function of T and H. Fig. 3 shows the prediction of f_M at the lowest temperature when J_c is approximated as $J_c = A(1- T/T_C)^n$ where A and n are the parameters to fit (Matsushita *et al* submitted). The predicted curve shows a saturation towards $f_M=1$ in the low H region but shows a decrease in the high H region. It is notable that the saturation region extends more towards the higher H as the thickness D decreases. The slope of the curve towards higher H is nearly linear and can be much larger than predicted by the process 1).

Figure 4 shows the f_M-H relations observed for single-, poly-crystalline and powder specimens. It is understood that the results for single crystals and powders can be well explained as the large and small D cases, respectively in Fig. 3. The polycrystalline sintered specimen exhibits a little more complex behaviour. Its curve is like a powder specimen under the high H region but like a single-crystal in the low H region with a nearly flat transition region in the intermediate field range. This indicates that the polycrystalline grain boundaries are essentially the reversible paths of the flux lines under the higher field, attaching the sintered body with the powder-like behaviour. On the othe hand, under the lower field they become like the bulk, hence letting the sintered body to behave like the single-crystal of the same size. Bontemps *et al* (1990) has shown that the transition region in the polycrystalline samples, in some cases, has even a small positive slope with H over a narrow range of field. It is notable that the flat transition region is often observed at nearly 1/3 of the full Meissner state. Wohlleben *et al* (1990) has recently advocated that this is due to the high anisotropy of the cuprate, assuming that the grains oriented with the c-axis parallel with H can expel the flux lines while those perpendicular to it cannot. Although this can explain why the flat region is located at 1/3 of the perfect diamagnetism elegantly, we cannot adopt the idea because the powder should then also show the flat dependence on H at 1/3 level, which clearly is not the case in Fig. 4.

They have also reported quite an interesting result. Namely, even the doped poly-crystalline system $YBa_2(Cu_{1-x}Fe_x)_3O_y$ ($x<0.08$) attains the full Meissner state if measured under low enough field of 2 Oe (Ruppel *et al* preprint). This clearly tells us that the low Meissner fraction of the dopted specimens, which has frequently been attributed to the inhomogeneous nature of the doped specimens, may be rather due to the relatively high pinning force in the doped specimens.We have experienced a significantly small Meissner fraction in a QMG(quench-melt-growth) specimen of YBCO in which the pinning force is much stronger than in the usual single crystal due to the presence of tiny precipitates (Murakami *et al* 1990). This further justifies the conclusions above.

From the above arguments it is clear that the Meissner signal should be the smaller for the larger sized specimen under the same field and hence that the comparison of the values needs a great precaution as to evaluate the sample homogeneity. Nevertheless, the Meissner curve provides us with useful informations about the equilibrium magnetization and the pinning irreversibilty if the dependence of it on the field is examined.

3. Lower Critical Field, H_{c1}

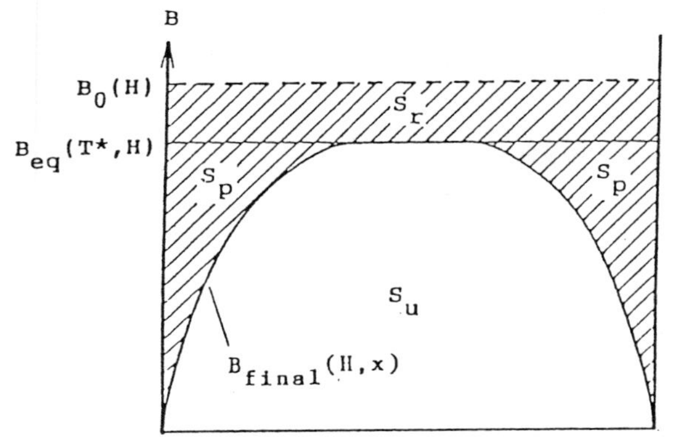

Fig.1 The final flux density distribution after field cooling. The unshaded area corresponds to flux lines trapped

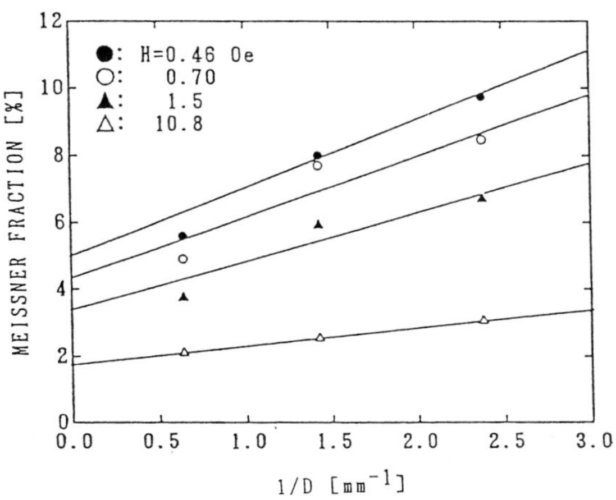

Fig.2 Meissner fraction vs. sample thickness for $(La_{0.94}Sr_{0.06})_2CuO_4$ single crystals (H⊥c) with ac-plane on sample surface

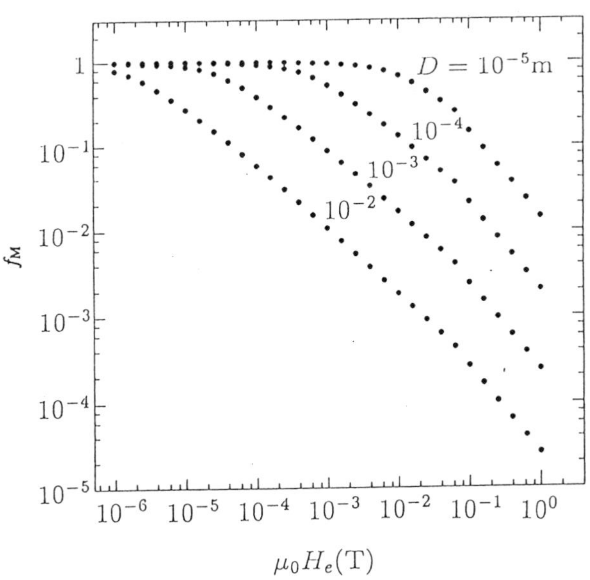

Fig.3 Predicted Meissner fraction as a function of field for specimens with various thicknesses ($A=1x10^6 A/cm^2$, n=1.8, T*=93K)

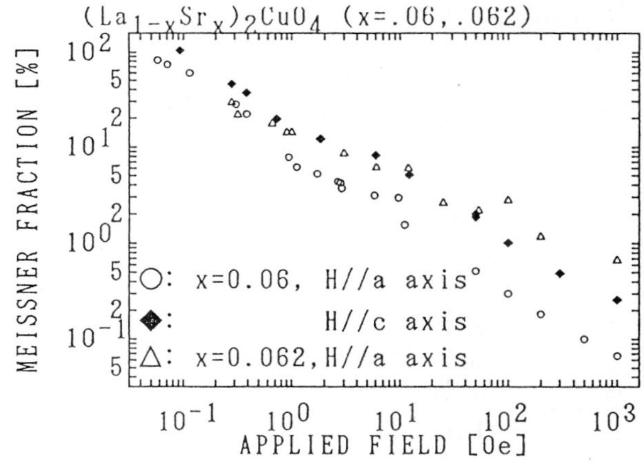

Fig.4a Meissner fraction of single- crystalline $(La_{0.94}Sr_{0.06})_2CuO_4$ (D= 0.4mm) for H⊥c (ab plane on sample face, 0.2mm) and H//c (ac plane on sample face, 0.4mm)

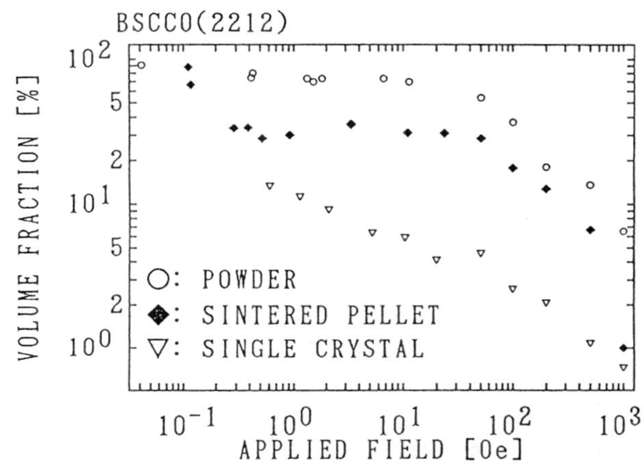

Fig.4b Meissner fraction of powder (1-20μm), single-(H⊥c,0.2mm) and poly-crystalline (5-10μm) $Bi_2Sr_2CaCu_2O_y$

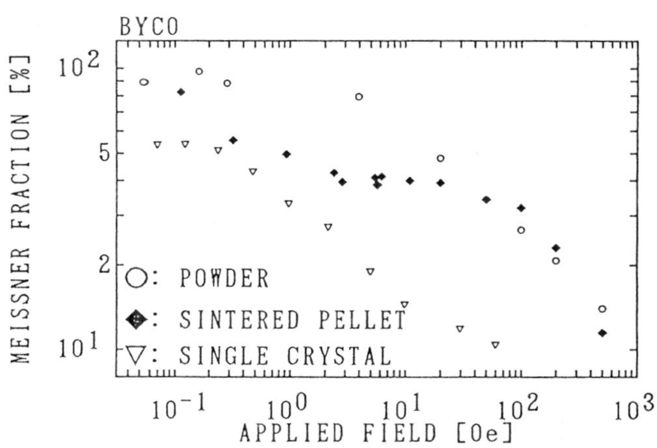

Fig.4c Meissner fraction of powder (~5μm), single-(H⊥c,0.24mm) and poly-crystalline (~10μm) $YBa_2Cu_3O_y$

Several reports so far have claimed the anomalous $H_{c1}(T)$ curve in cuprate superconductors, finding a sharp uprise of the curve in the low temperature region near 1/2 T_c or lower (Strobel *et al* 1988, Adrian *et al* 1989, Umezawa *et al* 1989, Kopylov *et al* 1989, Wacenovsky *et al* 1989, Sato *et al* 1989). Whether this is true or not, is a matter of great importance in order to assess the mechanism of superconductivity (Hirashima *et al* 1990, Ye *et al* preprint, Koyama *et al* 1990).

Our conclusions are (Naito *et al* 1990) :

1) There is no evidence for the H_{c1}-T curve to be abnormal.

2) The reason why such a sharp uprise has been observed is due to the nearly exponential increase in the pinning force with lowering of the temperature. The H_{c1}values in principle should be obtainable as the lowest field where zero-field cooled magnetization curve deviates from the perfect diamagnetism line. But because of the large pinning force in the lower temperature range, it can so happen that flux lines do not penetrate easily into the specimen even as the field exceeds H_{c1}. Then H_{c1} value may be erroneously taken to be larger than the true value.

3) The difficulty can be avoided if the magnetization curve above H_{c1} is back-extrapolated from the higher fields ($H>H_{c1}$) to H_{c1} according to the Bean's critical state model.

In order for the quantitative extrapolation to be performed, we followed a relationship in the intermediate field region $H_{c1}<H<H^*$:

$$B_{eq}(H)=(4\pi J_c D/10)^{1/2}(H+4\pi M_{obs})^{1/2} \qquad (1)$$

where $B_{eq}(H)$ is the equilibrium flux density, J_c is the pinning critical current, D is the thickness of the specimen, M_{obs} is the magnetization, and H^* is the field above which the flux distribution becomes finite at the center of the specimen.

According to this equation, $B_{eq}(H)$ is proportional to $(H+4\pi M_{obs})^{1/2}$ as far as J_c is constant. Since $B_{eq}(H)$ is nearly linear with H above H_{c1} according to the GL theory, the plot H vs. $(H+4\pi M_{obs})^{1/2}$ should give nearly a straight line above H_{c1}, hence providing means to determine H_{c1} by extrapolating the straight line portion to $H+4\pi M_{obs}=0$ and by multiplying an assigned factor according to the GL analysis (Naito *et al* 1990). Fig. 5 shows an example of H vs. $(H+4\pi M_{obs})^{1/2}$ plot, indicating that H_{c1} can be assigned with more confidence in this plot than in H vs. M plot. From the slope of the straight line portion of the curve, one can also obtain J_c value.

Figures 6 and 7 illustrate thus obtained H_{c1} and J_c as functions of temperature. As seen in the figure, no abrupt upturns are observed in the H_{c1}-T curves obtained by the novel method proposed. It is notable that a significantly larger value of H_{c1} than those in Fig. 6 would be estimated from the same results by solely assigning the H value where M-H curve starts deviating from the perfect diamagnetism line as H_{c1}. This discrepancy becomes significant below certain temperatures depending upon the orientation where J_c values exceed 10^4A/cm^2 in Fig. 7.

The results obtained here indicate that anomalies so far reported are due to the contribution from the large pinning force which increase nearly exponentially with lowering of temperature in the cuprate superconductors. Our current conclusion is that H_{c1}-T curve is rather normal in accordance with the framework of the BCS type pairing mechanism.

4. Problems Associated with Upper Critical Field, H_{c2}

The upper critical field is another important quantity to evaluate the characteristic lengths of superconductivity. The standard method to determine H_{c2} has been to observe the resistive transition T_c under the magnetic field H. However, it has been known for the cuprate that the transition region is broadened quite significantly in the resistivity-T curve under magnetic

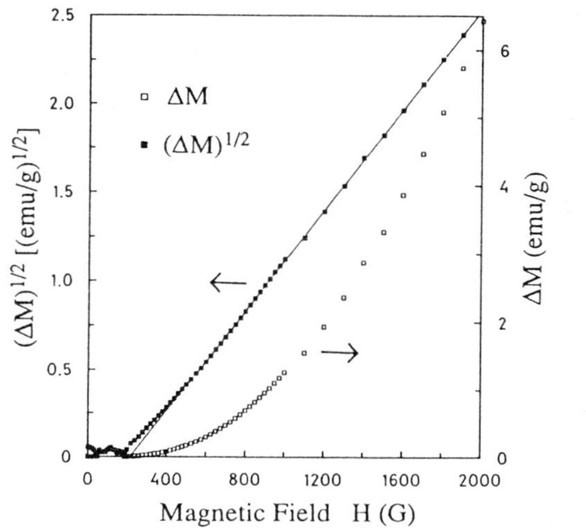

Fig.5 $\Delta M(=H+4\pi M_{obs})$ vs. H and $\Delta M^{1/2}$ vs. H curves at 7.5K for H⊥c (ac plane on the sample face)

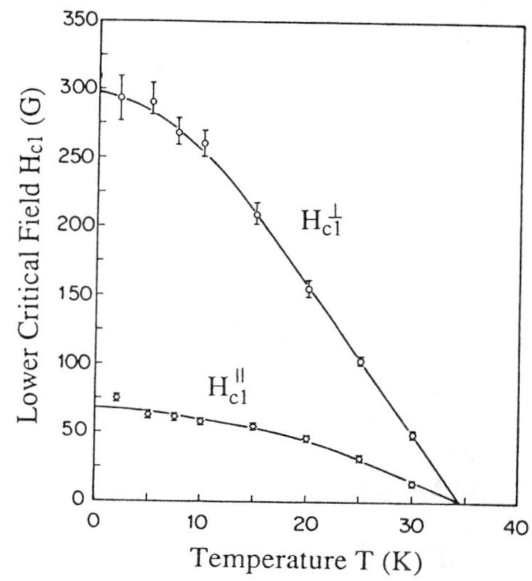

Fig.6 Temperature dependence of H_{c1} for $(La_{0.93}Sr_{0.07})_2CuO_4$ single crystals

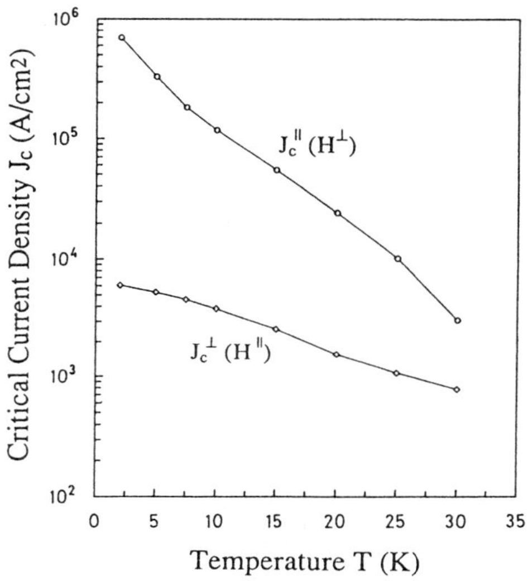

Fig.7 Dependence of J_c as obtained in the H_{c1} measurements for the same specimens for Fig.6

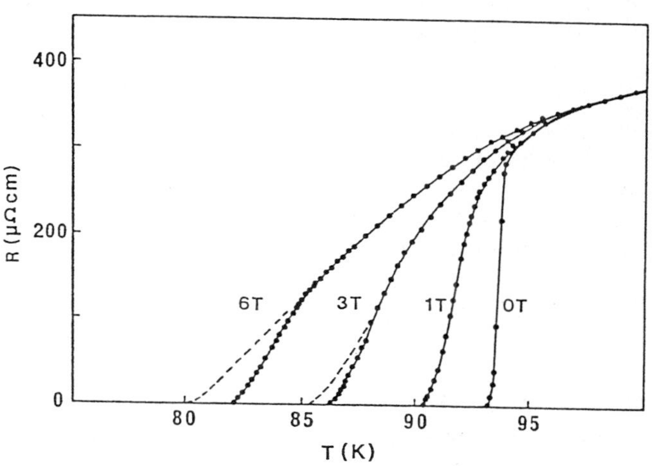

Fig.8 Difference in resistivity-T curves for high J_c (QMG method, solid line) and low J_c (single crystal, dotted line) samples of YBCO (schematic)

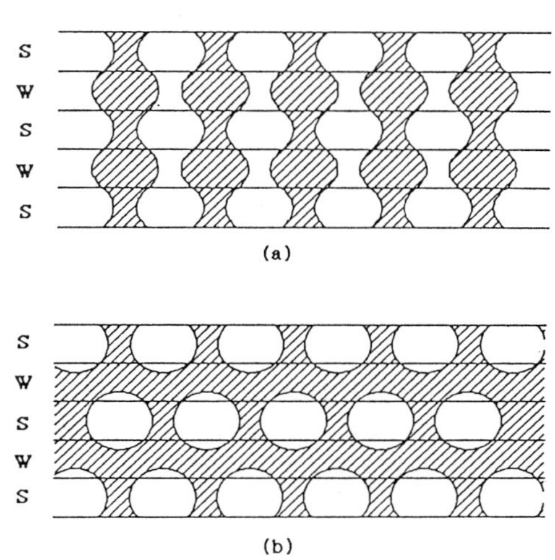

Fig.9 Schematic illustration of vortex cores where superconductivity is destroyed (shaded area) in the stack of weak (W) and strong (S) superconductors. The case (b) is more exaggerated than in (a).

fields, thus making the H_{c2} values ambiguous. Unless one understands the mechanism behind this broadening phenomenon, one can not obtain the reliable value of H_{c2}.

So far, T_c under the magnetic field has been defined frequently as the temperature where the resistive transition attains 10, 50, 90 or 100% of the normal resistivity value ρ_n. The H_{c2}-T curves thus obtained have raised anomalous upturn curvatures below T_c (H=0) when the latter three criteria are adopted. Oh *et al* (1988) has suggested that the 0.9 ρ_n criterion give the straight H_{c2}-T relation which obeys the mean field GL prediction. In order to assess whether the H_{c2}-T curve is anomalous or not, however, comparisons among these criteria are not fruitful.

Examples of mechanisms proposed to interpret the field-induced broadening of the superconductivity transition are superconducting glassy state (Muller *et al* 1987), giant flux creep (Malozemoff e*t al* 1988), Kosterlitz-Thouless transition (Martin *et al* 1989), and enhanced critical fluctuation (Ikeda *et al* 1989). Here discussions are made advocating that such a concept as " giant fluctuation" proposed by the auther should be introduced in order to systematically explain the unique experimental facts observed with cuprates (Kitazawa *et al* 1989b, Kitazawa *et al* 1989c).

The former three mechanisms may be ruled out because of the following observations.

1) The width of the broadening of the transition over the temperature does not depend on the relative orientation between the field H and current I, indicating the nonsignificant involvement of the Lorentz force. This was first reported in LSCO by the authors (Kitazawa *et al* 1989d, Kambe *et al* 1989) and confirmed in YBCO (Woo *et al* 1989) and BSCCO (Kadowaki *et al* 1989) as well.

2) Instead, the width is dominantly determined by the relative orientation between H and the c-axis (Kitazawa *et al* 1989d, Kambe *et al* 1989, Woo *et al* 1989, Kadowaki *et al* 1989).

3) The resistivity in the transition region is completely independent of the relative orientation between H and I but is significantly dependent on H vs. c-axis orientation. This was shown by Iye *et al* (1989) in BSCCO over the whole transition range and YBCO in the higher T region of the transition.

The above three findings clearly indicate that the broadening is not governed by the mechanism which assumes vortex motion because it is motivated by the Lorentz force, the vector product of H and I. There may be a possibility that the microscopic shapes of vortex and current lines are kinked or jogged. But this should still give some difference for the two extreme cases H//I and H⊥I, which has not been observed experimentally except in the lower foot of the ρ-T curve. Therefore, the superconducting fluctuation mechanism should be favoured.

Furthermore, Salamon *et al* (1988) reported that the specific heat jump at T_c is largely suppressed under H of about 2T when H is parallel to c-axis (Salamon *et al* 1988). This means that the contribution to the specific heat due to superconductivity is spread over a wide range of temperature if the entropy conservation rule is taken into account. Therefore, this also indicates the significant broadening of the temperature range over which the superconducting fluctuation takes place.

Based on the above argument the authors have introduced the term "giant fluctuation" in order for the appropriate expression for the extraordinary enhancement of the superconducting fluctuation (Kitazawa *et al* 1989b, Kitazawa *et al* 1989c). The same proposal has also been made by Kadowaki *et al* (preprint). Although, theoretical approach has been attempted based on fluctuation (Ikeda *et al* 1989), the H_{c2}-T relationship has not yet been quantitatively clarified.

The contribution from superconducting fluctuation is significant even in the absence of magnetic field as observed by the excess conductivity in the normal state, being sustainable even up above $1.5T_c$ (Freitas *et al* 1987). It's enchancement becomes enormous if the field is applied along c-axis, while it is minimal along ab-plane. When the magnetic field is applied, the superconducting correlation length perpendicular to the field is limited to the Landau radius $r_0=(\Phi_0/2\pi H)^{1/2}$ where Φ_0 is the quantized flux. If the field is along c-axis, the correlation is also limited along c-axis by the extremely short coherence length ξ_c. Therefore the correlation volume V_c is roughly equal to $V_c\sim \Phi_0\xi_c/2H$. Because ξ_c in YBCO is estimated to be shorter than a few angstroms, and even shorter in BSCCO, the superconducting condensation energy associated with this volume becomes comparable with the thermal energy at the temperature near T_c. Hence, both the superconducting or non-superconducting regions can readily be created by the thermal excitation over a wide range of temperature around T_c as H is intensified. This roughly explains why "giant fluctuation" can manifest itself in such an anisotropic superconductors as cuprates.

It is not obvious how the fluctuation view switches to the other mechanism such as flux creep as the temperature is lowered. It has been noticed that high crirical current YBCO specimens with numerous tiny precipitates (QMG-Quench and Melt Growth-specimens) (Murakami *et al* 1990) show a sharper approach to zero resistance in the ρ-T curve than the ordinary single crystal with a weaker pinning force, but only in the foot of the resistive transition curve. The higher temperature region seems to remain essentially the same as shown in Fig. 8. This cross-over of the two mechanisms from fluctuation to flux creep seems to provide the knee structure frequently seen in the ρ-T curves and to suggest that the fluctuation can be suppressed by introducing the pinning centers at least in the lower temperature side.

It has been known that the cuprate superconductors are the alternative stacks of metallic conducting layers and nearly insulating layers as depicted by scanning tunneling microscope in the case of BSCCO (Hasegawa *et al* 1990). This seems to be the main reason why such short coherence length is assigned in cuprates along c-axis; ξ_c is sometimes estimated even less than 1Å for BSCCO which is shorter than the inter-layer spacing of adjacent CuO_2 layers across the intervening $(BiO_2)_2$ layer. The authors therefore would like to adopt a view that the cuprate superconductors are a stack of superconducting sheets which are just weakly coupled to each other. In the state of giant fluctuation, the weakly superconducting region should be influenced the most.

Although deformation of vortex lines in a complex manner have been frequently discussed in the HTSC, the connectivity of them has been thought to be secured. But as shown in Fig.9, we would rather like to raise a question whether the connectivity of a single vortex line must always be kept in the weak region. The figure illustrates the two extreme cases in which the connectivity is sustained(a) and is completely destroyed across the weaker superconducting layers (b). In case (b), each of the superconducting $(CuO_2)_2$ layers is decoupled from the others by the intervening layers. We presume the state of the "giant fluctuation" to be situated between (a) and (b), where the connectivity is broken locally and hence the quantized vortex "particles", instead of lines, on each of the strongly superconducting layers can move independently of those on the other layers at least locally due to thermal activation. A further elaboration is needed for the quantitative arguments.

5. Conclusions

Some anomalous magnetic properties of the superconducting state of cuprates reported should not be attributed to the anomalous superconductivity but rather to the unusual contributions from the vortex state or to the exaggerated superconducting fluctuation. The Meissner curves can be understood by considering how the pinning force develops as the temperature is lowered while the equilibrium diamagnetic magnetization is maintained only at the specimen surface, which brings about the dependence of the Meissner fraction on the size of the specimens and on the intensity of the field. H_{c1} can be obtained correctly not by judging the first point of deviation of the magnetization curve from the perfect diamagnetic line but rather by analyzing the curve above H_{c1} and by extrapolating it back to H_{c1} according to the critical state model. The reason for the H_{c2}-T line to be obscured in the HTSC can be attributed to the "giant fluctuation" of superconductivity.

References

Adrian H, Assmus W, Hohr A, Koalewski J, Spille H and Steglich F 1989 *Physica* **C162-164** 329
Bloom S H, Kuric M V, Yao Y S, Guertin R P, Nichols D, Jee C, Kebede A, Crow J E, Mihlisin T, Myer G N and Shulottmann P *High-Temperature Superconductors* Eds Brodsky M B, Dynes R C,Kitazawa K and Tuller H L, MRS 1988 p19
Bontemps N, Monod P, Regnier P, *European MRS Meet. Proc.*, 1990 Strousbourg, Germany Preprint.
van Dover R B, Cava R J, Batlogg B and Rietman E A 1987 *Phys. Rev.* B**35** 5337
Freitas P P, Tsuei C C and Plasket T S 1987 *Phys. Rev.* B**36** 833
Hirashima D and Matsuura T 1990 *J. Phys. Soc. Jpn.* **59** 24,
Hasegawa T and Kiatzawa K 1990 *Jpn. J. Appl. Phys.* **29** L434
Ikeda R, Ohmi T and Tsuneto T 1989 *J. Phys. Soc. Jpn.* **58** 1377, 1990 *J. Phys. Soc. Jpn.* **59** 1397, 1990 *J. Phys. Soc. Jpn.* **59** 1740
Iye Y, Nakamura S and Tamegai T 1989 *Physica* **C159** 433
Junod A 1990 *Physical Properties of High Temperature Super-conductors* Ed Ginsberg D M 1990 (World Scientific Publication)
Kadowaki K, Li J N, Menken M J V, Menovsky A A and Franse J J M 1989 *Physica* **C161** 313
Kadowaki K, Li J N, Menken M J V and Franse J J M preprint for *Physica* C
Kambe S, Naito M, Kitazawa K, Tanaka N and Kojima H 1989 *Physica* **C160** 243
Kitazawa K, Nakamura O, Matsushita T, Tomioka Y, Motohira N, Murakami M and Takei H 1989a *Adv. in Superconductivity* II, Eds. Ishiguro T and Kajimura K *Proc. Int. Symp. Superconductivity* p 609
Kitazawa K, Kambe S and Okazaki N, 1989b *Superconductivity and Applications, Proc.Taiwan Int. Symp. Superconductivity* Eds. Wu P T, Ku H C, Lee W H and Liu R S (World Scientific Publication) p 79
Kitazawa K, Kambe S and Naito M, 1989c *Strong Correlation and Superconductivity*, Eds.Fukuyama H, Maekawa S and Malozemoff A P (Springer-Verlag) p 361
Kitazawa K, Kambe S, Naito M, Tanaka I and Kojima H, 1989d *Jpn. J. Appl. Phys.* **28** L555
Kitazawa K, Matsushita T, Nakamura O, Tomioka Y, Motohira N,Tamura T, Hasegawa T, Kishio K, Tanaka I and Kojima H 1990 *Proc. Int. Conf. Superconductivity* Ed. Joshi S K
Kopylov V N and Schegolev I F 1989 *Physica* **C162-164** 1143
Koyama T, Takezawa N and Tachiki N 1990 *Physica* **C168** 69
Krusin-Elbaum L, Malozemoff A P, Yeshurun Y, Cronemeyer D C and Holtzberg F 1988 *Physica* C **153-155** 1469
Krusin-Elbaum L, Malozemoff A P, Yeshurun Y, Cronemeyer D C and Holtzberg F, 1989 *Phys. Rev.* B**39** 2936
Malozemoff A P, Yeshurun Y, Krusin-Elbaum L, Worthington T K, Cronemeyer D C, Dinger T, Holtzberg F, Mcguier T R and Kes P 1988 *Proc. First Latin American Conference of High Temperature Superconductivity* (World Scientific Publication) p112
Malozemoff A P 1989 *Physical Properties of High Temperature Superconductors*, Ed.Ginsberg D M (World Scientific Publication)
Martin S, Fiory A T, Fleming R M, Espinosa G P and Cooper A S 1989 *Phys. Rev. Lett.* **62** 677
Matsushita T, Otabe E S, Matsuno T, Murakami M and Kitazawa K submitted to*Physica* C
Muller K A, Takashige M and Bednortz J G 1987 *Phys. Rev. Lett.* **58** 1143
Murakami M, Gotoh S, Koshizuka N, Tanaka S, Matsushita T, Kambe S and Kitazawa K 1990 *Cryogenics* **30** in press
Naito M, Matsuda A, Kitazawa K, Kambe S, Tanaka I and Kojima H 1990 *Phys. Rev.* B**41** 4823
Oh B, Char K, Kent A D, Naito M, Beasley M, Geballe T H, Hammond R H, Kapitulunik and Graybeal J M 1988 *Phys. Rev.* B**37** 7861
Ruppel S, Michels G, Geus H, Kalenborn J, Schlabitz W and Wohlleben D preprint
Salamon M B, Inderhees S E, Rice J P, Pazol B G, Ginsberg D M and Goldenfeld N 1988 *Phys. Rev.* B**38** 885,
Sato M, Shamoto S, Sera M and Fujishita H 1989 *Solid State Commun.* **72** 689
Strobel J P, Thoma A, Hensel B, Adrian H and Saeman-Ischenco 1988 *Phisica* **C153-155** 1537
Tinkham M 1988 *Phys. Rev. Lett.* **61** 1658
Umezawa A, Crabtree G W, Vandervoort K G, Welp U, Kwok W K and Liu J Z 1989 *Physica* **C162-164** 733
Wacenovsky M, Weber H W, Hyun O B and Finnemore D K 1989 *Physica* **C162-164** 1629
Wohlleben S, Michels G, Ruppel S, Braunisch W, Domel R, Roden B and Schlabitz W 1990 *Some Aspects of Superconductivity* Eds. Gupta L C and Malik S K (Nova Scientific Pub.) to be published
Woo K C, Gray K E, Kampwirth R T, Kang J H, Stein S J, East R and McKay 1989*Phys. Rev. Lett.* **63**1877
Ye Z, Umezawa H and Teshima R preprint

Supercond. Sci. Technol. **4** (1991) S43–50. Printed in the UK

Flux pinning and critical currents in melt processed YBaCuO superconductors

M. Murakami, S. Gotoh, H. Fujimoto, K. Yamaguchi, N. Koshizuka and S. Tanaka
ISTEC Superconductivity Research Laboratory, 1-10-13, Shinonome, Koto-ku, Tokyo 135 Japan

Abstract
We report flux pinning behavior of melt processed YBaCuO containing Y_2BaCuO_5(211) inclusions. Although the size of the 211 inclusions is relatively large compared to the coherence length, they can work as pinning centers. We present the mechanism for pinning due to large normal precipitates. While large flux creep is observed in thin films exhibiting high J_c values, flux creep rate is very small in melt processed YBaCuO having a magnitude lower J_c values, which is understandable by considering that the pinning potential of large normal precipitates is much larger than that of small pinning centers and thereby the flux motion is less affected by thermal activation.

1. Introduction

Critical current density(J_c) has the key to major applications of superconductors. Oxide superconductors belong to type II, where a magnetic field whose strength is greater than lower critical field(H_{c1}) can penetrate the superconductor in the form of quantized fluxoid.

When the currents(J) are passed through the superconductor in the mixed state, the Lorentz force($\mathbf{F_L=JxB}$) acts on the fluxoids. If the crystal is perfect, the fluxoid will move and leading to the appearance of some resistance, which indicates that the clean type II superconductors cannot carry resistanceless current in the mixed state. On the other hand, if the crystal contains imperfections such as normal precipitate, the flux motion is prevented, which is known as flux pinning. Resistanceless current can flow through the superconductor so long as the flux pinning force(F_p) is greater than the Lorentz force. Consequently, J_c is considered to be the current density when F_L becomes equal to F_p and is given by the following relation

$$F_p=F_L=J_cB$$

In order to achieve high J_c values, we need to increase F_p, which will be accomplished by introducing effective pinning centers into the superconductor.

Fig. 1 is schematic illustration for the flux pinning mechanism by a normal particle. The region where the fluxoid penetrates the superconductor loses the condensation energy which is $H_c^2/8\pi$ per unit volume where H_c is the thermodynamical critical field. When the fluxoid intersects with a normal particle, it can lower its energy by that amount per unit penetrated volume; viz., by $U_p=(H_c^2/8\pi)(\pi\xi^2d)$, where d is the size of a normal particle. U_p is the pinning energy and the corresponding pinning force f_p is given by

$$f_p=U_p/2\xi \; =(H_c^2/16)\xi d$$

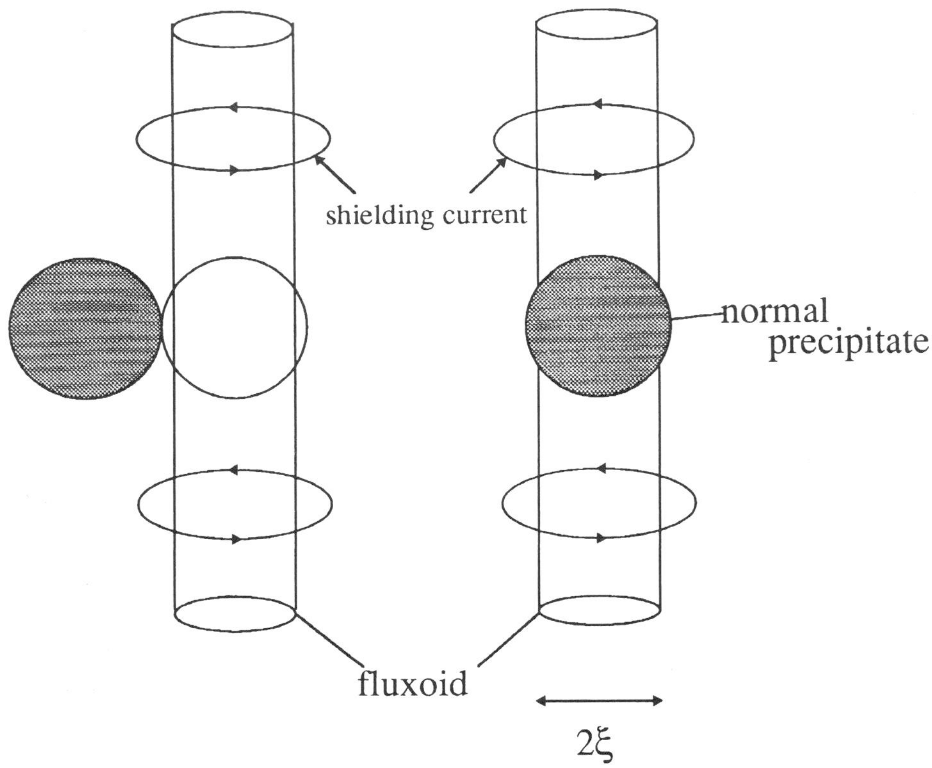

Fig. 1. Schematic illustration for the flux pinning mechanism. The fluxoid penetrating a normal particle can lower its energy, since it does not lose the condensation energy($H_c^2/8\pi$ per unit volume) in the penetrated volume($\pi\xi d^2$).

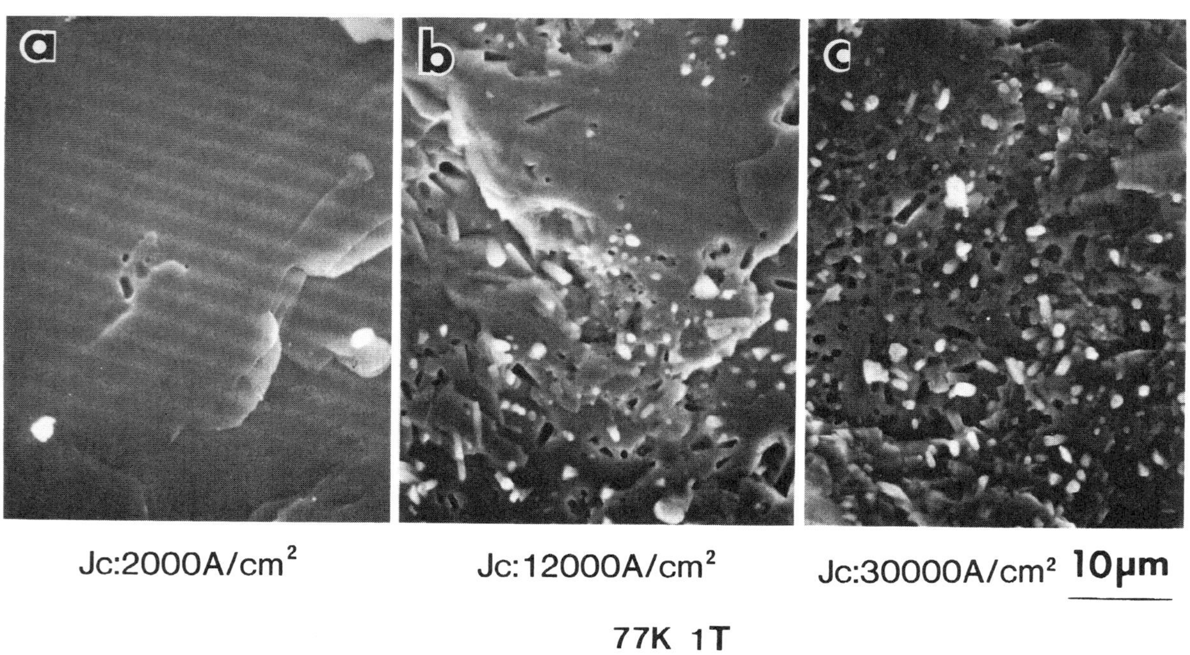

Fig. 2. Scanning electron micrographs for the fractured surfaces of the samples (a) Y1.0; (b) Y1.2; (c) Y1.5.

In order to obtain bulk pinning force we need to add up f_p. In the simplest case, the bulk pinning force per unit volume F_p is given by

$$F_p=Nf_p$$

where N is the number of fluxoid-pinning center interactions per unit volume. Therefore, large F_p or large J_c values can be obtained by either increasing the density of pinning centers or increasing f_p.

We have found that the introduction of non-superconducting particles into the superconducting matrix is possible in the Y-Ba-Cu-O system through Melt-Powder-Melt-Growth (MPMG) process[1)]. In this paper, we present how the normal particles affect J_c and flux creep and then present possible mechanism for flux pinning due to normal particles.

2. Experimental

YBaCuO samples were prepared by the MPMG process whose details have been reported elsewhere[1-4)]. Using the process, we can dope Y_2BaCuO_5(211) inclusions into the $YBa_2Cu_3O_x$ matrix simply by changing the starting composition toward the 211 rich region from the stoichiometry. In the present experiments we prepared samples with the compositions of Y:Ba:Cu=1:2:3; 1.2:2.1:3.1; 1.5:2.25:3.25. (Hereafter we name the samples Y1.0, Y1.2, and Y1.5 respectively.) These samples contains 211 inclusions approximately 0%, 10%, and 25% in volume, respectively. Fig. 2 shows scanning electron micrographs for fractured surface of the samples. Small white particles about 1 micron in diameter are the 211. It is clear that the 211's concentration increases with changing the composition toward the 211 rich region.

Critical current density was determined from magnetization measurements using a vibrating sample magnetometer. Flux creep was measured by monitoring magnetization decay during increasing and decreasing field process using a SQUID magnetometer.

3. Results

3.1. Critical current density

Fig. 3 shows magnetic field dependence of J_c for samples Y1.0, Y1.2, and Y1.5 with magnetic field parallel and perpendicular to the c axis. The data were calculated from magnetization hysteresis using the following relation[5)]

$$4\pi(M^+-M^-) = (4\pi/10)\, J_c\, d/2$$

where M^+ and M^- are the magnetization(emu/cm^3) in increasing and decreasing field process, d is the sample thickness(d) and Jc is the critical current density(A/cm^2). We confirmed that magnetic hysteresis scales with sample thickness, which guarantees that the shielding currents are not localized[6)]. In the case of $H\perp c$ axis, we prepared the sample as schematically which has a large aspect ration along the c axis so that the shielding currents flow mainly along the c axis, otherwise we cannot observe the anisotropy.

It is clear that J_c can be increased with increasing the volume fraction of 211 inclusions for both H//c and $H\perp c$ directions, indicating that the dispersion of nonsuperconducting particles contribute to the improvement in flux pinning force even in high temperature superconductors and their contribution is isotropic. According to the microscopic observation 211's distribution is homogeneous and their crystal orientation is random, which may be the source of isotropic contribution to flux pinning.

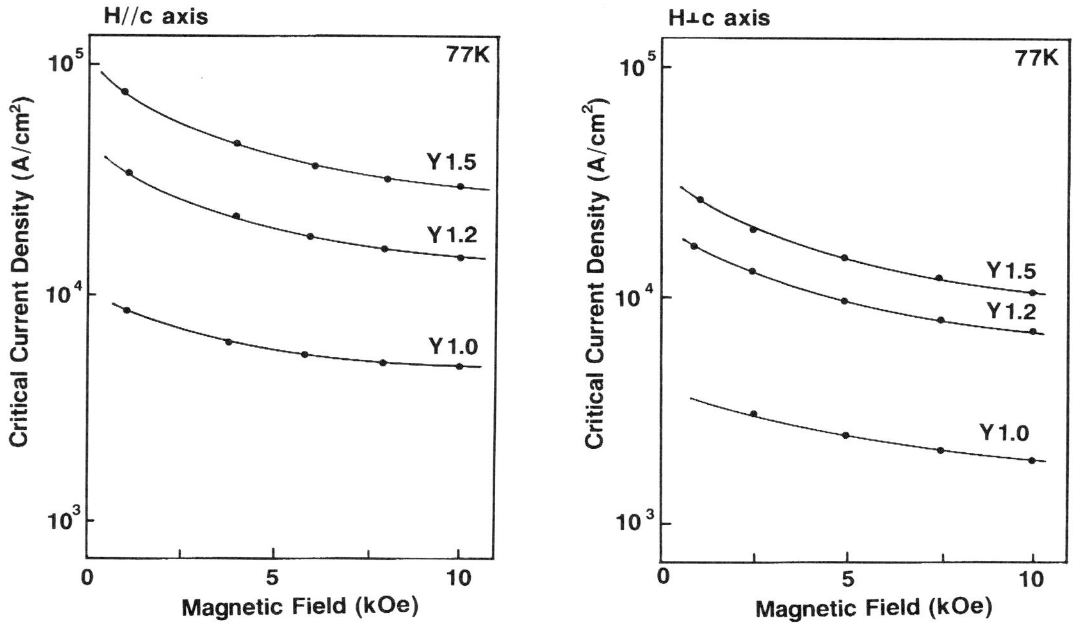

Fig. 3. Magnetic field dependence of J_c for samples Y1.0, Y1.2, and Y1.5 with magnetic field parallel and perpendicular to the c axis.

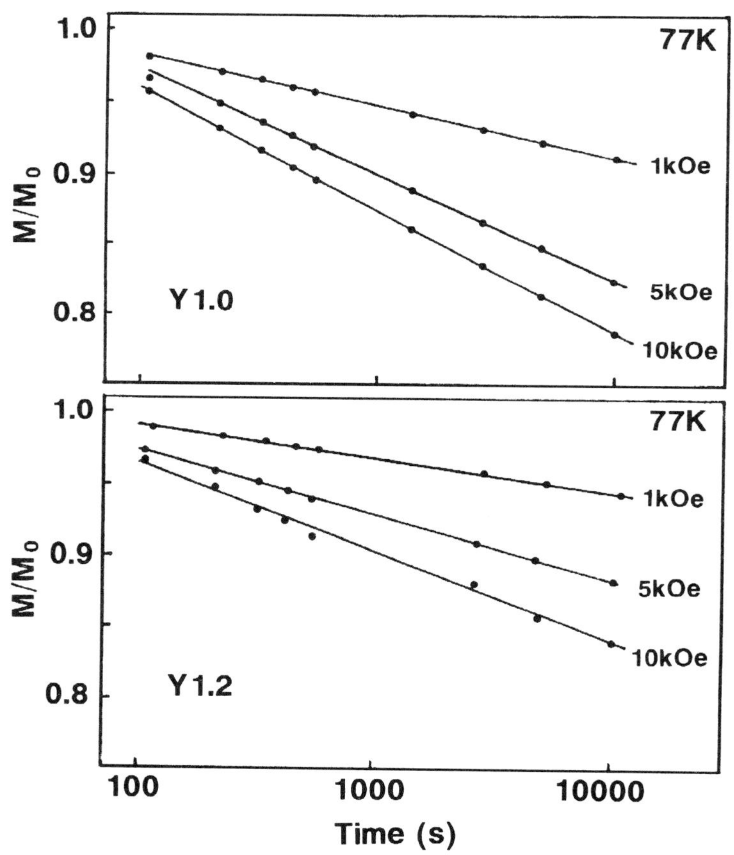

Fig. 4. Time decay of magnetization for the samples Y1.0 and Y1.2 at 77K and various magnetic fields.

It was also found that J_c values have anisotropy about 3:1 between the magnetic field parallel and perpendicular to the c axis. This anisotropy is considered to be intrinsic and mainly due to anisotropy in the coherence length as will be discussed later.

3.2. Flux creep

Fig. 4 shows flux creep for Y1.0 and Y1.2 at 77K and various magnetic fields. The data are the average of decay in magnetization between increasing and decreasing field process. By this treatment we can remove the effect of reversible magnetization.

The sample with 211 inclusions exhibits lower flux creep rate, which suggests that flux creep can be reduced by increasing flux pinning force or J_c. However, such dependence cannot be generalized as shown below. If we obtain the pinning energy by the following relation

$$dM/M_0 = -(U/kT)\,d\ln t$$

we get U values of 1 eV for Y1.2 and 0.6 eV for Y1.0 at 77K and 1kOe. These values are much larger than the U values of thin films having one order of magnitude larger J_c values. We believe that the size of pinning center may affect flux creep.

4. Discussion

4.1. Flux pinning due to pinning centers in various sizes

From the data presented above it is clear that 211 inclusions can contribute to flux pinning. However, it has been pointed out that a normal particle itself does not work as a pinning center, since the size is too large compared to the coherence length. However, as already reported by Coote et al.[7] large normal precipitates do act as pinning centers in conventional superconductors.

Here we consider how the defects of various sizes can affect flux pinning. Fig. 5 shows schematic illustrations for fluxoid-pinning center interaction for the defects in various sizes. In the case of point defects of the order of coherence length(Fig. 5(a)), which is believed to be dominant pinning centers in single crystal or thin film, the maximum pinning energy is obtained by

$$U_p = (H_c^2/8\pi)(4/3)\,\pi\xi^3$$

If we choose H_c=3000 G and ξ =2nm at 77K, U_p becomes 0.01 eV which is on the order of thermal energy kT. This may drive point defects less effective as pinning site owing to the thermal activation. Extremely large flux creep in high J_c thin film may then be attributed to this small pinning energy. This suggests that we need to increase the elementary pinning energy to avoid gigantic flux creep.

In the case of defects d in length (Fig 5(b)), the pinning energy is obtained by

$$U_p = (H_c^2/8\pi)\pi\xi^2 d$$

When d is 1 micron which is a typical size of the 211 inclusions, U_p becomes 500 times larger than that of a point defect. Therefore, the fluxoid pinned by the large precipitate will be less affected by thermal activation. We believe that smaller flux creep rate in MPMG processed YBaCuO is attributable to large pinning energy due to large pinning centers.

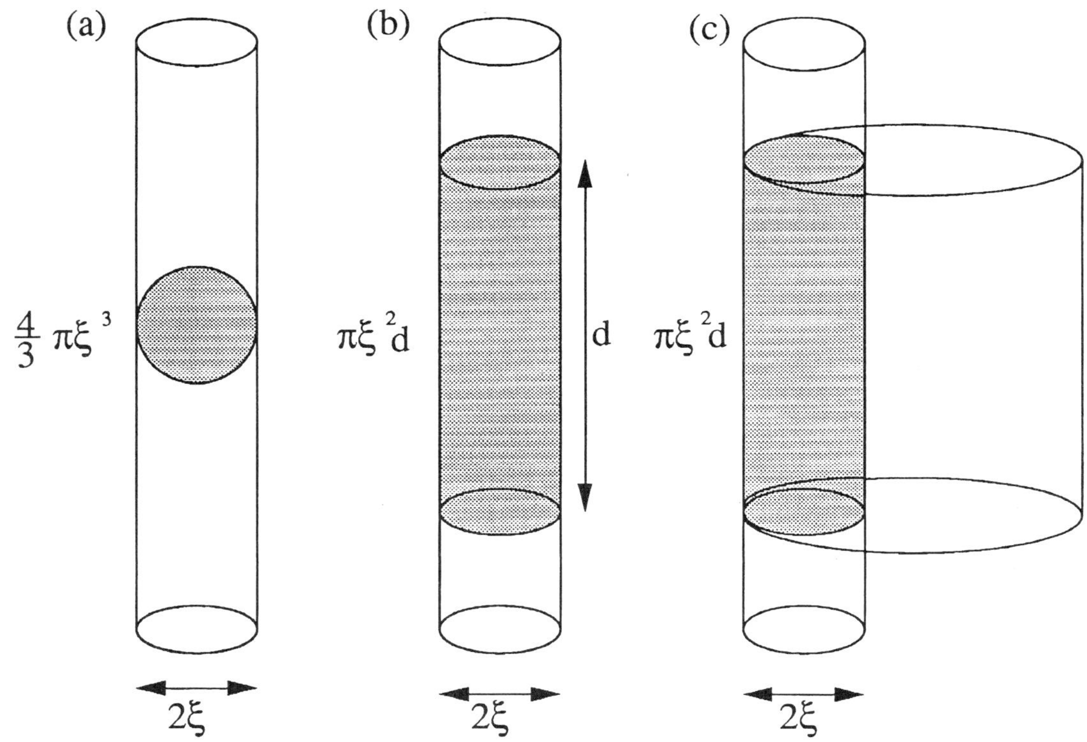

Fig. 5. Schematic illustrations for fluxoid-pinning center interaction for the defects in various sizes: (a) a point defect 2ξ in size; (b) a defect 2ξ in diameter and d in length; (c) a large defect d in diameter and d in length.

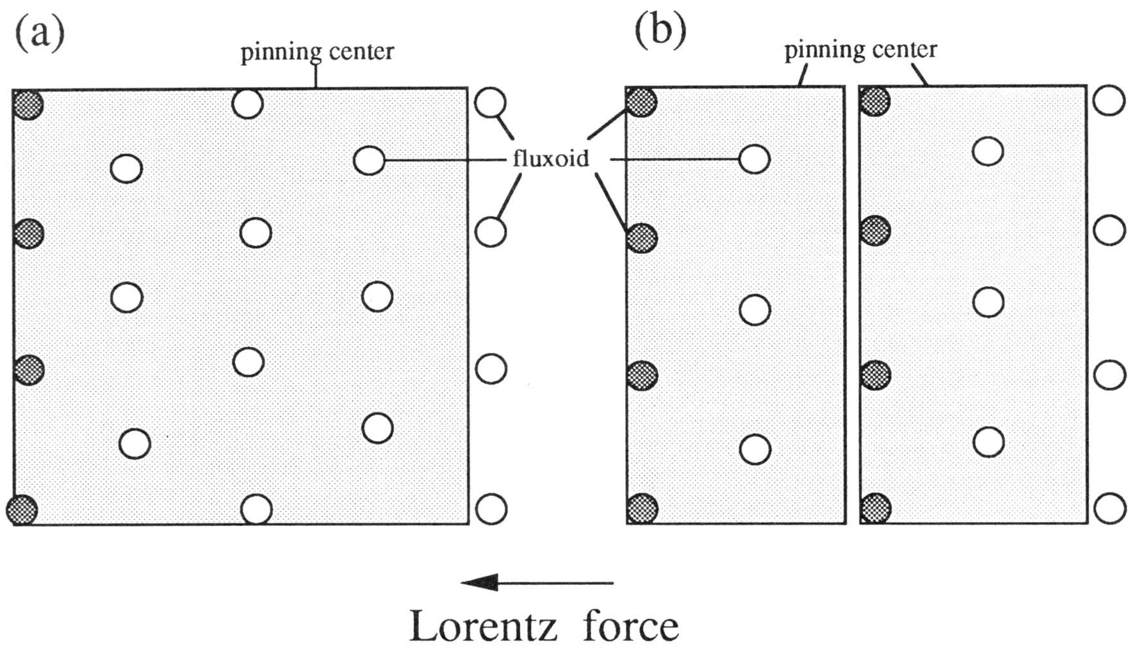

Fig. 6. Schematic illustrations for interaction between the fluxoids and large pinning centers. The shaded fluxoids are pinned at the interface. The number of pinned fluxoids can be increased by deviding the pinning center.

However, large normal precipitates can contribute to macroscopic bulk pinning force less effectively than smaller precipitates provided that the volume fraction of the precipitates is constant. As schematically shown in Fig. 5(c), in the case of large normal precipitate, only the interface can work as pinning center, because no energy is needed for motion of the fluxoids within the precipitate. When the precipitate is divided more pinning sites are produced as shown in Fig. 6. Therefore, bulk pinning force F_p or J_c varies as 1/d, which has been confirmed by Morimoto et al.[8] in melt processed (Y, Ho)BaCuO.

4.2. Estimation of pinning force

In high T_c superconductors, the coherence length is small and the penetration depth is large, which indicates that the flux line lattice is soft. In such a case, each element of the flux lattice could be individually pinned by the pinning site. Bulk pinning force density is then obtained by $F_p=Nf_p$. The number of fluxoids pinned by one normal particle d in size is d/a_f where a_f is the flux lattice spacing and given by $a_f=1.075(\phi_0/B)^{1/2}$ where ϕ_0 is the flux quantum. N becomes $N_p(d/a_f)$ where N_p is the density of 211 inclusions. Therefore

$$F_p = N_p(d/a_f)\, f_p$$

N_p is related to the volume fraction (V_f) by $V_f=N_pd^3$. From relation $F_p=J_cB/10$ (in practical units, e. g., dyn,A/cm^2, G) , J_c becomes

$$J_c = N_p(d^2/a_f)\, \xi(10H_c^2/16)B^{-1}$$

or

$$J_c = (V_f/d)\, \xi(10H_c^2/16)\phi_0^{-1/2}B^{-1/2}$$

This relation shows that J_c is proportional to V_f, 1/d, and $B^{-1/2}$, which does not conflict with the present results. J_c values at 77K and 10kOe can be calculated to be 25000A/cm^2 and 10000A/cm^2 for Y1.5 and Y1.2, respectively(H//c axis and ξ=2nm). These values are in good agreement with empirical values. While, these calculations are based on the flux pinning due to the 211 inclusions, the fact that the sample Y1.0 without the 211 inclusions exhibit flux pinning indicates that other defects such as twin planes or stacking faults can also work as pinning centers, which may explain why the calculated J_c values are somewhat lower than actual J_c values for Y1.2 and Y1.5.

We can also estimate how large anisotropy exists in J_c. It is clear that anisotropy in J_c is ascribed to the anisotropy in the coherence length, which affects U_p and f_p. Fig. 7 shows schematically how the anisotropy in ξ affects f_p. When magnetic field is applied perpendicular to the c axis as shown in Fig. 7(b), vortex will become elliptic. In such a case U_p becomes $(H_c^2/8\pi)\pi\xi_{ab}\xi_c d$ which is smaller than U_p with H//c axis(Fig. 7(a)) by a factor of ξ_{ab}/ξ_c. However, the anisotorpy in pinning force can be observed only when the force is directed perpendicular to the c axis, because in the other direction f_p becomes the same as in the case H//c axis since $f_p=U_p/2\xi_c=(H_c^2/16)\xi_{ab}d$. In this study observed anisotropy is about 3:1, which may indicate that ξ_{ab}/ξ_c is about 3, though more careful microstructural observation is needed to draw definite conclusions.

5. Conclusions

In this paper, we showed that relatively large normal inclusions can work as pinning centers in high temperature superconductors. Small flux creep rate is attributed to relatively large individual pinning energy due to large normal particles. Anisotropy in J_c was about 3:1 which may reflect the anisotropy in the coherence length.

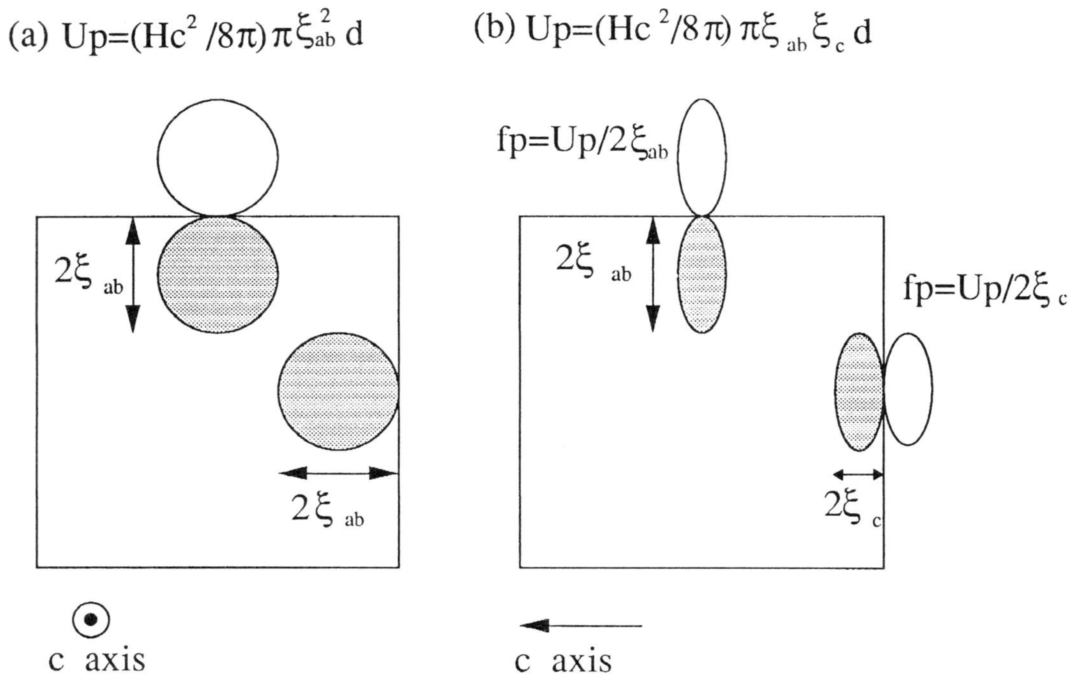

Fig. 7 Shematic illustration showing how the anisotropy in ξ affects U_p and f_p.

Acknowledgment

We are grateful to T. Oyama of Superconductivity Research Laboratory for sample preparation. We are also grateful to Prof. T. Matsushita of Kyushu Institute of Technology for his useful discussion.

References

1) H. Fujimoto, M. Murakami, S. Gotoh, T. Oyama, Y. Shiohara, N. Koshizuka and S. Tanaka: *Applied Superconductivity II* (Springer-Verlag, Tokyo, 1990) p. 285.
2) M. Murakami: Mod. Phys. Lett. **4** (1990) 163.
3) M. Murakami, S. Gotoh, H. Fujimoto, N. Koshizuka and S. Tanaka: *Proc. AMSAHTS 90* to be published.
4) M. Murakami, T. Oyama, H. Fujimoto, S. Gotoh, Y. Shiohara, N. Koshizuka and S. Tanaka: *Proc. ICMC 90* to be published.
5) C. P. Bean: Phys. Rev. Lett. **8** (1962) 250.
6) M. Murakami, M. Morita and N. Koyama: Jpn. J. Appl. Phys. **28** (1989) L1125.
7) R. I. Coote, J. E. Evetts and A. M. Campbell: Canad. J. Phys. **50** (1972) 421.
8) T. Morimoto et al.: *Proc. ICMC 90* to be published.

Supercond. Sci. Technol. **4** (1991) S51–58. Printed in the UK

Research Advances and Needs Related to HTSC Electronics

John M. Rowell
Conductus, Inc.
969 W. Maude Avenue
Sunnyvale, California, 94086, USA

Abstract

The creation of an electronics technology using the HTSC oxide materials requires at the very least mastery of the synthesis of thin films of these materials and understanding of their properties. This paper presents a personal evaluation of the major advances that have been made in film preparation since the discovery of these materials in 1986 and an outline of the outstanding needs that must be addressed before HTSC electronics is widespread. These advances and needs will be described for each of the steps required to create such a technology, beginning with substrates and films and ending with the products themselves. These materials have presented challenges that could not have been anticipated in 1986, but the challenges are being overcome with surprising rapidity.

Substrates

Advances

Early in 1987 the first YBCO films were deposited on $SrTiO_3$, MgO, and YSZ. Of these, $SrTiO_3$ has by far the best lattice match, but it has a very large dielectric constant and loss. MgO and YSZ are reasonably stable to YBCO chemically but have a poor lattice match.

A significant advance is the commercial availability of $LaAlO_3$ (Simon R W 1989), a decent microwave substrate with good epitaxy but a fairly high dielectric constant. It is likely, however, that $LaAlO_3$ will not be the final substrate of choice because it is extensively twinned. Another substrate claimed to be lattice matched and twin free is $CaNdAlO_3$.

Needs

The field needs a lattice matched, non–reactive, smooth, strong, twin–free, low–dielectric–constant, low–loss, large area (2–4") substrate that is also cheap. All the currently available substrates represent some kind of compromise. They either offer a good lattice match and a high dielectric constant, or a low dielectric constant and a poorer lattice match, or twinning, or a large divergence in thermal expansion coefficient.

The cost of substrates varies from about $200 for a 1 cm x 1 cm $SrTiO_3$ substrate to $25 for YSZ of the same size. The cost of substrates can be high when, in a typical deposition, nine to sixteen substrates are assembled on the heater block. Many people have discussed the technological importance of growing films directly on large area substrates, say 2" or 4" in

0953-2048/91/01S051 + 08 $03.50

diameter. At present, substrates larger than 2" in diameter are restricted to sapphire and silicon only. In both these cases, it is necessary to use a buffer layer between the substrate and YBCO film to improve the epitaxial match and reduce chemical reactions between substrate and film.

Buffered Substrates

Advances

Films of YBCO grown directly on sapphire are not of comparable quality to those on $SrTiO_3$ and $LaAlO_3$, for example. However, if a thin buffer layer of $SrTiO_3$ (Char K et al 1990) or MgO (Berezin A B et al 1990) is deposited between the sapphire and the YBCO, the films have all the hallmarks of good films, namely high T_c, sharp transitions, and low normal state resistivity (below 100 $\mu\Omega$ cm at 100K). In addition, the resistance vs. temperature is close to linear and extrapolates close to zero.

Such buffer layers also result in YBCO films with surface resistance markedly lower, by a factor of about 20 at 4.2K and 10 GHz. This improvement in surface resistance occurs because the buffer layer allows better in–plane epitaxial growth of the films.

YBCO films on silicon have proved difficult to grow due to chemical interactions and the large difference in rates of thermal contraction. Recently two groups have achieved significant improvements in film quality by the use of YSZ buffer layers. In the first case (Fork D K et al 1990), YSZ alone is used; in the second case, a thin layer of Y_2O_3 is inserted between the YSZ and the YBCO. In the latter case (Myoren et al 1990), critical currents of the films are extremely high, over 10^7 amps/cm^2 at 4.2K and over 10^6 amps/cm^2 at 77K. To date, these YBCO films on YSZ on Si are kept thin so they do not crack on cooling from the growth temperature.

Needs

It is clear that buffer layers offer significant advantages in the growth of YBCO (and probably other superconducting films) on cheap and readily available substrates like sapphire and silicon. To make further progress, it will be necessary to understand epitaxial relationships in numerous buffer layer systems. Some buffers are already multiple layers, as mentioned above; e.g., YSZ / Y_2O_3. It will be necessary to control growth of both the buffer layer and the superconducting film over large areas. In some cases we will have to learn how to deposit single crystal buffer layers on polycrystalline materials. In other cases, substrates of interest will be either amorphous or extremely fine–grained polycrystalline materials. It has even been suggested by some people that for interconnects it is desirable to grow buffer layers and YBCO on plastics such as polyimide. For growth on silicon, it will be necessary to learn further how to control the strain caused by differences in thermal contraction rates between substrate and film.

Film Synthesis

Many deposition techniques have been used to create very high quality superconducting films. I will emphasize here the methods which appear to be most popular, at least in the United States, and which offer advantages of simplicity or low equipment costs.

Achievements: Post Annealed Films

The barium fluoride process (Mankiewich P et al 1988) produces very good films by any criterion, including surface resistance. High–quality thallium films have been made by laser ablation and post annealing in thallium vapor (Bourne L C et al 1990). In both cases, the high temperature post–anneal limits the applicability of the process to single–layer films. However, the process is ideal for large–area substrates and for deposition on both sides of a substrate.

Advances: In Situ Films

Laser ablation (Wu X D, Venkatesan T, et al 1990) and off–axis sputtering (Eom C B et al 1990) have the advantage that the film composition is very close to that of the target. Laser deposition is a versatile, rapid process. The use of multiple targets sequentially moved in front of the laser beam allows multilayers of some complexity to be fabricated. An interesting area of research is the creation of multilayers or superlattice structures. In YBCO / NdCeCuO superlattices (Gross R et al 1990), the critical current of the YBCO layer reaches over 10^7 amps/cm^2 at 4.2K, the largest number reported at this temperature. Averaged over the total thickness of the film, the J_c is not extraordinary, merely comparable to that of the best single YBCO films. This result, however, does indicate that pinning in YBCO can be enhanced, by some kind of differential strain, beyond values commonly observed using many kinds of deposition techniques.

A disadvantage of laser deposition is that generally the highest quality film is only deposited over a relatively small area, say about 1 cm x 1 cm. This area can be increased by rastering the laser beam across a large target or by rastering the substrate through the "sweet spot" in the laser deposition plume.

A typical laser deposition rate for YBCO films is about 1 ångstrom per second, but recently it has been shown that rates of 50 or 100 ångstroms per second are not detrimental (Wu X D, Muenchausen R E et al 1990). Although the crystalline quality of the film suffers somewhat, as reflected in channeling experiments, the critical current density stays high up to deposition rates of 150 Å/sec. Obviously the flux pinning in these films is not correlated with this structural measurement. It is also interesting that so many deposition techniques yield films with similarly high critical currents at 77K, say values of 2 to 4 x 10^6 amps/cm^2. These techniques range from sequential rapidly–quenched deposition by pulsed lasers to continuous, very slow deposition by off–axis sputtering.

Off–axis sputtering has become a widely used technique. Although the deposition rate is quite slow (films of a few thousand ångstroms thickness take a few hours to deposit), the process appears to have the potential to make large–area microwave films with uniform properties.

Control of Orientation

A major and important difference between low–T_c materials such as Nb and the high–T_c oxides is the anisotropic properties of the latter. Generally, films are grown with c axis normal to the substrate, namely the more highly conductive copper oxide planes in the plane of the substrate. Recently several groups have grown films with the a axis perpendicular to the substrate. This is significant for some devices (for example, tunnel junctions) because the a–axis coherence length is longer than the c–axis coherence length.

Choices of Materials

Many oxide materials can be considered for applications. YBCO is the most commonly used material in the United States and Europe. Only a few groups work with thallium films, which at present require a post anneal in thallium vapor. The bismuth and the lead–doped bismuth materials are popular in Japan. The only other oxide which seems to have the potential to be commercially important is $BaKBiO_3$.

All the higher–T_c materials exist as multiple phases, and it is difficult to grow single phase films of the bismuth and thallium materials. Although these two families of compounds have higher transition temperatures, up to 125K, generally T_c's are in the range 100 to 110K. At present, YBCO at 77K has critical current densities as high as either the bismuth or thallium materials. So on the basis purely of J_c, there is no advantage yet for bismuth or thallium films over YBCO. For microwave applications, surface resistance is a more valued criterion, and YBCO films fall comfortably below copper up to frequencies of about 100 GHz. Recent data shows an advantage in surface resistance for thallium films in a small temperature window from roughly 70K to 90K (Bourne L C et al 1990).

In addition to choice of material, there is, of course, flexibility of choice in operating temperature. Comparison of reduced temperatures indicates that using YBCO at 77K is equivalent to using NbN at 13K or Nb at 8K, neither of which has been considered seriously. 77K might be a viable operating temperature for thallium films if they could be made with the maximum T_c of 125K. $BaKBiO_3$ operating at 15K is equivalent to operating Nb near 4K, whereas YBCO operating at 27K is better than NbN operating at 4.2K. Thus it appears that liquid neon refrigeration should receive more serious attention for YBCO applications.

Needs

As the demand grows for films on larger areas and in multilayer device structures, it will be important to develop an in situ multilayer process for large area deposition. There are increasing signs that CVD growth might satisfy this need.

Sputtering also has the potential for large–area deposition. At present, 2" diameter films are being made by off–axis sputtering. For some kinds of microwave devices, it will be necessary to deposit high–quality superconducting films on both sides of the substrate. This is most easily done by CVD, but in principle, all the common deposition methods can be used. Large–area deposition will also require uniform heating of the substrate, say to ±5°C.

For the growth of films on substrates containing other devices, such as GaAs or Si components, reduction in the growth temperature of the superconducting film will have obvious advantages. Despite some success in this direction, it still appears that the highest quality films are grown at temperatures near 700°C.

Any multilayer technology will require the growth of very smooth films. Study of the available literature shows that by changing deposition conditions and substrate materials, films have been grown with surface textures that are occasionally smooth but often very rough. This problem will have to receive more attention.

An important need is to control the epitaxy of the films, particularly in the plane of the substrate. This fact is illustrated, for example, by the observation of the reduction in critical current across grain boundaries in YBCO films as a function of the misalignment of the adjacent grains (Dimos et al 1990). This work implies that any polycrystalline film with grain boundary misalignments greater than, say, 10 degrees, will have significantly reduced critical current densities. Other reports show that the surface resistance of the films increases with the number of high–angle grain boundaries in the plane of the film (Laderman SS et al 1990). So it appears that HTSC electronics, in contrast to LTSC, is faced with the necessity to use single crystalline epitaxial films in most cases. We also have to face the additional complication of their anisotropy.

Processing

Advances

In general, processing of the HTSC oxides has not presented the unusual challenges that their synthesis has created. Lithography, ion milling, and chemical etching of these oxides appear to be relatively straightforward. The method of choice, at least below 1 micron, is ion milling or focused ion milling. Small structures (down to about 0.1 micron) have been demonstrated without any serious degradation of superconducting properties.

The most complex circuit made in HTSC to date, and therefore the most complex processing procedure, appears to be the flux transformer fabricated for use in a SQUID (Wellstood F C et al 1990). In this circuit, several interesting challenges had to be overcome — for example, a contact between two superconducting films made at different times. An insulating epitaxial layer of $SrTiO_3$ separates the two YBCO films in other areas. The top film of the structure was patterned into a spiral using a lithographic process. TEM cross sections of the YBCO / $SrTiO_3$ / YBCO structure indicates fully epitaxial growth, which is probably a necessity for a high transition temperature in the top film of the structure. This example illustrates the need for an epitaxial technology in the future.

It would be a mistake to assume that there are no inherent problems in the processing of HTSC materials. The oxides are much more sensitive to damage, e.g. from alpha particles, than niobium. Particle damage introduces structural changes and increases in resistance; eventually superconductivity is destroyed altogether. In some reports, the critical current density is affected even more rapidly than the critical temperature.

Ohmic contacts are a necessity in all electronic technologies. Silver and gold have been used to prepare very low resistance contacts (Ekin J et al 1988). In bulk materials, values of 10^{-10} ohms for 1 square centimeter contact area have been achieved, and values almost as small have been reported for films.

Another difference between the high–T_c and low–T_c materials is the sensitivity of the former to changes in oxygen content and to chemical doping. In fact, a number of device structures have been created by local doping techniques; for example, by extraction of oxygen or by poisoning with materials such as silicon.

Needs

In multilayer structures, it will be of great value to develop etches which preferentially remove one layer but not others. There has been one report of such a case (Shokoohi F K et al 1989). Little work has been done in the study of the storage life of high–quality films — for example, a measurement of surface resistance over a period of months with exposure to room atmosphere. On the other hand, the early reports of severe degradation of films exposed to air and humidity were probably exaggerated because of the low quality of these early films.

Low–temperature processing will be desirable in any hybrid structure that utilizes HTSC materials because of the large changes in properties that occur with oxygen loss at temperatures greater than 300°C.

Characterization

Although no new characterization techniques have been developed specifically for oxides, all the conventional structural, chemical, and electrical instruments are successfully being used to characterize these materials.

Needs

There is a significant need to develop a standard measurement of surface resistance that can be applied to unpatterned films from 4.2 to 100K. This technique should eventually be recognized by practitioners in the field as the standard measurement.

As films are deposited on larger area substrates, it will also be necessary to develop techniques to measure them nondestructively at low temperatures. Film uniformity over these large areas will also be an issue. It is becoming appreciated that film properties, especially critical current density, must be determined by local structure and chemistry over small length scales. Characterization tools are not available which allow determination, say, of local oxygen content or anti–site defects over these distances.

Junctions

Achievements

It is a major disappointment that hysteretic Josephson junctions of the conventional type have not been made with YBCO. Nevertheless, significant progress is being made in the demonstration of other types of devices. Trilayer processing has been demonstrated in both YBCO / PrBCO / YBCO (Wu X D, Venkatesan T et al 1990) and in the bismuth system (Mizuno K et al 1990). Even more encouraging progress has been made with weak links of various types. One is the grain boundary introduced into the film by growing it on a substrate made of a bicrystal with misalignment greater than 10 degrees. Recently an impressive SQUID was demonstrated in which the two junctions were produced over such a substrate (Gross R, Chaudhari P, et al 1990). The rest of the film presumably had no grain boundaries at all. These SQUIDs operated at temperatures near 90K. The grain boundary seems to be nature's way of providing a weak link device with dimensions comparable to the very short coherence lengths of these materials. Other interesting weak link devices are the step junction, in which an oriented film is made to follow a sharp step in the substrate material, and edge junctions, in which two c–axis oreinted films are made to touch each other in the plane. The availability of these three types of weak link device gives hope that their reproducibility will be improved and circuits of some complexity will be made from them.

Very promising results have been reported for the highest–T_c isotropic material, $BaKBiO_3$ (Dynes R C et al 1990). A tunnel junction made by pressing a single crystal against a thin film exhibits a well–defined energy gap of about 3.7 kT_c and a small Josephson current, which is reported to modulate with magnetic field. Thus the difficulties associated with preparation of tunnel junctions with the oxides seems to be associated with their anisotropy. The anisotropy of the higher–T_c oxides has also been illustrated in SNS junctions. In structures of the type YBCO / Au / Nb, it was reported that no supercurrent flowed when the c axis was normal to the gold surface (Akoh H et al 1990). However, when the ab planes were at an angle or normal to the gold surface, a supercurrent was observed.

Needs

Although the progress with individual weak link devices of various types is promising, any circuit technology will require devices that can be created reproducibly anywhere on the substrate with uniformity of critical current and I_cR products. The strong likelihood of hysteretic junctions using $BaKBiO_3$ indicates that 15K will be a desirable operating temperature, well within the range of closed–cycle refrigerators. Flux transfer logic circuits appear likely at higher temperatures. The critical role that will be played by anisotropy in these materials has already been discussed. The understanding of anisotropy constraints on device performance will be important.

Devices / Components

Despite the challenges of the oxide materials, some signficant advances are being made at the device / component level. Microwave components have been delivered to the U.S. Navy for a space experiment, and good progress is being made in fabricating various microwave components from thallium films. High–T_c SQUIDs have demonstrated performance approaching that of today's commercial SQUIDs operating at 4.2K.

Almost all the devices that have been built from high–T_c superconductors were proposed earlier for low–T_c materials However, a novel flux flow type of cryotron (Hohenwarter G K G et al 1989) has been reported in high–T_c in which the field created by a control line is coupled into a weak link in a second film. This device has demonstrated quite a high frequency response.

Needs

Although significant reductions in surface resistance values have been achieved as film quality has improved, the present values are still well above theoretical limits; hence, further improvement is required, especially at high temperatures and in the presence of high microwave powers. This will require improved control of film microstructure. In many applications, such as the infrared detectors, noise in the films is an important parameter. Thus it is required that low–noise films be made on substrates that are not always ideal lattice–matched materials. This, too, will require control of the film microstructure. To date, the focus of attention in the oxide materials has been on their superconducting properties. However, oxides exhibit many other interesting physical behaviors, some of which have been used in applications of completely different types. It appears likely that eventually new devices will be created using the wide variety of properties exhibited by such oxides.

Products

Although it cannot be claimed that a significant HTSC technology exists, nevertheless HTSC products are appearing in the marketplace. As noted above, substrates and sputtering targets are widely available commercially. Microwave films and microwave filters are on the market. It seems reasonable that infrared detectors will be available before the end of 1991 and that high–T_c magnetometers will be demonstrated, at least as protoytpes, in the same time frame.

On favorable substrates, interconnects could be made available in two years or so. Some simple digital circuits operating at 15K with hysteretic $BaKBiO_3$ junctions will probably be demonstrated in less than 3 years. Progress with nonhysteretic junctions gives some hope that digital circuits at 77K of the flux transfer type will be fabricated quite soon.

Summary

Given the most unusual properties of these oxide superconductors and their differences from the more familiar superconductors, progress in many areas of thin film synthesis and electronic applications has been rapid. Films of high quality can now be made by many processes. The progress in weak link devices is very encouraging. The recent demonstration of a BCS–like tunnel junction in $BaKBiO_3$ creates the possibility of Josephson circuits operating with closed–cycle refrigerators.

References

Akoh H, Camerlingo C and Takada S 1990 *Appl. Phys. Lett.* 56 (15)
Berezin A B, Yuan C W and de Lozanne L L 1990 *Appl. Phys. Lett.* 57 (90)
Bourne L E, Hammond R B, Robinson McD, Eddy M M, Olson W L and James T W 1990 *Appl. Phys. Lett.* 56 (2233)
Char K, Newman N, Garrison S M, Barton R W, Taber R C, Laderman S S and Jacowitz R D 1990 *Appl. Phys. Lett.* 57 (4) pp 409–411
Dimos et al 1990 *Physical Review B* Volume 41 #7
Dynes et al 1990, Proceedings of LT 19, to be published
Ekin J W, Larson T M, Bergren N F, Nelson H A, Swartzlander A B, Kazmerski L, Panson H A and Blankenship B A 1988 Mankiewich P *Appl. Phys. Lett.* 52 p. 1819
Eom C B, Sun J, Yamamoto K, Marshall A F, Luther K E, Geballe T H and Laderman S S 1989 *Appl. Phys. Lett.* 55 (6)
Fork D K, Fenner D B, Barton R W, Phillips J M, Connell G A N, Boyce J B and Geballe T H 1990 to be published
Gross R et al 1990 *Appl. Phys. Lett.* 57 (2)
Gross R, Chaudhari P, Kawasaki M, Metchen M B and Gupta A 1990 to be published
Gupta A, Gross R, Olsson E, Segmüller A, Koren G and Tsuei C C 1990 *Physical Review Letters* Volume 64 Number 26 pp 3191–3194
Hammond R B, Negrete G V, Bourne L C, Strother D D, Cardon A H and Eddy M M 1990 *Appl. Phys. Lett.* 57 (8) p 825
Hohenwarter G K G, Martens J S, McGinnis D P, Beyer J B, Nordman J E and Ginley D S 1989 Proce IEEE MAG 25 954
Laderman S S, Taber R C, Jacowitz R D, Eom C B, Hylton T L, Marshall A F, Beasley M R and Geballe T H 1990 to be published
Mankiewich P, Schwartz D B, Howard R E, Jackel L L, Straunghn B L, Burkhardt E G and Dayem A H 1988 SPIE 948 37
Mizuno K, Higashino K, Setsune K and Wada K 1990 *Appl. Phys. Lett.* 56 (15) p 1470
Myoren H, Nishiyama Y, Naokazu M, Kai Y, Yamanaka Y, Osaka Y and Nishiyama F 1990 *Japanese Journal of Applied Physics* Vol. 29, No. 6, pp. L955–L957
Shokoohi F K, Schiavone L M, Rogers C T, Inam A, Wu X D, Nazar L and Venkatesan T 1989 *Appl. Phys. Lett.* 55 (2661)
Simon R W 1989, SPIE 1187 2
Wellstood F C, Kingston J J and Clarke J 1990 *Appl. Phys. Lett.* 56 (23) pp 2336–2338
Wu X D, Muenchausen R E, Foltyn S, Estler R C, Dye R C, Flamme C, Nogar N S, Garcia A R, Martin J and Tesmer J 1990 *Appl. Phys. Lett.* 56 (15)
Wu X D, Venkatesan T et al 1990 *Appl. Phys. Lett.* 56 (4)

Supercond. Sci. Technol. 4 (1991) S59–66. Printed in the UK

Real space bosons in the bismuth oxide and copper oxide superconductors

Nevill Mott

Cavendish Laboratory and IRC in Superconductivity, Madingley Road, Cambridge

ABSTRACT: It is suggested, following many other investigators, that, both in the new high temperature superconductors and the cubic materials such as $Ba_{1-x}K_xBiO_3$ ($T_c = 34K$), the phenomenon is caused by real space bosons, the phonons playing no major role. In the latter, Bi^{3+} moving through heavily screened Bi^{5+} are the bosons; T_c is the temperature at which the boson gas becomes non-degenerate, so bosons carry the current above T_c. In the quasi-two dimensional copper oxide materials it is maintained, following previous papers, that spin bipolarons are formed, but there is now strong evidence that only a fraction of the carriers (oxygen 2p holes hybridised with $Cu3d^8$) form bosons, and that both above and below T_c a condensed gas of fermions exists. The fermions are heavy; a new description of the role of the copper moments is given. The fermions appear not to be responsible for the conductivity above T_c, being probably localized by disorder, so here again the current is carried by bosons.

In superconductors described by the BCS theory, the critical temperature T_c is that at which the "bosons", the strongly overlapping gas of Cooper pairs, disappears; above T_c the electric properties are those of normal metals. Before the discovery of the high-temperature superconductors there have been suggestions (see for instance the review of Micnas et al 1990) that materials could exist in which superconductivity is caused by a degenerate gas of real space bosons which do *not* overlap, and that T_c is the temperature at which the gas of bosons becomes wholly non-degenerate. For such materials bosons will exist above T_c, and can have a major effect on the electrical properties (Alexandrov et al 1986). In this lecture I shall suggest that both the cubic bismuth compounds such as $Ba_{1-x}K_xBiO_3$ and the quasi-two-dimensional copper oxide superconductors ($La_{2-x}Sr_xCuO_4$, $YBa_2Cu_3O_{7-\delta}$) are in this category. In the former all the carriers form bosons, so that the analysis is similar to that for superfluid He^4. In the latter probably only a proportion of order 10% do so, but none the less the evidence suggests that above T_c the bosons carry the current.

I will begin with the cubic bismuth compounds, discussed in detail by Micnas et al (1990). In for instance $Ba_{1-x}K_xBiO_3$, with $T_c \sim 34K$, if Ba gives up two electrons, K one and oxygen receives six, Bi must give up $6–2(1–x)–x = 4+x$, leaving $1–x$ electrons. It is supposed that these form $\frac{1}{2}(1–x)$ Bi^{3+} ions, moving as bosons through an ordered lattice of Bi^{5+}. Naturally the empty Bi^{5+} site will be strongly screened by overlap from the oxygens; no site will be strongly charged. The Bi^{3+}, with 2 electrons is s states, are spinless bosons. If we use for T_c the formula for the degeneracy temperature of a gas of non-interacting bosons

0953-2048/91/01S059 + 08 $03.50

$$kT_c = 3.3\,\hbar^2\, n^{2/3} / m \qquad (1)$$

where m is the effective mass of a boson and n their number for unit volume, then the prediction that T_c varies as $n^{2/3}$ is verified experimentally up to x = 0.3 (de Jongh 1988) where a charge-density wave is set up; that is, the Bi^{3+} sites order. By comparison with the observed T_c, we find $m \sim 20\, m_e$.

The model predicts that above T_c the current is carried wholly by bosons. The evidence for this quoted by Micnas et al is that above T_c, and in the state with or without charge density waves, the material is diamagnetic. The conductivity above T_c is very low, or order $10\Omega^{-1}\,cm^{-1}$ with a temperature dependence depending on composition that can have either sign. For a non-degenerate Bose gas we may write

$$\sigma = n\, q^2\, \ell/(mkT)^{1/2} \qquad q = 2e$$

and if ℓ is caused by boson-boson collisions

$$\ell = 1/nA$$

where **A** is the collision cross section. A value of **A** of about 10^{-17} cm^2 is required to give the observed resistivity. For coulomb scattering **A** should *decrease* with increasing T, giving an increase in σ; the term 1/kT gives the opposite effect. So either positive or negative values of $d\rho/dT$ can be understood.

We turn now to $La_{2-x}\, Sr_x\, CuO_4$ and $YBa_2Cu_3O_{7-\delta}$ (Mott 1990d). For x = 0 and $\delta <$ 0.5 these are antiferromagnetic insulators. The carriers are thought to be oxygen 2p holes, hybridised with $Cu(3d^8)$; without this hybridisation the effective mass would be very large, which is why Cu, rather than another metal, is necessary. In LSCO these holes move in an impurity band caused by the Sr (much as holes move in Si : B); in YBCO with $\delta = 0$ there is one carrier (hole) per formula unit; the corresponding positive charge is not present as point defects but in a change of the structure which locates it in the region of a few atomic radii; so we have, so to speak, the equivalent of an *ordered* impurity band. In either case we believe the Hubbard U splits the band into upper and lower Hubbard bands, for reasons to be discussed below.

For superconduction we need a mechanism of pairing, and one which involves the apparent disappearance of the moments in the metallic superconducting state. We believe – following many authors, that the carrier forms some kind of "spin polaron", orienting copper $3d^9$ spins in an area round it, so that they resonate with a frequency of – say $\sim 10^{14}\ s^{-1}$ as the carrier moves. A degenerate gas of such entities is envisaged. In the author's previous papers a spin polaron of the kind identified in $Gd_{3-x}V_xS_4$ (Von Molnar et al 1983) is envisaged (V denoting a vacancy) as illustrated in fig. 1. This is heavy, suffers Anderson localization in the random field of the vacancies, and metallic conduction at low temperatures occurs only in a strong magnetic field, when these polarons cannot form. More recently however in the lanthanum copper oxide neutron scattering has shown (Jurgens 1990) that the very strong antiferromagnetic coupling between neighbouring copper spins in the planes persists, so that perhaps an extension of the tJ model is appropriate, copper ions in the neighbourhood of the oxygen hole retaining their antiferromagnetic local order which dies away gradually – so that all spins are resonating but keep short range order (fig. 2). The heavy mass of such a system has its origin in the fact that it must tunnel to the next nearest Cu ion. That coupling between two such polarons with opposite moments can exist is easy to see through the anitferromagnetic attraction A, B in fig. 3. We think this system, rather than the fermion, behaves like a spin polaron, the diffusion coefficient being determined by the spin flip frequency $kT_N/\hbar$, so that the diffusion coefficient is independent of temperatures. Any such charged pseudoparticle will distort the lattice round it, so phonons may well play a role in bonding and in determining the effective mass.

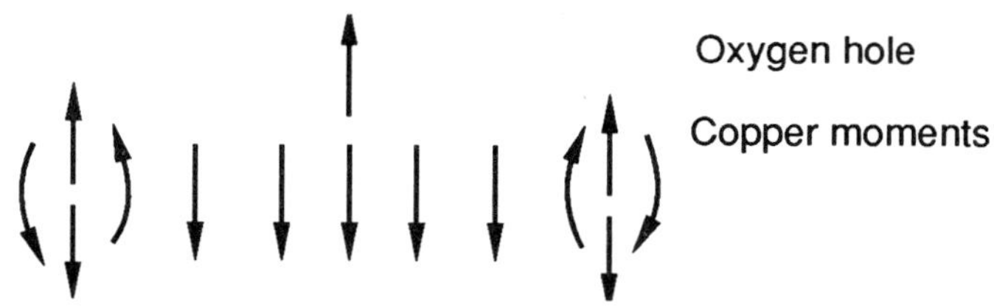

Figure 1

Form of spin polaron envisaged in previous papers; at the extremity of the polaron moments are resonating.

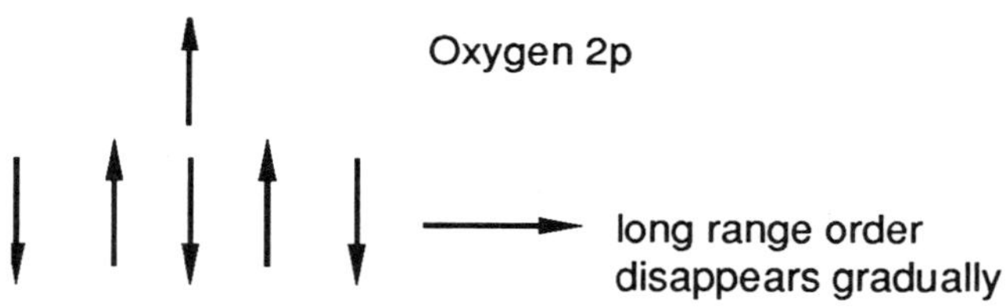

Figure 2

Suggested form of spin polaron, the moment being of order 1 Bohr magneton; the copper ions contribute no moment.

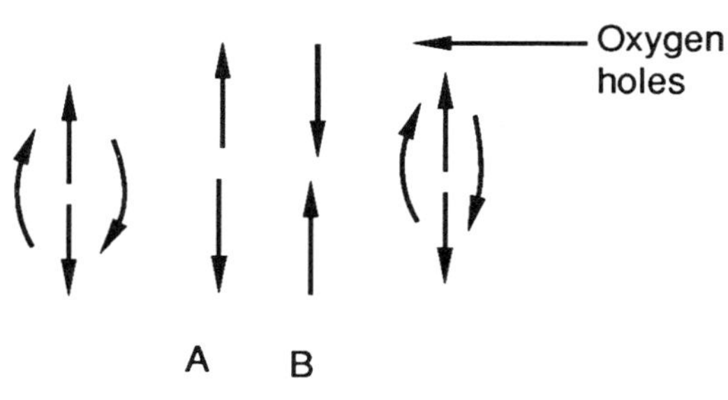

Figure 3

Showing coupling between two spin polarons of the form suggested in fig. 2.

Estimates of the effective mass of such polarons and bipolarons have not been made; we follow Kamimura et al (1989) in believing that the polaron is heavier. The reason Kamimura gives is that a large effective mass results from the necessity of reversing spin directions as the polaron moves, and in a bipolaron the effects of the two positive holes on the copper spins partially cancel each other. For the bipolaron (not the fermion) we believe that Vigren's (1973) prediction that diffusion is activated by spin flip, with a diffusion coefficient D of the order

$$D \sim \frac{1}{4}\omega a_o^2 \qquad (2)$$

and independent of temperature; here $\hbar\omega = kT_N$ where T_N is a Néel temperature, and a_o the distance the polaron diffuses as a consequence of spin flip.

With such a model, and no interaction between the bosons, T_c is given by eq. (1) and we expect for the specific heat at low temperatures $c_v \sim T^{3/2}$. We believe however that interaction is essential for superconductivity (Mott 1990d) and leads to c_v T^3.

An issue to be discussed in this paper is whether all the fermions form bosons, so that above T_c a non-degenerate gas of bosons is alone available to carry the current as in the cubic bismuth oxides, or whether only a fraction of them does so. In the author's papers it was supposed that the former was the case. A deduction was made about the thermopower S, supposing that conduction is in an impurity band. The Heikes formula was used (Heikes and Ure 1961)

$$S = (k/q)\, \ell n\, \{(1-z)/z\}$$

where q = 2e for bosons and z is the ratio of carriers to sites. In an impurity band in a material such as LSCO the strontium atoms are the sites, and without compensation no conduction is to be expected for large values of the Hubbard U, but if bosons are formed then clearly $z = {}^1/_2$ and according to (2) S should vanish. Actually several measurements (e.g. Howson et al 1989) show that S is small (4 μ V/k in contrast to k/e ~ 57) independent of T up to 300K and of either sign depending on heat treatment and oxygen content. It seems to us difficult to account for this in any other way.

Also a spin polaron hypothesis can account for the observation that in "good" specimens the resistivity ρ above T_c is proportional to T, and rises above the Ioffe-Regel value. This follows from eqn. (2), using the Einstein relation

$$\sigma = n\, q^2\, D/kT$$

if D is independent of temperature. We note that a diffusing spin polaron cannot be scattered by phonons or point defects; only macroscopic strains, producing changes of density and thus a change in the energy of the pseudoparticle over a volume comparable with its size might introduce a small activation energy, giving a resistivity of the form

$$T \exp (W/kT) \sim T + W/k$$

showing the appearance of an apparent residual resistivity. It is observed (Loram et al 1990a) that the addition of cobalt to YBCO, which is located outside the planes and is thought to produce internal strains, does produce a large residual resistance when ρ is extrapolated to zero temperature: zinc which goes in on the planes, does not.

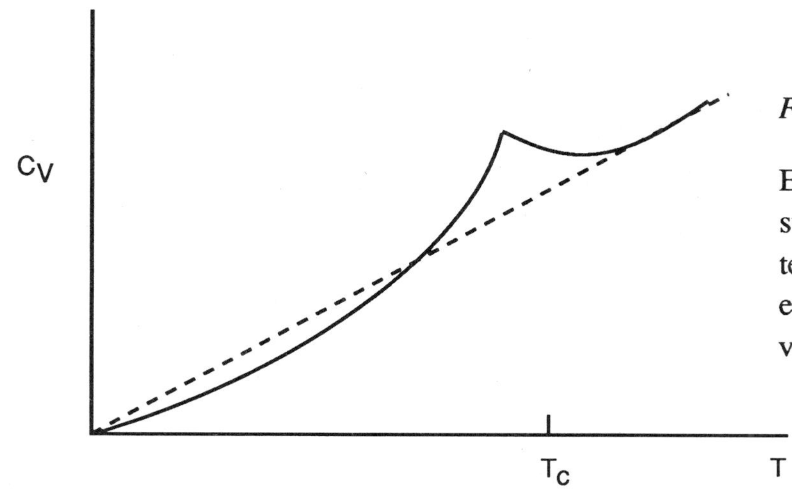

Figure 4

Electronic specific heat of high T_c superconductor as a function of temperature. The dotted line shows the extrapolation of the high temperature value to T–v.

The evidence, however, that only a fraction of the fermions form bosons comes from the work of Loram et al (1990a,b) on the specific heat, which particularly for YBCO appears as in fig. 4. Above T_c a specific heat which increases as T suggests a degenerate gas of fermions with degeneracy temperature $T_F \sim 1500K$. The magnetic susceptibility is independent of temperature and corresponds to the Pauli form

$$X = n\, \mu^2/kT_F$$

where μ is one Bohr magnetron; if this is caused by spin polarons, their moment must be about this, and as already indicated this rules out the kind of polaron shown in fig. 1. For the type shown in fig. 2 it appears that the copper moments contribute zero moment. The width of the peak increases when the planes are less strong coupled, illustrating the strong superconducting fluctuations expected in two dimensional systems. However for our purpose the important point is that the entropy corresponding to the disappearance of the condensed phase of the bosons is only about 1/10 of nk $\ell n2$, the value it should have for a non-interacting gas. Though disorder may account for some of this (Mott 1990b), it seems strong evidence that only a fraction of the carriers form bosons.

The evidence from the thermopower, however, suggests that it cannot be the fermions that carry the current; as remarked by Micnas et al, the thermopower for fermions should be of a totally different form from that observed and much larger. Also, as emphasized by Yu et al

(1989), the thermopower is unaffected by a magnetic field up to 30 Tesla. The current must then be carried by spinless particles, and the fermions must be Anderson localized, as they are in gadolinium sulphide at low temperature.

We believe the reason why only a fraction of the fermions form bosons must be that a boson takes up *more* volume in real space than two fermions – which is perhaps surprising. It may be that the fermions do not form extended polarons and are rather centres orienting just one extra copper spin and are unable to become any bigger because of overcrowding. The overcrowding effect which prevents the formation of more than a small number of (dielectric) bipolarons is shown by the work of Ruscher et al (1988a,b) on $NbO_{2.5-x}$ and ternary Nb-W oxides; these authors introduce this idea.

An estimate of the number of current carriers would be of great interest. Unfortunately the Hall coefficient, if our model is correct, cannot be used; according to the present author (Mott 1990c), we expect for YBCO

$$R_H = (T_o/T)/nec$$

and T_o is difficult to estimate.

The model with 10% of bosons has certain other advantages. The density of states in YBCO at T=0 will be as in fig. 5 (cf. Wheatley 1990); in the shaded region of energies the (hole) fermions form bosons, so no states are available just above E_F, and no linear specific heat is expected. Wheatley shows that if the fermions are heavier than the bosons, a sharp energy should separate the ranges as in the fermion spectrum where the latter form bosons from that where they remain fermions. Above T_c the density of bosons above E_F is small, and a linear specific heat expected, which can be responsible for the Korringa broadening of the oxygen Knight shift (Friedel 1990) and for the observations of a Fermi energy by angle resolved photo-emission. The bosons will have little effect on the specific heat above T_c, *only* on the electrical properties through the assumption that the fermions are immobile.

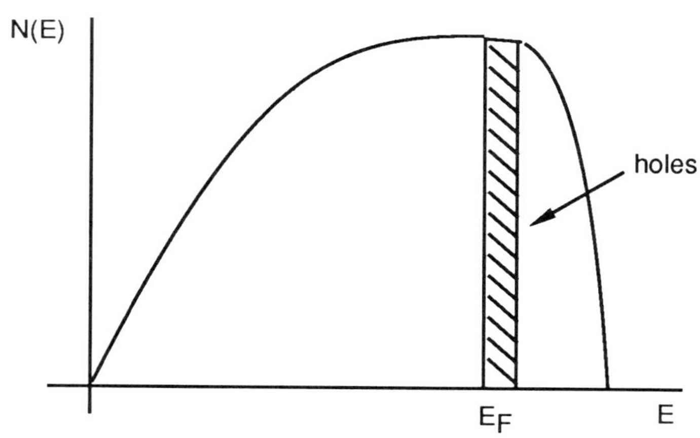

Figure 5

Density of states in YBCO (schematic). The energy range on the fermion spectrum occupied by bosons at T = 0 is shaded.

The drop in T_c resulting from adding zinc (which is incorporated in the planes in YBCO) fits this hypothesis too; with some of the copper moments put out of action the number of bosons formed will decrease, so T_c drops according to eqn (1).

An alternative explanation of the small values of the entropy has been described by de Jongh (1980a,b) and earlier papers, and quantified by a comparison with the critical point in antiferromagnetics (Matsubara and Matsuda 1956). Here it is thought that fluctuations, giving rise to local superconductivity about T_c, can account for the missing entropy; the apparent linear specific heat above T_c shown in fig 4 is thus thought to be due to such fluctuations. An argument against this interpretation is that the renormalisation of phonon frequency by superconductivity shows a sharp cut-off at T_c. (Salje 1990)

References

Alexandrov A S, Ranninger J and Robaszkiewics S 1986 Phys. Rev. B **33** 4526

de Jongh L J 1988 Physica C **152** 171

Friedel J 1989 J Phys: Condensed Matter **1** 7757

Heikes R R and Ure R W 1961 Thermoelectricity, p81, Interscience N.Y.

Howson M A, Salaman M B, Friedmann T A, Inderless S E, Rice J P, Ginsberg D M, Ghoron K 1989 J Phys. Condens Matter **1** 465

Jurgens M J G M 1990 Doctor Thesis University of Leiden "Neutron scattering studies on the high T_c superconductors $YBa_2Cu_3O_{6+}$"

Kamimura H, Matsuno S and Saito R 1988 Solid State Commun. **67** 363

Loram J W, Mirza K A and Freeman P F 1990 Physica C (in press)

Loram J W and Mirza K A 1990 Properties of High T_c Superconductors, Springer-Verlag

Matsubura T and Matsuda H 1956 Prog. Theor. Phys. **16** 596

Micnas R, Ranninger J and Robaszkiewics S 1990 Rev. Mod. Phys. **62** 113

Mott N F 1990a Phil. Mag. Lett. **61** 217; 1990b ibid **62** 37; 1990c ibid (in press); 1990d Adv. in Phys. **39** 55

Ruscher C, Salje E and Hussain A 1988 J. Phys. C **21** 3737 and 4465

Salje E 1990 Phil Mag Lett (in press); see also Salje, this volume.

Vigren P T 1973 J Phys C **6** 967

Von Molnar S, Briggs A, Floquet J and Remenyi G 1983 Phys. Rev. Lett. **57** 706

Wheatley J M 1990 Phys. Rev. B **42** 4946

Yu R C, Naughton M J, Yan X, Chaikin P M, Holtzberg F, Green R L, Stuart J and Davies P 1988 Phys Rev B **37** 7963

Supercond. Sci. Technol. **4** (1991) S67–69. Printed in the UK

The Metal-Insulator Transition in $La_{1.85}Sr_{0.15}CuO_4$ with Various Substitutions for Cu†

S. Guha*, Marta Z. Cieplak+, Gang Xiao**, J. Q. Xiao++, C. L. Chien++, H. Kojima*, and P. Lindenfeld*

*Serin Physics Laboratory, Rutgers University, Piscataway, NJ 08855, +Polish Academy of Sciences, **Brown University, ++Johns Hopkins University.

We have determined the concentrations x of Zn, Ga, Fe, and Ni impurities which lead to a metal-insulator transition when substituted for Cu in $La_{1.85}Sr_{0.15}CuO_4$. In each case these values of x are larger than those which are associated with the disappearance of superconductivity.

One of the special features of the perovskite superconductors is that relatively small changes in composition may cause them to become insulators. The approach to the metal-insulator (M-I) transition also drastically affects the transition temperature and other superconducting properties. In $La_{1.85}Sr_{0.15}CuO_4$ substitution of between about 2 and 4% of Ga, Zn, Fe, Ni, or Co for Cu causes T_c to go to zero (Cieplak et al. 1989, Xiao et al. 1989, Cieplak et al. 1990).

The investigation of the composition at which the M-I transition takes place has received much less attention, even though measurements of resistance as a function of temperature indicate that there is a close correlation between the disappearance of superconductivity and the M-I transition. Part of the reason is that unless the measurements are carried below 1K in magnetic fields sufficient to destroy any remaining superconductivity, estimates of the composition at which the M-I transition occurs are subject to considerable uncertainty.

We have made measurements on polycrystalline specimens with impurities of Zn, Ga, Fe, and Ni prepared by the same methods as those of previous investigations (Cieplak et al. 1989, Xiao et al. 1989) down to 50 mK and in fields up to 8T.

In fields sufficient to reach the normal state we usually observe the negative magnetoresistance characteristic of electron localization. We extrapolate the normal-state magnetoresistance R(H,T) to H=0 to determine R(0,T), the presumed zero-field resistance in the normal state. From the results at different temperatures we estimate $\sigma(0,0)$, the normal-state conductivity at T=0.

0953-2048/91/01S067 + 03 $03.50

Figure 1 shows the zero-field variation of resistivity with temperature for a series of specimens with Zn impurity, plotted as log ρ against $T^{-1/2}$.

Fig. 2 shows $\sigma(0,0)$ as a function of the impurity fraction x, showing an M-I transition for Zn close to x=0.12. The figure also includes the results for Ga, Fe, and Ni.

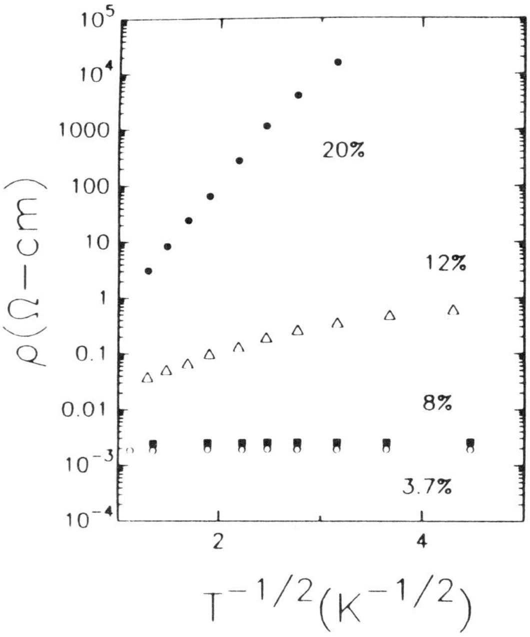

Fig. 1 The resistivity measured in zero magnetic field for four specimens with Zn, plotted as log ρ against $T^{-1/2}$.

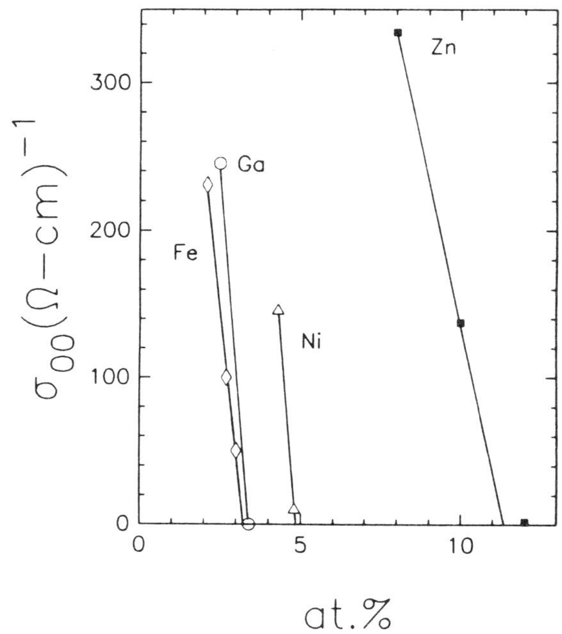

Fig. 2 The normal-state conductivity as extrapolated to H=0 and T=0.

We ascribe the difference between different impurities to the expectation that there are two distinct mechanisms which lead to the M-I transition, and a third which, in addition, affects the disappearance of superconductivity.

The M-I transition is affected firstly by the disorder which results from the introduction of impurities, and secondly by the change in carrier concentration. Thirdly the impurities may lead to the appearance of localized magnetic moments with an associated strong effect on superconductivity (Xiao et al. 1990a).

The change in carrier concentration n has been investigated by studies of the resistivity (Cieplak et al. 1990) and the Hall effect (Xiao et al. 1990b). The results indicate that n is increased by about 5% for each at. % of Zn, decreased by the same amount for Ni, and decreased more (by about a factor of two) for Fe and Ga.

We see that there is a strong correlation between the values of n at the M-I transition and the values x_{MI} at which the transition takes place for the various impurity elements. Zn, which increases n, has the largest x_{MI}. Ni decreases n and depresses x_{MI}, and for Fe both parameters are decreased even further.

For each of the impurities which we have used there is a metallic but non-superconducting range of values of x. This was also true in another system (granular aluminum) with an M-I transition entirely determined by disorder (Miller et al. 1988). In the present series there is a substantial intermediate range, which we ascribe primarily to the strong effect on T_c of the localized moments which are associated with each of the impurities. It has been shown (Xiao et al. 1990a, Kataev et al. 1990) that even non-magnetic impurities lead to localized magnetic moments when they are substituted for Cu so as to produce unpaired copper spins.

The work of Miller et al. (1988) already suggested the general presence of a metallic, non-superconducting phase in superconductors with a disorder-induced M-I transition. The present work shows the existence of this phase in the LaSrCuO system, with the strong inference of its presence also in the other superconductors which share this material's essential structural features.

We would like to thank Andrew Covalesky, Deborah Doatch, and Han Truong for their help.

†Supported by the National Science Foundation, Grant Nos. DMR85-11982 and DMR89-17027.

Cieplak M Z, Xiao G, Bakhshai A and Chien C L 1989 Phys. Rev. B39 4222.
Cieplak M Z, Xiao G and Chien C L, 1990 Proc. Europ. Conf. on High-T_c Thin Films and Single Crystals, Ustron, Poland; ed. W. Gorzkowski, M. Gutowski, A. Reich, H. Szymczak, (Singapore: World Scientific).
Kataev V E, Kukovitskii E F, Teitelbaum G B and Finkelstein A M, 1990 JETP Lett. 51 129.
Miller T A, Kunchur M, Zhang Y Z, Lindenfeld P and McLean W L, 1988 Phys. Rev 61 217.
Xiao G, Bakhshai A, Cieplak M Z, Tesanovic Z and Chien C L, 1989 Phys. Rev. B39 315.
Xiao G, Cieplak M Z, Xiao J Q and Chien C L, 1990a to be published.
Xiao G, et al. 1990b to be published.

Supercond. Sci. Technol. **4** (1991) S70–72. Printed in the UK

Phonon-order parameter coupling in $YBa_2Cu_3O_{7-\delta}$ as observed by infrared absorption spectroscopy

B.Güttler
Physikalisch-Technische Bundesanstalt, Bundesallee 100, 3300 Braunschweig, Germany
E.Salje, P.Freeman, J.Blunt, M.Harris
Interdisciplinary Research Centre in Superconductivity, University of Cambridge, Madingley Road, Cambridge, U.K.
T.Duffield, C.Ager, H.Hughes
Department of Physics, Cavendish Laboratories, University of Cambridge, Madingley Road, Cambridge, U.K.

Abstract. Infrared absorption spectoscopy has been used to investigate the complete phonon frequency range ($20cm^{-1}$ - $700cm^{-1}$) at temperatures between 10K and 300K. A direct coupling of bulk phonon modes in $YBa_2Cu_3O_{7-\delta}$ with the superconductor order parameter is deduced from the critical temperature dependence in the superconducting low temperature phase ($T_c \sim 92K$). The excess intensity and the frequency shift of the mode near $580cm^{-1}$ are shown to be proportional to the square of a BCS-like order parameter in agreement with the predictions of Hard Mode Spectroscopy.

1. Introduction

Phonon mediated correlations have been discussed as one of the main processes which lead to high temperature superconductivity (see e.g. Phillips 1989). Studies of the isotope effect revealed, however, controversial experimental observations (Batlogg *et al* 1987, Franck *et al* 1989). It is thus the temperature dependence of the phonon properties in the superconducting state which may elucidate the relevance of the phonon involvement in superconductivity (Feile 1989, Genzel *et al* 1989). In this paper we confirm the existence of such temperature effects using infrared spectroscopy (i.e. bulk phonon spectroscopy) and, furthermore, we shall argue that a quantitative evaluation of the phonon spectra in the low temperature phase allows for the correlation with the thermodynamic order parameter of the superconducting phase.

2. Results and Discussion

Absorption spectra at 10K and 300K are shown in Fig.1. They exhibit an electronic background increasing almost linearly towards higher energies at room temperature. This background is overlain by phonon bands near $152cm^{-1}$, $193cm^{-1}$, and $580cm^{-1}$, and a complex double peak structure near $300cm^{-1}$. These features are in good agreement with the data from reflection spectroscopy concerning their number and position (Genzel *et al* 1989). All phonon bands reveal a distinct temperature dependence exhibiting the most pronounced appearance at low temperatures. The $580cm^{-1}$ mode is of particular interest because its energy is significantly larger than the highest estimates made for the superconducting gap energy ($8kT_c \sim 500cm^{-1}$, Collins et al 1989). This is important as it excludes electron phonon interaction effects related to the excitation of charge carriers across the energy gap (Axe and Shirane 1973). We use the temperature effect on the $580cm^{-1}$ mode in the vicinity of the phase transition around 92K for an evaluation of the transition behaviour on a

0953-2048/91/01S070 + 03 $03.50

thermodynamic basis. The excess free energy related to any phase transition should lead to a phonon renormalisation provided a physical coupling mechanism between the mode susceptibility and the order parameter exists (Bruce *et al* 1980, Bismayer 1988, Güttler *et al* 1989). Following the general treatment a nonlinear coupling of the type $\xi Q_i^2 Q^2$ between the phonon and the normal coordinate is always compatible with the symmetry requirements of the phonon. The strength ξ depends on the physical mechanism of the coupling. Our results show that the strength is non-zero even from modes with phonon energies well above the superconducting gap energy.

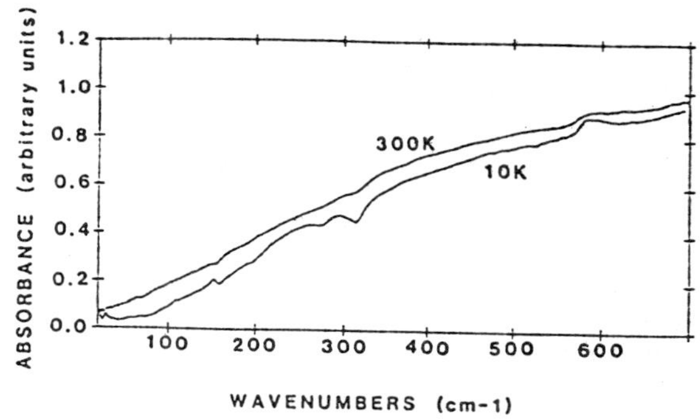

Fig.1 Absorption spectra of $YBa_2Cu_3O_{7-\delta}$ at 10K and 300K

A quantitative determination follows from the excess intensity of the 580cm^{-1} mode related to the low temperature phase. The excess intensity is revealed by the different spectra shown in Fig.2. The spectrum at 150K is subtracted from the spectrum at low temperatures. The absorption intensity of the mode *i* at any temperature T in the low temperature phase is

$$A_i\,(\mathrm{T}) = A_i\,(150\mathrm{K})\,\mathrm{x}\,\left(1 + \alpha\, Q\ (\mathrm{T})^2\right)$$

with $\alpha = 0.9$ +/- 0.1 for the 580cm^{-1} mode where Q is the normalised order parameter. The intensity increases almost linearly for $T_c > T > 1/2T_c$ on lowering the temperature while a mode softening is observed at the same time (~ 5cm^{-1}). The temperature effect flattens out at lower temperatures. We find a linear relation between the excess intensity ΔA_i and frequency shift $\Delta\omega_i$ of the mode *i* in the low temperature phase which is indeed expected from Hard Mode Spectroscopy.(Güttler *et al* 1989) for all temperature $T \leq T_c$. The most accurate information is revealed by the intensity changes. Fig.3 shows a comparison between the temperature dependence of $\Delta A_{580\mathrm{cm}^{-1}}$ and the square of the gap energy as predicted by BCS-theory. Hence we find a reasonable agreement between our data and a BCS-type temperature dependence of the order parameter. This, of course, does not imply that BCS-theory is correct for $YBa_2Cu_3O_{7-\delta}$ but only shows that the temperature dependence of the order parameter is BCS-like (or, indeed , follows a saturated Landau-Ginzburg behaviour,Salje *et al* 1990). The magnitude of the observed frequency changes is typical for phase transitions with excess Gibbs free energies of some 2-5 kJ/mole (see e.g. Bismayer 1988). The scaling of the frequency shift with the order parameter was also found by Salje (1990) who showed that the temperature evolution of the excess specific heat due to phonon anomalies in $YBa_2Cu_3O_7$ is proportional to the measured total excess heat capacity. Obhi and Salje (1990) have demonstrated that the relative changes of the phonon frequencies and the excess specific heat are also proportional to each other for Co-doped $YBa_2Cu_3O_7$.

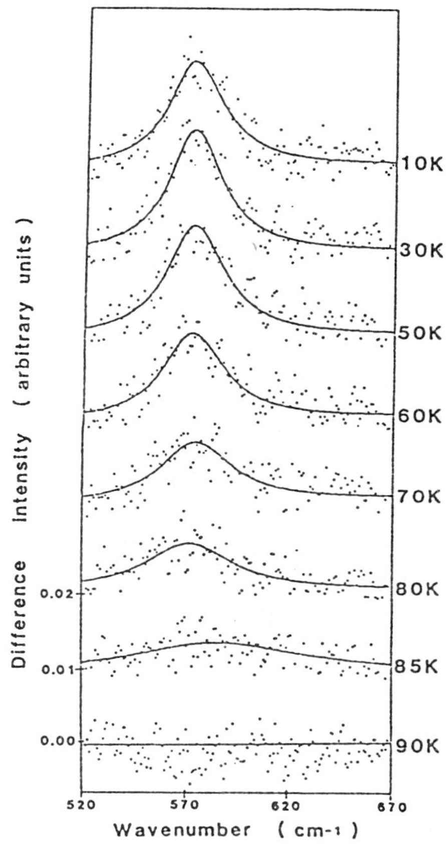

Fig.2 Difference spectra of the mode near 580cm^{-1}. Spectra at various temperatures were subtracted by the absorption spectra at 150K.

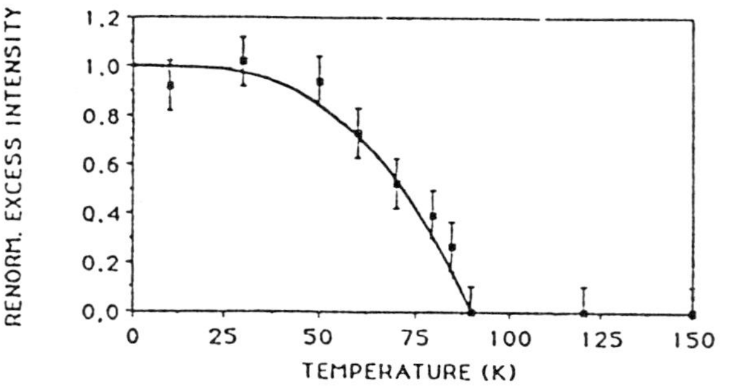

Fig.3 Temperature dependence of the excess intensity of the mode near 580cm^{-1} (squares) and the square of a BCS-like gap energy (line)

3. References

Axe JD and Shirane G (1973) Phys. Rev. Lett. **30**, 214

Batlogg B, Cava RJ, Jarayaman A, van Dover RB, Kourouklis GA, Sunshine S, Murphy DW, Rupp LW, Chen HS, White A, Mujsce AM and Rietman A (1987) Phys. Rev. **58**, 2333

Bismayer U (1988) In: Physical properties and thermodynamic behaviour of minerals. NATO ASI (ed. E.Salje) 143

Bruce AD, Taylor W, Murray AF (1980) J Phys C **13**: 483

Collins RT, Schlesinger Z, Holtzberg F and Feild C (1989) Phys. Rev. Lett. **63**, 422

Feile R (1989) Physica C **159**, 1

Franck JP, Jung J, Salomons GJ, Miner WA, Mohamed M A-K, Chrzanowski J, Gygax S, Irwin JC, Mitchell DF and Sproule I (1989) Physica C, **162-164**, 753

Genzel L, Wittlin A, Bauer M, Cardona M, Schönherr E and Simon A (1989) Phys. Rev. B **40**, 2170.

Güttler B, Salje E and Putnis A (1989) Phys Chem Minerals **16**, 365

Obhi S and Salje E (1990) Physica C submitted

Phillips JC (1989) "Physics of high-T_c superconductors", Academic Press, Inc. Boston, San Diego, New York, Berkeley, London, Sydney, Toyko, Toronto

Salje E, Wruck B and Thomas H (1990) Z.Phys. submitted

Salje E (1990) Phil.Mag.Lett. in press

Supercond. Sci. Technol. **4** (1991) S73–75. Printed in the UK

Anisotropic electronic structures in cuprate superconductors investigated by STM/STS

T. Hasegawa, M. Nantoh, N. Motohira, K. Kishio and K. Kitazawa

Department of Industrial Chemistry, University of Tokyo, Hongo, Tokyo 113, Japan

Abstract. The cleaved (ab plane) and edge (bc plane) surfaces of $Bi_2Sr_2CaCu_2O_y$ (BSCCO) single crystal was observed by scanning tunneling microscopy/spectroscopy (STM/STS) at room temperature. The STM image and STS conductance images at various bias voltages have revealed that the $(BiO)_2$ unit layer is insulative with an energy gap of 0.1 eV, while the $(CuO_2)_2$ unit layer is essentially metallic but with anomalous DOS profile around E_F.

1. Introduction

It is generally believed that cuprate superconductors can be regarded as the stacking of superconducting CuO_2 plane and non-superconducting or "weakly" superconducting layers composed of relatively inert ions such as alkaline earth, Bi, Tl and/or oxygen. This seems to bring about the high anisotropy in the superconductivity which is raising the various difficulties for the practical application. The reduction of critical current under a magnetic field especially along the c-axis is thought to be related with the extremely short coherence length along this direction. Difficulties in fabricating a Josephson junction are mainly caused by the same origin. It is therefore of a vital importance to understand the anisotropy of the electronic structure by knowing the contribution from each component layer. For this purpose STM/STS seems to be best suited. This paper describes the difference between the $(BiO)_2$ and $(CuO_2)_2$ layers in BSCCO by achieving an atomic resolution on the cross section of the layered structure.

2. Experimental

The specimen used was a floating-zone solidified boule composed of stacks of single-crystals slightly misoriented to each other. The typical dimension of the specimen is $10x5x5$ mm^3, which is sufficiently large for the STM observation on both cleaved and edge surfaces. Whole process of surface preparation and STM measurement were performed at room temperature under dry nitrogen atmosphere. The details of the STM/STS instrument and surface preparation procedure were described elsewhere (Hasegawa *et al* 1990).

3. Results and Discussion

The cleaved specimen of single crystalline BSCCO is known to expose the BiO mono-layer on surface. The STM image of the cleaved surface given in Fig. 1 shows the atomic structure corresponding to Bi atoms. The average interval between the corrugation is 0.37 nm, which is quite consistent with the atomic spacing between Bi atoms in the quasi-tetragonal unit cell. However, the locations of Bi atoms are modulated along the b-axis, similar to the observation in a ultra high vacuum (Shih *et al* 1989). The tunneling spectrum on the cleaved surface was almost independent of the measurement location. Fig. 2 plots the normalized conductance value (dI/dV)/(I/V), which is though to be a good measure of local density of states N(eV), averaged over the 5x5 nm^2 area against tip voltage V. The tunneling spectrum in Fig. 2 indicates a flat bottom around V=0, implying that the surface BiO mono-layer is a semiconductor with an energy gap of ~0.2 eV as has been reported by Tanaka *et al* (1989).

0953-2048/91/01S073 + 03 $03.50

Fig. 3(a) shows the STM image on the edge surface taken at the tip bias voltage of -0.4 V. In Fig. 3(a), one can clearly see the layered structure along the direction labeled "c" with a periodicity of 1.5 nm, which agrees well with a half of c-spacing. The brighter and darker layers in Fig. 3(a) were assigned to be the $(CuO_2)_2$ and $(BiO)_2$ layers, respectively (Hasegawa *et al* 1990). Figs. 3(b)-(e) are the normalized conductance maps observed at various tip voltages. It should be stressed that the conductance image on the edge surface strongly depends on the bias voltage. At relatively lower bias voltages (Figs. 3(b)-(c)), the $(CuO_2)_2$ layer is more conductive relative to the $(BiO)_2$ layer, while both layers are almost comparable in Figs 3(d)-(e). This result indicates that the density of states around the Fermi level (E_F) is contributed from the $(CuO_2)_2$ layer dominantly than from the $(BiO)_2$ layer.

Figs. 4(a) and 4(b) compare the (dI/dV)/(I/V)-V curves summed over the $(BiO)_2$ layer and the $(CuO_2)_2$ layer regions in Fig. 3(a), respectively. The tunneling spectrum on the edge of $(BiO)_2$ layer shows similar feature to that on the BiO mono-layer in Fig. 2, although the energy gap in the bulk $(BiO)_2$ layer is smaller (0.1 eV) than the surface (BiO) layer (0.2 eV). The conductance at lower bias voltage on the edge of CuO_2 layer is much larger than that on the edge of BiO layer, indicative of the essentially metallic and semiconducting natures of the two layers, respectively. The tunneling spectrum on the $(CuO_2)_2$ layer exhibits a quite unique

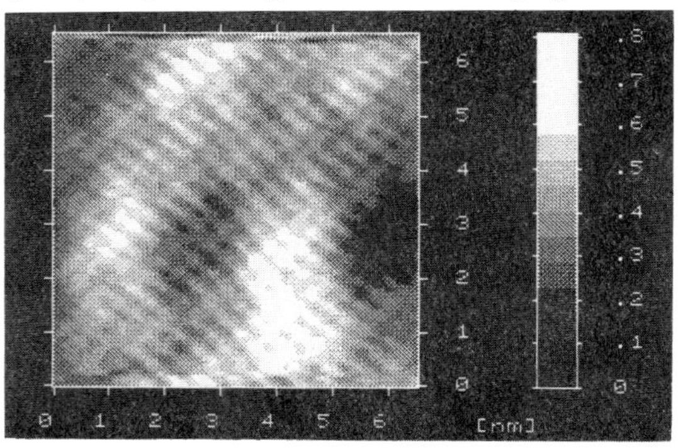

Fig. 1 STM image of a cleaved surface of BSCCO observed at a tip voltage of -1.5 V

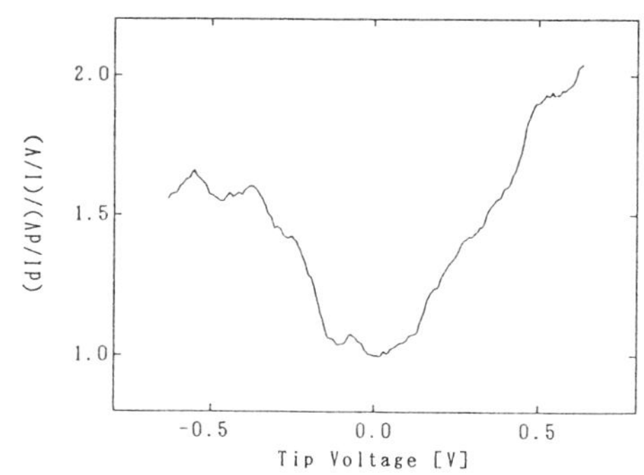

Fig. 2 Tunneling spectrum on a cleaves surface of BSCCO

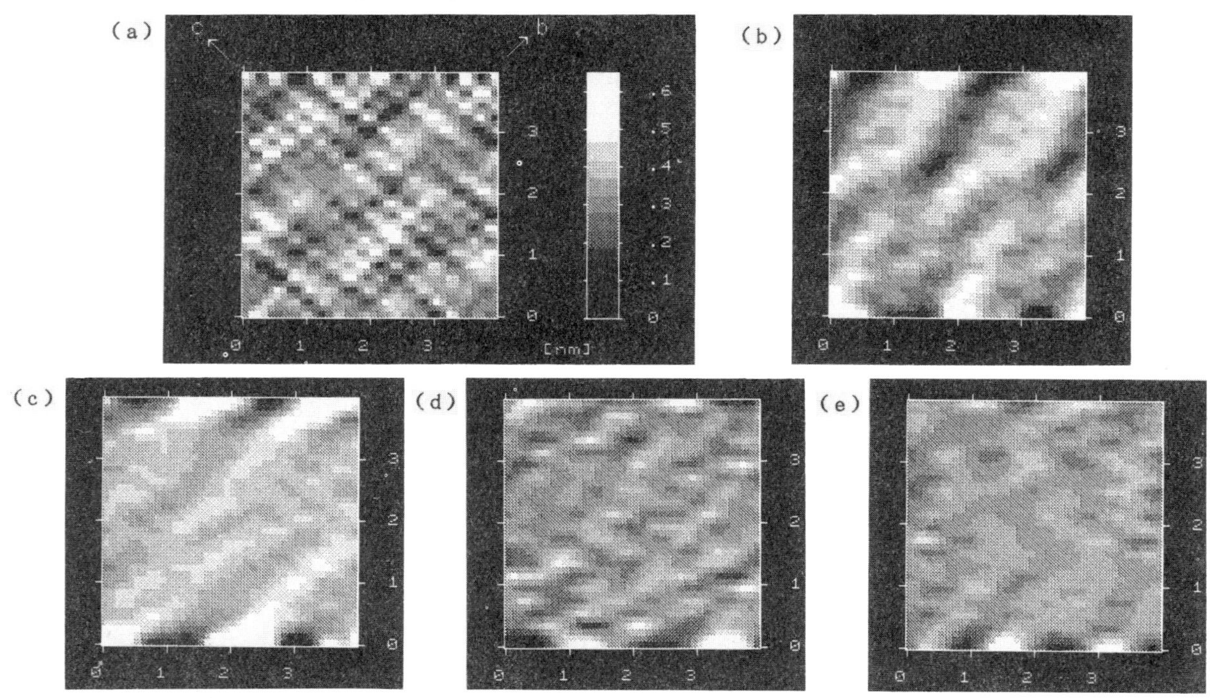

Fig. 3 STM image (a) and conductance maps taken at the tip voltage of -0.1 (b), -0.3 (c) -0.5 (d) and -0.7 V (e) on the bc plane of BSCCO

behavior that the normalized conductance is almost linear with respect to |V|. In the normal metal, however, N(E) is expected to be independent of E near the Fermi level. Actually, the tunneling spectrum on the fresh surface of Au film shows a flat behavior, as seen in Fig. 5.

Consequently,we may conclude that the $(CuO_2)_2$ layer is metallic but that it is a strange metal to possess a sharp minimum of N(E) at E_F. There have been studies report the same anomaly around E_F in the HTSC's from tunnel junctions (Ikuta *et al* 1988). The observation here tells us that this anomaly is originated from the individual nature of each CuO_2 layer.

Kambe *et al* (1989) measured the anisotropy of the normal resistivity in the single crystalline La-Sr-Cu-O, and advocated that the conduction along the c-axis is dominantly semiconductive above T_c, but each of the CuO_2 layer is weakly coupled to give the zero resistance a little below T_c. This indicates that the transfer probability is even less than the thermally assisted electron hopping across the adjacent $(CuO_2)_2$ layers. They estimated the upper limit of J_c along the c-direction based on this model. The present observation is presumably in accordance with this idea: the weakly coupled Josephson junctions along the c-axis.

The observation here clearly demonstrates that the electronic structure in the cuprate superconductor is not completely itinerant but it is nearly tow dimensional. From the practical point of views, the fact that tunneling conductance is very small on the BiO surface layer for bias voltage -0.1~+0.1 V suggests that the Josephson junction is difficult to be fabricated along this direction. A further insulating film, if applied on BiO layer, would reduce the conductance and hence no Josephson current should be obtained. Therefore, based on this observation, we believe that efforts to fabricate the junction on the non-basal plane should be more promising.

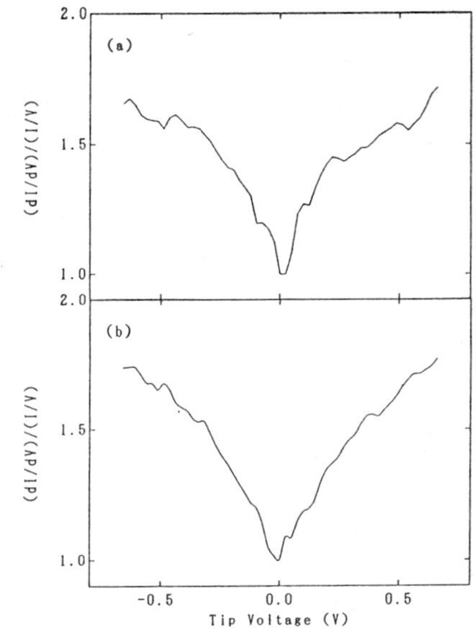

Fig. 4 Tunneling spectra on the $(BiO)_2$ layer (a) and the $(CuO_2)_2$ layer (b)

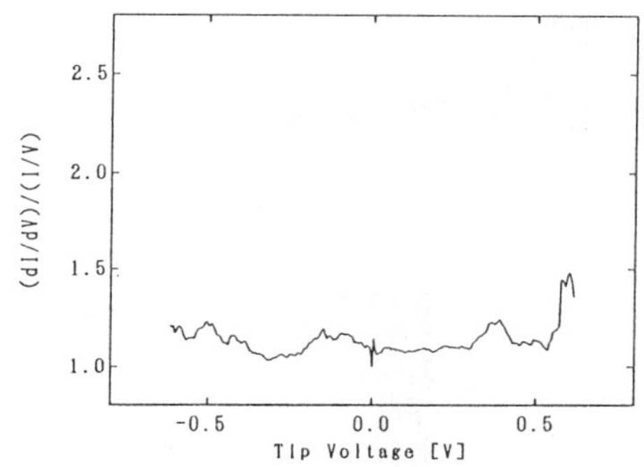

Fig. 5 Tunneling spectrum on the Au film

4. Conclusion

Based on the atomic resolution STM/STS on the cleaved and edge surface of BSCCO, we proposed the anisotropic electronic scheme that the $(BiO)_2$ layer is semiconducting, while $(CuO_2)_2$ layer is metallic. Practical applications of oxide superconductors should take the layered electronic structure into account.

References

Hasegawa T and Kitazawa K 1990 *Jpn. J. Appl. Phys.* **29** L434
Ikuta H, Maeda A, Uchinokura K and Tanaka S 1988 *Jpn. J. Appl. Phys.* **27** L1038
Kambe S, Kitazawa K, Naito M, Fukuoka A, Tanaka I and Kojima H 1989 *Physica* **C160** 35
Tanaka M, Takahashi T, Katayama-Yoshida H, Yamazaki S, Fujinami M, Okabe Y, Mizutani W, Ono M and Kajimura K 1989 *Nature* **339** 691
Shih C K, Feenstra R M, Kirtley J R and Chandrashekhar G V 1989 *Phys. Rev.* **B40** 2682

Supercond. Sci. Technol. 4 (1991) S76–78. Printed in the UK

THE CRYSTALLINE ELECTRIC FIELD AS A DIRECT PROBE FOR CHARGE TRANSFER PROCESSES IN HIGH TEMPERATURE SUPERCONDUCTORS

P. ALLENSPACH, J. MESOT, U. STAUB and A. FURRER, Laboratory for Neutron Scattering, ETH Zürich, CH-5232 Villigen PSI, Switzerland

H. BLANK, H. MUTKA and C. VETTIER, Institut Laue-Langevin, 156X, F-38042 Grenoble Cedex, France

E. KALDIS, J. KARPINSKI and S. RUSIECKI, Laboratory for Solid State Physics, ETH Zürich, CH-8093 Zürich, Switzerland

Abstract. We describe the inelastic-neutron-scattering technique to observe the crystal field in high T_c superconductors as a direct probe of the local symmetry and the charge distribution at the rare-earth site. For $ErBa_2Cu_3O_x$ ($6<x<7$) we succeeded to directly prove the oxygen-vacancy induced charge redistribution in the CuO_2 planes. An empirical relation between T_c and the observed charge transfer $\Delta\rho$ was derived which is highly nonlinear close to T_c=90 K.

1. Introduction

There has been a considerable effort to study the crystalline electric field (CEF) interaction in the perovskite-type compounds $RBa_2Cu_3O_x$ (R denotes rare earth) by the inelastic neutron scattering (INS) technique [1]. The R^{3+} ions are sandwiched between the CuO_2 planes and experience the CEF potential

$$V_{CEF} = \sum_i \rho(\vec{R}_i)/|\vec{r}-\vec{R}_i|, \qquad (1)$$

where $\rho(\vec{R}_i)$ denotes the charge distribution at the site $\vec{R}_i$ of the ith ligand ion. The R^{3+} ions thus constitute local probes of both the structure and the charge distribution in the CuO_2 planes, since for metallic systems eq.(1) is a short-range potential. This property is particularly useful to directly check the mechanism of electronic charge transfer from the CuO chains to the CuO_2 planes, which has recently been suggested to play a crucial role in the oxygen-vacancy induced suppression of superconductivity in $RBa_2Cu_3O_x$ ($6<x<7$) [2]. Based on our results we have put forward the conjecture that the large differences in the pressure dependence of T_c versus oxygen content may be consistently understood in terms of pressure induced charge transfer to the planes [3].

2. Experimental

Details of the experimental procedure are described elsewhere [3]. The INS experiments were performed for $ErBa_2Cu_3O_x$ and $ErBa_2Cu_4O_8$ with use of the time-of-flight (TOF) spectrometers IN4 and IN5 at the high-flux reactor of the Institut Laue-Langevin (ILL) in Grenoble. The powdered samples were

0953-2048/91/01S076 + 03 $03.50

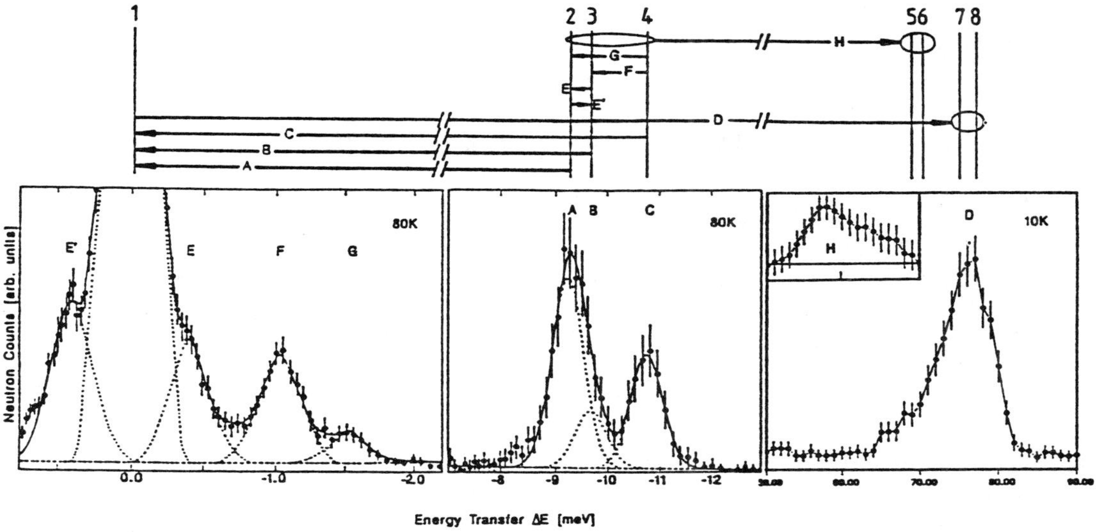

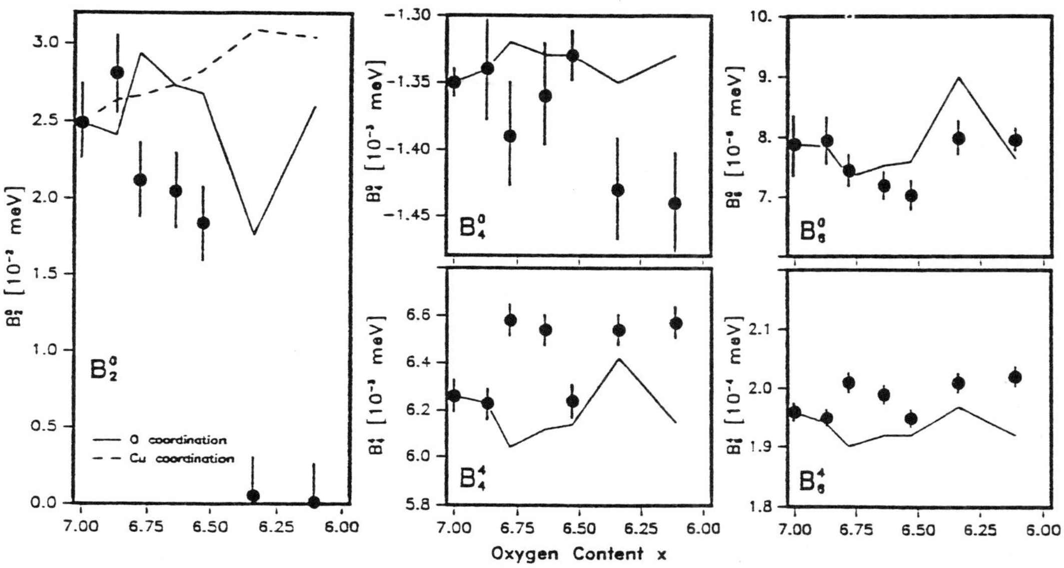

Fig. 1: Energy spectra of neutrons scattered from $ErBa_2Cu_4O_7$,

Fig. 2: CEF parameters (dots) and extrapolated structural behavior (lines)

filled into cylindrical aluminum containers of 10 mm diameter and 50 mm height under helium atmosphere. For the pressure experiments we used a steel clamp with a sample chamber of 6 mm diameter and 15 mm height.

3. Analysis of results

We base the analysis of the observed energy spectra on the CEF Hamiltonian for orthorhombic symmetry D_{2h}:

$$H_{CEF} = \sum_{n=1}^{3} \sum_{m=0}^{n} B_{2n}^{2m} O_{2n}^{2m} , \qquad (2)$$

where the B_{2n}^{2m} denote the CEF parameters and the O_{2n}^{2m} are operator equivalents [7]. The CEF Hamiltonian gives rise to a decomposition of the ground-state J-multiplet $^4I_{15/2}$ of Er^{3+} in $ErBa_2Cu_3O_x$ and $ErBa_2Cu_4O_8$ into eight doublet states Γ_5, see Fig. 1.

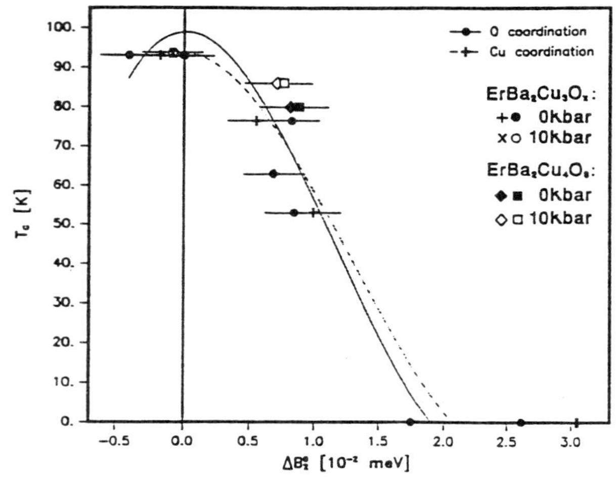

Fig. 3: Empirical relation between T_c and ΔB_2^0 ($\sim \Delta\rho$).

4. Discussion

The CEF parameters determined for $ErBa_2Cu_3O_x$ were found to vary considerably as a function of the oxygen content x, see Fig. 2. According to eq.(1) this variation may be due to structural modifications of the coordinating ligand ions as well as to charge redistributions at the ligand sites. We have been able to separate the structural and charge effects by an extrapolation procedure described in detail in Refs.[3] and [4]. These extrapolated values due to structural changes are shown as lines in Fig. 2. The fourth- and sixth-order CEF parameters are mainly governed by the nearest neighbouring oxygen O(2) and O(3) ions, whereas the second-order CEF parameter is influenced by the next-nearest neighbouring copper Cu(2) ions as well. We interpret the differences $\Delta B_{2n}^{2m} = B_{2n}^{2m}$(exrapolated)-$B_{2n}^{2m}$(observed) in terms of charge redistributions in the superconducting CuO_2 planes, which are particularly pronounced for the second order CEF parameter. ΔB_{2n}^{2m} is proportional to the actual charge $\Delta\rho$ being transferred to the planes. In Fig. 3 we plot T_c versus $\Delta B_2^0 \sim \Delta\rho$. In addition, the results obtained for $ErBa_2Cu_4O_8$ nicely follow the behavior derived for $ErBa_2Cu_3O_x$, and our present pressure studies indicate that the drastic differences in dT_c/dp may well be understood in terms of our empirical relation T_c versus $\Delta\rho$. In particular, we anticipate dT_c/dp to be directly related to the slope of our T_c versus $\Delta\rho$-curve. For $ErBa_2Cu_3O_7$ and $ErBa_2Cu_4O_8$, application of 10 kbar pressure raises T_c by approximately 1 K and 6 K, respecively, thus we predict the change of the 2nd order CEF parameter to be $\Delta B_2^0 = -10^{-3}$meV as verified in the present experiments, see Fig. 3.

[1] ALLENSPACH P et al (1989a) , Phys. Rev. B 39 2226.
[2] CAVA R J et al (1988), Physica C 156 523.
[3] FURRER A et al (1989), J. Phys.: Condens. Matter. 1 3715.
[4] ALLENSPACH P et al (1989b), Physica C 161 671.

Supercond. Sci. Technol. 4 (1991) S79–81. Printed in the UK

The Effect of Cu/M (M=Mn,Co,Ni,Zn) Substitution upon Superconductivity in $Nd_{1.85}Ce_{0.15}CuO_4$

G Hilscher, M Forsthuber, S Pöllinger, A Hanninger*, and P Rogl*

Institut für Experimentalphysik T.U. Wien, A-1040 Wien, Austria
*Institut für Physikalische Chemie, Univ.Wien, A-1090 Wien, Austria

Abstract. We report on the structural chemistry, ac susceptibility-, and resistivity measurements of $Nd_{1.85}Ce_{0.15}(Cu_{1-x}M_x)O_{4}4$ where Cu/M substitution reduces superconductivity with an initial rate of -4.5K/mol% Zn while 1% Co or Ni suppresses superconductivity below 4.2 K.

The new class of electron doped copper oxide superconductors $(R_{1-y}Ce_y)_2CuO_4$ (with R = Pr, Nd, Sm and Eu) are unique in that the charge carriers are electrons. They crystallize in the tetragonal T' - Nd_2CuO_4 structure where the copper atoms are in square planar coordination with oxygens forming the Cu-O sheets [1,2]. Since this structure exhibits only one Cu site in the Cu-O plane, Cu/M substitution can be explored without the complication of two different lattice sites for Cu, as is the case in YBCO. $(Nd,Ce)_2CuO_4$ is a magnetic superconductor where antiferromagnetism of the R - ions coexists with superconductivity at low temperatures [2]. This and the electron hole symmetry in the systems NCCO and LSCO [3] makes the system attractive to study the mechanisms of superconductivity in ceramic cuprates where the magnetic and superconducting ordering temperatures are close to each other.

Sample preparation and experimental methods are described in detail in reference [4]. Room temperature X-ray powder analysis of the samples $Nd_{1.85}Ce_{0.15}Cu_{1-x}M_xO_4$ proved the existence of homogeneous regions whose range of extension heavily depends on the nature of the solute M. Whereas rather limited solid solubilities smaller than $x \simeq 0.025$ have been observed for the large early 3d metals V, Cr as well as for the small, late 3d metals Fe and Co, slightly extended solid solution ranges, however, seem to exist for Mn and for Zn ($x \simeq 0.05$). X-ray analysis in all cases is consistent with a random distribution of (Nd,Ce) and of (Cu,M) on their crystallographic sites. No deviations from the *I4/mmm* crystal symmetry are observed on heavily exposed Guinier powder photographs. The variation of the unit cell dimensions versus composition is shown in Fig.1.

The superconducting fraction estimated from the comparison of the diamagnetic ac susceptibility signal with that of a Nb superconductor of the same geometry in fields of 80 A/m is 48 % , 43 % and 3-4 % for Zn_x = 0.0, 0.01 and 0.025, respectively. To determine the superconducting fraction the ac field of 80 A/m was used, since both types of samples (the pellets with 2.5 mm diameter and bar shaped samples cut from cylindrical samples of 8 mm diameter) show that at this ac field, the diamagnetic response is mainly due to the screening of the grains. At very low ac fields (0.8 A/m) a nearly complete shielding is achieved for x=0.0, which is drastically reduced with increasing ac field (for a more detailed discussion concerning the inter- and intagranular losses of the ac susceptibility we refer to reference [4]). In agreement with susceptibility data the resistivity measurements show that Zn supresses superconductivity at an initial rate of 4.5 K/mol% Zn.

0953-2048/91/01S079 + 03 $03.50

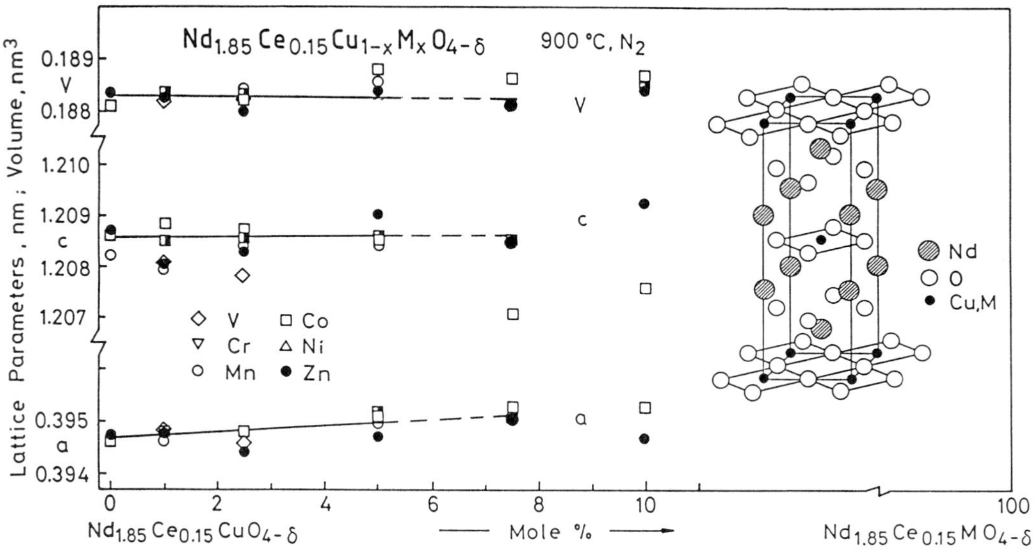

Fig.1: Unit cell dimensions as a function of x.

The variation of the superconducting transition temperatures (Tc-midpoint) is shown in Fig.2 as a function of the Mn,Co,Ni and Zn content. We omitted in this graph the Tc values of V and Cr, because we are uncertain about the solid solubility of these elements in NCCO. These specimens exhibit even at a doping level of 1 % V or Cr a rather broad transition of about 10 K with a Tc-midpoint of 16 K. For V the transition temperature increases again up to 19.5 K at x = 0.05 while in the case of Cu/Cr substitution the Tc values scatter between 11 and 15 K for a Cr content up to 7.5%.

Zn substitution has obviously a less dramatic effect upon the depression of superconductivity (-4.5 K/mol% Zn) than in YBCO [5] and LSCO [6] which is in good agreement with results recently obtained by Felner et al. [7]. Moreover we stress that the experiments on Cu/Co and Cu/Ni substitution yield a dramatic depression of superconductivity; for doping levels as low as 1 mole % superconductivity is already

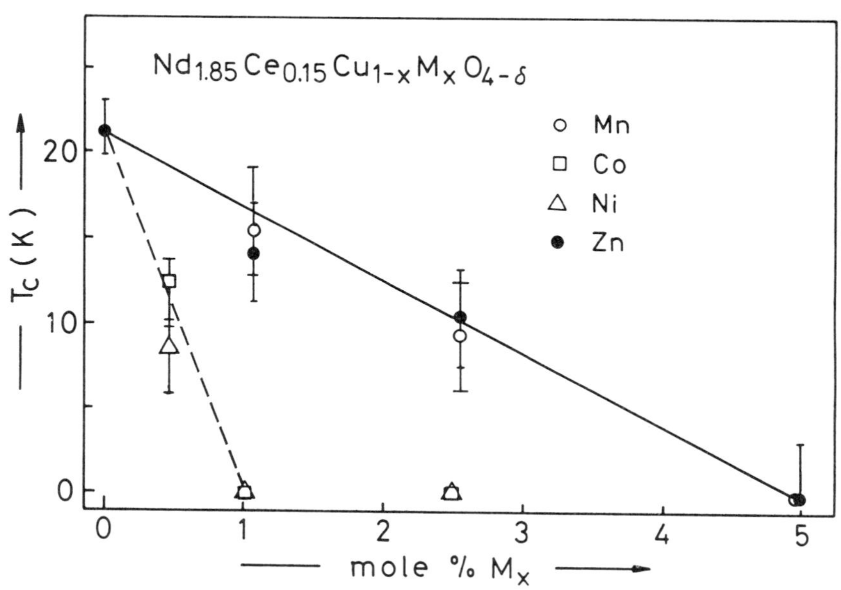

Fig.2: Superconducting transition temperatures for various substituents M as a function of x.

suppressed. This appears to be the reversed situation in comparison with the hole doped LSCO with respect to Zn and 3d metal substitution. In the latter system the magnetic 3d metal substitutions have nearly the same or a minor impact upon the depression of T_c than Zn.

The different behaviour of the electron and hole doped superconductors with respect to substitutions indicate that either the larger coherence volume of NCCO in comparison with LSCO or the recently proposed phase separation [8] in NCCO plays a significant role. In this context it is of importance to note that the coherence length ξ of ceramic NCCO with approximately 90 Å [9] and ξ_{ab} = 80Å, 70Å for a single crystal and a c axis oriented thin film[3,10] is in between those of conventional superconductors (generally larger than 400 Å) and the hole doped superconductors (smaller than 20 Å). In the case of a small coherence volume non-magnetic scattering should have a similar effect upon the T_c suppression as magnetic impurity scattering , while a considerably distinct effect is to be expected for a larger coherence volume. In this view Mn is expected to suppress T_c as effectively as Co and Ni. Surprisingly, however, Mn is found to have nearly the same effect upon the variation of T_c as Zn. In view of the model of phase separation [8], according to which samples over a wide range of Ce content y ($0.0 < y < 0.2$) contain only one superconducting phase with y=0.165 while the other phase with variable Ce content is not sperconducting, a physical interpretation of the compositional dependence of T_c on Cu/M substitution becomes problematic because of the uncertainty of the Cu/M ratio in the superconducting phase.

Acknowledgement.This work was supported by the Austrian Science Foundation under projects 7056 and 7060.

References

1) Y Tokura, H Tagaki and S Uchida , Nature 337 (1989) 345
2) C L Seaman, N Y Ayoub, T Bornholm, E A Early, S Ghamaty, B W Lee, J T Markert,J J Neumeier, T S Tsai and M B Maple Physica C 159 (1989) 391
3) M B Maple MRS Bulletin 15 (1990) 60
4) G Hilscher, S Pöllinger, M Forsthuber, N Pillmayr, K Remschnig, P Rogl, M Reissner, W Steiner and P Knoll Physica C 167(1990) 472
5) G Hilscher, N Pillmayr, R Eibler, E Bauer, K Remschnig and P Rogl, Z Phys B72 (1988) 461
6) K Remschnig, P Rogl, E Bauer, R Eibler, G Hilscher, H Kirchmayr and N Pillmayr in High T_c Superconductors (Ed.H.W.Weber) Plenum (1988) p 99
7) I Felner, D Hechel and U Yaron, Physica C 165 (1990) 247
8) P Lightfoot, D R Richards, B Darbowski, D G Hinks, S Pei, D T Marx, A W Mitchell, Y Zheng and J D Jorgensen Physica C 168 (1990) 627d
9) J P Sröbel, M Klauda, M Lippert, B Hensel, G Saemann-Ischenko, W Gerhäuser, H W Neumüller, W Ose and K F Renk Physica C 162-164 (1989) 1209
10) M Suzuki and M Hikata Phys. Rev. B 41 (1990) 9566

Supercond. Sci. Technol. 4 (1991) S82–84. Printed in the UK

Oxygen contents, crystal structures and physical properties of $Tl_2Ba_2CuO_{6+\delta}$.

Y. Kubo, Y. Shimakawa, T. Manako, H. Igarashi, F. Izumi* and H. Asano**

Fundamental Research Laboratories, NEC Corporation,
4-1-1 Miyazaki, Miyamae-ku, Kawasaki 213, Japan
*National Institute for Ressearch in Inorganic Materials, Tsukuba 305, Japan
**Institute of Materials Science, University of Tsukuba, Tsukuba 305, Japan

Abstract. $Tl_2Ba_2CuO_{6+\delta}$ shows a gradual transition from a 85-K superconductor to a nonsuperconducting normal metal, which arises from over-doping of hole carriers by excess oxygen. Neutron diffraction study reveals that excess oxygen is incorporated at an interstitial site between double TlO layers. No remarkable changes are observed at the superconductor-metal phase boundary in both Hall coefficient and spin susceptibility, suggesting the very similarity and continuity between both the electronic structures. This is in sharp contrast with the previous observations for $La_{2-x}Sr_xCuO_4$.

1. Introduction

$Tl_2Ba_2CuO_{6+\delta}$ (Tl-2201) shows a gradual transition from a 85-K superconductor to a nonsuperconducting normal metal, as excess oxygen content is increased from ~0 to ~0.1 (Shimakawa et al 1989a, 1989b). Hall measurement demonstrates that this transition arises from over-doping of hole carriers by excess oxygen (Kubo et al 1989). These facts strongly supports the idea that superconductivity is suppressed if the carrier concentration exceeds an optimum value, which was first pointed out by Torrance et al (1988) for $La_{2-x}Sr_xCuO_4$.

Until now, most studies of High T_c superconductivity have focused on an insulator-to-superconductor transition which takes place at lower carrier density region than an optimum value. However, considering that superconductivity appears in an intermediate region between an antiferromagnetic insulator and a normal metal, studies from a metallic side should be as important as those from an insulating side. In particular, it is crucial to clarify whether any singularities exist at the superconductor-normal metal phase boundary, because studies on $La_{2-x}Sr_xCuO_4$ have suggested some singularities in both Hall coefficient R_H (Takagi et al 1989) and spin susceptibility χ_s (Torrance et al 1989); R_H changes its sign from positive to negative and χ_s shows a maximum at the phase boundary.

2. Experimental Results

Neutron diffraction study was performed on four single phase samples with T_c's of 73K, 58K, 48K and 0K(metallic) using a high resolution TOF neutron powder diffractometer at the National Laboratory for High Energy Physics. The obtained data were analyzed using a Rietveld refinement program RIETAN (Izumi et al 1987). An interstitial oxygen site O(4) is confirmed between double TlO layers as shown in Fig.1. The amount of O(4) atom, which corresponds to the excess oxygen content δ, is increased from ~0 to ~0.1 per formula unit with decreasing T_c, showing an excellent agreement with the change in oxygen content determined by weight measurement (Shimakawa et al 1989).

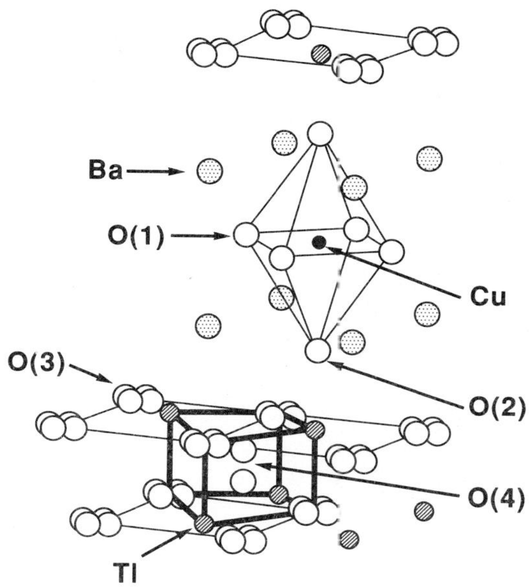

Fig. 1 Crystal structure model for $Tl_2Ba_2CuO_{6+\delta}$.

Temperature dependence of the resistivity is well fitted to a power-law dependence $\rho = \rho_0 + T^n$ for all T_c's as shown in Fig.2. High T_c samples ($\delta \sim 0$) exhibit a linear T dependence (n=1) which is usually observed in high T_c Cu oxides, while excess hole-doping gradualy evolves the T dependence toward quadratic (n=2) and lowers T_c down to 0 K. It is noted that such power-law dependences are difficult to be explained in terms of electron-phonon scattering. Figure 3 shows δ dependense of R_H at 120 K and 300 K. It should be noted that R_H is always positive and gradually decreased with δ, showing no remarkable change at the superconductor-metal phase boundary.
The magnetic susceptibility in the normal state shows nearly temperature independent behavior after subtracting the Curie term which seems to be due to Cu^{2+} ions substituting for the Tl site. The obtained nearly T independent spin susceptibility χ_s is linearly increased with δ, showing no indication of singularity at the superconductor-metal phase boundary.

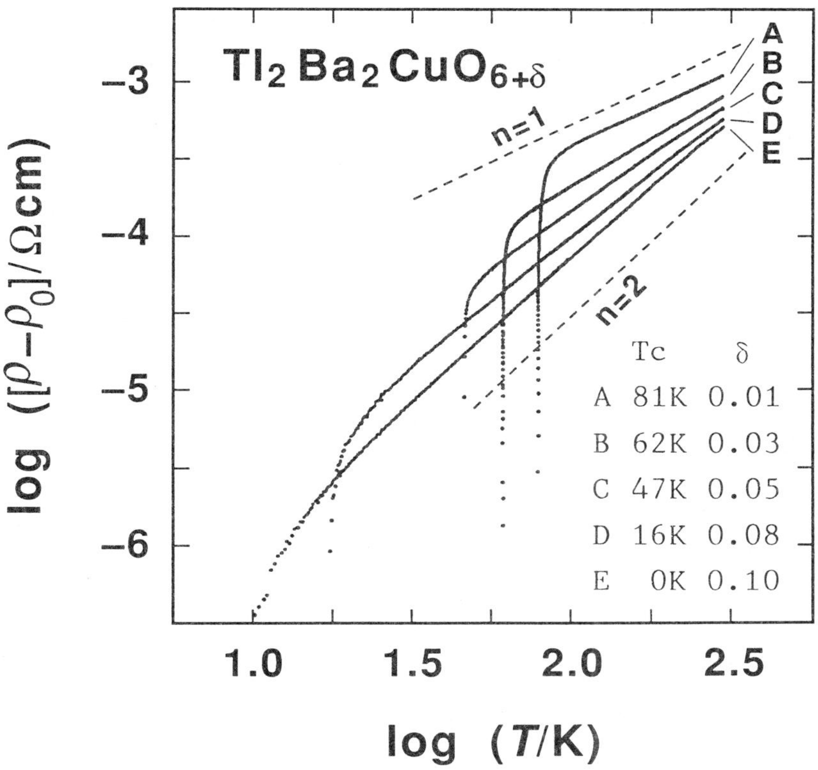

Fig.2. Fit of the resistivity data to a power-law temperature dependence $\rho = \rho_0 + AT^n$. Dashed lines indicate the slopes for n=1 and n=2.

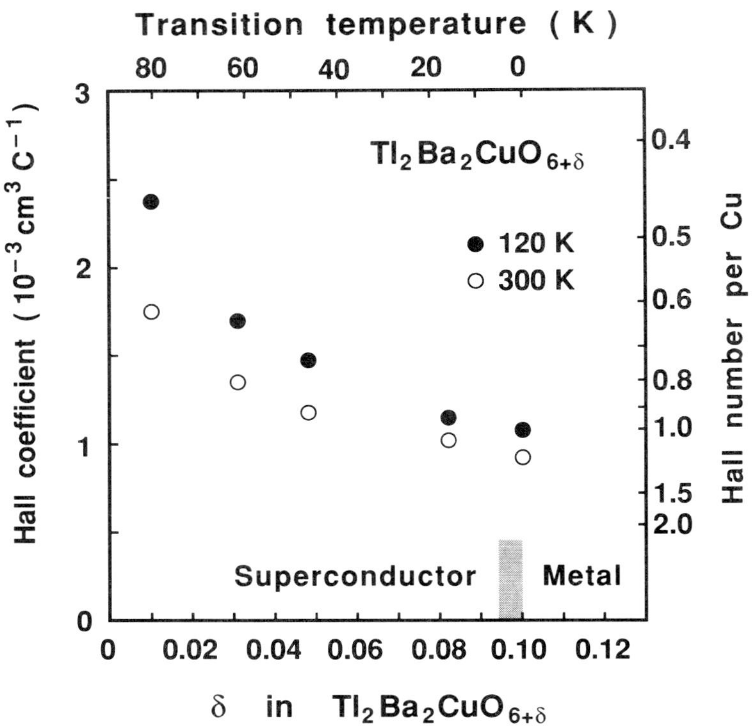

Fig.3. R_H at 120 K and 300 K plotted against the excess oxygen content δ.

3. Discussion

The most important point to be noted is that the 85-K superconductivity is gradually connected to the normal metal phase without any indication of discontinuity in both transport and magnetic properties, which suggests the very similarity and continuity between both the electronic structures. This might imply that the electronic structures of high T_c materials are basically described as a Fermi liquid. The reason why the present results are in sharp contrast to those for the La system might be related to the difference in the hole-doping mechanism. In Tl-2201, the Cu-O octahedron layer is scarcely disturbed because the dopant of excess oxygen is incorporated in the TlO double layer. Whereas in the La system, holes are doped by Sr-substitution for La, making the Cu-O octahedron layer essentially a solid-solution.

References

Izumi F, Asano H, Murata H and Watanabe N 1987 *J. Appl. Crystallogr.* **20** 441
Kubo Y, Shimakawa Y, Manako T, Satoh T, Iijima S, Ichihashi T and Igarashi H 1989 *Physica C* **162-164** 991
Shimakawa Y, Kubo Y, Manako T, Satoh T, Iijima S, Ichihashi T and Igarashi H 1989a *Physica C* **157** 279
Shimakawa Y, Kubo Y, Manako T and Igarashi H 1989b *Phys. Rev B* **40** 11400.
Takagi H, Ido T, Ishibashi S, Uota M, Uchida S and Tokura Y 1989 *Phys. Rev.B* **40** 2254
Torrance J B, Tokura Y, Nazzal A I, Bezinge A, Huang T C and Parkin S S P 1988 *Phys. Rev. Lett.* **61** 1127
Torrance J B, Bezinge A, Nazaal A I, Huang T C, Parkin S S P, Keane D T, LaPlaca S J, Horn P M and Held GA 1989 *Phys. Rev. B* **40** 8872

Supercond. Sci. Technol. **4** (1991) S85–87. Printed in the UK

Determination of the Angular Dependence of H_{c2} in High T_c Single Crystals by a Microwave Technique.

D. Shaltiel[a,b,c], H. Bill[b], A. Grayevsky[a], A. Junod[c], D. Lovy[b], S. Sadowski[c] and E. Walker[c].

a. Racah Institute of Physics, Hebrew University, Givat Ram, Jerusalem 91904, Israel.
b. Groupe Physico-Chimie du Solide, Sciences II, Univ. de Genève, quai E. Ansermet, CH1211 Geneva 4, Switzerland.
c. Dept. de Physique de la Matière Condensée, as b above.

Abstract. It is shown that using an ESR spectrometer with magnetic field modulation and sweeping the temperature across T_c (at a constant and a very low magnetic field), is equivalent to temperature modulation. The signal intensity obtained when crossing T_c is proportional to $1/(\partial H_{c2}/\partial T)$ at $T = T_c$. Using the WHH relation $H_{c2(T=0)} = 0.7\, T_c(\partial H_{c2}/\partial T)_{T=T_c}$ it enabled the measurement of the relative angular variation of H_{c2} in single crystals of YBaCuO with T_c~85K. The data fit the Ginzberg-Landau theory. This very sensitive technique can be used to characterize properties of high T_c superconducting materials. Results on thin films will also be presented.

2. Introduction

Microwave absorption as a function of the temperature can measure the surface impedance in super conductors, and has been used to investigate the penetration depth λ(T). Usually the complex microwave resistivity was studied from changes in the quality factor and the frequency of a cavity with and without the material under study[1]. An alternative method to measure the microwave resistivity is to use an ESR spectrometer at a fixed magnetic field, to apply temperature modulation instead of magnetic modulation, and to sweep the temperature. Such measurements will yield the derivative of the microwave resistivity as shown in fig. 1a and 1b. Although such a system can be built and is expected to have a higher sensitivity than the first method, to the best of our knowledge it has not yet been reported. In addition to the difficulties in devising such a modulation, the drawback is, that only low frequency modulation, of the order of a few Hz can be achieved, and therefore its sensitivity due to 1/f noise is reduced in comparison to conventional spectrometers where a magnetic field modulation up to 10^5 Hz is used. In this work we demonstrate that in high Tc superconductors and close to the transition temperature magnetic field modulation is

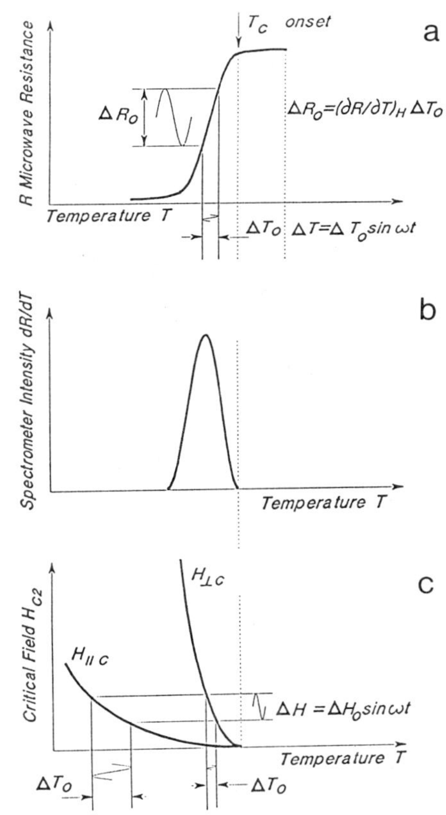

Fig. 1 Schematic presentation of temperature modulation via magnetic modulation

equivalent to temperature modulation when sweeping the temperature across T_c at constant and low magnetic field. Using this property, we have measured the relative angular variation of H_{c2} in single crystals. The method also enables us to characterize properties of high T_c superconductors in polycrystaline materials, single crystals and thin films.

2. Theoretical considerations

Fig. 1c schematically shows the $H_{c2}(T)$ curve as a function of temperature. When the modulating magnetic field $\Delta H = \Delta H_o \sin\omega t$, crosses the H_{c2} curve, it shifts the superconducting material into the normal state or the superconducting state whenever it adds to, or is subtracted from the constant $H_{c2}(T)$ field respectively. As seen in fig. 1c this magnetic field modulation is equivalent to temperature modulation and the amplitude ΔT_o is given by $\Delta T_o = \Delta H_o/(\partial H_{c2}/\partial T)$. For constant and close to zero magnetic fields $[\partial H_{c2}/\partial T]$ at $T = T_c$ is $H_{c2(T=0)}/(0.7T_c)$[2]. Hence

$$\Delta T_o = \Delta H_o(0.7\ T_c/H_{c2})_{(T=0)} \text{ at } H \approx 0 \qquad (1)$$

Thus, in single crystals with anisotropic properties, ΔT_o will depend on the orientation of the crystal with respect to the magnetic field. Therefore, as disussed above, sweeping the temperature will show a signal proportional to the derivative of the microwave surface resistivity. Its maximum intensity, I, will be proportional to ΔT_o and therefore inversely proportional to $\partial H_{c2}/\partial T_{(T=Tc)}$ or to $H_{c2(T=0)}/T_c$.

In uniaxial superconductors it has been confirmed that the mass tensor formulation of Ginzburg-Landau theory holds[3]. The upper critical fields is given by $H_{c2}(\theta) = m_3\cos^2\theta + m_1\sin^2\theta$ where $m_1 = m_{xx} = m_{yy}$ and $m_3 = m_{zz}$, $H_{c2}(n)$ and $H_{c2}(p)$ are the critical fields normal or parallel to the CuO layers respectively and θ is the angle between normal to the CuO layers and the magnetic field. Hence:

$$I(\theta)/I(n) = H_{c2}(n)/H_{c2}(\theta) = [\cos^2\theta + (H_{c2}(n)/H_{c2}(p))^2\sin^2\theta]^{1/2} \qquad (2)$$

and for $\theta = 90°$, $\Delta T_o(n)/\Delta T_o(p) = I(n)/I(p) = H_{c2}(p)/H_{c2}(n)$. Thus, measuring the ratio of the microwave absorption intensities normal and parallel to the CuO planes in a single crystal should give the ratio of the critical fields in these two directions. Furthermore the angular dependence of the critical field should follow eq. 2.

Here, effects of demagnetization, due to shielding currents induced by the magnetic modulation, have been neglected. They will depend on the shape of the crystal and its anisotropy properties and hence on the orientation with respect to the magnetic field. The calculation of the demagnetization effects is complex, however it can be argued that in high H_{c2} superconductors such as the high T_c compounds, the contribution of the demagnetising field is negligible.

3. Experimental

Single crystals and oriented thin films of $YBa_2Cu_3O_{(6+x)}$ x=0.9 have been used in this experiment. They were prepared by the flux method and evaporation techniques respectively then treated to maximize the oxygen content. A Varian spectrometer with low frequency (150Hz) modulation coils attached to the magnetic poles has been used. The sample was introduced on the side wall of a cavity. The angular dependence was obtained by rotating the magnet and therefore the microwave field around the sample did not vary during the measurements. The variation of the temperature was always obtained by slowly cooling the cavity from above to below T_c. Fig. 2 shows typical microwave absorption spectra of two single crystals in the form of platelets and an oriented thin film where the c axis was perpendicular to their plane (The spectra were taken for $\theta = 90$. Only minute changes in the line shape were observed for different angles.). One single crystal shows a narrow line with maximum at T = 88.5K and whose width is about 1K. The other crystal shows a structure composed of several peaks indicatng that this crystal contains regions with slightly different transition temperatures. The microwave absorption spectrum of the thin film shows a broad transition of 20K though its resistivity curve has indicated a sharp transition of a few degrees. Part of this width may be attributed to inhomogeneities in the edges of the film.

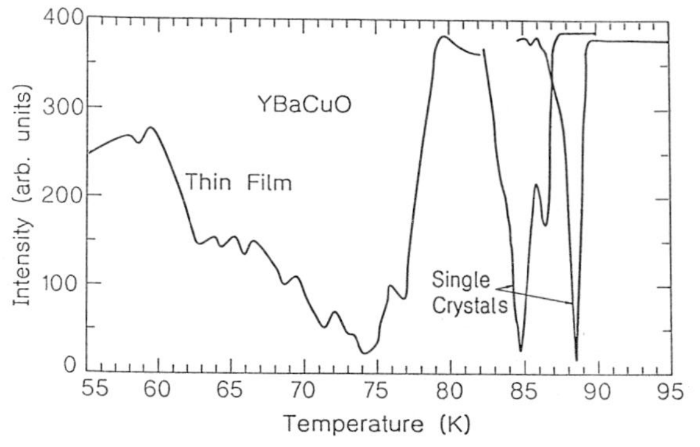

Fig. 2 Microwave absorbtion vs. temperature of two single crystals and a thin film

Fig. 3 shows the angular dependence of the maximum intensity for the single crystal with one peak whose spectrum was shown in fig. 2. The full line is the presentation of eq. 2 and therefore the results are in good agreement with the Ginzburg Landau theory. The ratio of $H_{c2}(p)/H_{c2}(n) \approx 3$ is in agreement with measurements obtained from the slope of $\partial H_{c2}/\partial T$ using direct susceptibility methods.(4)

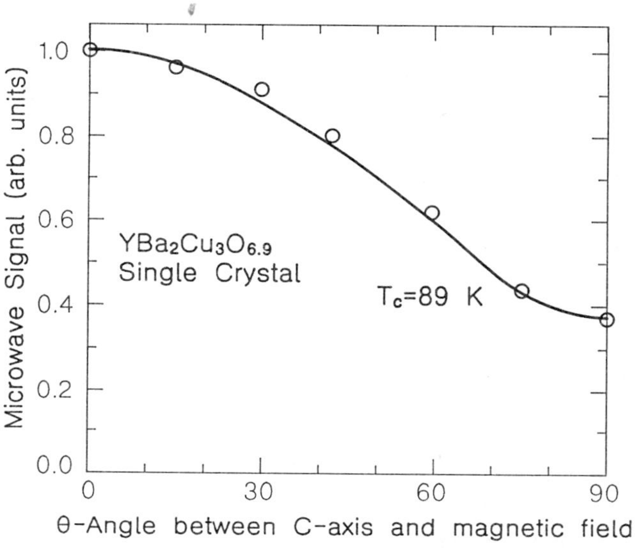

Fig. 3 Angular dependence of the microwave absorption intensity. The full line presents eq. 2.

4. Conclusion

Using the microwave absorption technique we were able to measure the relative angular behavior of $H_{c2}(\theta)$ in high quality single crystals of $YBa_2Cu_3O_{6+x}$ $x \approx 0.9$. We found an agreement with the Ginzburg-Landau theory and results obtained by other methods on the numerical values $H_{c2}(p)/H_{c2}(n)$ for single crystals were in agreement. Oriented thin films show very broad transition temperatures, in variance with results obtained by susceptibility and resistivity measurements. The method also enables us to determine transition temperatures and to characterize the quality of superconducting materials. It has the advantage of being contactless and very sensitive (around 10^{-8} gram of YBaCuO).

References

1. Carini J.P. et al (1988) Phys. Rev. **B 37** 6.
2. Werthamer N.R. et al (1966) Phys. Rev. **147**, 295.
3. Kogan V.G. & Clam J.R. (1987) Jap. J. App. Phys Suppl. **26-3**, 1159.
4. Bauhofer W. et al (1989) Phys. Rev. Lett. **63**, 2520.

Acknowledgement - This work was partly supported by BSF Grant No. 87-00275

Supercond. Sci. Technol. **4** (1991) S88–90. Printed in the UK

Flux-line lattice dynamics and irreversibility line in single crystalline $Bi_2Sr_2CaCu_2O_{8+\delta}$

K. Kadowaki, N. J. Li, F. R. de Boer, P. H. Frings and J. J. M. Franse

Natuurkundig Laboratorium der Universiteit van Amsterdam,
Valckenierstraat 65, 1018-XE Amsterdam, The Netherlands

Abstract. High-field magnetisation measurements up to 38 T have been performed on single-crystalline $Bi_2Sr_2CaCu_2O_{8+\delta}$ in order to investigate the irreversible behaviour associated with the flux line dynamics. It was found that the irreversible magnetisation vanishes asymptotically at high fields well below B_{c2}. As temperature is increased, above approximately 40 K, the magnetisation essentially becomes reversible. Critical currents J_c and volume pinning forces F_p were calculated and compared with the results of $YBa_2Cu_3O_{7-\delta}$. The irreversibility line deduced from the present measurements as well as from other techniques is discussed in terms of a possible boundary, above which superconducting phase fluctuations play a dominant role for the dissipation.

1. Introduction

A number of recent experiments on single crystals of high T_c superconductors such as $YBa_2Cu_3O_{7-\delta}$ and $Bi_2Sr_2CaCu_2O_{8+\delta}$ (Malozemoff et al. 1988, Gammel et al. 1988, Kadowaki et al. 1990a, Li et al. 1990a) have revealed the existence of a phase boundary in the superconducting mixed state, named the irreversibility line. This line separates the B-T phase diagram into two regions. Crossing this irreversibility line, physical parameters change drastically: irreversible magnetisation becomes reversible, consequently the critical current becomes essentially zero. Because of this the resistivity becomes appreciable and dissipation which seems to be ohmic becomes markedly even in a relatively low magnetic field. These rather unusual features are most significant in the single-crystalline $Bi_2Sr_2CaCu_2O_{8+\delta}$, which is known as a highly two dimensional superconductor with T_c of 85 K. There have been a number of attempts to explain the phenomena associated with the irreversibility line, based on a classical approach in terms of flux motion due to a Lorentz-force-type of driving force (Kes and van der Beek 1990, Kes 1990, Palstra et al. 1990). Although these approaches are considered to be rather successful in accounting for phenomena such as magnetisation relaxation and resistivity broadening by simply assuming small pinning forces in these compounds, the fundamental interaction between the flux lines and the transport current has not been established yet. Recent investigations of the precise angular dependence of the resistivity near T_c in magnetic fields in single-crystalline $Bi_2Sr_2CaCu_2O_{8+\delta}$ (Kadowaki et al. 1990b) and in highly oriented thin films (Iye et al. 1990) strongly suggests that there is no Lorentz-force-type of interaction in this regime. The peculiar angular dependence of the resistivity in magnetic fields, especially along the c-direction, is also inconsistent with the flux-flow model and casts some doubt about the traditional interpretation.

2. Experiment

Magnetisation measurements have been performed in a semi-continuous high magnetic fields up to 38 T at various temperatures in the high field installation at the University of Amsterdam. The single crystal used in this experiment has been grown by the travelling solvent floating zone method (Kadowaki, et al. 1990d). The single-crystal sample with a total mass of 94.3 mg consists of a collection of twenty-one thin disks of 3 mm in diameter, which were aligned along the common c-axis, stacked together and cast into stycast. After each pulse, the sample is warmed up above T_c to eliminate the trapped flux from the sample. The temperature of the sample is monitored by the silicon diode sensor with an accuracy of ± 0.5 K.

3. Results and Discussion

Figure 1 shows a set of irreversible magnetisation curves obtained at various temperatures. A number of interesting features are summarised: 1. The diamagnetic magnetisation is much

0953-2048/91/01S088 + 03 $03.50

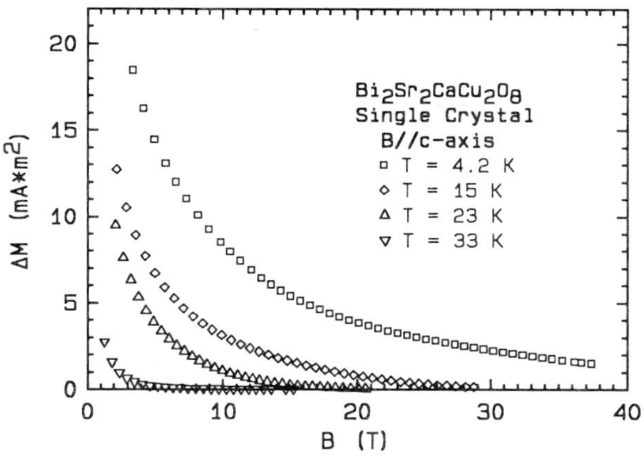

fig. 1. Irreversible magnetisation curves ΔM (=|M⁺ −M⁻|)at various temperatures.

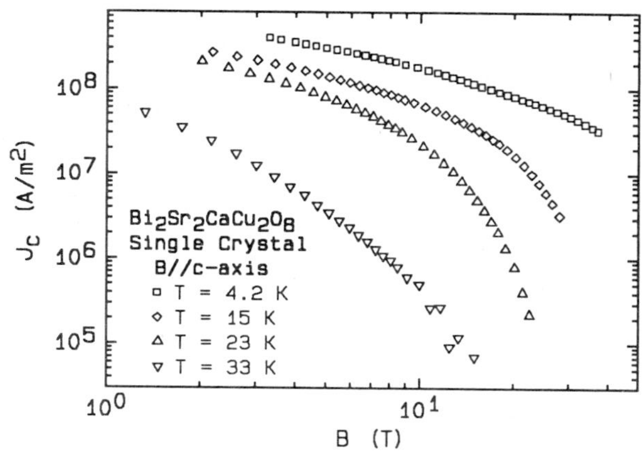

fig. 2. Critical current densities at various temperatures.

smaller compared with that of $YBa_2Cu_3O_{7-\delta}$ (Li et al. 1990b). 2. The magnetisation diminishes asymptotically in the higher field region even for fields well below the estimated upper critical field $B_{c2}(T)$ line. 3. This feature is more dramatic at higher temperatures: above about 40 K, the irreversibility is below the detection limit in the present system. 4. For magnetic fields crossing the irreversibility line, there is no sudden disappearance of irreversibility in the magnetisation, in strong contrast to the results of $YBa_2Cu_3O_{7-\delta}$ (Li et al. 1990). This gradual disappearance of the irreversibility hinders a precise determination of the irreversibility line $B_{irr}(T)$.

In fig. 2 the calculated critical current, J_c, is shown based on the data in fig. 1. A dramatic reduction of J_c is seen in the higher-field and higher-temperature region. The J_c values at higher temperatures are compared with the values obtained from low field ac-susceptibility measurements and found to be fairly consistent (Li et al. 1990a).

Knowing J_c, the volume pinning force $F_p = J_c \times B$ can be calculated. This is given in fig. 3 for the data shown in figs. 1 and 2. In correspondence with the weak critical current, the pinning force is also much weaker than that of $YBa_2Cu_3O_{7-\delta}$. The broad peak in the F_p curves found at all temperatures in a rather low-field region ($B/B_{irr} \approx 0.1$) suggests an unconventional pinning mechanism. It seems that the individual pinning force is already broken at the low field region. At higher fields above $b \approx 0.2$ the pinning force is flat and the flux line lattice is governed by the shear stress of the flux-line lattice. The weak and long tail in higher field regime ($B/B_{irr} \approx 0.7$ - 1.0) also indicates unusual-flux-line lattice dynamics. It appears that the scaling law of Kramer (1973) is still valid in spite of the large uncertainty of the B_{irr}-value. This implies that the physical parameters which describe the pinning force scale with $b = B/B_{irr}$, indicating that a classical flux-pinning picture is valid in this region. Based on the classical model, the elementary pinning force due to oxygen vacancies has been evaluated by Kes (1990). Although the pinning force calculated contains ambiguous parameters, reasonable agreement with the experimental results can be obtained. It is noted, however, that the strong magnetic field dependence of the critical current is in contradiction to this model.

The irreversibility field B_{irr} obtained by the present measurements at various temperatures is plotted in the B-T phase diagram in fig. 4 together with the data obtained by other techniques. Although the B_{irr}-value at 4.2 K could not be reached within our field range, a field range from 60 to 80 T can crudely be estimated by a natural extrapolation of the pinning force curve to zero. This result constrains the lower limit of the $B_{c2}(0)$-value to be 60 - 80 T, since $B_{irr} \leq B_{c2}$. As a consequence, the upper limit of the in-plane coherence length $\xi_{ab}(0) \leq 20$ - 23 Å can be set, values that are considerably shorter than the value previously obtained (Palstra et al. 1988). Above 15 K, B_{irr} comes into our attainable field range. As seen in fig. 4, those points follow the trend of the line obtained from low field measurements by other techniques. The data above 30 K, however, deviate significantly. The reason is not obvious but it seems that the time scale of the measurements plays a significant role in this region.

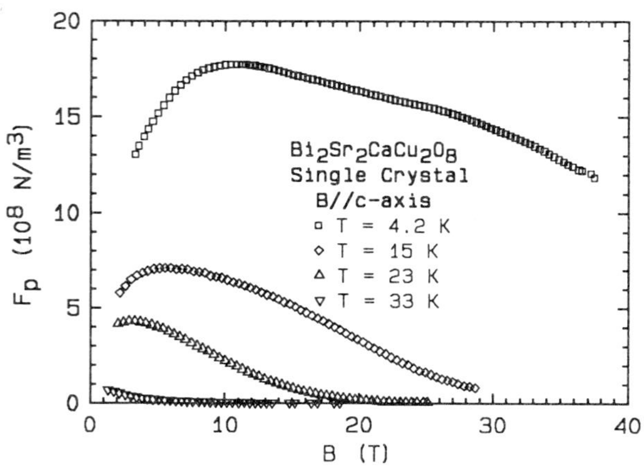

fig. 3. Pinning force densities at various temperatures.

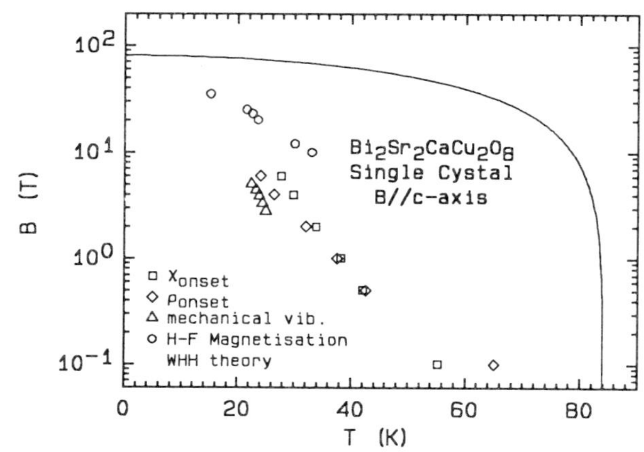

fig. 4. The B-T phase diagram in magnetic fields. The irreversible lines obtained by various techniques are compared: (○)-present measurements, (□)-onset of ac-susceptibility, (◇)-onset of resistivity, (−)-WHH theory calculated using a value $dB_{c2}(T)/dT|_{T=Tc} = -1.36$ T/K.

The phase enclosed by the $B_{c2}(T)$ and $B_{irr}(T)$ lines in fig. 4 is most of concern here. This phase region is highly dissipative with the ohmic law and has been interpreted by the traditional flux creep and flux flow model. However, the recent precise resistivity studies on single-crystalline $Bi_2Sr_2CaCu_2O_{8+\delta}$ in magnetic fields (Kadowaki et al. 1990a, 1990b, 1990c) has revealed evidence that the Lorentz-force-type of interactions are absent. Furthermore, the fact that the resistivity in the ab-plane is not influenced by magnetic fields up to 5 T sets the lower limit for $dB_{c2}(T)/dT$ to - 100 T/K, resulting in a value of the coherence length $\xi_c(0) \approx 0.3$ Å at most. These unrealistic values for the physical parameters cast a serious doubt whether or not the conventional anisotropic three dimensional Ginzburg-Landau approach is still valid for this compound. The weak pinning forces as well as the extremely short coherence length in this compound lead us to the speculative superconducting-phase-fluctuation model (Kadowaki et al. 1990b), which is based on weakly coupled Josephson links and which is an extension of Tinkham's model (1988) without invoking special dissipation mechanisms, special configuration of the flux lines or meandering of the current path.

References

Iye Y et al. 1990 *Physica* **C166** 62.

Kadowaki K et al. 1990a *Proceedings of the Int. Conf. on Magnetic Phase Transition (YAMADA Conference XXV), April 13 -16, Osaka, Japan, to be published in J. Magn. Magn. Mater.*

Kadowaki K et al. 1990b *to be published.*

Kadowaki K et al. 1990c *Proceedings of the Int. Conf. on LT19, Aug. 16 - 22, Sussex, Brighton, UK.*

Kadowaki K et al. 1990d *to be published.*

Kes P H and van der Beek C V 1990 *Proceedings of the ICMC'90, Topical Conference on High Temperature Superconductors, Materials Aspects, May 9 - 11, Garmisch-Partenkirchen, FRG.*

Kes P H 1990 *Proceedings of the Int. Conf. on Transport Properties of Superconductors, April 29 - May 4, Rio de Janeiro, Brazil.*

Kramer E J 1973 *J. Appl. Phys.* **44** 1360.

Gammel P L et al. 1988 *Phys. Rev.* **61** 1666.

Li J N et al. 1990a *Proceedings of the Int. Conf. on LT19, Sussex, Brighton, UK.*

Li J N et al. 1990b *Physica* **C169** 81.

Malozemoff A P et al. 1988 *Phys. Rev.* **B38** 7203.

Palstra T T M et al. 1988 *Phys. Rev.* **B38** 5102.

Palstra T T M et al. 1990 *Phys. Rev.* **B41** 6621.

Tinkham M 1988 *Phys. Rev. Lett.* **61** 1658.

Supercond. Sci. Technol. **4** (1991) S91–93. Printed in the UK

Magnetic flux creep as a percolation phenomenon.

A Gurevich*, H Küpfer and C Keller

KFK, Institut für Technische Physik, Postfach 3640, D - 7500, Karlsruhe, FRG
*also Institute of High Temperatures, 127412, Moscow, USSR

Abstract. Flux creep in macroinhomogeneous superconductors has been formulated as a vortex percolation. The model proposed is in good agreement with our experimental data and some recent experiments on magnetic relaxation in oriented-grained $YBa_2Cu_3O_{7-x}$.

1. Introduction

In this paper we consider flux creep as a percolation of vortices in a random pinning potential, where the flux motion becomes essentially nonuniform due to macroscopic modulations of a local pinning potential by material inhomogeneities, say, inherent stresses, nonuniform oxygen distribution in high-T_c oxides, their granular crystalline structure etc. Similar granularity may arise in the vortex structure itself due to the destruction of its long-range order by pinning, thermal fluctuations, effects of anisotropy, magnetic pre-history of the sample etc. In both cases the vortices begin moving along a network of resistive channels, which changes considerably the dependences of the flux-creep parameters on temperature T and magnetic induction B as compared to a uniform vortex motion.

2. Theoretical model

Let us consider a superconductor in which the local flux-creep activation energy, $W(\mathbf{r},T,B)$ is a random function of the coordinates, $\mathbf{r}$ (Figure 1). The resistivity,

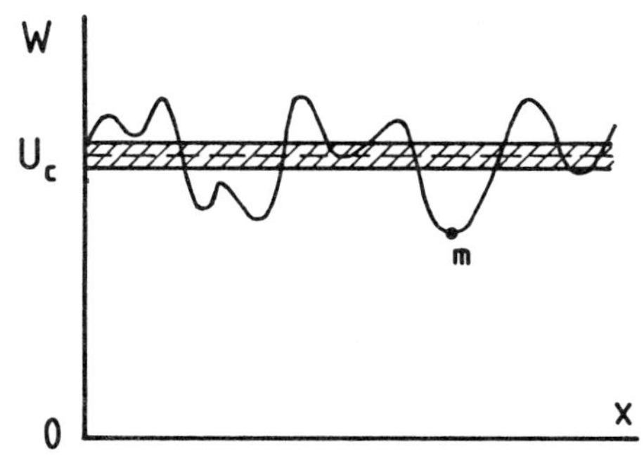

Fig.1. Sketch of W(x) (the domain with $|W - yU_c| < k_BT$ is hatched)

Fig. 2. Percolation channels (hatched) in a random potential (Zallen and Scher 1971)

$\rho(\mathbf{r}) = \rho_0 \exp(-W/k_BT)$, of such a sample can be highly nonuniform even in the case of weak inhomogeneities in $W(\mathbf{r})$. This implies that the specimen is separated into domains of considerably different local flux-creep rates, with only some part of the low-W domains contributing to the decay of macroscopic currents. Indeed, this process results from a macroscopic flux motion through the whole cross-section of the superconductor, hence the domains with a fixed value of W contribute to ρ only if their concentration, p, exceeds the percolation threshold, p_c, i.e. $p(U_c)=p_c$. At $k_BT \ll W$ the main contribution comes from domains with $W \simeq U_c$ which form a network of interlaced channels for the vortex motion (Figure 2). Such an infinite percolation cluster has a complicated topological structure, and its main part consists of the domains which do not contribute to ρ (so-called 'dead ends') (Last and Thouless 1971). Notice, these features lead to different distribution functions F(W) obtained by means of the electric noise, resistive and magnetic methods. For instance, the 'dead ends' do not affect the average resistivity, $<\rho>$, but contribute to the noise spectrum due to a local redistribution of vortices.

Let us now examine the dependence of the macroscopic activation energy, U, on T and B provided that the average resistivity $<\rho> \propto \exp(-U/k_BT)$ can be calculated within the framework of the effective medium theory (Gurevich et al 1990). Then

$$U(T,B) = U_c y(T,B) - \pi^2 k_B^2 T^2 F'(U_c)/6F(U_c) y(T,B), \qquad k_BT \ll U \tag{1}$$

Here U_c is the percolation threshold for the potential $W(\mathbf{r})$ at $T=0$, the prime denotes the differentiation with respect to W and the function y(T,B) determines the dependence of W on T and B which is assumed to be $W(\mathbf{r},T,B)=W_o(r)y(T,B)$ with $y(0,0)=1$ and $y(T_c,B)=0$. Formula (1) implies that U is determined by the vicinity of an energy 'belt' of width $\sim k_BT$ at $W \simeq U_c y$ (see Figure 1), and the condition of the vortex percolation yields $U_c\, y = U + <\delta U>$, where $<\delta U>$ is an average dispersion of U coming from domains with $|W - U_c y| \simeq k_BT$. At low T the superconducting parameters are independent of T, if there are no weak links in the specimen. Then $y(T)=$const, hence $U(T,B)=U_o(B)+AT^2$, where U_o and A are determined by formula (1). The energy U(T) increases with T if $F'(U_c)<0$, that is, if the function F(W) has a pronounced high-energy tail. Just this case usually corresponds to F(W) obtained from magnetic relaxation (Hagen and Griessen 1989), or electric noise measurements (Ferrari et al 1990), the details of F(W) being inessential for U(T). Therefore, the activation energy U(T) can exhibit qualitatively different behaviour as compared to the local values W(T), in particular, U(T) can increase with T even if W(T) always decreases with T.

The increase of U(T) at low T implies that U(T) passes through a maximum, U_m, at some $T=T_m(B)$. The values T_m and U_m depend on the shape of F(W), but at $T_m \simeq T_c$ their field dependences can be established by the scaling $U(T,B)= =k_BTG(z)$, $z=T/y(T,B)$ (Gurevich et al 1990). Indeed, at $T_c - T_m \ll T_c$ the equation $dU/dT=0$ for T_m reduces to $dG(z_m)/dz_m = 0$, which yields $T_m = y(T_m,B)z_m$, $z_m =$ const, or $T_c - T_m = (z_m/y_o)^{1/a}(B/B_{c2})^{b/a}T_c$, if one assumes $y=y_o(1-T/T_c)^a(B/B_{c2})^b$. Thus, $T_m(B)$ decreases with B and $U_m(B)=T_m(B)G(z_m)$, $G(z_m)=$const. The value of T_m is connected to the irreversibility line T_{irr} defined as $U(T_{irr}) \simeq k_BT_{irr}$. At $(T_m,T_{irr}) \simeq T_c$, that is, $B \ll B_{c2}$ the field dependences of the values $T_c - T_m(B)$ and $T_c - T_{irr}(B)$ are the same, which yields $[T_c - T_m(B)]/[T_c - T_{irr}(B)] =$ const.

Therefore, the results independent of the specific form of F(W) are : 1. The increase of U(T) at low T and the maximum in U(T) at $T=T_m(B)$; 2. The independence of the product U_oA of B (see formula (1)); 3. The linear proportionality of U_m versus T_m at $T_c - T_m \ll T_c$; 4. $[T_c - T_m(B)]/[T_c - T_{irr}(B)]=$ const at $(T_m,T_{irr}) \simeq T_c$.

3. Experimental results.

The above features of U(T,B) have been observed in our magnetic relaxation experiments carried out on the oriented-grained $YBa_2Cu_3O_{7-x}$. The decay of the magnetic moment M(t) was found to be logarithmic within all our experimental time window $t_i<t<t_f$ with $t_i \simeq 15$s, $t_f \simeq (1 \div 5)10^4$s and $M(t_f) \simeq (0.9 \div 0.7)M(t_i)$.

The activation energy $U_{ex} = -k_B T M(t_i) d\ln t/dM$ was measured at $4.2K < T < 90K$ and $0 < B < 12T$. The details were published elsewhere (Keller et al 1990).

Our experimental data for $\mathbf{B} \perp \mathbf{c}$ are shown in Figure 3, where the solid lines in the

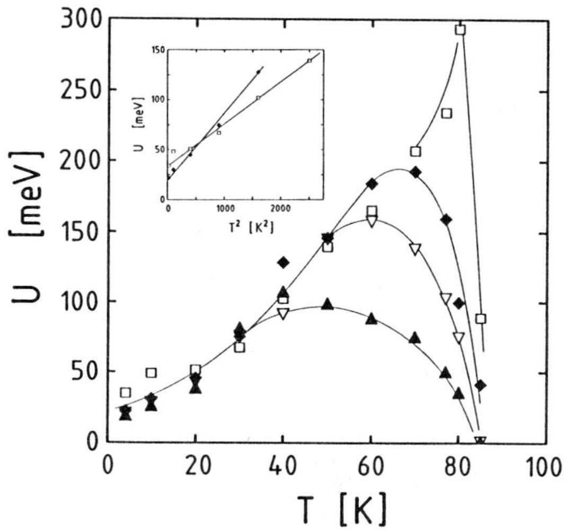

Fig.3.U versus T for different B: □ - 1T; ◆ - 4T; ∇ - 8T; ▲ - 12T. Insert: Examples of the fit to $U = U_0 + AT^2$ at B=1T and 4T.

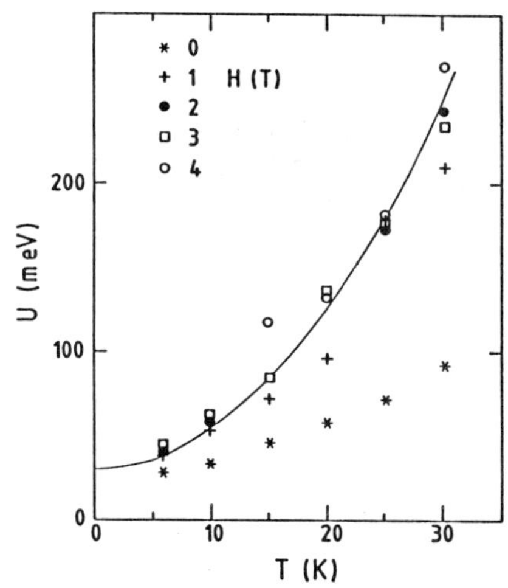

Fig.4. U versus T for different $\mathbf{B} \parallel \mathbf{c}$ (Campbell et. al., 1990).The solid curve corresponds to $U = U_0 + AT^2$ with $U_0 =$ 30 meV and $A = 0.244$ meV/K².

insert correspond to $U = U_0 + AT^2$ with fit parameters U_0 and A. This formula is in good agreement with our low-T data. At $\mathbf{B} \parallel \mathbf{c}$ we observed a similar U(T,B) but a rather linear dependence of U(T) at low T. This may result from granularity (twin structures etc). In Figure 4 the recent data by Campbell et al (1990) for the oriented - grained $YBa_2Cu_3O_{7-x}$ and $\mathbf{B} \parallel \mathbf{c}$ are shown.

As follows from Figures 3,5, the value $T_m(B)$ decreases with B, and $U_m(T_m)$ depends linearly on T_m in accordance with the above arguments. We have also measured $T_{irr}(B)$ and found that both the ratio $[T_c - T_m]/[T_c - T_{irr}]$ and the product AU_0 obtained from the low-T data are independent of B with an accuracy $\simeq 10\%$.

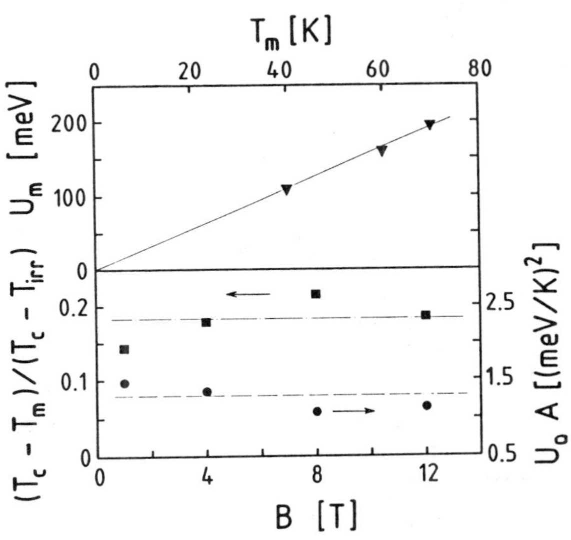

Fig. 5. U_m versus T_m (upper half). Field dependences of $[T_c - T_m]/[T_c - T_{irr}]$ and AU_0. The lines are the best fit.

Campbell I A, Fruchter L and Cabanel R 1990 Phys. Rev. Lett. **64** 1561
Ferrari M J et al 1990, Phys. Rev. Lett. **64** 72
Gurevich A, Küpfer H and Keller C 1990 (submitted to Phys. Rev. Lett.)
Hagen C W and Griessen R 1989 Phys. Rev. Lett. **62** 2857
Keller C et al 1990 Cryogenics **30** 401; **30** 411
Last B D and Thouless D J 1971 Phys. Rev. Lett. **27** 1971
Zallen R and Scher H 1971 Phys. Rev. **B4** 4471

Supercond. Sci. Technol. **4** (1991) S94–96. Printed in the UK

Dynamic Mixed State of High T_c Epitaxial Thin Films

Pan V M, Prokhorov V G, Kaminsky G G, Kasatkin A L, Kuznetsov M A and Tretiatchenko C G

Institute of Metal Physics, Kiev 252142, USSR

Abstract. The resistive state behavior has been studied for the $YBa_2Cu_3O_{7-\delta}$ epitaxial films deposited by the laser ablation onto $SrTiO_3$ (100) single crystals. The I-V curves and R(H,T) have been measured. The dissipation processes, temperature and field dependencies of the elementary pinning force and pinning potential have been shown to be adequately described by the spatially inhomogeneous superconducting order parameter model and the existence of easy flux slip channels.

1. Introduction

The investigation of vortex motion in dynamic mixed state could give a lot of information about pinning processes and thermally assisted flux creep (TAFC) in high temperature superconductors (HTSC). Nevertheless, the dissipation processes at weak links, which could obscure the viscous flux flow (VFF) resistivity, confuse correct interpretation of the experimental data, obtained from I-V curves. On the other hand, most of the theoretical models do not take into account an influence of the spatial inhomogeneity of the superconducting order parameter, that, as to our mind, is essential. In the present work we used the specially developed pulse technique for the investigation of dissipation processes which are connected with flux creep and VFF in epitaxial HTSC films. We interpret the results taking into account the specimen inhomogeneity.

2. Experimental results

The epitaxial $YBa_2Cu_3O_{7-x}$ films prepared by laser ablation were investigated (Gaponov et al. 1989). $SrTiO_3$ and ZrO_2 single crystals were used as substrates. The film thickness was 0.15 - 0.20 μm. The dc measured superconducting transition temperature was T_c = 89 - 90 K; the transition width was $\leq$ 1 K in the absence of applied field. The residual resistivity was $\rho_b = \rho(100\ K) = 10^{-4}\ \Omega$ cm. The specimens were finished for measurements by photolitography. The sample sizes were as follows: 20 μm width, 100 μm distance between potential probes. The silver contacts were vacuum deposited.

To avoid an influence of the thermal electric and contact effects the I-V curves were measured by dc as well as by pulse lock in device. In the last case samples were supplied by rectangular double polar pulses, which after relaxation of transition processes were integrated and stood in computer memory. The difference signal from both devices was monitored with X-Y recorder. The pulse duration was about 36 μs and the sequence frequency was approximately 870 Hz.

Investigating I-V curves for $YBa_2Cu_3O_7$ films at high temperatures by the pulse technique, which reduces a role of the sample heating, the VFF regions have been revealed.

The field dependence of the VFF resistance is shown in Fig.1. It is seen that this dependence could be divided into two parts with different ρ_f (B) behavior. The initial part is well described by the $\rho_f \propto B^{1/2}$ law (see the insert in Fig.1). At higher fields the VFF $\rho_f \propto B$ being usual for conventional type II superconductors. It should be also noted that the linear dependence does not start from the zero ρ_f.

It is seen (Fig.2, Insert) that the temperature dependence, $f_p(t)$, where $t = T/T_c$, is not monotonous. At $t \le 0.9$ $f_p \propto (1-t)$ and at $t \ge 0.9$ $f_p \propto (1-t)^{5/2}$.

3. Interpretation

The experimental data on the field dependence of the VFF and the temperature behavior of the elementary pinning force cannot be reasonably explained by the existing theories of dynamic mixed state in HTSC (e.g. Dew-Hughes 1988). We suppose that because of the extremely small coherence length in such materials, each type crystal defects (misfit dislocations, twin domain and grain boundaries, etc.) could result in significant spatial variation of the superconducting order parameter, $\Delta(\mathbf{r})$, and its strong suppression near the defect. Thus, the vortex motion under the Lorentz force (creep as well as VFF) at relatively low fields and high temperatures could occur only along one-dimensional "easy slip" channels, where the local pinning is weaker, rather than in the entire film volume. Taking into account this assumption, the well known expression for the field dependence of the VFF (Stephen 1965) can be modified:

$$\rho_f(B) \simeq \alpha \rho_n \frac{(\phi_o B)^{1/2}}{L B_{c2}},$$

where ρ_n is the normal resistivity, ϕ_o is the flux quantum, L is the mean distance between channels, and α is the factor about 1. As the field increases the vortex densities in channels and in strongly pinned regions are equalized, the VFF becomes two-dimensional, and, therefore, ρ_f (B) becomes linear.

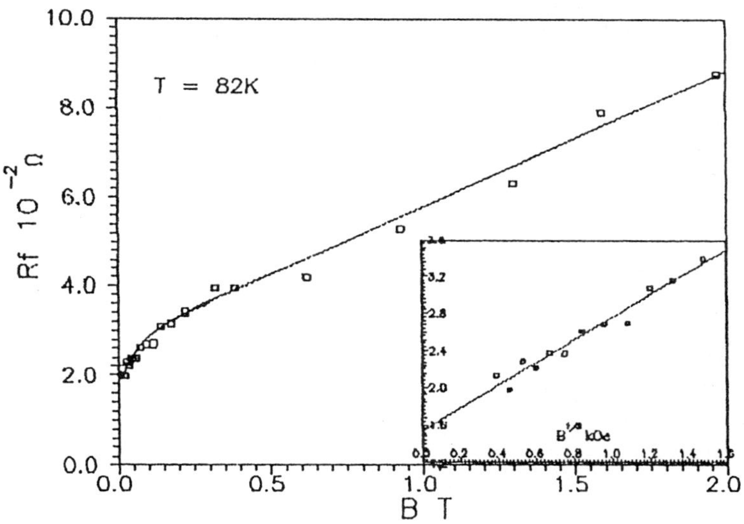

Fig.1.

The revealed temperature dependence of the pinning force $f_p \propto (1-t)^{5/2}$ at $t \to 1$ (Fig.2) agrees also well with the concept of the vortex motion along easy slip channels.

Indeed, for the elementary pinning force in the channel the following expression could be written:

$$f_p \propto U_o = \alpha \frac{H_c^2(t)}{8\pi}\left[\frac{\Delta_i}{\Delta_o}\right]^2 V_c ,$$

where U_o is the vortex activation energy in the channel, $V_c \propto \xi d a_o \propto (1-t)^{-1/2}$ is the correlation volume of the flux line lattice in a single thermal activation process, ξ is the coherence length, d is the film thickness, a_o is the flux line lattice parameter, Δ_i/Δ_o is the ratio of the superconducting order parameters in channels and in shores, which obeys the $(\Delta_i/\Delta_o)^2 \propto (1-t)$ law (Deutscher and Müller 1987), and $H_c(t) \propto (1-t^2)$ is the thermodynamic critical field. Thus, in the frameworks of present concept $f_p \propto U \propto (1-t)^{3/2}(1-t^2)$, and near T_c correspondingly $f_p \propto (1-t)^{5/2}$ that well agrees with the experimental data.

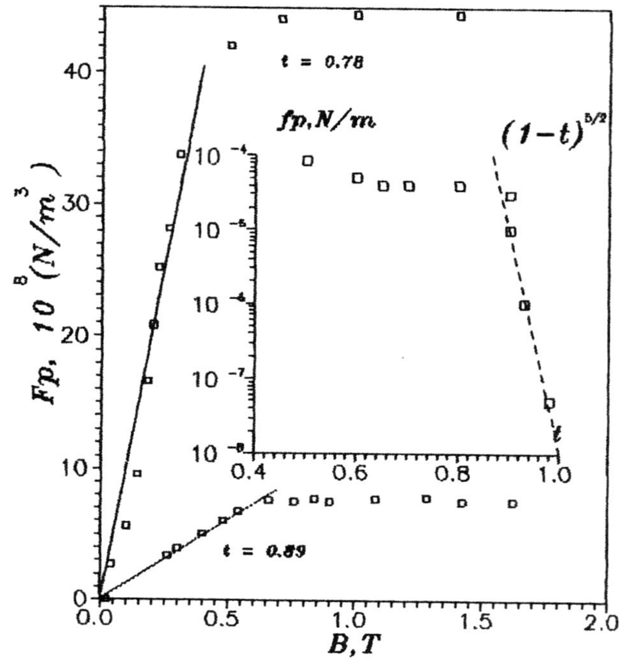

Fig.2.

Conclusion

The main peculiarity of the HTSC film electrodynamics in the mixed state is the existence of easy slip channels (ESC), where flux lines could move along planar crystal defects. The developed ESC theory taking into account the spatial inhomogeneity of superconducting order parameter enables to describe reasonably the whole collection of experimental results.

Acknowledgement

The authors are indebted to Prof. S.V.Gaponov and his colleagues from the Institute of Applied Physics, Gorki, for the high quality HTSC films.

References

Bardeen J and Stephen M J 1965 Phys.Rev. **140** 1197.
Deutscher G and Müller K A 1987 Phys.Rev.Lett. **59** 1745.
Dew-Hughes D 1988 Cryogenics **28** 674.
Gaponov S V et al. 1989 ZhETF (Sov.) **95** 2191.

Supercond. Sci. Technol. 4 (1991) S97–99. Printed in the UK

Critical Currents in Tl-2122 and Y-124 Sinters under High Hydrostatic Pressure

J Diederichs[a], W Reith[a], B Sundqvist[b], J Niska[c], K E Easterling[c], and J S Schilling[a]

[a] Sektion Physik der Universität München, D-8000 München 40, FRG
[b] Department of Experimental Physics, Umeå University, S-90187 Umeå, Sweden
[c] Department of Engineering Materials, Luleå Univ. of Techn., S-95187 Luleå, Sweden

Abstract. Utilizing ac susceptibility measurements we have investigated the *inter*- and *intra*granular properties of selected bulk materials under hydrostatic pressures to 3.5 GPa. A single phase $Tl_2CaBa_2Cu_2O_8$ sample with superconducting transition temperature $T_c \simeq 107\,K$ shows a clear maximum in $T_c(P)$ at $\sim 2.2\,GPa$ coupled with a nearly twofold increase to 3 GPa of the *inter*granular critical current density J_{cJ} near T_c. The second sample, a $YBa_2Cu_4O_8$ sinter, shows a large linear increase in T_c with pressure (+5.5 K/GPa) as well as a fivefold increase J_{cJ} to 3 GPa. Applying a critical state model, we find that $T_c(P) \propto \lambda_g^{-2}(P)$, where λ_g is the *intra*granular (*London*) penetration depth.

1. Introduction

The nature of the contacts between grains in sintered high-T_c superconductors which leads to weak-link behaviour and, therefore, to severely reduced transport critical currents is poorly understood. We follow Clem (1988) in modeling the sinter as an array of *Josephson*-coupled strongly superconducting grains and in distinguishing between *inter*- and *intra*granular properties. We find ac susceptibility measurements to be a valuable tool for studying these properties and use Müller's (1989) critical state model to extract parameters such as the *inter*granular critical current density J_{cJ} and the *intra*granular London penetration depth λ_g from the data. An external ac field H allows *inter*granular flux to sweep in and out of the sample. If pinning centers are present, a flux gradient is set up which is proportional to J_{cJ},

$$\left|\frac{dH}{dx}\right| = J_{cJ}(H). \quad (1)$$

Müller (1989) assumes a power-law temperature dependence $J_{cJ}(T) \propto (1 - T/T_c)^n$ which is valid only near T_c. As discribed in Goßner *et al* (1990) we derive the exponent n with the help of the *Bean* model and use only $J_{cJ}(\text{“}T = 0K\text{”})$ and λ_g as fit prameters to calculate $\chi_{ac}(T)$ utilizing a fortran program which K-H Müller kindly provided us. This model has proven to be very successful in describing the ac and dc field dependence of $\chi_{ac}(T)$ for the samples studied here (Goßner *et al* 1990, Diederichs *et al* 1990).

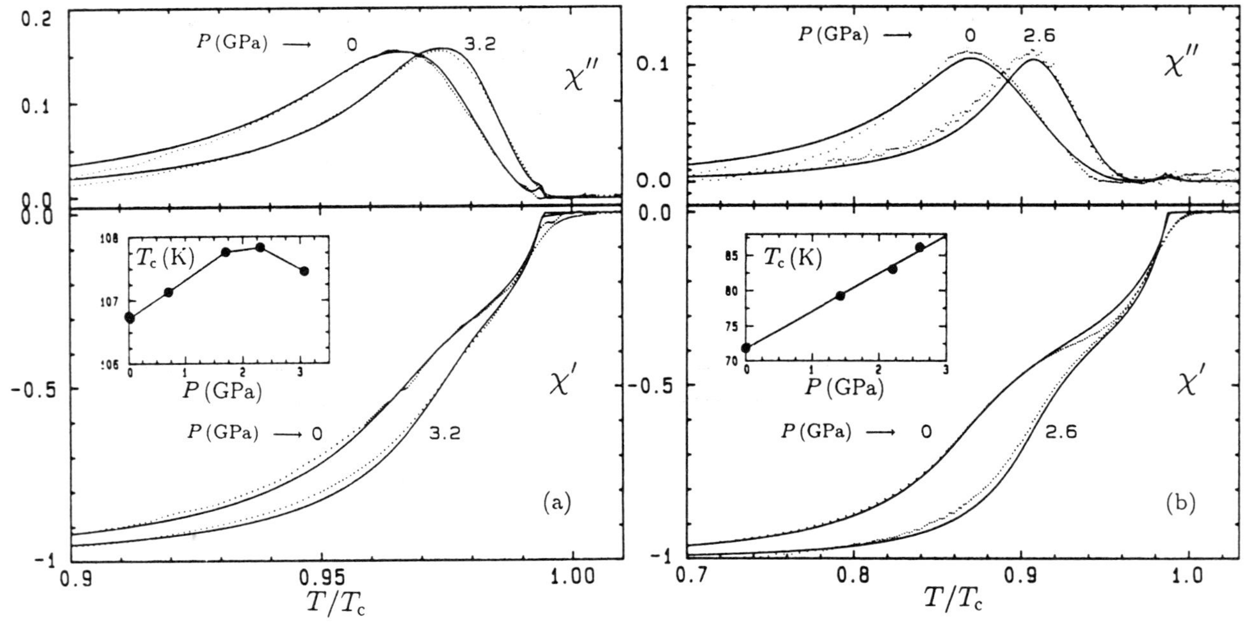

Figure 1: **Real χ' and imaginary χ'' parts of the ac susceptibility versus T/T_c for (a) $Tl_2CaBa_2Cu_2O_8$ at 2 Oe ac field and (b) $YBa_2Cu_4O_8$ at 0.25 Oe ac field for two pressures. Ambient dc field is the earth's field. The points are the data while the solid lines are fit's using Müller's model. Insets show pressure dependence of T_c.**

2. Experimental

The $Tl_2CaBa_2Cu_2O_8$ sample was prepared by normal sinter reaction, whereas the $YBa_2Cu_4O_8$ sinter was hot isostatic pressed. From optical microscopy we estimate the superconducting volume fraction to be $f_s = 60\%$ and 98% and the mean grain size $R_g = 4\mu$m and 4μm for the above samples, respectively (more details in Andersson *et al* 1990 and Diederichs *et al* 1990). Hydrostatic pressure to 4.0 GPa is generated using a metal-gasket cell developed by Fasol and Schilling (1978). Ac susceptibility is measured at 5 kHz using a miniature coil system (0.3 mm I.D. x 1 mm O.D.) and a two-phase lock-in amplifier.

3. Results

In Figure 1 we compare the ac susceptibility at two pressures for the samples studied. The two-step structure seen in $\chi'(T)$ separates *inter*grain screening at the lower temperature from *intra*grain screening at the higher temperature. The weaker the *inter*grain contact, the further the lower step is shifted to lower temperature.

Fits of the ambient pressure data on $Tl_2CaBa_2Cu_2O_8$ to Müller's model yield $J_{cJ} = 340$ A/cm^2 ($n = 1.7$) at 5 K below T_c, an unusually high value for a sinter. At the same reduced tempeature T/T_c the $YBa_2Cu_4O_8$ sample exhibits a $60\times$ smaller value of J_{cJ} ($n = 2.5$). Under pressure the lower step shifts into the upper *intra*grain step, implying a strong improvement of the contacts. This conclusion is supported

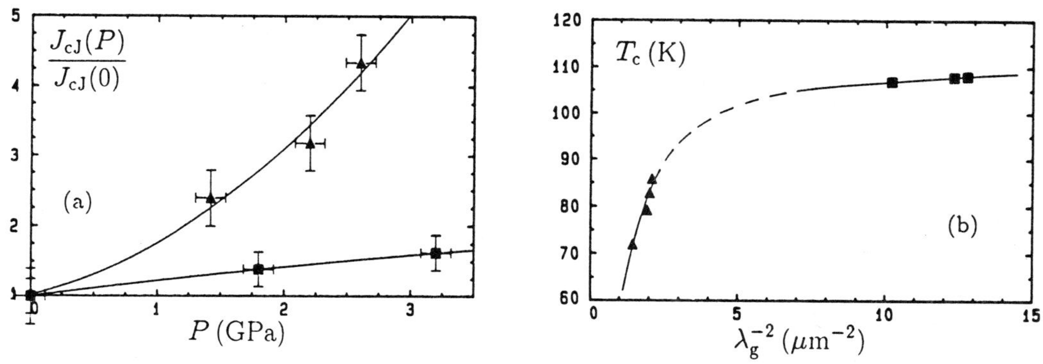

Figure 2: (a) Relative increase of the critical current density under pressure and (b) λ_g^{-2} versus T_c for $\triangle$: $YBa_2Cu_4O_8$ and $\square$: $Tl_2CaBa_2Cu_2O_8$.

by the results of fits of the data to Müller's model where a strong increase in the critical current density is found under pressure, as shown in Figure 2. According to Clem (1988) increasing critical current density can be understood by a decrease of the normal barrier resistance (and thus increasing the coupling energy). The barrier resistance could be changed by variation of the intrinsic or the barrier charge density or by mechanical improvement of the contacts.

The London penetration depth λ_g is found to initially decrease under pressure for both samples and, additionally, $\lambda_g^{-2}(P)$ is approximately proportional to $T_c(P)$, as shown in Figure 2b. Taken together, the data in Figure 2b would suggest that T_c saturates at $\sim$ 110 K for large λ_g^{-2}. It is interesting to note that under high pressures to nearly 15 GPa van Eenige *et al* (1990) find that $T_c(P)$ for $YBa_2Cu_4O_8$ indeed saturates at $\sim$ 108 K. These results are consistent with the systematics found in the muon-spin-resonance experiments of Uemura *et al* (1989) on two layer compounds.

Acknowledgments

This work is supported by the *Bundesministerium für Forschung und Technologie.*

References

Clem J R 1988 *Physica C* **153-155** 50
Müller K H 1989 *Physica C* **159** 717
Goßner H and Schilling J S 1990 *in preparation*
Diederichs J, Reith W, Schilling J S, Andersson K, Sundqvist B, Niska J and Easterling K E 1990 *in preperation*
Andersson K, Sundqvist B, Niska J and Easterling K E 1990 *submitted to Physica C*
Fasol G and Schilling J S 1978 *Rev. Sci. Instr.* **49** 1722
Van Eenige E N, Griessen R, Wijngaarden R J, Karpinski J, Kaldis E, Rusiecki S and Jilek E 1990 *Physica C* **168** 482
Uemura Y J, Luke G M, Sternlieb B J, Brewer J H, Carolan J F, Hardy W N, Kadono R, Kempton J R, Kiefl R F, Kreitzman S R, Mulhern P, Riseman T M, Williams D Ll, Yang B X, Uchidy S, Takagi H, Gopalakrishnan J, Sleight A, Subramanian M A, Chien C L, Cieplak M Z, Xiao G, Lee V Y, Statt B W, Stronach C, Kossler W J and Yu X H 1989 *Phys. Rev. Lett.* **62** 2317

Supercond. Sci. Technol. **4** (1991) S100–102. Printed in the UK

Critical current anisotropy of BiSrCaCuO thin films under magnetic fields.

H. Raffy[x], S. Labdi[x], O. Laborde[+*], P. Monceau*.

x Laboratoire de Physique des Solides, Bât.510, Université Paris-Sud, 91405 Orsay France
+ Service National des Champs Intenses, CNRS, 38042 Grenoble France.
* Centre de Recherches sur les Très Basses Températures, CNRS, 38042 Grenoble France

Abstract . We report on the strong angular dependence under magnetic field of the critical currents of c-axis oriented $Bi_2Sr_2CaCu_2O_8$ thin films, measured as a function of the angle θ between the c-axis and the applied magnetic field ($H \leq 20$ T). The curves $J_c(\theta)$ at a given field H present a peaked maximum occurring when H is parallel to the CuO_2 planes ($\theta = 90°$). At $\theta \neq 90°$, J_c appears to depend only on the perpendicular field component $H\perp$ (along the c axis). In particular at 4.2 K and for $H\perp > 0.02$ T, it is shown that $J_c(H, \theta)$ behaves as $|H\cos\theta|^{-\alpha}$ with $0.44 \leq \alpha \leq 0.50$.

1 - Introduction.

High Tc oxide superconductors have a very anisotropic crystalline structure, which can be schematically described as a layered structure. $Bi_2Sr_2CaCu_2O_8$ superconductors are characterized by a very weak coupling between superconducting CuO_2 layers separated by a distance of about 12 Å and a very short coherence length ξ_c along the c axis, implying a well marked 2D character for this system. In this paper we report on some results demonstrating the strong anisotropy of the basal plane critical currents as a function of the orientation of the applied magnetic field, and we discuss the origin of the observed dissipation effect.

2 - Experimental.

The films were deposited by DC single target sputtering on MgO (100) single crystal substrate at room temperature and crystallized by a post-deposition annealing in flowing O_2-N_2 gas at 810°C-820°C (H. Raffy *et al* 1988). X-ray diffraction studies showed that the films are constituted of the 2212 phase and are highly c-axis oriented (misorientation ± 1°). The films used in this study were 1000 Å thick and had a critical temperature ranging from 85 K to 88 K.

Critical current measurements were performed on laser-patterned strip lines 10-100 µm wide, 500-1000 µm long, using a four probe technique, with a voltage criterion of 0.5-2.5 µV. The magnetic field was applied with a rotatable electromagnet ($H \leq 1$ T) or with a Bitter coil ($H \leq 20$ T) and a rotatable sample holder. In these measurements the field was perpendicular to the current if not otherwise specified.

3 - Results and discussion

We first consider some typical results to illustrate the anisotropic

0953-2048/91/01S100 + 03 $03.50

behaviour of the critical current density $J_c(H,T)$ in the two extreme situations H //(θ=90°) or H⊥(θ=0) to the CuO_2 planes.

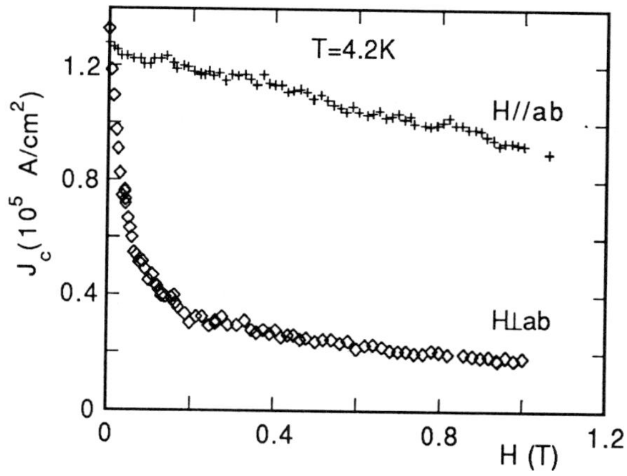

Fig.1 Dependence of J_c on magnetic field H// or H⊥ to the CuO_2 plane for a 1000 Å Bi oxide (2212) film.

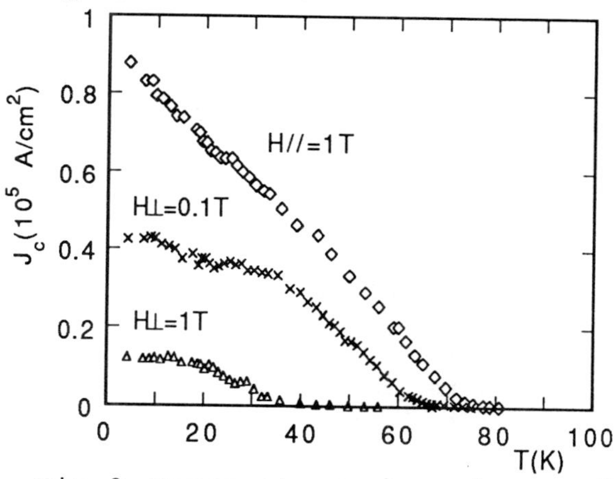

Fig.2 Temperature dependence of J_c in an applied magnetic H// or H⊥ for the same film with T_c=85K.

Figure 1 shows that, at 4.2 K, J_c (H//) exhibits a slow linear decrease, with a slope very sensitive to field orientation (as explained below). J_c (H⊥) decreases steeply as soon as a field is applied and is found to follow the law $(H\perp)^{-\alpha}$ for H ≥0.02 T, with α =0.45. This anisotropy is consistent with an easy motion due to the Lorentz force along the CuO_2 planes for the vortices perpendicular to them and implies a much more difficult motion along the c axis for flux parallel to the CuO_2 planes. The latter might be ascribed to an intrinsic pinning effect of the crystalline structure in which the flux will be trapped between the CuO_2 layers due to a modulation of the superconducting order parameter along the c axis as described by Tachiki *et al* (1989); however, Tachiki *et al* predict that J_c(H//) would stay constant for $H \ll H_{c2}$, whereas a decrease is observed (Fig.1). On the contrary, a motion of the vortices along the CuO_2 plane is most likely impeded by extrinsic pinning centers.

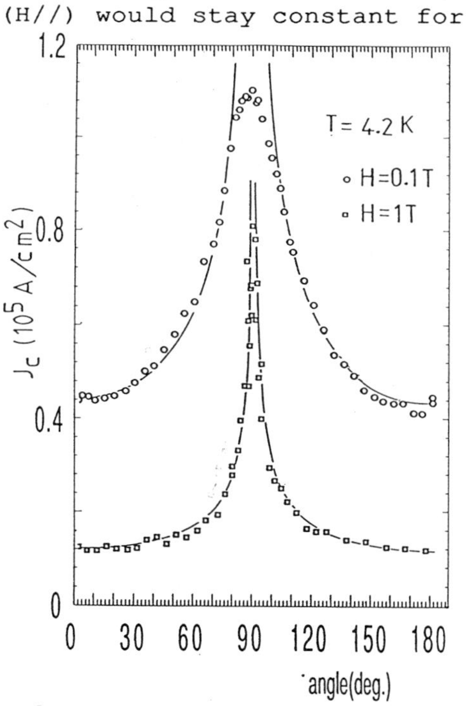

Fig.3 Angular dependence at 4.2K of J_c for a Bi oxide (2212) film as a function of field orientation (H parallel to the CuO_2 planes at θ=90°). The solid curves represent the function: $J_c(H\perp)/|\cos\theta|^{1/2}$.

Figure 2 displays the temperature dependence of J_c measured at constant field H// or H⊥. With increasing temperature, J_c(H//,T) decreases smoothly and vanishes rather close to T_c. The effect of temperature is much more dramatic in the H⊥ case: J_c(H⊥,T) after a moderate decrease at low T, decreases abruptly to very low J_c values in a temperature range well below T_c. In fact, the higher the field H⊥, the farther below Tc, J_c vanishes . This fact may be the sign of an abrupt increase of the flux mobility with growing temperature depending on changes in the vortex structure (3D to 2D vortices) and/or in the flux line distribution (lattice or liquid) (Doniach 1990 and ref.therein). It may be the sign of a progressive decoupling between the CuO_2 planes.

We consider now the angular dependence of the critical current, measured at a

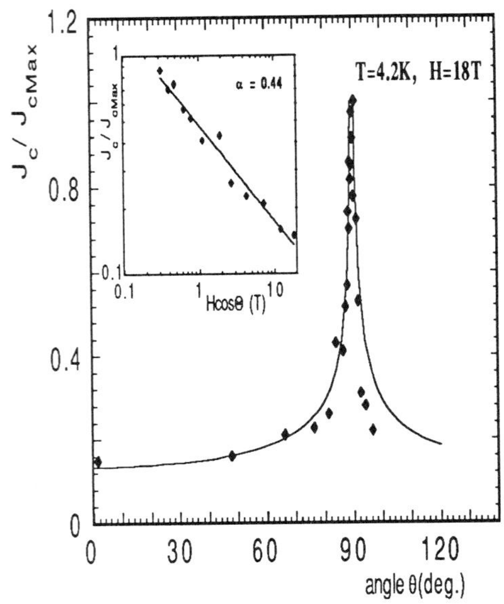

Fig.4 Dependence of J_c on sample orientation in the applied field H=18T. The solid curve represents $J_c(H\perp)/|\cos\theta|^{0.44}$. The insert shows that J_c plotted as a function of $H\cos\theta$ ($0\leq\theta\leq 90°$) behaves as $(H\cos\theta)^{-\alpha}$ with $\alpha=0.44$.

constant field, as a function of the tilt angle θ of the field with respect to the c axis. The results are presented at 4.2 K, at low field in Figure 3 and one example at high field, H =18 T, is given in Figure 4. It is shown that in all cases the critical currents increase when the parallel field situation is approached (θ=90°). The width of the maximum is decreasing with increasing H up to 12 T, with Δθ(HMHW) ~1.5° and saturates. The solid curves are fits to the function $J_c(H\perp)/|\cos\theta|^{\alpha}$ with $0.44 \leq \alpha \leq 0.50$. The same type of behaviour was proposed by Tachiki et al.(1990) for YBaCuO results of Roas *et al* (1990). The insert of figure 4 show that $J_c(\theta)$ (H fixed, θ varying from 0 to 90°) plotted as a function of Hcosθ follows the same functional dependence $|H\cos\theta|^{-\alpha}$ as the one observed for $J_c(H\perp)$ (θ=0, H varying) indicated above. This behaviour holds for all the results obtained at different H values, for Hcosθ ≥0.02, i.e. the closer to θ=0, the higher the applied field H. More generally, measurements performed at higher temperature (T ≥30 K) show that for θ≠90°, $J_c(\theta)$ as a function of Hcosθ is also identical to $J_c(H\perp)$ measured at θ=0, although its functional dependence is different from that at low temperature.

In conclusion, these results indicate that at θ ≠90° the critical current is essentially determined by the perpendicular component of the magnetic field Hcosθ in agreement with ref.(P.Kes *et al*). In the experimental parallel case, most of the dissipation, especially at fields H >1 T, can be attributed to the existence of a small perpendicular component due to a non perfect alignment of the flux with the CuO_2 planes or to the small mosaïcity of the films. The behaviour of $J_c(H//)$ is found to scale that of $J_c(H\perp)$ with a scaling factor cosθ with θ =1–3°. It may explain that the values of $J_c(H//)$ with H// parallel or perpendicular to the current are very close (Raffy *et al* 1990, Fig.2). We have also shown that the values of the magnetoresistance measured on the same films are essentially governed by the perpendicular component of the applied field, Hcosθ, as well (to be published) . These results confirm the 2D behaviour of the $Bi_2Sr_2CaCu_2O_8$ compound.

Acknowledgments HR and SL thank A.Vaurès for technical assistance in sample preparation, S.Megtert for X-Ray diffraction studies and D.Beaupère, Laboratoires de Marcoussis CR-CGE, for laser patterning.

References.

Doniach S.,*preprint*

Kes P.H., Aarts J., Vinokur V.M., Van der Beek C.J. 1990, *Phys. Rev. Lett*, **64**, 108.

Raffy H., Vaurès A., Arabski J., Megtert S., Rochet F., Perrière J. 1988, *Sol. State Comm.* **68**,235

Raffy H., Labdi S., Laborde O. and Monceau P.1990, *LT 19 proceedings*.

Roas, Schultz L., Saemann-Ischenko G. 1990, *Phys.Rev.Lett.***64** 479

Tachiki M. and Takahashi S. 1989, *Sol. State Comm.* **70**, 291.

Tachiki M. and Takahashi S. 1990, *Sol. State Comm.* **72**, 1083.

Supercond. Sci. Technol. **4** (1991) S103–105. Printed in the UK

Critical Currents in Neutron Irradiated YBCO and BiSCCO Single Crystals

H. W. Weber, H. P. Wiesinger, W. Kritscha, and F. M. Sauerzopf
Atominstitut der Österreichischen Universitäten, A-1020 Wien, Austria
G. W. Crabtree,[1] J. Z. Liu,[1,*], Y. C. Chang,[2] and P. Z. Jiang[1]
[2]Science and Technology Center for Superconductivity and [1]Materials Science Division
Argonne National Laboratory, Argonne, IL 60439, U.S.A.

Abstract. Neutron irradiation of high-T_c single crystals in the fluence range up to $2x10^{22}$ m^2 (E > 0.1 MeV) has been found to lead to the following property changes: 1. The critical temperature decreases only moderately at a rate of approximately 4 K per 10^{22} neutrons per m^2; 2. The critical current densities J_c as deduced from magnetization measurements on the basis of an extended Bean model increase significantly; 3. These J_c-enhancements are larger at higher temperatures; 4. The anisotropy of J_c decreases significantly with neutron fluence; and 5. The irreversibility line in BiSCCO-2212 is shifted to much higher fields and temperatures.

Neutron irradiation has proved to be a valuable tool for studying flux pinning effects in superconductors, because well defined defects are introduced into the material, while the basic superconductive properties of the sample remain nearly unchanged, at least at relatively low fluences (~10^{22} m^{-2}, E>0.1 MeV). Based on these considerations, neutron irradiation experiments have been made on several high-quality single crystals of $YBa_2Cu_3O_{7-\delta}$ and $Bi_2Sr_2CaCu_2O_8$. Dc measurements of the transition temperature confirmed again that the changes of T_c remained relatively small (~4 K at $1x10^{22}$ m^{-2}) and that the broadening of the shielding curves (from 5 K to 7 K) was only moderate (Sauerzopf *et al.* 1990).

The main emphasis of the present work has been placed on the investigation of magnetization curves as a function of temperature, crystal orientation with respect to the external magnetic field, and neutron fluence. All the measurements were done using an 8 T SQUID-magnetometer. Because of the highly anisotropic nature of critical current densities in the single crystals prior to irradiation, the evaluation of J_c from the hysteretic magnetization curve in terms of the Bean model (Bean 1962) requires special attention (Gyorgy *et al.* 1989, Sauerzopf *et al.* 1990), in order to account for anisotropic current flow (especially when H is parallel to the basal plane) and for demagnetization corrections (especially when H is parallel to the c-axis). These extensions of the Bean model have been used throughout the present analysis, where we will denote (isotropic) currents flowing in the basal plane, i.e. H//c, by $J_c{}^a$, and currents flowing along the c-direction as deduced from an experiment with H parallel to the basal plane, by $J_c{}^c$, respectively.

Results on the fluence dependence of J_c are shown in Figure 1 for two YBCO single crystals (S 71, S 84), at fixed low temperature and at a fixed magnetic field of 2 T. In both cases, J_c is found to increase with neutron fluence; the data on crystal S 71 which was irradiated sequentially in small fluence steps, indicate, however, that the initial rapid increase of J_c at low fluences is followed by a slower dose dependence, with occasional kinks appearing, especially at higher fields and temperatures. In addition, the anisotropy of critical current densities is

0953-2048/91/01S103 + 03 $03.50

drastically reduced by irradiation, as shown in Figure 1b. Both of these observations indicate that the pre-irradiation defect structure and, hence, the flux pinning mechanisms prevailing prior to irradiation, play a major role in determining the radiation response of a particular single crystal. (Note, e.g., the large difference in critical current anisotropies between the two single crystals prior to irradiation.)

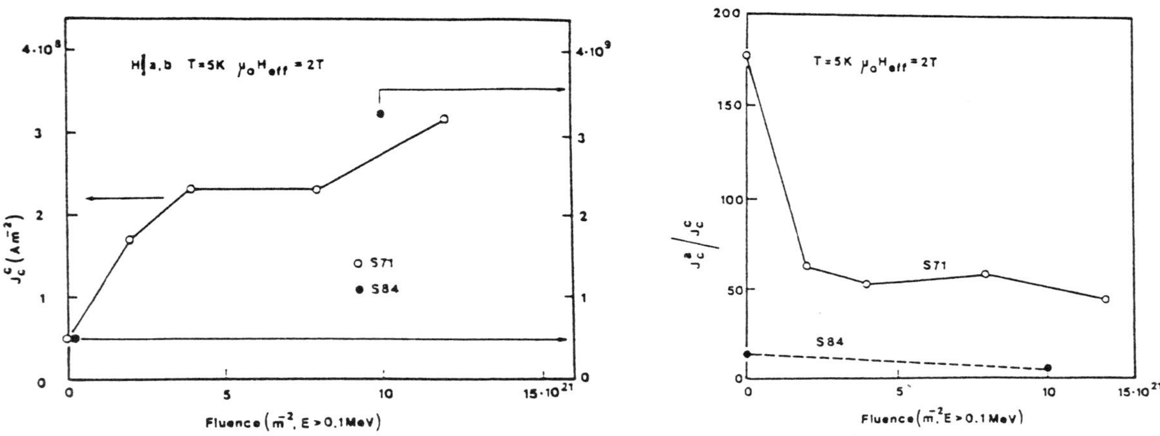

Fig. 1: Critical current densities as a function of fast neutron fluence for single crystals S 71 and S 84. a) Critical current density flowing along the c-direction, b) ratio of critical current densities J_c^a and J_c^c as a function of neutron fluence.

The *temperature dependence* of J_c prior to and following irradiation to $\sim 10^{22}$ m^{-2} (Sauerzopf *et al.* 1990) demonstrates for both crystals the beneficial effect of radiation-induced defects in a most remarkable way. Whereas the J_c-enhancement factors at 5 K (remnant state after a full magnetization cycle to 8 T) amount to 2.9 and 3.3 for crystals S 84 and S 71, respectively, the corresponding figures are 15 and 18 at liquid nitrogen temperature. From this *change of slope* in the temperature dependence of J_c following neutron irradiation we infer, that the radiation-induced defects have *deeper* pinning potentials than the pre-irradiation defect structure, since they are obviously capable of exerting stronger pinning forces at higher temperatures.

The nature of these radiation-induced defects has been investigated only recently and still in a somewhat preliminary way by TEM (Kirk *et al.* 1990, 1990a). First results obtained upon low-fluence 50 keV Xe-irradiation revealed the formation of relatively large (~5 nm) collision cascades, which in some cases consisted of recrystallized superconducting material, but misoriented by about 30°. These defects are certainly able to act as strong additional pinning centers. Secondly, first results on a crystal neutron-irradiated to high fluences, showed large areas of highly defective or amorphous material imbedded in 10-30 nm regions of relatively undisturbed superconductor. Although this defect structure is still found to provide considerable flux pinning, the distribution of these defects and their interaction with the pre-irradiation defect structure may provide clues for a more detailed understanding of the complicated fluence dependence of J_c at higher fields, temperatures and fluence levels.

The effectiveness of radiation-induced defects as flux pinning centers has also been proved by neutron irradiation experiments on BiSCCO-2212 single crystals (Kritscha *et al.* 1990). In these materials, the "irreversibility line" is usually found at relatively low temperatures and magnetic fields (Yeshurun *et al.* 1989, diamonds in Figure 2). Following neutron irradiation to a fluence of 6×10^{21} m^{-2}, a drastic broadening of the magnetization curves has been observed, especially also at fields and temperatures far above the pre-irradiation irreversibility line. A summary of results pertaining to this shift of the irreversibility line is presented in Figure 2, where the fields and temperatures separating hysteretic from reversible magnetization behavior as deduced from the dc magnetization measurements are shown. Clearly, significant flux pinning can be observed at fields and temperatures, at which the unirradiated sample had shown complete reversibility. Since the basic physical properties of the sample have changed only very little under irradiation, significant alterations of the intrinsic properties of the flux line lattice, such as a phase transition into a liquid state, seem very unlikely.

In summary, fast neutron irradiation has been shown to significantly enhance critical current densities in high-T_c single crystals. Based on TEM results, relatively large collision cascades as well as highly defective or amorphous areas have been identified to be the radiation-induced defect structure, which is certainly capable to provide strong flux pinning. In the case of YBCO, we conclude from the complicated J_c-dependence on neutron fluence that the radiation-induced defects also interact with the pre-irradiation defect structure. For BiSCCO, flux pinning has been observed in the previously reversible field and temperature range, which is considered as evidence for the existence of a flux line lattice also *above* the irreversibility line.

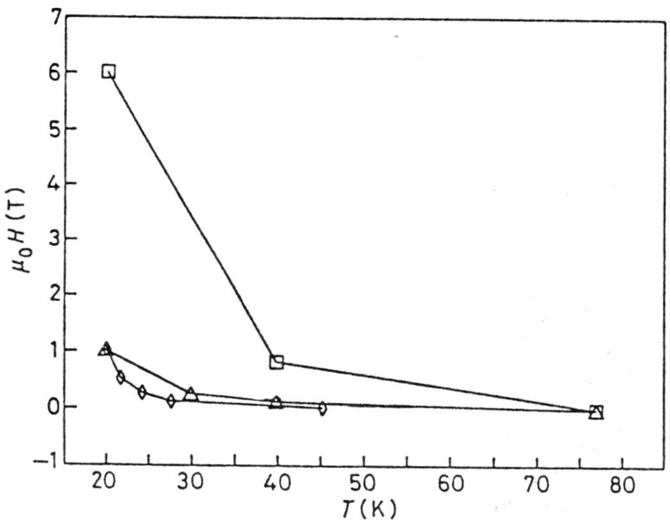

Fig. 2: Irreversibility line in BiSCCO-2212 prior to (triangles) and following (squares) neutron irradiation. Also shown are results obtained by dc measurements (diamonds) on another single crystal (Yeshurun *et al.* 1989), which are in excellent agreement with our pre-irradiation data.

Acknowledgements

This work has been supported in part by Fonds zur Förderung der Wissenschaftlichen Forschung, Wien, under contract # 6837 (HWW, HPW, WK, FMS), by the U.S. Department of Energy, Basic Energy Sciences - Materials Sciences, under contract # W-31-109-ENG-38 (GWC, JZL, PZJ), and by the National Science Foundation--Office of Science and Technology Centers under contract # STC-8809854 (YCC).

References

Bean C P 1964 *Rev.Mod.Phys.* **36**, 31

Gyorgy E M, van Dover R B, Jackson K A, Schneemeyer LF and Waszczak J V 1989 *Appl.Phys.Lett.* **55**, 283

Kirk M A, Frischherz M C, Liu J Z, Greenwood L R and Weber H W 1990 *Phil.Mag.Lett.* **62**, 41

Kirk M A, Frischherz M C, Zhang J P, Liu J Z and Weber H W 1990a *(submitted to Phys.Rev.Lett.)*

Kritscha W, Sauerzopf F M, Weber H W, Crabtree G W, Chang Y C and Jiang P Z 1990 *Europhys. Lett.* **12**, 179

Sauerzopf F M, Wiesinger H P, Kritscha W, Weber H W, Crabtree G W and Liu J Z 1990 *(submitted to Phys.Rev. B)*

Sauerzopf F M, Wiesinger H P,and Weber H W 1990a *Cryogenics* **30**, 650

Yeshurun Y, Malozemoff A P, Worthington T K, Yandrofski R M, Krusin-Elbaum L, Holtzberg F H, Dinger T R and Chandrasekhar G V 1989 *Cryogenics* **29**, 258

*Present address: Department of Physics, University of California, Davis, CA 95616, U.S.A.

Supercond. Sci. Technol. 4 (1991) S106–108. Printed in the UK

Pinning in Twin Boundaries of $YBa_2Cu_3O_{7-\delta}$ Single Crystals

W. K. Kwok,[2] U. Welp,[1,2] S. Fleshler,[2,*] K. G. Vandervoort,[1,2,+] G. W. Crabtree,[2]
J. Z. Liu,[3] J. Brooks,[4] J. Hettinger,[4] S. T. Hannahs[4]

(1)Science and Technology Center for Superconductivity, Argonne, IL 60439 USA
(2)Materials Science Division, Argonne National Laboratory,Argonne, IL 60439 USA
(3)Department of Physics, University of California-Davis, Davis, CA 95616 USA
(4)Department of Physics, Boston University, Boston, MA 02215 USA
*Also at Purdue University, West Lafayette, IN 47907 USA
+Also at University of Illinois at Chicago, Chicago, IL 60680 USA

Abstract. We present direct observation of Lorentz force induced flux motion and pinning by twin boundaries in single crystal $YBa_2Cu_3O_{7-\delta}$. From AC magnetoresistance measurements, we derive the effective pinning potential attributed to twin boundaries.

1. Introduction

The broad resistive transition in magnetic fields of the high T_c superconductors remains an intriguing subject of research. Several explanations, including thermally activated flux creep and flux flow (Malozemoff et al., 1989), Josephson junctions (Tinkham, 1988), superconducting fluctuations (T. Tsuneto, 1988), and flux line melting (Nelson, 1988, Fisher, 1989) have been proposed, but fail to quantitatively fit the experimental results satisfactorily. In a recent Letter (Kwok et al., 1990), we demonstrated that flux flow exists in the form of an excess angular dependent resistivity superimposed on an angular independent resistivity in a single crystal of $YBa_2Cu_3O_{7-\delta}$ with twin boundaries oriented only along one direction. In this paper, we report magnetoresistance measurements up to 8 Tesla on a high quality single crystal of $YBa_2Cu_3O_{7-\delta}$ which contains twin boundaries oriented along both the <110> and <1$\bar{1}$0> directions. We see the effects of flux flow as an excess angular dependent resistance which follows a $\sin^2(\theta)$ dependence. In addition, sharp decreases in resistances are found whenever the magnetic field is aligned along either the <110> or <1$\bar{1}$0> twin boundaries (TB). In order to separate the intrinsic pinning from pinning due to twin boundaries, we measured the temperature dependence of the resistance at several fields for H || TB where the sharp decrease in resistance was observed. From an Arrhenius plot representation of the data, we obtain the effective pinning potential U_{eff} of the twin boundaries.

2. Results and Discussion

The sample was prepared by a self flux method described elsewhere (Kaiser et al., 1987) with dimensions 1.2 x 0.5 x 0.04 mm^3. AC resistivity was measured by a four probe technique using gold wires attached to the sample with silver epoxy. Typical measuring current was 0.1 mA at a frequency of 17 Hz in the **ab** plane of the crystal. Angular dependent resistivity measurements at fixed temperatures and low fields H || **ab** were measured in a 1.5 Tesla transverse superconducting magnet. Measurements in higher fields up to 8 T (H || **ab**) were performed in the radial access transverse 8 T Bitter magnet at the Francis Bitter National Magnet Laboratory.

0953-2048/91/01S106 + 03 $03.50

Figure 1 shows the resistive transition in a magnetic field of 1.5T in the **ab** plane for fields parallel (zero Lorentz force) and perpendicular (maximum Lorentz force) to the measuring

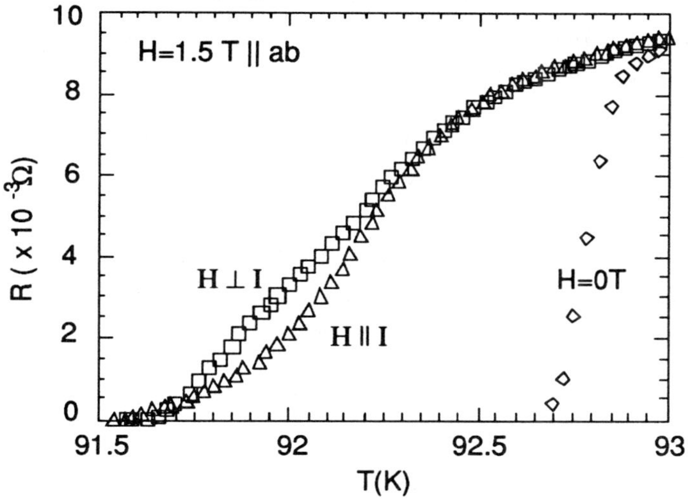

Fig. 1 Resistance versus temperature in a field of 1.5 Tesla for H parallel and perpendicular to the transport current. Both the current and magnetic field are parallel to the **ab** plane of the crystal.

current. The zero field transition is shown for comparison. In zero field, the resistive transition is very sharp with T_{onset} ~ 92.9 K and T_{zero} ~ 92.6 K. In the presence of a magnetic field, the anomalous broadening is evident even in the zero Lorentz force configuration (H || I). The sharp kink previously observed in the resistive tail of the transition for H ⊥ I (Kwok et al. 1990) is absent in this crystal, implying the strong sample dependence origin of the kink (Kwok et al. 1990b). In the maximum Lorentz force configuration (H ⊥ I) there is an 'excess' resistance which follows a $\sin^2(\theta)$ behavior, as shown in Figure 2 for H=1.5T and T=91.89K.

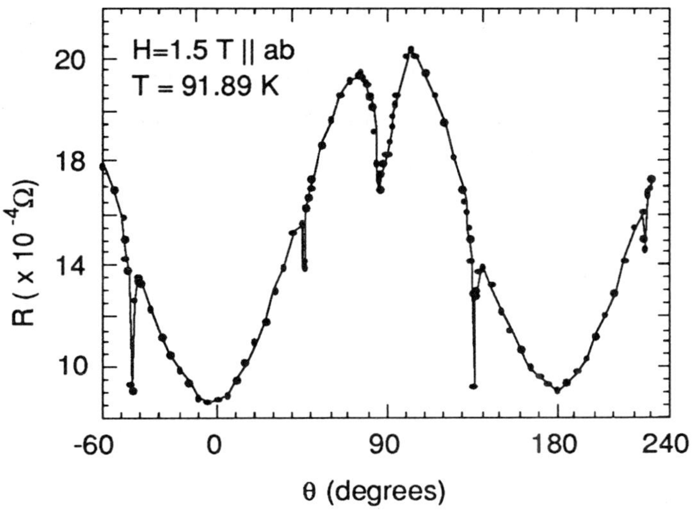

Fig. 2 Angular dependence of the resistance at H=1.5 T and 91.89 K. θ represents the angle between the magnetic field and transport current.

Minima are observed at 0 degrees and 180 degrees at the zero Lorentz force configuration (H || I) and a maximum is observed near 90 degrees. The pinning due to twin boundaries of both

orientations $\langle 110\rangle$ and $\langle 1\bar{1}0\rangle$ is shown by sharp decreases, about 2° wide in resistance at -45°, 45°, 135° and 225°. Another sharp decrease at $\sim 90^o$ may be due to pinning by oxygen vacancy cluster defects along the crystallographic a or b axis.
In order to isolate the effect of pinning due to twin boundaries alone, we oriented the magnetic field parallel to the twin boundaries at 135 degrees. The Arrhenius plot of the temperature dependence of the resistance for this orientation in fields up to 7.35 T is shown in Figure 3. The slopes of the Arrhenius curves at low temperatures remain virtually constant in fields up to 7 Tesla. We obtain an effective $U_o/k = 8 \times 10^4$ K from fitting the low temperature linear region of the Arrhenius curves. This value is 5 times larger than that obtained from the Arrhenius curve for H=5.88 T, H II I (zero Lorentz force) configuration, consistent with our earlier measurements (Kwok et al., 1990) on a single crystal with twin boundaries only along one direction. A substantial gain in pinning strength may be achieved, even in the presence of high magnetic fields with the introduction of more twin boundaries.

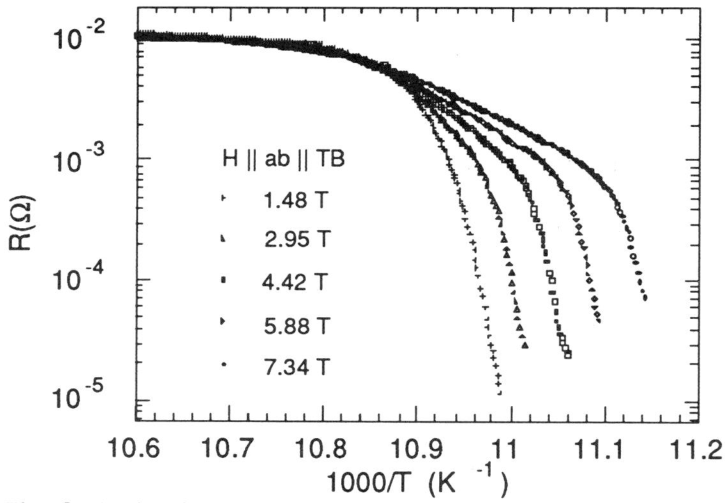

Fig. 3 Arrhenius plot of resistance versus 1000/T in different magnetic fields for H parallel to the twin boundaries.

3. Acknowledgements

We wish to thank the staff of the Francis Bitter National Magnet Laboratory for their hospitality and technical assistance. This work was supported by the U. S. Department of Energy, BES--Materials Science under contract #W-31-109-ENG-38 (WKK, SF, JZL, GWC) and the NSF--Office of Science and Technology Centers under contract #STC8809854 (UW, KGV) Science and Technology Center for Superconductivity). SF and KV acknowledge partial support from the Division of Educational Programs, Argonne National Laboratory.

4. References

Fisher M P A 1989, Phys. Rev. Lett. **62**, 1415
Kaiser D L, Holtzberg F, Chisholm M F, Worthington T K, 1987 J. Cryst. Growth **85**, 593
Kwok W K, Welp U, Crabtree G W, Vandervoort K G, Hulscher R, and Liu J Z, 1990 Phys. Rev. Lett. **64**, 966
Kwok W K, Welp U, Vandervoort K G, Fang Y, Crabtree G W, Liu J Z, 1990 Appl. Phys. Lett. **57**(12), 1
Malozemoff A P , Worthington T K, Zeldov E, Yeh N C, McElfresh M W, Holtzberg F, 1989 in Springer Series in Physics: Strong Correlations and Superconductivity, edited by Fukuyama H, Maekawa S, Malozemoff A P, eds. (Springer-Verlag, Heidelberg) pp. 349
Nelson D R 1988, Phys. Rev. Lett. **60**, 1973
Tinkham M 1988, Phys. Rev. Lett. **61**, 1658
Tsuneto T 1988, J. Phys. Soc. Jpn. **57**, 3499

Supercond. Sci. Technol. **4** (1991) S109–111. Printed in the UK

Non-logarithmic Magnetic Relaxation in Bi-Sr-Ca-Cu-O Single Crystals

K Niskanen*, P Norling*, P Svedlindh*, P Nordblad*, L Lundgren*, C Rossel+ and G V Chandrashekhar #

*Uppsala University, Department of Technology, Box 534, S-751 21 Uppsala, Sweden
+IBM Research Division, Zürich Research Laboratory, 8803 Rüschlikon, Switzerland
#IBM Research Division, T J Watson Research Center, Yorktown Heights, N Y 10598, USA

The time dependence of the low field susceptibility of a $Bi_{2.2}Sr_{1.7}CaCu_2O_8$ single crystal has been investigated in the field range 0.1-300 gauss and at temperatures $0.25 \leq T/T_c \leq 0.9$. The relaxation of the magnetization is found to change from a logarithmic to a non-logarithmic form as the applied field is increased, implying that a conventional flux creep model cannot sufficiently describe the behaviour.

1. Introduction

It has been observed that high temperature superconductors, both sintered materials and single crystals, show large relaxation effects in their magnetization. The time decay of the magnetization is usually considered as being logarithmic in time, and is therefore described in terms of a conventional flux creep model with thermally activated flux motion (Yeshurun et al 1988). The unusually large magnetic relaxation in high temperature superconductors is then explained by small values of the activation energies, due to short coherence length, and high measurement temperatures.

In this paper we present results from measurements of the zero-field-cooled and remanent magnetization in a $Bi_{2.2}Sr_{1.7}CaCu_2O_8$ single crystal. Our measurements show that the relaxation changes distinctly from a logarithmic to a non-logarithmic behaviour within a temperature dependent field region, as the applied field is increased. Similar results have been obtained from measurements on single crystals of the high temperature superconductor YBCO and the isotropic extreme type-II superconductor $PbMo_6S_8$ (Svedlindh et al 1990), which similar to the high temperature superconductors has a very short coherence length. Safar et al (1989) report on non-logarithmic relaxation in sintered Bi-Sr-Ca-Cu-O.

2. Experimental

The single crystal used in our measurements is approximately 0.5x0.5x0.01 mm^3, with the shortest edge along the c-axis. The critical temperature is 80 K, and the width of the transition is about 2 K in an applied field of 0.1 G. The effective mass anisotropy ratio for this material has previously been estimated to $m_c/m_{ab} \approx 3000$ (Farrell et al 1989).

A non-commercial dc SQUID magnetometer was used to study the time decay of the zero-field-cooled (ZFC) and remanent (RM) magnetization for fields 0.1-300 gauss and at temperatures $0.25 \leq T/T_c \leq 0.9$. The experimental time window was 0.3-3000 sec. The magnetic field was oriented parallel to the c-axis of the sample.

0953-2048/91/01S109 + 03 $03.50

3. Results and discussion

Figure 1 shows the ZFC magnetization (M_{ZFC}) and the RM magnetization (M_{RM}) as a function of field at 50 K. The different curves have been obtained from measurements of the time decay of the magnetization, and each curve illustrates the field dependence of the magnetization at a specific observation time. At low fields the relaxation is logarithmic in time and the relaxation rate $\partial M/\partial \ln(t)$ depends on field as $\partial M/\partial \ln(t) \propto H^2$ for both the ZFC and the RM magnetization. In a field region slightly below the field region where ($-M_{ZFC}$) has its maximum, the relaxation of the magnetization changes gradually from a logarithmic to a non-logarithmic form. The magnitude of the decay in the field region with a logarithmic decay is, within our experimental time window, at most a few percent of the magnetization value, while in the field region with a non-logarithmic decay it reaches about 30%.

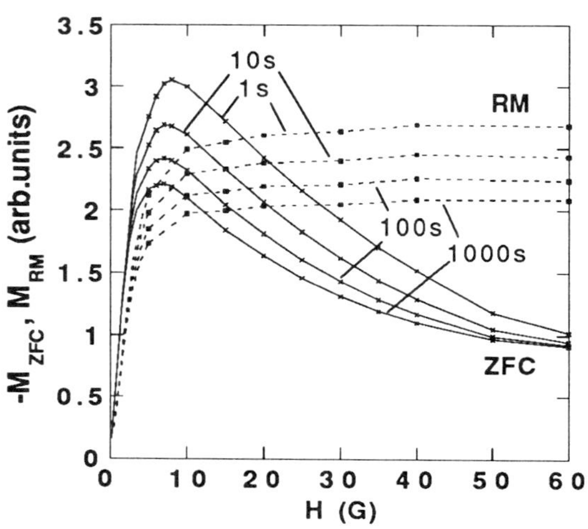

Figure 1. Magnetization as a function of field at 50 K.

Figure 2 shows the time dependence of the ZFC magnetization (Figure 2a) and the relaxation rate (Figure 2b) at 50 K for some different fields in the field region where there is a crossover from logarithmic to non-logarithmic decay. The non-logarithmic character of the decay is particularly evident in the relaxation rate, which exhibits a maximum at a field dependent observation time.

With increasing field this maximum is pushed towards shorter observation times. For lower temperatures, the crossover between logarithmic and non-logarithmic behaviour occurs at higher fields.

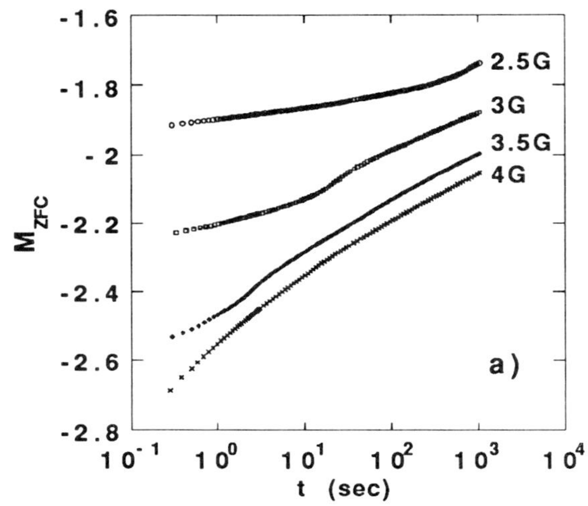

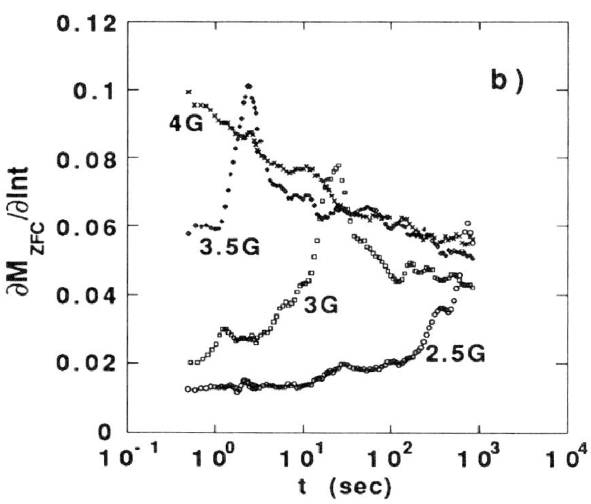

Figure 2. ZFC magnetization (a) and relaxation rate (b) at 50 K. The same relative units as in Figure 1 are used.

Using the extended Bean model (Yeshurun et al 1988), or some other flux creep model with a local field dependent critical current, it is possible to show that the relaxation rate of the ZFC magnetization should have a maximum when a fully penetrated flux state is reached. Thus we propose that the observation time where the ZFC relaxation rate at a given field exhibits a maximum corresponds to the time when full flux penetration in the sample is achieved.

The important implication of the results shown in Figure 2b is then that full flux penetration can be achieved both with increasing observation time and with increasing field.

Also the decay of the RM magnetization at 50 K shows a gradual change from a logarithmic to a non-logarithmic time dependence in the same field region as the ZFC magnetization, but in this case the relaxation rate curves do not show any maxima as a function of observation time. This is because a partially penetrated flux state does not occur in RM measurements. Since the field cooled magnetization value, the Meissner signal, is only about 20% of its ideal value at 3G and is strongly suppressed with increasing field, the sample is in a fully penetrated flux state prior to the removal of the applied field in the RM measurements.

The concept of a critical state as described by Anderson (1962) is only well defined if the reduction of the magnetization is small and if the time decay is logarithmic, which occurs when the screening current density (j) associated with the flux gradient in the sample is of the same order as the critical current density, i.e. when $(j_c-j)<<j_c$. Only in the low field region of our measurements, with logarithmic relaxation, it is possible that a critical state is well established and the flux creep model by Bean (1962) is applicable. The theory of collective flux creep (Feigel'man et al 1989), which discusses the nature of flux creep phenomena in the case of collective pinning by weak disorder, also includes the case when $j<<j_c$ and can therefore be applied in a wider field range. One result of this theory is that the screening currents depend on observation time as $j(t) \propto [T \ln(t/\tau_0)]^{-\alpha}$. This decay of the screening currents is then reflected in the time dependence of the magnetization so that $\partial M \propto [T \ln(t/\tau_0)]^{-\alpha}$. This expression gives a good description of the observed time dependence of the RM magnetization at high fields. Our data yield $\alpha \approx 0.6$-0.7 for $T/T_c \approx 0.6$, decreasing to $\alpha \approx 0.3$-0.4 for $T/T_c \approx 0.9$.

At lower temperatures, $T/T_c \leq 0.4$, some strange new features in the relaxation occur. The decay of both the ZFC and RM magnetization within our experimental time window can be as large as 80% of the magnetization value. Furthermore, the RM relaxation rate curves exhibit similar maxima as a function of observation time as the ZFC relaxation rate curves, although less pronounced. Moreover, the observed decay does not follow an expression of the type $\partial M \propto [T \ln(t/\tau_0)]^{-\alpha}$. A more detailed investigation of this temperature region will be made.

Our results indicate a complicated H-T (and t) diagram for the high temperature superconductor Bi-Sr-Ca-Cu-O. Conventional flux creep models are not sufficient to describe its complex magnetic relaxation behaviour, but the collective flux creep model seems applicable in a wider region of the H-T diagram.

Financial support from the Swedish Natural Science Research Council is acknowledged.

References

Anderson P W 1962 *Phys. Rev. Lett.* **9** 309

Bean C B 1962 *Phys. Rev. Lett.* **8** 250

Farrell D E, Bonham S, Foster J, Chang Y C, Jiang P Z, Vandervoort K G, Lam D J and Kogan V J 1989 *Phys. Rev. Lett.* **63** 782

Feigel'man M V, Geshkenbein V B, Larkin A I and Vinokur V M 1989 *Phys. Rev. Lett.* **63** 2303

Safar H, Durán C, Guimpel J, Civale L, Luzuriaga J, Rodriguez E, de la Cruz F, Fainstein C, Schneemeyer L F and Waszczak J V 1989 *Phys. Rev. B* **40** 7380

Svedlindh P, Niskanen K, Norling P, Nordblad P, Lundgren L, Rossel C, Sergent M, Chevrel R and Potel M 1990 preprint

Yeshurun Y, Malozemoff A P, Holzberg F and Dinger T R 1988 *Phys. Rev. B* **38** 11828

Supercond. Sci. Technol. **4** (1991) S112–114. Printed in the UK

High Resolution Patterning of High-T_c Supercconducting Thin Films

Z.G. Ivanov, P-Å. Nilsson, E-K. Andersson, and T. Claeson

Chalmers University of Technology, S-412 96 Göteborg, Sweden

Abstract

Micron and sub-micron size microbridges have been patterned in high T_c films. The degradation of superconductivity caused by ion beam etching and subsequent processing was minimized by using a thin silver buffer layer. Lift-off technology and post-annealing gave narrow microbridges with good Josephson response.

1. Introduction

Patterning of high T_c superconducting (HTS) films to micron and sub-micron size structures without seriously degrading the superconducting properties is needed for several electronic components. Good progress has been made using photolithography and wet etching, allowing dimensions down to 3μm with less than 1K decrease of T_c (Shokoohi et al 1989). Even smaller structures have been made with dry etching, in particular ion beam etching (IBE) (Enokihara et al 1988 and Tsuge et al 1988). Maskless techniques like focussed ion beam lithography (Harriot et al 1989) and pulsed laser evaporation (Zhang et al 1989) have produced sub-micron strips. Those techniques, however, are best suitable for not too complicated patterns and cause considerable degradation at strip widths less than 0.6 μm.
We have developed an Ar IBE based process and resist masks to pattern HTS bridges down to a 1 μm level. Even some of the sub-micron patterns (0.3-1 μm) were produced by a lift-off process followed by post-annealing of ex-situ deposited YBCO films.

2. Micron Level Patterns with Ar Ion Beam Etch

Y-Ba-Cu-O (YBCO) and Bi-Sr-Cu-O (BSCO) films were deposited in-situ on MgO substrates by pulsed laser ablation (Ivanov et al 1989 and Brorsson et al 1990). YBCO films were also deposited in-situ on sapphire substrates by dc magnetron sputtering (Zarembinski et al 1990). Resist masks were made by photo or electron lithography exposures using Shipley S1813 (photo resist, also used as an electron resist) and Shipley SAL601 negative electron resist. Resist lines down to 1 μm were obtained. A Kaufmann type Ar ion source was used (beam voltage 500V, current density 0.2 mA/cm^2, Ar pressure $2\cdot10^{-4}$ torr). The etch rate for YBCO, BSCO, SAL601 and S1813 were determined to be 7±1, 7±1, 12±2 and 8±1 nm/min respectively.
The resistance, as measured by a four-probe method, of a 4 μm wide and 20 μm YBCO bridge showed approximately the same temperature dependence above T_c as the film did before patterning, see Figure 1. Below T_c, a low resistance tail appeared and its extension depended upon the film quality. For a high quality laser deposited YBCO film on MgO (T_c=92K, $j_c=8\cdot10^5$ A/cm^2 at 77K) the tail was minimal and did not exceed 1.5±0.5K (dashed curve in 1a). For YBCO films on sapphire with T_c=74-82K and relatively poor j_c, the tail length was about 4±1K (see 1b). It varied from 5 to 10K for T_c=40-60K (1c). Similar results were obtained for BSCO. The large changes in T_c after the IBE of lower T_c films are probably due to inhomogeneities in the films.

0953-2048/91/01S112 + 03 $03.50

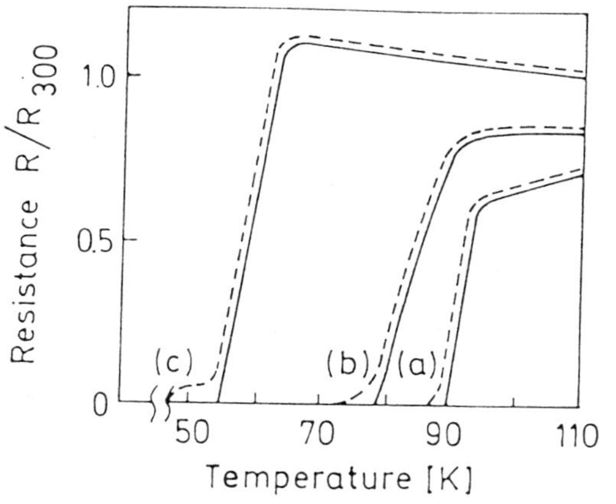

Fig.1 Normalized resistance vs temperature for YBCO films: (a) in-situ laser deposited on MgO (100) substrates; (b)-(c) in-situ magnetron sputtered on sapphire (solid line - before patterning, dashed line - after patterning bridges 4 μm wide and 20 μm long.

Of great nuisance, it was found that a subsequent anneal or processing step which did not raise the temperature of the bridge to more than 170-300°C (in an oxygen atmosphere) had a detrimental influence on the superconducting properties in an unpredictable way. This may have been caused by a resist after the ion etch and resist removal. To circumvent the problem, a 20-100 nm thick Ag layer was deposited on the YBCO film after the HTS deposition . Then there was no change in the superconducting properties at subsequent etching or at moderate anneal steps. Hence we infer that the T_c degradation, at least down to the 1 μm level, is not caused by ion damage but rather by environmental effects. The Ag buffer layer introduces an additional step of $Ar/0_2$ IBE to remove the Ag but it is worth the extra effort as it minimizes the degradation of the superconductivity.

3. Sub-Micron Patterns by Lift-Off and Post-Anneal

300 nm thick YBCO films were co-evaporated from Cu, Y, and BaF_2 sources at relatively low substrate temperature (100°C). They were deposited on a lift-off three layer stencil: 400 nm co-polymer resist was spun onto $SrTiO_3$ substrates, it was covered by a 20 nm thermally evaporated NiCr layer, and 80 nm PMMA was, finally, spun on top. The resists were exposed by an e-beam at 50 kV and the PMMA layer, determining the pattern size, was developed. The NiCr layer, which was used to avoid charging effects, was etched through the PMMA mask and, finally, the co-polymer was developed to form an under-cut for the lift-off. After lift-off, the microbridges were post-annealed at 890 °C for 15 min (Nilsson et al 1990).

The low substrate temperature during the deposition meant that T_c, after anneal, was no higher than 70K, but the same, in the patterned and the non-patterned films. Two types of bridges were fabricated: short bridges with length L=0.8-1.6 μm and width w=0.3-1.6 μm and long bridges with L=5-20μm and W=0.3-1-6 μm. A SEM picture of a short bridge is shown in Figure 2a. The lift-off process resulted in well defined structures with vertical walls.

The heat treatment gave small deformations: protrusions at the edges and small islands of sizes 0.1-0.3 μm outside the bridge-probably caused by thermal migration during the anneal.

I-V curves for the same short bridge are shown in Figure 2b. The critical current, I_c, was 1mA, corresponding to $j_c \approx 4 \cdot 10^5$ A/cm^2 at 4.2K. The I-V curve displays an excess current making a correct assessment of the normal state resistance, R_n, difficult. By choosing an R_n value just above I_c we obtain $I_cR_n \approx 1$ mV. Applying microwave radiation at f=4.35 GHz, I_c was depressed and Shapiro steps appeared at $V=\hbar f/2e$, see Fig. 2b. They were not distinguishable at such high voltages as implied by the I_cR_n product. The voltage separation indicates that the properties of the bridge are governed by one native grain boundary weak link localised in the bridge region.

The I-V curves of long bridges with L=20μm, W=0.8μm were sharper close to I_c while instabilities and hysteresis were encountered at higher voltages. An estimated I_cR_n as high as 10mV could be due to several grain boundary weak links connected in series.

(a) (b)

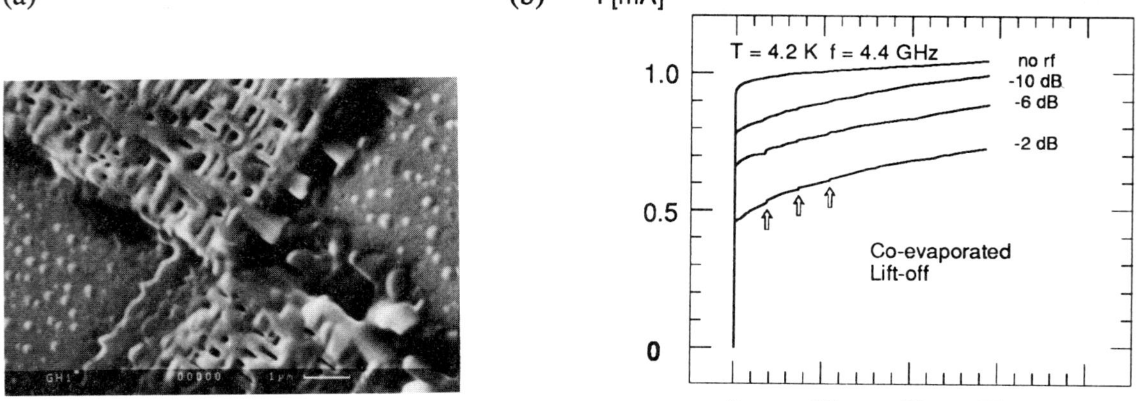

Fig. 2 (a) SEM micrography and (b) I-V curves measured at 4.2K for a microbridge with L=0.8 μm and W=0.8 μm under microwave radiation at 4.36 GHz.

4. Conclusions

It was found that the quality of the HTS film is determining the degradation caused by an ion etch down to the micron level. The effect of inhomogeneities is enhanced. Of great importance is the protection that a thin Ag buffer layer offers toward resist contamination that may seriously degrade the superconducting properties during subsequent processing.

A lift-off process may produce fine lines. It inhibits in-situ HTS deposition as the resist does not survive a high substrate temperature. However, an ex-situ anneal, after lift-off may restore superconductivity. Our most narrow lines (0.3μm) had a strongly depressed T_c but gave good microwave response.

Acknowledgement

The work utilized the Swedish Nanometer Facility supported by the Swedish Board of Technical Development and Natural Science Research Council. Help by J. Alarco in the SEM work is acknowledged.

References

Brorsson G, Ivanov Z and Nilsson P-Å 1990 Proceedings of the Conference on the Science and Technology of Thin-Film Superconductors April 30-May 4 Denver , USA

Enokihara A, Higashino H, Setsune K, Mitsuyu T and Wasa K 1988 Jpn. J. Appl. Phys. **27** L1521

Harriot L R, Polakos P A and Rice C E 1989 Appl. Phys. Lett. **55** 495

Ivanov Z, Brorsson G and Claeson T 1989 Appl. Phys. Lett **55** 2123

Nilsson P-Å, Andersson E-K and Brorsson G 1990 Proceedings of the Conference on the Science and Technology of Thin-Film Superconductors April 30-May 4 Denver , USA

Shokoohi F K, Schiavone L M, Rogers C T, Inam A, Wu X D, Nazar L and Venkatesan T 1989 Appl. Phys. Lett. **55** 2661

Tsuge H, Matsui S, Matsukura N, Kojima Y. and Wada Y 1988 Jpn. J. Appl. Phys **27** L2237

Zarembinski S, Ramos J, Ivanov Z and Claeson T 1990 Proceedings of the E-MRS Spring Conference, Symposium A, HT_c SuperconductorMaterials May 29-June 1 Strassbourg

Zheng J P, Kim H S, Ying Q Y, Barone R, Bush P, Shaw D T and Kwok H S 1989 Appl. Phys. Lett. 55 1044

Supercond. Sci. Technol. 4 (1991) S115–117. Printed in the UK

Spatially resolved measurements of the current transport in $Y_1Ba_2Cu_3O_7$-films

D.Koelle, F.Kober, R.Gross, R.P.Huebener, A.Gupta[+], J.Mannhart[o], B.Roas[*#]

Phys. Institut, Lehrstuhl ExperimentalphysikII, Universität Tübingen, D-7400Tübingen, FRG
+IBM T.J.Watson Research Center Yorktown Heights, NY10598-0218, USA
°IBM, Zürich Research Laboratory, 8803Rüschlikon, Switzerland
*Physikalisches Institut, Universität Erlangen, D-8520 Erlangen, FRG
#Siemens AG, Forschungslaboratorien, D-8520 Erlangen, FRG

The spatial variation of the resistive transition and critical current density j_c in $Y_1Ba_2Cu_3O_7$- films, grown on singlecrystalline and polycrystalline $SrTiO_3$, has been imaged by low temperature scanning electron microscopy (LTSEM) with a spatial resolution of about 1μm. In polycrystalline samples the limitation of j_c due to the weak link behavior of the grainboundary network has been directly imaged. Our studies on epitaxial YBCO-films with j_c(77K) above $10^6 A/cm^2$ reveal temperature variations in the local resistive transition curves up to 200mK. Possible mechanisms for the observed spatial variations are discussed.

1.Introduction

A detailed understanding of the transport properties of hTc thin films is important for fundamental reasons and for possible technological applications. For an unambiguous interpretation of the experimental observations like the shape of the transition curve or the I-V-curves, information about the spatial homogeneity of the samples is necessary, because local effects can be important.
Low temperature scanning electron microscopy (LTSEM) offers the possibility to image the spatial distribution of the critical current density j_c and the resistive transition in superconducting films (Gross etal 1988, Gross etal 1989a, Koelle etal 1990). This provides further insight to the weak link behavior of grain boundaries and their effect on supercurrent flow in bicrystalline and polycrystalline films. We report on LTSEM-measurements on polycrystalline YBCO-films with grain sizes above the LTSEM resolution limit. We also present results on c-axis oriented epitaxial YBCO- films. Here we have investigated defects limiting the critical current density and the resistive transition of high quality samples with j_c(77K) above $10^6 A/cm^2$.

2.LTSEM- Principle of measurement

One needs a superconducting bridge patterned into a thin film, which is mounted on a temperature-controlled sample holder connected to the cryostage of a SEM (Gross etal 1989b). At $T<T_c$ the sample is biased with a constant current I_B and scanned by the pulsed electron beam (repetition rate:10-20kHz). The beam induced voltage change δV along the sample is measured in dependence of the position (x,y) of the beam focus on the sample surface with lock-in technique. The electron beam causes a local temperature rise (typically 0.1-1K) in the irradiated area. This

can lead to a local resistivity change, detected as a voltage change $\delta V(x,y)$. The voltage signal is always a function of a whole cross sectional area (perpendicular to I_B) containing the irradiated location. Hence upon increasing I_B, the first signal appears when j_c of the weakest cross sectional area is reduced below the bias current density j_B due to the heating by the electron beam. By further increasing I_B more and more cross sectional areas become resistive, and imaging of the distibution $j_c(x,y)$ is possible (for details see Hartmann etal 1989).

3.Polycrystalline $Y_1Ba_2Cu_3O_7$-films

LTSEM revealed a strongly inhomogeneous j_c-distribution in polycystalline YBCO-films (Gross etal 1988). Experiments on bicrystalline YBCO-films (Chaudhari etal 1988) showed that single grain boundaries reduce j_c and this was directly imaged by Mannhart etal (1989). We also have investigated polycrystalline YBCO-films with large grains (epitaxially grown on polycrystalline $SrTiO_3$ substrates consisting of randomly oriented grains with grain sizes between 5μm and 30μm). Superconducting lines containing only a few grains in parallel with sizes above the LTSEM resolution limit have been patterned by laser ablation. Fig.1a shows a sketch of the grain boundary configuration of a bridge 30μm wide and 180μm long. The corresponding LTSEM images in brightness modulation are shown below (Fig1b,c). Bright regions indicate resistive locations on the sample. At low temperatures only grain boundaries contribute to the voltage signal (Fig.1b). With increasing bias current I_B more and more grain boundaries become resistive according to their different j_c-values. The inhomogeneity perpendicular to the current flow reflects spatial variations of j_c within single grain boundaries or between different grain boundaries connected in parallel. The resistive behavior of the grain boundaries corresponds to the typical foot structure in the R(T)-curve down to low T. At temperatures near T_c (steep portion of the R(T)-curve) also the grains become resistive, as shown in Fig.1c.

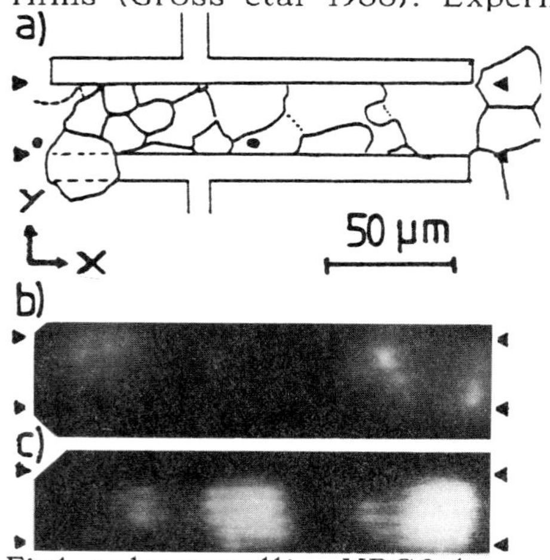

Fig1: polycrystalline YBCO-line
a) grainboundary configuration
b) $\delta V(x,y)$ at T=10K, I_B=0.6mA
c) $\delta V(x,y)$ at T=91K, I_B=0.1mA

Fig.2 shows LTSEM images in y-modulation for a single linescan running along the bridge. The voltage signals change in a complex way with increasing bias current I_B. This suggests, that the IV-curves of various grain boundaries are quite different leading to current redistribution by changing I_B.

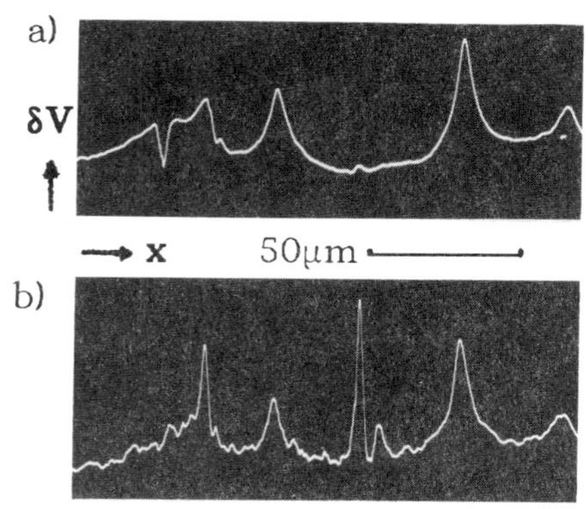

Fig2 $\delta V(x)$ for a linescan along a polycrystalline YBCO-line at T=10K a) I_B=377μA b) I_B=390μA

4.Epitaxial $Y_1Ba_2Cu_3O_7$-films

We have investigated two samples grown epitaxially by laser deposition on <100> $SrTiO_3$, prepared under the same conditions and patterned by chemically wet etching as narrow lines ($10 \times 180\ \mu m^2$). j_c(77K) of sample II was $5\times10^6 A/cm^2$, whereas in sample I it was 20 times lower. The R(T)-curve of this sample showed a foot structure, shifting T_c^0 by about 700mK. This j_c- and T_c-reduction was found to be caused by a single weak link, and all other parts had values of $j_c(77K)$ above $10^6 A/cm^2$. We did not observe significant vaiations in the voltage signals perpendicular to the current flow, so we can use the one-dimensional approximation.

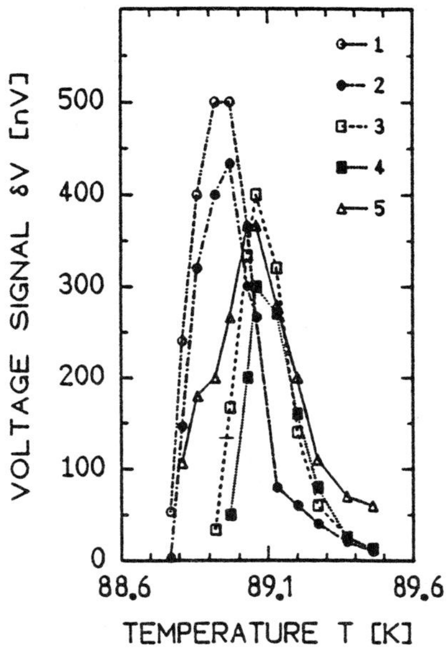

Fig3 LTSEM-signals δV(T) for five locations on an epitaxial YBCO-film

In this case we have the proportionality

$$\delta V(x,T) \sim d\rho(x,T)/dT \quad (x \parallel I_B)$$

In Fig.3 δV(x,T) is plotted versus the temperature T for five locations on the bridge. The inhomogeneous behavior is reflected by the relative shifts of these local transition curves by up to about 200mK, whereas the widths of the curves do not vary significantly. We found that the observed inhomogeneous behavior is current independent and variations in the sizes of different cross sectional areas (e.g. caused by precipitates) as a reason for the observed inhomogeneous behavior can be excluded.

For this behavior we can offer two explanations: first: we have identical transition curves, but different T_c-values, for example caused by variations in the oxygen content. Second: weak links with varying coupling energy exist in the bridge, acting as highly damped Josephson Junctions. In the limit of zero current, the transition curves are broadened due to thermally activated phase slippage, as shown by Gross etal (1990) on bicrystalline YBCO-films.

This work was supported by the Bundesministerium für Forschung und Technologie (Project No.13N5482)

5.References

Chaudhari P, Mannhart J, Dimos D, Tsuei CC, Chi J, Oprysko MM and Scheuermann M 1988 Phys.Rev.Lett.**60** 1653

Gross R, Bosch J, Huebener RP, Mannhart J, Tsuei CC, Scheuermann M, Oprysko MM and Chi CC 1988 Nature (London) **332** 818

Gross R, Hipler K, Mannhart J, Huebener RP, Chaudhari P, Dimos D, Tsuei CC Schubert J and Poppe U 1989a Appl.Phys.Lett.**55** 2132

Gross R, Bosch J, Wener H-G, Fischer J, Huebener RP 1989b Cryogenics**29** 716

Gross R, Chaudhari P, Dimos D, Gupta A, Koren G 1990 Phys.Rev.Lett.**64** 228

Hartmann M, Hipler K, Koelle D, Kober F, Bernhardt K, Sermet T, Gross R and Huebener RP 1989 Z.Phys.B **75** 423

Koelle D, Kober F, Hartmann M, Gross R, Huebener RP, Roas B, Schultz L and Saemann-Ischenko G 1990 PhysicaC **167** 79

Mannhart J, Gross R, Hipler K, Huebener RP, Tsuei CC, Dimos D, Chaudhari P 1989 Science **245** 839

Supercond. Sci. Technol. 4 (1991) S118–120. Printed in the UK

In-situ growth of Bi-Sr-Ca-Cu-O films by shutter-controlled molecular beam epitaxy

Ichiro Tsukada, Yoshimi Nakayama,† Ichiro Terasaki, Atsutaka Maeda and Kunimitsu Uchinokura

Department of Applied Physics, The University of Tokyo, 7-3-1 Hongo, Bunkyo-ku, Tokyo 113, Japan.

Abstract. Thin films of $Bi_2Sr_2Ca_{n-1}Cu_nO_y$ (n=1, 2, 3 and 4) were prepared *in situ* by molecular beam epitaxy technique. Among them n=2, 3 and 4 phases were superconducting. Layer-by-layer deposition technique allowed us to construct the structure of n=4. Reflection electron energy loss spectroscopy of these films were investigated. These measurements showed that thin films with compositions of n=2 have almost the same electronic structures above 1.5 eV as the corresponding single crystals. An ultrathin film of n=2 phase was also prepared *in situ*, whose typical thickness was about 62 Å. This film showed superconducting transition at 38 K.

1. Introduction

Recent developements of superconducting thin-film processes are remarkable. Among them, molecular beam epitaxy (MBE) technique has advanced rapidly in recent years. Since pure ozone was found very effective for oxidation (Berkley *et al* 1988), several groups have applied this technique to prepare thin films of $Bi_2Sr_2Ca_{n-1}Cu_nO_y$ (Nakayama *et al* 1989a, Schlom *et al* 1990). To obtain high-quality films, it is necessary to prepare thin films without post-annealing. First report of obtaining superconducting $Bi_2Sr_2CuO_y$ films without post-annealing was made by Nakayama *et al* (1989a), in which films were prepared by co-evaporation technique. After this report, we have tried to prepare $Bi_2Sr_2Ca_{n-1}Cu_nO_y$ films with $n \geq 2$ by sequential deposition technique (Nakayama *et al* 1989b, Uchinokura *et al* 1990). In this paper, $Bi_2Sr_2Ca_{n-1}Cu_nO_y$ thin films with n=1, 2, 3 and 4 were prepared systematically and their physical properties were studied.

The combination of MBE and pure ozone has an interesting possibility of obtaining ultrathin films. Several reports of preparing ultrathin films have been made, however, only on $YBa_2Cu_3O_y$ or the compounds with the same structure (Michikami *et al* 1990, Xi *et al* 1989, Terashima *et al* 1990). We have tried to prepare ultrathin films of n=2 phase, whose thicknesses are below 100 Å, in order to investigate the minimum *c*-axis length of $Bi_2Sr_2CaCu_2O_y$ for exhibiting superconductivity.

2. Sample preparations

The method to prepare thin films is the same as previously reported ones (Nakayama *et al* 1989b, Uchinokura *et al* 1990). Films are deposited in a modified MBE machine. Each metal sources, i.e., Bi, Sr, Ca and Cu, are provided by Knudsen cells.

Pure ozone gas providing system is almost the same as before but a tight protective shield is added around this system for safe operation. Ozone gas is introduced into the epitaxial chamber when the evaporation started and is turned-off when substrates are cooled down to 200°C. During evaporation, the pressure in the epitaxial chamber is kept to 5×10^{-4} Pa.

---

†Present address:Department of Industrial Chemistry, The University of Tokyo.

0953-2048/91/01S118 + 03 $03.50

MgO (001) and $SrTiO_3$ (001) are used for substrates. Typical substrate temperature is 770°C, which is monitored by an optical pyrometer. The substrate holder is made of Inconel. This material is hardly oxidized by ozone.

Molecular beams of metal sources are controlled by shutters at the top of each Knudsen cells. Only one type of molecular beams is permitted to reach the substrates at any time. A sequence of the deposition for n=2 phase is (Bi-Sr-Cu-Ca-Cu-Sr-Bi), which corresponds to half of the c-axis length, i.e., 15.4 Å, and this sequence are repeated several times. When we intend to make n=1, 3 and 4 phases the sequence of (-Cu-Ca-) is subtracted or added. Usually we have repeated 60 cycles of this sequence. Ultrathin films were deposited by the repetition of 4 cycles of (Bi-Sr-Cu-Ca-Cu-Sr-Bi) sequence.

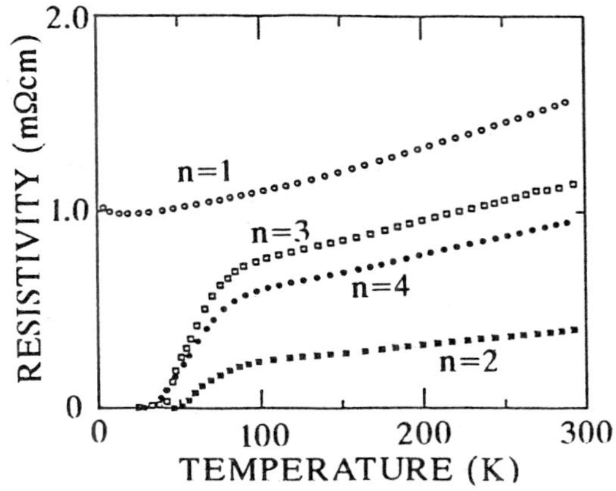

Fig. 1 Temperature dependence of resistivity of $Bi_2Sr_2Ca_{n-1}Cu_nO_y$ films of n=1, 2, 3 and 4 phases.

3. Experimental results and discussion

Obtained films are highly c-axis oriented and the c-axis lengths are calculated as 24.6, 30.9, 37.1 and 43.0 Å for n=1, 2, 3 and 4 phases, respectively. Figure 1 shows the temperature dependence of the resistivity. All films show the metallic behavior of the resistivity above T_c and n=2, 3 and 4 phases show superconducting transition. These films have broad transition width and show zero resistivity at 50, 40 and 33 K, respectively. The film of n=1 phase did not show the superconductivity above 4.2 K. The chemical compositions of these films are not stoichiometric. For example, the film of n=2 phase had the following chemical composition: Bi:Sr:Ca:Cu=1.70:1.93:1.93:2.35 obtained by Inductively Coupled Plasma Atomic Emission Spectroscopy.

The reflection electron energy loss spectroscopy (EELS) of the n=2 phase film was studied. The energy of primary electron is 1 keV and the energy resolution is 0.55 eV. The loss function is shown in Fig. 2 together with those of the single crystals with n=1 and n=2 phases. In contrast to the fact that there are quite large difference between the energy loss functions of the single crystals of the two phases, it is clearly seen that the loss function of n=2 phase film is very similar to that of the corresponding single crystals. So we may conclude that the n=2 phase film has almost the same electronic structure as the corresponding single crystal at least above 1.5 eV.

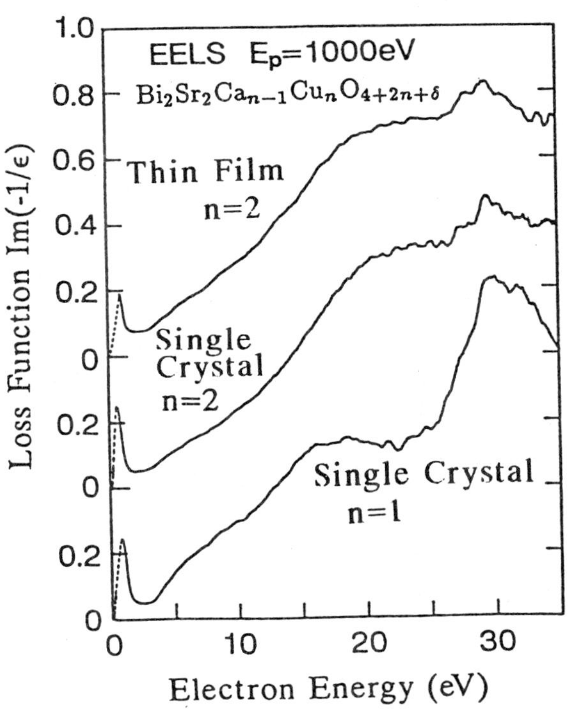

Fig. 2 Loss functions $Im(-1/\epsilon(\omega))$ of a thin film of n=2 phase and single crystals of n=1 and 2 phases. Dashed lines below 1.5 eV were extrapolations to zero at 0 eV.

Next we tried to prepare ultrathin films of n=2 phase. The x-ray diffraction pattern and the resistivity of one of the ultrathin films on MgO (001) are shown in Fig. 3. These films were prepared by almost the same method as mentioned above. There was, however, a little difference in the deposition sequence. To enhance oxidation of Bi and Cu, intermissions of 1 min were placed after the deposition of Bi and Cu, and the sequence was repeated 4 times. Only (002) peak (around 5.8°) in the typical diffraction pattern of the n=2 phase is observed in the inset of Fig. 3, but additional two peaks appear below 5.8°, i.e., 4.2° and 3.2°. We can not identify the origin of these peaks. Further experiments to improve crystallinity of ultrathin films are necessary to observe Laue peaks of ultrathin films.

This film showed superconducting transition at 38 K. But the resistivity at room temperature was about 2.7 mΩcm, which is a little higher than expected. We think that the quality of this film as a single crystal may not be good enough. It is important to adjust the chemical compositions of films to ideal stoichiometric values.

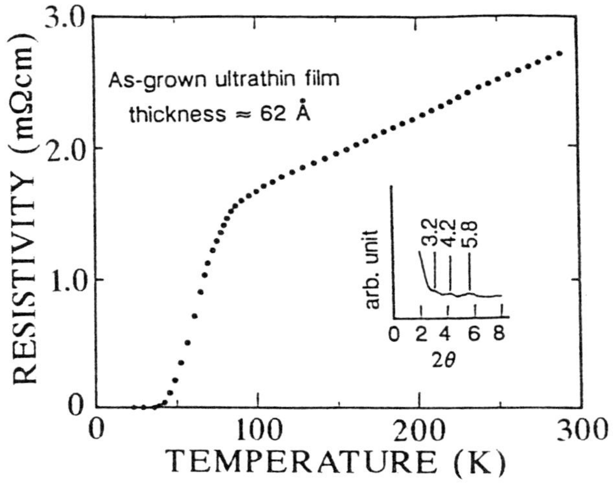

Fig. 3 Temperature dependence of resistivity of $Bi_2Sr_2CaCu_2O_y$ ultrathin film. The inset shows Cu-K_α x-ray diffraction pattern of this film in $2\theta = 2° \sim 8°$ region.

4. Summary

In summary, thin films of $Bi_2Sr_2Ca_{n-1}Cu_nO_y$ with n=1, 2, 3 and 4 were prepared by MBE using pure ozone. Films with n=2, 3 and 4 phases show superconductivity below 50, 40 and 33 K. The study of EELS shows that the film of n=2 phase has almost the same electronic structure as the corresponding single crystal. Ultrathin films of n=2 phase have been prepared by the same method. One of them show superconducting transition at 38 K.

We thank the members of the Superconductivity Research Laboratory, International Superconductivity Technology Center for the chemical analysis. This work was supported in part by the Grant-in-Aid for Scientific Research from the Ministry of Education, Science and Culture.

References

Berkley D D, Johnson B R, Anand N, Beauchamp K M, Conroy L E, Goldman A M, Maps J, Mauersberger K, Mecartney M L, Morton J, Tuominen M and Zhang Y-J 1988 *Appl. Phys. Lett.* **53** 1973

Nakayama Y, Ochimizu H, Maeda A, Kawazu A, Uchinokura K and Tanaka S 1989a *Jpn. J. Appl. Phys.* **28** L1217

Michikami O, Asahi M and Asano H 1990 *Jpn. J. Appl. Phys.* **29** L298

Nakayama Y, Tsukada I, Maeda A and Uchinokura K 1989b *Jpn. J. Appl. Phys.* **28** L1809

Schlom D G, Marshall A F, Sizemore J T, Chen Z J, Eckstein J N, Bozovic I, von Dessonneck K E, Harris J S Jr and Bravman J C 1990 *J. Cryst. Growth* **102** 361

Terashima T, Bando Y, Iijima K, Yamamoto K, Hirata K, Hayashi K, Kamigaki K and Terauchi H 1990 in *Abstracts of The Second ISTEC Workshop on Superconductivity* May 28-30 Kagoshima Japan pp 101-104

Uchinokura K, Nakayama Y, Tsukada I, Terasaki I and Maeda A 1990 To be published in *Proceedings of the Conference on the Science and Technology of Thin-Film Superconductors* Denver Colorado USA April 30-May 4 (New York: Plenum)

Xi X X, Geerk J, Linker G, Li Q and Meyer O 1989 *Appl. Phys. Lett.* **54** 2367

Supercond. Sci. Technol. **4** (1991) S121–123. Printed in the UK

Electronic States and Origin of Superconductivity in $La_{2-x}Sr_xCuO_4$

H. Kamimura, M. Eto, K. Ishida, S. Matsuno, T. Schimizu and H. Ushio*

Dept. of Physics, Univ. of Tokyo, Bunkyo-ku, Tokyo, Japan 113

Abstract. The ground state of CuO_6 in La-compounds is $^3B_{1g}$ in the superconducting phase, supporting Kamimura's two-story house model. Based on this model the Sr concentration dependence of T_c and the normal state Hall effect and the temperature dependence of the normal state resistivity are calculated, and all the results are in good agreement with experiments.

Recently Kamimura and Eto have showed that the ground state of CuO_6 octahedra with dopant holes changes from $^1A_{1g}$ state to $^3B_{1g}$ state when the Sr concentration x increases in $La_{2-x}Sr_xCuO_4$[1], as shown in Fig. 1. This supports the "two-story house model" proposed by Kamimura[2,3]. Identifying the x_c in Fig.1 to the onset of superconductivity, they have predicted two kinds of phase diagram, which are shown in Fig.2.[1]

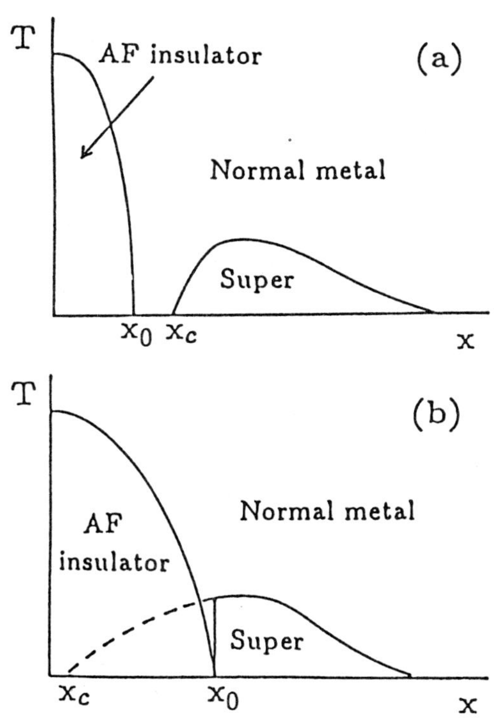

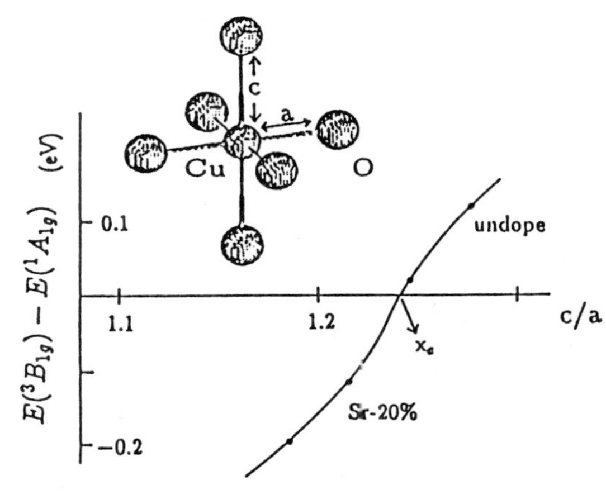

Fig.1. The energy difference between $^3B_{1g}$ and $^1A_{1g}$ states vs. Cu-apical O distance c in CuO_6 cluster. The ground state changes at x_c.

Fig.2. Two kinds of phase diagrams in the temperature vs. dopant concentration plane. (a) corresponds to the case $x_c > x_0$ (hole doping case) while (b) to the case of $x_c < x_0$ (electron doping case), where x_c and x_0 are the concentrations at which $^3B_{1g}$ ground state appears and AF ordering disappears, respectively.

According to his model, dopant holes occupy the lower Hubbard band constructed from an a_{1g} antibonding orbital of Cu dz^2 main character, and their spins become parallel to those of the localized holes occupying the lower Hubbard level of the Cu dx^2-y^2 antibonding orbital due to the large intra-atomic exchange interaction with a ferromagnetic sign (about 2 eV), like the Hund's coupling in atoms. This model is treated by the following Hamiltonian of the coupled fermion-spin system[4];

$$\mathcal{H}_{eff} = \sum_{i,j,\sigma} t_{ij} c^{\dagger}_{i\sigma} c_{j\sigma} + J \sum_{i,j} \vec{s}_i \cdot \vec{s}_j - K \sum_i \vec{\sigma}_i \cdot \vec{s}_i + U_d \sum_i \tilde{n}_{i\uparrow} \tilde{n}_{i\downarrow}, \qquad (J, K > 0) \ . \quad (1)$$

Ishida and Kamimura[4] have showed in the computer simulation work of two-dimensional square lattice system with 16 sites that a dopant hole creates spin-polarization in the first story, i.e. a spin-polaron, and that two dopant holes on different sublattices interact with a long-ranged antiferromagnetic and attractive interaction. With use of this interaction shown in Fig.3-(a) and assuming the formation of Cooper pairs for spin-polarons, Matsuno and Kamimura[5] have calculated the Sr concentration dependence of T_c. In doing so they have chosen the energy dispersion of the a_{1g} hole in the lower Hubbard band so as to reproduce the a_{1g} band in the LSDA band structure calculation by Shiraishi *et al.*[6] By choosing the strength of the interaction V_0 in Fig. 3-(a) so as to obtain the maximum value of T_c (40K), they have obtained the good agreement with experiments by Torrance *et al.*[7], as shown in Fig.3-(b).

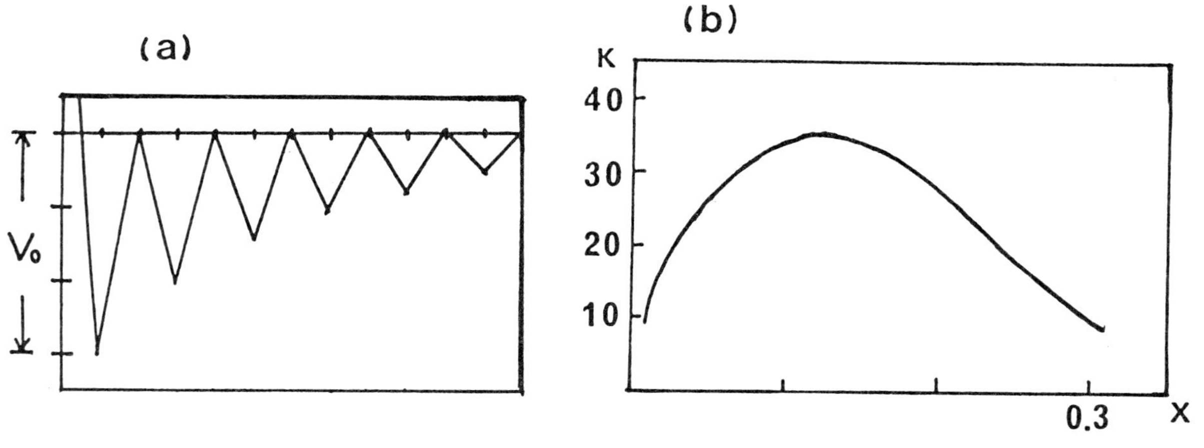

Fig.3. (a) Schematic picture of attractive potential which acts only between spin-polarons on different sublattices. (b) Calculated Sr concentration dependence of T_c. V_0 has been taken as 0.3J.

Based on the present model Schimizu and Kamimura[8] and Ushio and Kamimura[9] have calculated the normal state Hall effect and resistivity, respectively, whose results are shown in Figs.4-(a) and -(b). In Fig.4-(a) the calculated concentration dependence of the Hall coefficient for T = 0K is compared with the experimental results by Takagi *et al.*[10] for T = 80K(circles) and 300K(squares). The discrepancy in the temperature dependence is resolved by taking into account the contribution of spin-polarization in the first story to the Hall current. As a result we find the good agreement with experiments as for both the concentration and temperature dependences. On the other hand, the calculated temperature dependence of resistivity due to the electron-phonon interaction above T_c is shown in Fig.4-(b) for various values of x. Resistivity above T_c is proportional to T. This is found to be consistent with experimental results.

(a) (b)

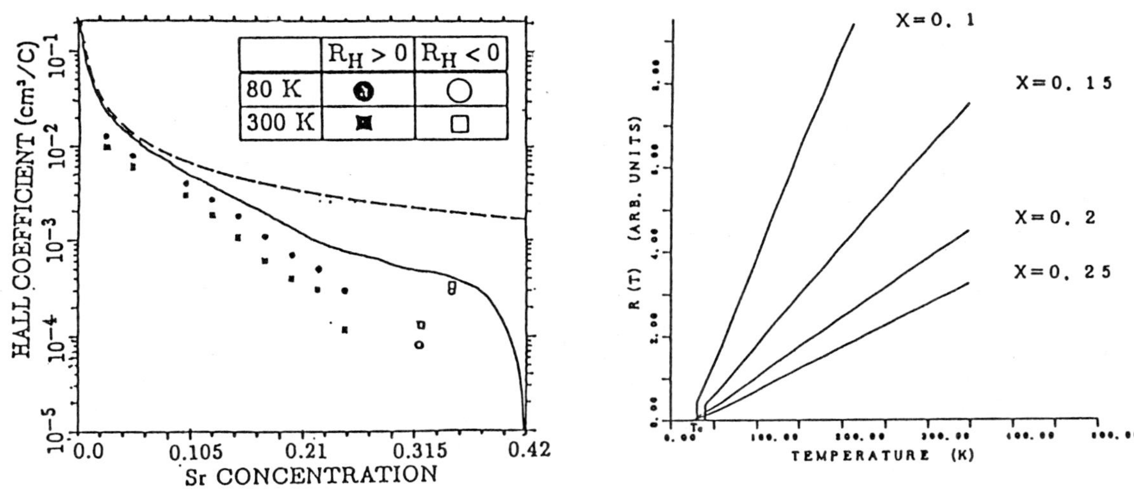

Fig.4. (a) The Hall coefficient of $La_{2-x}Sr_xCuO_4$ as a function of x. The broken line represents the 1/x line, and the solid line the theoretical result. The circles and squares are the experimental data for T=80 and 300 by Takagi *et al.*,respectively. (b) Calculated T dependence of resistivity for various x.

References

*Permanent address: Tokyo National College of Technology, Hachioji-shi, Japan 193

[1] H.Kamimura and M.Eto: J. Phys. Soc. Jpn. Vol.59, No.9 (1990) (to be published)

[2] H.Kamimura: in *Proc. Adriatico Conf. on Theoretical Understanding of High Temperature Superconductivity* p.163 (1988) (World Scientific, Singapore)

[3] H.Kamimura, S.Matsuno, and R.Saito: in *Mechanism of High Temperature Superconductivity*, eds. H.Kamimura and A.Oshiyama (Springer, Heidelberg, 1989) Springer Series in Materials Science Vol.11, p.8

[4] K.Ishida and H.Kamimura: Phys. Rev. B (1990) (to be published)

[5] S.Matsuno and H.Kamimura: in preparation (1990), see also in Physica C 162-164 (1989) 781-782

[6] K.Shiraishi, A.Oshiyama, N.Shima, T.Nakayama, and H.Kamimura: Solid State Commun. 66 (1988) 629

[7] J.B.Torrance, A.Bezinge, A.I.Nazzal, and S.S.P.Parkin: Physica C 162-164 (1989) 291-295

[8] T.Schimizu and H.Kamimura: J. Phys. Soc. Jpn. Vol.59 No.10 (1990) (to be published)

[9] H.Ushio and H.Kamimura: in preparation (1990)

[10] H.Takagi, Y.Tokura, and S.Uchida: in *Mechanism of High Temperature Superconductivity*, eds. H.Kamimura and A.Oshiyama (Springer, Heidelberg, 1989) Springer Series in Materials Science Vol.11, p.238

Supercond. Sci. Technol. **4** (1991) S124–126. Printed in the UK

Progress in High T_c Superconducting Ceramic Antennas

Y.S. He*, A.S. He+, X.X. Zhang*, C.S. Shi*, N. Lu*, B. Tian*, D.R. Lu*, M.L. Zhou# and W.G. Wang#

*Tsinghua University, Beijing 100084, P.R. CHINA
+North University of Technology, Beijing 100041, P.R. CHINA
#Department of Material Science, Zhongnan University of Technology, Changsha 410083, P.R. CHINA

Abstract. Two sets of new type high T_c superconducting antennas have been designed and fabricated. When compared with the corresponding copper antennas under the liquid nitrogen temperature, the "U" type superconducting dipole antenna made from Y-based material obtained a 3 dB relative gain enhancement(RGE) at the frequency of 740 MHz and a 7 dB RGE at 5.8 GHz for the Bi-based superconducting microstrip patch antenna was achieved and reported for the first time, which indicated that the newly discovered high T_c superconducting materials do have great potential in improving the performance of the conventional metal antennas.

1. Introduction

Superconducting antenna is an important application of superconductivity in the field of microwave. By using the conventional superconductors, the research about superconducting antenna began in 1960s(Walker 1969). However, until recent years, the development of the high T_c superconducting oxide ceramics made the practical utilization of the superconducting antenna possible. From 1988 to 1989, the research groups in Birmingham University and in Tsinghua University both successfully fabricated the high T_c superconducting antennas and obtained the satisfactory results(Gough 1988, Khamas 1988, He 1989). In this paper, the results of a series of the newly designed "U" type dipole antennas made from Y-Ba-Cu-O and the microstrip patch antennas from Bi(Pb)-Sr-Ca-Cu-O will be presented and discussed.

2. Antenna Design and Fabrication

Under the condition that the power input for the antenna is the same, the signal magnitude of the antenna will be only dependent upon its efficiency. The efficiency of antenna can be written as:

$$\eta = R_r/[R_r+(R_l+R_m)]=1/[1+(R_l+R_m)/R_r] \qquad (1)$$

where R_r is the radiation resistance, (R_l+R_m) is the loss resistance of the antenna and its matching unit. Evidently, if not only antenna itself, but also its matching unit is superconducting, the total loss resistance (R_l+R_m) will be very small and thus the efficiency of the antenna could be much higher than that of the normal metal one.

0953-2048/91/01S124 + 03 $03.50

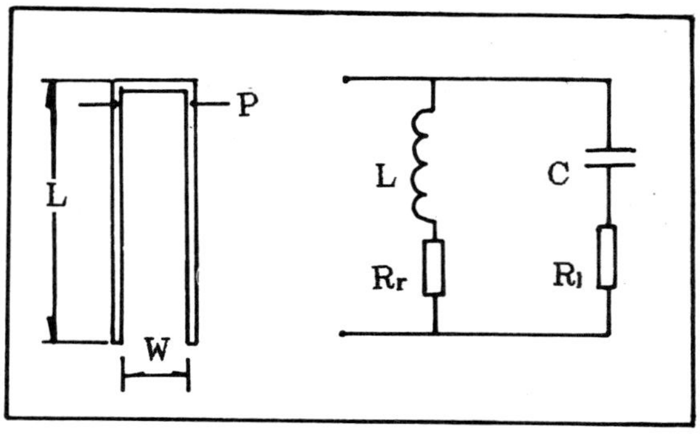

Figure 1 "U" type dipole antenna and its equivalent circuit.

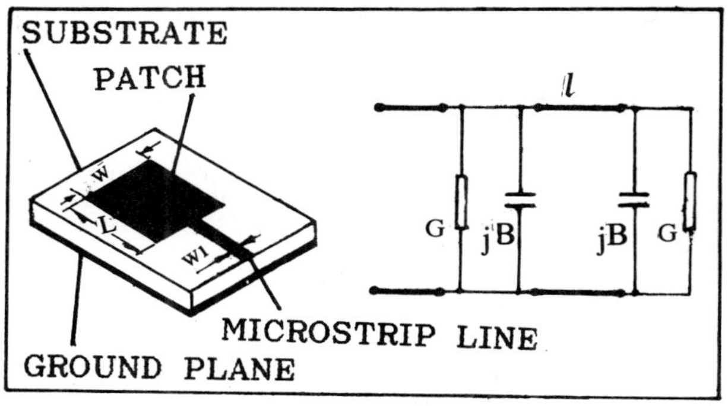

Figure 2 Microstrip antenna and its equivalent circuit.

The "U" type dipole antenna was designed by the present group, in which, a short bar was used as the antenna and two parallel wires were added as matching unit, forming a reciprocal "U" as shown in Figure 1. The equivalent circuit is also given. The optimum matching point P will be chosen experimentally. The antennas were shaped to the desired geometry by using extruded plastic wire which was a mixture of $YBa_2Cu_3O_{7-y}$ powder and some polymer solution. The polymer carrier was burnt out and the antennas were sintered again in flowing oxygen to achieve better superconductivity.

The microstrip patch antenna shown in Figure 2 was formed as follows: its microstrip patch and ground plane were all made from the slabs of $Bi_{2-x}Pb_xSr_2Ca_2Cu_3O_{10-y}$ bulk material(T_{c0}=110 K) with the thickness less than 0.5 mm, and a Teflon plate of 0.5 mm thickness was chosen as the substrate. The 50 Ω feed lines will be connected to the microstrip line via a special transformer.

3. Results and Discussions

The basic method in experiment is to compare the superconducting antenna with the same size and same form copper antenna under the liquid nitrogen temperature. The measured antenna was used as the transmitting antenna which with its matching network was put into a liquid nitrogen container, while a standard metal antenna was a receiving antenna. The signals both received and transmitted were probed by a HP 8505A NETWORK ANALYZER for the "U" type dipole antenna and a HP MODEL 560A SCALE NETWORK ANALYZER or a HP 8410C NETWORK ANALYZER for the microstrip patch antenna. The relative gain enhancement(REG) of the superconducting antenna is defined as the difference between the logarithm output values(S_{21}) of the superconducting and normal metal antennas.

Table 1 The results of "U" type dipole antennas

				S_{21} (dB)		REG(dB)	
No.	W(mm)	L(mm)	f_o(MHz)	Cu	Supercond.	Expt.	Calc.
1	12	130	462.0	-16.7	-15.8	0.9	1.2
2	8	130	490.5	-18.1	-17.1	1.0	1.8
3	4	130	492.0	-26.4	-24.0	2.4	4.4
4	4	85	740.0	-35.0	-32.0	3.0	4.4

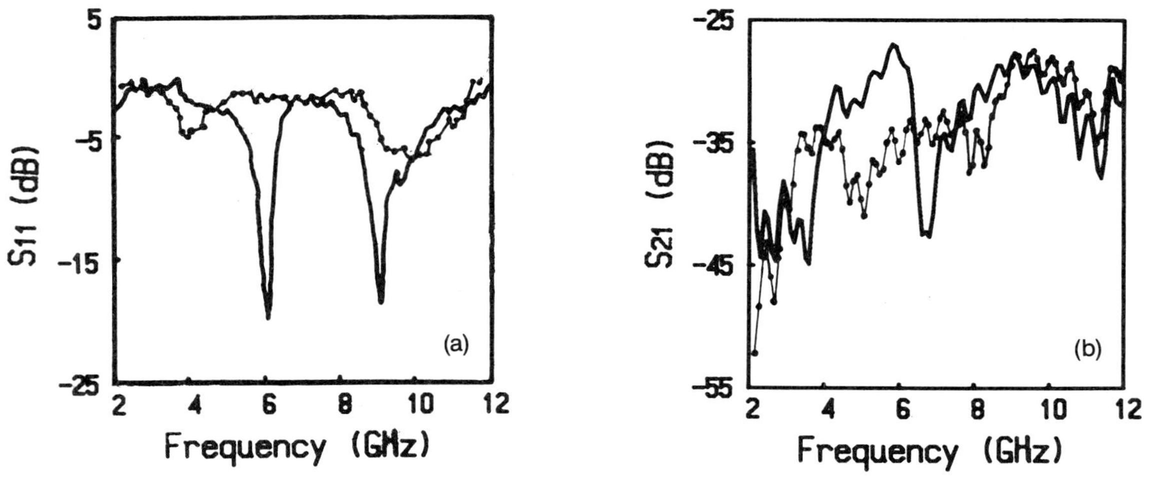

Figure 3. Frequency dependence of measured return loss S_{11}(a) and transmission coefficient S_{21}(b) of copper (dots and short lines) and superconducting (solid line) antennas at 77 K.

The experimental results of the "U" type superconducting dipole antennas from Y-Ba-Cu-O and the corresponding copper ones were summarized in Table 1, i.e., the geometry parameters(W,L), the center frequency f_o of the Y-Ba-Cu-O antennas, the transmission coefficient S_{21} and the REG. The calculated theoretical REG values are also presented. It can be found that, in the frequency range from 400 to 1000 MHz, the "U" type superconducting dipole antennas could radiate more electromagnetic energy than the corresponding copper dipole antennas and the RGE could reach up to 3 dB at the frequency of 740 MHz. The results also showed the consistency with theoretical predictions.

The results of the microstrip patch antennas were more attractive, especially those of No.3 antenna (W=11.94 mm, W_1=2.14 mm, L=9.88 mm). In comparison to the same kind copper antenna, the return loss(S_{11}) and transmission coefficient(S_{21}) of superconducting antenna were shown in Figure 3. The two minima of the return loss of superconducting antenna at 6 and 9 GHz respectively, are very close to the maxima of its transmission coefficient, showing excellent consistency. It can be seen that the highest RGE was 7 dB at 5.8 GHz. In a even larger frequency range, say 4-8 GHz, where the highest peak of the transmission coefficient curve of copper antenna is at 7.1 GHz, the RGE will still be 5.5 dB. Near 9 GHz, the RGE was near zero, which is reasonable since the surface resistance of Bi-Sr-Ca-Cu-O is very close to that of copper at such high frequency. The above results were the best and first reported for microstrip antenna at such high frequency. Considered the frequency band used in the satellite communication is 4-6 GHz, it showed us the prospect of the practical application of the high T_c superconducting antennas. It is believed that by using high quality superconducting films, better results will be achieved and such work was also being carried out in the present group.

References

Gough C E, Khamas S K, et al, 1988 Proc. Int. Conf. ASc 88,

He Y S, et al, 1989 Proc. Int. Conf. on High T_c Supercond. ed Z.X. Zhao (Singapore: World Scientific) pp 118

Khamas S K, Mehler M J, et al, 1988 Electronic Letters, **24** 460

Walker G B, Haden C R, 1969 J. App. Phys., **40** 2035

Supercond. Sci. Technol. 4 (1991) S127–129. Printed in the UK

Surface Impedance and Granularity in Superconducting Cuprates

J. Halbritter

Institut für Material- und Festkörperforschung, Kernforschungszentrum Karlsruhe, Postfach 7500 Karlsruhe 1, Federal Republic of Germany

Abstract. The rf residual surface resistance R_{res} is an important figure of merit for superconducting microwave devices and for establishing the homogeneity of superconducting surfaces. In granular superconductors not only are high residual surface resistance values R_{res} observed, but also strong field dependencies of R_{res} (T, ω, H) and the penetration depth λ_{res} (T, ω, H) are encountered even at very low field levels (of order 1 Oe). These observations can be explained in terms of granularity consisting of weak links and of insulating interfaces, which provide additional degrees of freedom due to fluxoid motion and electromagnetic strip line modes. These degrees of freedom cause additional interaction mechanisms with applied fields.
These mechanisms allow a coherent description of the observed T, ω, H_{rf} and H_{dc} dependence of R_{res} and λ_{res} for the granular cuprates and for similar effects in Pb, Nb, Nb_3Sn and NbN. In addition, on the basis of such rf data, it is possible to quantify the density of weak links and their critical fields and currents.

It is well accepted that weak links dominate the residual rf surface impedance achieved with current surface preparations. But the actual mechanism operating at these weak links is usually not stated and unambiguous experimental evidence is hard to find. But recent rf measurement of grain-aligned $YBa_2Cu_3O_{7-x}$ bulk material fabricated using a liquid phase process [1] shows this in Fig. 1 [2] clearly. Describing the quality of this material by a geometric factor $p > 1$ with $\rho(T) = \Sigma\rho_b + p\rho^i(T)$, a weak-link boundary resistivity ρ_b and the intrinsic resistivity $\rho^i(T)$, the results in Fig. 1 can be described as follows:

ab-plane with its "single crystal type" behavior ($p \approx 2.5$) has in the normal state the normal skin effect of $R_{nc} \approx p\, R_{nc}^i$. The steep decrease of R below T_c fits to $R(T) \simeq p\, R_{BCS}^i(T)$ with $2\Delta/kT_c \approx 8$. The ac-cut has many weak links and complicated percolation as indicated by $p_{ac} > 100$. Thus in the normal state the normal skin effect cannot be applied ($R_{nc} \approx 10\, R_{nc}^i \neq p\, R_{nc}^i$). In cooling in these rf-dc-fields, the weak links become superconducting more than 2 K below T_c(ab) with a smeared-out field dependent R(T,B)-decay with no resemblance to the BCS-theory.

Such decays are commonly measured by ac methodes being related to the impedance of fluxoids at weak links. The field dependence of this "weak link surface impedance" shown in Fig. 2 [4] gives insight into the fluxoid related mechanism [5]. The superconducting weak link parameters are summarized in Table 1 where consequences of the Josephson critical currents as, e.g., penetration depths λ and H_{C1}-fields are given. $H_{C1_{J/G}}$ is the field where fluxoids first penetrate into inter ("J")- or intragrain ("G") weak links for H parallel to the surface. For the often used films of thickness t smaller than $\lambda_{J/G}$ rf field components $H_\perp$ exists reducing H_{C1} very much yielding thus an enhanced residual rf surface impedance [5].

0953-2048/91/01S127 + 03 $03.50

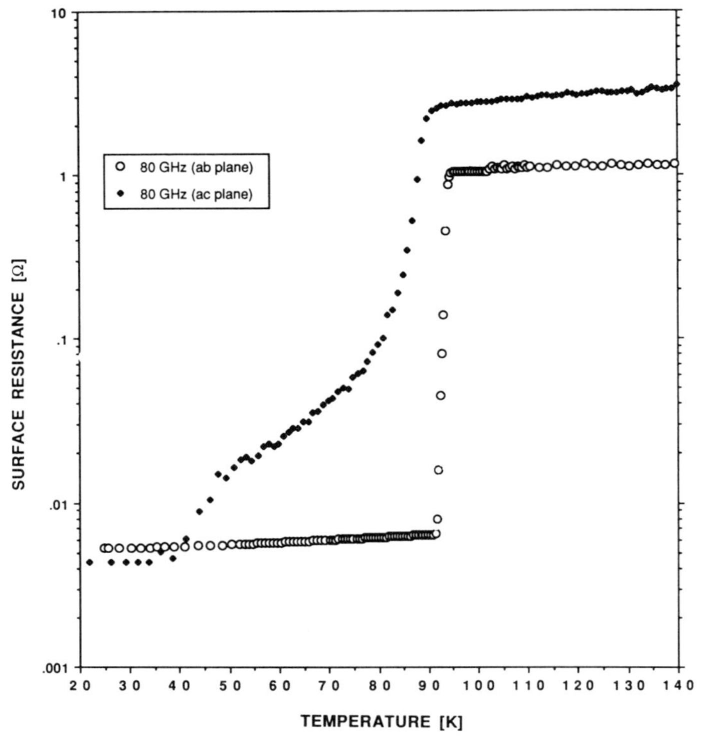

Fig. 1: Surface resistance R(T) obtained by endplate measurements in a Cu cavity at 80 GHz. The dc conduction of samples shows a percolation factor $p_{ab} \approx 2.5$ and $p_{ac} > 200$. This indicates many weak links hindering the current in the ac-cut. The intragrain weak links crossed in this case are strong, i.e. not detectable by $j_c(T,B)$-measurements [1], but quite obvious in the surface resistance R_{ac}.

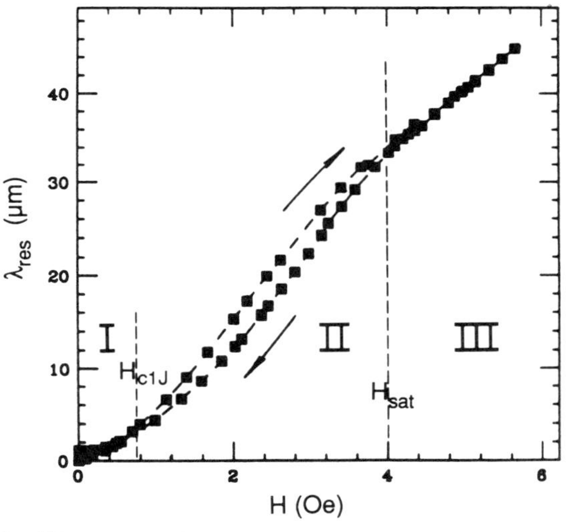

Fig. 2: "High"-field effects on the penetration depth at 85 K and 6 GHz, displayed as $\Delta\lambda_{eff}$ vs. B The dashed lines are guides to the eye. The

<u>regime I</u> is given by a $(H/H_{CJ})^2$- increase due to pair-breaking at weak links ($H_{CJ} \approx 1$ Oe). The

<u>regime II</u> is due to flux entry above H_{C1J} with a hysteresis due to flux pinning in the weak links. The linear increase of λ is due to flux reducing the mean Josephson shielding current. This change to a sublinear

<u>regime III</u> above H_{sat}, where λ increases further due to an increased density (∝ H) of oscillating flux.

All these effects are proportional to µ (≈ %) as areal ratio of weak links.

Table I: Parameters characterizing weak links of $YBa_2Cu_3O_{7-x}$. The abbreviations "J" or "G" are subscripts to be added to j_c, λ_J or H_{c1} as needed for clarity. $R_b = \rho_b d_b$ is the weak link boundary resistance between the grains of size d_b.

Weak link	abbr.	$R_b/\Omega cm^2$	$j_c/A/cm^2$	$\lambda_J/\mu m$	Hc1/Oe	Hc2/100 Oe
insulator		∞	0	∞	-	-
intergrain	"J"	$\geq 10^{-6}$	$\sim 10^2$	~ 30	≈ 1	1
intragrain	"G"	$\approx 5 \cdot 10^{-8}$	$\geq 10^4 - 10^5$	~ 3	≈ 100	> 100

We now turn back to the surface impedance addressed above in Fig. 2. In order for the surface impedance of a superconductor to have its simple form $Z = R + i\omega\mu_0\lambda$ independent of H, it is necessary that the superconductor remains in the Meissner state, i.e. $H \ll H_{cJ}$ or H_{cG} or H_c. Low weak link H_{C1}-fields ($\approx$ Oe) may exist in bulk YBCO, NbN or Nb_3Sn weakened further at the surface by corrosion or chemical attack. The latter mechanisms dominate for bulk Nb and Pb. Moreover, in materials containing a network of weak links, cooling through T_c creates trapped flux which cannot leave the network and which can be described by an equivalent magnetic field $H_{res} \leq H_{c1J}$ or H_{c1G}. The flux flow losses associated with this residual trapped flux cause rf residual losses that are proprotional to $R_{res} \propto \omega^2 H_{res}$.

The weak superconductivity at the weak links also leads to field dependences of the rf properties. Pair breaking (e.g., as described by the Ginzburg-Landau theory) predicts that the surface resistance and penetration depth increase like $(H/H_{cJ})^2$ -- see regime I in Fig. 1. But at sufficiently high fields ($\approx H_{c1J}$), the weak links of granular superconductors allow fluxoids to enter between grains as a specific degree of freedom. These are nucleated quickly ($f_N{}^{-1} \sim 10^{10}$/sec) and at high level of rf excitation can flow relatively easily. At low frequencies ($\leq 10^4$ Hz) flux pinning is encountered between the grains and can be described by a critical state model. Also for $H > H_{c1J}$, the penetrating flux enhances the Josephson penetration depth to $\lambda_J(H)$ as shown in Fig. 2 as regime II. The flux flow dissipation in this range yields an irreversible, nonlinear surface impedance for frequencies below the nucleation frequency f_N. Evidence for such behavior has actually been observed in intergrain weak linksμ even at 10 GHz and for intragrain weak link above H_{c1G} at 6 MHz. The data permit identification of H_{c1J} and H_{c1G} and thus estimates of j_{cJ} and j_{cG} or areal weak link ratio μ. At a field $H_{sat} = cj_c/\omega > H_{c1}$ the steep linear increase of R_{res} or λ_{res} with H crosses over to a flux flow shielding regime in which $\Delta R_{res} \propto \sqrt{\omega}$ slowly increases with H -- see regime III in Fig. 2. For $H_{sat} < H_{c1J}$ or frequencies above f_N no flux is penetrating and thus only H_{res} is causing $R_{res} \propto \omega^2 H_{res} \mu$.

References

1. K. Salama, V. Selvamanickam, Appl. Phys. Lett. (1990); C. Keller, H. Küpfer, R. Meier-Hirmer, U. Wiech, V. Selvamanickam, K. Salama, Cryogenics 30, 401 and 410 (1990).
2. J. Wosik, R. A. Kranenburg, J. C. Wolfe, V. Selvamanickam, K. Salama, J. Appl. Phys. (1990)
3. J. Halbritter, Int. J. Mod. Phys. B3, 719 (1989) and J. Mat. Research, submitted
4. Dong-Ho Wu, C.A. Shiffman, S. Sridhar, Phys. Rev. B38, 9311 (1988)

Supercond. Sci. Technol. **4** (1991) S130–132. Printed in the UK

Properties of rf-SQUIDs Fabricated from Epitaxial $YBa_2Cu_3O_7$ Films

G. Cui, Y. Zhang, K. Herrmann, Ch. Buchal, J. Schubert, W. Zander, A. I. Braginski and C. Heiden, Institute for Film- and Ion Technology, Forschungszentrum Jülich (KFA), D-5170 Jülich, FRG.

Abstract. Radio-frequency (rf) SQUIDs were evaluated with weak links in epitaxial $YBa_2Cu_3O_7$ films created by either local ion-implantation or step-edges in the substrate. It is shown that such weak links permit one to fabricate rf-SQUIDs with low 1/f noise. The step-edge SQUIDs show a promise of reproducible I_c control.

1. Introduction

The purpose of this work was to evaluate the performance of $YBa_2Cu_3O_7$ (YBCO) thin-film rf-SQUIDs with engineered and controllable weak-link microbridges. First reports on thin film YBCO one-hole rf-SQUIDs described results obtained with post-annealed, polycrystalline or textured films (Betts et al. 1989a, Daly et al. 1989b). These SQUIDs utilized the natural grain boundary (GB) junctions present in microbridges (30 and 10 μm wide) and exhibited temperatures of operation, T_{op}, up to 55 and 65K, respectively. The low-frequency 1/f noise was high, 10^{-3} to 10^{-2} $\Phi_0 Hz^{-1/2}$ at 10 Hz. The authors attributed it to random fluxoid motion at grain boundaries. Similar results were obtained in this laboratory with textured YBCO SQUID microbridges 0.5 to 10 μm wide (Tinchev 1990a).

Weak links engineered in epitaxial, GB-free YBCO films hold the promise of controllable properties and lower 1/f noise. Recently, Simon et al. (1990b) demonstrated operating rf-SQUIDs with weak links fabricated by several different means in epitaxial, *in-situ* grown YBCO . A narrow disordered stripe across a patterned microbridge was created by local in-diffusion of Al, a sharp step-edge in the substrate or by focussed ion-beam irradiation. In each case, the film's J_c was locally reduced by two or three orders of magnitude and, consequently, rf-SQUIDs operating between 4.2 and 60K were obtained.

We have been investigating similar rf-SQUIDs with step-edge weak links, and with links which were locally ion implanted. One of us (G.C.) experimented earlier with this approach to weak link formation (Shao et al. 1990c).

2. Fabrication and Measurements

Epitaxial, c-axis oriented YBCO films were *in-situ* deposited on (100) $SrTiO_3$ substrates by pulsed laser deposition (PLD) and magnetron sputtering. Film thicknesses were 100 and 200 nm. One-hole SQUID structures were patterned by optical lithography and Ar ion-beam etching (Lehmann et al. 1990d). Chemical etching in diluted H_3PO_4 or in EDTA (Shokoohi et al. 1989c) has also been used.

The examined SQUID microbridges were 10 μm long and nominally 2 μm wide. The one-hole rf-SQUID structure consisted of a 1.4 x 1.4 mm^2 flux concentrator with a central hole of 100×100^2 μm corresponding to an inductance of approximately 150 pH. The SQUID was loosely coupled to a 40-turn, 1 μH tank-circuit coil (coupling coefficient $\alpha < 0.2$). The tank circuit frequency was between 13 and 20 MHz. The SQUID noise was measured between 1 and 10^5 Hz using a HP 3562 spectrum analyser. The dewars used for measurements had a Mumetal or Mumetal and Pb shielding.

3. SQUIDs with Ion-Implanted Microbridges

The rf-SQUID structures were patterned by wet etching from magnetron sputtered, 200 nm thick, epitaxial YBCO films on $SrTiO_3$ having critical temperatures, T_c, of 86 to 89K. The microbridges were locally implanted using 100 KeV Ar-ions. After determining the dependence of the film resistivity upon the dose, the latter was set at 5×10^{12} cm^{-2}. The PMMA 950K resist serving as an implantation mask had 100-200 nm wide open slits written by e-beam lithography across the 2 μm wide microbridges. The PMMA was 300 nm thick, sufficient for protection of the underlayer (Biersack 1967). Prior to implantation, no SQUID response could be observed.

The locally implanted rf-SQUIDs exhibited a well-defined, periodic and nearly triangular $V(\Phi)$ response on up to 10 steps of the rf I-V curve. These SQUIDs operated over a wide temperature range, between 4.2 and 60K, in the dissipative or dispersive mode, as defined by the temperature-dependent critical current $I_c(T)$. At 4.2K, the estimated I_c was of the order of 10 μA and the SQUID-parameter β was approximately 3. Noise characteristics were measured over the temperature range of 4.2 to 50K. The lowest white noise voltage level above $f = 3$ Hz corresponded to 2×10^{-4} $\Phi_0 Hz^{1/2}$ at 4.2K and 8×10^{-4} $\Phi_0 Hz^{1/2}$ at 38K with a crossover to 1/f noise still observed near 3 Hz, as shown in Fig. 1. The presence of strong 50 Hz and harmonic signals in the spectrum and also the much lower intrinsic SQUID noise calculated (Jackel and Buhrman 1975) indicated that our shielding was inadequate.

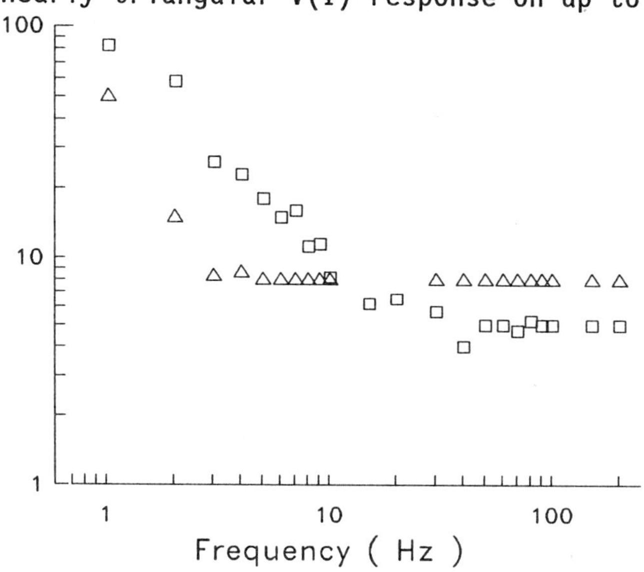

Fig. 1. Low-frequency SQUID noise spectra, in units of $10^{-4} \Phi_0 Hz^{-1/2}$: triangles - ion-beam implanted link (1) at T = 38K, squares - step-edge link (2) at 47K (comparable β's). The 1/f noise in (1) did not change significantly between 30 and 50K.

One of the SQUIDs was thermally cycled 8 times between 4.2 and 300K, with intermittent exposures to moisture, without any change in characteristics.

4. SQUIDs with Step-Edge Microbridges

Steps 20 to 200 nm high were Ar-ion beam milled in $SrTiO_3$ substrates through photolithographically defined Nb masks. Epitaxial YBCO films, 100 nm thick, were subsequently fabricated by PLD and exhibited a T_c = 87 to 89K. In such films, the J_c at 77K was typically in the 10^6 A/cm^2 range. The ion-beam patterned SQUID microbridges were 2 μm wide. In the absence of the step-edge, no SQUID response was observed.

At low step heights, h, a well-defined, triangular and periodic V(Φ) response was observed only over a narrow temperature range near T_c, e.g. 75 to 82K at h = 20 nm, indicating that at lower temperatures the I_c was too high. With the increasing step height, the temperature range of operation broadened and extended down to 4.2K. At h = 70 nm, the range extended from 4.2 to 55K. At 200 nm, the range of periodic response was typically 4.2 to 30K while between 65K and T_c a much stronger nonperiodic response was observed. This nonperiodic response could be tentatively associated with the random array of junctions in the 0.65 mm wide step in the SQUID loop located opposite to the weak link.

The grain boundaries which nucleate at both edges of a step, (where the film orientation changes from c- to a-axis and back), the reduction of film thickness at the step and the I_c anisotropy should contribute to the h-independent I_c reduction. Boundaries and other defects in the a-axis step plane will make I_c to depend upon h. The uniformity of I_c distribution on the chip was high at low step heights: up to 50% of the 16 SQUIDs had nearly identical T_{op} ranges. The spread in T_{op} increased with h, thus confirming the role of random defects and percolation paths in the step.

The best step-edge SQUID sample operated between 20 and 80K in the dissipative and dispersive mode. The inferred low-temperature I_c was of the order of 10 μA. The 1/f noise was much higher than in SQUIDs with implanted microbridges (Fig. 1) while the white noise level was comparable. At 74K, β = 1 and f > 60 Hz the noise was $5 \times 10^{-4}\ \Phi_0 Hz^{-1/2}$. However, even the step-edge SQUIDs exhibited lower 1/f than the bulk (Zhang 1990d) and thin film SQUIDs with grain-boundary junctions.

5. Conclusion

We confirmed that weak links created in epitaxial YBCO films permit one to fabricate rf-SQUIDs with 1/f noise lower than in grain-boundary devices. The step-edge SQUIDs show a promise of reproducible I_c control.

Acknowledgement

We thank Drs. H.-M. Mück (KFA) for the e-beam lithography, S.S. Tinchev (FUBA) for his input while at KFA and R.W. Simon (TRW) for valuable suggestions on step-edge fabrication.

References

Betts, K. et al. 1989a Trans. IEEE on Magnetics **25** 965

Biersack, J.P. 1967 in *Ion Beam Modification of Insulators*, ed. Mazzoldi, P. and Arnold G. (Amsterdam: Elsevier) pp 1-56 and p A11

Daly, K.P. et al. 1989b Trans IEEE on Magnetics **25** 1305

Jackel, L.D. and Buhrman, R.A. 1975 J. Low Temp. Phys. **19** 201

Lehmann, R., et al. 1990c Proc. ICMC '90 (in print)

Shao, K. et al., 1990c Progr. in HTS **22**, ed. Zhao Z.X., Cui G.J. and Han, R.S. (Singapore: World Scientific) pp 61-3

Shokoohi, F.K., et al. 1989c Appl. Phys. Lett. **55** 2661

Simon, R.W. 1990b Proc. 2nd Conf. Sci & Tech. of Thin Film Superconductors (in print)

Tinchev, S.S., 1990a (unpublished data)

Zhang Y. 1990d Ph.D. Dissertation, Univ. of Giessen.

Supercond. Sci. Technol. **4** (1991) S133–135. Printed in the UK

Proximity effect based $YBa_2Cu_3O_{7-\delta}$-Ag-Al – Al_2O_3 – Pb Josephson and tunnel junctions

M.A.M. Gijs, A.M. Gerrits, D. Scholten and Th. van Rooy

Philips Research Laboratories, 5600 JA Eindhoven, The Netherlands

Abstract. Two types of $YBa_2Cu_3O_{7-\delta}$-Ag-Al – Al_2O_3 – Pb proximity structures have been fabricated. The first junction type contains metallic shorts in the Al_2O_3 insulating layer and exhibits a Josephson current. Sharp Shapiro steps appear in a microwave field. The other junction type has a closed Al_2O_3 barrier and is a tunnel junction with high subgap resistance. At low temperatures an energy gap of 9-10 meV is found to be induced in the Al-layer by the $YBa_2Cu_3O_{7-\delta}$.

1. Introduction

A high-T_c superconductor based Josephson junction may be realised using a noble metal buffer layer between the oxide superconductor (S) and the insulating material (I) of the tunnel barrier. A normal metal (N) like Ag or Au has a low contact resistance with the superconducting oxide and is characterized by a large coherence length (typically 100 nm at low temperatures), while the superconducting coherence length of an oxide superconductor is only of the order of 1 nm, posing severe demands on the S-I interface. Obviously the study of S-N-S′ and S-N-I-S′ junctions, where S′ is a classical superconductor, is relevant for the Josephson junction objective and provides important information on the superconducting proximity effect (de Gennes 1964) between an oxide and a classical superconductor. We already reported supercurrents in $YBa_2Cu_3O_{7-\delta}$-Ag-Pb and $YBa_2Cu_3O_{7-\delta}$-Ag-Al-Pb thin film proximity structures (Gijs *et al.* 1989 1990a). Hereby the microwave induced appearance of Shapiro steps in the current-voltage (I-V) characteristic is an important confirmation that the proximity effect in a sandwich consisting of an oxide and classical superconductor separated by a normal metal really exists and that one is not measuring some very small Ohmic junction resistance. In subsequent experiments (Gijs *et al.* 1990b) a thin and closed Al_2O_3 layer was formed on the Al, giving rise to quasiparticle tunneling behaviour.

2. Experimental details

Junction fabrication (see Figure 1) is as follows. A 1 mm wide $YBa_2Cu_3O_{7-\delta}$ strip is sputtered on a (100) $SrTiO_3$ substrate with an Ar pressure of 0.003 Torr. Superconductivity is obtained after annealing in flowing O_2 at 850°C for 1/2 hour . Then a Ag strip of 20 nm thickness is sputtered after which the sample is heated up to 450°C in flowing O_2 during 1/2 hour to obtain a good superconducting contact between $YBa_2Cu_3O_{7-\delta}$ and Ag. After evaporation of a 25 nm thick Al strip the substrate is completely covered with SiO_2. Using a CF_4/O_2 plasma a $10 \times 10\ \mu m^2$ hole is etched in the quartz layer. At the end of this process the Al is to the larger part

0953-2048/91/01S133 + 03 $03.50

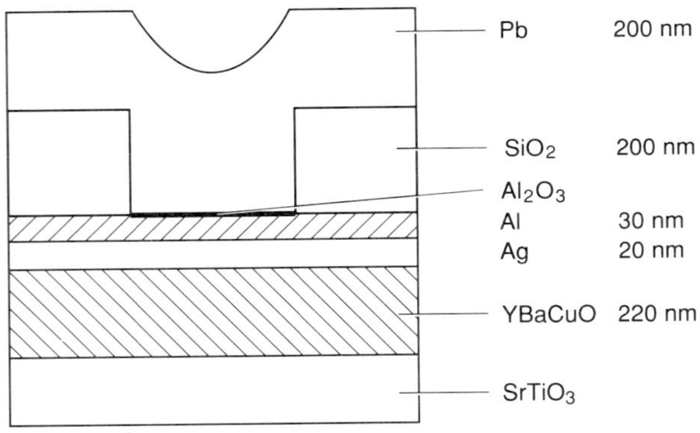

Figure 1 : Schematic diagram of the proximity junction geometry

covered with an Al_2O_3 layer, in which however pinholes exist which form the actual S-N-S′ junctions we study. To obtain a closed Al_2O_3 layer subsequent Ar-etching and re-oxidation is necessary. As a final step a crossed Pb strip is evaporated to form the second junction electrode. After contacting the sample with In, it was mounted in a continuous flow cryostat. Electrical measurements were done using a standard four probe technique. 6-12 GHz microwave radiation can be generated by a HP86245A source and is fed in a coaxial line which ends close to the sample.

3. Experimental results and discussion

In Figure 2 we show the I-V characteristics for a junction with pinholes at 3.9 K in the presence of 9.8 GHz microwaves at different levels of microwave source power. Only a small fraction of this power is coupled into the junction. The Josephson critical current I_c without microwaves, I_0 , is 70 μA and the normal junction resistance $R_n = 1\Omega$. As we only expect influence of microwaves when the microwave current amplitude in the junction I_{mw} is of the order of I_0 a rough estimate for the effective microwave power coupled into the junction is $R_n I_0^2 \simeq$ 5nW. With increasing power I_c is modulated and Shapiro steps are generated at voltages $V = n(h/2e)\nu = n \times 20.2\,\mu$V , with n integer and ν the microwave frequency. Important information on the nature of the proximity effect is obtained by studying the dependence of step amplitude on microwave power. In the case of a S-N-S′ Josephson junction the 'Resistively Shunted Junction' model is expected to apply (Russer 1972). We compared the experimental step height as a function of microwave current with the predictions of the model, using the parameter $h\nu/(2eR_nI_0) = 0.45$. We obtain a good agreement and find that the Josephson current obeys $I_c \propto \sin\varphi$, where φ is the superconducting phase difference across the junction.

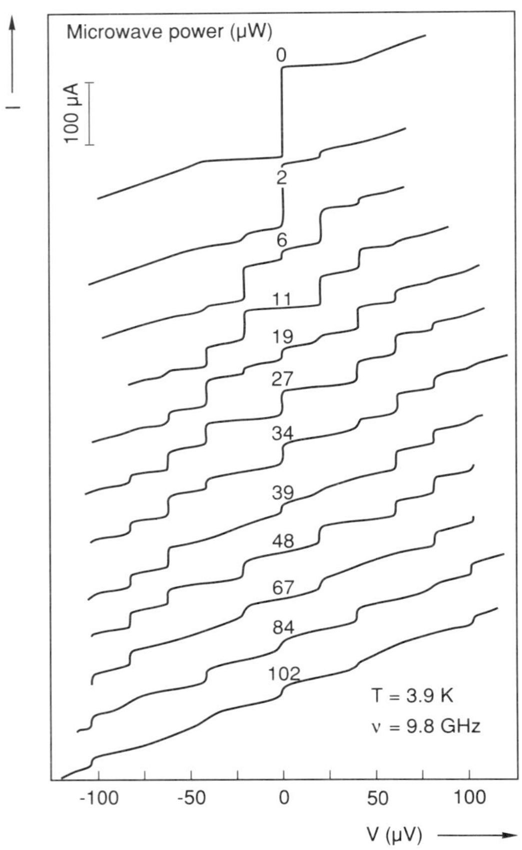

Figure 2 : I-V characteristic of a $YBa_2Cu_3O_{7-\delta}$-Ag-Al-Pb S-N-S′ junction at different levels of microwave source power.

In Figure 3 we show the dynamic conductance dI/dV for a junction of the second type, i.e. a tunnel junction with a closed Al_2O_3 barrier, at different temperatures. The derivative is ob-

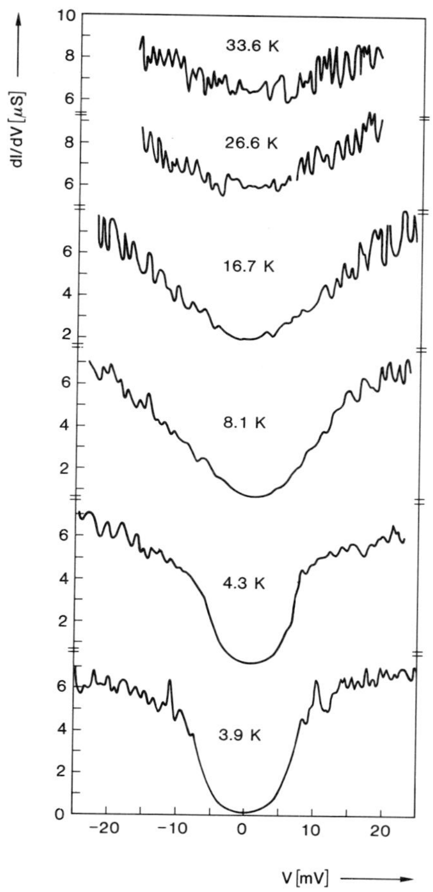

Figure 3 : Dynamic conductance dI/dV-V for a proximity effect based tunnel junction at different temperatures.

tained by numerical differentiation. At 3.9 K we observe a pronounced energy gap with very low subgap conductance. The irregular structure in dI/dV at higher voltages is probably due to noise, induced by inelastic tunneling events, which becomes very pronounced in the numerical derivative. At V=10-11 meV larger peaks appear which presumably are related to the peak in the tunneling density of states of the Pb. Indeed for a tunnel junction consisting of two superconducting electrodes with a different gap (in this case $\Delta_{1:2:3}$ and Δ_{Pb}) one would expect the appearance of a peak in the dynamic conductance dI/dV at a voltage $|V| = (\Delta_{1:2:3} + \Delta_{Pb})/e$. Although our data are not perfectly symmetric in voltage we can estimate, taking $\Delta_{Pb} = 1.2$ meV, that $\Delta_{1:2:3} = 9\text{-}10$ meV at this temperature. Note that $\Delta_{1:2:3}$ is a proximity superconducting gap, induced in the Al by the outdiffusion of superconductive carriers from the $YBa_2Cu_3O_{7-\delta}$ film. The characteristic length over which this diffusion takes place, is defined as the normal metal coherence length $\xi_N(T)$; from previous work (Gijs *et al.* 1989 1990a) we found at low temperatures that $\xi_N(T) \simeq (93\ \text{nm})/\sqrt{T}$ for our proximity systems. It is this decreasing length which makes the energy gap $\Delta_{1:2:3}$ disappear at higher temperatures.

4. Conclusion

We have fabricated $YBa_2Cu_3O_{7-\delta}$-Ag-Al-Pb junctions in which we studied the superconducting proximity effect using microwave irradiation. We find that the Josephson current is well described by a $\sin\varphi$ relation. Moreover we have made $YBa_2Cu_3O_{7-\delta}$-Ag-Al-Al_2O_3-Pb tunnel junctions with very low subgap conductance. The energy gap induced in the Al layer by the $YBa_2Cu_3O_{7-\delta}$ is 9-10 meV.

5. References

de Gennes P G 1964 *Rev. Mod. Phys* **36** 225

Gijs M A M, Scholten D, van Rooy Th and IJsselsteijn R 1989 *Solid State Commun.* **71** 575

Gijs M A M, Scholten D, van Rooy Th and Gerrits A M 1990a *Phys. Rev. B* **41** 11627

Gijs M A M, Scholten D, van Rooy Th and Gerrits A M 1990b *submitted to Appl. Phys. Lett.*

Russer P 1972 *J. Appl. Phys.* **43** 2008

Supercond. Sci. Technol. 4 (1991) S136–138. Printed in the UK

Window-Type Tunnel Devices on $YBa_2Cu_3O_{7-x}$ Thin Films

J.H.James, B.Dwir, M.Affronte, A.Münzner, T.Naucler, B.J.Kellett and D.Pavuna.

Institute of Micro & Optoelectronics, Department of Physics,
Swiss Federal Institute of Technology, (EPFL), Lausanne, CH-1015, Switzerland.

Abstract We present the first $YBa_2Cu_3O_{7-x}$-Pb tunnelling measurements using window-type tunnel devices. $YBa_2Cu_3O_{7-x}$ (YBCO) thin films are grown in-situ by dual Ar^+ ion beam sputtering. Tunnel barriers are formed on a chemically etched YBCO surface. Tunnelling dV/dI(V) characteristics show reproducible weak gap-like structure at 4.4±0.4meV and 22±2meV, the latter giving a $2\Delta/k_BT_c$ value of about 6.5.

Tunnelling data has been reported on $YBa_2Cu_3O_{7-x}$ (YBCO) single crystals (Gurvitch *et al*, 1989a) and on thin films by Geerk *et al*, (1989) and several other groups. However, almost all of the thin film tunnel structures have been simple cross-strip geometries of macroscopic dimensions (≈0.1-1mm^2). In the present work we describe tunnelling measurements over much smaller areas on YBCO films using window-type tunnel devices that avoid film-edge effects.

The YBCO films (100-200nm) are grown in-situ by dual Ar^+ ion beam sputtering from $YBa_2Cu_3O_{7-x}$ and $BaCuO_2$ targets. The deposition system (figure 1) has been described in detail elsewhere (James *et al*, 1989). During deposition, O_2 is directed at the growing film, the O_2 pressure is about 8.0 x 10^{-3} torr at the film surface. Films are grown at $0.1nms^{-1}$ at a substrate temperature of 690±40°C. Ar^+ beam conditions are typically about 600eV, 25mA. YBCO films grown in-situ on (100) $SrTiO_3$ are c-axis oriented and show T_{conset}=92K and T_{co}=89K with J_c(77K) in excess of 10^5Acm^{-2}. YBCO films grown in-situ under similar conditions on 200nm ITO (In_2O_3 +10mol% SnO_2) buffer layers on Si and GaAs show T_{co}=68K (Kellett *et al*, 1990) and 60K, respectively. The tunnel devices are fabricated using YBCO films on $SrTiO_3$.

Figure 2 is a schematic of the completed device structure. The YBCO base electrode layer is patterned with photoresist and etched using $1HNO_3:16H_3PO_4(85\%):3H_2O$ solution (1.3$nmsec^{-1}$ etch rate) or Ar^+ ion beam milled to form the base electrode. Acetone or photoresist do not degrade T_c, ΔT_c or J_c. Photoresist developer (Shipley MF319) depresses T_c by about 3K. To define the device windows, a 200nm Y_2O_3 isolation layer is deposited by ion beam sputtering and patterned by photoresist lift-off. Y_2O_3 is chosen for its low reactivity with YBCO. This is followed by a 100nm SiO_2 layer to protect the Y_2O_3 from HNO_3 attack during the patterning of the top Pb layer. At this stage the YBCO Tc is depressed by about 5K. The structure is therefore annealed in O_2 for up to two hours at 815°C followed by a cool down (200°Ch^{-1}) and 2 hour plateau at 450°C. Such post annealed films are c-axis textured with T_{co} up to 89K. The oxygen-deficient YBCO surface layer is removed by a 15sec etch in freshly prepared 1% (vol) Br in CH_3OH, (Phillips *et al*, 1989 and Cuculo *et al*, 1989) and dried with N_2 gas, the YBCO etch rate is about 4$nmsec^{-1}$. This etchant exposes any a-b plane surfaces (Gurvitch et al, 1989b). The etched YBCO surface is then exposed to ambient

0953-2048/91/01S136 + 03 $03.50

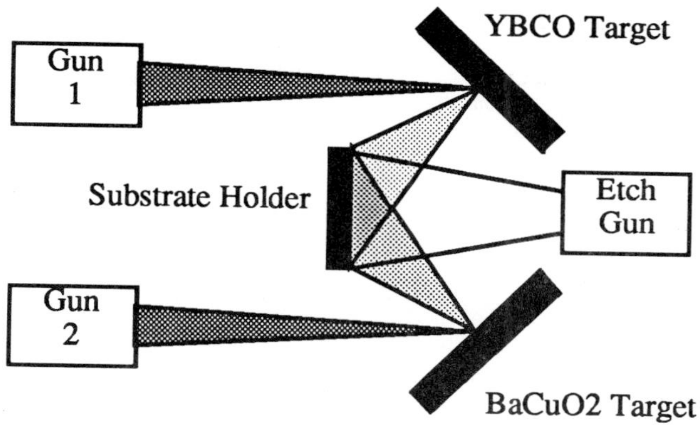

Figure 1. A schematic of the ion beam deposition system.

air for 15min. A 200nm Pb film is then evaporated and patterned with $1HNO_3(65\%):20H_2O$ etch solution ($0.7nmsec^{-1}$ etch rate) to form the counterelectrode. A natural tunnel barrier of unknown composition (possibly PbO) forms at the interface between the YBCO and the Pb. The tunnel barrier is fully encapsulated by the Pb and therefore not contaminated by photoresist during the all-important tunnel barrier formation stage. Device dimensions range from $100x100\mu m^2$ to $3x3\mu m^2$. Room temperature $80x80\mu m^2$ device resistances are typically 2 to 5kΩ. Devices formed on un-etched YBCO show resistances greater than 2MΩ.

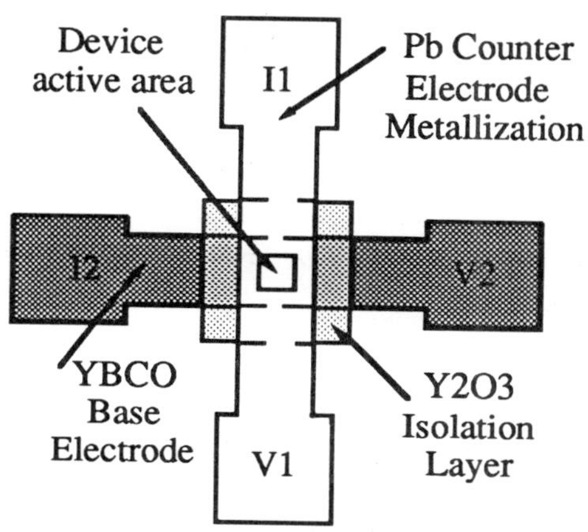

Figure.2. A schematic of the completed device structure.

Below the T_c of Pb (7.2K), the tunnelling I(V) characteristics are dominated by the Pb gap (1.3meV), as shown in figure 3a. Derivatives of the tunnelling I(V) characteristics are recorded using standard lock-in techniques. Reproducible weak gap-like structure in dV/dI(V) is observed at 4.4±0.4meV and 22±2meV (marked by the arrows in figure 3b), this structure exists above the T_c of Pb and is attributed to the YBCO gap in the c-axis and a-b axis directions respectively (Gurvitch *et al* 1989b). The upper gap value gives a $2\Delta/k_BT_c$ value of about 6.5. Figure 4 shows $d^2V/dI^2(V)$ at various temperatures and highlights the non-symmetric nature of the tunnelling characteristics about the Fermi level. The origin of this asymmetry is not clearly understood at present but may be due to strong variation of the quasiparticle density of states near the Fermi level. The characteristic energies of the tunnelling structure are in agreement with data reported by Geerk *et al*, (1989), Lanping *et al* (1989), Furuyama *et al* (1990) and Iguchi *et al* (1990) and others.

In conclusion we have fabricated simple window-type YBCO-Pb tunnel devices on YBCO thin films grown in-situ by dual Ar^+ ion beam sputtering. The tunnelling dV/dI(V) characteristics show reproducible weak gap-like structure at 4.4±0.4 and 22±2meV, the latter giving a $2\Delta/k_BT_C$ value of about 6.5.

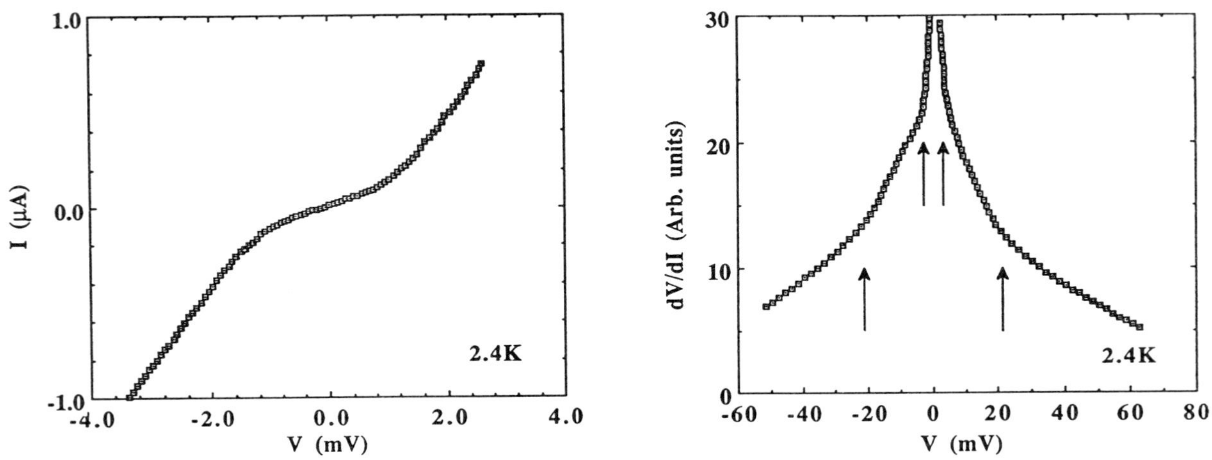

Figure 3. (a) Tunnelling I(V) characteristics and (b) dV/dI(V), at 2.4K.

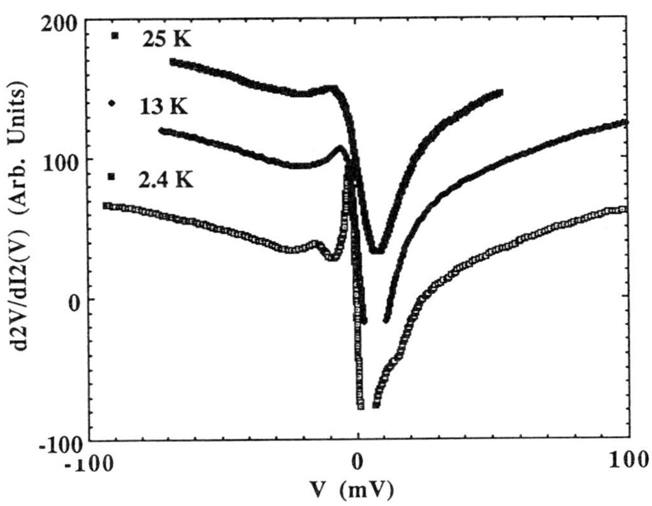

Figure 4. Variation of $d^2V/dI^2(V)$ tunnelling characteristics with temperature.

We are indebted to Swiss PTT, Bern for financial support and gratefully acknowledge stimulating discussions with Prof. F-K Reinhart and H.J. Scheel and the invaluable technical assistance of J.Faulkner and C.Faure.

Cucolo A M, Valles J M, Dynes R C, Gurvitch M, Phillips J M and Garno J P, 1989 *Physica C* **161** 351

Furuyama M, Iguchi I, Shirai K, Kusumori T, Ohtake H, Tomura S and Nasu M, 1990 *Jap. J. Appl. Phys.* **29**(3) L459

Geerk J, G.Linker, Meyer O, Li Q, Wang R.L and Xi X.X, 1989 *Physica C* **162** 837

Gurvitch M, Valles J.M, Cucolo A.M, Dynes R.C, Garno J.P, Schneemeyer L.F and Waszczak J.V, 1989a *Phys. Rev. Lett.* **63** 1008

Gurvitch M, Valles J.M, Dynes R.C, Cucolo A.,Schneemeyer L, 1989b *Physica C***162** 1067

Iguchi I, Furuyama M, Kusumori T, Shirai K, Tomura S, Nasu M and Ohtake H, 1990 *Jap. J. Appl. Phys.* **29**(4) L614

James J H, Kellett B J, Dwir B, Gauzzi A and Pavuna D, 1989 *Appl. Surf. Sci.* **43** 393

Kellett B J, James J H, Gauzzi A, Dwir B and Pavuna D, 1990 *Appl. Phys. Lett.* (10 Sept) in print

Lanping W.K, Jian H and Guowen W, 1989 *Phys. Rev. B* **40**(16) 10954

Phillips J P, Valles J M, Cucolo A M, Dynes R C, Gurvitch M, Garno J P and Marshall J H, 1989 *Proc. MRS Fall Meeting*

Supercond. Sci. Technol. 4 (1991) S139–141. Printed in the UK

Ordering of interstitial oxygen atoms in $La_2NiO_{4+\delta}$

Zenji Hiroi, Mikio Takano, and Yoshichika Bando

Institute for Chemical Research, Kyoto University, Uji, Kyoto 611, Japan

Abstract. Two types of homologous series of superstructures have been found for $La_2NiO_{4+\delta}$ containing excess oxygen, i.e. $\delta>0$. They are $2a\times ka\times lc$ (k,l = n or $2n$ depending upon n) and $\sqrt{2}a\times p\sqrt{2}a\times qc$ (most frequently p=2 and q=1), where $a\times a\times c$ stands for the K_2NiF_4-type cell. The origin of these superstructures has been assigned to the ordering of excess oxygen atoms located interstitially between two adjacent LaO layers. The most probable ordering schemes are proposed. Phases with discrete oxygen contents of δ=1/2n seem to exist.

Oxygen nonstoichiometry and its influences on structural and physical properties of $La_2MO_{4+\delta}$ (M=Cu, Ni, Co) have recently been studied extensively. Interest in these compounds has been renewed by a finding that excess oxygen atoms inject holes into CuO_2 sheets of the M=Cu compound and render it superconducting (Jorgensen et al 1988). According to powder and single crystal neutron diffraction studies (Jorgensen et al 1989 and Chaillout et al 1989), the excess oxygen atoms are located between two adjacent LaO layers, at interstitial sites tetrahedrally coordinated by La atoms, and this leads to a serious distortion of the lattice around them. Moreover, it has been suggested that a miscibility gap exists in the temperature vs. oxygen-content (4+δ) phase diagram, which is $0<\delta<0.08$ for $La_2CuO_{4+\delta}$, $0.02<\delta<0.12$ for $La_2NiO_{4+\delta}$, and $0<\delta<0.15$ for $La_2CoO_{4+\delta}$ at room temperature (Dabrowski et al 1989).

In perovskites and related compounds showing oxygen nonstoichiometry, ordering of oxygen vacancies has long been an interesting topic of solid state chemistry. However, ordering of interstitial oxygen atoms has been relatively unknown. Very recently, the modulated structure found in superconducting oxides of the Bi-Sr-Ca-Cu-O system has been attributed to ordering of excess oxygen atoms in $BiO_{1+\delta}$ sheets (Hewat et al 1989). It is interesting to notice that there is a structural similarity between the Bi and the La systems: The BiO and LaO double layers of the NaCl-type are sandwiched by perovskite-like layers which are $SrO/Ca_{m-1}Cu_mO_{2m}/SrO$ in the former and CuO_2 in the latter. The lattice mismatch between these two different types of layers must be the origin of the oxygen incorporation. We have expected that the excess oxygen atoms in $La_2MO_{4+\delta}$ also become ordered to minimize the increase in elastic energy caused by the incorporation. However, the previous diffraction studies did not find such ordering. We have examined $La_2NiO_{4+\delta}$ with various oxygen contents by means of electron diffraction and transmission electron microscopy (Hiroi et al 1990).

A wide variety of superlattice reflections have been observed in ED patterns from various samples except δ=0, revealing that at least two homologous series of superstructures exist in nonstoichiometric $La_2NiO_{4+\delta}$. One of them is characterized by one-dimensional arrays of superlattice spots along the [011]* direction with an interval of one-nth of the [011]*

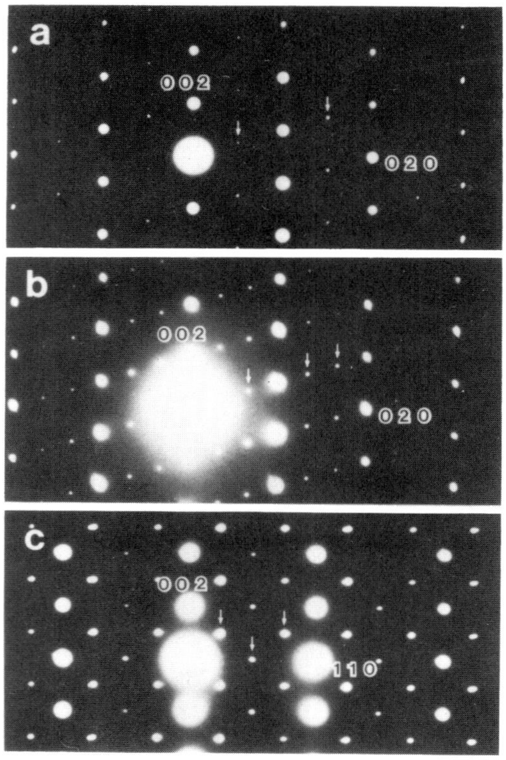

Fig. 1 [100] zone axis ED patterns of the $2a\times ka\times lc$ type superstructures with $k/2=l/2=n=2$ (a) and $k=l/2=n=3$ (b), and the $[1\bar{1}0]$ ED pattern from the $\sqrt{2}a\times2\sqrt{2}a\times c$ structure (c).

length. The n=2 and n=3 patterns are shown in Fig.1. Corresponding lattice images show bright (011) lattice fringes in every second and third places, respectively. These superstructures with small n values are formed over relatively wide ranges of 10^4 nm^2, while longer periodicities up to n=12 appear very locally. Examination of several zone axes ED patterns has revealed that this type of superstructure is described as $2a\times ka\times lc$ ($k,l=n$ or $2n$ depending upon n) on the basis of the ideal K_2NiF_4-type unit cell; for example, $2a\times 4a\times 4c$ for n=2 and $2a\times 3a\times 6c$ for n=3. Another type of superstructure is $\sqrt{2}a\times p\sqrt{2}a\times qc$ as typically seen in the $[1\bar{1}0]$ zone axis ED pattern of Fig.1(c) for $(p,q)=(2,1)$. Incommensurate periodicities of $2<p<3$ were frequently observed.

The origin of these superstructures is naturally attributed to ordering of the excess oxygen atoms: They were found only in samples with $\delta>0$, and moreover, high resolution lattice images, not shown here, have strongly suggested that the periodically arranged structural singular points which define their superlattice unit cell always lie within the LaO double layer containing excess oxygen. The most probable structural models for the $2a\times ka\times lc$ structures with $k/2=l/2=n=2$ and $k=l/2=n=3$ that consistently explain the observed reciprocal lattices and the high-resolution images are depicted in Fig.2. The interstitial oxygen atoms coordinated tetrahedrally by La atoms are aligned along the $[1\bar{1}1]$ direction within a (011) plane. The rows are separated from each other by $2a$ along the a axis. These special (011) planes are separated from each other by na along the b axis as shown in Figs.2(b) and (c). In Fig.2(b) these planes are further shifted from each other by a along the a axis. The resulting unit cell size is $2a\times 4a\times 4c$ in Fig.2(b) and $2a\times 3a\times 6c$ in Fig.2(c) as experimentally found. These models give excess oxygen contents of $1/(2n)$ for arbitrary n; for example, $\delta=0.25$ for n=2 and $\delta=0.17$ for n=3. There can exist a series of interstitial-oxygen ordered phases with discrete oxygen contents of $La_2NiO_{4+(1/(2n))}$ (n=2, 3, 4, ...).

Concerning the $\sqrt{2}a\times p\sqrt{2}a\times qc$ type superstructure, the excess oxygen atoms are separated from each other by $\sqrt{2}a$ along the $[1\bar{1}0]$ direction and by $p\sqrt{2}a$ along the [110] direction within a basal plane as shown in Fig.3. For p=2 the $[1\bar{1}0]$ excess oxygen rows are shifted by $\sqrt{2}a$ along the [110] direction in adjacent basal planes to form a body-centered-orthorhombic lattice sized to $\sqrt{2}a\times2\sqrt{2}a\times c$ as shown in Fig.3. For $p>2$ a monoclinic superlattice unit cell seems to be formed. Excess oxygen content is again generally expressed as $\delta=1/(2p)$. It is noted that the alignment of excess oxygen atoms along the $[1\bar{1}0]$ direction is similar to that in the Bi compounds.

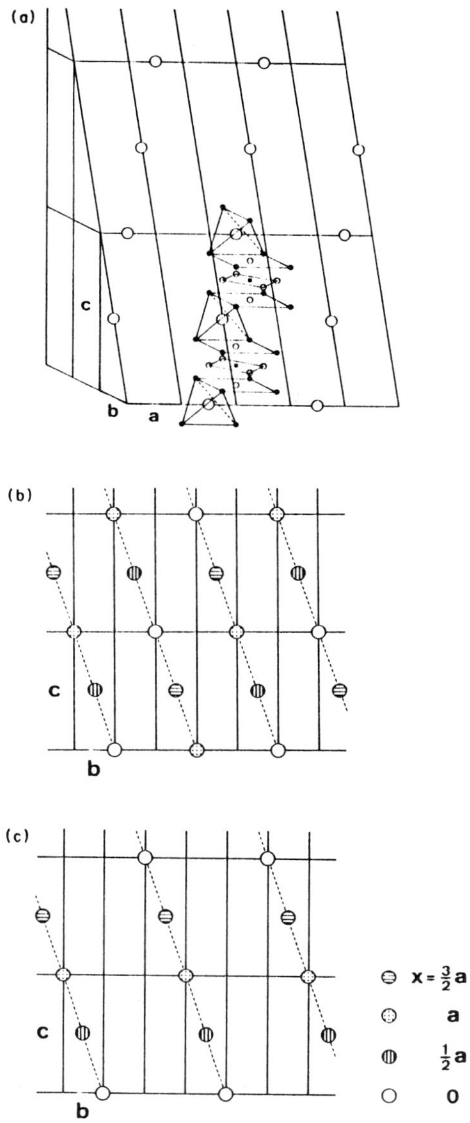

Fig. 2 Most probable structure models for the 2a×ka×lc type superstructures. Large and small open circles in (a) represent interstitial and normal oxygen atoms, respectively.

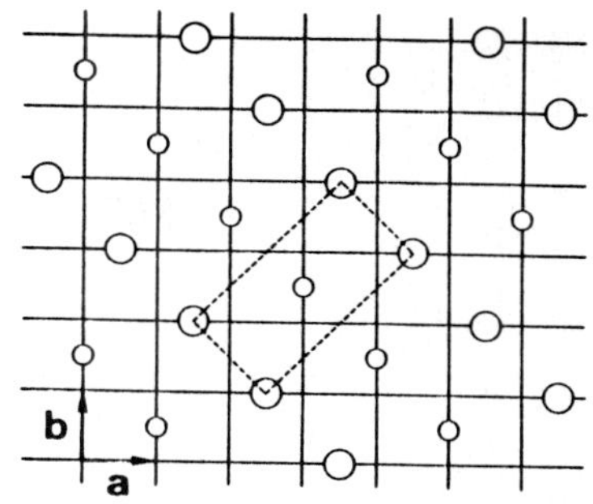

Fig.3 Schematic illustration of the most probable basal plane structure of the $\sqrt{2}a\times2\sqrt{2}a\times c$ superlattice. Large and small circles represent interstitial oxygen atoms at z=0 and z=c/2, respectively.

We encountered experimental difficulties in confirming the relation between these superstructures and excess oxygen content. Various superstructures appeared or disappeared as they were illuminated by the electron beam even in the normal observation mode. It also happened that superlattice spots, once disappeared on heating for a few seconds by the electron beam intensified by removing the condenser aperture, reappeared just as before in a few tens of seconds as continuously monitored using the aperture again. We have interpreted these results as indicating that excess oxygen atoms are rather mobile and become localized in an ordered way only at low temperatures, not much higher than room temperature. This point should be considered in preparing single ordered phases.

References

Chaillout C, Cheong S W, Fisk Z, Lehmann M S, Marezio M, Morosin B and Schirber J E 1989 *Phys. Scr.* T **29** 97

Dabrowski B, Jorgensen J D, Hinks D G, Pei S, Richards D R, Vanfleet H B and Decker D L 1989 *Physica* C **162-164** 99

Hewat E A, Capponi J J and Marezio M 1898 *Physica* C **157** 502

Hiroi Z, Obata T, Takano M, Bando Y, Takeda Y and Yamamoto O 1990 *Phys. Rev.* B **41** 11665

Jorgensen J D, Dabrowski B, Shiyou Pei, Hinks D G and Soderholm L 1988 *Phys. Rev.* B **38** 11337

Jorgensen J D, Dabrowski B, Shiyou Pei, Richards D R and Hinks D G 1989 *Phys. Rev.* B **40** 2187

Supercond. Sci. Technol. **4** (1991) S142–144. Printed in the UK

LATTICE DYNAMICS AND FAR INFRARED SPECTRA OF $LnBa_2Cu_3O_{7-y}$

S. Onari, A. Ono, H. Yamamoto, T. Arai and +T. Mori.

Institute of Applied Physics, University of Tsukuba, Tsukuba 305, Japan. +Department of Electrical Engineering, Ibaraki National College of Technology, Katsuta 312. Japan.

Abstract. Far infrared spectra of $LnBa_2Cu_3O_{7-y}$ (Ln = Y, Ho, Sm, Gd, Eu) are studied by changing the oxygen concentration. Lattice dynamical calculations by shell model are performed. For y=0.5, a large unit cell of $(LnBa_2Cu_3O_{7-y})_2$ doubled in the b axis direction is considered. The space group for y=0.5 belongs to the same space group for y=0.0 and the zone edge optical phonon modes are added by folding back to the zone center. Results of the lattice dynamics reproduce well the observed infrared and Raman scattering.

1. Introduction

Studies of the lattice vibrations and the free carriers are important for understanding the physical nature of the high Tc superconductors. In this paper infrared reflection spectra of $LnBa_2Cu_3O_{7-y}$ for various oxygen concentration are reported. Calculation of the lattice dynamics was performed by the use of the shell model and the calculated results are shown to be in good agreement with the experimental data.

2. Experimental

Properly weighted mixed powder of Ln_2O_3, $BaCO_3$ and CuO is sintered at 900℃ for 24 h in the air. After it is pulverized, pellets of 10 mm diameter with a thickness of ca. 1 mm are made by a 4 ton press. As an example, the oxygen concentrations x for $EuBa_2Cu_3O_x$ were 6.25, 6.44, 6.54, 6.66 and 6.88 for the sample pellets obtained by quenching from the temperatures of 930, 800, 700, 600 and 500° C, respectively. These oxygen concentrations x were determined by the method of the conventional iodometry.

Infrared reflection spectra from the surfaces which were polished with carborandom #1200 and finished with 1 μm alumina were measured by Digilab Fourier transform spectrometer FTS 16B.

Lattice dynamical calculation was performed with minicomputer.

3. Experimental Results

In Fig. 1 and 2 , the infrared reflection spectra of $SmBa_2Cu_3O_{7-y}$ and $GdBa_2Cu_3O_{7-y}$ quenched from various temperatures are shown. The tetragonal polycrystalline samples have the structures in the reflection spectra near 640, 590, 530, 355 and 250 cm^{-1}, and the orthorhombic samples have rather smaller structures in the reflection spectra near 617, 570, 317 and 285 cm^{-1}.

The difference between tetragonal and orthorhombic $LnBa_2Cu_3O_{7-y}$ is strongly correlated with the oxygen concentration 7-y. The occupation probability of oxygenn on (0,1/2,0) site is reported for $YBa_2Cu_3O_{7-y}$

to be ca. 80% at 400° C and ca. 65% at 600° C, and that on (1/2,0,0) site is almost zero below 600° C. The oxygen occupation probabilities of both (1/2,0,0) and (0,1/2,0) sites are ca.13% at the temperature above 900° C (Jorgensen et al. 1987). The oxygen occupation probability in $LnBa_2Cu_3O_{7-y}$ may be similar to that of $YBa_2Cu_3O_{7-y}$.

The infrared active vibration bands near 590, 355 and 250 cm^{-1}, which are observed in the tetragonal structure disappear in the orthorhombic structure with the increase of the oxygen concentration, and new vibration bands appear near 317 and 285 cm^{-1} in the orthorhombic structure.

As the orthorhombic-tetragonal structure change is related largely with the oxygen concentration , these experimental facts suggest that the infrared active bands at 590, 355 and also at 250 cm^{-1} are strongly influenced by the oxygens around the Cu atom which is surrounded by four Ba atoms.

4. Lattice Dymanical Calculation

Calculation of the lattice dynamical vibration frequencies (Thomsen 1988, Kress et al. 1988, Liu et al. 1988) of $YBa_2Cu_3O_7$, and $YBa_2Cu_3O_6$ are performed with the use of the shell model. In the shell model calculation the force constants between two atoms are defined as shown in Table 1. (Onari et al. 1989). The vibration frequencies of each modes for $YBa_2Cu_3O_6$ and $GdBa_2Cu_3O_6$ obtained by the calculation are shown in Table 2. The mass of Gd atom is about 1.7 times larger than that of Y atom, however, the reflection spectra of the tetragonal and also orthorhombic $YBa_2Cu_3O_{7-y}$ and $GdBa_2Cu_3O_{7-y}$ are quite similar in shape and wavenumber of the vibration modes to each other . These quite interesting experimental results suggest that the vibration of Y and Gd atoms may play not much role in the infrared spectra in the wave number range of 210-700 cm^{-1}. These results are consistent with the shell model calculation.

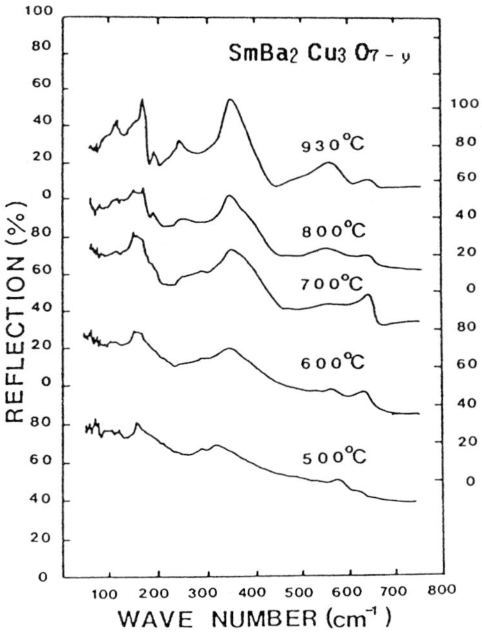

Fig. 1
Infrared reflection spectra of $SmBa_2Cu_3O_{7-y}$.

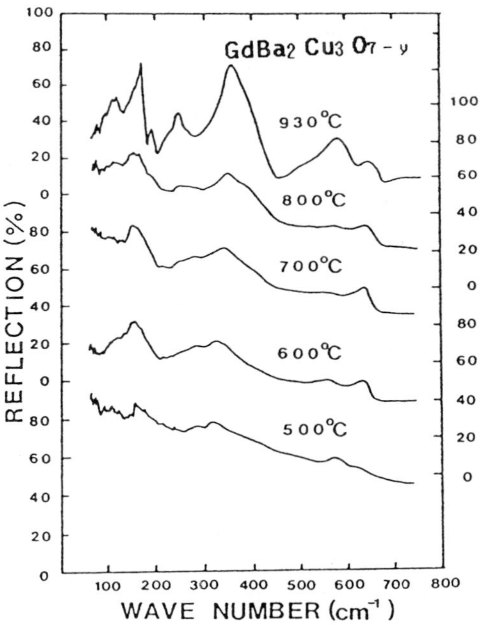

Fig. 2
Infrared reflection spectra of $GdBa_2Cu_3O_{7-y}$.

Table 1

Parameters of Shell mopdel for $Y(Gd)Ba_2Cu_3O_7$

1) Born-Mayer Potential

$V=a_{ij}\exp(-b_{ij}r)$

	a_{ij} (eV)	b_{ij} (A^{-1})
Y-O	3010	2.90
Ba-O	3225	2.90
Cu(1)-O	1260	3.45
Cu(2)-O	1260	3.29
O-O	1000	3.00

2) Charge and force constants

Ion	Z(e)	Y(e)	k(e^2/v)
Y	2.85	-1.42	872
Ba	1.90	2.32	165
Cu(1)	2.00	3.22	1000
Cu(2)	2.00	3.22	1000
O(II III)	-1.81	-2.70	252
O(I IV)	-1.81	-2.70	252(k//) 1675(k⊥)

Table 2

Vibration frequencies by shell model calculation. Raman active A_{1g}, B_{1g} and E_g modes have the same common vibration frequencies under the present assumption.

	$YBa_2Cu_3O_6$		$GdBa_2Cu_3O_6$		$Y(Gd)Ba_2Cu_3O_6$
	$65cm^{-1}$		$64cm^{-1}$		$106cm^{-1}$
	137		133	A_{1g}	175
A_{2u}	164	A_{2u}	164		417
	304		254		467
	492		484		
				B_{1g}	369
B_{2u}	228	B_{2u}	228		
					82
	108		98		164
	137		137	E_g	405
E_u	227	E_u	200		422
	323		322		521
	399		399		
	521		521		

For $y=0.5$, a large unit cell of $(LnBa_2Cu_3O_{7-y})_2$ doubled in the b axis direction is considered. The space group of this unit cell for $y=0.5$ belongs to the same space group for $y=0.0$ and the optical phonon modes at Γ point are expressed as follows.

$$11A_g + 3B_{1g} + 8B_{2g} + 11B_{3g} + 2A_u + 3B_{1u} + 13B_{2u} + 11B_{3u}$$

These modes correspond to the sum of the phonons at the zone center and the phonons of the zone edge of $LnBa_2Cu_3O_7$ added by folding back to the zone center. The results of the shell model calculation have one to one correspondence between the phonon frequencies at Γ point and zone edge in the b axis direction. Raman scattering data are explained well by this calculation.

From these calculations, we can notice that the high frequency infrared active modes are rather insensitive to the oxygen concentration.

This work is partially supported by Murata Science Foundation.

REFERENCES

D.J.Jorgensen, M.A.Beno, D.G.Hinks, L.Soderholm, J.K.Volin, R.L.Hitterman, J.D.Grace, I.K.Schuller, C.U.Serge, K.Zang & M.S.Kleefish, Phys. Rev. B36 (1987) 3608

W.Kress, U.Schroder, J.Prade, A.D.Kulkarni and F.W.de Wette, Phys. Rev. B38 (1988) 2906

R.Liu, C.Thomsen, W.Kress, M.Cardona, B.Gegenheimer, F.W.deWette, J.Prade, D.Kulkarni and U.Schroder, Phys. Rev. B37 (1988) 7971

S.Onari, S.Hidaka, T.Arai and T.Mori, Solid State Commun.71(1989)195

C.Thomsen, M.Cardona, W.Kress, R.Liu, L.Genzel, M.Bauer, E.Schonherr & U.Schroder, Solid State Commun. 65 (1988) 1139

Supercond. Sci. Technol. **4** (1991) S145–147. Printed in the UK

Thick film YBCO microstrip resonators

R Mossavati, J C Gallop, W J Radcliffe, T Button+ and N McN Alford+

National Physical Laboratory, Teddington TW11 0LW, UK & +ICI Advanced Materials, Runcorn WA, UK

Abstract We present results on a microstrip resonator made from thick film YBCO. A number of resonant modes of this structure have been measured in the frequency range 4-12GHz and results for the centre frequency and Q of these modes are presented.

1. Introduction

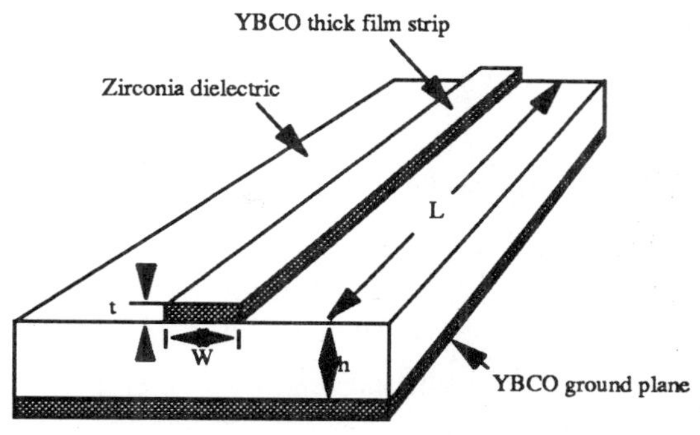

Fig.1. Schematic of HTS microstrip resonator

A transmission line acts as a resonator at those frequencies for which it is an integer number of half wavelengths long. The microstrip geometry consists of a thin dielectric slab of thickness h and complex permittivity $\varepsilon = \varepsilon' + j\varepsilon''$. On top of it is a strip of conductor of width W and complex conductivity $\sigma = \sigma_1 + j\sigma_2$, and underneath a large ground plane, effectively infinite in size which, for simplicity, we assume is made of the same material as the strip (see fig.1 for schematic view of resonator). For a line of length L the resonant frequencies ω_n are found from the condition that

$$2L/\lambda=n \text{ (n an integer)}$$

where $\lambda=2\pi v/\omega=2\pi/(\mu\varepsilon_{eff})^{1/2}$, μ being the permeability of the dielectric medium which we assume to be equal to μ_o, the permeability of free space. ε_{eff} is the effective dielectric constant for the resonator which is not exactly equal to that of the dielectric due to the presence of fringing field in the $\varepsilon = 1$ region above the dielectric. Thus

$$\omega_n = n\pi/(L^2\mu_o\varepsilon_{eff})^{1/2}$$

The complex surface impedance of a strip of width W made from a superconductor with penetration depth λ and complex conductivity σ at frequency ω is

$$Z_s = (i\omega\mu_o/\sigma)^{1/2}$$

which is valid except very close to T_c when the film thickness is comparable with or less than the penetration depth of the electromagnetic fields into the thick film. The real and imaginary parts of this impedance R_s and X_s are given by

$$R_s = [(\omega\mu_o/2\sigma_r^2)(\sigma_r - \sigma_2)]^{1/2} \;\&\; X_s = [(\omega\mu_o/2\sigma_r^2)(\sigma_r + \sigma_2)]^{1/2}$$

where $\sigma_r^2 = \sigma_1^2 + \sigma_2^2$ and the contribution to the total Q of a stripline resonator arising from losses in the superconductor is

$$Q_s = \Gamma_n/R_s$$

where Γ_n is a geometry factor for the nth mode of the line and is defined by

$$\Gamma = \frac{2\omega\int[\mu_r(r)\mu_o H^2 + \varepsilon_r(r)\varepsilon_o E^2]dV}{\int H^2 dS}$$

Here the numerator represents the stored energy in the resonator, and the integral is taken over all space, including the vacuum region of the circuit, wherever the electromagnetic fields are not negligible. The integral in the denominator is taken over the entire surface of the conductors.

Since analytical expressions are not available for the field distributions the value of Γ will inevitably be approximate. The simplest approximation assumes that there are no fringing fields, and in this case it is straightforward to show that

$$\Gamma_n < \omega_n \mu_o h$$

2. Microstrip Measurements

A microstrip resonator consisting of a thick film YBCO strip (23 mm long and 1.3 mm wide) deposited on one side of a 0.9 mm thick zirconia substrate, the other side of which is covered with a thick film YBCO groundplane. The resonator was mounted in a coppper box clamped to the cold-head of a closed cycle refrigerator capable of remaining at any temperature between 8K and 300K. Coaxial cables connected to the copper box and coupling probes protrude into it to within about 2 mm of the strip. The microwave power transmitted through the box was measured as a function of frequency and temperature. A number of microstrip resonances were identified when $T < 93$ K which were not visible above that temperature. This is because the normal state resistance of YBCO is rather high, some hundreds of times worse than copper, so that the linewidths are too large and the amplitudes too low to be detected in the normal state.

A comparison with an identical resonator made from copper foil shows that the cross-over frequency, that is the frequency below which the HTS material performs better than copper at the same temperature (77 K) is higher than 12 GHz. Already this combination of substrate and thick film process looks promising for microwave applications. We present results on resonator Q and centre frequency of the modes as a function of temperature, interpret this data in terms of loss mechanisms in the HTS material and in the substrate dielectric. Suggestions for further improvement in performance are also put forward.

3. Discussion

The Q factor of any resonance observed may be defined as

$$Q = \omega_n/\Delta\omega$$

where $\Delta\omega$ is the full width at the half power points. Fig.2 shows the total Q (Q_t) as a function of temperature for a resonance of the above microstrip circuit at 12.0 GHz. Q_t is the sum of various contributions to the loss, from radiated power Q_r, loss in the dielectric Q_e as well as ohmic loss in the superconductor surfaces Q_s.

$$1/Q_t = 1/Q_r + 1/Q_e + 1/Q_s$$

However we may use Q_t to set an upper limit on the superconductor surface resistance, using the approximate analysis for Γ_n given above. Thus from the data we know that at 77K and 12.0 GHz $Q_s > Q_t = 720$, and

$$R_s < \Gamma_n/Q_t < \omega\mu_o h/Q_t < 118\mathrm{m}\Omega$$

This must be regarded as almost certainly a serious overestimate of the surface resistance of the YBCO thick film since the loss tangent for zirconia has been measured as 2×10^{-3} at 77K. If the loss tangent for the microstrip resonator described here were as great as this, dielectric losses alone would account for the value of Q_t observed. Q_r consists of two contributions, the first due to dissipation in the copper box which encloses the microstrip while the second arises from the coupling of the circuit to external components. An estimate of the latter contribution may be obtained from the insertion loss which in this case is >40dB at resonance, sufficient to make this mechanism quite negligible as a source of loss compared with other contributions.

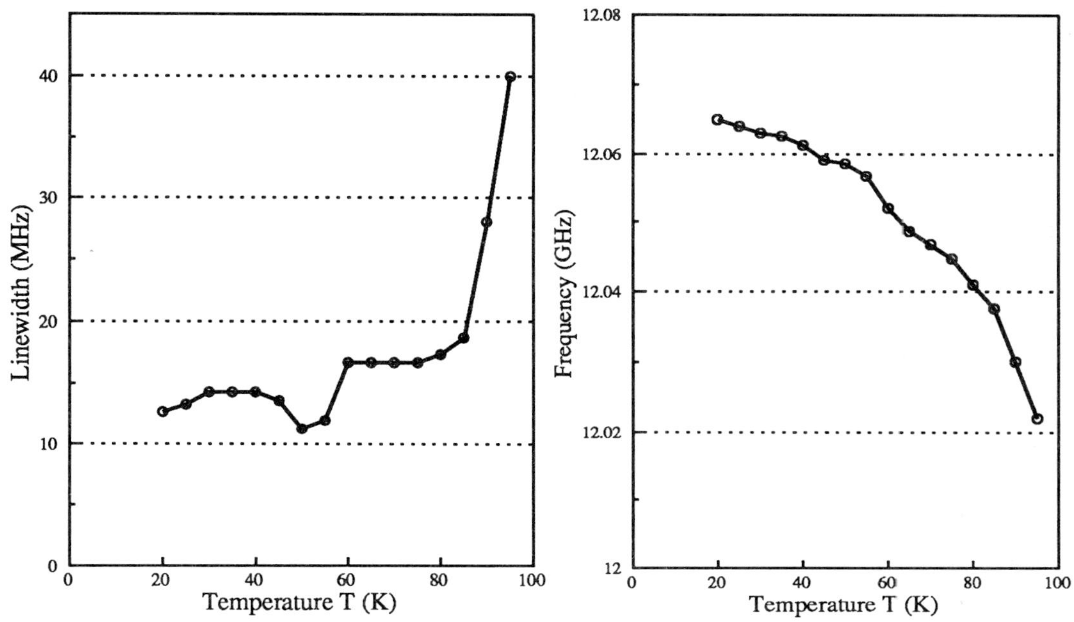

Fig. 2a) Linewidth variation and b) centre frequency variation with temperature for the YBCO thick film on zirconia resonator, as described above.

Conclusion

It may be concluded from these preliminary experiments that thick film YBCO resonant circuits have superior characteristics to identical copper circuits operated at 77 K up to at least 12 GHz, and therefore appear to have useful potential for microwave applications. Further, the simplicity of the strip-line resonator technique presents interesting possibilites for use in material characterisation. The method complements the microwave cavity perturbation technique[1] in that the HTS surface which is sampled is primarily the one in contact with the substrate.

References

1. W J Radcliffe, J C Gallop, C D Langham, M Gee & M Stewart, Physica C 153-5 635-6 (1988)

Supercond. Sci. Technol. **4** (1991) S148–150. Printed in the UK

Noise in $YBa_2Cu_3O_{7-\delta}$ Grain Boundary Junction dc SQUIDs

R. Gross
Lehrstuhl Experimentalphysik II, Universität Tübingen, D-7400 Tübingen, FRG

P. Chaudhari, M. Kawasaki, M. B. Ketchen, and A. Gupta
IBM Thomas J. Watson Research Center, P. O. Box 218, Yorktown Heights, NY 10598, USA

Abstract: We have fabricated low noise $YBa_2Cu_3O_{7-\delta}$ grain boundary junction dc-SQUIDs with a square washer geometry design. The SQUIDs show perfectly periodic voltage-flux characteristics from 4.2 to 87K. The noise characteristics of these SQUIDs have been measured as a function of temperature, bias current, and applied magnetic flux. Above about 82K the noise is dominated by flux noise presumably caused by thermally assisted flux motion in the film. Below this temperature the SQUID noise is dominated by voltage noise across the junctions.

The key features necessary for the fabrication of useful, reliable, and reproducible high-T_c dc-SQUIDs are a superconducting loop consisting of a high quality epitaxial film intersected by just two controllable Josephson elements, and a low noise input coil. The first key element can be obtained by using epitaxial $YBa_2Cu_3O_{7-\delta}$ films grown on a $SrTiO_3$ bicrystal.[1–3] Here grain boundary junctions (GBJs) can be used as weak links in an otherwise fully epitaxial SQUID loop structure. The method of fabricating $YBa_2Cu_3O_{7-\delta}$ GBJs has been described in a number of papers.[4] The GBJs show Josephson like behavior and can be fabricated in a controlled way.[2,3,5] For our SQUID experiments thin films of $YBa_2Cu_3O_{7-\delta}$ were deposited by laser ablation[6] on [001] tilt $SrTiO_3$ bicrystals. Using laser ablation patterning[7] a number of dc-SQUID devices were made with the two GBJs $5 - 10\mu m$ wide in a nominally $0.35\mu m$ thick film. The SQUIDs had a square washer geometry design [2,8] with values of the loop inductance L ranging between 60 and 265pH. Due to the relatively small spread of the critical current density J_c of the GBJs prepared under the same conditions, the critical current $2I_c$ of the SQUIDs could be selected within a quite narrow range by the patterning process. In order to make the coupling energy $I_c\Phi_0/2\pi$ much larger than the thermal energy k_BT at 77K, $2I_c \sim 50 - 200\mu A$ was chosen. The current-voltage characteristics (IVCs) of the GBJs with misorientation angles larger than about 15° are close to those calculated according to the resistively shunted junction (RSJ) model.[2,9] The I_cR_N products of the GBJs typically range between 50 and $300\mu V$ at 77K. Here, in accordance with the RSJ-model, R_N is denoted as the normal resistance of the GBJ. Details of the IVCs and the resistive transition of the GBJ dc-SQUIDs are given in Ref.2 and 3.

Fig.1 shows the flux-voltage characteristics of a 60pH GBJ dc-SQUID at 81K for different values of the bias current. The characteristics are perfectly periodic from 4.2 to 87K. The observed hysteresis parameter $h = \delta\Phi/\Delta\Phi_e$, i.e. the ratio of the shift $\delta\Phi$ of the flux-voltage curves caused by sweeping the applied flux with an amplitude $\Delta\Phi_e$ in opposite directions, was smaller than 1×10^{-4} at 77K for $\Delta\Phi_e \leq 100\Phi_0$. The maximum value of the flux responsivity $H = \partial V/\partial\Phi$ for the SQUID shown in Fig.1 is about $10\mu V/\Phi_0$ at 81K. By now, H ranged between 1 and $50\mu V/\Phi_0$ at 77K for the investigated GBJ dc-SQUIDs.

As shown in Ref.2, for a 60 and 110pH SQUID, an intrinsic energy resolution, measured above 1kHz in the small signal or "direct" mode, of 1.5×10^{-30} and $3.0 \times 10^{-30} J/\mathrm{Hz}$ was obtained. At 10Hz, the intrinsic energy resolution, limited by 1/f noise, was 1.2×10^{-28} and $5.5 \times 10^{-28} J/\mathrm{Hz}$, respectively.[2] For most investigated GBJ SQUIDs, above about 82-85K

0953-2048/91/01S148 + 03 $03.50

the 1/f noise has been found to be dominated by flux noise presumably caused by thermally assisted flux motion in the epitaxial film, in agreement with recent measurements by Ferrari et al.[10] At these temperatures the level of noise had a clear maximum where the transfer function is a maximum and a minimum where H=0.[3] Below about 82-85K, the measured voltage noise power was found to be, within the scatter of our data, independent of the applied flux, i.e. independent of H. At these temperatures, the voltage noise power spectrum $S_V(f)$ of the SQUID was found to be very similar to that of one of its junctions suggesting that the SQUID noise is dominated by voltage noise across the junctions. In Fig.2 $S_V(f)$ of a 140pH SQUID and one of its junctions is shown for T=77K. Here, the same reduced bias currents $I_b/2I_c$ and I_b/I_c were used for the SQUID and the individual junction. The noise characteristics of the single junctions were obtained by cutting the SQUID loop and measuring $S_V(f)$ of the individual junctions. Beyond $S_V(f)$ also $S_V(I_b)$ was found to be very similar for the GBJ SQUIDs and one of their junctions. This is shown in in Fig.3, where $S_V(100Hz)$ and $S_V(1000Hz)$ of a 60pH SQUID and one of its junctions is plotted versus the reduced bias current $I_b/2I_c$ and I_b/I_c, respectively.

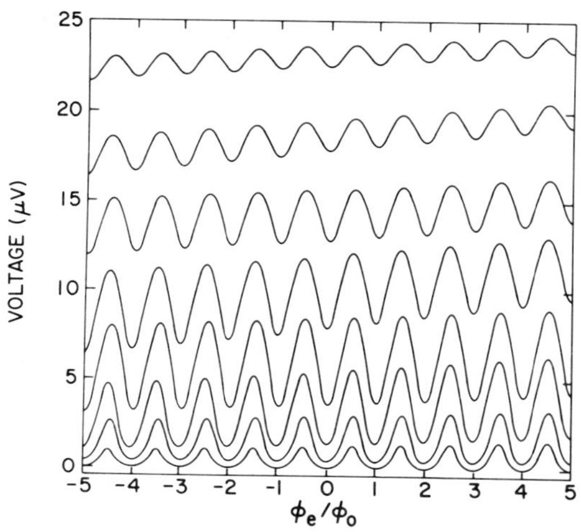

Fig.1: Voltage-flux characteristics of a 60pH GBJ dc-SQUID at 81K. The bias current is increased from 60 to 95μA in 5μA steps.

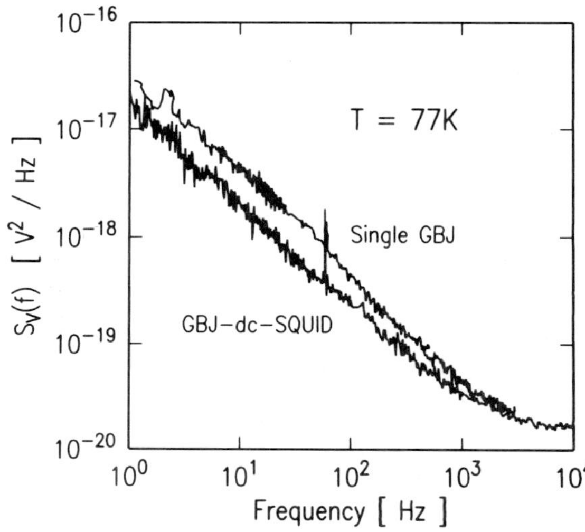

Fig.2: Voltage noise power spectrum $S_V(f)$ for a 140pH GBJ dc-SQUID and one of its junctions at T=77K.

The observed noise behavior suggests that the measured voltage noise at temperatures below about 82-85K is caused by the fluctuation of the GBJ parameters such as the fluctuation of the critical current. In the temperature regime where the junction noise is dominant, S_V is found to increase slightly with decreasing temperature. Since H is about constant if the dimensionless screening parameter $\beta = 2LI_c/\Phi_0 \gg 1$, the corresponding flux noise $S_\Phi = S_V/H^2$ shows about the same temperature dependence. Going to higher temperatures, S_Φ starts to increase at the temperature where either H decreases due to the decrease of the critical current or the flux noise due to flux motion in the film becomes the dominant noise source. This results in a minimum of S_Φ that could be observed for all GBJ SQUIDs at a temperature ranging between about 77 and 85K.[3] Note that for the SQUID shown in Fig.3, the flux noise power $S_\Phi = S_V/H^2$ that would correspond to the measured voltage noise power at 77K is about $2 \times 10^{-9}\Phi_0^2/\mathrm{Hz}$ at 100Hz. Since we do not see usual flux noise at this temperature, the flux noise power due to thermally assisted flux motion at 77K must

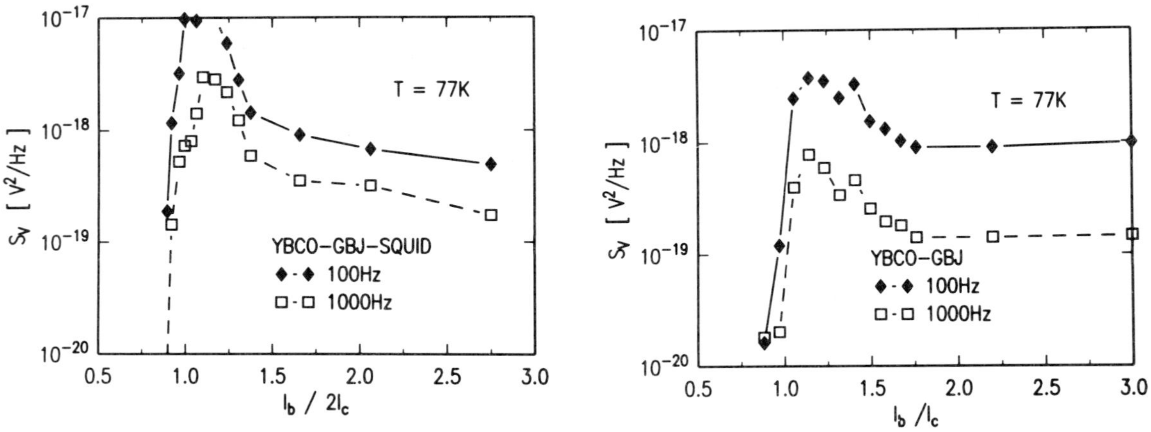

Fig.3: Voltage noise power S_V at 100Hz and 1kHz plotted versus the reduced bias current for a 60pH GBJ dc-SQUID (a) and for one of its junctions (b).

be smaller than this value. This is in agreement with the results of Ferrari et al.[10] These authors point out that the flux noise in epitaxial $YBa_2Cu_3O_{7-\delta}$ films at 77K is already close to the noise level in commercially available SQUIDs operating at 4.2K. To improve the low frequency noise properties of the GBJ SQUIDs at 77K the 1/f noise, now directly attributed to the junction noise, needs to be improved further.

References:

1. P. Chaudhari, J. Mannhart, D. Dimos, C.C. Tsuei, C.C. Chi, M.M. Oprysko, M. Scheuermann, Phys. Rev. Lett. 60, 1653 (1988)
2. R. Gross, P. Chaudhari, M. Kawasaki, M. B. Ketchen, A. Gupta, Appl. Phys. Lett., to be published
3. R. Gross, P. Chaudhari, M. Kawasaki, M. B. Ketchen, A. Gupta, Physica C, to be published
4. D. Dimos, P. Chaudhari, J. Mannhart, Phys. Rev. B41, 4038 (1990); D. Dimos, P. Chaudhari, J. Mannhart, F.K. LeGoues, Phys. Rev. Lett. 61, 219 (1988)
5. R. Gross, P. Chaudhari, D. Dimos, A. Gupta, G. Koren, Phys.Rev. Lett. 64, 228 (1990)
6. G. Koren, A. Gupta, A. Giess, E.A. Segmüller, T.R. McGuire, Appl. Phys.Lett. 54, 1055 (1989)
7. J. Mannhart, M. Scheuermann, C.C. Tsuei, M.M. Oprysko, C.C. Chi, C.P. Umbach, R. Koch, C. Miller, Appl. Phys. Lett. 52, 1271 (1988)
8. M.B. Ketchen, J.M. Jaycox, Appl. Phys. Lett. 40, 736 (1982)
9. W.C. Stewart, Appl. Phys. Lett. 12, 277 (1968); D.E. McCumber, J. Appl. Phys. 39, 3113 (1968)
10. M.J. Ferrari et al., Nature 341, 723, (1989)

Supercond. Sci. Technol. **4** (1991) S151–153. Printed in the UK

Magnetic Properties of $YBa_2Cu_3O_{7-x}$ Pellets Made at 100 kbar

S Gygax*, G Soerensen*, J P Franck†, J F Zhou†, M Kanzaki†, S C Cheng†, J Jung†

*Simon Fraser University, Burnaby, B.C. V5A 1S6, Canada
†University of Alberta, Edmonton, Alberta T6G 2J1, Canada

Abstract. $YBa_2Cu_3O_{7-x}$ powder was heat treated in hermetically sealed gold capsules at 100 kbar and 945°C. The resulting product is superconducting without further oxygen treatment. Magnetically determined transition temperatures are near 92.6 K. The product consists of fine particles with diameters in the range 50 to 1,000 nm and much reduced density of twins. At low fields the magnetization is reversible from T_c to 10 K. Results for ^{18}O-substituted samples treated by this process show an isotope shift of the transition temperature of about 0.6 K.

We describe the magnetic properties of specially prepared YBCO samples. Fully oxygenated $YBa_2Cu_3O_7$ was prepared by standard processes. The resulting pellets were ground and the powder sealed in a gold capsule of diameter 3mm and length 3mm and thickness 0.1mm. The capsule was backed up by an octahedral assembly, containing a local heater, and pressurized at 100 kbar. The capsule was then, under pressure, heated to 945°C and maintained at that temperature for 5 hours. The samples were then slowly cooled under pressure to room temperature. Occasionally we had to quench from 945°C to room temperature (due to apparatus failure). The resulting samples showed no differences from the slowly cooled ones. The resulting pellets are quite brittle; they exhibit superconductivity with T_c near 92.6 K without further oxygen treatment. The high transition temperature, as well as the result on ^{18}O substituted samples, show that oxygen was not lost during pressure treatment.

The magnetic properties of these samples were measured in two SQUID magnetometers: a commercial one for high fields (Quantum Design), and a specially constructed low field instrument. They overlap in the 1 to 10 mT range.

A typical set of low field transitions (1 mT) is shown in the bottom part of Fig. 1. It is immediately apparent that the shape of the zero-field-cooled (ZFC) curve does not show the sharp transition of a typical pressed ceramic pellet. Instead, it resembles the magnetic transitions one finds for fine powders. There is, in fact, very little difference in the magnetic behaviour between a pressure-treated pellet and a fine powder obtained by grinding this pellet under acetone. A typical transition for such a powder is shown in the upper part of Fig. 1. Also shown is the predicted transition for a collection of small spherical particles whose radius is comparable with the magnetic penetration depth λ. The shape arises from the temperature dependence of λ assumed to be BCS-like. The powder consists of a range of different particle sizes, and the fit between the two curves could be much improved by taking into account such a particle distribution. However, the main point is the close similarity in magnetic response between pressed pellet and powder. It strongly suggests that our pellets obtained under high pressure have an extreme granular morphology. The extremely small particle size in these pellets was also directly observed by transmission electron microscopy, grains from 500 Å diameter up were found. Powder x-ray spectroscopy shows only the usual $YBa_2Cu_3O_7$ structure, but with extreme broadening due to small grain size. The samples probably are broken up into small grains by evolving oxygen during the heating process. The small grain size in the samples is also responsible for the fast oxygen take-up on cooling.

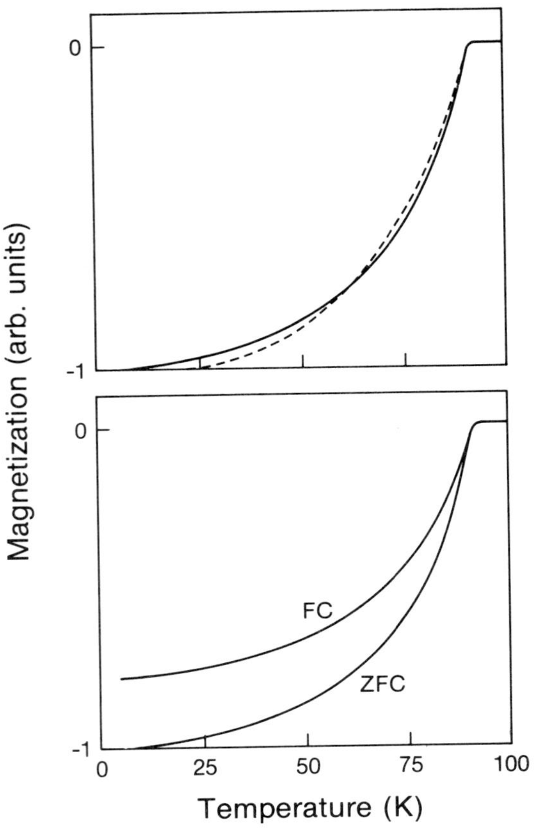

Fig. 1. Bottom: ZFC and FC magnetization of a pressure-treated pellet in a field of 0.1 mT. Top: Solid line: magnetization of the powdered pellet, H = 0.01 mT. Broken line: transition for spherical particles with $\lambda/r = 3.5$, λ = penetration depth, r = radius.

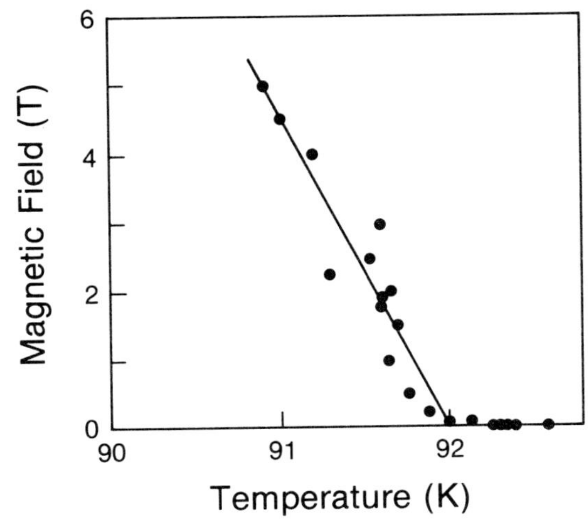

Fig. 2. Upper critical field for a pressure-treated pellet.

Magnetic determination of the critical temperatures in zero field gives $T_c = 92.6$ K. The shift of the critical temperature T with field is shown in Fig. 2. The slope of this curve, -4.53 T/K, falls between the values found for a de-twinned single crystal for the applied field directions perpendicular and parallel to the crystalline c-axis, Welp *et al.* (1989). Our data show the familiar negative curvature in low fields near T_c which seems to be absent in de-twinned single crystals.

Fig. 3 shows the high temperature part of the magnetization in a field of 0.1 mT for zero-field and field-cooled conditions. The transition is nearly reversible in a field of 0.1 mT and will become entirely reversible as the field is lowered further. We have studied the field dependence of T_r, the reversibility temperature, which separates the reversible and irreversible regions of the phase diagram. It is defined as the temperature above which zero-field and field-cooled magnetization coincide. The results for a pressurized pellet are presented as a semi-log plot in Fig. 4 which clearly shows the rapid transition in low fields to complete reversibility.

Considering the granular nature of our samples, complete reversibility in low fields is expected. Individual grains will show no irreversibility until at least one flux line can be nucleated and trapped. This field is of the order of $\phi_o/(\pi\lambda^2)$ Assuming that the grain size is comparable to λ, the observed transition field of 0.1 mT for complete reversibility implies a 1μm grain size. This is indeed the upper range of the particle size observed in our pressed pellets using an SEM. It is the larger particles which can trap flux first.

We have also investigated the reversibility line for a powder obtained by grinding a pressed pellet as well as for the reacted starting powder for our pellets. The results are also shown in Fig. 4. Not only does the powder have a similar shape for the transition curve as the pressed pellet, it also makes a sharp transition to complete reversibility, but at a somewhat lower field. This confirms our idea that the morphology is responsible for the unusual reversibility of the pressed pellets. In contrast, the starting powder has always some trapped field.

At high fields the reversibility line becomes increasingly difficult to determine since the ZFC curve approaches the FC one tangentially. The reversibility line seems to be somewhat sample dependent. For fields up to 0.2 T, we are able to fit the irreversibility line to a $H^{2/3}$ dependence, but above this field, deviations occur which are outside experimental error and also sample dependent.

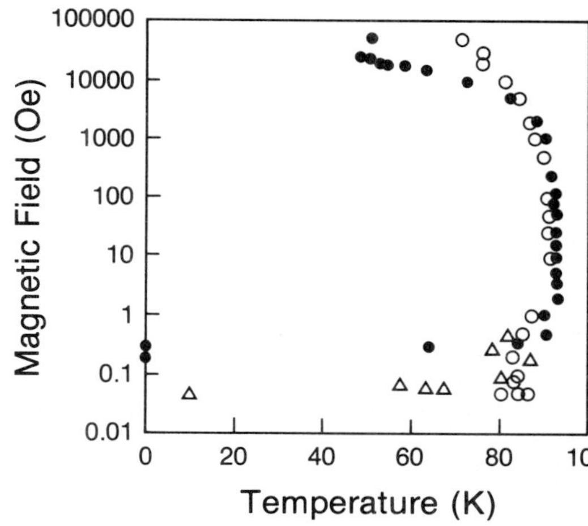

Fig. 4. Reversibility temperature for a pressure-treated pellet (filled dots), the powdered pellet (open triangles), and the starting material as powder (open circles).\

We finally report on a pressure-treated sample which was prepared from YBCO powder with a concentration of 85 at -% ^{18}O. This material had previously been used in an isotope effect study and showed an isotopic shift in T_c of 0.25 K, Franck *et al.* (1989). The results are shown in Fig. 5. The pellets show a lowered transition temperature by about 0.6 K. Back-exchanging this sample at normal pressure and a temperature of 900°C with ^{16}O raised T_c to values observed in ^{16}O pellets. This shows that a small, but finite isotope effect exists in these compounds.

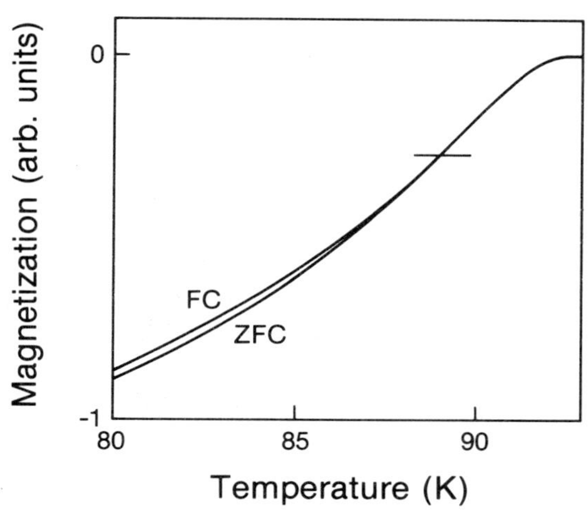

Fig. 3. Magnetization versus temperature for a pressure-treated pellet in a field of 0.1 mT.

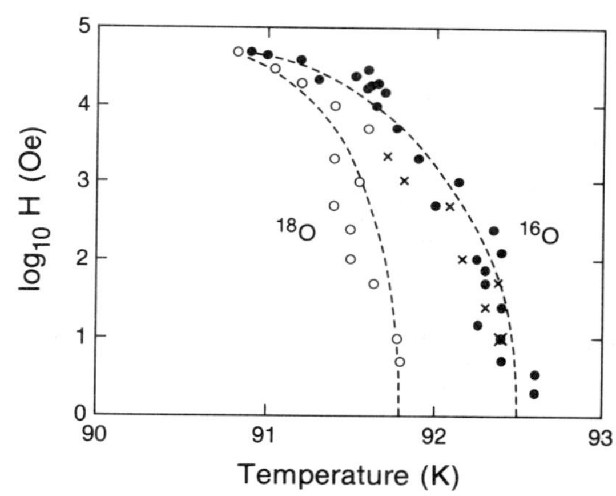

Fig. 5. Upper critical field for pressure-treated pellets. Black dots: made with ^{16}O. Open circles: made with ^{18}O (85%). Crosses: after back-exchange with ^{16}O.

References:

Franck J P, Jung J, Salomons G J, Miner W A, Mohamed M A-K, Chrzanowski J, Gygax S, Irwin J C, Mitchell D F and Sproule G I 1989 *Physica* C **162-163** 753

Welp U, Grimsditch M, You H, Kwok W K, Fang M M, Crabtree G W and Liu J Z *Physica* C **161**, 1

Supercond. Sci. Technol. 4 (1991) S154–156. Printed in the UK

Bi-based superconducting fibers prepared from amorphous materials

He Yusheng, Zhang Jincang*, He Aisheng#, Wang Jinsong and Huo Yujing

Physics Department, Tsinghua University, Beijing, China

Abstract. Bi-based superconducting fibers have been prepared from amorphous materials by the LHPG method. The onset transition temperature reached is 102 K, and the zero resistance temperature reached is 86 K. The critical current density J_c (77 K, 0 T) is greater than 5000 A/cm^2. Structure analysis indicates that the superconducting fibers have highly aligned morphology with a-axis parallel to the growth direction.

1. Introduction

Since the discovery of high temperature oxide superconductors by Bednorz and Muller (1986), many techniques for preparing superconducting materials have been explored, among which the vitrification and crystallization process (Tohge et al 1989), melt and textured growth method (Jin et al 1987) and quench and melt growth process (Murakami et al 1989) are quite noticeable. The common ground of these methods is to prepare high density materials through melting and achieve their superconductivity by subsequent thermal treatment. Similarly, by means of the laser-heated pedestal growth (LHPG) method (Feigelson et al 1988), Bi-based superconducting fibers have been fabricated from amorphous materials for the first time. The onset transition temperature reached is 102 K, and the zero resistance temperature reached is 86 K. The critical current density J_c (77 K, 0 T) is greater than 5000 A/cm^2.

2. Experimental

The starting raw materials were Bi_2O_3, $SrCO_3$, $CaCO_3$, CuO and PbO powders with analytic purity. After careful weighing , mixing and grinding, the fine powders were calcined in air at 820°C for 24 h. The calcined materials were ground again, rapidly heated to 1150-1250°C, and then quenched to room temperature at the cooling rate of more than 300°C/sec. The density of the amorphous materials prepared ranges from 6.0 to 6.2 g/cm^3, which is much greater than that of the sintered polycrystalline materials (4.8-5.5 g/cm^3) and close to the theoretical density of Bi-based superconducting materials. This is obviously beneficial for fabricating high quality superconducting fibers by the LHPG method.
The growth condition of the fibers is as follows. Firstly, the source rod was pulled at the rate of about 2 mm/min to become round, smooth and dense. Then the fibers were grown at the rate of 0.2-1 mm/min. The diameters of the resulting fibers were 100-340 μm.

---

* Visiting scholar from Henan Normal University, Xinxiang, China
# Visiting scholar from North University of Technology, Beijing, China

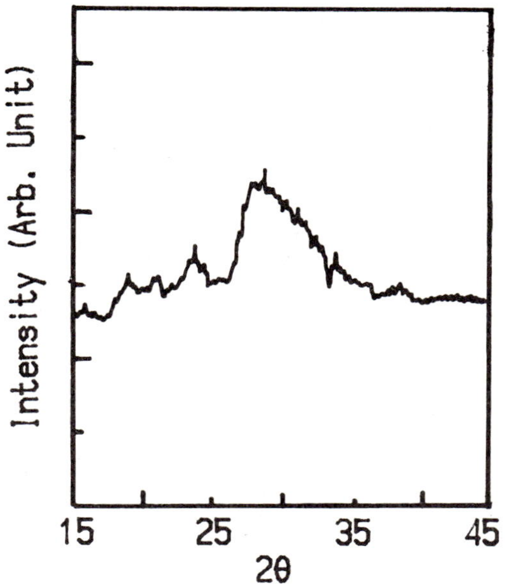

Figure 1 XRD spectrum of the source material

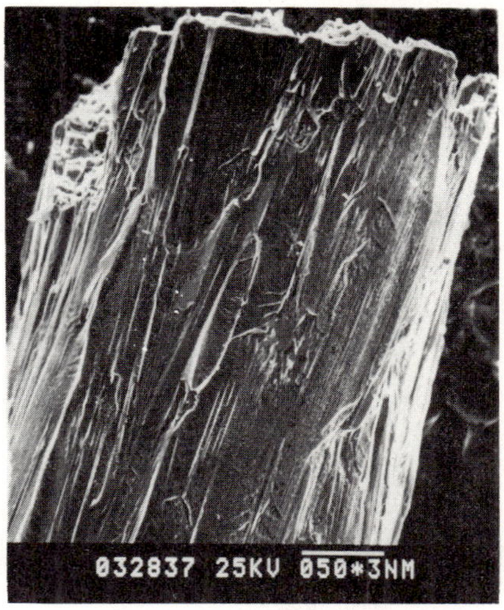

Figure 2 SEM photograph of the resulting fiber

The resistance of the fibers was measured by the standard four-probe technique. Measurements of the critical current density J_c were conducted at liquid nitrogen temperature (77 K) and under zero magnetic field (0 T) by using the conventional steady current method, with zero resistance criterion 1 μV/cm. Post-growth heat treatment was still necessary for the fibers to achieve their superconductivity.
The structures of the source materials and the fibers were determined by X-ray micro-diffraction (XRD). Scanning electron microscope (SEM) was used to study the morphology of the fibers and the composition analysis was carried out by EDAX.

3. Results and discussions

As shown in Figure 1, the XRD spectrum indicates that the source material prepared by the melt and quench method was amorphous. The main results of Bi-based superconducting fibers prepared from amorphous materials are given in Table 1. For the nominal composition of sample M40 and M45, Ca and Cu are excessive compared with 2223. Such a deviation from stoichiometry is in accordance with the EDAX analysis results of the composition changes during the growth of the fibers made from polycrystalline materials. For example, in the case using polycrystalline source materials, the as-grown fiber with similar nominal composition of M45 was superconducting without post-growth heat treatment and has comparatively high critical current

Table 1 Main results of the fibers prepared from amorphous materials

No.	nominal composition Bi(Pb)-Sr-Ca-Cu	Growth rate (mm/min)	T_{c0} (K)	J_c(77K, 0T) (A/cm^2)
M40	2-2-2.8-7.5	2+1	86	2250
M45	2(0.4)-2-2.8-7.5	2+0.5	81	2520
M67	2(0.4)-2-2-4	2+0.5	84	>5000

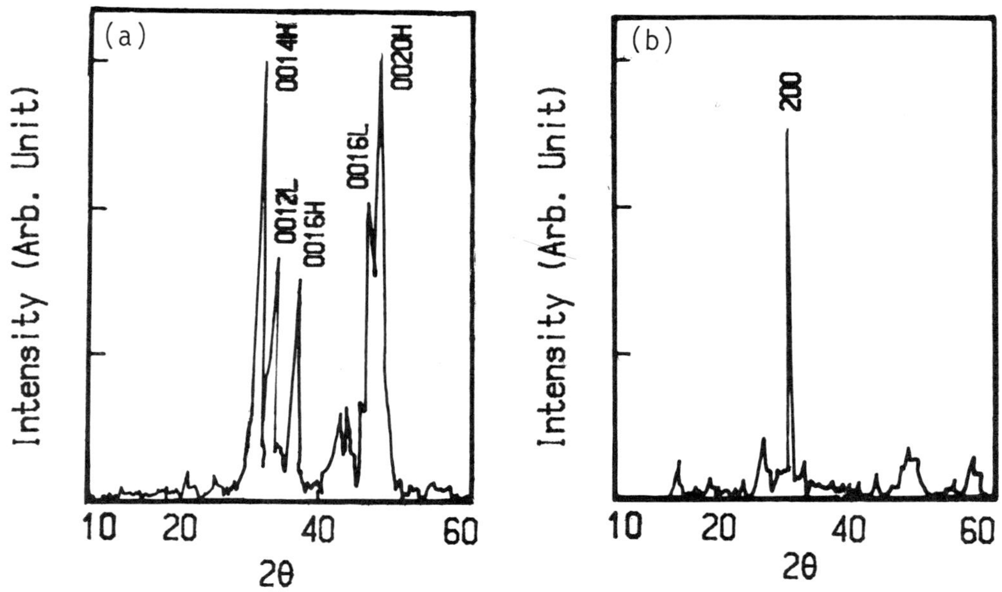

Figure 3 XRD results of the fiber (a) side surface (b) cross section

density (Huo Yujing et al 1990). In the present case, however, where amorphous source materials were used (Table 1), the highest J_c (greater than 5000 A/cm^2) is obtained from sample M67, which has no excess of Ca and only small amount of excess CuO as flux. It shows that the best nominal composition of amorphous materials may be different from that of polycrystalline materials, since comparatively short period high temperature treatment in the melt and quench method will give rise to small composition deviation.

Figure 2 gives a micrograph of the morphology of the side surface of the fiber. Among a number of growth wrinkles, several large pieces of single crystals with their a-b plane aligned parallel to the growth direction can be clearly seen from the fractured surface.

XRD results also show that the textured structure of the fiber is quite strong. As shown in Figure 3a, only the (00L) peaks appear in the XRD spectrum of the side surface. In the XRD spectrum of the cross section (Figure 3b), the (200) peak is most intense , all of the (00L) peaks, even the (220) peak almost disappear. It indicates that the a-axis (not just the a-b plane) is parallel to the growth direction. The highly aligned morphology with a-axis parallel to the growth direction will provide an optimized path for superconducting current along the fibers, which is closely related to high critical current density.

References

Bednorz J and Muller K A 1986 Z. Phys. B64 189

Feigelson R S, Gazit D, Fork D K and Geballe T H 1988 Science 240 1642

Huo Yujing et al 1990 Proc. C-MRS International '90 ed Hengde Li (Amsterdam: Elsevier) to be published

Jin S, Tiefel T H, Sherwood R C, Kammlott G W and Zahurak S M 1987 Appl. Phys. Lett. 51 943-945

Murakami M, Morita M and Koyama N 1989 Jpn. J. Appl. Phys. 28 1125

Tohge N, Tsuboi S, Akamatsu Y, Tatsumisago M and Minami T 1989 J. Ceram. Soc. Jpn. 97 334

Supercond. Sci. Technol. **4** (1991) S157–159. Printed in the UK

BiSrCaCuO/Ag Superconducting Composite Tapes

H. Kumakura, K. Togano, D.R. Dietderich, H. Maeda, J. Kase and T. Morimoto

National Research Institute for Metals, 1-2-1, Sengen, Tsukuba, Ibaraki 305 JAPAN
Asahi Glass Co. Ltd., Research Center, 1150 Hazawa, Kanagawa-ku, Yokohama 221 JAPAN

Abstract. $Bi_2Sr_2Ca_1Cu_2O_x$/Ag composite tapes were prepared by the combined process of doctor blade casting and crystal growth from the partially molten state. A highly textured microstructure was obtained for oxide layer thicknesses of 15µm or less by solidifying from 890°C. The textured tapes have excellent transport J_c values greater than 10^5 A/cm^2 at 4.2K in magnetic fields above 25T. The results indicate that textured Bi-based tapes are promising candidates for high field superconducting magnets.

1. Introduction

The application of superconducting materials to magnets requires the winding of tape or wire conductors with sufficient flexibility and sufficiently high transport critical current density J_c into a solenoid. Since oxide superconductors are intrinsically brittle, the developement of special techniques are essential for tape or wire fabrication. Recently, Kase et al.(1990a,b) have succeeded in fabricating flexible $Bi_2Sr_2Ca_1Cu_2O_x$/Ag composite tapes with a textured microstructure and excellent J_c by doctor blade casting and partial melting. In this paper, we report on the relationship between the oxide microstructure and current carrying capacity with oxide layer thickness.

2. Microstructure

The conductors were prepared by the doctor blade casting process.
A slurry consisting of $Bi_2Sr_2Ca_1Cu_2O_x$(2212) powder and various organic materials was cast under a doctor blade into green tapes 30-200µm in thickness. The green tapes were mounted on 50µm-thick Ag tapes and heat treated at 500°C for 2h to remove the organic formulation. The subsequent high temperature heat treatment had a strong effect on the microstructure and superconducting properties of the oxide (Kase et al. 1990b). The optimum high temperature heat treatment of the composite tapes was 890°C for 10m, slow cooling to 870°C at a rate of 5-10°C/h, holding at 870°C for 3h, and lastly, cooling to room temperature at a rate of 30-100°C/h.
During heat treatment the thickness of the oxide layer decreases from 1/3 to 1/5 of the green tape thickness. The oxide layer also sticks well to the Ag substrate providing good electrical and thermal contact. This suggests that Ag would be a good stabilizing material.
Figure 1(a), (b) and (c) show the fractured cross-section of the composite tapes with various oxide thicknesses. When the oxide thickness is less than 15µm, a well-aligned microstructure with plate-like grains is observed (Fig. 1(a) and (b)). Most of the peaks observed in X-ray

0953-2048/91/01S157 + 03 $03.50

diffraction patterns of the surface of these tapes can be indexed as (00ℓ) reflections of the 2212 phase, indicating a strong preferred orientation with the c-axes aligned perpendicular to the tape surface. This textured microstructure was obtained by solidifying from the partially molten state during slow cooling from 890°C to 870°C.
When the oxide layer is thicker than 30μm, however, a highly textured microstructure was only observed at the surface of the oxide layer as shown in Fig. 1(c). However, textured microstructures were also observed near the oxide/Ag interface but with no regularity. These results suggest that the texturing proceeds from the surface or possibly from the Ag/oxide interface inward during slow cooling from 890°C.
Some portion of the Ag substrate dissolves into the liquid oxide during the heat treatment lowering the melting point of the oxide (Dietderich et al. 1990).
Furthermore, the Ag substrate seems to play an important role in texturing. Preliminary observations show that highly textured 2212 microstructures are not produced when Pt or Au are used as substrate materials.

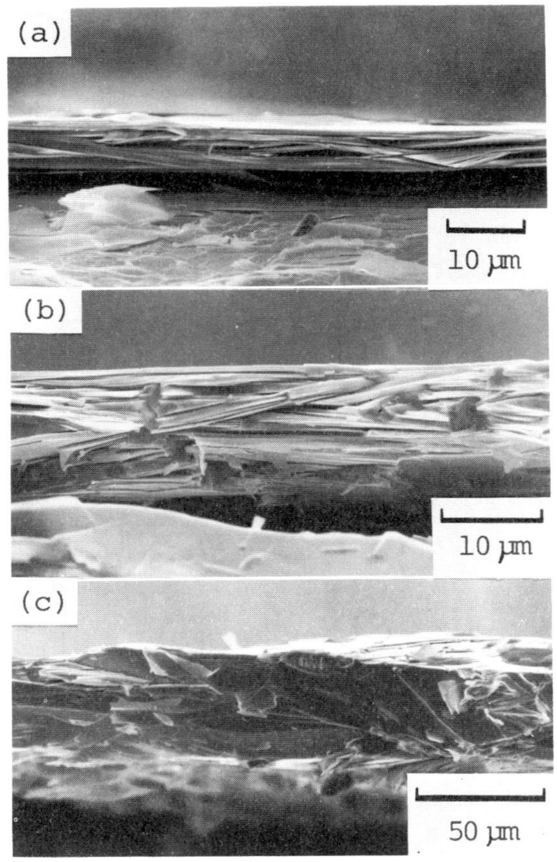

Fig. 1 Scanning electron micrographs of the fractured cross section of the 2212 oxide layers of several thicknesses.

3. Critical Current Properties

Transport critical current I_c was measured by the resistive method in the temperature range 4.2K to 77K in magnetic fields up to 27T. Measurements at 4.2K and 18-27T were carried out at the High Field Laboratory for Superconducting Materials at Tohoku University. The potential criterion for I_c determination was 1μV/cm and the cross-sectional area of the oxide layer was used to calculate J_c.
The textured tapes have a much higher J_c (at least 10 times greater) than non-textured tapes due to better coupling between grains. At 77K, J_c values greater than 10^4A/cm^2 were easily obtained in zero magnetic field. However, this high J_c was substantially reduced by magnetic fields as weak as 0.01-0.1T, with the sharpest decrease in J_c occuring for fields applied perpendicular to the tape surface (parallel to the c-axes).
This large magnetic field dependence of J_c significantly reduced at low temperature such that at 4.2K excellent J_c values were obtained even in high magnetic fields (Kumakura et al. 1990).
Figure 2 shows I_c and J_c at 10T and 4.2K as a function of oxide layer thickness. The critical current I_c increases with increasing thickness, while J_c decreases with increasing thickness. The I_c vs thickness curve appears to have two regions, for thicknesses less than 15μm, I_c rapidly

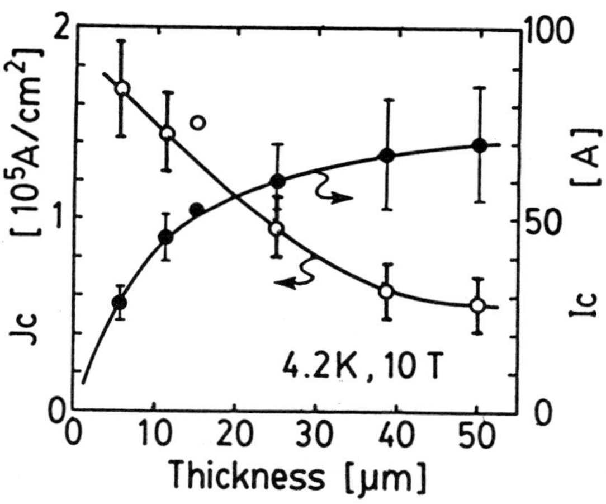

Fig. 2 I_c and J_c at 10T and 4.2K as a function of 2212 oxide layer thickness.

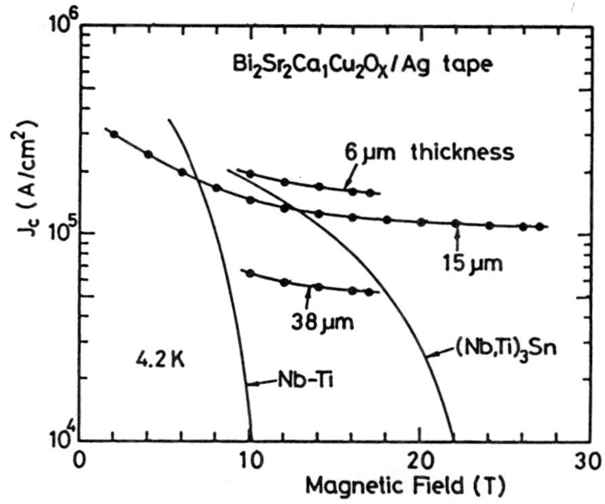

Fig. 3 J_c-B curves at 4.2K for the 2212 oxide/Ag composite tapes and conventional metallic superconductors Nb-Ti and $(Nb,Ti)_3Sn$.

increases, while for larger thicknesses it increases slowly. This I_c behavior can be understood by observing the oxide microstructure. As the oxide layer thickness increases, more non-textured regions appear in the oxide layer.

Figure 3 shows J_c-B curves at 4.2K in magnetic fields parallel to the tape surface for the 2212 tapes for several oxide layer thicknesses. The tapes with oxide layers of 6μm and 15μm thick have grain-oriented microstructures through the entire layer, as shown in Fig. 1(a) and (b). These tapes have excellent J_c values, higher than 10^5A/cm^2 in fields up to 27T. Due to the extraordinarily high B_{c2} value at 4.2K, the magnetic field dependence of J_c for the 2212 tapes is much smaller than that of the conventional metallic superconductors such as $(Nb,Ti)_3Sn$. These results indicate that textured 2212 tapes are very promising for high magnetic field application above 20T for which no practical superconducting tape or wire is currently available.

Acknowledgement

The authors are grateful to Profs. Y. Muto and K. Watanabe of Tohoku University for providing them the opportunity to use the high field testing facilities.

References

Kase J, Irisawa N, Morimoto T, Togano K, Kumakura H, Dietderich D R, and Maeda H 1990a Appl. Phys. Lett. 56 970.

Kase J, Togano K, Kumakura H, Dietderich D R, Irisawa N, Morimoto T and Maeda H 1990b Jpn. J. Appl. Phys. 29 L1096.

Dietderich D R, Ullmann B, Freyhardt H C, Kase J, Kumakura H, Togano K, Maeda H 1990 Jpn. J. Appl. Phys. 29 L1100.

Kumakura, Togano K, Kase J, Morimoto T, and Maeda H 1990 Cryogenics (in press).

Supercond. Sci. Technol. 4 (1991) S160–162. Printed in the UK

CRITICAL CURRENTS IN SUPERCONDUCTING TETRAGONAL SINGLE CRYSTAL $HoBa_2Cu_3O_{7-\delta}$

H. Arabi and J. W. Ross

Physics Department, Manchester University, Manchester, M13 9PL, U.K.

Magnetisation measurements on a high quality superconducting tetragonal single crystal of Ho-123 show that anisotropy of both H_{c1} and the critical current is much stronger than in orthorhombic Ho-123. From the temperature dependence of the magnetic relaxation we show that pinning within the Cu-O planes is the same as in Y-123 but is much greater when the field is applied parallel to the planes.

1. Introduction

Early in the history of high temperature superconductivity it was established that only the orthorhombic form of RE-123 compounds was superconducting although a wide variation in the lattice parameters was observed at that time. This was attributed to oxygen deficiency and was accompanied by a lowering and widening of the superconducting transition (Beyers and Shaw (1989)). We have prepared by a specialised technique single crystals of tetragonal Ho-123 which have a sharp transition at 92±1K.

2. Experimental Details

The partial melting procedure of Schneemeyer et al. (1987) was used but homogeneous mixing of the components was achieved via the citrate gel process. This method has the added advantages of small oxide particle size (<1μm) and the exclusion of grinding processes which might introduce impurities. Stoichiometric amounts of holmium, barium and copper were mixed in an aqueous solution of their nitrates in the respective anion proportions of 1:4:10 or 1:4:6. Citric acid was then added and the solution evaporated at 78-80°C to yield a solid precursor which was fired in air at ~250°C. At this stage X-ray analysis confirmed the presence of barium carbonate, metallic copper and holmium oxide. This finely divided powder, contained in a high purity alumina crucible, was held at 1040°C for 12 hours, cooled at 0.4°C hr^{-1} from 1010°C to 910°C then at 20°C hr^{-1} to 700°C and finally at 10°C hr^{-1} to room temperature. The heating and cooling operations were carried out under a constant stream of high purity oxygen.

The crystals formed in the BaO and CuO rich flux were of varying sizes ranging from $0.3x0.3x0.05mm^3$ to $3x3x0.5mm^3$ with the smaller crystals showing better morphology. Impurities of Si and Al were less than the detection limits of SEMA (<0.5%) while chemical analysis showed carbon incorporation to be <0.25%. Laue, Buerger precession and crystal diffraction X-ray methods showed the crystals to be tetragonal. Neutron diffraction on a crystal with a mosaic spread of <1° gave a=3.858±.001Å and c=11.650±.003Å at 20K. No evidence for a superlattice (Balestrino et al. (1988)) was found: the oxygen atoms showed no axial preference within the Cu(I)-O planes. TEM confirmed the tetragonal symmetry and showed a complete absence of twinning (Arabi et al. (1990)).

Magnetisation data were taken in fields up to 6T using a home-made vibrating sample magnetometer with a sensitivity of $10^{-7}Am^2$. Relaxation rates were determined from measurements taken from 5 to $5x10^3$ sec after a field of 0.1T was applied to the sample cooled in zero field.

3. Results

Magnetic measurements were made on a 2.4mg crystal with sides 1 x 1.05 x $0.34mm^3$ where the c-axis is along the shortest dimension: Laue patterns show the Cu-O planes were parallel to the $1x1.05mm^2$ faces to within <1°.

H_{c1} was determined by carefully noting the deviation from linearity of the magnetic moment v. field plots when the field was applied along the c-axis ($H^{\parallel}c$) and along an a-axis ($H^{\perp}c$). The respective values are 0.17±0.04T and 3.5±0.7mT, the large uncertainties reflecting the well known difficulty of this test.

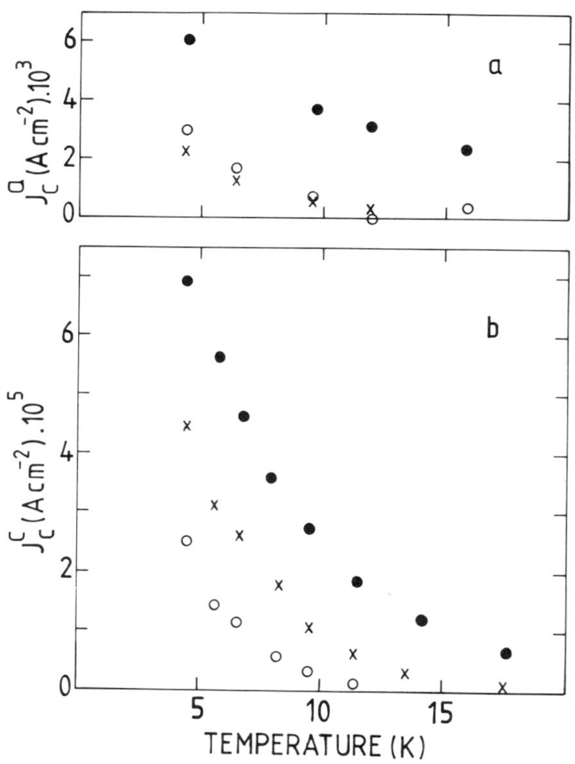

Fig. 1 The temperature variation of J_c for a) $H^{\perp}c$. b) $H^{\parallel}c$ in the remanent state (●) and fields of 2T (X) and 4T (O).

Critical current densities (J_c) for the two field directions at 2T and 4T were deduced via the modified Bean model (Gyorgy et al. (1989)) from the difference in magnetisation on increasing and decreasing the magnetic field up to 6T. Remanent magnetisation in the residual field of the superconducting solenoid (<3mT) was also used to calculate J_c. The temperature variation of the critical current densities is shown in figure 1. J_c^c/J_c^a is approximately 20 times greater than that of orthorhombic Ho-123 (Tholence et al. (1988)) with our value of J_c^c a factor of 4 smaller. The critical current densities of tetragonal Ho-123 decrease more rapidly with temperature than orthorhombic Ho-123 and untwinned Tm-123 (Moshchalkov et al. (1990)).

At temperatures up to 37K a linear logarithmic time dependence of the magnetisation was observed after 0.1T was suddenly applied along the c-axis of the crystal. Figure 2 shows that the creep rate variation with temperature has its characteristic form (Yeshurun and Malozemoff (1988)). Yeshurun et al. (1988) have extended the Bean model for creep by incorporating the dependence of J_c on field and time. At low temperatures where $H_{c1}<H<H^*$ we have fitted our results to the expression

$$\frac{dM}{d\ln t} = \frac{1}{4\pi H^*} [H^2 - H_{c1}{}^2] \frac{kT}{U_o} \qquad (1)$$

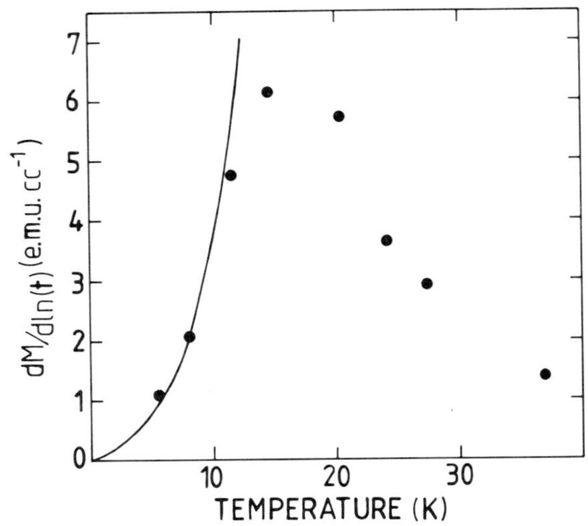

Fig. 2 Creep rate v. temperature for 0.1T ($H \parallel c$). The line is a fit to equation 1 with U_o=0.013eV (see text).

where H is the applied field corrected for demagnetisation, (N=0.64 for H parallel to the short dimension of the crystal), U_o is the pinning activation energy. $H^*=4\pi J_c^c(T)R/10$ is the field for which currents first flow through the entire volume of the sample. We have used 0.05cm for the radius of the cylinder approximating our sample and $J_c^c(T)$ values are taken from Figure 1. Yeshurun et al. (1989) find U_o~0.015eV from a similar fit to their data on Y-123. We were not able to detect any relaxation when $H \perp c$ indicating stronger pinning for this direction than observed in Y-123 (Yeshurun et al. (1989)).

4. Discussion

Even though H_{c1} ratios are not the most accurate way to deduce effective mass ratios (see for example Kogan (1989)) we are confident that our results show that the mass tensor in tetragonal Ho-123 is much more anisotropic than in orthorhombic Ho-123 or Y-123. Complications due to intergrowth suggested by Krusin-Elbaum et al. (1990) as a factor in the apparent high effective mass anisotropy of BISSCO should be less likely to occur in the Y-123 structure where the Cu-O planes are much closer together. Kogan (1989) has pointed out that greater mass anistropy results in an increase of the $C_{66,hard}$ shear modulus of the flux line lattice and reduced creep when $H \perp c$ as we have observed. The random distribution of oxygen atoms within the Cu-O planes, however, has not changed the pinning energy for vortex motion within the plane.

5. Acknowledgements

The authors thank R. H. Fenn, P. Midgley, O. S. Mills, M. E. Clamp and R. Vincent for invaluable contributions to the diffraction experiments.

6. References

Arabi H et al. XVth Congress of Int. Union of Crystallography, Bordeaux 1990 Acta Cryst. (to be published)
Balestrino G, Barbanera S, Paoletti A and Paoli P 1988 Phys Rev B 38 6609
Beyers R and Shaw T M 1989 Solid State Phys 42 135
Gyorgy E M et al. 1989 App Phys Lett 55 283
Kogan V G 1989 Physica C 162-164 1689
Krusin-Elbaum L et al. 1990 Physica A 163 125
Moshchalkov V V et al. 1990 Physica C 166 185
Schneemeyer L F et al. 1987 Nature 328 601
Tholence J L et al. 1988 Physica C 153-155 1479
Yeshurun Y and Malozemoff A P 1988 Phys Rev Lett 60 2202
Yeshurun Y et al. 1988 Phys Rev B 38 11828
Yeshurun Y et al. 1989 Cryogenics 29 258

Supercond. Sci. Technol. 4 (1991) S163–165. Printed in the UK

Magnetic field dependence of critical current in $Bi_2Sr_2CaCu_2O_{8+y}$ films prepared by laser ablation

Takeshi Fukami, Terukazu Nishizaki*, Takashi Yamamoto and Takafumi Aomine

Department of Physics, Kyushu University, Fukuoka 812, Japan
*Department of Information Systems, Kyushu University, Kasuga 816, Japan

Abstract: We measured the magnetic field H dependence of critical current I_c in $Bi_2Sr_2CaCu_2O_{8+y}$ films varying the temperature T. A hysteresis of I_c versus H was observed in the low T, became weaker as T increased and vanished at about 60 K. This hysteresis comes from the nonreversibility of I_c due to the flux pinning near bridges. The field dependence of current density J_c in high H had an abrupt change above about 30 K at a threshold field H_T, which decreased as the T increased. The plots of H_T versus T may show a phase transition from a rigid flux lattice to a flux-melting state.

1. Introduction

Oxide superconductors have exhibited a pronounced broadening of the resistivity versus temperature T curves in the magnetic field H (Palstra et al 1988, Kitazawa et al 1989, Fukami et al 1989a,b). This broadening has been also observed not only under **I**⊥**H** but also under **I**//**H** (Kitazawa et al 1989, Fukami et al 1989b), here **I** is current vector and **H** magnetic field vector. This result shows that the macroscopic Lorentz force does not play any important role for the broadening. Recently, for explanation of this broad transition, a flux-lattice melting state has been researched. And to search the origin of the small critical current density J_c in $Bi_2Sr_2CaCu_2O_{8+y}$ (BSCC) films, we certified the existence of bridges as one of weak links using the same method as one by Aomine and coworkers (Mizuno and Aomine 1983, Aomine and Yonekura 1986, Yamamoto and Aomine 1988). Our aims in this investigation are to confirm some clear sign of the transition on curves of J_c versus H and whether bridges exist or not in our films.

2. Experimental Results

The thin films of BSCC were deposited on single crystalline MgO(100) substrates by the laser ablation technique (Fukami et al 1989a). The c-axis of crystal almost orients perpendicular to the surface of the substrate. The critical current I_c was taken to be the current at which the sample voltage reached 3 μV.

Figure 1 shows typical H dependences of I_c in low fields at five points of temperature for **H**//c-axis and Figure 2 shows typical H dependences of I_c at three points of temperature for **H**⊥c-axis. The characteristic behaviors are summarized as follows. (1) A hysteresis was observed for **H**//c-axis, became weaker as the temperature increased and disappeared at about 68 K. It could not be, however, observed for **H**⊥c-axis between 4.2 K and 78 K. (2) The field dependence of I_c for **H**⊥c-axis is smaller than one for **H**//c-axis. Figure 3 shows the plots of log J_c versus log H as function of T. In low field region, these plots in Figure 3 had a similar behavior that J_c decreased linearly, but in high field region very different behaviors were observed with increasing T.

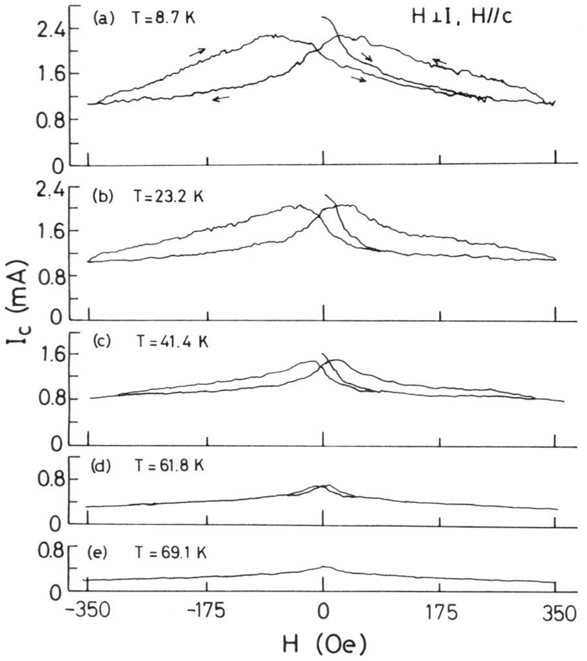

Figure 1. Magnetic field dependence of I_c for **H**//c-axis.

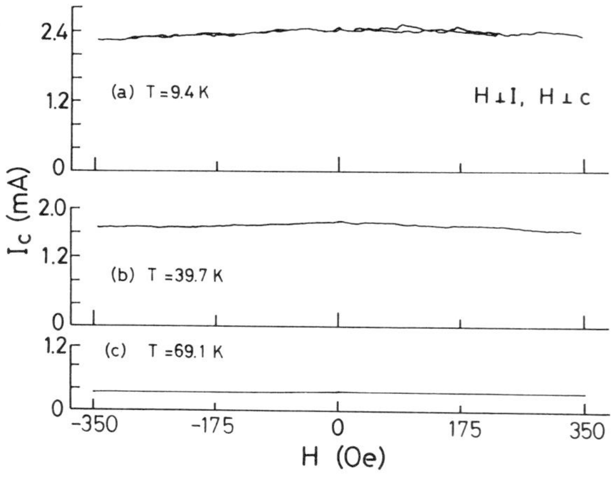

Figure 2. Field dependence of I_c for **H**⊥c-axis.

At 10 K and 15 K the dependences were almost linear from 0.05 T to 11 T. The change became stronger for T higher than 27.5 K. Here let us define a threshold field as a crossing point of a straight line in low field region and one in high field region in log J_c versus log H for **H**//c-axis. This point shifts to the weak field side as the temperature increases. For the case of **H**⊥c-axis the clear bending could not be observed at temperatures not higher than within 11 T. Using the values of H_T determined thus, plots of H_T versus T are shown in Figure 4. For **H**//c-axis these plots are limited within very narrow region of T. The plots for **H**⊥c-axis seem to shift to higher side of T though only two points are obtained.

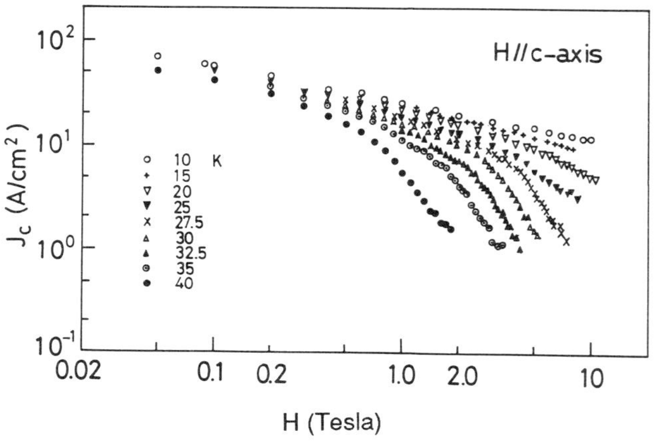

Figure 3. Field dependence of J_c for **H**// c-axis in high field region.

3. Discussion

It has been shown that the hysteresis of I_c of superconducting granular aluminum bridges prepared by photolithography in low values of H disappears as the temperature increases (Aomine and Yonekura (1986)). And the cases for hysteresis to disappear are in the following: the constricted part is too long for the electrical influence of the wide part adjacent to the constricted one to reach and the bulk pinning is dominant over the edge pinning (Aomine et al 1989). As for our films, for **H**//c-axis an appropriate length of the constricted part would exist and it would enter a state in which at least one of the conditions mentioned above would realize effectively as the temperature increases. For **H**⊥c-axis, the bulk pinning may be dominant over the edge pinning. The anisotropic behaviors of I_c versus H would be ascribed to the anisotropy of pinning force near bridges arising from the preferred orientation of c-axis of the film. In addition to the above fact, the J_c was very small compared with the values of $J_c = 8 \times 10^4$ A/cm^2 at 77 K (Steibeck et al 1989) and $J_c = 10^5$ A/cm^2 (Hong et al 1989). Then, the J_c of our films was considered to be determined by such a weak link as a bridge.

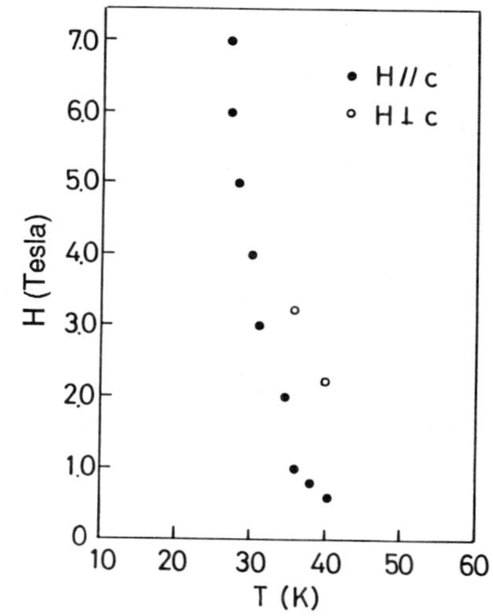

Figure 4. H_T versus T.

Next let us consider the origin of the abrupt change of the J_c versus H plots at high temperature region. One of the most hopeful candidates to explain the field dependence of J_c may be a transition from a rigid flux lattice to a flux-lattice melting state as the temperature increases. Gammel et al (1988) used this concept to explain the mechanical properties in $YBa_2Cu_3O_7$ and $Bi_{2.2}Sr_2Ca_{0.8}Cu_2O_8$ in magnetic field. They obtained a phase diagram in a T-H plane. This is very similar to one shown in Figure 4. And this diagram is also very similar to one expected from the theoretical investigation by Houghton et al (1989). Thus, the abrupt change of J_c in J_c versus H plots as a function of T may be ascribed to the transition to the flux-lattice melting state.

References

Aomine T and Yonekura A 1986 *Phys. Lett.* **114A** 16.

Aomine T, Tanaka E, Yamasaki S, Tani K and Yonekura A 1989 *J. Low Temp. Phys.* **74** 263.

Fukami T, Kamura T, Youssef A A A, Horie Y and Mase S 1989a *Physica C* **159** 422.

Fukami T, Kamura T, Youssef A A A, Horie Y and Mase S 1989b *Physica C* **159** 427.

Gammel P L, Scneemeyer L F, Waszczak J V and Bishop D J 1988 *Phys. Rev. Lett.* **61** 1666.

Hidaka H, Oda M, Suzuki M, Katsui A, Murakami T, Kobayashi N and Muto Y 1987 *Physica B* **148** 329.

Hong M, Kwo J and Yeh J -J 1989 *J. Cryst. Growth* **91** 382.

Houghton A, Pelcovits R A and Sudbo A 1989 *Phys. Rev. B* **40** 6763.

Iye Y, Tamegai T, Takeya H and Takei H 1987 *Jpn. J. Appl. Phys.* **26** L1057.

Kitazawa K, Kambe S, Naito M, Tanaka I and Kojima H 1989 *Jpn. J. Appl. Phys.* **28** L555.

Mizuno K and Aomine T 1983 *J. Phys. Soc. Jpn.* **52** 4311.

Palstra T T M, Batlogg B, Schneemeyer L F and Waszczak J V 1988 *Phys. Rev. Lett.* **61** 1662.

Steibeck J, Tsaur B Y, Anderson A C and Strauss A J 1989 *Appl. Phys. Lett.* **54** 466.

Yamamoto T and Aomine T 1988 *Jpn. J. Appl. Phys.* **27** L1917.

Supercond. Sci. Technol. 4 (1991) S166–168. Printed in the UK

Thin Film Preparation, Transport Properties and Superconductivity of $YBa_2Cu_3O_7$

H. Adrian[1], C. Tomé-Rosa[1], G. Jakob[1], A. Walkenhorst[1], M. Maul[1], M. Paulson[1], M. Schmitt[1], P. Przysłupski[1,2], G. Adrian[3], M. Huth[1], and Th. Becherer[1]

[1]Inst. für Festkörperphysik, Technische Hochschule Darmstadt, FRG; [2]Inst. of Physics, Polish Academy of Sciences, Warszawa, Poland; [3]Daimler-Benz AG, Forschungsinst. Frankfurt, FRG.

Abstract: We report the preparation of high-T_c $YBa_2Cu_3O_7$ films on single crystalline $SrTiO_3$, $LaGaO_3$, $NdAlCaO_4$, MgO, ZrO_2, and Al_2O_3 substrates by sputtering and laser ablation. In all cases c-axis oriented films with T_c's near 90 K and narrow transitions were obtained. The films grow epitaxially on $SrTiO_3$ and $LaGaO_3$, highly oriented on MgO, ZrO_2 and $NdAlCaO_4$ and less oriented on Al_2O_3. On patterned films we obtained $j_c(77K) = 4 \cdot 10^6 A/cm^2$ for films on $SrTiO_3$, $4 \cdot 10^4 A/cm^2$ on $NdAlCaO_4$ and $4 \cdot 10^3 A/cm^2$ on Al_2O_3 revealing the importance of epitaxial growth. In magnetic fields of 12 Tesla with B || c-axis we find $j_c(12T, 20K) = 2 \cdot 10^6 A/cm^2$.

1. Introduction

The dependence of the normal state and superconductive transport properties of high-T_c-superconductors on the microstructure of the material is an important aspect for future applications. In order to obtain a variation of the microstructure we prepared $YBa_2Cu_3O_7$-films on $SrTiO_3$, $LaGaO_3$, MgO, $NdAlCaO_4$, ZrO_2, and Al_2O_3 substrates. The films were wet-chemically patterned into an 8-probe structure which allows measurements of resistivity $\rho(T)$, Hall-effect $R_H(T)$ and upper critical field $B_{c2}(T)$ on a 2 mm long and 0.2 mm wide strip and measurements of the critical current densities $j_c(B,T)$ on a 100 μm long and 10 μm wide strip of the same film.

2. Experimental

Films on $SrTiO_3$ (100), $LaGaO_3$ (001), MgO (100), ZrO_2 (random), and $NdAlCaO_4$ (001) were prepared by high pressure (3 mbar) dc-sputtering in pure oxygen (Poppe et al. 1988) from a water-cooled $YBa_2Cu_3O_7$ target onto heated substrates. Additionally films on ZrO_2 (random) and Al_2O_3 (1120) were prepared by laser ablation using an excimer laser (LPX 301 with KrF, $\lambda = 248$ nm) and a rotating ceramic target. Further details of the thin film preparation, the patterning and the fabrication of low resistance contacts have been reported earlier (Tomé-Rosa et al. 1990). The structure of the films was studied by X-ray diffraction using Cu-K_α-radiation. Measurements in magnetic fields up to 12 Tesla were made in a $NbTi/Nb_3Sn$-hybrid magnet. In all experiments reported in this paper the field was oriented perpendicular to the film surface, i. e. parallel to the c-axis of the $YBa_2Cu_3O_7$ crystal structure. The criterion for the $j_c(B,T)$-measurements was 1 μV across a 100 μm long strip.

3. Results and Discussion

On all substrate materials used c-axis oriented $YBa_2Cu_3O_7$ films were obtained. The inductively measured $T_{c,mid}$-values and transition widths ΔT_c (10% to 90%) were 90.8 K and 0.4 K

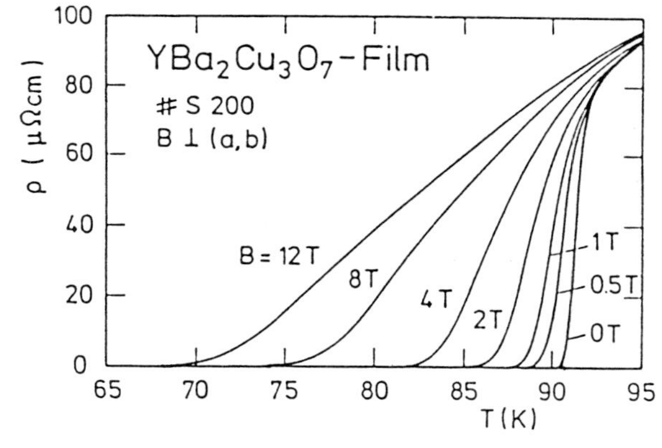

Fig. 1: Resistive transitions in external magnetic fields up to 12 Tesla

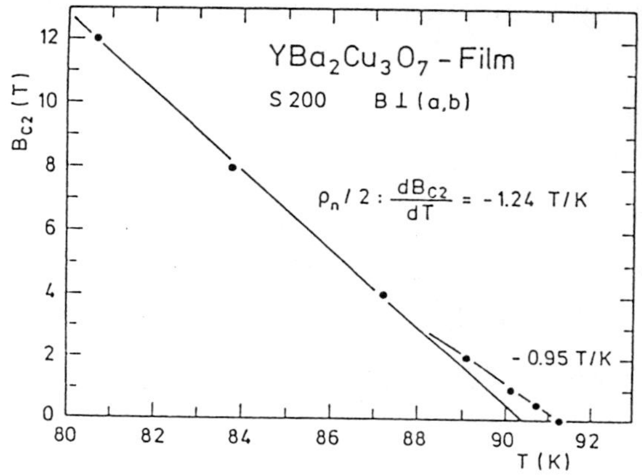

Fig. 2: Temperature dependence of the upper critical field $B_{c2}(T)$

for films on $SrTiO_3$, 89.5 K and 0.5 K ($LaGaO_3$), 88 K and 5 K ($NdAlCaO_4$), 87.5 K and 1.5 K (ZrO_2), 87 K and 2 K (Al_2O_3), and 86 K and 1.5 K (MgO). In the X-ray diffraction diagrams additional to the substrate peaks only (00ℓ)-peaks of the $YBa_2Cu_3O_7$ structure are present except for extremely small (200)-peaks in case of MgO, $NdAlCaO_4$, and ZrO_2. In order to examine the quality of the film growth we measured ϑ-scans of the (005)-peaks. The extremely small FWHM-value of 0.4° for films on $SrTiO_3$ and $LaGaO_3$ demonstrates epitaxial growth. For the other substrates we find FWHM-values of 0.75° (MgO), 1.0° (ZrO_2), 1.2° ($NdAlCaO_4$), and 2.1° (Al_2O_3). In case of the latter substrates the films grow only with c-axis orientation due to growth anisotropy rather than due to epitaxy.

Measurements in magnetic fields have been carried out so far on epitaxial films on $SrTiO_3$ and c-axis oriented films on Al_2O_3 (Adrian et al. 1990). The resistive transitions for a film on $SrTiO_3$ in fields up to 12 T are displayed in Figure 1, which clearly shows the broadening of the transition well known from bulk samples. From these data we obtain $B_{c2}(T)$ presented in Figure 2. The initial slope of the critical field depends on the definition of B_{c2}. Evaluating the midpoints we obtain $dB_{c2}/dT = -1.24T/K$.

Figure 3 shows the temperature dependence of the same film's carrier concentration $n_H(T) = (R_H(T) \cdot e)^{-1}$ determined from Hall-effect measurements. The data can be fitted by a straight line which nearly extrapolates through zero with $n_H(240K) = 6 \cdot 10^{21}cm^{-3} = 1V_{cell}^{-1}$ as expected by valence considerations. The critical current density in zero field can be described down to 40 K by $j_c(T) = j_c^*(0)(1 - T/T_c)^{1.4}$ with $j_c^*(0) = 5.5 \cdot 10^7 A/cm^2$ as shown in Figure 4. The films

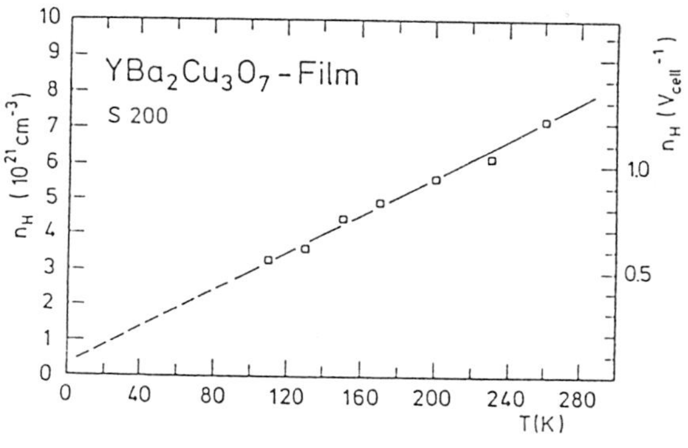

Fig. 3: Temperature dependence of the Hall concentration $n_H(T)$

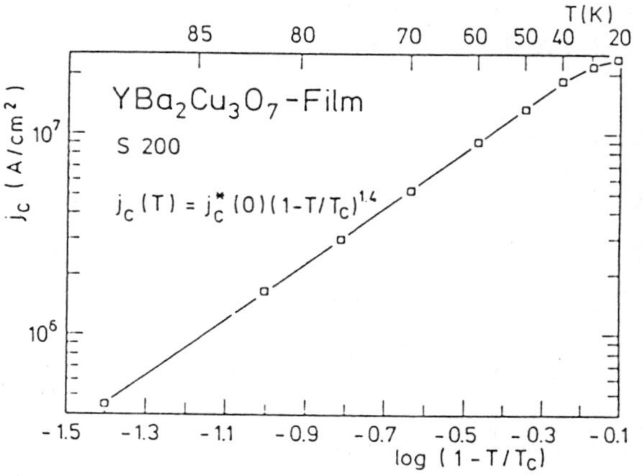

Fig. 4: Critical current density $j_c(T)$

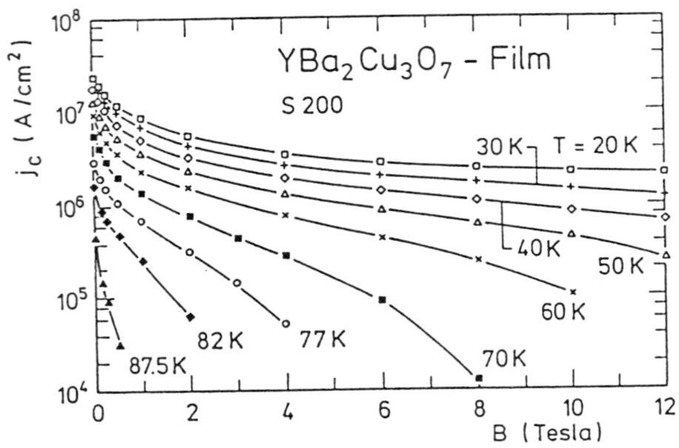

Fig. 5: Critical current density $j_c(B)$ at fixed temperatures

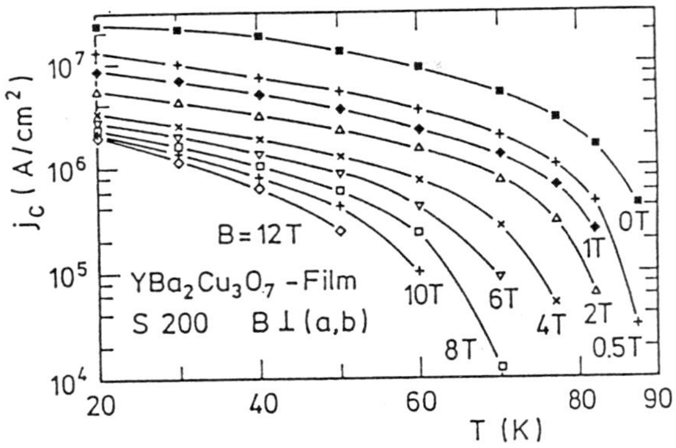

Fig. 6: Critical current density $j_c(T)$ for fixed magnetic fields

on $SrTiO_3$ are characterized by $j_c(77K) = 4 \cdot 10^6 A/cm^2$, whereas the corresponding values for films on $NdAlCaO_4$ are $4 \cdot 10^4 A/cm^2$ and for films on sapphire $4 \cdot 10^3 A/cm^2$.

The full dependence $j_c(B, T)$ for $B \| c$ is presented in Figures 5 and 6. The measured values exceed those of Hettinger et al. (1989) for coevaporated films and are virtually identical with the data of Roas et al. (1990) for laser ablated films for the same orientation of the external field.

It is interesting to note that the intersection of the $j_c = 10^4 A/cm^2$ plane with the $j_c(B, T)$-surface corresponds to a $B^*_{c2}(T)$ curve with $dB^*_{c2}/dT = -0.4T/K$. Since $\rho = E/j$ this criterion corresponds for $E = 1\mu V/100\mu m$ and $j = 10^4 A/cm^2$ to $\rho^* = 10^{-2}\mu\Omega cm \approx 10^{-4}\rho_n$.

Conclusions

High pressure dc-sputtering is suitable to prepare high-T_c c-axis oriented films on a large variety of oxide substrates. The critical current density $j_c(B, T)$ is comparable to the best films produced by laser ablation. For substrates with significant interdiffusion, such as sapphire, laser ablation is superior to sputtering due to higher deposition rates. Our experiments show that c-axis orientation is not sufficient for high j_c, but epitaxial growth is needed.

Acknowledgements

The authors acknowledge financial support by the Bundesminister für Forschung und Technologie and the Deutsche Forschungsgemeinschaft through SFB 252. Two of us were financially supported by the Conselho Nacional de Desenvolvimento Científico e Tecnológico do Brasil (C. T.-R.) and the Alexander von Humboldt-Stiftung (P. P.) respectively.

References

Adrian G, Wilkens W, Adrian H and Maul M 1990, this conference.

Hettinger J D, Swanson A G, Skocpol W J, Brooks J S, Craybead J M, Mankiewich P M, Howard R E, Stroughn B L and Burkhardt E G 1989 *Phys. Rev. Lett.* **62** 2044

Poppe U, Schubert J, Arons R R, Evers W, Freiburg C H, Reichert W, Schmidt K, Sybertz W and Urban K 1988 *Solid State Commun.* **66** 661

Roas B, Schultz L and Saemann-Ischenko G 1990 *Phys. Rev. Lett.* **64** 479

Tomé-Rosa C, Walkenhorst A, Maul M, Jakob G, Adrian H, Haberle K, Przysłupski P and Adrian G 1990 *Proc. of the XIXth Int. Conf. on Low Temp. Physics*, Brighton, UK, to appear in *Physica C*

Supercond. Sci. Technol. 4 (1991) S169–171. Printed in the UK

Superconductive and Normal State Transport Properties of $YBa_2Cu_3O_7$ Films on Sapphire in High Magnetic Fields

G.Adrian, W. Wilkens, H. Adrian*, and M. Maul*

Daimler Benz AG, Research Institute Frankfurt, FRG; * Institute for Solid State Physics, Technical University Darmstadt, FRG

Thin films of $YBa_2Cu_3O_7$ were deposited on (1120) Al_2O_3 by laser ablation. X-ray diffraction analysis revealed that the films grow c-axis oriented. The FWHM of the rocking curves were determined to 2.1^o for films prepared at substrate temperatures of $T_S \geq 830^oC$, exhibiting typical inductively measured T_c -values around 87 K (transition width around 1 K). By means of ESCA depth profiles the diffusion of the Al of the substrate into the superconducting film was analysed. The resistivity $\rho(T)$ is ≈ 500 $\mu\Omega$cm at 300 K and exhibits the typical linear temperature dependence with $\rho(300K)/\rho(100K) \approx 2$. In magnetic fields B up to 12 T with the field direction perpendicular to the film surface $\rho(T)$ as well as the critical current density $j_c(T)$ were monitored. The resistive transition shows only little broadening at B = 12 T and $T_{c,zero}$ is shifted to 75 K, whereas for j_c (T) a strong reduction in magnetic fields is observed.

1. Introduction

While it is generally possible to produce epitaxial $YBa_2Cu_3O_7$ films with prominent superconducting properties on proper substrate materials such as (100) $SrTiO_3$ with various deposition techniques – e. g. high pressure sputtering (Poppe et al. 1989), ICM sputtering (Li et al. 1989), laser ablation (Venkatesan et al. 1989) and MOCVD (Watanabe et al. 1990) – the application of substrates customary for microelectronic devices such as Si or Al_2O_3 is still a problem. The difficulty consists at first in the diffusion processes occurring at the high substrate temperatures T_S required for the formation of the 1-2-3 phase. However because of the availability, the price, and other advantages of these usual substrates it is important to undertake special efforts to develope the high–T_c superconductor thin film technology with these substrates. As high deposition rates should be profitable, we choose the method of deposition by laser ablation for preparing $YBa_2Cu_3O_7$ films on sapphire.

2. Experimental

The $YBa_2Cu_3O_7$ films were prepared by ablation from a sintered bulk target using a KrF excimer laser (Lambda Physik LPX 301). An applied energy density of $\approx$ 1 J/cm^2 leads at a substrate–target distance of 5 cm to a deposition rate of 0.2 nm/pulse and 2 nm/s, respectively. The oxygen pressure during the ablation process was maintained at $5{\cdot}10^{-2}$ to $2{\cdot}10^{-1}$ mbar. At these low pressures the resulting film exhibits an extremely small density of "droplets" (≤ 2 mm^{-2}) as can be seen in Fig. 1. During the deposition T_S was held constant at 730^oC to 880^oC; T_S was determined by an optical pyrometer after a dense $YBa_2Cu_3O_7$ film covered the substrate. For homogenization the substrates were rotated during the deposition. Afterwards the films were cooled down to 500^oC in 1 bar O_2 and held for 10 min. Stoichiometry distributions were investigated by ESCA depth profiles, structural properties

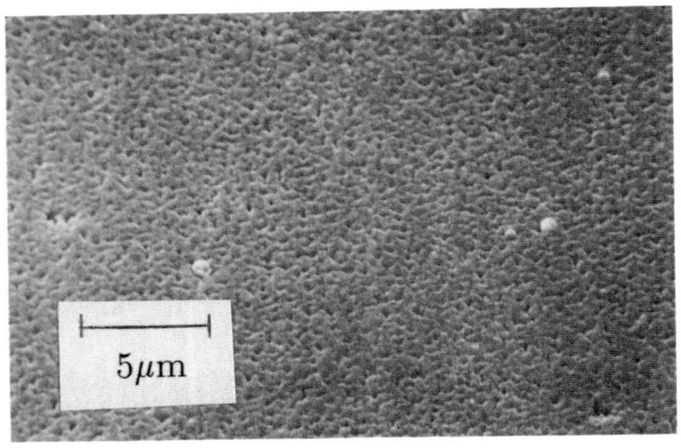

Fig. 1 REM photograph of a $YBa_2Cu_3O_7$ film prepared at 830°C and 0.1 mbar O_2.

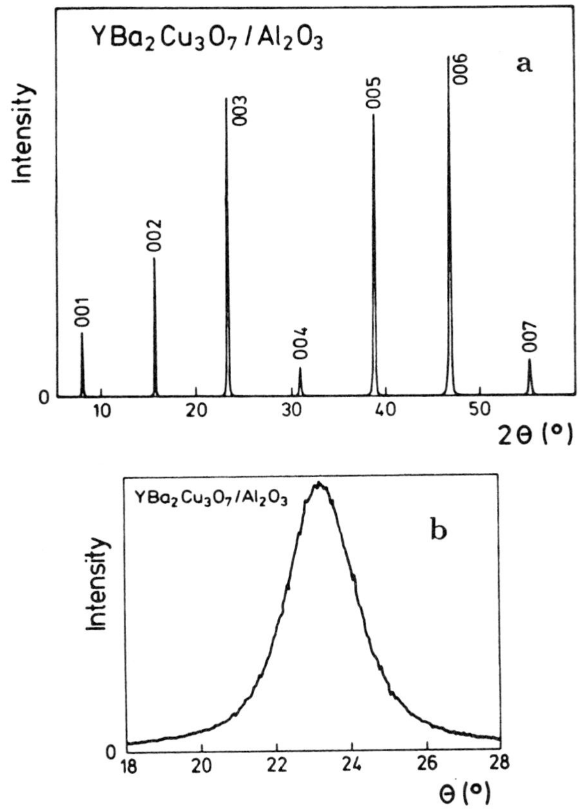

Fig. 2 x-ray diffraction pattern (a) and rocking curve (b) of a film on Al_2O_3.

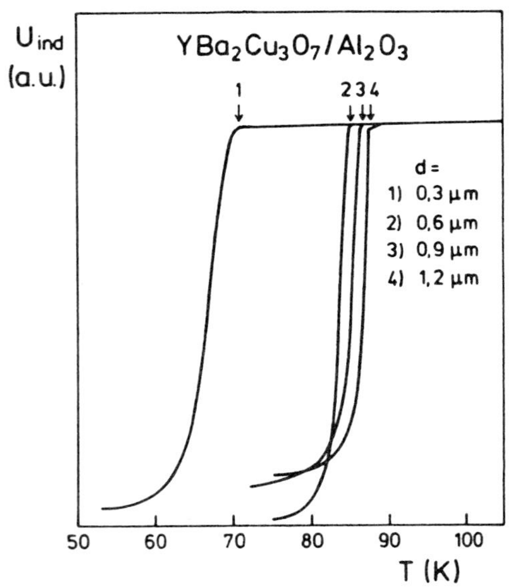

Fig. 3 Inductive transitions of films with the indicated thicknesses.

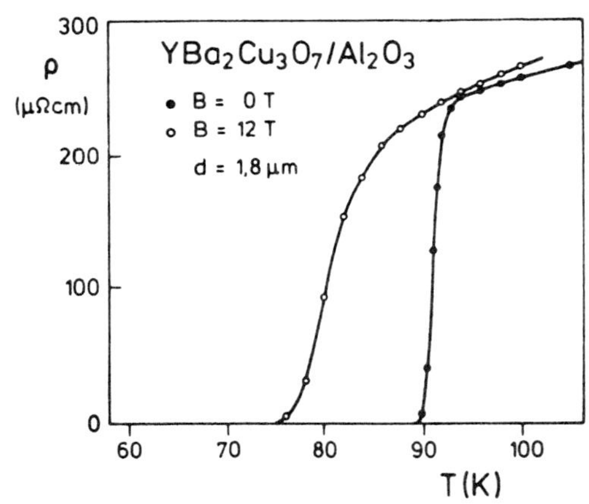

Fig. 4 Resistive transition of a film at the indicated fields.

were checked by X-ray diffraction in Bragg-Brentano (BB) and Seeman-Bohlin (SB) geometry.

3. Results and Discussion

All films grow single phase with the c-axis oriented perpendicular to the film surface (see Fig. 2a, where a X-ray diffraction pattern in BB geometry is shown). From analyzing X-ray diffraction in SB geometry a small polycrystalline portion could be detected with its intensity increasing with decreasing T_S. Best results i. e. narrowest rocking curves were obtained at $T_S \geq 830°C$ with FWHM = 2.1°, as can be seen in Fig. 2b for the (006) line. Similar X-ray-data have been reported from Char et al. (1990) for significant lower T_S and a FWHM of 8° has been found for sputtered films on random oriented Al_2O_3 by Schauer et al. (1990). At such high substrate temperatures a reaction between the superconductor and the Al_2O_3 occurs at the interface. Consequently a dependence of the superconducting properties on the film thickness d appears. As can be seen in Fig. 3, where the inductive transitions for samples with different d are shown, a drastic reduction of T_c is observed for d = 0.3 μm. The resistivity curves of corresponding samples show contributions from superconducting and semiconducting phases. These observations are consistent with the results from ESCA depth profiles from which a thickness of $\approx$ 200 nm for the disturbed interface layer can be estimated. For film thicknesses of 1 ... 2 μm typical inductive T_c values of 85 to 88 K are measured. These films exhibit a resistivity of $\approx$ 500 $\mu\Omega$cm at ambient temperature and a linear ρ(T) behavior with $\rho(300K)/\rho(100K) \approx 2$.

In Fig. 4 the resistive transition of a film at B = 12 T and in zero field is displayed. From the shift of $T_{c,zero}$ one can deduce a slope of the upper critical field of $(dB_{c2}/dT)_{T_c} \approx -0.86$ T/K for B || c-axis. The transition at 12 T is broadened about a factor of 6 compared to the zero field data. For films with d $\leq$ 0.6 μm this broadening and the reduction in field is slightly stronger.

Critical current measurements were carried out with laser patterned microbridges of 100 μm width and 2 mm length applying the 1-μV-criterion. In the temperature region between 80 K and 20 K a linear j_c(T) behavior was detected with $j_c(20K) \approx 6 \cdot 10^4$ A/cm^2 for films with d $\geq$ 0.9μm and $dj_c(T)/dT \approx 10^3$ A/(cm^2·K). These values are about a factor of 10 larger than those reported by Schauer et al. (1990). The field dependence of the critical current exhibits two destinct regions: In the low field region up to $\approx$ 1 T a steep decrease of j_c(T) is observed, whereby a scaling of the slope n = d(ln j_c(T))/d(ln B) with the temperature T can be recognized. At 50 K a slope of n = 1 was found as can be seen in Fig. 5. In the high field region 1 T $\leq$ B $\leq$ 12 T especially at low temperatures j_c(B) is almost constant. Whereas in lower fields j_c is independent of the film thickness, the high field "saturation value" is drastically influenced by d. This difference is also enhanced with increasing temperature. Simultaneous with the lowered j_c a broadening of the I–V curve appears. This is illustrated by the bars in Fig. 5, where the lower end-point corresponds to a 0.1-μV-criterion, the upper to a 1-μV-criterion. Thus it might be possible by growing even thicker films to reach higher critical currents.

Although we have shown that it is possible to prepare $YBa_2Cu_3O_7$ films with good structural and superconducting properties directly on sapphire, the comparatively low j_c values might force us to the application of buffer layers. These layers must grow epitaxial on Al_2O_3 and serve as proper underlayer for the epitaxy of $YBa_2Cu_3O_7$. Wiener-Avnear et al. (1990) realized an improvement of the superconductor-film quality with $SrTiO_3$ underlayer, but even their T_c values are still not perfect. However a progress in this direction will enable us to utilize both the properties of epitaxial superconductor films and those of the customary substrates for microelectronic devices.

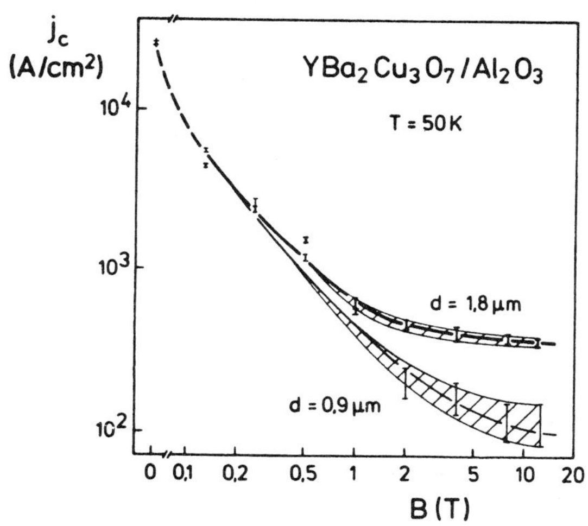

Fig. 5 Field dependence of j_c(50K) of 2 films with different thicknesses

The authors thank I. Büttner, G. Grabe, and H.-W. Grueninger for experimental assistance and the Bundesminister für Forschung und Technologie for financial support.

References

Char K et al 1990 *Appl. Phys. Lett.* **56** 785

Li Q et al 1989 *Appl. Phys. Lett.* **55** 1792

Poppe U et al 1989 *Solid State Commun.* **71** 569

Schauer W et al 1990 *Cryogenics* **30** 589

Venkatesan T et al 1989 *Appl. Phys. Lett.* **54** 581

Watanabe K et al 1990 *Appl. Phys. Lett.* **56** 1490

Wiener-Avnear E et al 1990 *Appl. Phys. Lett.* **56** 1802

Supercond. Sci. Technol. 4 (1991) S172–174. Printed in the UK

Oxygenation of $YBa_2Cu_3O_x$ Thin Films and Multilayers

R.G.Humphreys, N.G.Chew, J.A.Edwards, J.S.Satchell, S.W.Goodyear and S.E.Blenkinsop

Royal Signals and Radar Establishment, Malvern, Worcs. WR14 3PS, UK

Abstract. The oxygen requirements for the successful growth of $YBa_2Cu_3O_7$ thin films have been studied. Good quality films can be produced at much lower pressures than suggested by the data of Bormann and Nolting. The main benefit from the use of atomic oxygen is during cooling after growth rather than during growth itself. Epitaxial Y_2O_3 overlayers inhibit oxygen diffusion into smooth $YBa_2Cu_3O_x$ films.

Thin films of $YBa_2Cu_3O_7$ have been grown by evaporation of the metals from electron beam heated sources onto MgO substrates in the presence of $\sim 10^{-4}$ mbar of oxygen, 10% of which consists of atoms rather than molecules. More details of the growth process and the calibration procedures have been given elsewhere[1–3]. We have previously shown that the composition must be carefully controlled to obtain high critical current density and smooth surfaces, at least at the low oxygen pressures permissible during evaporation. Fully epitaxial multilayers of $YBa_2Cu_3O_7$ and Y_2O_3 have been grown. The growth conditions for the Y_2O_3 were the same as those used for the superconductor (690C, 2 10^{-4} mbar oxygen with 10% atomic). RHEED observations during growth showed that the <100> directions of the Y_2O_3 were aligned with the <110> directions of the MgO and $YBa_2Cu_3O_7$. X-ray measurements showed that both materials grew (001) oriented. In this paper we concentrate on the oxygen requirements of thin films and multilayers.

An adequate oxygen supply is clearly necessary in order to grow $YBa_2Cu_3O_x$ thin films in situ from the metals. The pressure required is a function of the growth temperature, and

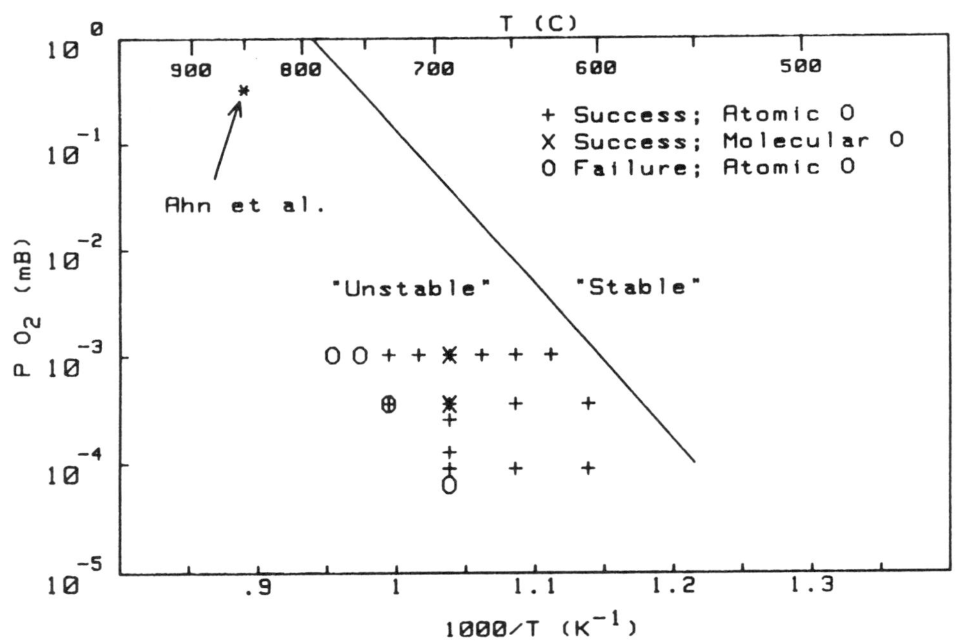

Fig.1 Temperature–pressure diagram showing conditions for film growth.

0953-2048/91/01S172 + 03 $03.50

has been deduced from measurements of the stability of bulk material by Bormann and Nolting[4]. Our results are summarised in Fig. 1. All the films indicated were grown under conditions of lower oxygen pressure than would be required by the Bormann and Nolting line, (solid line in Fig. 1). Most of the results shown were obtained using atomic oxygen, which might be expected to improve the stability range of the film, but two films were successfully grown using only O_2. Furthermore, it appears that the high temperature limit set by our data is pessimistic: growth at 740C and 2 10^{-4} mbar was successful when the layer was prenucleated at 690C, but not when nucleation was attempted at the higher temperature (this led to the growth of a transparent, non-epitaxial insulating film). This suggests that it is nucleation rather than the stability of the $YBa_2Cu_3O_x$ film which is limiting our other results.

This evidence indicates clearly that the Bormann and Nolting line does not give the practical stability limit for the growth of $YBa_2Cu_3O_x$ films. Although thin film growth is not an equilibrium process, it seems probable that the reason is indicated by the work of Ahn et al.[5], who showed that a number of oxygen pressure plateaus can be observed in oxygen titration, depending on the exact composition. They found the absolute stability limit of $YBa_2Cu_3O_x$ was at much higher temperature, as indicated by the asterisk in Fig. 1, which is consistent with our data. A corollary of this explanation is that the properties of material grown at these low pressures may be more sensitive to composition errors than that grown at high pressure.

Active oxygen is used by many groups growing high temperature superconductor thin films, but its precise role remains uncertain. We have carried out experiments in which the only intentional variable was whether the microwave power to the oxygen plasma[6] was switched on. Table 1 shows the critical temperatures of films grown and cooled in different combinations of atomic and molecular oxygen, together with the effect of post-annealing at 500C in one bar of oxygen. These films were grown at a substrate temperature of 690C and a pressure of 2 10^{-4} mbar, and cooled in ~10^{-2} mbar. There is remarkably little difference between the critical temperatures of films grown with and without activated oxygen. There is, however, a clear benefit in cooling the films from the growth temperature in the presence of atomic oxygen, and this remains true even if the layer is subsequently annealed in 1 bar of oxygen. Although the use of atomic oxygen is clearly necessary to produce the best films in our growth regime, it is possible to grow films with reasonably good electrical properties (T_c=84.5K, J_c=1.2 10^6A/cm^2 at 77K) without any atomic oxygen being used.

The effect of atomic oxygen was corroborated in a further experiment. The sample grown with, but cooled without, atomic oxygen was cleaved into two pieces. One was replaced in the evaporator, heated to 690C, and cooled following our post-growth procedure. Its T_c increased from 86.5 to 91.8K. The other was annealed at 700C in oxygen at atmospheric pressure, and its T_c only increased by about a degree. This underlines the importance of oxygen activity during cooldown, and is supported by the following results on bilayers.

	Atomic Oxygen during growth			
	As grown		Annealed 500C	
Atomic Oxygen during cool down	on	off	on	off
on	88.5	88.9	91.5	89
off	68.2	71.5	86.5	84.5

Table 1. Effect of atomic oxygen on the zero resistance critical temperature

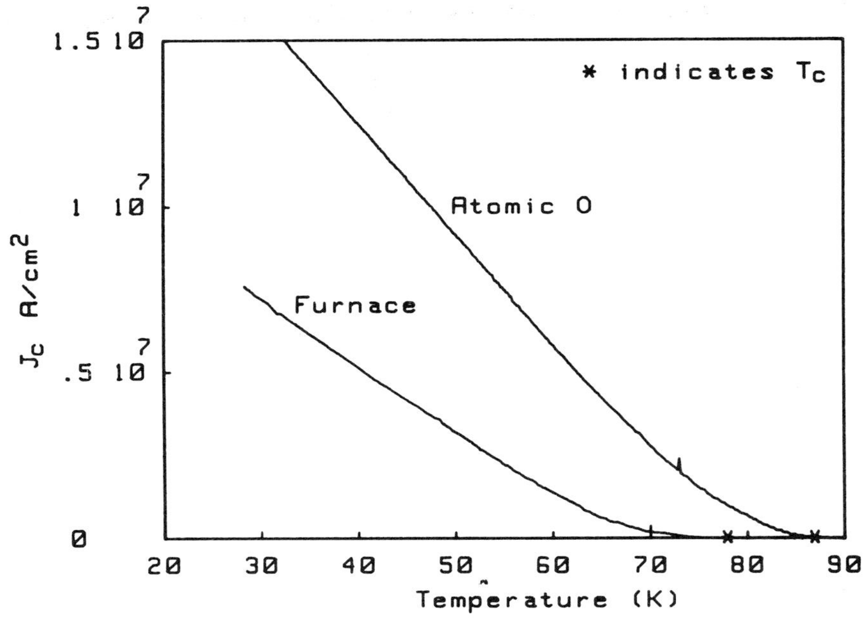

Fig. 2 Critical current density vs temperature for a film coated with 200A of epitaxial Y_2O_3, after furnace annealing, followed by atomic oxygen.

It is found that the T_c of $YBa_2Cu_3O_x$ coated with 200Aof Y_2O_3 is markedly reduced as grown, and a remarkable spread was observed in the annealing behaviour. This has been correlated with the surface morphology of the films. Films with lumpy surfaces (containing excess Cu) are found to oxygenate rapidly, but smooth films (slightly Y rich and Cu deficient) are slow to oxygenate at 500C and have depressed T_c and J_c. This is attributed to the Y_2O_3 acting as an effective diffusion barrier, even when only 200A thick. One such film was extensively annealed at 500 (4 hours) and 550C (0.5 hours) without improving its T_c beyond that achieved in a short anneal. It was cleaved into two pieces, one of which was ion milled to remove the Y_2O_3. Subsequent furnace annealing still did not give a high T_c. However, on reinserting in the growth chamber, and annealing in atomic oxygen T_c, the T_c of both halves rose to 87K (Fig. 2).

These results can be understood if oxygen sites other than the chain ones are incompletely populated in the film at the growth temperature. Subsequent treatment with molecular oxygen fills the chains, but the much higher activity of oxygen atoms is required to fill the other sites. It is encouraging that the Y_2O_3 is of sufficient integrity to act as a diffusion barrier, and that oxygenation was eventually possible through it using atomic oxygen annealing. These data show the importance of using the correct post-growth cooling procedure for smooth multilayers.

1. R.G.Humphreys, J.S.Satchell, N.G.Chew, J.A.Edwards, S.W.Goodyear, S.E.Blenkinsop and A.G.Cullis Supercond. Sci. Technol. 3, 38 (1990)

2. N.G.Chew, S.W.Goodyear, J.A.Edwards, J.S.Satchell, S.E.Blenkinsop and R.G.Humphreys To be published.

3. J.A.Edwards, N.G.Chew, S.W.Goodyear, J.S.Satchell, S.E.Blenkinsop and R.G.Humphreys. Presented at the E-MRS, Strasbourg, May 1990.

4. R.Bormann and J.Nolting App.Phys.Lett. 54, 2148 (1989)

5. B.T.Ahn, V.Y.Lee, R.Beyers, T.M.Gur and R.A.Huggins Physica C 167, 529 (1990)

6. N.G.Chew, R.G.Humphreys and J.S.Satchell UK patent application 9009319 (1990)

Supercond. Sci. Technol. **4** (1991) S175–177. Printed in the UK

Preparation and properties of in-situ sputtered YBCO films‡

A.M. Kadin*, P.H. Ballentine†, D.S. Mallory*, and J.P. Allen†

* University of Rochester, Rochester, New York 14627 USA
† CVC Products, Inc., Rochester, New York 14603 USA

Abstract. We have used rf magnetron sputtering from a large near-stoichiometric YBCO target onto heated substrates to produce c-axis oriented films with T_c=87 K *in situ*, without a subsequent post-anneal. Effects of negative ion bombardment were avoided by placing substrates above the center of the target, directly above a grounded ion shield. Film properties are rather sensitive to small changes in target composition. At 66 K, critical currents of optimum films are greater than 1MA/cm^2, and microwave surface resistance at 2 GHz is less than 10 mΩ.

Sputtering is a well-controlled method for depositing thin films of multi-element materials, which will produce stoichiometric films from a stoichimetric target under appropriate conditions. However, the process of sputtering from a target of $YBa_2Cu_3O_7$ (YBCO) can be plagued by serious problems associated with negative ion bombardment of the growing film. This bombardment produces resputtering of the film, which causes compositional shifts and nonuniformity. We have developed a process using a large-area target in a configuration that avoids these negative ion effects, enabling us to obtain reproducible, uniform YBCO films at a reasonable deposition rate.

Many potential applications of YBCO films require large in-plane currents, so that the crystalline c-axis must lie perpendicular to the substrate. This can be achieved on a wide variety of substrates if the films grow as deposited in the tetragonal crystal structure, and this in turn requires substrate temperatures of order 600°C or above. If the films are cooled in the presence of oxygen, they convert to the proper orthorhombic structure *in situ*, without the need for a subsequent post-anneal. In this way we have obtained films with zero resistance at 87 K and critical current density at 77 K of order 10^6 A/cm^2. We also find, however, that the microstructure and superconducting properties of these in-situ films can be very sensitive to small changes in film composition and to defects resulting from ion bombardment.

The YBCO films discussed here were deposited using an 8-inch diameter oxide target in the rf magnetron mode, with most of the sputtering occuring from a "magnetron racetrack" about 5-7 inches in diameter. As we showed earlier (Kadin *et al.* 1989), resputtering effects from negative ion bombardment (from oxygen ions formed at the target) are concentrated directly above this racetrack region. In this region, films are much thinner, and dramatically depleted in Ba (and to a lesser extent, Cu) compared to films deposited above the center of the target. In a central region a few inches in diameter, however, there is very little ion bombardment and the composition of the film is close to that of the target. We have therefore chosen this center region for the placement of our substrates (see Figure 1). This configuration is essentially equivalent to off-axis sputtering (Eom *et al.* 1989), except that the large target and cylindrical symmetry yield a higher deposition rate (≈1 Å/sec) and better spatial uniformity.

---

‡ Supported in part by the National Science Foundation, NASA Jet Propulsion Laboratory, and the New York State Institute on Superconductivity.

For most of this work, the target material was made from the constituent oxides to the proper (nearly stoichiometric) composition and reacted at high temperatures, then reground to form a fine powder. This unsintered loose powder, spread in a tray about 2 mm deep, formed the target for this sputter-up configuration. Although some care must be taken to degas the target during initial power-up, targets of this type have yielded reproducible, high quality films for dozens of runs. The results are comparable to those obtained using a solid sintered target of similar size and composition, which is much more expensive.

Recent efforts have concentrated on the in-situ production of c-axis oriented YBCO thin films by heating the substrate (single-crystal MgO, cubic ZrO_2, $LaAlO_3$, or $SrTiO_3$) to 650-750°C during deposition, and providing a sufficient supply of oxygen during and particularly after deposition. Recently, we have been sputtering in an atmosphere of 10 mTorr oxygen and 20 mTorr Ar, and cooling down afterwards in ≈100 Torr oxygen. We found, however, that one additional modification was needed to obtain high-quality superconducting films. This is the "negative ion shield", a grounded disk placed below the substrate and above the center of the target (see Figure 1). Without this shield, we still obtained a good c-axis oriented film, but it exhibited an extended c-axis with a substantially depressed T_c (below 77 K). We suspect this is due to frozen-in defects produced by the small amount of negative ion bombardment still present above the center of the target. Using the shield to eliminate any possibility of ion damage, we obtained films with T_c=87 K and a proper c-axis spacing of 11.67 Å.

We would like to obtain c-axis oriented epitaxial films with proper 1-2-3 composition, smooth surfaces, high critical temperature, large critical currents, and low microwave losses. However, although our results are very reproducible, we have had difficulties in trying to attain all of these properties at once. In particular, we were surprised to find that rather small changes in film composition translate into rather large changes in some of these properties. In addition, we have identified a small but significant shift from the target composition to that of the film. For a stoichiometric target with metallic concentrations Y/Ba/Cu of 17/33/50, careful measurements of the film composition with inductively-coupled plasma spectroscopy (ICP) gave a composition of 14/36/50. We suspect that the Ba enrichment is the counterpart of the Ba depletion taking place above the plasma ring; in steady-state, those Ba atoms have to go somewhere. Films with this composition ("Ba-rich") were very smooth and highly c-axis ori-

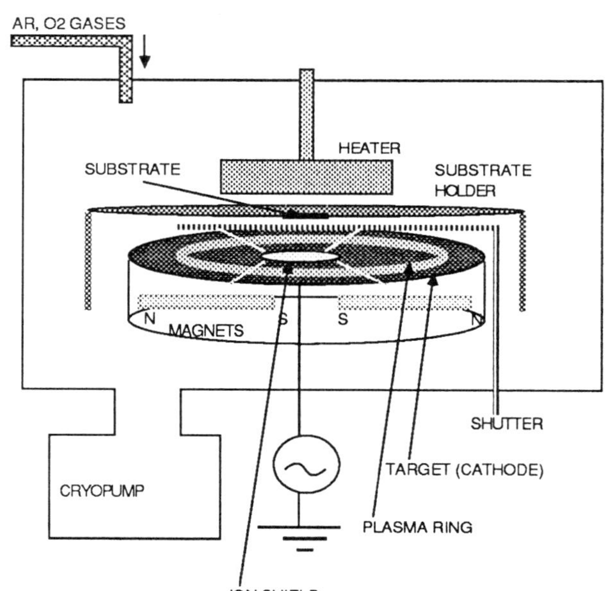

Figure 1. Schematic of UR/CVC magnetron sputtering system for YBCO films, showing placement of ion shield relative to target and substrate.

ented, but unfortunately with depressed values of T_c. Altering the target composition slightly, films that were Cu-rich with composition 14/29/57 exhibited very good dc super- conducting properties as well as orientation, but were very rough. Microscopic examination showed that the extra Cu led to a high density of CuO crystallites sticking out of the top of the film, some of which could even be wiped off. In contrast, films that were Y-rich (19/32/49) were smooth and had good superconducting properties, but seemed to have a high-resistance surface layer that sometimes made electrical contract difficult. Finally, films with ideal 1-2-3 composition (to better than 1%) showed very good superconducting properties, good electrical contact, and relatively smooth films. The normalized dc resistance curves for these films are shown in Figure 2.

We measured the critical current density of these films using transport measurements on samples patterned into narrow strips 20 μm wide. For stoichiometric films, J_c was ≈1 MA/cm^2 at 77 K, increasing to ≈10 MA/cm^2 at 4 K. Off-stoichiometric films were somewhat lower. We also carried out measurements of the microwave surface resistance R_s in the 1 - 3 GHz range, using a stripline resonator (Kadin *et al.* 1990). These results indicate that R_s at 2 GHz in the best films is less than 10 mΩ at 66 K, but goes down to <<1 mΩ at 4 K.

In summary, we have developed a method for preparing *in-situ* c-axis oriented films on heated MgO, ZrO_2, $LaAlO_3$, and $SrTiO_3$ substrates. Careful control of film composition to within 1% is necessary to combine excellent superconducting properties with proper film morphology.

References

Eom C.B., Sun J.Z., Yamamoto K., Marshall A.F., Luther K.E., Geballe T.H., and Laderman S.S. 1989, *Applied Physics Letters* **55** 595.

Kadin A.M., Ballentine P.H., Argana J., and Rath R.C. 1989 *IEEE Transactions on Magnetics* **25** 2437.

Kadin A.M., Mallory D.S., Ballentine P.H., Rottersman M. and Rath R.C. 1990, in *Superconductivity and Applications*, ed. H.S. Kwok, D.T. Shaw, and Y.H. Kao (New York: Plenum) pp. 757-766.

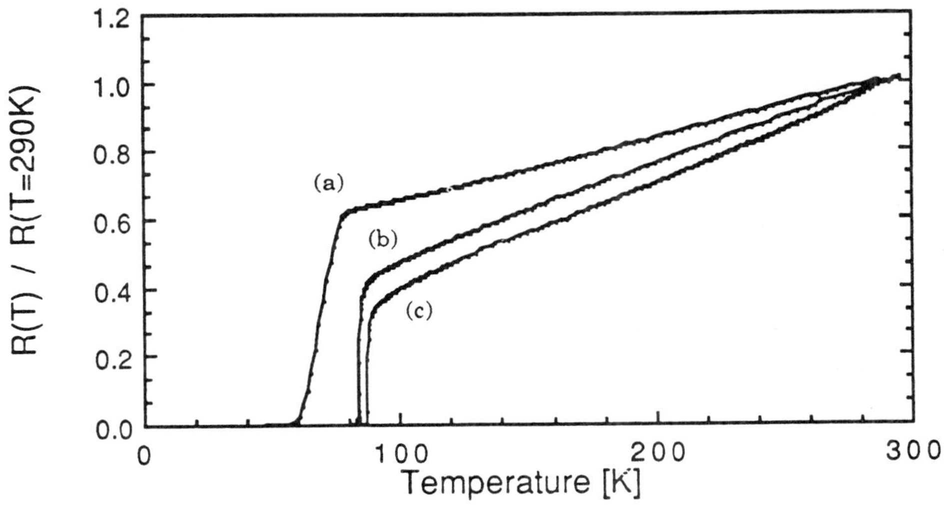

Figure 2. Normalized R(T) for several *in-situ* YBCO films of slightly different composition, ≈0.5μm thick on (100) MgO substrates. (a) Ba-rich (Y/Ba/Cu = 14/36/50), (b) Y-rich (19/32/49), and (c) Cu-rich (14/29/57), as well as stoichiometric films (17/33/50).

Supercond. Sci. Technol. **4** (1991) S178–180. Printed in the UK

Coevaporated Y-Ba-Cu-O thin films on different substrates

H Ohlsén, J Hudner* and L Stolt

Swedish Institute of Microelectronics, P.O. Box 1084,
S - 164 21 Kista, Sweden

* Royal Institute of Technology, Department of Solid State
Electronics, P.O. Box 1298, S - 164 28 Kista, Sweden

Substrates of $SrTiO_3$ are generally recognized as the most suitable for growing epitaxial superconducting thin films of $YBa_2Cu_3O_7$ (YBCO) with high transition temperature and high critical current density. Regardless if different deposition methods such as sputtering, coevaporation, or laser ablation are used, superconducting films with T_c above 90 K and I_c exceeding 10^6 A/cm^2 at 77 K have been prepared on this substrate material. However, $SrTiO_3$ is not suitible for high frequency applications due to its high losses in this regime. It is therefore of interest to study film preparation on other substrates which are better suited for applications. Several alternatives to $SrTiO_3$ have recently been proposed. In this work a comparison is made between coevaporated YBCO layers using the BaF_2 method on seven different, commercially available, single crystal substrate materials.

The layers of YBCO were fabricated using an UHV system equipped with two electron beam guns (Y and Cu) and one effusion source (BaF_2 powder). During the evaporation process oxygen was bleeded into the chamber through a needle valve directed towards the substrate. The evaporation rates and inlet of oxygen were controlled by a quadrupole mass-spectrometer system (Hudner et al 1990). No intentional heating of the substrates was used and the temperature did not exceed 150°C during deposition. The base pressure was 10^{-9} mbar and the oxygen pressure during deposition was $5x10^{-6}$ mbar. The total $YBa_2F_4Cu_3O_x$-rate was approximately 5 Å/s.

The single crystal substrates ($SrTiO_3$, $LaAlO_3$, $LaGaO_3$, Al_2O_3, $CaNdAlO_4$, $LiNbO_3$ and MgO), all of the size 1x1 cm^2, were positioned close together in a sample holder. The configuration was such that composition deviation as compared to each other was minimized and estimated to be smaller than 5%.

After evaporation the as-deposited films were amorphous (from XRD) and insulating. The films of each deposition run were annealed together in flowing wet oxygen (2 l/min) at 800-900°C in order to exchange the flourine for oxygen and form the tetragonal non-superconducting phase (Mankiewich et al 1987). The oxygen was bubbled through high purity water heated to 95°C and then let into the furnace. This was followed by a slow temperature ramp down of 2.5^{o}C/min in dry oxygen during

0953-2048/91/01S178 + 03 $03.50

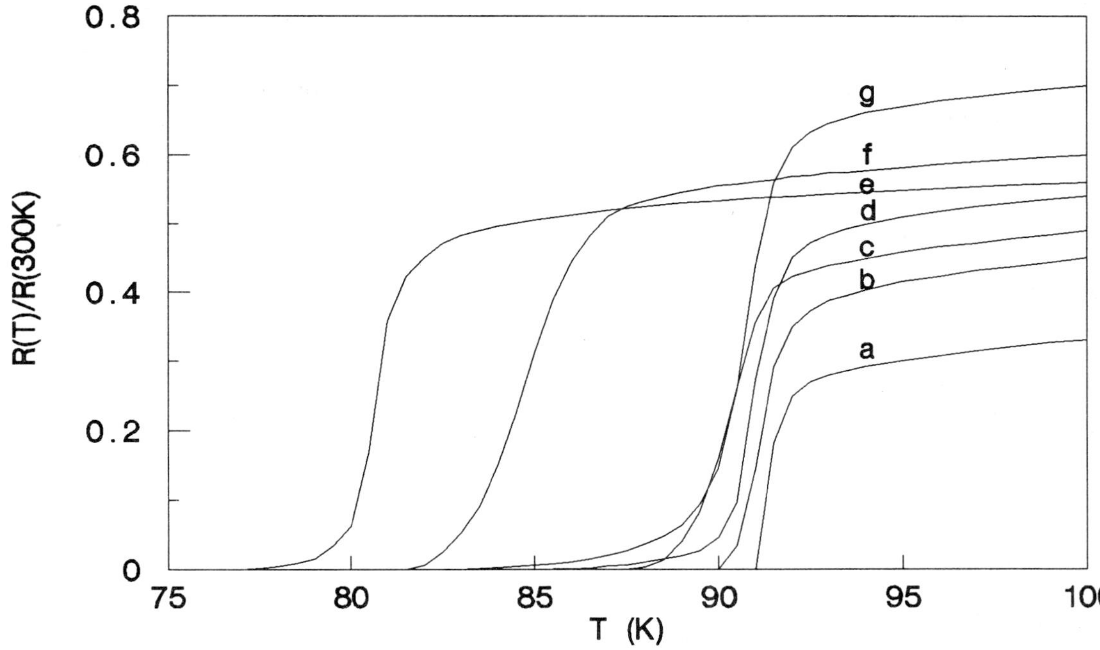

Fig. 1. Normalized resistivity at 300 K versus temperature for: a-$SrTiO_3$, b-$LaAlO_3$, c-$LaGaO_3$, d-Al_2O_3, e-MgO, f-$CaNdAlO_4$, and g-$LiNbO_3$.

which the superconducting orthorombic YBCO phase is formed. The annealing time was counted from the moment the samples are pushed into the center of the hot furnace until the ramp down procedure is initiated.

In Fig. 1 the resistivity, normalized at 300 K, versus temperature is shown for a one micron thick YBCO film on the different substrates. The annealing time at 900°C was 10 min. It can be seen that the superconducting onset temperature is the same for the films on $SrTiO_3$, $LaAlO_3$, Al_2O_3, and $LiNbO_3$ substrates (around 94 K). The onset is only 1-2 K lower for the film on $LaGaO_3$ while it is about 5 and 10 K lower for $CaNdAlO_4$ and MgO, respectively. The YBCO film on Al_2O_3 and $LiNbO_3$ exhibit a tail in resistivity leading to a low T_c value despite the higher onset temperature.

In Table 1 results from three depositions with the same composition denoted **A**, **B** and **C** are presented. **A** represents a deposition of a one micron thick YBCO film on all the seven different substrates and an annealing at 900°C for 10 min (see Fig. 1). **B** represents a deposition (one micron thick) without $CaNdAlO_4$ and an annealing at 840°C for 25 min. **C** represents a deposition of a 0.8 micron thick YBCO film on $SrTiO_3$, $LaAlO_3$, $LaGaO_3$, and $CaNdAlO_4$ annealed at 840°C for 25 min. It should be pointed out that the heat treatments used in these experiments are not optimized for YBCO films on each single substrate (see e.g. Ohlsén et al 1989 for the case of Al_2O_3). None of the prepared films showed a semiconducting behaviour in the normal state as shown by the R(100K)/R(300K) ratio, and for the $SrTiO_3$ films the resistivity is linear in temperature and extrapolates to zero at zero temperature. The heat treatment at 840°C results in longer tails in resistivity leading to lower T_c values for the films on Al_2O_3, $LiNbO_3$ and MgO indicating a higher degree of interdiffusion between film and substrate. It can also be seen that for T_c values lower than

Table 1

Substrate		$SrTiO_3$	$LaAlO_3$	$LaGaO_3$	$CaNdAlO_4$	Al_2O_3	$LiNbO_3$	MgO
Orientation		(100)	(100)	(100)	(001)	$(1\bar{1}02)$	x-cut	(100)
$T_c(R=0)$ (K)	A	91.0	90.0	87.5	81.5	84.5	82.5	77.0
	B	91.0	90.0	87.5	-	72.0	77.0	76.0
	C	90.5	90.0	87.0	85.5	-	-	-
$\frac{R(100K)}{R(300K)}$	A	0.33	0.45	0.49	0.59	0.54	0.70	0.55
	B	0.34	0.42	0.46	-	0.66	0.66	0.62
	C	0.33	0.43	0.49	0.47	-	-	-
$\frac{R(Subst.)}{R(SrTiO_3)}$ (300K)	A	1.0	2.3	28.4	3.0	6.8	10.6	2.2
	B	1.0	1.5	9.8	-	4.4	4.6	4.2
	C	1.0	1.9	0.9	2.2	-	-	-

85 K the resistance ratio between the value at 300 K and 100 K is higher than 0.5. The resistance ratio at 300 K for the layers on the different substrates as compared to $SrTiO_3$ show that the resistance is higher in all cases except for $LaGaO_3$ in case **C**. Measurements at a current density of 1600 A/cm^2 of the films **C** in Table 1 resulted in a T_c of 89.5, 88.5, 74, and 79 K for $SrTiO_3$, $LaAlO_3$, $LaGaO_3$ and $CaNdAlO_4$, respectively.

An analysis using X-ray diffraction showed that the YBCO layer grows with the a-axis perpendicular to the surface on the $LaAlO_3$ and $CaNdAlO_4$ substrates. The film on $SrTiO_3$ was found to have a mixture of a- and c-axis orientation while it was completely (110)-axis oriented for $LaGaO_3$. The YBCO layers on Al_2O_3, $LiNbO_3$ and MgO were preferentially c-axis oriented.

Using the BaF_2 method with a high temperature post-annealing procedure, these results indicate that especially the $LaAlO_3$ substrate seems to be a promising substitute for $SrTiO_3$. An investigation leading to an optimized heat treatment specifically for the $LaAlO_3$ substrate to reduce interdiffusion will most probably increase both zero resistivity temperature and the critical current density.

This work has been supported by the Swedish Natural Science Research Council, the Swedish Board for Technical Development, the National Energy Administration and the Swedish Industrial Superconductivity Initiative.

References

Hudner J, Ohlsén H, Stolt L and Östling M 1990 Proc. EMRS Spring Conf. May 29 - June 1, Strasbourg, France

Mankiewich P M, Scofield J H, Skocpol W J, Howard R E, Dayem A H and Good E 1987 Appl. Phys. Lett. **51** 1753

Ohlsén H, Hudner J and Stolt L 1989 J. of the Less-Common Metals **151** 317

Supercond. Sci. Technol. 4 (1991) S181–183. Printed in the UK

Normal State Magnetic Susceptibility of Some High T_c Oxides.

J.R. Cooper, IRC in Superconductivity, University of Cambridge, Madingley Road, Cambridge CB3 OHE, U.K.

Abstract. The magnetic susceptibility (χ) of several sintered samples of $YBa_2(Cu_{1-x}M_x)_3O_7$ with M = Co or Zn has been measured in the normal state. Co doping introduces localised magnetic moments corresponding to 3.7 Bohr magnetons per Co atom and also gives a strong increase in the spin susceptibility of YBCO. In contrast the intrinsic effects of Zn doping seem to be small. The effect of oxygen loading on $\chi(T)$ of $Bi_2Sr_2CaCu_2O_{8+\delta}$ has also been studied.

1. Introduction

Systematic measurements of the normal state magnetic susceptibility (χ) of high T_c oxides may help towards understanding their electronic properties. They are a useful complement to NMR, ESR and specific heat studies.

2. Experimental

$\chi(T)$ measurements were made on the same sintered samples of $YBa_2(Cu_{1-x}M_x)_3O_7$ with M = Co or Zn, which have been extensively studied in our laboratory, by structural (Schmahl et al 1989), high resolution specific heat (Loram, Mirza and Freeman 1990) critical current (Blunt et al 1989) electrical resistivity (Porch et al 1990, Cooper et al 1990) and microwave experiments (Porch et al 1990). Data were taken using a Quantum Design SQUID magnetometer from just below T_c to 330K in fields of 50,000 gauss and linearity checks were made at lower fields for several fixed temperatures. Corrections were made for the gelatin or tubular quartz sample holders used. These corresponded to a temperature independent signal which was 20 or 60% respectively of that from 0.2gms of pure YBCO. The estimated accuracy of the χ_o values quoted here is ± 3%.

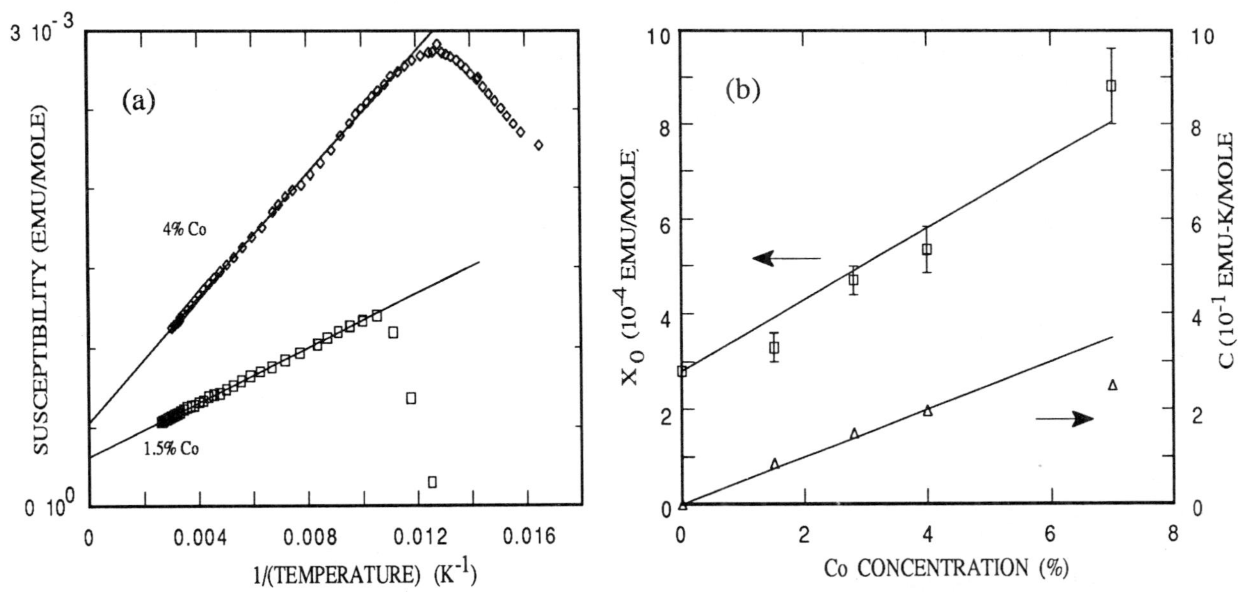

Fig.1a *Curie plots for two Co doped YBCO samples with x = 1.5 and 4%* . Fig.1b *Curie parameters (χ_0 and C)versus Co concentration x (%)*

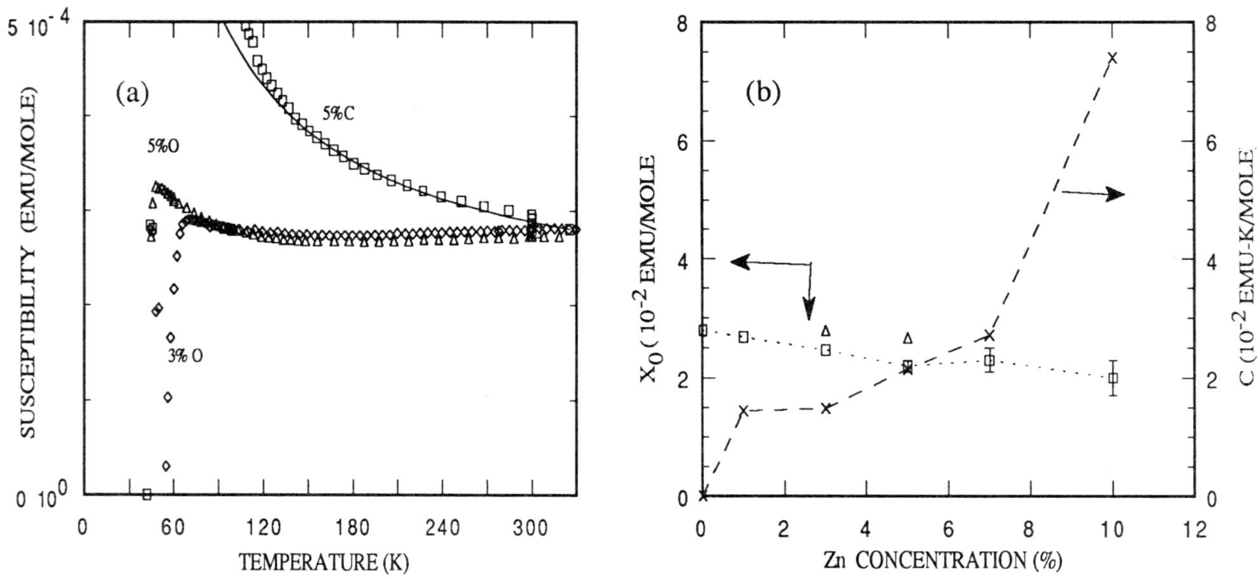

Fig.2a *χ(T) for three Zn doped samples prepared simultaneously. O using barium oxalate and C using the carbonate. Solid line is a fit to equation 1.*

Fig.2b *Plot of Curie parameters versus Zn concentration x(%). Open squares* $-\chi_0$ *and crosses - C for "carbonate" samples, triangles* $-\chi_0$ *for "oxalate" samples.*

3. Results and Discussion

As shown in Figure 1a the $\chi(T)$ data for the Co alloys fit a Curie law:

$$\chi = \chi_0 + C/T \qquad (1)$$

down to temperatures 10 to 20% higher than the zero-field value of T_c. The Curie constant C increases linearly with cobalt concentration (x) up to at least 4% (Figure 1b). Therefore in this temperature range the interactions between the magnetic moments on Co atoms are negligible and from Figure 1b a magnetic moment of 3.7±0.2 Bohr magnetons (μ_B) per Co atom can be calculated. This is closer to the spin only value expected for Co^{2+} (3.9 μ_B) than for Co^{3+}(4.7μ_B). Surprisingly χ_o also increases substantially with Co doping. At 4% doping the increase is nearly equal to the measured susceptibility of pure YBCO (Figure 1b).

Most of the Zn alloys studied here were also prepared from a mixture of metal oxides and barium carbonate. For these samples $\chi(T)$ could be fitted to equation (1) from 330 to ca. 200K but there were positive deviations at lower temperature (for example curve 5%C in Figure 2a). The results of these fits are shown in Figure 2b. Although C increases systematically with x , two samples prepared using barium oxalate showed a much smaller upturn in χ (Figure 2a) so the Curie term probably arises from impurity phases. A typical value of C in Figure 2b (2x10^{-2} emu-K/mole) corresponds to 1.8% of the Cu atoms having spin s=1/2. The χ_o values are nearly constant. However, a 20% decrease can be detected at 10% doping, which corresponds to 15% of the planar Cu(2) atoms being substituted by non-magnetic Zn atoms, and a T_c of only 10K.

Finally in Figures 3a and 3b are shown some results of a $\chi(T)$ study for single phase $Bi_2Sr_2CaCu_2O_{8+\delta}$, for which T_c can be varied from 90 to 60K by oxygen treatment (Namgung et al 1989). The oxygen content δ was varied by annealing the sample for 12 - 36 hours at various temperatures between 300 and 850°C in a gas flow of pure O_2, air or 2% O_2/N_2 mixture and then quenching into liquid nitrogen. Resistivity and AC susceptibility measurements have also been made. The samples used up to now were too small to determine the changes in δ by weighing, so δ was estimated by comparing our resistivity data with those of Namgung et al (1989). The three representative samples in Figure 3, Q4, Q7 and Q10, have resistivity midpoints of 62, 90 and 81K respectively corresponding to δ= 0.195, 0.18 and 0.17(A R West personal communication 1990). The AC susceptibility data gives sharper

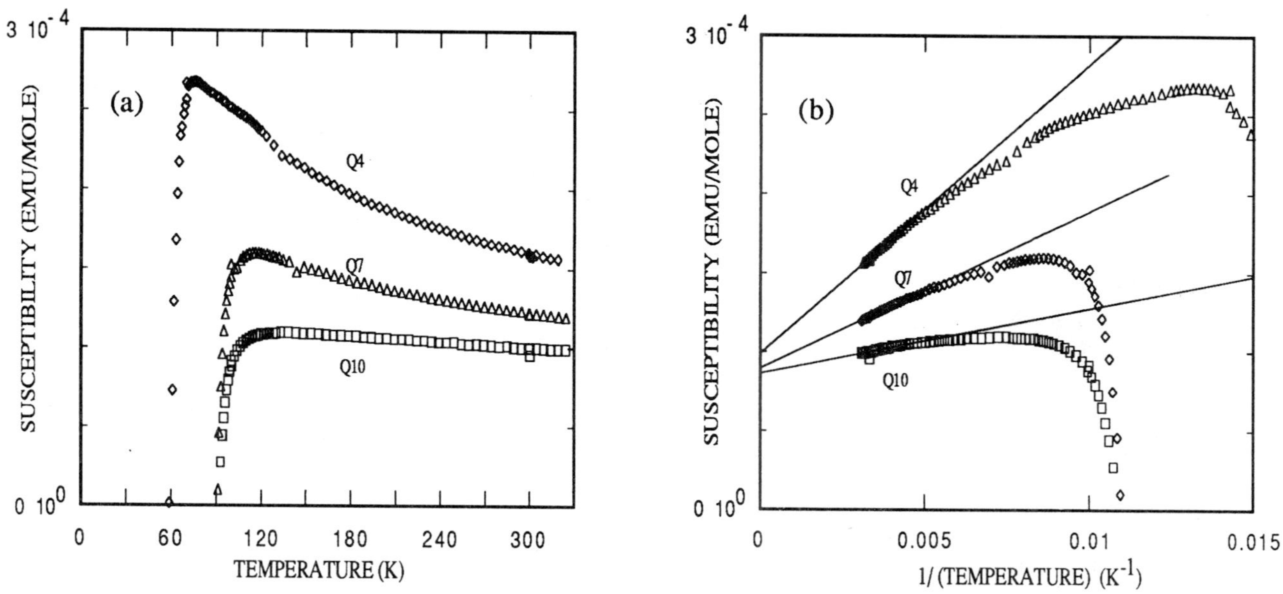

Fig.3a *$\chi(T)$ plots for a quenched $Bi_2Sr_2CaCu_2O_{8+\delta}$ sample. Q4 from 306C in O_2, Q7 from 606°C and Q10 from 850°C both in 2% O_2.* Fig. 3b *Curie plots of the data in Fig.3a. Solid lines are least squares fits above 200K.*

onsets with T_c= 62, 90 and 88K respectively. There are large changes in $\chi(T)$ with δ (Figure 3a). As δ is increased (Q10 to Q4) $\chi(T)$ becomes more Curie like and less "metallic", in contrast to the resistivity (not shown here) which becomes <u>more</u> metallic as δ is increased and T_c falls to 60K.

Curie plots are shown in Figure 3b. Good fits to Equation (1) are only obtained down to 200K which may possibly be due to the onset of diamagnetism from superconducting fluctuations in this extremely anisotropic compound. χ_o changes by ca. 15% while the number of localised spins (taken to be S = $^1/_2$) increases substantially with δ. Actually the numbers of spins per formula unit derived from Figure 3b, 0.049 (Q4), 0.027 (Q7) and 0.011 (Q10) closely follow the changes in δ given above. Thus it seems that most of the oxygen atoms introduced by the present oxygen treatment introduce a localised spin and it would be interesting to study this system by ESR. AC susceptibility measurements (not shown) indicate that for all values of T_c the intergrain screening currents develop below 60K.

In summary, we have made magnetic susceptibility measurements on a number of high T_c oxides in the normal state and have begun to correlate them with other physical properties.

Acknowledgements

P A Freeman (IRC, Cambridge) prepared most of the doped YBCO samples and C Namgung, J T S Irvine and A R West (University of Aberdeen) the single phase 2212 BISCO sample. J D Johnson and A R Perry helped with operation of the SQUID and data analysis. I am grateful to J S Schilling for a preprint of similar work on BISCO (Allgeier and Schilling 1990) and to W Y Liang, J W Loram and S D Obertelli for helpful discussions.

References

Allgeier C and Schilling J S 1990 to appear in Physica C
Blunt F J et al 1989 Physica C **162** 1605
Cooper J R et al 1990 These proceedings
Loram J W, Mirza K A and P A Freeman 1990 Submitted to Physica C
Porch A, Cooper J R, Cheah H M, Edmends J S and Waldram J R 1990 Physica **B165-166** 1199
Namgung C, Irvine J T S, Binks J H, Lachowski E E and West A R 1989 Supercond.Sci. Technol. **2** 181-184
Schmahl W W et al 1989 Phil. Mag. Lett. **60** 241

Supercond. Sci. Technol. **4** (1991) S184–186. Printed in the UK

Specific Heat Evidence for Chain Superconductivity in $YBa_2(Cu_{1-x}Co_x)_3O_{7-\delta}$

J.W.Loram, K.A.Mirza, P.F.Freeman and J.J.Tallon*

IRC in Superconductivity, Madingley Road, Cambridge CB3 OHE, UK.
* Physics and Engineering Department ,DSIR,Lower Hutt,New Zealand.

Abstract We have investigated the superconducting specific heat anomalies in $YBa_2(Cu_{1-x}Co_x)_3O_7$ for $0 \leq x \leq .07$. An initial small increase in T_c with x is accompanied by a rapid decrease in size of the anomaly. We suggest that this anomalous behaviour results from the suppression of superconducting order on the CuO chains . For x>.03 T_c decreases linearly with x due presumably to magnetic pair breaking with little further change in anomaly height. At low temperatures we find evidence of magnetic order.

1. Introduction

Metal ion substitutions for Cu in $YBa_2(Cu_{1-x}M_x)_3O_{7-\delta}$ have been extensively investigated and are reviewed by Markhert et al (1989). Two distinct patterns emerge from these studies. Divalent ions such as non-magnetic Zn or magnetic Ni substitute partially or wholly into Cu(2) plane sites and induce an immediate and rapid decrease in T_c due to magnetic pair breaking (Loram et al 1990a). The structure of such systems remains orthorhombic to large x. In contrast trivalent ions such as magnetic Co or Fe or non-magnetic Ga or Al substitute primarily for chain Cu(1). In these systems T_c remains constant up to x ~ .01 to .03 before decreasing at higher x rather slowly for non-magnetic dopants and more rapidly for magnetic ions. A macroscopic orthorhombic to tetragonal transition is observed around x ~ .025 (for Co) though there is evidence for an orthorhombic microdomain structure to higher x (Schmall et al 1989). No concensus has been reached on the origin of the initial plateau in T_c for trivalent substitutions.

2. Experimental

In this paper we present measurements of superconducting specific heat anomalies in $YBa_2(Cu_{1-x}Co_x)_3O_7$ for $0 \leq x \leq .07$. To resolve the small specific heat anomalies we use a differential technique (Loram 1983) to measure directly the difference in specific heat ΔC = C(x,T)-C(.07,T) between each sample and a $YBa_2(Cu_{.93}Co_{.07})_3O_7$ reference thus eliminating most of the large phonon term. Samples were prepared in the form of 5g buttons by a standard powder technique, the final stages being an anneal in flowing oxygen at 650 °C for 6 hours followed by slow cooling in oxygen (1 °C/min) to room temperature. Structural studies of these samples have been published (Schmall et al 1989) and resistivity data is presented at this conference (Porch et al 1990).

3. Results

The raw data for ΔC reveals a softening of low frequency phonons (~ 15 meV) and hardening of high frequency phonons (~ 60 meV) with increasing x, similar to that observed in $YBa_2Cu_3O_{7-\delta}$ with increasing δ. After subtraction of a phonon term we observe the low temperature anomalies suggestive of spin glass order shown in Fig 1. If the total spin glass entropy is ~ k_B /Co^{3+} , most of the anomaly must lie above 14K for x>.015. The decrease in magnitude with x (for x>.015) of the anomaly in the He temperature range shows that the mean Co-Co interaction energy increases more rapidly than linearly with x and is ~ 50-100K

0953-2048/91/01S184 + 03 $03.50

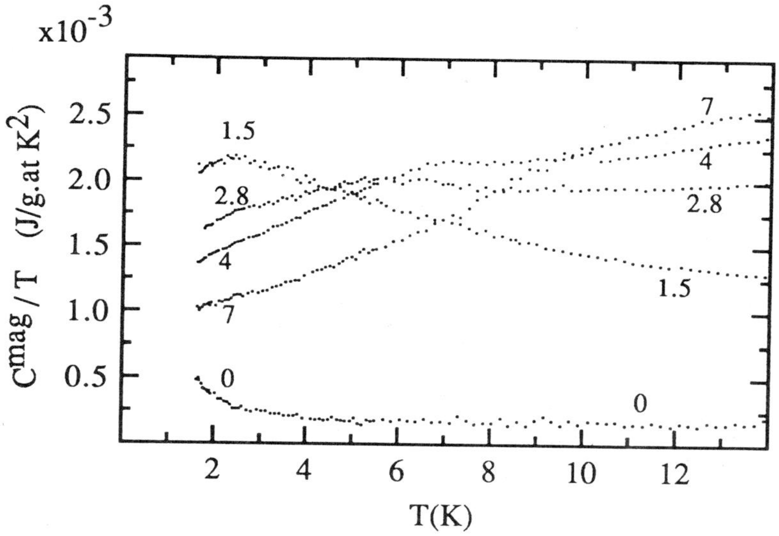

Fig 1. Low temperature magnetic anomalies in $YBa_2(Cu_{1-x}Co_x)_3O_7$. Labels show 100x.

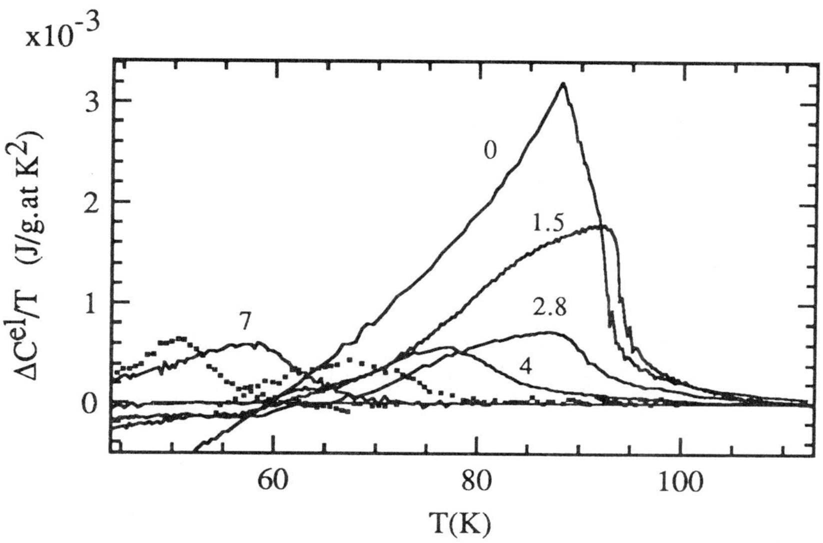

Fig 2.
Connected points: Specific heat anomalies in $YBa_2(Cu_{1-x}Co_x)_3O_7$. Labels show 100x.
Disconnected points: Anomalies in $YBa_2Cu_3O_{7-\delta}$ for δ=.2 (T_c~74K) and δ=.5 (T_c~54K)

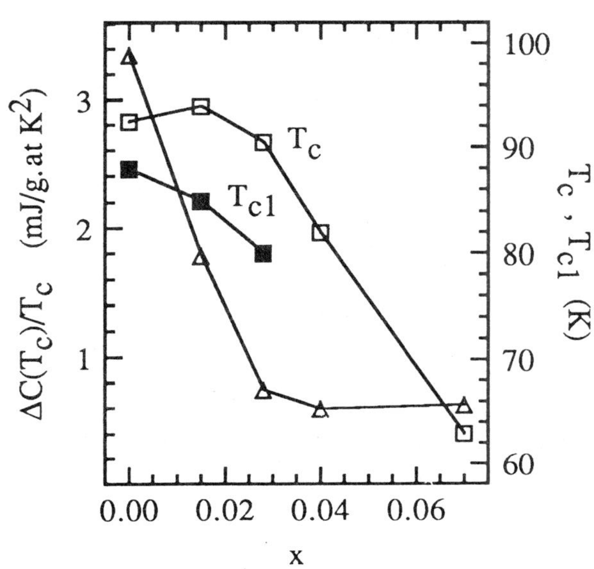

Fig 3. Anomaly height $\Delta C(T_c)/T_c$ (Δ), and transition temperatures T_c and T_{c1} against x for $YBa_2(Cu_{1-x}Co_x)_3O_{7-\delta}$.

for x=.07. Unless Co clustering is significant the Co-Co interaction may be of RKKY origin via unpaired chain electrons.

In Fig 2 we show the superconducting anomalies after subtraction of a smooth background term from ΔC. T_c, defined as the temperature of maximum negative slope of the anomaly, is shown in Fig 3. T_c shows an initial **rise** from 92.5K at x=0 to 94K at x ~ .015 followed by a linear decrease above x=.028, in agreement with transport and magnetic determinations. The most striking feature of the results is the immediate rapid decrease in size of the anomaly (and hence superconducting pair density) with x as shown in Fig 2. A similar effect has been previously reported for Fe^{3+} (Junod et al 1988) and Cr^{3+} (Kim et al 1989) substitution. The initial rise in T_c is especially surprising in view of this sharp fall in pair density and the very rapid decrease observed in the Hall carrier density n_H with x (Clayhold et al 1989). The specific heat behaviour on Co doping closely mirrors the effect of oxygen depletion in $YBa_2Cu_3O_{7-\delta}$. Junod et al (1989) find that the anomaly exhibits a rapid decrease in magnitude with little change in T_c for $0 \leq \delta \leq 0.15$ similar to our results for x≤.028 and this is confirmed by our results for $0 \leq \delta \leq .175$ presented in these Proceedings (Loram et al 1990b). Our measurements for δ=.2 (T_c~74K) and δ=.5 (T_c~54K) are shown in Fig 2 and are similar to Co doping with x > .04. In view of this striking similarity in behaviour for Co (and other trivalent ion) doping and oxygen depletion, it is interesting to speculate whether there is a common explanation.

We suggest that in undoped fully oxygenated YBCO the superconducting order parameter is large on the chains as well as on the planes, accounting for the uniquely large specific heat anomaly (pair density) and lower than usual anisotropy. Disrupting the chains by trivalent ion substitution or oxygen depletion suppresses the order parameter on the chains, thus reducing the size of the anomaly and increasing the anisotropy. Koch et al (1989) find that the carrier density on the chains is at least as large as that on the planes, and thus our observation (Loram et al 1990a) that the electronic specific heat coefficient at T=0 in YBCO is close to zero requires an energy gap for both chains and planes. Possible evidence for two distinct but coupled order parameters may be seen in the shape of the anomaly for the x=0 sample in Fig 2 which shows a sharp peak at T_{c1}~88K and a second order step with mid-point at T_c =92.5K (Loram et al 1990b). Doping the chains with Co strongly suppresses the 88K peak leaving small broad bumps at T_{c1}~85K and 80K for x=.015 and .028 respectively, but has a much weaker effect on the step at T_c. For this reason we tentatively attribute T_c and T_{c1} to the onset of superconducting order on the planes and chains respectively. It is significant that the rather constant anomaly size for x≥.04 or $\delta \geq .2$ is similar in magnitude to that observed in the more 2-dimensional Bi or Tl superconductors and we suggest that in this regime superconductivity is limited to the planes. If this interpretation of the specific heat results is correct it is clear that chain superconductivity in YBCO more than doubles the superconducting condensation energy.

References

Clayhold J.et al 1989 Phys. Rev **B39** , 77

Dunlap B.D.et al. 1988 Physica **C153-155**, 1100

Junod A. et al 1988 Physica **C152**, 495

Junod A, Eckert D, Graf T, Triscone G, Muller J.1989 Physica **C162-164**, 482

Kim S, Fisher R.A, Phillips N.E, Gordon J.E. 1989 Physica **C162-164** , 494

Koch B, Geserich H.P, Th. Wolf 1989 Solid State Comms.**71**, 495

Loram J.W. 1983 J.Phys **E16** , 367

Loram J.W, Mirza K.A, Freeman P.F 1990a. Submitted to Physica C

Loram J.W, Cooper J.R, Mirza K.A. 1990b These proceedings, **PS** 3.23

Markhert J.T,Dalichaouch Y,Maple M.B. 1990 "Physical Properties of High Temperature Superconductors 1", Ed D.M.Ginsberg(World Scientific)

Porch A, Cooper J.R, CheahH.M, Edmends J.S, Waldram J.R. 1990 LT19.

Schmall W.W et al. 1989 Phil.Mag. Lett. **60**, 241.

Supercond. Sci. Technol. **4** (1991) S187–189. Printed in the UK

Critical Fields And Superconducting Properties Of Single Crystal Bi-Ca-Sr-Cu-O

L.J.Chang, H.E.Horng and H.C.Yang[+], H.C.Chen[+]

Department of Physics, National Taiwan Normal University, Taipei, Taiwan, R.O.C.
+Department of Physics, National Taiwan University, Taipei, Taiwan, R.O.C.

Abstract. The magnetic data associated with the perpendicular and parallel magnetic fields to the ab-plane of Bi-Ca-Sr-Cu-O single crystal will be shown to study the effects of the magnetic field. It is found that all lower and upper critical fields are upturn curves in both orientations of magnetic fields. Comparing to H_{c2}'s , H_{c1}'s are extremely low. The parallel critical fields are about twice larger than those for perpendicular field in all temperatures. Finally, the temperature dependence of magnetic penetration depths and coherence lengths of this sample were calculated.

1. Introduction

One of the important superconducting oxides is BiCaSrCuO system, which was first discovered by Akimitsu et al(1988). After then, Maeda et al(1988) found superconducting oxides with T_c higher than 100K in this Bi-system. Since it is very hard to obtain a single phase from a bulk sample of this system, large and high-quality single crystals are required to measure physical properties and to elucidate the superconducting mechanism. Although a few papers have been reported by Shishido et al(1989), Katsui et al(1988) and Schneemeyer et al(1988) on single crystal works, a detailed study on single crystal is absent. Therefore, in this report, the systematical data in whole temperature range for magnetic fields both parallel and perpendicular to ab-plane will be presented.

2. Sample Growth

The starting materials were prepared by mixing high purity powders of Bi_2O_3, $SrCO_3$, $CaCO_3$ and CuO with the appropriate ratio (Bi:Ca:Sr:Cu=2:1:2:2). Then the 20g starting material was reground with alkali chloride solvent of KCl(20g) and NaCl(20g). The mixed powder was annealed in a alumina crucible by the following steps: (1) The powder was heated to 1200°C with the heating rate 300°C/hr and then kept at 1200°C for 2 hours. (2) The temperature was decreased down to 950°C at a rate of 300°C/hr and stayed at this temperature for 10 hours. (3) The followed cooling rates were 3°C to 800°C, 5°C to 700°C and 10°C to 300°C. And then, (4) furnace cooling to room temperature. After then, thin and irregular crystal plates about 10μm thick were found on the solvent surface. One of those was cut into a rectangular shape (1.2x1.0 mm^2) and its plane was proved being ab-plane with lattice constants a=5.33Å and b=5.25Å by x-ray precession method.

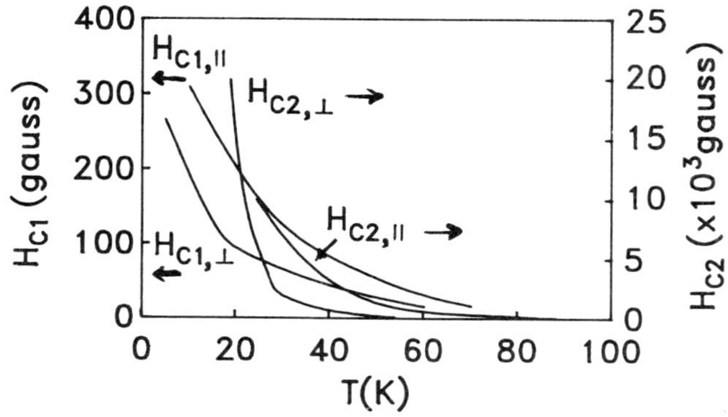

Fig. 1 Temperature dependence of critical fields.

3. Results And Discussion

The magnetic data taken from SQUID magnetometer (Quantum Design) will be shown to study the effects of the orientation of the magnetic field. In Fig. 1, the lower and upper critical fields(expressed as H_{c1} and H_{c2} respectively) for the magnetic fields both parallel and perpendicular to the ab-plane were plotted as a function of temperature. H_{c1} was defined as the magnetic field at the maximum of the -M vs H curve, while H_{c2} was determined by ZFC susceptibility measurement. It is found that all critical fields are upturn curves and the values for parallel magnetic field are larger than those for perpendicular case. Comparing to H_{c2} at a certain temperature, H_{c1} is extremely low. This phenomenon seems very common for most oxide superconducting systems.

One of the basic lengths of superconductivity is the magnetic penetration depth $\lambda(T)$, which is related fundamentally to the symmetry of the superconducting state and thus to the mechanism of pairing. By using ZFC data, $\lambda(T)$ at 5G parallel low-field was calculated by the thin plate equation:

$$\chi/\chi_o = 1- (\lambda/a)\tanh(a/\lambda) \quad ,$$

where a is one half of the thickness of the plate single crystal. In our calculation, Which is shown in Fig.2, the zero temperature susceptibility for a Nb plate was considered as χ_o in order to eliminate the demagnetization factor and χ/χ_o is 0.98 at T~0 . The value of $\lambda(0)$ was obtained as 1086Å, which was comparable to $\lambda(0)$=900~1400Å for YBCO single crystal work done by Krusin-Elbaum et al(1989). Taking $\lambda(0)$ to be 1086Å, the experimental $\lambda(T)$ is compared with two theoretical predictions: two-fluid relation (T.F.) and Ginzburg-Landau theory (GL). Unlike Krusin-Elbaum's YBCO work, which was explained by both GL and TF theories quite well, our results cannot be fitteed by either theory near the transition temperature.

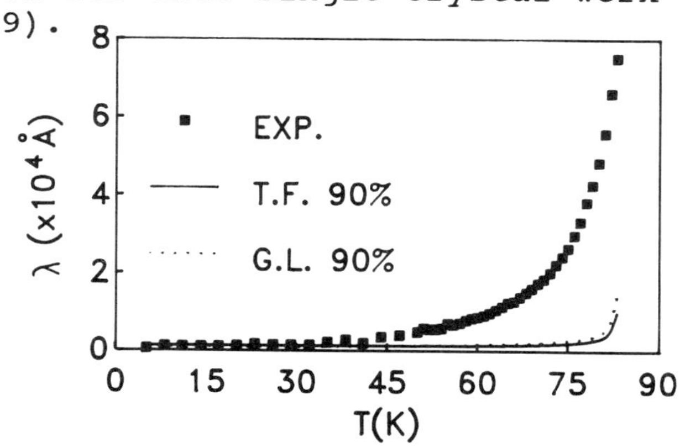

Fig. 2 The calculated magnetic penetration depth and the theoretical fitting.

This could be caused by the

 defect of the sample and

it may be solved by improving crystal growth process. It is noted that T_c's in our fitting are taken as the transition temperature at 90% height of the dc magnetic susceptibility drop.

Rather than penetration depth, the other important superconducting length is coherence length. The coherence lengths parallel ($\xi_\perp$) and perpendicular ($\xi_\parallel$) to ab-palne were calculated according to Lawrence and Doniach model based on an anisotropic Ginzburg-Landau theory for layered superconductors. The coherence lengths in this model are given by

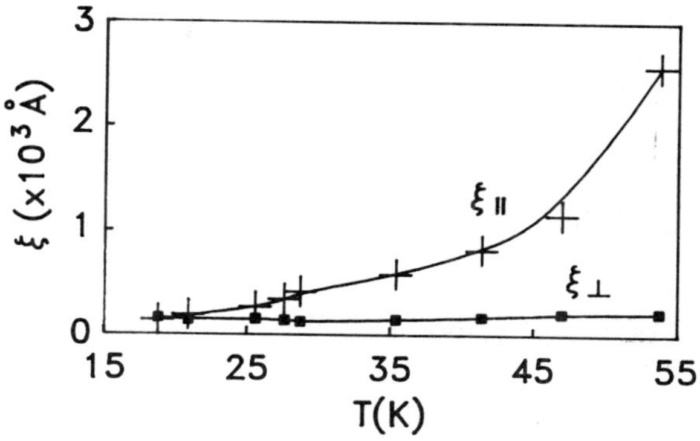

Fig. 3 Temperature dependence coherence lengths calculated according to L-D model.

$$H_{c2,\parallel}(T) = \Phi_o/2\pi\xi_\parallel(T)\xi_\perp(T)$$

and

$$H_{c2,\perp}(T) = \Phi_o/2\pi\xi_\parallel(T)^2 \quad ,$$

here Φ_o is the fluxoid. Dc magnetic susceptibility results of temperature dependent parallel and perpendicular upper critical fields in Fig. 1 were used in this model. The calculated coherence lengths for parallel and perpendicular fields are shown in Fig. 3 as a function of temperature. The absence of data in this figure for higher temperature below T_c is due to the missing of critical fields in Fig. 1. It is observed that the parallel coherence length is much larger than that of perpendicular one at higher temperature range and these two values meet at low temperature . We also obtained that $\xi_\perp$ kept about constant as decreasing temperature.

4. Summary

In summary, critical fields and coherence lengths were observed to be angular dependent. The zero-temperature parallel coherence length (=1086Å) is comparable to that of YBCO system. The mismatch of experimental and theoretical coherence lengths may not be intrinsic for Bi-system and worth an advanced study.

References

Akimitsu J, Yamazaki A, Sawa H and Fujiki H 1988 Jpn. J. Appl. Phys. **26** L2080

Katsui A and Ohtsuka H 1988 J. of Crystal Growth **91** 261

Krusin-Elbaum L,Greene R L, Holtzberg F, Malozemoff A P and Yeshurum Y 1989 Phys. Rev. Lett. **62** 217

Maeda H, Tanaka Y, Fukutomi M and Asano T 1988 Jpn. J. Appl. Phys. **27** L209

Schneemeyer L F et al. 1988 Nature **332** 422

Shishido T, Shindo D, Ukei K, Sasaki T, Toyota N and Fukuda T 1989 Jpn. J. Appl. Phys. **28** L791

Supercond. Sci. Technol. **4** (1991) S190–192. Printed in the UK

Point-Contact Tunnelling Measurements of Tl-Pb-Ca-Pr-Sr-Cu-O

S J Chandler and C J Adkins

Cavendish Laboratory, Madingley Road, Cambridge, CB3 OHE

Abstract. We report point-contact tunnelling measurements of the high-T_c superconductor $(Tl_{0.5}Pb_{0.5})(Ca_{0.8}Pr_{0.2})Sr_2Cu_2O_y$. The *I-V* characteristics clearly show an energy gap and are qualitatively similar to those of a classical S-I-S tunnel junction with $2\Delta/k_bT_c$ of 5.1.

1. Introduction

Electron tunnelling is a powerful technique for studying the density of states of a superconductor. The point-contact method of tunnel junction fabrication developed by Levinstein and Kunzler (1966) uses a sharpened metal tip, heavily anodised to form an insulating barrier, in contact with a small region on the surface of a bulk sample. Anodising of the tip is not necessary for point-contact junctions formed on high-temperature superconductors since these have insulating surface layers possibly caused by carbonate contamination due to reaction with atmospheric carbon dioxide (Egdell *et al* 1990) or by atomic segregation during annealing (Gavaler *et al* 1988). Their surface layers are much thicker than their short coherence lengths (Wonltungton *et al* 1987) and so impede attempts to vacuum-tunnel using a scanning tunnelling microscope (STM) but allows a point-contact tunnel junction to be formed by carefully pushing the tip into the surface layer until the layer is thin enough for a measurable tunnel current to flow.

2. Experiment

The necessary fine-control of a point-contact has been obtained by manipulating a tip with three-degrees of freedom using a single, cylindrical, piezoelectric transducer, the design being similar to that of many contemporary STMs (Binnig and Smith 1986). Course approach of the tip to the sample is achieved at 4.2K by using a 7μm/turn differential screw and differential spring system. Despite vibration isolation, the cryostat contacts are found to be rather unstable; a particular current-voltage characteristic may remain stable for several minutes and then change unpredictably. For this reason four-terminal measurements of the tip-sample current are made using a microcomputer controlled voltage-biased source and data acquisition system. Sets of identical *I-V* characteristics may later be averaged and differentiated numerically to obtain the dynamic conductance. The conductance of more stable junctions may be measured directly using a lock-in amplifier technique in which the bias is slowly swept over the range of interest.

3. Results

We report measurements on a sample of high-quality, single-phase, polycrystaline ceramic with composition $(Tl_{0.5}Pb_{0.5})(Ca_{0.8}Pr_{0.2})Sr_2Cu_2O_y$ and T_c (zero resistance) of 106 K (Liu *et al* 1989) using an electrochemically etched tantalum tip. The conductance often varied smoothly with voltage with no evidence of a gap-like structure. However, fine adjustment of the contact produced *I-V* characteristics with the attributes of a good superconducting tunnel junction, i.e.

0953-2048/91/01S190 + 03 $03.50

• low-conductance in a region about zero-bias
• conductance overshoots at the edge of this region
• symmetry in the voltage at which the overshoots occur

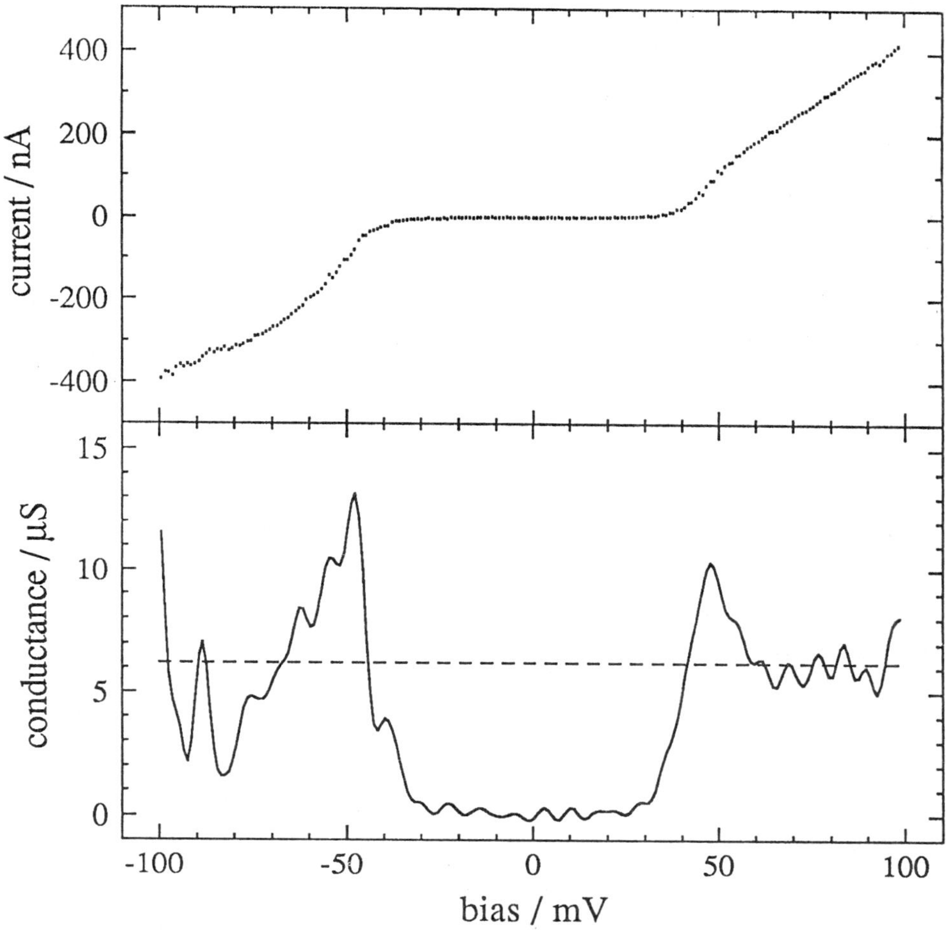

Figure 1 The *I-V* and conductivity (*dI*/*dV*) curves of a Ta point-contact onto $(Tl_{0.5}Pb_{0.5})(Ca_{0.8}Pr_{0.2})Sr_2Cu_2O_y$. The dashed line shows the estimated background conductance.

The above curves are carefully calibrated and the zero-bias conductance was unmeasurable – less than 2% of the beyond-gap conductance. Features both inside and outside the gap region are attributed to noise. The conductance maxima occur at ±48±1 mV and the conductance crosses the estimated background value (dotted line in Figure 1) at ±43±1 mV. We use the latter as a measure of $\Delta_1+\Delta_2$. Comparison with the BCS density of states shows a slight broadening though this is much smaller than in many reports (Wnuk *et al.*1990).

There are two possible interpretations :

(a) The structure results from S-I-S tunnelling between superconducting tantalum and the high-T_c material. In this case we obtain $\Delta \approx 43\pm1$ meV and $2\Delta/k_bT_c \approx 9.4\pm0.3$.

(b) We are observing tunnelling between two grains of the high-temperature superconductor, one of which is in good contact with the tip and the other is part of the bulk ceramic. The tunnelling barrier is between the grains. This model gives an energy gap of 21.5 meV and $2\Delta/k_bT_c$ of 4.7±0.3, a value which is in agreement with our earlier work on $Y_1Ba_2Cu_3O_y$ (Edgar et al 1988) where we find $2\Delta/k_bT_c \approx 5$.

A similar model was used by Viera *et al* (1989) to explain the observation of S-I-S characteristics when tunnelling from a normal Pt-Rh into $Tl_2Ba_2Ca_2Cu_3O_{10+\delta}$. This explanation is also supported by the majority of good tunnelling experiments which report values of $2\Delta/k_bT_c$ in the range 4-6 for the high-T_c superconductors (Kirtley 1990).

We therefore conclude that $(Tl_{0.5}Pb_{0.5})(Ca_{0.8}Pr_{0.2})Sr_2Cu_2O_y$ has an energy gap of 21.5 meV and $2\Delta/k_bT_c \approx 4.7\pm0.3$.

Acknowledgments

Samples prepared by R.S.Liu, Interdisciplinary Research Centre in Superconductivity, University of Cambridge.

Levinstein H.J. and Kunzler J.E. 1966 *Phys. Lett.* **20** 581

Egdell R.G., Flavell W.R. and Golden M.S. 1990 *Supercond. Sci. Technol.* **3** 8

Gavaler J.R. and Braginski A.I. 1988 *Physica C* **153-155** 1435

Wonltungton T.R., Gallagher W.J.and Dinger T.R. 1987 *Phys. Rev. Lett.* **59** 1160

Binnig G. and Smith D.P.E. 1986 *Rev. Sci. Instrum.* **57** 1688

Liu R.S., Liang J.M., Wu S.F., Huang Y.T. and Wu P.T. 1989 *Physica C* **159** 385

Wnuk J.J., Smokers R.T.M., Nolden F.W., Schreurs L.W.M. Wang Y.S. and van Kempen H. 1990 LT-19 satellite conference on high temperature superconductivity. To be published in *Supercond. Sci. Technol.*

Edgar A., Adkins C.J. and Chandler S.J. 1987 *J. Phys. C : Solid State Phys.* **20** L1009

Viera S., Rodrigo J.G., Ramos M.A., Rao K.V. and Makino Y. 1989 *Phys .Rev.* **B40** 11,403

Kirtley J.R. 1990 Preprint to be published in *Intl. J. Mod. Phys.* **B**

Supercond. Sci. Technol. **4** (1991) S193–195. Printed in the UK

Study of electronic structure in single Tl-O sheet superconductors by EELS

J Yuan, L M Brown*, W Y Liang, R S Liu and P P Edwards

IRC in superconductivity, Madingley Road, Cambridge CB3 0HE
*Cavendish Laboratory, Madingley Road, Cambridge CB3 0HE

Abstract O 1s absorption spectra for a series of metal oxide samples based on a common structure unit $(Tl,A)(Ca,Y)Sr_2Cu_2O_{7-\delta}$ (A=Pb,V) have been measured using high energy electron energy loss spectroscopy. The pre-edge feature contains transition from the O 1s core level to the states both in the Cu-O_2 sheets, which can be correlated to the transition temeprature, and to states in Tl(A)-O square planner sheets. Preliminary vanadium 2p spectroscopy suggests that V occupies a well defined atomic site.

1. Introduction

The highest values of the superconducting transition temperature (T_c) have been realized in the thallium containing copper oxides (Sleight, Subramanian and Toraddi, 1989). Two homologous series, $Tl_2Ba_2Ca_{n-1}Cu_2O_{2n+4}$ (n=1,2,3 and 4) and $TlBa_2Ca_{n-1}Cu_2O_{2n+3}$ (n=1,2,3,4 and 5) have been identified. In common with other high tempreature copper oxide superconductors, the structure of these compounds contains CuO_2 sheets, separated by Ca ions when n is greater than one. These sheets, together with the Ba ions sitting above and below them form an essentially oxygen deficient perovskite structure. They are interconnected by one or two sock-salt related Tl-O layer(s). The charge balance between different layers in the structure can be influenced by chemical substitution of chastens in the Tl-O and Ca layers by ions of different valency. This chemical 'doping' by replacing Ca with Y has been very successful since Ca and Y can form solid solution. Substitution of Tl by Bi and Pb has only been partially successful since it is subjected to a greater structure constraint. In this respect, a recent discovery of superconducting Tl-V-Sr-Ca-Cu-O compound (T_c ~ 87 K) is very interesting (Liu *et al*, 1990). In this short paper, we are going to report an study of the O 1s absorption spectra of $Tl_{0.5}Pb_{0.5}Ca_{1-x}Y_xCu_2O_7$ (0<x<1), $TlSr_2YCu_2O_7$, $TlVSr_2CaCu_2O_7$, all derived from $TlBa_2CaCu_2O_7$ (1212) compound by various chemical substitution at Ca or Tl rich layer. For the V compound, the spectra also include the absorption by V 2p core electrons, which enable us to derive information about the valence state of V in the compounds.

2. Experimental

The ceramic samples were prepared with usually solid state reaction method and details have been discussed elsewhere (Liu *et al*, 1990a,b). For the electron energy loss spectroscopy (EELS) investigation, the sintered sample was crushed between glass microslides. The resulting powder on a holy carbon film covered grid was quickly transferred into the vacuum chamber of a VG-HB501 Scanning Transmission Electron Microscope (STEM)where EELS measurements were carried out. Spectra were collected from electron transparent particles, using a slight defocused electron probe. The incident electron has an energy of 100 keV (k_0 = 27 $Å^{-1}$) and a convergence angle of about 8 mrads ($\Delta k \sim 0.2 Å^{-1}$). The collection angle was set to about 32 mrads ($\Delta k \sim 0.9$ $Å^{-1}$). Given that the mean value of the core-orbital radii of O 1s is about 0.1 Å, the core electron transitions obeys the dipole selection rule. The possible anisotropic effect is strongly suppressed with the use of large collection angle because Δk can sample a substantial fraction of the Brillouin Zone. The data was collected with a resolution of about 0.7 eV. The probable error in absolute energy determination is about 0.5 eV.

3. Results

O-K absorption edges of $(Tl_{0.5}Pb_{0.5})(Ca_{1-x}Y_x)Sr_2Cu_2O_{7-\delta}$ are shown in Fig. 1, together with the O-K edge from $TlSr_2YCu_2O_7$. For later comparison, we also presented the O K absorption edges of $YBa_2Cu_3O_7$ and $YBa_2Cu_3O_{6.1}$. The spectra are all normalized to have the same height at the main peak at 536 eV. The most prominent result is the existence of a well defined pre-edge peak at 528.7 eV, for the semiconducting $Tl_{0.5}Pb_{0.5}Sr_2YCu_2O_7$ and $TlSr_2YCu_2O_7$ samples. As the Ca/Y ratio in the sample increases, the pre-edge feature broadens but is initially still centred at the same energy. At x < 0.7, however the edge appears to shift towards at a lower energy and there is a clear evidence of an additional low energy feature appearing at 527 eV. Throughout the energy position of the main peak remains the same, at about 536 eV. In the V doped 1212 compound, the feature in the O 1s absorption edge is not very well resolved. The V 2p core level transition is observed just about 15 eV before the O-K edge and show well defined cheracteristic peaks.

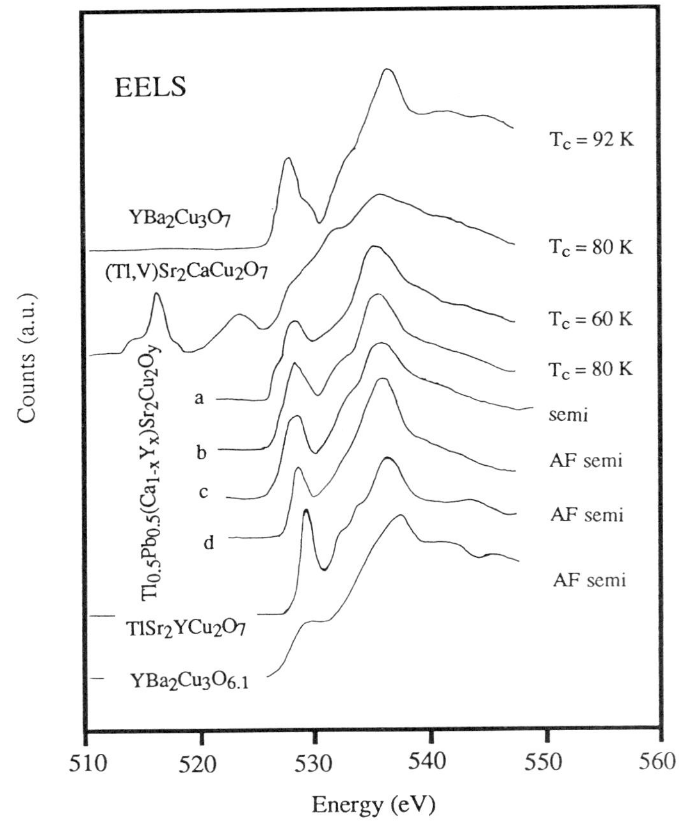

Fig. 1 The O K-absorption edge of various single Tl-O sheet superconducting and related compounds. a,b,c and d refers to $Tl_{0.5}Pb_{0.5}(Ca_{1-x}Y_x)Sr_2Cu_2O_7$ compounds with x=1,0.5,0.2 and 0.0 respec-tively. Also presented for comparion iare the O K-edge of $YBa_2Cu_3O_7$ and $O_{6.1}$ samples. AF = antiferromagnetic; semi=semiconducting; T_C = superconducting transition temperature.

4. Discussion and conclusions

Various interpretations of the pre-edge structure are possible: (i) the pre-peak is related to the holes in the valence band as in other p-type superconductors; (ii) the peak maybe caused by O 2p states hybridized with Tl 5d/6s states; (iii) the peak is related to the transition into the upper Hubbered band which is predominantly Cu 3d in character, but with considerable O 2p admixture. The most striking feature of the O-K edges in the Tl compounds is the existence of the pre-edge feature throughout the composition range spanning from metals to insulators. This demonstrates that not all pre-edge feature in the O-K absorption edge is necessarily associated with doping induced holes in CuO_2 sheets. To test the possibility of (iii), we compared the pre-edge feature of both $TlSr_2YCu_2O_7$ and $Tl_{0.5}Pb_{0.5}Sr_2YCu_2O_7$ with $YBa_2Cu_3O_{6.1}$. All are antiferromagnetic and are different only in the replacement of Tl-O layer by the Cu-O chain structure. As the copper in the chain site in $YBa_2Cu_3O_{6.1}$ has a d^{10} configuration, then the threshold in the O-K absorption edge at 529 eV may be assigned to the transition into the two dimensional conducting band at the CuO_2 sheets. However, the intensity of this transition is much smaller than we have observed in the Tl compound. Since the degree of covalency in two CuO_2 sheets of similar size should be approximately the same, this rules out the transition into the upper Hubbered band being solely responsible for the pre-edge peak in the semiconducting phases. We believe that the pre-edge peak reflects predominantly the transition into the valence band in the Tl-O layers. The band structure calculation (Meshkov *et al*, 1990) shows that the two energy bands in the CuO_2 and TlO layer cross the Fermi surfaces of the Ca rich phase, forming hole pockets in both TlO and CuO_2 layers. According to the band calculation the electronic pockets in the Tl-O layer will be filled as Y replaces Ca. This is confirmed by a steady growth of the spectral weight of the main pre-edge peak (as displayed in fig. 2), as a function of Ca doping, not by increasing the spectra height, but by broadening into the low energy region. With real Tl containing oxide possessing many Tl-O short range order, the predicted closure of these electronic pockets with complement replacement by Y cannot be realized. Still any charge carrier in the Tl-O layer will be localized by the short range order themselves. Consistent with this interpretation, the pre-edge peak intensity of the $TlSr_2YCu_2O_7$ is stronger than that of $Tl_{0.5}Pb_{0.5}Sr_2YCu_2O_7$.

The effect of replacing Ca by Y in $Tl_{0.5}Pb_{0.5}Sr_2(Ca_{1-x}Y_x)Cu_2O_7$ is reflected in the fine structure of the O1s absorption edge (fig. 1). This mainly shows up as broadening of the pre-edge feature. It can also be interpreted in terms of a growth of a new peak below 528.2 eV as a distinct feature is observed in Ca rich sample. At the moment, we lack the photoemission data for this series of compounds to obtain the binding energies of various oxygen atoms in the unit cell. However, for most CuO_2 sheet based superconductors, the lowest oxygen 1s binding energies have been found to be between 528.5-529.0 eV (Lindberg *et al*, 1990). It is reasonable therefore to assume the binding energy for O 1s in these compounds are also within that range. Thus the new peak reflects the holes in the important CuO_2 sheet. In this respect, the compounds are typical of p-type CuO sheet based superconductors. Another evidence supporting such assignment is the reasonable agreement between the composition dependence of the spectra weight of this new low energy peak and the superconducting transition temperature (fig. 2). Compared with the first pre-edge peak in $YBa_2Cu_3O_7$, the hole related peak in Tl containing compound is much smaller. However, one has to remember the copper at the chain site also contributes as well now.

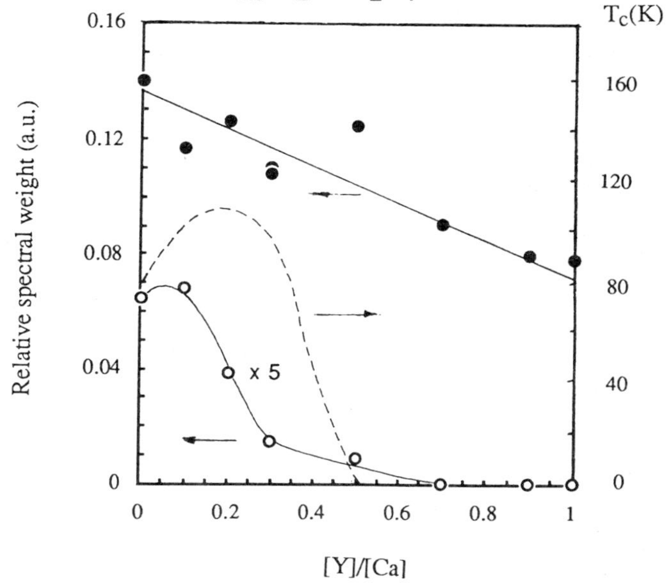

Fig. 2 The composition dependence of the relative spectra weights of the two pre-edge peaks (open circle: at 527.5 eV; full circle: at 528.5 eV), all normalized to the main peak at 536 eV, in O -K edge of $Tl_{0.5}Pb_{0.5}(Ca_{1-x}Y_x)Sr_2Cu_2O_7$ compounds, compared with that of superconducting transition temperature (T_c, dashed line).

In the V doped sample, the pre-edge feature of the O-1s absorption edge is not very defined. However, the transition from the V 2p core level is very sharp and is dominated by the excitonic line. Because of this the density of states near the V ions cannot be deduced easily. But the intensity distribution indicates that the vanadium ions of high valency occupy a well defined atomic site with pseud-octahedral point symmetry (Groot *et al*, 1990). This suggests that the V ions is most probably substituted for Tl in the matrix. It will introduce a large amount of empty 3d states at the Fermi level and may explain the lack of feature in the O 1s absorption edge.

Reference

de Groot F M F, Fuggle J C, Thole B T and Sawatzky G A 1990 *Phys. Rev.* **B41**, 928

Lindberg P A P, Shen Z X, Spicer W E and Lindau I 1990 *Surf. Sci. Rep.* **11** 137

Liu R S, Edwards P P, Huang Y T, Wu S F and Wu P T 1990, *J Sol. State Chem.*, **86**, 334-339

Liu R S, Zhou W, Janes R and Edwards P P 1990 Submitted to the *J. Sol. State Chem.*

Meshkov S V, Molotkov S N, Nazin S S, Smirnova I S and TatarskII V V 1990 *Phys. C* **166** 476-482

Sleight A W, Subramanian M A and Torardi C C 1989 *Mater. Res. Soc. Bull.* **XIV**, 45

Supercond. Sci. Technol. 4 (1991) S196–198. Printed in the UK

DYNAMIC YOUNGS MODULUS RELAXATION AND SOFTENING IN $YBa_2Cu_3O_{7-x}$

E.BONETTI[1], E.G.CAMPARI[1], P.CAMMAROTA[2], A.CASAGRANDE[2] and S.MANTOVANI[3].

1 Dipartimento di Fisica Universita' di Bologna, Consorzio INFM, via Irnerio 46- 40126 Bologna ITALY

2 Istituto di Metallurgia, Universita' di Bologna, CIRM v.le Risorgimento 4 - 40122 Bologna ITALY.

3 Dipartimento di Fisica Universita' di Modena,Consorzio INFM, via Campi 213/A- 41100 Modena ITALY.

Abstract Measurements of the dynamic Youngs modulus were performed in polycrystalline YBCO by the vibrating reed technique (10^2-10^4 Hz) at different temperatures between 300 and 850 K. An abrupt and strong modulus softening, indicative of a phase transition, occurs with increasing x at about 0.50. Isothermal modulus measurements suggest that the orthogonal phase first transforms into a tetragonal matrix with embedded micro-ortho domains. Successively, a first order kinetics yields the tetragonal phase.

1. Introduction

It is now well established that the superconducting properties in YBCO are strongly correlated to the content and degree of order of the oxygen in the system (Cava 1990,Jorgensen 1987).

Measurements of the elastic behaviour of superconductors by internal friction and ultrasonic attenuation techniques (Dominec 1989) have indicated that strong stiffening of the elastic constants occur near 210 K indicating, according to Laegreid (1988), the development of a structural instability at that temperature.

In a previous paper (Bonetti 1990a) dealing with the anelastic properties of polycrystalline YBCO at high temperatures as a function of the oxygen content, a modulus softening was reported similar to that observed at higher frequencies by other authors (Tallon 1988). This softening is indicative of a phase transition.

As anelasticity measurements can reveal structural transformations with high sensitivity, these techniques have been employed to obtain detailed information on structural instability phenomena in YBCO, due to the loss and disordering of oxygen. The temperature range investigated was selected so as to allow easy diffusion of oxygen.

2 Experimental

Polycrystalline YBCO specimens were prepared as described previously (Bonetti 1990b). Youngs modulus measurements were performed by a vibrating reed technique in an oxygen partial pressure of $5x10^{-3}$ mmHg. The heating rate was 2 K/min. The isotherms, at selected temperatures in the 650-850 K range lasted up to 10^5 s. The oxygen stoichiometry was determined by an iodometric method.

3 Results

In fig 1, the dynamic Youngs modulus (M) typical of initially orthorombic sample (x=0.08) is reported as a function of temperature, together with oxygen deficiency. A strong and abrupt softening of the modulus is evident. It occurs for x growing towards 0.5 i.e. at the oxygen deficiency which characterize the ortho-tetra phase transition.

0953-2048/91/01S196 + 03 $03.50

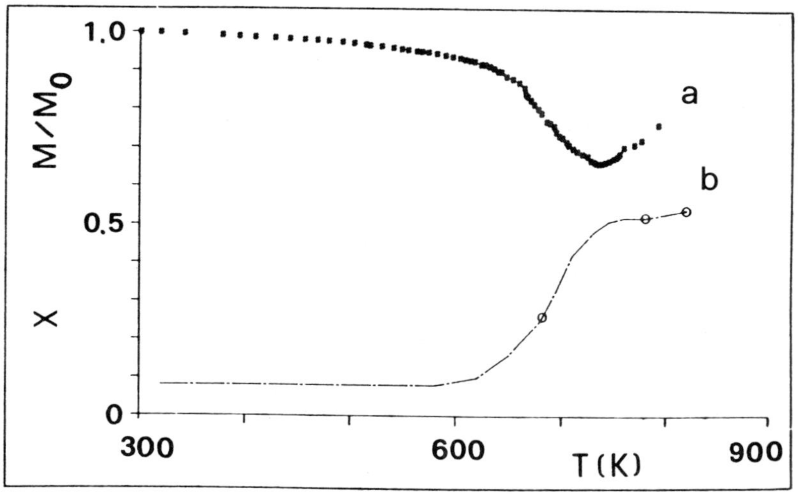

Fig.1 Temperature dependence of the dynamic Youngs modulus M (a), and oxygen deficiency x (b),for an as prepared sample. Heating rate 2 K/min, vibration frequency at 293 K 1032 Hz. M_0 is the dynamic modulus at room temperature.

The thermal treatment of the specimens submitted to isothermal measurements is the same as the one of the specimen of Fig.1, up to the isotherm temperatures. These, were selected so as to cover a range below and above the softening temperature and are indicated by the points in Fig.1.

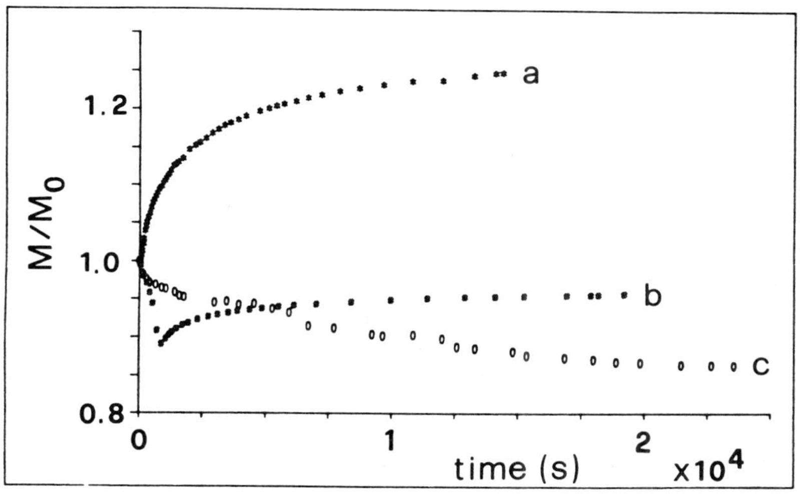

Fig.2 Time dependent dynamic Youngs modulus behaviour at the following temperatures: a:820 K, b:780 K, c:680 K. M_0'is the modulus value at the beginning of each isotherm.

As can be seen in Fig.2 the isotherms display two different trends, that is, a modulus increase, whose strenght depends on annealing temperature (curves a, b) and a decrease, whose duration also depends on temperature (curves b, c). Assuming that the oxygen mobility in the basal plane is responsible for the trends of curves a and b, a first order kinetics:

$$(dm'/dt) = -\frac{1}{\tau}(m'-1) \qquad 1)$$

should be employed to analise the process. Here m' is the normalised modulus and τ a proportionality constant. τ is assumed to follow an Arrhenian equation:

$$\tau = \tau_0 \exp(H/kT) \qquad 2)$$

which is usually employed for atoms or defects jumping from one site to another over a potential barrier (Nowick,1972). But while equation 1 corresponds to an exponential behaviour, experimental data are better fitted by a logarithmic curve such as:

$$M(t) = A+B \log(t+1) \qquad 3).$$

This suggests a distribution in activation energies (Primak 1955, Oguz 1980) that is, many processes of the type given by equation 1, with different energies, whose sum yields the observed modulus behaviour.

The decreasing trend, if is connected with the oxygen outflow, should also follow a first order kinetics as stated by Ottaviani (1990) and Strauven (1988). Further measurements to test the order of the kinetics are still in progress.

4 Discussion

From the reported results about the modulus measurements and oxygen content determinations, it can be stated that the observed modulus softening corresponds to the orthorombic to tetragonal phase transition. This transition can be observed at different temperatures by changing the oxygen partial pressure in ramping temperature measurements, or after different times by performing isothermal measurements (e.g. curve b fig. 2). The modulus trend observed in isothermal measurements such as that of curve a and the rising part of curve b of fig. 2 should not be caused by oxygen outflow. In fact, Ottaviani (1990) measured the oxygen outflow of YBCO in isothermal tests at temperatures up to 926 K and found that no outflow occurred after 200 min. We also measured the oxygen content of samples (initial value of x=0.08) brought to temperatures in the 650-850 K range and immediately cooled to room temperature and of samples kept for different times at various annealing temperatures. The two groups showed x value differences less than 10 %. We think these differences cannot account for the observed modulus increase, expecially since the modulus variation timescale is much longer than the oxygen one.

So we suggest that an order-disorder oxygen redistribution process might be responsible of the observed relaxation. Such a conservative process (that is, with a constant oxygen content) was also proposed by Tu(1989), who speaks of a "twinning" type relaxation. The process consists of two steps: first a collapse of the orthorombic twinned structure giving rise to a tetragonal arrangement with embedded small ortho domains. Successively a first order kinetics controlled by long range oxygen diffusion leads to a complete order-disorder transformation of the structure.

References

Bonetti E,Campari E G,Cammarota P,Casagrande A,1990 Proc. EMRS Symp. on High T_c Superc. Strasbourg.

Bonetti E,Campari E G,Casagrande A,Mantovani S,1990 High Temp. Superc. Third Nat. Meeting Genova (Italy) World Sc.Publishing Co.

Cava R J,Hewat A,Hewat E,Batlogg B,Marezio M,Rabe K,Krajewsky J,Peck W,Rupp L,1990 Physica C 165, 419.

Dominec L,1989 Superc. Sci. Technol. 2,91

Jorgensen J,Beno M, Hinks D,Soderholm L,Volin K,Grace J,Hitterman R,Schuller I,Serge C,Zhang K,Kleefish M,1978 Phys. Rev. B 36,3608

Laegreid T,Fossheim K,1988 Europhys. Lett. 6,81

Nowick A,Berry B,1972 Anelastic Relaxation In Crystalline Solids

Oguz S,Paesler M,1980 Phys. Rev. B22,6213

Ottaviani C,Nobili C,Rossi M.,Sparpaglione M.,1990 Physica C in press.

Primak W,1955 Phys. Rev. 100,1677

Strauven H,Loquet J,Verbeke O,Bruynseraede Y,1988 Solid State Comm. 65,293

Tallon J L,Schuitema A H,Tapp N E,1988 Appl. Phys. Lett. 52,507

Tu K N,Yeh N C,Park S I,Tsuei C,1989 Phys Rev. B 39,304

Supercond. Sci. Technol. 4 (1991) S199–201. Printed in the UK

Elastic behaviour of $Nd_{1.85}Ce_{0.15}CuO_{4-y}$ and Nd_2CuO_{4-y}

Chang Fanggao, A. Al-Kheffaji, P.J. Ford, D.P. Almond, B. Chapman, J. Freestone, D.A. Ladds, Ll. Mañosa, M. Cankurtaran and G.A. Saunders

Schools of Physics and Materials Science, University of Bath, Claverton Down, Bath BA2 7AY

Abstract. Ultrasonic measurements on the electron doped high T_c superconductor $Nd_{1.85}Ce_{0.15}CuO_{4-y}$ and Nd_2CuO_{4-y} its non superconducting parent compound are presented. Although the superconductor shows no evidence of acoustic mode softening below room temperature, there is a clear change in the temperature dependence of the longitudinal mode velocity around 220K. This effect is far less pronounced for the shear mode and is not observed in the parent compound, indicating that the intermediate valence state of the Ce ions plays an important role in the elastic behaviour of the superconductor. The first results for the pressure dependences of the elastic constants are reported.

1. Introduction

In the class of high temperature superconductors with the general formula $Ln_{2-x}Ce_xCuO_{4-y}$ (Ln = Pr, Nd and Sm) conduction takes place by electrons (Tokura et al 1989; Takagi et al 1989), in contrast to all the previously known high T_c oxides where conduction is by holes. The most studied of these electron doped superconductors is $Nd_{1.85}Ce_{0.15}CuO_{4-y}$ with x = 0.15 which has a superconducting transition temperature of about 23K. Because of the complicated nature of these materials, an understanding of the properties of the superconductor requires a knowledge of those of the parent compound Nd_2CuO_{4-y}. Here we present ultrasonic wave velocity measurements in polycrystalline $Nd_{1.85}Ce_{0.15}CuO_{4-y}$ and Nd_2CuO_{4-y}. Our initial studies (Al-Kheffaji et al 1989) showed that sound velocity measurements provide evidence for mode softening in $La_{1.8}Sr_{0.2}CuO_4$ but not in $Nd_{1.85}Ce_{0.15}CuO_{4-y}$.

2. Results and Discussion

Details of the preparation and characterisation of these samples, as well as the experimental techniques involved, are described elsewhere (Chang Fanggao et al 1990).The longitudinal and shear mode velocities of $Nd_{1.85}Ce_{0.15}CuO_{4-y}$ and Nd_2CuO_{4-y} all increase

0953-2048/91/01S199 + 03 $03.50

in a continuous manner as the temperature decreases (fig. 1). There is no reduction in the velocity as the temperature is reduced as was observed for $La_{2-x}Sr_xCuO_4$ (Fossheim et al 1987) and attributed to mode softening associated with the orthorhombic to tetragonal transition (Al-Kheffaji et al 1989). However, Fig. 1 shows that the temperature dependence of the longitudinal mode velocity of $Nd_{1.85}Ce_{0.15}CuO_{4-y}$ has a pronounced change of gradient near 220K. This is also observed, but much less strongly, in the shear mode velocity. The parent material Nd_2CuO_{4-y} does not show a change in slope in ultrasonic wave velocity around 220K. However it does show quite marked hysteresis on warming and cooling between 220K and 290K. Such behaviour is often associated with a structural transition, although our x-ray and differential scanning calorimetry investigations (Chang Fanggao et al 1990) did not indicate any evidence for this in Nd_2CuO_{4-y}. The bulk B_o and shear C_{44} moduli data are shown in Fig. 2. In each case there is a clear change in slope around 220K in the $Nd_{1.85}Ce_{0.15}CuO_{4-y}$ which is not apparent in the parent material. In addition, the elastic constants of the superconductor vary much more strongly with temperature than do those of Nd_2CuO_{4-y}. This marked contrast between the two materials suggests that the elastic behaviour of $Nd_{1.85}Ce_{0.15}CuO_{4-y}$ is markedly influenced by the presence of the Ce ions. On cooling to low temperatures a number of cerium compounds undergo a valence change from the Ce^{3+} state towards the Ce^{4+} state. However, the transition does not take place from one integral valence state to another: the Ce ions in the resultant phase have an intermediate valence. An important feature of intermediate valence compounds (Tu Hailing et al 1984) is a softening of the longitudinal acoustic modes because they change the interatomic distance directly and therefore alter the volume and are strongly coupled to the valence instability. This is not the case for transverse modes. The change of gradient observed in the temperature dependence of the longitudinal mode velocity around 220K is consistent with enhanced coupling of this mode to

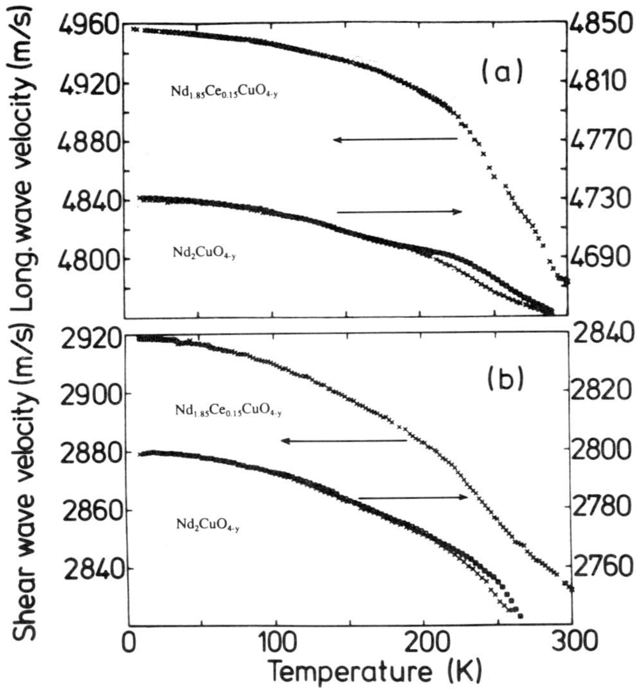

Fig. 1. The temperature dependences of the velocities (ms^{-1}) of (a) longitudinal and (b) shear ultrasonic waves for $Nd_{1.85}Ce_{0.15}CuO_{4-y}$ and Nd_2CuO_{4-y} samples.

the Ce ions as their net valence increases below this temperature. Such effects would be far less important for the volume independent shear modes in line with our observations. Thus the elastic properties of $Nd_{1.85}Ce_{0.15}CuO_{4-y}$ depend strongly on the intermediate valence state of the cerium ion. An examination of other experimental data on this material (Chang Fanggao et al 1990) shows that intermediate valency plays an important role in many properties of this electron superconductor.

The velocities of longitudinal and shear 5MHz ultrasonic waves propagated in the superconductor and host materials were studied under pressure. In all cases the velocities are found to increase linearly with pressure the effects being much more marked in the longitudinal than the shear velocity. Values of 12.4 and 36.2 obtained for $(\partial B/\partial P)_{P=0}$ for $Nd_{1.85}Ce_{0.15}CuO_{4-y}$ and Nd_2CuO_{4-y} respectively are substantially smaller than that for $YBa_2Cu_3O_{7-x}$. An important outcome of this study is to lend support to the use of the bulk modulus B_o measured ultrasonically rather than that $B^T(P)$ from x-ray determinations at very high pressures. This will be discussed in a future publication.

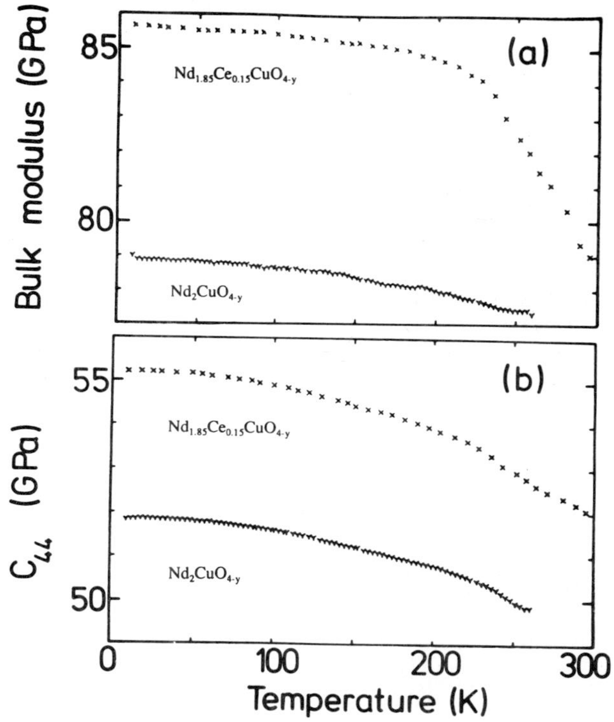

Fig. 2. Temperature dependences of (a) bulk B_o and (b) shear C_{44} moduli for $Nd_{1.85}Ce_{0.15}CuO_{4-y}$ and Nd_2CuO_{4-y} ceramic samples.

References

Al-Kheffaji A, Freestone J, Almond D P, Saunders G A and Wang Q 1989 Phys. Condens. Matter **1**, 5993-6.

Chang Fanggao, Al-Kheffaji A, Ford P J, Ladds D A, Freestone J, Chapman B, Mañosa Ll, Almond D P and Saunders G A 1990 Supercond. Sci. Technol. **3**, 422-8.

Fossheim K, Laegreid T, Sandvold E, Vassenden F, Müller K A and Bednorz J G 1987 Solid State Commun. **63**, 531-3

Hailing Tu, Saunders G A, Yogurtcu Y K, Bach H and Methfessel S 1984 J. Phys. C: Solid State Phys. **17** 4559-73.

Takagi H, Uchida S and Tokura Y 1989 Phys. Rev. Lett. **62**, 1197-200.

Tokura Y, Takagi H and Uchida S 1989 Nature **337** 345-7.

Supercond. Sci. Technol. **4** (1991) S202–204. Printed in the UK

Phonon properties and the specific heat of $YBa_2Cu_3O_7$ and $Bi_2Sr_2CaCu_2O_{8+x}$

S. J. Collocott and J. G. Collins

CSIRO Division of Applied Physics, Lindfield, Australia 2070.

Abstract. The lattice vibrational spectra, $g(\nu)$, of $YBa_2Cu_3O_7$ and $Bi_2Sr_2CaCu_2O_{8+x}$, are used to predict qualitatively the observed behaviour of their lattice specific heat. The temperature dependence of the Debye temperature, $\theta_D(T)$, is calculated from published $g(\nu)$ spectra for both $YBa_2Cu_3O_7$ and $Bi_2Sr_2CaCu_2O_{8+x}$, and compared with the experimentally derived values of $\theta_D(T)$. For both ceramics there is poor quantitative agreement between the calculation and experiment, which we attribute to difficulties in accurately determining $g(\nu)$ at low frequencies.

1. Introduction

The low temperature specific heat, C_p, of both $YBa_2Cu_3O_7$ and $Bi_2Sr_2CaCu_2O_{8+x}$ is characterised by an upturn in C_p/T below ≈3 K, a finite value (typically 7 mJ/(mol K^2)) for the coefficient, γ_0, of the linear term in C_p in $YBa_2Cu_3O_7$, and a zero or near zero value of γ_0 in $Bi_2Sr_2CaCu_2O_{8+x}$ (see the review by Fisher *et al.* 1988). The C_p/T upturn and finite value of γ_0 in $YBa_2Cu_3O_7$ are thought to be due primarily to small amounts of $BaCuO_2$, which is present as an impurity phase, whilst the origin of the C_p/T upturn in $Bi_2Sr_2CaCu_2O_{8+x}$ is unclear (Fisher *et al.* 1988). As a part of a larger specific heat study of the C_p/T upturn and γ_0 in both $YBa_2Cu_3O_7$ and $Bi_2Sr_2CaCu_2O_{8+x}$, and therefore of how impurities contribute to C_p, we decided on an approach whereby the lattice C_p was calculated from the spectrum of lattice vibrations (phonon density of states). The contribution of impurities to C_p in these materials can then be estimated by subtracting the calculated lattice C_p from the experimental C_p values.

2. Phonon Properties and the Experimental C_p

The lattice vibrational spectra of $YBa_2Cu_3O_7$ and $Bi_2Sr_2CaCu_2O_{8+x}$ are very similar and can be characterised by a number of gross features: a small Debye-like ($g(\nu) \propto \nu^2$) region, a set of dispersive acoustic modes, a group of optic modes lying approximately 1.2 THz above the acoustic branch in $YBa_2Cu_3O_7$, and a zone-centre optic mode which lies approximately 1 THz below the acoustic branch at the zone edge in $Bi_2Sr_2CaCu_2O_{8+x}$ (Mase 1989). This information allows one to predict qualitatively the observed behaviour of the low temperature lattice specific heat in these materials, namely a departure from simple T^3 behaviour, particularly in $Bi_2Sr_2CaCu_2O_{8+x}$ (see Figure 1), due to dispersion and to the excitation of the low-lying optic modes. The low-lying optic modes can be represented as a group of independent oscillators which manifest themselves as an Einstein term in C_p.

These qualitative predictions are confirmed, with the C_p data for $YBa_2Cu_3O_7$ (specimen

0953-2048/91/01S202 + 03 $03.50

YB46, Collocott *et al.* 1988) from 2 to 15 K (fitted with $1/C_p^2$ weighting) being well represented by

$$C_p = 9.1T^{-2} + 5.67T + 0.3172T^3 + 4303(85.6/T)^2 \frac{\exp(85.6/T)}{[1-\exp(85.6/T)]^2} \quad \mathrm{mJ/(mol\ K)} \quad (1)$$

%rms deviation 2.2%, and that of $Bi_2Sr_2CaCu_2O_{8+x}$ (specimen BSC12, Collocott and Driver 1990) by

$$C_p = 24.9T^{-2} + 3.58T + 1.8093T^3 + 3591(52.2/T)^2 \frac{\exp(52.2/T)}{[1-\exp(52.2/T)]^2} \quad \mathrm{mJ/(mol\ K)} \quad (2)$$

%rms deviation 1.0%. The T^{-2} and T-terms are included to account for C_p/T^3 rising below 5 K (see Figure 1). The limiting value for the Debye temperature, θ_D^0, calculated from the coefficient of the T^3-term, is 430 K, cf. 370–500 K (Fisher *et al.* 1988), and 253 K, cf. 250 K (Fisher *et al.* 1988), for $YBa_2Cu_3O_7$ and $Bi_2Sr_2CaCu_2O_{8+x}$, respectively. The value of 52.2 K for the Einstein temperature, θ_E, for $Bi_2Sr_2CaCu_2O_{8+x}$ is consistent with the phonon dispersion curves which show a zone-centred optic mode at 1.8 THz (87 K). The upward curvature, and weak peak, in C_p/T^3 above 6 K ($\approx \theta_E/10$) shown by $Bi_2Sr_2CaCu_2O_{8+x}$ in Figure 1 is consistent with excitation of the optic mode at 1.8 THz. For $YBa_2Cu_3O_7$, which has an optic mode at 3.4 THz (162 K) and consequently a higher value of θ_E, there is not such a pronounced upward curvature or peak in C_p/T^3.

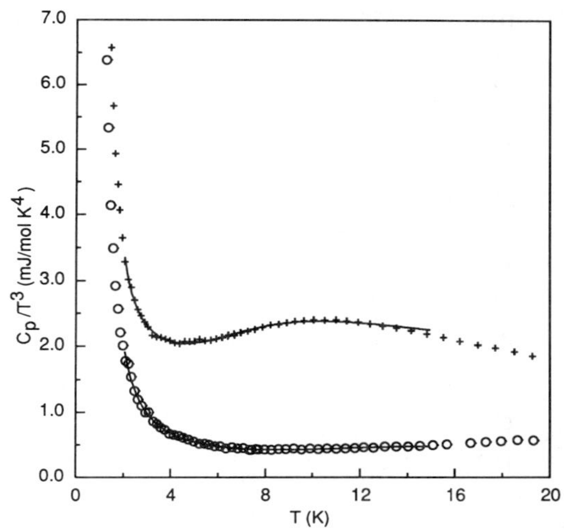

Fig. 1 C_p/T^3 versus T for $YBa_2Cu_3O_7$ (O) and $Bi_2Sr_2CaCu_2O_{8+x}$ (+). The solid curves are given by the equations in the text.

3. Calculation of C_p and $\theta_D(T)$

C_p was calculated by numerical integration of the formula

$$C_v/27R = \int_0^\infty x^2 \exp x(\exp x - 1)^{-2} g(\nu)\mathrm{d}\nu \quad (3)$$

and the Debye temperature, θ_D, by equating $C_v/27R$ to the Debye expression $J_4(x_D)x_D^{-3}$. Here $x = h\nu/kT$ and $x_D = \theta_D(T)/T$, where h is Planck's constant, k is Boltzmann's constant and ν is the lattice vibrational frequency. The Debye integral $J_4(x_D)$ is defined by

$$J_4(u) = \int_0^u x^4 \exp x(\exp x - 1)^{-2}\mathrm{d}x \quad (4)$$

and $g(\nu)$ is the normalised lattice phonon density of states function:

$$\int_0^\infty g(\nu)\mathrm{d}\nu = 1. \quad (5)$$

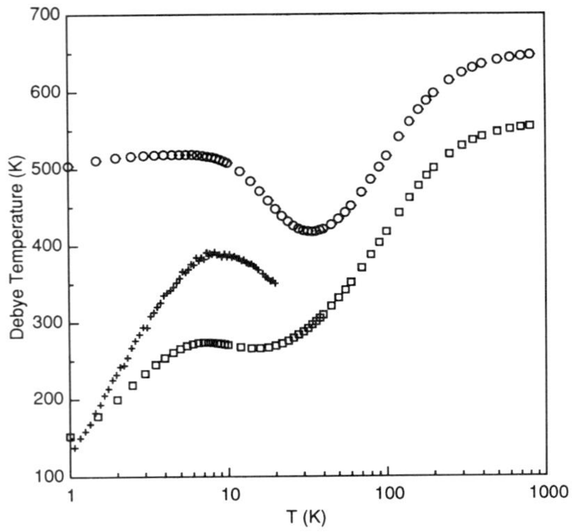

Fig. 2 Debye temperature versus T for $YBa_2Cu_3O_7$ calculated from Mase (1989) (○), Chaplot (□) and experiment (+).

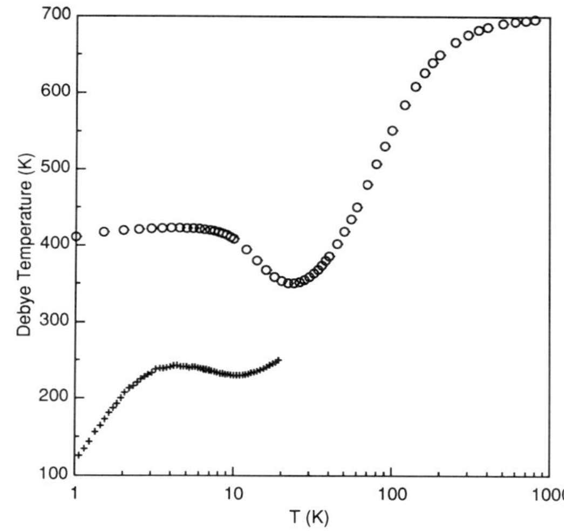

Fig. 3 Debye temperature versus T for $Bi_2Sr_2CaCu_2O_{8+x}$ calculated from Mase (1989) (○) and experiment (+).

The spectral function $g(\nu)$ was taken from the published data of Mase (1989) and Chaplot (1988). At low frequencies, below 1 THz, where sampling of $g(\nu)$ is unreliable, a smooth 'best estimate' transition was made to a Debye curve, $g(\nu) \propto \nu^2$.

Figure 2 shows $\theta_D(T)$ for $YBa_2Cu_3O_7$ calculated from the $g(\nu)$ spectra of Mase (1989) and Chaplot (1988) and from the experimental C_p values. It is seen that there is a marked difference in the $\theta_D(T)$ values calculated from the two independent measurements of $g(\nu)$, as well as poor quantitative agreement with experiment. Figure 3 shows $\theta_D(T)$ for $Bi_2Sr_2CaCu_2O_{8+x}$ calculated from the $g(\nu)$ spectrum of Mase (1989) and again it is seen there is poor agreement between the calculation and experiment. The poor quantitative agreement between the calculated and experimental values of $\theta_D(T)$ is attributed to difficulties in accurately determining $g(\nu)$ at low energies. It is the region of $g(\nu)$ where $\nu \to 0$ that has the most effect on the calculation of the low temperature C_p and θ_D^0, and it is in this region that we have had to make a judicious estimate of $g(\nu)$. To improve the calculation, we need more accurate values of $g(\nu)$ as $\nu \to 0$ from better neutron measurements on good quality, ideally single-crystal, specimens. Elastic moduli from ultrasonic measurements on single crystals would also assist in determining θ_D^0.

4. Conclusion

For both $YBa_2Cu_3O_7$ and $Bi_2Sr_2CaCu_2O_{8+x}$, knowledge of the lattice vibrational spectra allows one to predict qualitatively the behaviour of the low temperature lattice C_p. There is poor quantitative agreement between the calculated and experimental values of $\theta_D(T)$. This is attributed to difficulties in accurately determining $g(\nu)$ at low frequencies.

References

Chaplot S L 1988 *Phys. Rev. B* **37** 7435

Collocott S J, Driver R, Welsh H K and Andrikidis C 1988 *Physica C* **152** 401

Collocott S J and Driver R 1990 *Physica C* **167** 598

Fisher R A, Gordon J E and Phillips 1988 *J. of Superconductivity* **1** 231

Mase S. 1989 *Studies of High Temperature Superconductors* vol. 2, ed. A. Narlikar (New York: Nova Science Publishers) pp 129–70

Supercond. Sci. Technol. 4 (1991) S205–207. Printed in the UK

Magnetic Transition in Strong Field for Single Crystal $Bi_2Sr_2CaCu_2O_x$ *

Shiying Ding, Zheng Yu, Guang Yang[+], Zhongxian Zhao[+]

Physics Department and Solid Microstructure Lab., Nanjing University, Nanjing 210008, PR China.
[+]Institute of Physics, Academia Sinica, Beijing 100080, PR China.

Abstract. Susceptibility χ was measured in wide temperature range and strong applied field for single crystal $Bi_2Sr_2Cacu_2O_x$. Normal state diamagnetism was detected at 90K<T<320K. At 10K<T<90K two magnetic transition temperatures Tc $\approx$ 90K and Tc* $\approx$ 25K were found.

1. Introduction

It has been reported that some parameters including critical current density Jc, flux pinning energy U_0 and flux line lattice melting temperature T* of Bi system superconductor are smaller than the counterpart of Y system at high temperature, say 77K (for examples, see Palstra et al. 1989, Nelson 1989). Palstra et al. has attributed these difference to different intrinsic anisotropy for the both high Tc superconductors. However, Shi et al.(1989)has obtained Jc > 10^7 A/cm^2 at 10K for $Bi_2Sr_2CaCu_2O_x$ (B2212). This Jc is in the same order as the one of YBCO measured by Küpfer et al. (1988). Shi et al. explain the high Jc as strong flux pinning which is not intrinsic property. therefore it is not be clarified why Jc and relevant parameters of Bi system superconductor have more sensitive temperature dependence. In this contribution, we present the experiment result on the dependence of susceptibility of single crystal B2212 sample on temperature at strong applied field.

2. Experiment and Result

The single crystal B2212 sample suffered a final treatment in pure nitrogen atmosphere after grown. The sample preparation and structure and composition examination will be reported elsewhere (Zhao et al.1990). The magnetic measurement was carried out by a commercial SQUID. Fig.1 displays the mass susceptibilities χ_m in wide temperature range and two applied fields parallel to the crystal axis b. The data show that the B2212 single crystal sample has diamagnetism at temperatures up to the highest measured value of about 320K, in spite of the χ_m is rather small . Fig.1 shows also that a first magnetic transition occurs at Tc of around 90K. Fig.2 extends the χ_m(T) to quite low temperature of about 10K, and shows a second transition at temperature Tc* of around 25K. The second transition is sharper than the first one and shows a stronger diamagnetic state exists. The sample is nearly rectangular with a weight of 7.2mg. Mangetization measurements were caried out in field cooling (FC).

---

* This work is supported by the National Center for Research and Development on Superconductivity.

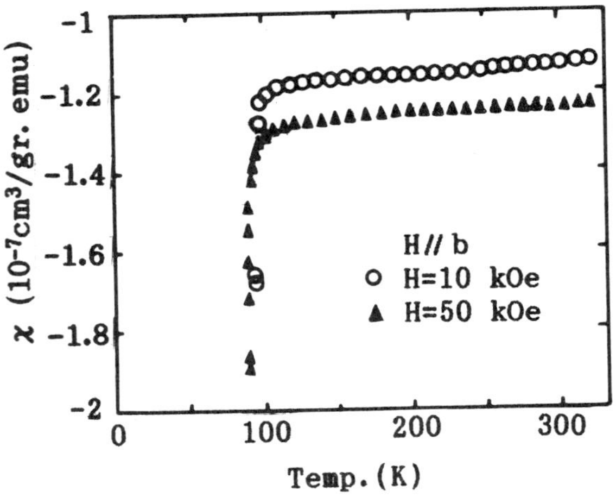

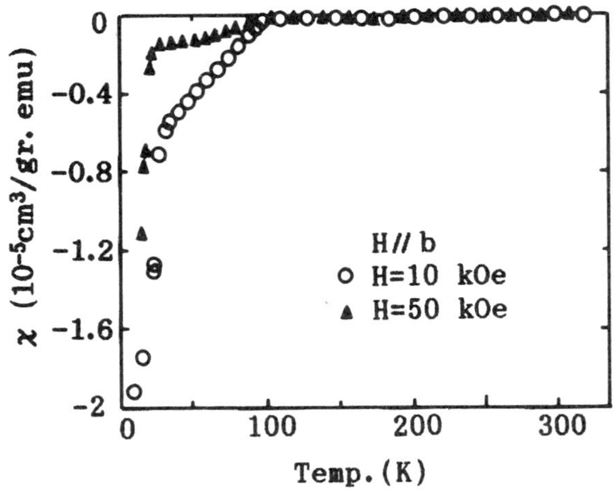

Fig.1 The $\chi_m(T)$ of B2212 single crystal shown normal state diamagnetism. H // b.

Fig.2 The $\chi_m(T)$ of B2212 single crystal shown two transitions. H // b.

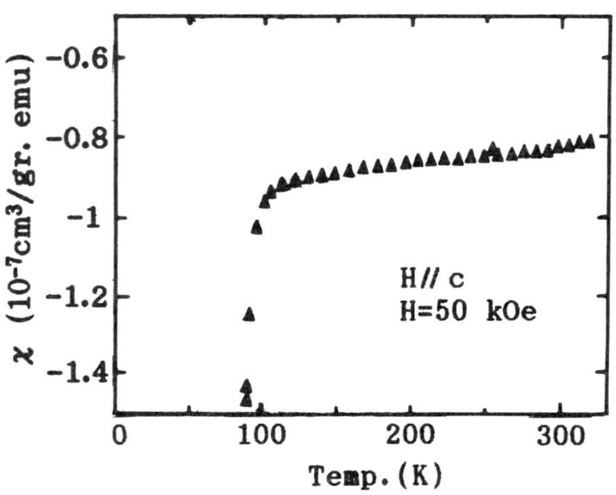

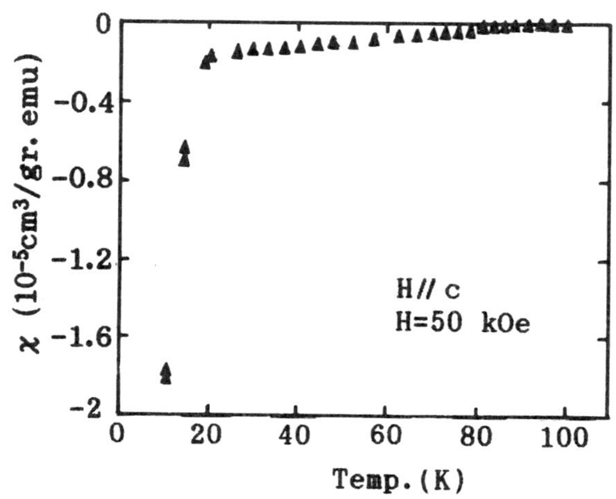

Fig.3 $\chi_m(T)$ of B2212 single crystal shown the different χ_m of normal state diamagnetism at H // c.

Fig.4 $\chi_m(T)$ of B2212 single crystal shown also two transitions at H // c.

Fig.3 illustrates the similar property as the one in fig.1 at H // c. It is worthy to note that the normal state diamagnetism is anisotropic because the χ_m are different from each other for H // b and H // c.

Fig.4 demonstrates there exist also two magnetic transitions for H // c with the same Tc and Tc* as those of in H // b and the strong diamagnetism at temperatures below Tc* $\approx$25K.

3. Discussion

The anisotropic susceptibility χ at H // b and H // c in normal state must not come from the diamagnetism of atoms or molecules of impurity or of any unreacted components such as Bi_2O_3 etc. which should spread statistically in the sample. Whereas the anisotropic χ at T>Tc must associates with the sample's anisotropic structure as well as superconductivity.

X-ray diffraction and TEM structure studies have shown that the sample are Ca deficient resulting from lattice vacancy of Ca, although the sample is still in single 2212 phase (Zhao et al. 1990). It is reasonable that the enhancement of diamagnetism at T< Tc* results from the superconducting transition of Ca deficient area which is non-superconducting at T>Tc*.

Motoi et al.(1989) and Ding et al.(1990) have pointed out that low oxygen pressure anneal treatment enhances Tc of B2212 up to about 90K. Our results support their conclusion.

It is very interest to point out that the Tc* is about the same temperature below which the Jc and flux pinning energy are becoming higher. The Tc* is also about the same temperature as flux lattice melting one in B2212. Therefore it is very important to confirm if our speculation is true as discussed in introduction.

In summary, We have shown the anisotropic diamagnetism associated with the structural as well as superconductive anisotropy in normal state B2212 single crystal. A new diamagnetic transition with Tc* $\approx$25K which is attributed to Ca deficient regime in B2212 phase has been found. It is not clear that if the Tc* is related to the low flux pinning energy at high temperatures and low flux line lattice melting temperature of B2212.

References

Ding S Y 1990 Submitted in Acta Physica Sinica.

Küpfer H, Apfelstedt I, Flükiger R, Keller C, Meier-Hirmer, Runtssch B, Touroski, Weiech U and Wolf T 1988 Cryogenics **25** 650

Motoi Y, Ikeda,Uwe H and Sakudo T 1989 Physica C **162-164** 929

Nelson D R,1989 Physica C **162-164** 1156

Palstra T T M, Batlog B, Van Dover R B, Schneemeyer L F and Waszszak J V to appear in Phys. Rev.B.

Shi D, Boley M S, Welp U, Cheng J G and Liao Y 1989 Physical Rev. B **40** 5255

Zhao et al. 1990 to be published.

Supercond. Sci. Technol. 4 (1991) S208–210. Printed in the UK

CARRIER LOCALIZATION IN HYDROGEN-DOPED $La_{1.9}A_{1.1}Cu_2O_{6+y}$ (A= Ca, Sr)

W. Ye, T. Takabatake, T. Ekino and H. Fujii
Faculty of Integrated Arts and Sciences, Hiroshima University
Hiroshima 730, Japan

The effect of hydrogen absorption on transport and magnetic properties was studied for $La_{1.9}A_{1.1}Cu_2O_{6+y}$ (A= Ca and Sr) having two-dimensional CuO_2 network. The results indicate that the introduction of hydrogen atom causes a strong localization of carriers and induces a metal-insulator transition in these systems.

1. Introduction

We reported previously that hole concentration in high-T_c cuprates can be controlled by introducing hydrogen into the system (Takabatake et al 1989). In $YBa_2Cu_3O_7$ and $Bi_2Sr_2CaCu_2O_{8+y}$, hydrogen atoms behave as protons and thus play a role of electron donor to the CuO_2 plane. The metallic compounds of $La_{1.9}A_{1.1}Cu_2O_{6+y}$ (A= Ca, Sr) have been considered as a good candidate for high-T_c superconductors because they have common structural features to the high-Tc cuprates above mentioned, i.e., pyramidal Cu-O clusters forming double two-dimensional CuO_2 planes separated by a cation plane (Torrance et al 1988, Doverspike et al 1989 and Tamegai and Iye 1989). For $La_{1.9}Ca_{1.1}Cu_2O_6$, the nominal hole concentration of 0.05 per CuO_2 may be too small to exhibit superconductivity, whereas that of 0.26 holes per CuO_2 in oxygen-annealed samples of $La_{1.9}Sr_{1.1}Cu_2O_{6.21}$ seems to be in excess for superconductivity. We expected therefore the latter system would show superconductivity if the hole concentration is reduced by hydrogenation. It should be noted that hole doping in $La_2CaCu_2O_6$ by partial substitution of Sr for La has been successful in converting the system to a superconductor with T_c= 60 K (Cava et al 1990).

2. Sample preparation

The host samples $La_{1.9}A_{1.1}Cu_2O_{6+y}$(A= Ca and Sr) were prepared by a standard solid-state reaction. Final heat treatment was performed in a flowing oxygen atmosphere at 1060 °C for 50 h, which was followed by furnace cooling. Hydrogenation was performed by exposing samples to 7N hydrogen gas of 0.9 bar at 350 °C. The amount of hydrogen incorporated into the sample was determined volumetrically from the overall decrease in pressure of hydrogen in the reaction cell. Powder X-ray diffraction analysis showed that samples with hydrogen mole fraction x up to 1.0 and 0.5 are single-phased for A= Ca and Sr, respectively. With increasing x in both systems, the lattice parameter a gradually increases but the c-parameter remains unchanged.

3. Results and discussion

For both the host compounds, the resistivity ρ is about 8 mΩcm near room temperature. Upon hydrogenation, ρ at 280 K increases exponentially with increasing x, as shown in the inset of Fig.1. Even at x= 0.1, the values increase by one and three order of magnitude for A= Ca and Sr, respectively. These effects are much drastic than that found in $Bi_2Sr_2CaCu_2O_{8+y}H_x$, where the increase in ρ at x= 0.1 is only 40 % (Takabatake et al 1989). With decreasing temperature, $\rho(T)$ of hydrogenized samples show semiconducting behavior. For semiconducting oxides, the resistivity is generally of the form of $\rho(T)= \rho_0 \exp(T_0/T)^{1/n}$. Figure 1 shows that data of $La_{1.9}Ca_{1.1}Cu_2O_6H_x$ obey the law for n=3 below 50 K, which is indicative of Mott variable range hopping in the two dimensional systems. In this case, the characteristic temperature T_0 is correlated with the localization length $1/\alpha$ as $T_0 = 3\alpha^2/k_B N(E_F)$, where $N(E_F)$ is the electronic density of states at the Fermi level (Knotek et al 1973). For the samples with A= Sr, T_0 was estimated in a similar way, regardless of a small deviation from the n= 3 law at low temperatures.

In an attempt to estimate the value of $N(E_F)$, we measured the magnetic susceptibility χ between 4.2 and 500 K. Figure 2 shows that $\chi(T)$ of the

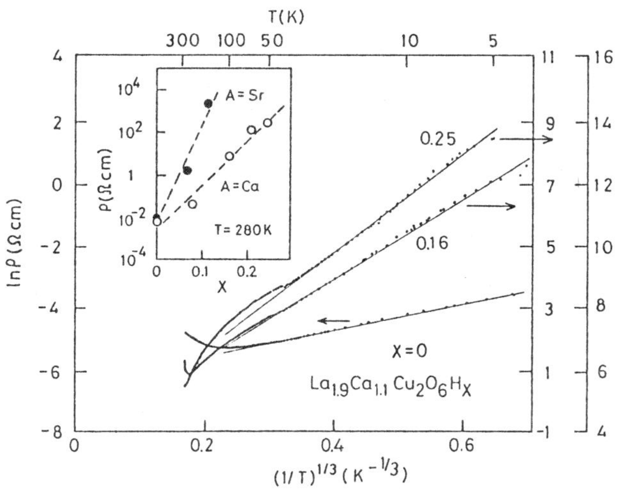

Fig.1 Log of electrical resitivity vs $T^{-1/3}$ of $La_{1.9}A_{1.1}Cu_2O_6H_x$ for A= Ca. Inset: log of resistivity at 280 K vs hydrogen concentration x for A= Ca and Sr.

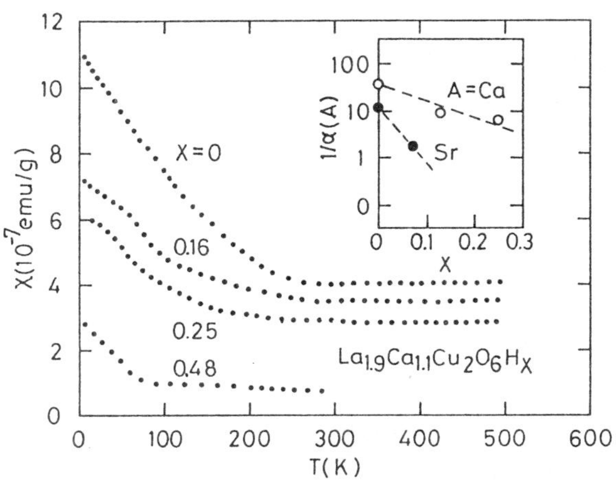

Fig.2 Magnetic susceptibility vs temperature of $La_{1.9}A_{1.1}Cu_2O_6H_x$ for A= Ca. Inset: log of localization length vs hydrogen concentration x for A= Ca and Sr.

system with A= Ca gradually decreases with increasing x over the temperature range. The upturn below 200 K may originate in small amount of isolated Cu^{2+} ions in defects and/or impurity phases. It should be noted that recent μ^+SR experiments on our sample of $La_{1.9}Ca_{1.1}Cu_2O_6H_{0.25}$ revealed the development of an antiferromagnetic ordering at temperatures below 300 K (Shiratake et al 1990). However, $\chi(T)$ of the sample of x= 0.25 gives no indication of magnetic order up to 500 K. Since $\chi(T)$ becomes independent of T above 250 K irrespective of x, we may assume that the contribution from the two-dimensional antiferromagnetic spin correlations does not strongly depend on x in the studied range. Then, the decrease in $\chi(T)$ with x can be ascribed to the decrease in the Pauli susceptibility χ_p. The absolute value of χ_p was determined by assuming χ_p to be zero for the insulatiang samples of x= 0.48 for A= Ca and of x= 0.52 for A= Sr. Using the value of $N(E_F)$ thus obtained, we evaluated $1/\alpha$ and plotted the results in the inset of Fig. 2. With increasing x, $1/\alpha$ decreases exponentially for both systems and the decrease is much stronger in the Sr system than in the Ca system. Nguyen et al (1980) reported that Ca^{2+} ions in the host compound preferentially occupy the 2a site, whereas Sr^{2+} and La^{3+} ions occupy randomly the 2a site and it results in an incorporation of excess oxygen into the 2b site neighbouring the La^{3+} ions. The holes derived from the excess oxygen may weakly localize near the cation layer sandwiched between the CuO_2 planes. The exponential decrease in localization length with x indicates that the holes are trapped by the hydrogen atom, as similar in the passivation of shallow acceptors in hydrogenized silicon (Pankove et al 1983). In the present compounds with rather localized hole carriers, the role of hydrogen is significantly different from that in $Bi_2Sr_2CaCu_2O_{8+y}H_x$, where the electrons transferred from hydrogen gradually fill the hole band derived from oxygen-holes and thus change the T_c (Takabatake et al 1989). Comparing these results, we may conclude that the mobile hole concentration can be controled by hydrogenation only when the hole band is well developed.

Ferefences

Cava R J et al 1990 Nature 345 602.

Doverspike K, Liu J-H, Dwight K and Wold A 1989 J.Solid State Chem. 82 30.

Knotek M L, Pollak M and Donovan T M 1973 Phy. Rev. Lett. 30 853.

Nguyen N, Er-Rakho L, Michel C, Choisnet J and Raveau B 1980 Mat. Res. Bull 15 891.

Pankove J I, Carlson D E, Berkeyheiser J E and Wance R O 1983 Phys. Rev. Lett. 51 2224.

Takabatake T, Ye W, Orimo S, Tamagai T and Fujii H 1989 Physica C 162-164 65.

Tamegai T and Iye Y, 1989 Physica C 159 181.

Torrance J B, Tokura Y, Nazzal A, Parkin S S P 1988 Phys. Rev. Lett. 60 542.

Shiratake S, Nishida N, Ye W, Takabatake T and Tamegai T, to be published.

Supercond. Sci. Technol. **4** (1991) S211–213. Printed in the UK

EFFECTS OF OXYGEN STOICHIOMETRY ON CRITICAL TEMPERATURE AND CRYSTAL STRUCTURE OF $Er_{1-x}Pr_xBa_2Cu_3O_{7-\delta}$ AND $ErBa_{2-x}La_xCu_3O_{7-\delta}$

D.Wagener, M.Buchgeister, W.Hiller, S.M.Hosseini and K.Kopitzki
Institut für Strahlen- und Kernphysik der Universität Bonn,
Nußallee 14-16, D-5300 Bonn 1, F.R.G.

Abstract. For ceramic samples of HTC-superconductor $ErBa_2Cu_3O_{7-\delta}$, in which erbium ions were partially substituted by praseodymium ions or barium ions by lanthanum ions, the critical temperature T_c and the crystal structure were determined in dependence of the oxygen content of the compound. T_c was inferred from standard four-probe resistivity measurements and the crystal structures from powder X-ray diffraction. The critical temperature versus oxygen content and versus total charge of the noncopper cations is discussed.

The critical temperature T_c of HTC-superconductors of the type $ABa_2Cu_3O_{7-\delta}$ (A = Y or Lanthanide) is strongly influenced by the oxygen deficiency δ of the compound. Generally the oxygen content dependence of T_c shows plateaus at about 90 K and about 60 K (Cava *et al* 1987), the length of which was found to be affected by the ionic radius of the substituted A-atom (Buchgeister *et al* 1990). T_c can also be influenced by partial substitution of the A-atoms or the Ba-atoms by atoms of lower or higher valence (Tokura *et al* 1988). In this work investigations were performed on the compound $ErBa_2Cu_3O_{7-\delta}$; Er-atoms were partially substituted by Pr-atoms and Ba-atoms by La-atoms. We studied the influence of these substitutions on T_c and crystal structure for varied oxygen content.

Samples of $Er_{1-x}Pr_xBa_2Cu_3O_{7-\delta}$ and $ErBa_{2-x}La_xCu_3O_{7-\delta}$ for x = 0.1, 0.2 and 0.3 were prepared by mixing appropriate amounts of Er_2O_3, Pr_6O_{11}, La_2O_3, $BaCO_3$ and CuO powders and subsequent calcination in air at 920 °C for doping with Pr and at 930 °C for doping with La. The ground material was annealed in flowing oxygen at the same temperatures for 12 h, pressed to pellets, annealed again and cooled down to room temperature. The oxygen reduction was performed as described by Buchgeister *et al* (1990). All samples had an initial oxygen content of about 7.00 ± 0.04 in agreement with the oxygen contents for the corresponding substitutions by Neumeier *et al* (1990) for Pr and Cava *et al* (1988) for La in the $YBa_2Cu_3O_{7-\delta}$-system. Cava *et al* found higher initial oxygen contents only for La concentrations x > 0.4. The stoichiometry of the cations was checked for each prepared sample set by electron microprobe. T_c of the pellets was inferred from standard four-probe resistivity measurements and the lattice parameters a, b and c were determined from X-ray powder diffraction with CuK_α-radiation. No additional peaks of impurity phases occured for doping with Pr. For doping with La the observed reflections of $BaCuO_2$ were smaller than two standard deviations of background intensity. The lattice constants a, b and c for the fully oxydized samples are listed in table I.

Fig. 1 shows for the system $Er_{1-x}Pr_xBa_2Cu_3O_{7-\delta}$ the dependence of the critical temperature T_c on the oxygen deficiency δ for x = 0.0, 0.1, 0.2 and 0.3. The value of T_c of the fully oxidized samples decreases rapidly with increasing percentage of Pr-atoms. No sample exhibits a clear

0953-2048/91/01S211 + 03 $03.50

Table I: Lattice parameters a, b and c of fully oxydized samples at room temperature

		$Er_{1-x}Pr_xBa_2Cu_3O_{7.00\pm0.04}$			$ErBa_{2-x}La_xCu_3O_{7.00\pm0.04}$			
	x = 0.0	0.1	0.2	0.3	0.1	0.2	0.3	
a:	3.816(1)	3.819(1)	3.823(1)	3.830(1)	3.820(1)	3.822(1)	3.835(2)	[Å]
b:	3.880(1)	3.888(1)	3.894(1)	3.892(1)	3.887(1)	3.882(1)	3.875(2)	[Å]
c:	11.672(1)	11.676(3)	11.676(3)	11.674(3)	11.652(2)	11.632(2)	11.672(6)	[Å]

90 K plateau. For increasing x the 60 K plateau becomes more extended and the value of δ for which superconductivity vanishes decreases only slightly. The plateau structure of $ErBa_2Cu_3O_{7-\delta}$ was related by Rupp *et al* (1989) to the bond length between the apex oxygen and the sheet copper atom. If the oxygen content is reduced this bond length increases. Neumeier *et al* (1989) found for the system $Y_{1-x}Pr_xBa_2Cu_3O_7$ for increasing x the opposite effect. So the extended 60 K plateau of $Er_{1-x}Pr_xBa_2Cu_3O_7$ could be explained by a compensation of the two effects mentioned.

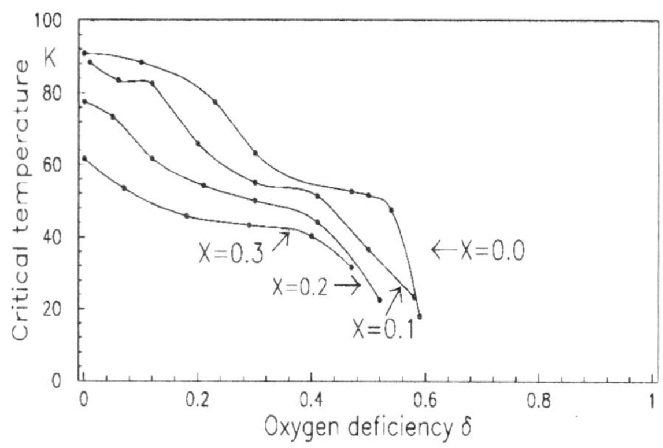

Fig. 1: Critical temperature T_c in dependence of oxygen deficiency δ for $Er_{1-x}Pr_xBa_2Cu_3O_{7-\delta}$.

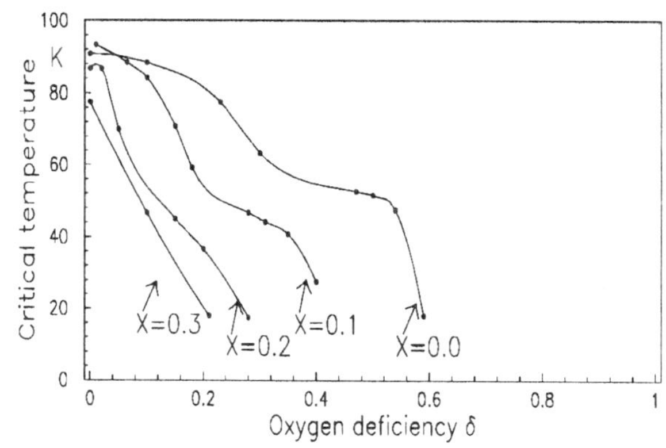

Fig. 2: Critical temperature T_c in dependence of oxygen deficiency δ for $ErBa_{2-x}La_xCu_3O_{7-\delta}$.

Analogous experimental data for the system $ErBa_{2-x}La_xCu_3O_{7-\delta}$ are shown in fig. 2. They are remarkably different from those for doping with Pr. The measured values of T_c of the fully oxydized samples for x < 0.3 are all around 90 K. With increasing concentration x of La the lengths of the 90 K and 60 K plateaus decrease in a similar way.

In the same manner as reported by Tokura *et al* (1988) for the systems $YBa_{2-x}La_xCu_3O_{7-\delta}$ and

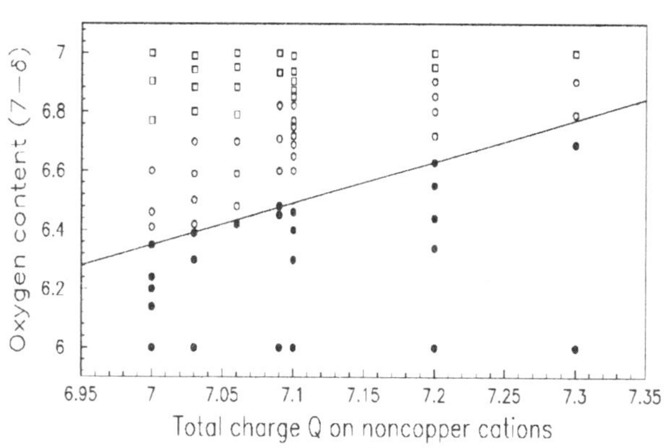

Fig. 3: Electrical characteristics of $Er_{1-x}Pr_xBa_2Cu_3O_{7-\delta}$ and $ErBa_{2-x}La_xCu_3O_{7-\delta}$ in dependence of oxygen content (7-δ) and total charge Q on the noncopper cations. The symbols are explained in the text.

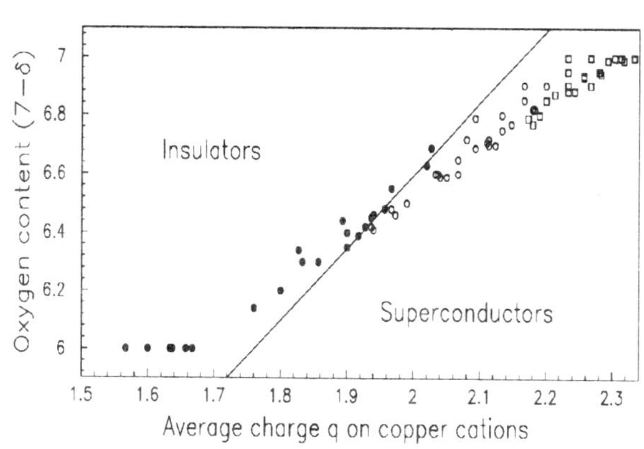

Fig. 4: Electrical characteristics of $Er_{1-x}Pr_xBa_2Cu_3O_{7-\delta}$ and $ErBa_{2-x}La_xCu_3O_{7-\delta}$ in dependence of oxygen content (7-δ) and the average charge q on the copper cations. The symbols are explained in the text.

$Y_{1-x}Ca_xBa_2Cu_3O_{7-\delta}$ we plotted our results as a function of the oxygen content (7-δ) and the total charge Q on the noncopper cations (fig. 3). The solid line seperates superconducting (open symbols) and insulating phases (closed symbols), phases with metallic-like resistivity are marked by open squares, those with semiconductor-like resistivity by open circles. If the valence of Pr is assumed to be +3.3 as determined by Neumeier *et al* (1990) the electrical characteristics of $Er_{1-x}Pr_xBa_2Cu_3O_{7-\delta}$ are reproduced rather well in the diagram.
Instead of characterizing the samples by the total charge Q on the noncopper cations we can also use the average charge q on the copper cations (fig. 4) determined by the condition of electrical neutrality $Q + 3q = 2(7-\delta)$.
The meaning of the solid line in fig. 4 is the same as in fig. 3. It defines q_c. T_c of superconducting samples is found to increase for increasing horizontal distance from that line. According to the model of Tokura *et al* (1988) the number of mobile holes per sheet Cu atom can be determined by the relation $p_{sh} = 3/2\ (q-q_c)$, assuming that the charge q is distributed over two sheets and one chain.

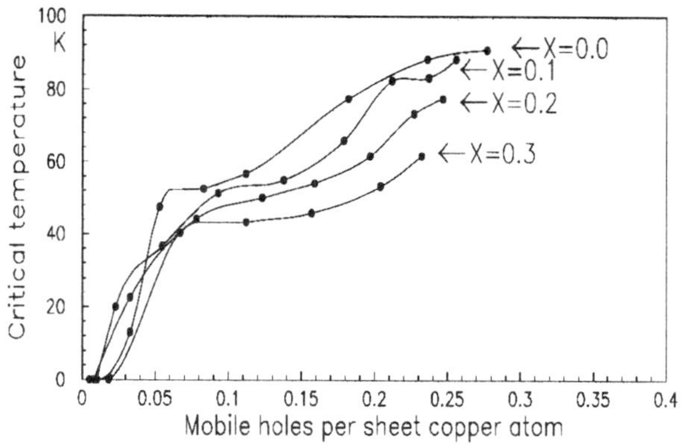

Fig. 5: T_c in dependence of mobile holes per sheet Cu atom for $Er_{1-x}Pr_xBa_2Cu_3O_{7-\delta}$.

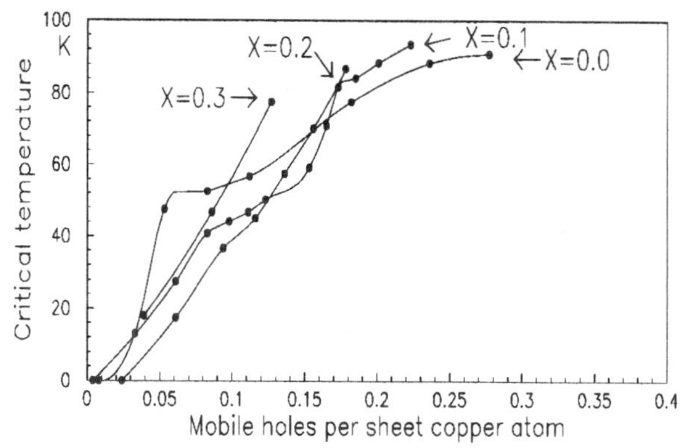

Fig. 6: T_c in dependence of mobile holes per sheet Cu atom for $ErBa_{2-x}La_xCu_3O_{7-\delta}$.

In fig. 5 and 6 T_c is plotted against p_{sh} for doping with Pr and La. T_c is found to increase for higher hole concentration per sheet Cu atom in the same way as reported by Tokura *et al* (1988). The number p_{sh} of mobile holes can also be determined by Hall effect measurements. The results should agree with the values of p_{sh} calculated according to the model of Tokura *et al* (1988) and should especially confirm the assumed valence of Pr. Appropriated experiments are in preparation.

We thank Dr.B.Spiering for the analysis of the prepared samples by electron microprobe. Three of the authors would like to acknowledge support from the Cusanuswerk (M.B.), the Friedrich-Naumann-Stiftung (W.H.) and the Alexander-von Humbold-Stiftung (S.M.H).

References

Buchgeister M, Hiller W, Hosseini S M, Kopitzki K and Wagener D 1990 *Proc.Int.Conf. on Transp. Prop. of Superconductors* ed R Nicolsky (Singapore: World Scientific) to be publ.

Cava R J, Batlogg B, Chen C H, Rietman E A, Zahurak S M and Werder D 1987 *Phys. Rev.* **B36** 5719

Cava R J, Batlogg B, Fleming R M, Sunshine S A, Ramirez A, Rietman E A, Zahurak S M and van Dover R B 1988 *Phys. Rev.* **B37** 5912

Neumeier J J, Bjørnholm T, Maple M B, Rhyne J J and Gotaas J A 1990 *Physica C* **166** 191

Rupp B, Pörschke E, Meuffels P, Fischer P and Allenspach P 1989 *Phys. Rev.* **B40** 4472

Tokura Y, Torrance J B, Huang T C and Nazzal A I 1988 *Phys. Rev.* **B38** 7156

Supercond. Sci. Technol. **4** (1991) S214–216. Printed in the UK

Microwave conductivity measurements of YBCO thin films

J R Waldram, M Robson, H-M Cheah and R E Somekh

IRC in Superconductivity, Madingley Road, Cambridge, UK CB3 OHE

Abstract The paper discusses techniques for determining the complex conductivities of thin films of high-T_c superconductors at microwave frequencies. Preliminary results suggest that measured microwave conductivities are sensitive to defects extended normal to the microwave current, which also affect the normal conductivity and J_c.

1 Methods of measurement

The complex conductivity $\sigma = \sigma_1 - i\sigma_2$ of a superconductor is usually measured at microwave frequencies by incorporating a sample into a resonator, and deducing from the measured bandwidth and frequency of resonance the *surface impedance* $Z_s = (i\omega\mu_o/\sigma)^{1/2}$ which is related to the *complex microwave skin depth* δ by the relation $Z_s = i\omega\mu_o\delta$. The same method may be used for thin films, but we must recognise that the effective surface impedance measured at the front of the film depends on its thickness and on the boundary condition at the back of the film. For instance, if the film forms part of the wall of the resonator, we may often take the magnetic field to be negligible at the back of the film. The effective surface impedance is then given by

$$Z_{eff} = Z_s \tanh(\delta/t) \qquad (1)$$

where t is the thickness of the film. In the *thin film limit* this reduces to the simple result $Z_{eff} = 1/\sigma t$, and diverges as t becomes small.

For a very thin film Z_{eff} may become comparable with Z_∞, the impedance of free space, when transmission through the film becomes important. One is unlikely to be working with films thin enough to make this an important issue when the film is appreciably resistive. There are, however, a number of reports in the literature of very small but finite intrinsic residual losses in good films at low temperatures which may well be explicable as radiation loss through the film, or through micro-cracks in the film.

The results reported below were obtained using two films back-to-back with the dielectric substrate inside so as to form a simple strip line resonator, which was placed inside a screening tube beyond cut-off to avoid radiation loss. No edge corrections were made.

2 Measurements for laser–ablated YBCO on MgO

Results were obtained using a 10mm x 8mm film of YBCO on MgO. The film was about 0.4μm thick. It was prepared by laser ablation at the University of California at Berkeley, and kindly supplied to us by Prof J. Clarke. The film had strong c-axis orientation normal to the film. The temperature dependence of dc resistivity is shown in Figure 1; the dc conductivity at 80K was measured as $1.7\times10^6\ \Omega^{-1}m^{-1} \pm 50\%$. Comparable films have J_c values of about $2.5\times10^{10} Am^{-2}$. Because only one film was available the strip line was completed using a copper foil as the other conductor. The line spacing was 0.5mm, and the fundamental resonance appeared at about 6.0GHz.

The apparent surface resistance of this film with the copper and MgO contributions subtracted is shown as a function of temperature in Figure 1. Here the surface resistance falls rapidly to a

0953-2048/91/01S214 + 03 $03.50

very low value, as in the best films known. This is clearly an excellent film showing surface resistance comparable to what is seen for single crystals in the same orientation.

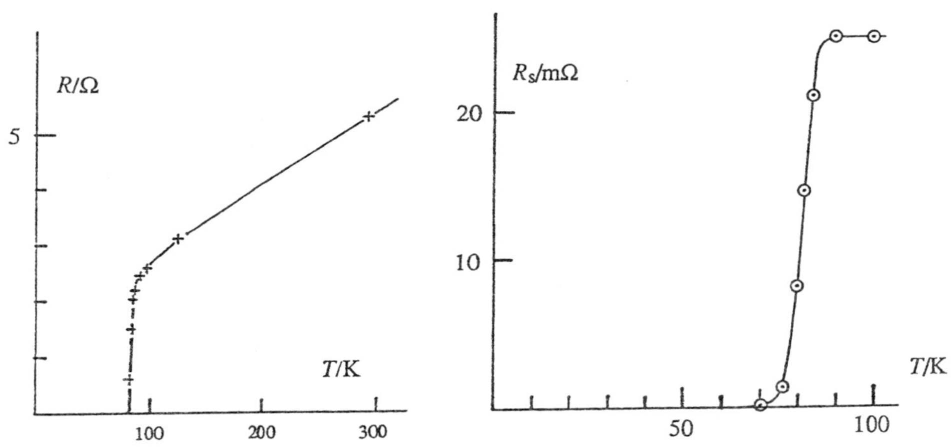

Figure 1 Laser-ablated YBCO on Mgo. Left: dc film resistance. Right: surface resistance

3 Measurements for sputtered YBCO on YSZ

Our second set of results was obtained using a pair of 10mm x 2mm films of sputtered YBCO on yttrium stabilised zirconia (YSZ). The films had rather variable thickness, averaging about 0.25μm. They were prepared early in the Cambridge sputtering programme and were by no means perfect. They had strong c-axis orientation normal to the film, but rocking curves on comparable films suggest that they contain grain boundaries with appreciable angles. The temperature dependences of dc resistivity are shown in Figure 2; the dc conductivity at 80K was measured as $1.6 \times 10^5\ \Omega^{-1}m^{-1} \pm 50\%$, but the critical current density of similar films was less than $10^{10}Am^{-2}$. The line spacing was 1mm, and the fundamental resonance appeared at about 7.3GHz.

The effects of the YSZ, thermal expansion etc were eliminated by repeating the experiment twice, once with copper foils in place of the YBCO and once with copper foils and air dielectric. Using the thin film limit the measured complex foil resistivity was computed and is plotted in the complex plane in Figure 2. (Near and above T_c the resonance became so broad that measurements were impossible.)

4 Comment

Figure 2 (left) shows that for the imperfect film of YBCO on YSZ the real part of the resistivity in the superconducting state is much greater than that observed in single crystals and powders, but still much smaller than the measured dc resistivity. This suggests strongly that the microwave current is flowing mainly through superconducting regions but occasionally crossing a narrow normal boundary, even at low temperatures. Why should this happen at microwave frequencies, when at low frequencies the films appear to be completely superconducting below about 80K?

At low frequencies, the supercurrent can loop around normal regions, and the sample will appear to be completely superconducting as soon as a percolation path through superconducting regions has been formed. But at microwave frequencies current paths which loop around normal regions may be blocked because they carry an inductive load. The most important normal regions will be those which have an appreciable extent a normal to the microwave current, such as, for instance, small-angle grain boundaries on which impurities have collected. Simple impedance arguments suggest that the current will flow through such a normal region rather than around it when a is greater than about $|\delta|(2\pi b/t)^{1/2}$ where δ is the normal state skin depth, and b is the width of the normal region. (We have assumed that the normal part of the film is in the thin film limit.) We estimate that in typical films this critical length will be about 10–25μm.

In film containing normal regions longer than this critical length we expect microwave currents to be flowing in series through superconducting and normal regions. In the thin film limit the effect should be that the measured complex resistivity is the sum of a real normal term, which depends on the geometry and resistivity of the normal grain boundaries, and an almost pure imaginary superconducting term. If the current flows uniformly this term should be about equal to the ideal superconducting complex resistivity $1/\sigma_s \approx i/\sigma_2$, but if the film contains some normal boundaries shorter than the critical length, these normal regions will force the supercurrent to follow a tortuous path, which would be expected to increase the superconducting contribution.

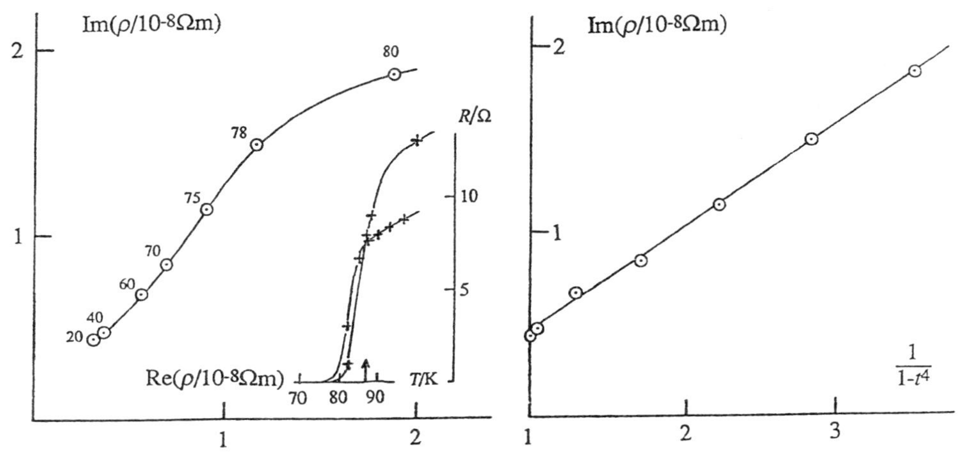

Figure 2 Sputtered YBCO on YSZ. Left: resistivity plotted in the complex plane, labelled by temperature T/K. Right: Im(ρ) as a function of $(1-t^4)^{-1}$; T_c is taken as 87K. Inset: dc resistance.

To test this model we plot in Figure 2 the imaginary part of the measured resistivity against the empirical function $[1-(T/T_c)^4]^{-1}$ which is known to fit well other measurements of $1/\sigma_2$ in YBCO. The fit is reasonable, and the penetration depth deduced from the slope is 4μm – about twice the accepted bulk value, a very plausible result if we assume that the film also contains some shorter normal regions which the supercurrent circulates around. The fall with temperature of the normal part of the measured resistivity is presumably due to some normal regions becoming superconducting and others becoming narrower because of the proximity effect.

Our work suggests that film quality as determined by normal state conductivity and J_c correlates to some extent with low microwave loss, presumably because the same defects, which may well be grain boundaries, limit all of these properties. But the dc transition width, which was about the same for all of our films, was not a good indicator of microwave performance, which may well be degraded by comparatively short defects even though the sample has percolation paths which give it a narrow transition in dc tests.

5 Summary

The complex conductivities of YBCO thin films may be conveniently measured when they are in the form of a strip line. Width of dc transition is not a good test of microwave quality, but the type of defect which produces excess microwave loss, probably a normal region extended 10μm or more normal to the current, may also degrade dc normal conductivity and J_c.

Supercond. Sci. Technol. **4** (1991) S217–219. Printed in the UK

Flux motion, proximity effect and critical currents in $Y_1Ba_2Cu_3O_7$/silver composites

J Jung, M A-K Mohamed and J P Franck

Department of Physics, University of Alberta, Edmonton, Alberta T6G 2J1, Canada

Abstract. We studied the magnetic and transport properties of pure $Y_1Ba_2Cu_3O_{7-\delta}$ and $Y_1Ba_2Cu_3O_{7-\delta}$/intergrain Ag (10 and 30 wt %) composites. The results show that the enhancement of the intergrain "transport" critical current density in the $Y_1Ba_2Cu_3O_{7-\delta}$/Ag (10 wt %) composite is caused by the SNS proximity junctions built up by intergrain silver and not by increased intergrain flux pinning.

We report on studies of flux pinning at low magnetic fields (0-1200 G) and critical current densities (both intergrain and intragrain ones) in the pure $Y_1Ba_2Cu_3O_{7-\delta}$ ceramics and the $Y_1Ba_2Cu_3O_{7-\delta}$/intergrain silver (10 and 30 wt %) composites. We investigated possible reasons for the enhancement of the transport intergrain critical current density in the $Y_1Ba_2Cu_3O_{7-\delta}$/silver (10 wt %) composite (Dwir *et al* 1989). It was suggested that this effect could be caused either by increased intergrain flux pinning or by improved coupling between superconducting grains (Ma *et al* 1989; Weinberger *et al* 1989).

The magnetic properties of the composites were measured using a SQUID magnetometer and a Hall probe device. The intergrain and intragrain critical current densities (J_{CT} and J_{CM}) were measured using dc current-voltage four-probe technique and high magnetic field (4.5 T) hysteresis loops. We also measured changes in the lattice constants and in the intragrain twin boundary structure.

We performed the following magnetic measurements:

-- the dc magnetization and the remanent magnetic moment versus temperature (10-95 K) for different applied magnetic fields over the range of 10 G-30 kG (Fig. 1);

-- the low-field (200 G) and high-field (4.5 T) hysteresis loops at 10 K and 77 K;

-- the dependence of the diamagnetic shielding, the Meissner effect, and the trapped magnetic field on an applied field (0-1200 G) at 77 K for both the zero-field-cooling (ZFC) and field-cooling (FC) cases;

-- the dependence of the decay rate of the trapped field on the initial trapped field at 77 K for the ZFC and FC cases.

The results of these measurements show that silver added to $Y_1Ba_2Cu_3O_{7-\delta}$ during the sintering process degrades the superconducting properties of $Y_1Ba_2Cu_3O_{7-\delta}$/intergrain silver composites: the diamagnetic shielding, the Meissner effect, and the trapped field. The activation energy E for intergranular flux creep in the $Y_1Ba_2Cu_3O_{7-\delta}$/Ag composites was found to be reduced in comparison to that for the pure $Y_1Ba_2Cu_3O_{7-\delta}$ sample (Table 1). This energy does not depend on the number of pinning centers and the amount of intergrain silver, and does not affect the intergrain J_{CT} or intragrain J_{CM}. The measurement of the transport critical current density revealed enhancement of the intergrain J_{CT} in the $Y_1Ba_2Cu_3O_{7-\delta}$/Ag (10 wt %) composite. On the other hand the magnetic intragrain critical current densities in the $Y_1Ba_2Cu_3O_{7-\delta}$/Ag composites were found to be reduced as compared to those in the pure $Y_1Ba_2Cu_3O_{7-\delta}$ (Table 1).

0953-2048/91/01S217 + 03 $03.50

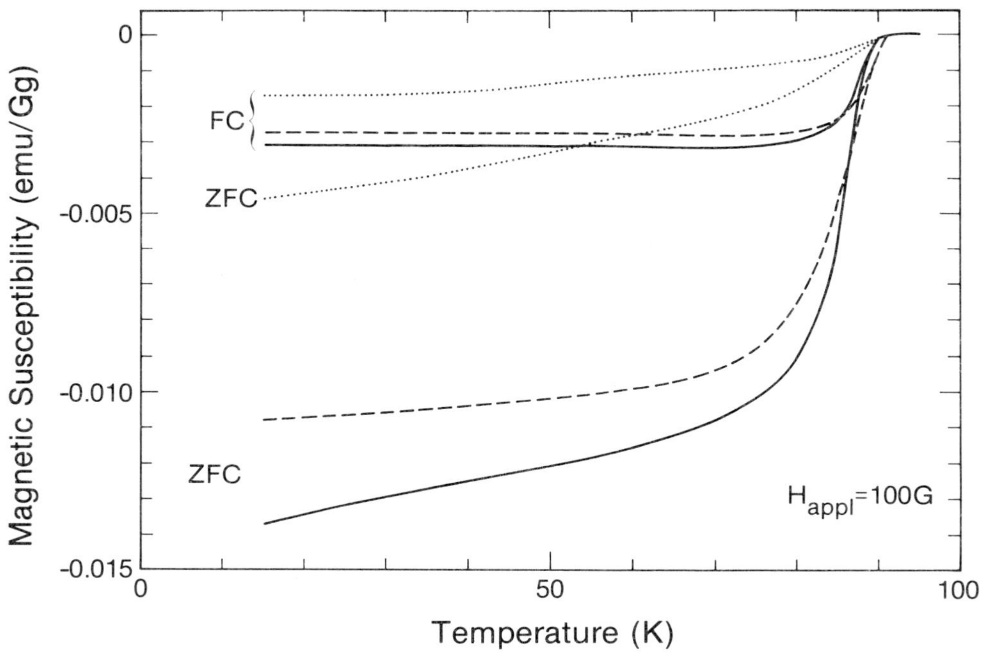

Fig. 1. Magnetic susceptibility vs. temperature, measured for the pure $Y_1Ba_2Cu_3O_{7-\delta}$ (solid curve), the $Y_1Ba_2Cu_3O_{7-\delta}$/Ag (10 wt %) composite (dashed curve) and the $Y_1Ba_2Cu_3O_{7-\delta}$/Ag (30 wt %) composite (dotted curve) at an applied magnetic field of 100 G. The difference in magnetic moments M_{ZFC}-M_{FC} decreases if silver is added to $Y_1Ba_2Cu_3O_{7-\delta}$.

Table 1: The intergrain (J_{CT}) and intragrain (J_{CM}) critical current densities and the activation energy for intergrain flux creep E in the pure $Y_1Ba_2Cu_3O_{7-\delta}$ and the $Y_1Ba_2Cu_3O_{7-\delta}$/Ag composites.

Sample (wt % Ag)	Average Grain Size (μm)	Oxygen Content (7-δ)	T_c (K)	J_{CT} (77K) (A/cm^2)	J_{CM} (77K) (A/cm^2)	E(ZFC,77K) (eV)
0	3.7	6.957	90.5(R=0)	68.6	3.6x10^5	1.6 ±0.3
10	6.1	6.915	91.0(R=0)	230.6	2.2x10^5	0.28 ±0.02
30	6.4	6.830	91.0(R=0)	0.78	3.8x10^4	0.26 ±0.03

The results suggest that the intergrain flux pinning is not responsible for the enhancement of J_{CT} in the $Y_1Ba_2Cu_3O_{7-\delta}$/Ag (10 wt %) composite. It appears that the superconductor-metal-superconductor (SNS) proximity junctions built up by intergrain silver are responsible for the increase of J_{CT}.

The reduction of the intragrain J_{CM} in the $Y_1Ba_2Cu_3O_{7-\delta}$/Ag composites was found to be related to the amount of oxygen inside the grains (Table 1). The oxygen content was estimated from the changes in the lattice constant c. The intragrain structure of twin boundaries or their density was not affected by silver as revealed by transmission electron microscopy.

References:

Dwir B, Affronte M and Pavuna D 1989 *Appl. Phys. Lett.* **55** 399

Ma H, Liang S, Feng Q, Sun Y, Yan S and Zhang H 1989 *Europhysics Lett.* **10** 375

Weinberger B R, Lynds L, Potrepka D M, Snow D B, Burila C T, Eaton Jr H E,Cipolli R, Tan Z and Budnick J I 1989 *Physica* **C 161** 91

Supercond. Sci. Technol. **4** (1991) S220–222. Printed in the UK

Vortex pinning and flux creep in $Bi_2Sr_2(Ca_{1-x}R_x)Cu_2O_{8+y}$ (R = Y,Gd)

PAJ de Groot, RA Rose, B Jayaram, GP Rapson and M Guillot[+]

Physics Department, University of Southampton, Southampton SO9 5NH, UK
[+]Service National des Champs Intenses, C.N.R.S, 38042 Grenoble Cedex, France

Abstract.

We present magnetisation measurements on powders of $Bi_2Sr_2(Ca_{1-x}R_x)Cu_2O_{8+y}$ (R = Y,Gd). The observed magnetic relaxations and the field dependence of the pinning force have been described in terms of a thermally activated flux-creep model without the need to consider phase transitions of the vortex system.
We have found that the intra-grain J_c decreases rapidly with rare earth or Y substitution. Even in the concentration range for which T_c remains constant, J_c decreases: $J_c(x{=}0)/J_c(x{=}.32) = 5.7$.

1. Introduction

Magnetic investigations of the Bi and Tl families of HTS have revealed unusual properties of the vortex state (e.g. large relaxations of the superconducting currents, large magneto resistances and a rapid decrease of intra-grain critical currents above about 30K). To explain these observations, several researchers have suggested the presence of a melting or glass-like transition of the vortex system in HTS. Measurements of the magnetic field dependence of the pinning force ($F_p = J_c \times B$) have been used to support the presence of a transition (Markiewicz 1988)). The pinning force follows surprisingly well the simple relation derived by Kramer (1973):

$$F_p = K\, b^{\frac{1}{2}}\, (1-b^2) \qquad (1)$$

where K is a constant and $b = B/B^*$. For conventional superconductors $B^* = B_{c2}$. In Kramer's theory B^* is the field for which the shear modulus of the vortex lattice goes to zero: hence the vortex lattice melts at B^*. For Bi and Tl based high temperature superconductors values for B^* are found which are an order of magnitude smaller than B_{c2}. Based on these observations suggestions have been made that $B^*(T)$ represents a melting line.

We have made detailed magnetic hysteresis measurements in fields up to 20T on powders of $Bi_2Sr_2CaCu_2O_{8+y}$ (BSCCO) and $Bi_2Sr_2(Ca_{1-x}R_x)Cu_2O_{8+y}$ (R = Y,Gd), to investigate these claims and to investigate the dependence of this behaviour upon systematic changes in the parameters of the superconducting state.

2. Experimentals

Powders of $Bi_2Sr_2(Ca_{1-x}R_x)Cu_2O_{8+y}$ (x = 0%, 5%, 10%, 20%, 32%, 45% and R = Y,Gd) have been prepared by standard solid state reaction methods. The powders were single phase as determined from X-ray diffraction data. Critical temperatures have been determined by measurements of resistivity (on sintered pellets) and a.c. susceptibility.

We have studied the magnetic properties of these samples using a 12T VSM (Oxford Instruments Ltd). Measurements in higher fields (up to 20T) have been performed with an extraction magnetometer at the S.N.C.I. The VSM measurements have been done with various sweep rates of the applied magnetic field. In the measurements with the extraction magnetometer, at each value of the field, a waiting time of 2 minutes was chosen to allow for relaxation of the screening currents.

3. Magnetic Hysteresis of $Bi_2Sr_2CaCu_2O_{8-y}$

Our magnetic studies reveal a strong dependence of the hysteresis width on the sweep rate of the applied magnetic field. At temperatures above about 30K the motion of the vortices has been found to be diffusive (Ota et al 1989). Our results also confirm the rapid decrease of (intra-grain) vortex pinning roughly at a line $B^*(T)$. However, the $F_p(B)$ curves deviate clearly from the Kramer relation (see fig.2). These deviations are particularly clear for high fields where F_p tails off and remains finite for $B > B^*$.

We have analyzed these results assuming thermally activated motion of the vortices, starting from the flux creep relation for the electrical field (e.g. Dew-Hughes 1988):

$$E = 2\, a_0\, B\, \Omega_0 \exp(-U/k_BT) \sinh(JBa_0/N_pk_BT) \qquad (2)$$

The electrical field will be proportional to the sweep rate (via some geometrical factor) in a magnetic hysteresis experiment. At a given temperature and magnetic field the flux creep equation gives for the hysteresis width as function of the sweep rate (β = dB/dt):

$$\Delta M/\Delta M^* = \sinh^{-1}(\beta/\beta^*) \qquad (3)$$

where ΔM^* and β^* are constants. A fit of our data using this relation gives good agreement (fig.1).

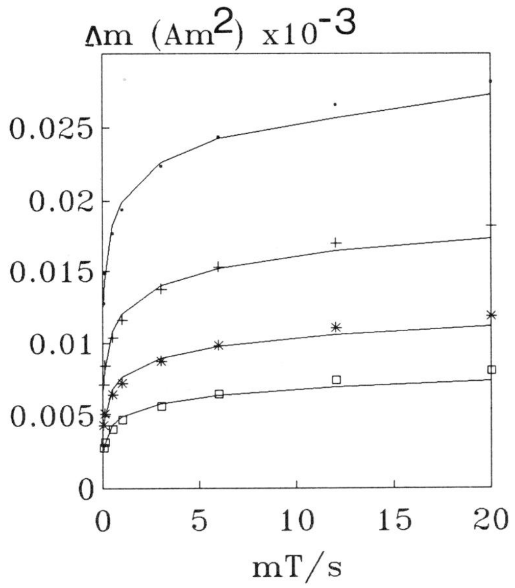

Fig.1 Sweep rate dependence of ΔM in BSCCO. T=25K, B = 0.75T (•), 1.05T (+), 1.35T (*), 1.65T (▫).

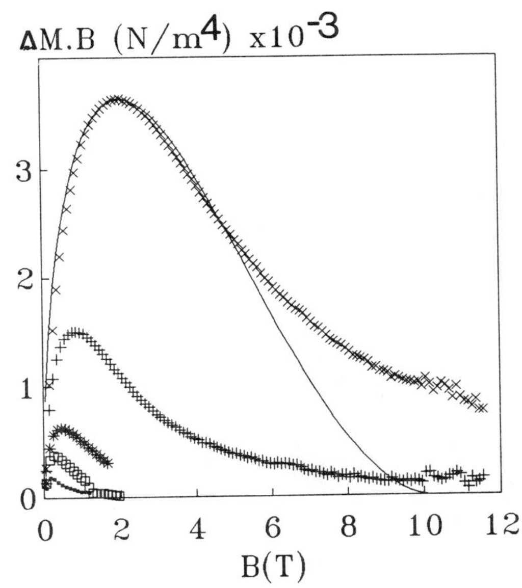

Fig.2 Pinning force for BSCCO. β=1mT/s and T = 15K (×), 20K (+), 25K (*), 30K (▫), 35K (▪).

To model our $F_p(B)$ data using the flux creep equation, we need to know the field dependence of the pinning potential. We have chosen a pinning potential as given by Dew-Hughes (1987) for pinning by normal precipitates: $U(b) \propto (1-b)$ (where $b=B/B_0$ with B_0 a constant). Combining this pinning potential with the flux creep equation gives:

$$F_p = C\, b^{\frac{1}{2}} \sinh^{-1}(Ab^{-\frac{1}{2}} \exp(u(1-b))) \qquad (4)$$

where C,A and u are field independent parameters. In figure 3 we have given an example of $F_p(B)$ data obtained from our hysteresis measurements fitted by eye to a curve obtained from eq. 4. The agreement of the flux creep model with our data is good. In contrast to comparisons using eq. 1, the high field region is well described. Therefore, the $B^*(T)$ (or irreversibility) line indicates the line where J_c falls below some small current criterion (e.g. the instrument resolution) and we find no evidence for a transition in the vortex system at this line.

4. Effects of Substitutions with Y and Gd

In the rare earth and Y substituted BSCCO superconductors T_c remains unchanged for x<0.3; above this concentration T_c decreases, becoming zero at x=0.6. The hole concentration, decreases with x over the whole concentration range: 0 < x < 0.6 (Jayaram et al. 1989). Our transverse field MuSR studies on Gd substituted BSCCO revealed that the penetration depth increases with x over the whole range: 0<x<0.6. The penetration depth roughly doubles on going from x = 0 to x = 0.3. The vortex interaction energy thus changes rapidly with x. Our magnetic hysteresis measurements reveal strong relaxations of the magnetisation similar to the pure (x = 0) compound. The pinning force decreases rapidly with Y or Gd concentration (fig.4) even in the range where T_c is constant (at x=0.32 the pinning force reaches only 18% of the value for pure $Bi_2Sr_2CaCu_2O_{8+y}$). However the shape of the $F_p(B)$ curve changes only slightly upon these substitutions. A vortex lattice melting line based on the decreasing part of the $F_p(B)$ curve (e.g. using Kramer's relation) would thus remain roughly unchanged. This appears in contrast to the melting model for which changes in vortex interaction energy would be expected to change the melting phase line.

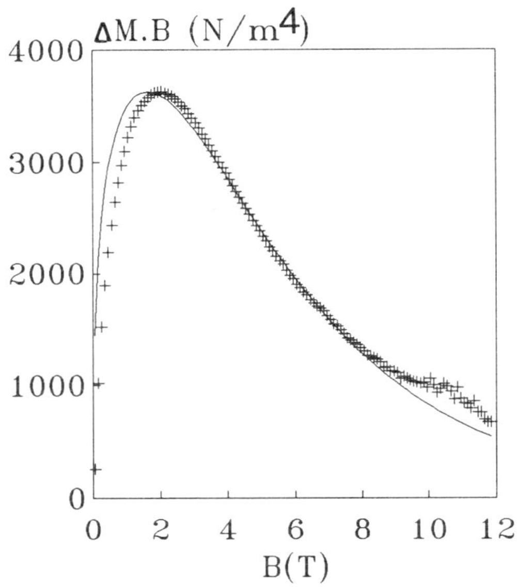

Fig.3 Pinning force for BSCCO at 15K and β=1mT/s. Solid line is a fit using eq.4.

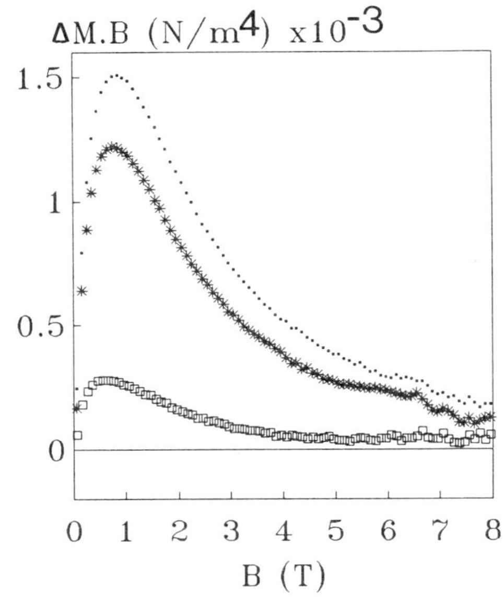

Fig.4 Pinning force for Y substituted BSCCO. T=25K, β=1mT/s and x = 0% (•), 10% (*), 32% (□).

References

Dew-Hughes D 1987 *Phil. Mag.* **B55** 459
Dew-Hughes D 1988 *Cryogenics* **28** 674
Ota SB, Rose RA, Jayaram B, de Groot PAJ and Lanchester PC 1989 *Physica C* **157** 520
Jayaram B, Lanchester PC and Weller MT 1989 *Physica C* **160** 17
Kramer EJ 1973 *J. Appl. Phys.* **44** 1360
Markiewicz RS 1988 *J. Phys. C* **21** L1173

Supercond. Sci. Technol. 4 (1991) S223–225. Printed in the UK

Relaxation effects in $Bi_2Sr_2Ca_1Cu_2O_{8+x}$ and $Bi_{1.7}Pb_{0.3}Sr_2Ca_1Cu_2O_{8+x}$ single crystals

E.Agostinelli, D.Fiorani, J.Tejada*, A.M.Testa

I.T.S.E., C.N.R., Area della Ricerca, P.B.10,00016 Rome (Italy)
* Depto. de Fisica Fonamental, Universidad de Barcelona (Spain)

Abstract Relaxation effects have been investigated in $Bi_2Sr_2Ca_1Cu_2O_{8+x}$ and $Bi_{1.7}Pb_{0.3}Sr_2Ca_1Cu_2O_{8+x}$ single crystals by critical current and magnetic relaxation measurements. Both crystals consisted of 2212 single phase material (T_C= 85K and 92K for the pure and the doped samples, respectively).

1. Introduction

A great deal of research is in progress in order to obtain high temperature superconducting materials able to carry high critical current densities in presence of large magnetic fields, without detectable dissipation. For example, critical current densities of 10^4 - 10^6 Acm^{-2} have been reported for single crystals by Enomoto et al.(1987) and for thin films by Watanabe et al.(1989), even in large magnetic fields. However, applications of high T_C materials at high temperatures are seriously affected by thermally activated flux creep, which causes a relaxation of persistent currents. A combination of a large thermal activation energy and a relatively small pinning energy is regarded as responsible of the observed large flux creep (see for example Yeshurun et al., 1989). The introduction of crystalline defects acting as pinning centers is therefore necessary to sustain resistanceless currents in magnetic fields. Various methods have been used to introduce defects into a high T_C superconductor, such as fast neutrons irradiation (Van Dover et al. 1989), mechanical stress (Kramer et al.,in press), quench and melt growth processes (Murakami et al .,1988). In this paper we discuss the effects of a partial substitution of lead on the critical current density and relaxation processes of a BSCCO single crystal.

2. Results and discussion

Single crystals of BSCCO (platelet of 2.2x1.7x0.3 mm^3) and $Bi_{1.7}Pb_{0.3}Sr_2Ca_1Cu_2O_{8+x}$ (disk with r=0.98 mm and thickness t=0.16 mm) have been selected for the measurements.
Both X-ray diffraction and magnetic characterization (Agostinelli et al., in press) showed the presence of single phase 2212, with a superconducting transition at 85 K for the pure-Bi compound and 92 K for the doped one.
The critical current density J_C has been estimated for both samples by using the critical state model (Bean, 1962), which relates J_C to the width of hysteresis cycles. J_C vs H plots at 4.2 K are reported in Figure 1. The doped sample exhibited a

0953-2048/91/01S223 + 03 $03.50

critical current density one order of magnitude higher than the pure-Bi compound, indicating an enhancement of the pinning strength. Such an enhancement could be ascribed to local fluctuations in the lead content which could, in turn, originate local changes in T_c, thus acting as more "effective" pinning centers than those present in the unsubstituted compound.

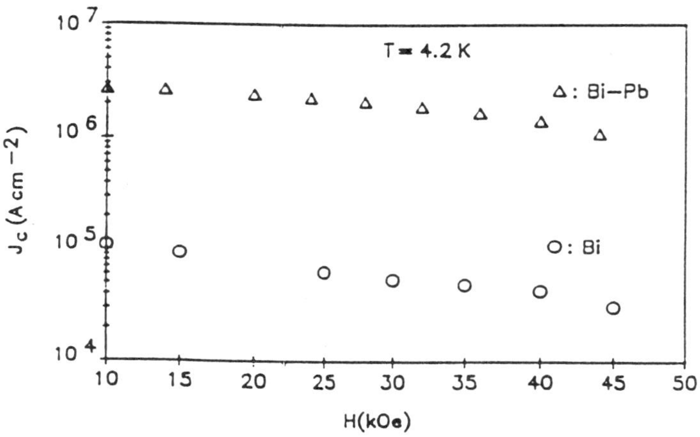

Fig. 1 Critical current density as a function of field (H//c) for BSCCO (o) and BPSCCO (Δ) single crystals.

Relaxation properties have been investigated by measuring the time decay of the zero field cooled magnetization (M_{zfc}). The sample has first been cooled down in zero field from above T_c and then, at different temperatures, a field H= 10 kOe has been applied parallel to the c-axis. In Figure 2 magnetization data, normalized to the first measured value, are reported as a function of time for the BSCCO (2a) and BPSCCO (2b) crystals. The time decay is logarithmic, in agreement with the classical flux creep model. The observed relaxation rates of the two compounds, though, differ significantly: for example, at T= 15K after 10^3 s, the relative decrease of M_{zfc} is ≈35% for the pure compound, but only ≈15% for the doped one. Relaxation data have been analyzed in the framework of the Hagen-Griessen model (1989) which assumes the existence of a distribution of energy barriers for, the thermally activated flux motion.

Following this model, an effective (temperature dependent) activation energy E_o= (18±2)meV has been deduced at 4.2K for the pure compound, consistenly with remanent magnetization data (Agostinelli et al.,1990). Furthermore, the observed E_o was found to increase with temperature (E_o≈36 meV at 15 K).

As implied in the H-G model, this behaviour is related to relaxations associated with the high energy tail of the activation energies distribution. For the Pb-doped crystal a higher E_o {E_o= (27±3)meV at 4.2K} has been observed.

Further experimental investigation of the field and temperature dependence of the elementary pinning force and relaxation measurements in a wider temperature range are in progress in order to get a better insight into the pinning mechanism, and to study the influence of lead doping on dissipative effects in Bi-based compounds.

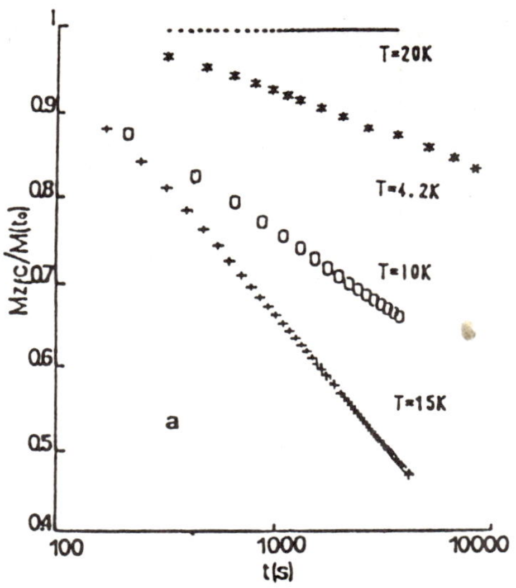

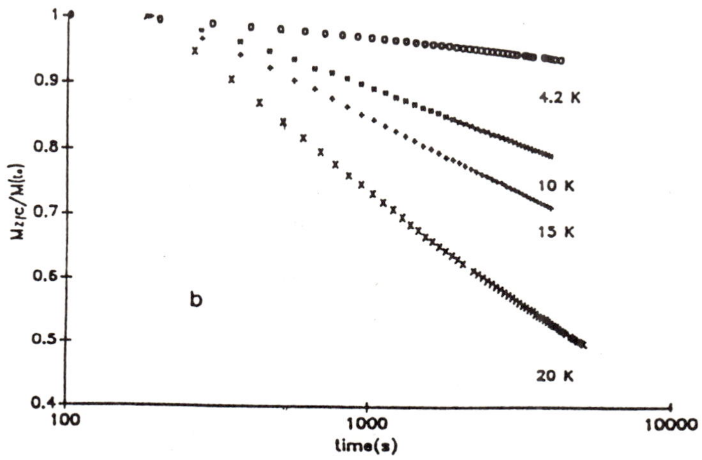

Fig. 2 Time dependence of M_{zfc} for BSCCO (a) and BPSCCO (b) single crystals, with H= 10 kOe // c-axis.

Agostinelli E., Filaci P., Fiorani D., Testa A.M., Tejada J., "Studies in high temperature superconductors" (New York: Nova Science), in press.

Agostinelli E., Balestrino G., Fillon G., Fiorani D., Muzi R., Paroli P., Tejada J., Testa A.M., 1990 J.Magn.Magn.Mat.**83** 509

Bean C.P., Phys.Rev.Lett. 1962 **8** 250.

Enomoto Y., Murakami T., Suzuki M., Moriwaki K., Jpn.J.Appl.Phys. 1987 **26** L1248

Hagen C.W., Griessen R., Phys.Rev.Lett. 1989 **62** 2857

Kramer M.J., Chumbley L.S., McCallum R.W., J.Mater.Sci. in press

Murakami M., Morita M., Doi K., Miyamoto K., Jpn.J.Appl.Phys. 1989 **28** 1189

Van Dover R., Gyorgy E., Schneemeyer L., Mitchell J., Rao K., Puzniak R., Waszczak J. Nature 1989 **342** 55

Watanabe K., Yamane H., Kurosawa H., Hirai T., Kobayashi N., Iwasaki H., Noto K., Mutop Y., Appl.Phys.Lett. 1989 **54** 575

Yeshurun Y. et al, Cryogenics 1989 **29** 258

Supercond. Sci. Technol. 4 (1991) S226–228. Printed in the UK

Direct Observation of Flux Behavior in Melt Processed YBaCuO Using The Faraday Effect of Iron Garnet Films

S. Gotoh, N. Koshizuka, M. Yoshida, M. Murakami, H. Fujimoto and S. Tanaka
Superconductivity Research Laboratory, ISTEC,
1-10-13, Shinonome, Koto-ku, Tokyo 135, Japan

Abstract. The dynamic magnetic flux behavior in $YBa_2Cu_3O_x$ crystals made by the melt process was observed by using the Faraday effect of an iron garnet film with stripe magnetic domains. It was found that the overall flux gradient was in good agreement with that predicted by the critical state model. However, some samples contained a region where flux could not easily penetrate, indicating a large flux pinning force.

1. Introduction

Since the magnetic flux distribution in type II superconductors is not directly observable by magnetization measurements, the flux density gradient is usually described by the critical state model. A more direct method using the decoration technique has been applied to observe the flux line lattice. However, it was difficult to observe the behavior of flux line lattice over the total volume of the sample, and dynamic observations were not possible (Träuble and Essmann 1968). Recently, Moser et al.(1989) observed the flux penetration into both single crystal and sintered $YBa_2Cu_3O_x$ samples using the Faraday effect of an evaporated magneto-optical EuS/EuF_2 film, and Indenbom et al.(1990) introduced a similar method using a ferromagnetic film with perpendicular magnetization.

We have also succeeded in the observation of flux behavior in high-Tc superconductors using the Faraday effect of iron garnet films (Gotoh et al. 1990). In this paper, we report on the observations of flux penetration and trapping in $YBa_2Cu_3O_x$ crystals made by the melt process.

2. Experimental

Samples of $YBa_2Cu_3O_x$ were prepared by the quench and melt growth (QMG) process (Murakami et al. 1989). Rectangular specimens with dimensions approximately $0.4x1x1.5mm^3$ were cut from the QMG-processed plates so that the c axis was parallel to the surface plane of the plate. Mechanical polishing was carried out on the plates to obtain a shiny surface. The magneto-optical thin film of $(BiGdYb)_3Fe_5O_{12}$ was prepared on a transparent gadolinium gallium garnet (GGG) substrate. The optical detection system with an electromagnet, an optical cryostat and a polarization microscope was used. The $YBa_2Cu_3O_x$ plate was fixed on a cooling holder inside the cryostat and then the film/substrate was placed onto the sample surface film-side down and in good thermal contact with the sample. The electromagnet was placed under the cryostat to apply a magnetic field perpendicular to the sample surface with a maximum field of 850 Oe.

0953-2048/91/01S226 + 03 $03.50

3. Results and Discussion

The changes in the stripe domain patterns of the iron garnet film on the $YBa_2Cu_3O_x$ sample zero-field cooled to 77 K and measured at 77 K with increasing applied fields are shown in Fig. 1. The principle of this method is to observe the image of the stripe domains in the iron garnet film, which reflect the flux density inside the sample. The stripe domains are caused by the perpendicular magnetization in the iron garnet film. The applied field can be estimated from changes in the domain width because the domain width with the magnetization parallel to the applied field expands, and the domain width with the opposite magnetization becomes narrower, when a magnetic field is applied perpendicular to the surface.

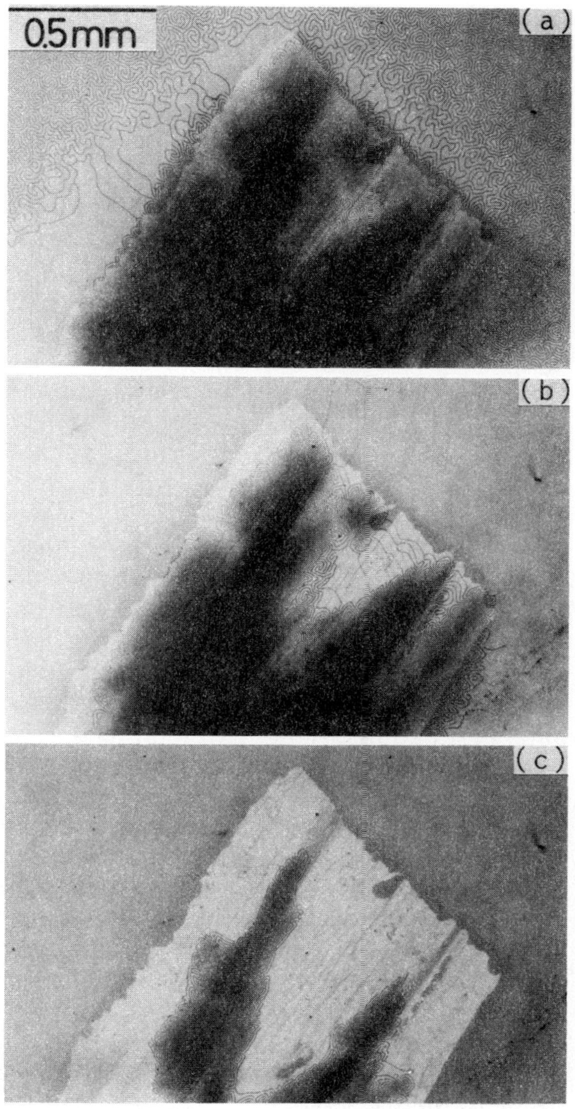

Fig. 1. Stripe domain structures of iron garnet film on $YBa_2Cu_3O_x$ for increasing fields at 77 K after zero-field cooling. (a)H=380 Oe; (b)H=510 Oe; (c)H=850 Oe.

Except for the superconducting sample itself and its vicinity, the white stripe domain reflects the homogeneity of the applied field, as shown in Fig. 1(a). A narrow domain structure remains in the interior of the sample because the applied field does not penetrate into the center. However, in the region surrounding the superconducting sample, the white domain width increases as flux is excluded from the sample due to the superconducting shielding effect by flux pinning. The applied flux penetrates into both the edges of the sample and the areas having a small pinning force. When the applied field increases, the flux penetrates into the sample, as shown in Figs. 1(b) and 1(c). The stripe domain structure disappears and a single domain appears as the applied field exceeds a temperature-dependent threshold value. Fig. 1(c) shows the nonuniform flux gradient. This indicates that a region with a large pinning force is located in the sample. We believe that this behavior is due to the nonuniform distribution of fine Y_2BaCuO_5 inclusions introduced in the QMG process. In fact optical microscopic observation revealed that this region contained that high density of Y_2BaCuO_5 inclusions.

Figure 2(a) shows the domain pattern at 77 K after field cooling under a field of 425 Oe. There is no change of the domain structure both inside and outside the sample. This indicates that almost all applied field is trapped in the sample below Tc due to the large pinning force.

When the applied field was removed, the trapped flux appears as positive remanent magnetization as shown in Fig. 2(b). Fig. 2(c) shows the magnetic flux density profile of the line indicated by arrows in Fig. 2(b), which is estimated by the change of the domain width. Although large trapped flux remains in the center of the sample, the flux abruptly decreases in the end of the sample with large flux gradients. Such a flux profile is in good agreement with that predicted by the critical state model.

4. Conclusions

Dynamic flux motion was directly observed in the QMG-processed $YBa_2Cu_3O_x$ using the Faraday effect of an iron garnet film with stripe domains. Since the domain pattern easily changes depending on the flux distribution, we could estimate the local gradient of the flux density from the domain width. It was found that in the particular sample we used here, the region with the largest pinning force was located inside the sample.

Acknowledgments

We would like to thank Dr.Y.Yokoyama of Electrotechnical Laboratory for the preparation of the iron garnet films used. This work was supported by New Energy and Industrial Technology Development Organization under the management of the R&D of Basic Technology for Future Industries.

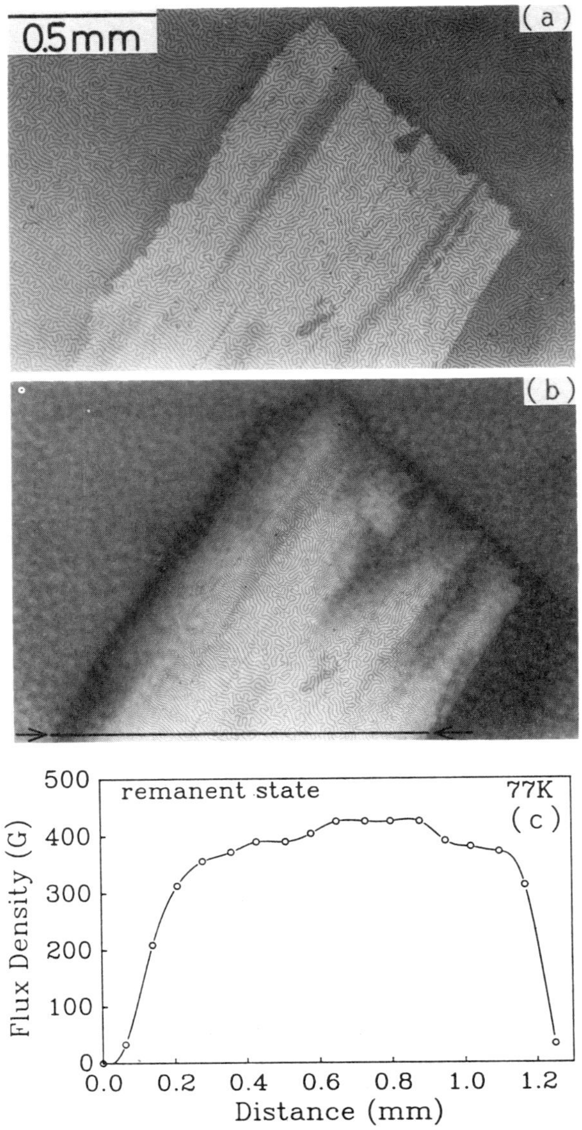

Fig. 2. Stripe domain structures at 77 K after field cooling under a magnetic field of 425 Oe. (a)H=425 Oe; (b)H=0 Oe. (c) shows the flux profile of the line indicated by arrows in (b).

References

Träuble H and Essmann U 1968 Phys. Stat. Sol. 25 373.

Moser N, Koblischka M R, Kronmüller H, Gegenheimer B and Theuss H 1989 Physica C 158 117.

Indenbom M V, Kolesnikov N N, Kulakov M P, Naumenko I G, Nikitenko V I, Polyanskii A A, Vershinin N F and Vlasko-Vlasov V K 1990 Physica C 166 486.

Gotoh S, Koshizuka N, Yoshida M, Murakami M, Fujimoto and Tanaka S 1990 Jpn. J. Appl. Phys. 29 L1083.

Murakami M, Morita M, Doi K and Miyamoto K 1989 Jpn. J.Appl.Phys. 28 1189.

Supercond. Sci. Technol. **4** (1991) S229–231. Printed in the UK

Theory of Magnetic Flux States in Oxide Superconductors

Takeo Izuyama, Satoshi Ishizaka and Ken Masuda

Institute of Physics, University of Tokyo at Komaba, Meguro-ku, Tokyo 153 , *Japan*

Statistical mechanics for the assembly of the flux lines is developed for oxide superconductors with many crystal imperfections. For perfect crystals we provide a phase diagram where molten flux phase appears but does not occupy large regions in the H-T plane. We find new phase which we call the spontaneous flux phase. For imperfect crystals, we adopt the replica technique. It is concluded that the intrinsic pinning can possibly exist.

An important feature of the oxide superconductors is that even in a single crystal the sample is seriously imperfect for the sake of the carrier doping. The question may be asked: Can the magnetic flux lines be pinned by crystal imperfections? Such intrinsic pinning is not obvious. It seems at the first sight that in a large sample the pinning potentials acting upon each flux line are averaged out. There exists a wide-spread sentiment that the pinning is weak in oxide superconductors. To consider this kind of problems, we give here a general formulation of statistical mechanics of the flux lines in these materials.

Consider the case where the external field is applied along the c-axis. We assume that the supercurrent is confined within a CuO_2 plane. The condensate wave function spreads only over very thin sheets of the CuO_2 layers. The magnetic field in the material can be regarded as a superposition of cylinders of fluxes, each of which carries one flux quantum ; hc/2e. We adopt $\kappa \equiv \lambda/\xi = 100$ in numerical calculations. For simplicity we assume that the center of the vortex core is located at a center of the unit square formed by four Cu ions- The configuration of each flux line is specified by the series of of the core locations. Each such string is traced from the bottom surface to the upper surface of the sample. On the top surface the vortex cores are located at $r_1, r_2, \cdots, r_n$. On the bottom they are found at $r_1', r_2', \cdots, r_n'$. Let the number of the CuO_2 layers be M+1. The partition function is given by

$$Z[\{v\}] = \frac{1}{n!} \sum_{\{\vec{r}\},\{\vec{r}'\}} \langle \vec{r}_1 \cdots \vec{r}_n | \mathcal{T} | \vec{r}_1{}' \cdots \vec{r}_n{}' \rangle \ .$$

Let $v_\nu(\vec{r})$ be the impurity potential that could appear at site $\vec{r}$ on the ν-th layer. The transfer matrix is given by

$$\mathcal{T}(\{v\}) = \mathcal{T}_1 \cdot \mathcal{T}_2 \cdots \mathcal{T}_{M-1} \ .$$

The transfer matrix connecting the ν th layer to the $(\nu+1)$th is expressed as

$$\begin{aligned} \mathcal{T}_\nu[\{v_\nu(\vec{r})\}] = \exp \Bigg[\sum_j \Bigg\{ & \ln \sum_{\vec{r}_j} \frac{1}{N^2} e^{-\beta v_\nu(\vec{r}_j)} \\ & + \ln \sum_{\vec{k}_j} |\vec{k}_j\rangle \Big\{ \eta + 2\zeta(\cos k_x + \cos k_y) \Big\} \langle \vec{k}_j | \Bigg\} \\ & + \sum_{j<k} \ln \sum_{\vec{r}_j, \vec{r}_k} |\vec{r}_j\rangle |\vec{r}_k\rangle e^{-\beta V(\vec{r}_j - \vec{r}_k)} \langle \vec{r}_j | \langle \vec{r}_k | \Bigg] \ . \end{aligned}$$

The impurity potentials $v_\nu(\vec{r})$ are regarded as stochastic variables with the Gaussian distribution. We take the sample average.

$$\overline{\ln Z} = \prod_{\vec{r}} \prod_{\nu=1}^{M} \left[\frac{1}{\sqrt{\pi\sigma^2}} \int dv_\nu(\vec{r}) \exp(-v_\nu(\vec{r})^2/\sigma^2) \right] \ln Z[\{v\}] \ .$$

The replica technique initiated by S.F.Edwards is adopted

$$\ln Z[\{v\}] = \lim_{m \to 0} \frac{1}{m} \left(Z[\{v\}]^m - 1 \right) \ .$$

Then the free energy is simply given by the ground state energy of the Hamiltonian $\overline{H}_m$:

$$\begin{aligned} \overline{H_m} = & -\beta \frac{\sigma^2}{4} \sum_{\vec{r}} \sum_{\mu \neq \nu}^{m} n_\mu(\vec{r}) n_\nu(\vec{r}) \\ & -\beta^{-1} \sum_{\vec{k}} \sum_{\mu}^{m} \ln \epsilon(\vec{k}) a_\mu^\dagger[\vec{k}] a_\mu[\vec{k}] \\ & + \frac{1}{2} \sum_{\vec{r} \neq \vec{r}'} \sum_{\mu}^{m} V(\vec{r} - \vec{r}') n_\mu(\vec{r}) n_\mu(\vec{r}') \ , \end{aligned}$$

$$a_\mu^\dagger(\vec{r}) a_\mu(\vec{r}) \equiv n_\mu(\vec{r})$$

$$V(\vec{r}_j - \vec{r}_k) = \frac{\phi_0^2}{8\pi^2\lambda^2} K_0\left(\frac{|\vec{r}_j - \vec{r}_k|}{\lambda}\right) \delta$$

The width of the CuO_2 layer is defined by δ.

$$\overline{H_m}\psi_0(m) = E_0(m)\psi_0(m)$$

$$F = -\beta^{-1}\overline{\ln Z} = \lim_{m\to 0}\left(\frac{1}{m}E_0(m)\right)\left(\sum_{\{\vec{r}\}}\psi_0(\{\vec{r}_j^{\,\mu}\})\right)^2$$

In the above, $a_\mu(\vec{r})$ is the Bose operator of the particle (string) with μ-th colour ($\mu = 1, \cdots, m$), $a_\mu[k]$ is its Fourier transform, $\phi_0 = hc/2e$, $\eta = \exp(-\beta\delta)$, and $\zeta = \exp(-\beta\varepsilon)$; δ being the energy of the field within a cylinder between two adjacent layers, when the cylinder is straight, and ε the field energy when the string connects a site to its nearest neighbor site during one "time step". $\Delta E \equiv \delta - \varepsilon$ is the increase of the field energy through bending of the string. This is rather large, and hence the bending is not very easy. The number of the strings is given by minimizing the Gibbs free energy; $G = F - (M/4\pi)\,\phi_0 H\, l\, n$. l is the separation between layers.

For the perfect crystals we have $\sigma = 0$. We obtain the Abrikosov lattice in a large region of the H-T plane. Following R.D.Nelson, we use the Lindeman criterion for the melting of the triangular lattice. The results are shown below in the figure.

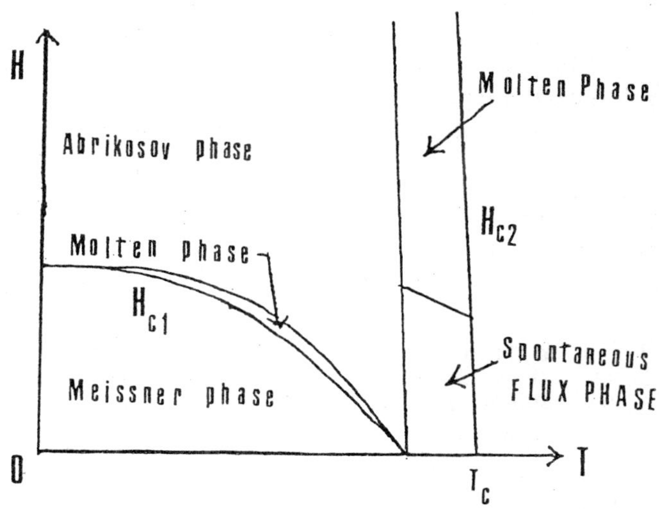

The spontaneous flux phase is characterized by the entangled strings running in the opposite directions even in the absence of the external field. For H = 0 this phase appears at $T_c > T > 0.9\,T_c$.

Consider the case of non-zero σ. We have neglected the Parisi replica symmetry break down. Hence we get no disclinations, no dislocations or any amorphous solid phase. In our case, two coloured lattices are bound, when the zero point amplitude of their relative motion is sufficiently smaller than the lattice spacing of the triangular lattice formed through $V(r_i - r_j)$. According to this criterion, we obtain the pinned states in addition to the depinned states.

References

Edwards S F 1971 *Proc. 3rd Int. Conf. on Amorphous Materials 1970* ed A J Compff and S Newman (New York: Plenum) p 83

Izuyama T 1982 *J. Phys. Soc. Japan* **51** 3449

Nelson R D 1988 *Phys. Rev. Lett.* **60** 1973

Supercond. Sci. Technol. **4** (1991) S232–234. Printed in the UK

Acoustic Emission in YBCO Ceramics

M P Staines and N E Flower

D.S.I.R. Physical Sciences, P O Box 31313, Lower Hutt, New Zealand

Abstract. We report measurements of the acoustic emission generated during sintering and heat treatment of $YBa_2Cu_3O_{7-y}$ for unsubstituted and Ca and La substituted material. Intense acoustic emission due to microcracking caused by anisotropic lattice contraction occurs in a relatively narrow temperature range on cooling after sintering. We see no evidence for microcracking associated with the tetragonal to orthorhombic phase transition in $YBa_2Cu_3O_{7-y}$. If microcracking is produced solely by anisotropic contraction it is not expected to greatly affect J_C since it will occur in the basal plane and it should be reduced in textured material.

1. Introduction

The current carrying capabilities of the cuprate superconductors, ceramic materials with short anisotropic coherence lengths, might be expected to depend critically on conditions at the grain boundaries. Anisotropic thermal expansion and, in $YBa_2Cu_3O_{7-y}$, large changes in oxygen stoichiometry during preparation make these materials susceptible to cracking. Instances of grain boundary and intragranular microcracking have been observed in $YBa_2Cu_3O_{7-y}$ (Shaw *et al* 1989) but the influence of microcracking on J_C has yet to be demonstrated.

Acoustic emission measurements during heating and cooling of $YBa_2Cu_3O_{7-y}$ (Richardson and De Jonghe 1989) in helium and oxygen atmospheres revealed acoustic emission which appeared to be correlated with the orthorhombic-tetragonal (OT) phase transition suggesting that this is an additional source of microcracking in $YBa_2Cu_3O_{7-y}$.

The phase transition is displaced to lower temperatures and the oxygen diffusion rate is enhanced by Ca substitution (Tallon *et al* 1990). We initially performed acoustic emission measurements with a view to seeing if microcracking at the phase transition was reduced by Ca substitution.

2. Experimental

Samples were produced by the conventional ceramic route. Unsintered discs 2mm thick were prepared by pressing single phase starting material in a 12.5 mm die at 200 MPa. Acoustic emission measurements were made using a transducer (DECI model SE 150-M) coupled to the sample in a furnace by a vertically mounted 250 mm long fused quartz rod. A 20 g zirconia weight held the sample in place with a thermocouple in contact with its top surface. The entire furnace and transducer assembly were mounted inside a gas-tight chamber to avoid the possibility of acoustic coupling to the transducer from vacuum feedthroughs. Acoustic emission counts were recorded at 20 second intervals during temperature cycling.

3. Results and Discussion

The initial measurements were made in air, heating and cooling at a rate of 200°C/hour. Figure 1a shows part of a sequence of acoustic emission measurements on a single $YBa_2Cu_3O_{7-y}$ sample recorded on cooling from progressively lower maximum temperatures. The sample was held at the maximum temperature for 12 minutes on each occasion. Cumulative counts for each cooling are plotted against temperature.

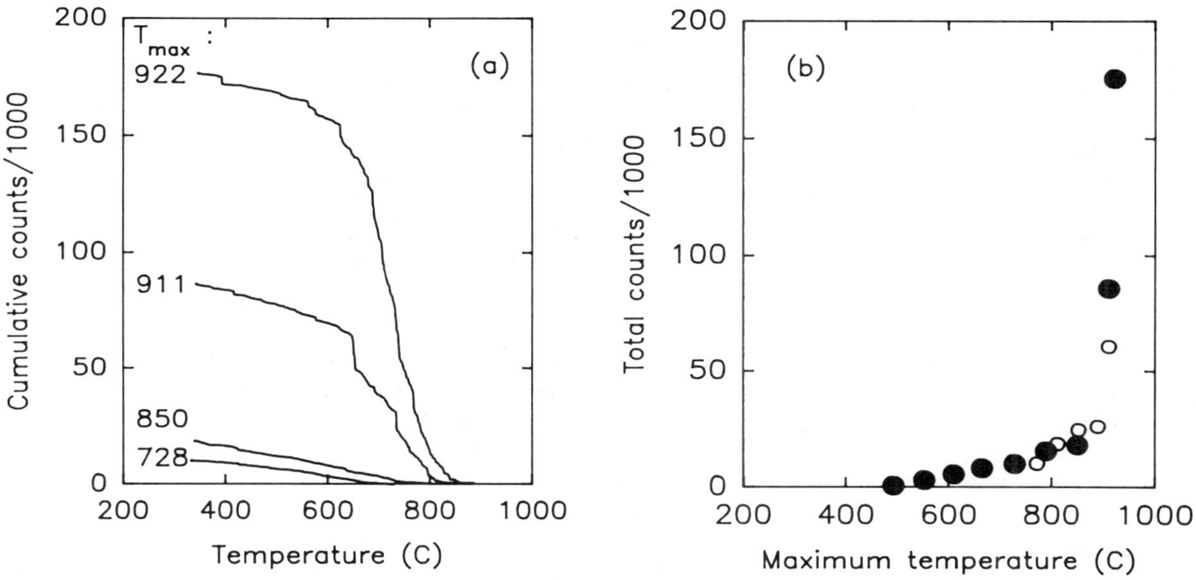

Figure 1. a) Cumulative acoustic emission counts for a $YBa_2Cu_3O_{7-y}$ sample during cooling runs in air from the temperatures indicated. b) Total counts in cooling to 350°C from the maximum temperature for the cycle. Solid circles: first sequence of cycles with temperatures decreasing. Hollow circles: second sequence.

The first cycle represents the first cooling after sintering at 922°C. Initially cooling is uneventful but starting at about 850°C there is an interval of intense acoustic emission which tails off to a lower level at about 650°C.
As the maximum temperature is reduced in subsequent cycles the level of emission in this band drops until below 890°C only a constant lower level emission remains. This behaviour is summarised in Figure 1b in which total counts recorded to 350°C are plotted against the maximum temperature for each cooling cycle. Following this set of measurements in which the maximum temperature was reduced a second sequence cycling to higher temperatures again confirmed the change in behaviour at 890°C.

These results are consistent with microcracking due to thermal expansion anisotropy. Sintering takes place at temperatures above 890°C and on cooling below the sintering temperature stress builds up until microcracking occurs at a temperature which depends on the anisotropy in expansion coefficient, grain size, Young's modulus, and fracture surface energy of the material (Cleveland and Bradt, 1978, Ohya *et al* 1987). For sufficiently fine grain size microcracking will be suppressed. This critical size has been estimated at about 1 micron in $YBa_2Cu_3O_{7-y}$ (Shaw *et al* 1989). The onset of sintering at 890°C was independently confirmed by thermal expansion measurements which showed shrinkage beginning at this temperature.

We made measurements on a range of samples produced from different material sintered at 920°C for 4 hours. Although the samples differed greatly in porosity, with final densities from 70 to 87 % of theoretical density, we estimated from SEM images of fracture surfaces that the average grain size was similar at around 5 microns. The intensity of acoustic emission in different samples cannot be reliably compared because of differences in coupling to the transducer but the onset temperature of emission was fairly constant at around 90°C below the sintering temperature. The most porous samples typically produced intense acoustic emission over a narrow temperature range - about 100 degrees, while in the denser samples high count rates were sustained for several hundred degrees below the onset.

A notable observation is the absence of any feature in the acoustic emission cooling spectrum at temperatures around the OT transition - 615°C in air. After initial sintering at 920°C we made careful measurements temperature cycling a $YBa_2Cu_3O_{7-y}$ sample through the OT transition in atmospheres of 1 and 0.02 atm. of oxygen with corresponding OT transition temperatures of 675°C and 550°C respectively. The upper temperature for these cycles was limited to 820°C by the need to avoid further sintering at low partial pressures of oxygen which would obscure any changes due to the shift in the transition temperature. There was no change in the level of acoustic emission with oxygen partial pressure nor any qualitative change with Ca or La substitution (15% Ca substituted 75:25 for Y and Ba, and 20% La substitution for Ba). The La substituted material is in fact tetragonal across the entire temperature range. Thus it appears that there is no measureable microcracking associated with the OT transition, which predominantly imposes shear stresses which can be accommodated by a network of twin planes parallel to the c axis without microcracking.

If anisotropic lattice contraction is the major cause of microcracking in ceramic cuprate superconductors this has some fortunate implications for processing. Because grains are placed in tensile stress in the c direction, due to its more rapid contraction, cracking will occur in the basal plane. Microcracks in this orientation do not create a barrier to current flow in the ab plane and should have little effect on J_C. Furthermore, in c axis textured polycrystalline material the fact that the lattice contraction in individual grains more nearly matches that of the aggregate of crystals should reduce the severity of microcracking. Measurements on samples which have been grain aligned using a magnetic field are currently underway to test this prediction.

Acknowledgement. This work was partially supported by Electricorp New Zealand. Thanks to Simon Gibson for technical assistance.

References.

Cleveland J J and Bradt R C 1978 *J. Am. Ceram. Soc.* 61 478

Ohya Y, Nakagawa Z and Hamano K 1987 *J. Am. Ceram. Soc.* 70 C-184

Richardson T J and De Jonghe L C 1989 *Proc. Symposium on High Temperature Superconducting Compounds: Processing and Related Properties* eds S H Whang and A DasGupta (Warrendale, PA: TMS) pp 633-7

Shaw T M, Shinde S L, Dimos D, Cook R F, Duncombe P R and Kroll C 1989 *J. Mater. Res.* 4 248

Tallon J L, Pooke D M, Staines M P, Bowden M E, Flower N E, Buckley R G, Presland M R, and Davis R L 1990 submitted to *Physica C*

Supercond. Sci. Technol. **4** (1991) S235–237. Printed in the UK

CORRELATION OF INFRARED ABSORPTION INTENSITY WITH T_c

G.A. Thomas, S.L. Cooper, J. Orenstein, D.H. Rapkine, A.J. Millis, J.V. Waszczak and L.F. Schneemeyer

AT&T Bell Labs, Murray Hill, NJ 07974 USA

Abstract. The magnitude of T_c in $Ba_2YCu_3O_{6+x}$ for 0<x<1 scales with the spectral weight of a contribution to the conductivity centered at a frequency (0.1 to 0.2 eV), near the magnetic exchange energy, J.

1. Introduction.

As carriers are added to insulating cuprates by changing the O-concentration, the initial electronic excitations occur at finite frequencies. As the carrier density increases, these excitations can be divided into two classes by their rate of increase with doping. Above the insulator-metal transition, a zero-frequency component emerges, but the finite-frequency spectral weight remains. Some remnant of the initial excitations (Varma 1989, Lee 1989, Anderson 1989) may interact in a significant way with the low frequency quasi-particles in the normal and superconducting states.

2. Observed Spectra.

Optical reflectivity studies (Cooper 1989, Orenstein 1990, Thomas 1990) with unpolarized light in the a-b plane of microtwinned crystals of $Ba_2YCu_3O_{6+x}$ for 0<x<1 yield the results for the real part of the conductivity shown in Figure 1. We divide the IR absorption into two parts based on the peak position and the way in which the contribution grows.

3. Total IR Spectral Weight.

We define the IR spectral region as having its low frequency edge at the minimum near 50 meV that is clearly seen for the samples with T_c = 30, 50, and 80K. We identify the high frequency edge as the point of the minimum in conductivity in the insulating sample near 1.25 eV. We use these minimum and maximum values to define the limits of integration of the conductivity which allows a calculation of the "total" IR spectral weight. We use these limits (with data at the lowest available T) even in samples which have no features at these frequencies, under the assumption that the spectral weights still provide reasonable estimates. The "total" IR spectral weight thus defined is shown by the solid circles in Figure 2.

0953-2048/91/01S235 + 03 $03.50

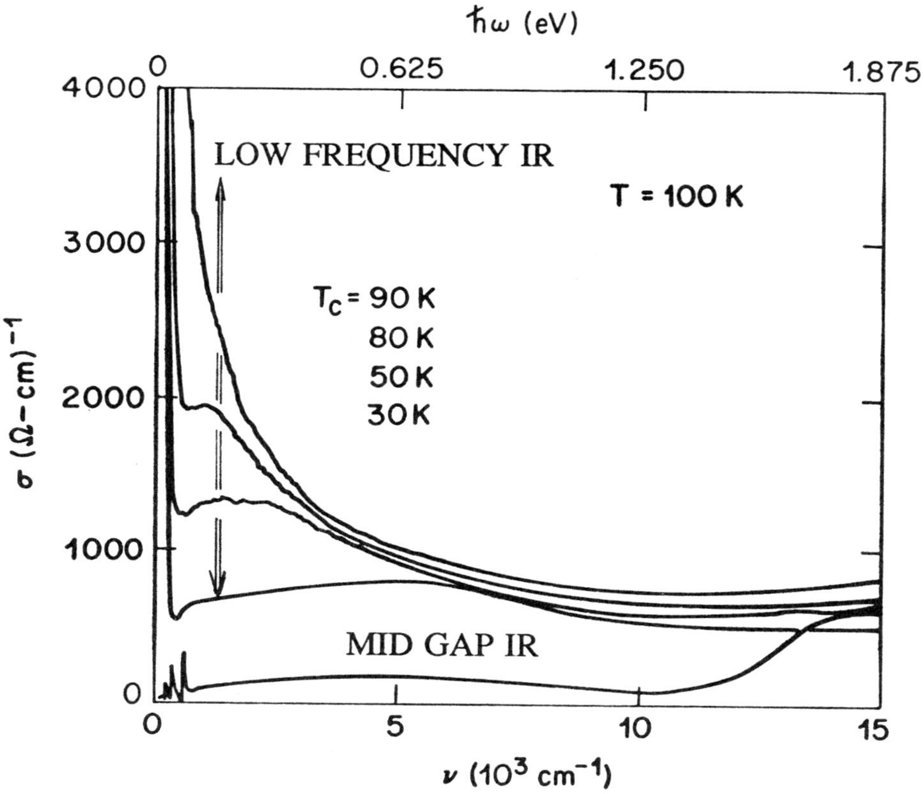

Figure 1. Real part of the conductivity as a function of frequency at T=100K for five crystals of $Ba_2YCu_3O_{6+x}$ with 0.1<x<1. The four superconducting samples are characterized by their values of T_c, and the insulating sample has an estimated value of x=0.1.

4. Low Frequency IR Spectral Weight.

There is a peak in the infrared region of the conductivity in Figure 1 just above .6 eV for the insulating sample and the sample with T_c = 30K. The spectral weight near this frequency grows only slightly above the concentration for T_c = 30K. However, the weight at lower frequency grows significantly at higher doping both within the IR range defined above and in the "Drude" peak at zero frequency. The appearance of a peak in the IR range just above 0.1 eV for the samples with T_c of 50 and 80K also suggests that it is reasonable to define a low frequency IR component as the part of the IR spectral weight that scales with the "Drude" part. Although the mid IR part and the low frequency IR part are not resolved in any of the samples shown, their relative contributions can be estimated using the above criteria. These low frequency IR spectral weights are plotted as the open circles in Figure 2.

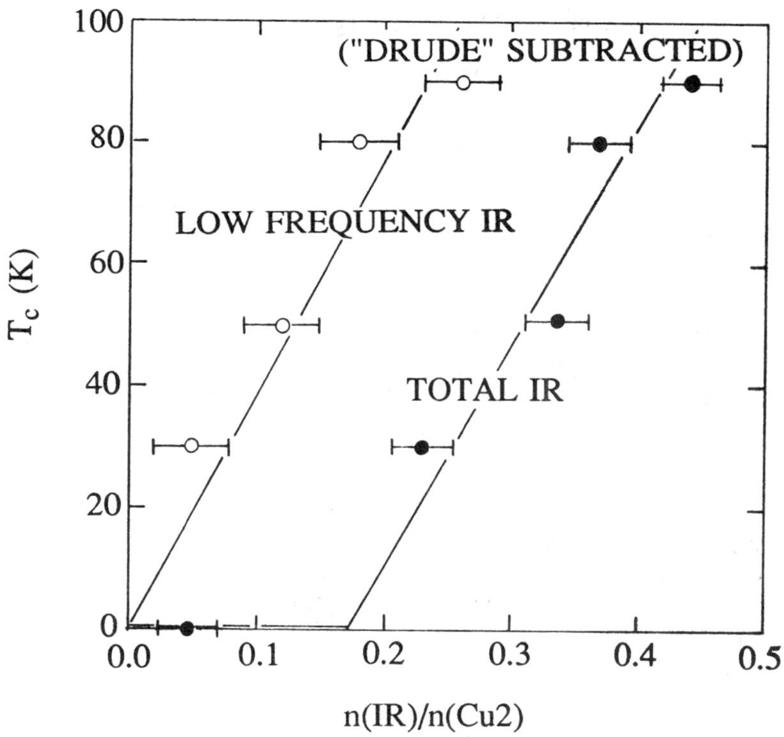

Figure 2. Values of T_c as a function of effective carrier density normalized to the density of planar Cu atoms.

5. Scaling of T_c.

Figure 2 shows the values of T_c for the five samples of Figure 1 as a function of the measurements of the IR spectral weights described above. If we assume that the appropriate effective masses are given by the band masses, then the spectral weights divided by the density of planar Cu atoms gives the relative number of holes in the states from which the IR absorption originates. Beyond a certain relative density of holes (just below 0.2), the superconducting phase begins (near where the metallic phase begins). At higher hole density, T_c scales with the excess density in the IR (solid points). The low frequency part of the IR spectrum itself scales with T_c (open circles), indicating the same effect from the same origin.

References.

Anderson, P.W. 1990 in "Strong Correlation and Superconductivity" H. Fukuyama et al., eds., (Spinger-Verlag, Berlin), p.2.

Cooper, S.L., et al. 1989 Phys. Rev. **B40,** 11385.

Lee, P.A. 1990, in "High Temperature Superconductivity", K.S. Bedell et al., Eds. (Addison Wesley, CA).

Orenstein, J., et al. 1990 Phys. Rev. B, in press.;

Thomas, G.A., et al. 1990 Proceedings of LT19, in press.

Varma, C.M. 1989 Int. J. of Mod. Phys. **B3,** 2083.

Supercond. Sci. Technol. 4 (1991) S238–240. Printed in the UK

Magnetisation measurements on melt-textured $YBa_2Cu_3O_{7-\delta}$

N.McN. Alford[a], T.W. Button[a], A.D. Caplin[b], S.M. Cassidy[b], J.R. Laverty[b], D.G. McCartney[c], R.J. Pollard[c] and Z.X. Shen[b]

a ICI Advanced Materials, Runcorn, U.K.
b Blackett Laboratory, Imperial College, London, U.K.
c Department of Materials Science and Engineering, University of Liverpool, Liverpool, U.K.

Abstract

Quasi-single crystal specimens were grown from rods of bulk $YBa_2Cu_3O_{7-\delta}$ by a melt-textured (MT) growth process. The material's magnetisation was highly anisotropic which supports the microstructural picture. While M-H data show there is a high bulk critical current density, the flux pinning in MT material differs little from that in single crystals or grains of sintered material.

1. Melt-Textured Growth and Sample Microstructure

Rods of bulk single phase $YBa_2Cu_3O_{7-\delta}$ were heated to 1050°C in air and then cooled slowly in a temperature gradient. The temperature for the peritectic formation of $YBa_2Cu_3O_{7-\delta}$ moves along the rod, leading to the growth of grains which extend fully across the 2 mm diameter rods and have the same c-axis alignment. To date there is no control over the angle θ between the rod axis and the crystal c-axis. For high transport currents the basal plane should ideally lie along the rod axis (θ=90°), but typically θ ranges from 40° to 50°.

Two main types of large scale defects are found. Firstly Y_2BaCuO_5 (211 phase) appears as spheroidal inclusions in the 123 matrix. There has been some speculation that the defect structures around such inclusions lead to enhanced pinning of flux lines in the superconducting state (Murakami et al. 1989), and some evidence for this in another material system (Shi et al. 1990). The second type of defect is a thin (<1μm) amorphous BaCuO impurity layer deposited on (001) planes at ~50 μm intervals, and extending for macroscopic distances along those planes. However the optical domain pattern is coherent along the rod, indicating that the impurity layers are not continuous. To obtain fully superconducting samples, the rods were annealed at 450°C in flowing O_2 for 48 hours. That this short oxygenation treatment suffices is probably due to rapid O_2 diffusion along the impurity layers. The growth and microstructure of melt-textured $YBa_2Cu_3O_{7-\delta}$ has been described in more detail by Pollard et al. (1989) and Ringnalda et al. (1990).

2. Experimental Details

Magnetisation measurements were performed with a Vibrating Sample Magnetometer (VSM), with the sample temperature controlled by pumping on either liquid nitrogen or helium. M-H loops were swept out to determine the critical current density of the material, with care being taken to ensure the sample was in the critical state. Flux creep was monitored in zero field (< 5 mT) for 3000 seconds after each M-H loop. Transport measurements were carried out using a four-probe method with voltage contacts separated by 1.5 mm (hence including about 30 of the amorphous BaCuO layers). A pulsed current technique was used for transport critical current measurements (Meisels et al. 1988).

0953-2048/91/01S238 + 03 $03.50

Figure 1

MT samples used for magnetisation data, with the field orientation shown. WS1 is used for in-plane currents and WS2 for out-of-plane currents (sample dimensions are 0.33 x 0.50 x 0.60 mm for WS1 and 0.95 x 0.57 x 0.58 mm for WS2).

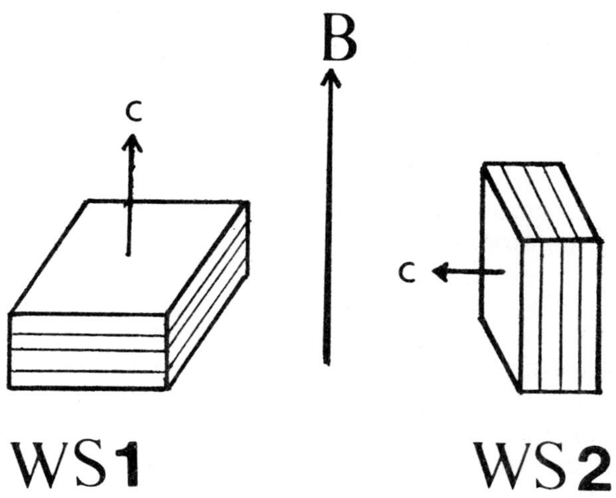

Figure 2

Magnetic Jc's for MT (sample WS1) and PX YBCO (intragranular contribution).

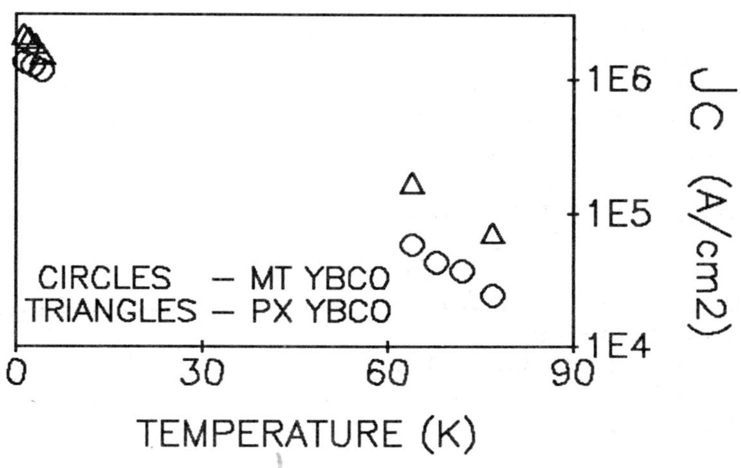

Figure 3

Normalised creep rates in PX and MT (sample WS1) material.

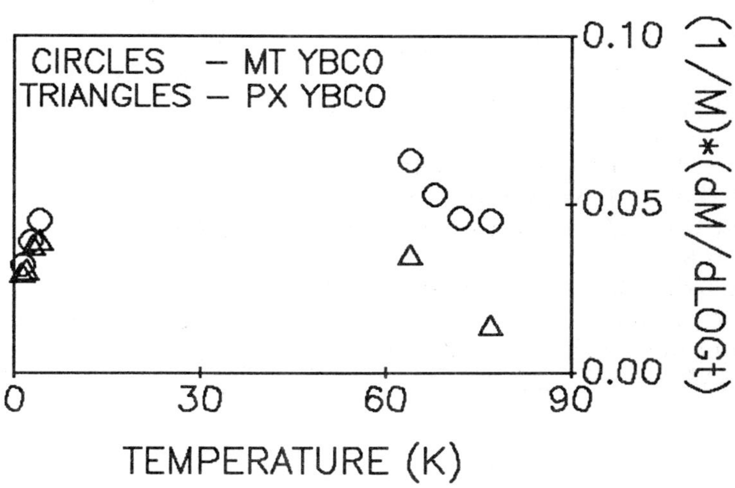

The melt-textured rods cleave easily along the amorphous layers and this was used to establish the sample orientation in the VSM. Two samples were cut out from cleaved disks using a wire saw. The samples with their field orientation are indicated in figure 1: WS1 was used to look at currents in the basal plane ($J_c^{(ab)}$), while WS2 provides information about both $J_c^{(ab)}$ and the out-of-plane currents ($J_c^{(c)}$). The anisotropy in J_c *and* the length scales on which the currents flow dictate which current direction dominates the magnetisation (Gyorgy et al. 1989). Magnetisation measurements were carried out also on a sample of bulk polycrystalline (PX) YBCO to compare the intra-grain currents in bulk and melt-textured material.

3. Results and Discussion

Resistance measurements show a 1 K wide superconducting transition with zero resistance at 88 K. Meissner effect measurements on the VSM confirm a T_c of 88 K, but with only 2% flux expulsion in a 1 mT field. This low ratio of M_{FC} to M_{ZFC} is peculiar to melt textured material, as the ratios for single crystal and bulk material are typically 50%. The melt-textured material is free of large scale non-superconducting regions (checked by measuring specimens cut from the interior of the rod), and so this Meissner anomaly indicates strong flux pinning at very low fields.

Magnetisation measurements of WS2 show that the BaCuO impurity layers limit the out-of-plane currents. The hysteresis in M-H data does not scale with sample dimension (from 2 to 0.5 mm). Assuming a c-axis dimension of 50 μm gives $J_c^{(c)}/J_c^{(ab)} \sim 10^{-2}$, which agrees reasonably with the single crystal data of Gyorgy et al. (1989). Transport data show the normal state resistivity of this material is large (~15 mΩcm at 300 K) and has a small temperature gradient, indicating that connection through the amorphous layers is poor. Despite this, the transport J_c along the rod is quite high, exceeding 400 A/cm^2 at 82 K (and too high at lower temperatures for us to measure), suggesting that although the amorphous layers do act as weak links, their critical current is high, perhaps because of the good crystalline alignment in MT material.

Clearly MT material shows much larger M_{REM} and absolute flux creep than does polycrystalline (PX) material. However, when translated into J_c, using the sample diameter for the former and grain size for the later (~5 μm), both give values of 10^6 A/cm^2 at 4 K (figure 2). Data from Crabtree et al. (1987) on single crystals give similar results. Likewise the normalised creep rates at 4.2 K differ by less than 10% between PX and MT (figure 3). We conclude that, other than at very low fields of the order of mT, flux pinning in MT, PX and single crystal YBaCuO is much the same. Therefore, in our MT material the large particles of 211 phase are unimportant as pinning centres.

Acknowledgments: Financial support from SERC and National Power TEC is acknowledged.

References

Crabtree G W, Liu J Z, Umezawa A, Kwok W K, Sowers C H, Malik S K, Veal B W, Lam D J, Brodsky M B and Downey J W 1987 *Phys Rev* **B 36** 4021

Gyorgy E M, van Dover R B, Jackson K A, Schneemeyer L F and Waszczak J V 1989 *App. Phys. Lett* **55** 283

Meisels R, Bungre S and Caplin A D 1988 *J. Less Common Metals* **151** 83

Murakami M, Morita M, Doi K and Miyamoto K 1989 *Jap. J. Appl. Phys.* **28** 1189

Pollard R J, McCartney D G, Alford N McN and Button T W 1990 *Supercond. Sci. Technol.* **2** 169

Ringnalda J, Kiely C J, Pollard R J, McCartney D G and Beanland R 1990 *Proc. XIIth International Congress for Electron Microscopy* (to be published)

Shi D, Xu M, Fang M M, Chen J G, Cornelius A L and Lanan S G 1990 *Phys. Rev.* **B 41** 8833

Supercond. Sci. Technol. **4** (1991) S241–243. Printed in the UK

Grain Boundary Carbon Contamination of HTSC-layers on Copper Based Ribbons

Š.Molokáč[1)], V.Hajko Jr.[1)], Ju.N.Ivaščenko[2)], S.V.Pan[2)]

[1)] Institute of Experimental Physics, Slovak Academy of Sciences, 043 53 Košice, Czechoslovakia

[2)] Institute for Problems of Materials Science, Ukrainian Acad. Sci., 252180 Kiev, USSR

Abstract. We examine the rise of the carbon contamination from atmospheric source during preparation of the thermooxidized HTSC-layers on microcrystalline Y-Ba-Cu ribbons. It was found some correlation between the pecularities of grain boundaries carbon contamination and the value of Jc. The results of the Auger scanning microscope microanalysis are correlated with resistivity and critical current value for our sample.

1. Introduction

One of the major problems with traditional ceramic high temperature superconductors (HTSC) is a very low magnitude of critical current density Jc ($\sim 10^3$ A cm^{-2}) (Cava 1987), although highly oriented thin films present about 2-3 order of magnitude Jc higher than bulk ceramics (Xi 1988). Grain boundary contamination influences flux pinning and thus acts as a weak link between grains, thereby reducing Jc.

Some experiments (Nakahara 1987) give special attention to the thickness and chemical status of C-contaminated layers in the $YBaCu_3O_{7-x}$ ceramic and thin films with grained structure.

2. Sample characterization and experimental details

In this paper we examine the rise of the carbon contamination from atmospheric source during preparation of thermooxidized HTSC-layers on microcrystalline Y-Ba-Cu ribbons. The sample were prepared by melt spinning method (Verkin 1988). Rapidly quenched (RQ) master alloy ribbons of 50 μm thickness, 2-3 mm width and some meters length contained 95 wt% of copper and 5 wt% of yttrium and barium with atomic ratio Y:Ba = 1:2. The 12 mm long samples were statistically selected from different sites of the ribbons and then annealed in the air flow (sample A) and oxygen flow (sample B) with the same 2-step time-temperature annealing conditions: 673 K (6 hours) and 743 K (4 hours).

"In situ" fracture surface was studied on Auger microprobe JAMP-10S with system pressure of 8.10^{-8} Pa and electron probe diameter 0.1 μm. Sputtering of fracture surface was performed with 1 kV Ar^+ ion beam.

Electrical dc resistance measurements were carried out by an automatic four-probes method.

3. Results and discussion

In recent papers (Verkin 1988, Hajko 1988, Hajko 1989) we show, mainly on the basis of magnetic measurements, that the RQ-copper ribbons doped with Y and Ba and annealed in air, show superconductive behaviour below $T < 90$ K.

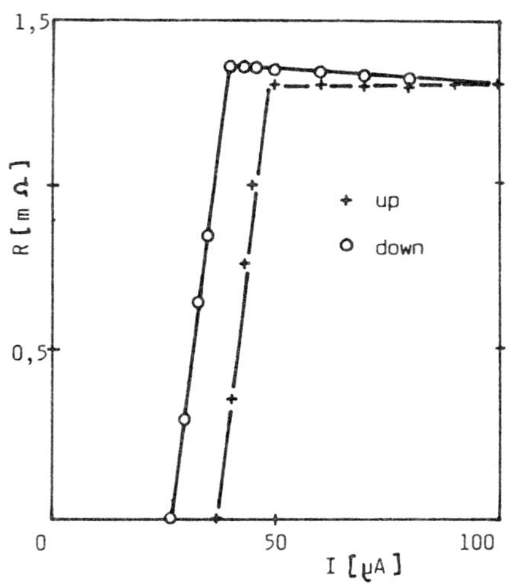

Fig.1. Current-resistance dependence at room temperature for sample A

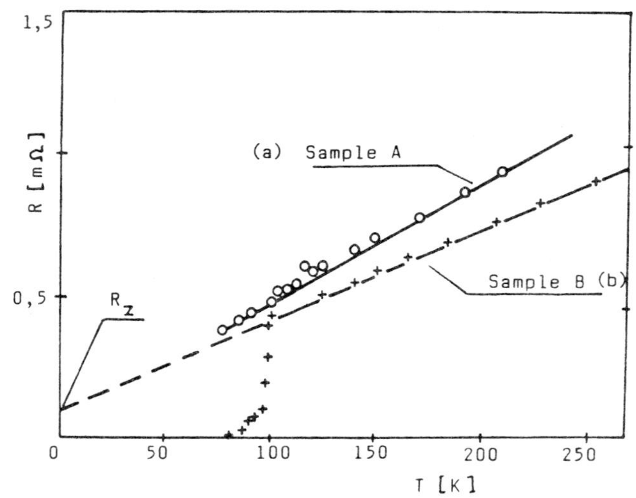

Fig.2. Resistance-temperature dependence of a/ sample A (air annealing) b/ sample B (oxygen annealing)

The same behaviour as in our previous works was found for sample A. The typical current versus dc resistance dependence at room temperature is shown in Figure 1. The jump from $R_L \lesssim 10^{-4}\ \Omega$ (sensitivity limit of our measuring system) to $R_H \sim 10^{-3}\ \Omega$ was registered at $I_{CA} \sim 3.10^{-5}$ A.

Both the decrease of the resistance value R_H with decreasing temperature (Figure 2a) and the sensitivity limit of the measuring system does not allow to measure R-I dependence below 220 K.

For the sample B the non-linear R-I dependence was not observed above 105 K. Below this temperature the sample showed non-linear R-I character with critical current value $I_{CB} \sim 5.10^{-4}$ A, but this value did not remarkably change with decreasing temperature. The sample B showed a superconducting critical temperature of $T_c \sim 93$ K. This midpoint temperature is defined as the temperature correspond to the half of extrapolated "residual resistance" $R_Z(T \rightarrow 0)$ (Figure 2b). The deviation from the linearity of the electrical resistance was observed below 105 K and "zero resistance" ($R_L < 10^{-5}\ \Omega$) was achieved below 90 K.

The detail structure studies were applied to explain the difference in the electrical characteristics of the samples A and B. The surface oxide layers thick about 6-8 μm have multiphase character (Hajko 1989) with a significantly higher concentration of Y, Ba and oxygen. Auger microprobe analysis of the "in situ" fracture surface were performed for four points of metallic Cu matrix and for six point of surface oxide layer.

The absence of Y-peak in Auger spectra (Figure 3,4) is caused by very low value of the sensitivity factor of yttrium, which represents about 18% of this factor for barium.

It can be seen that the main difference between samples A and B is that in the specimen A carbon was detected (air annealed sample) on the grain boundaries of the surface oxide layers as well as in the Cu matrix (Figure 3). The carbon peak is significantly higher for the surface oxide layer than for the metallic matrix. From the analysis of the shape of C-Auger peaks it is evident, that carbon is in the free state (graphite-like state).

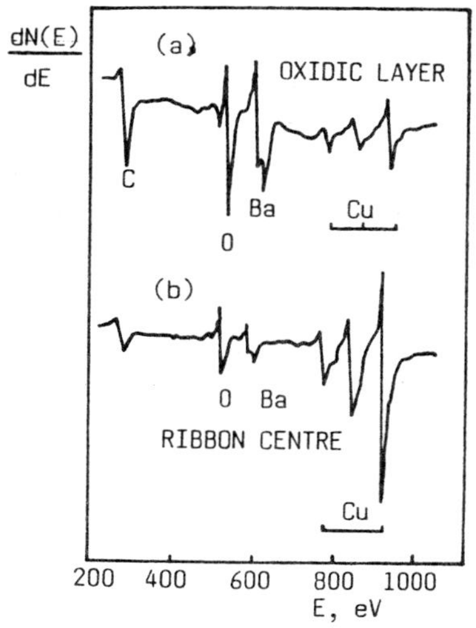

Fig.3. Auger microprobe analysis of sample A

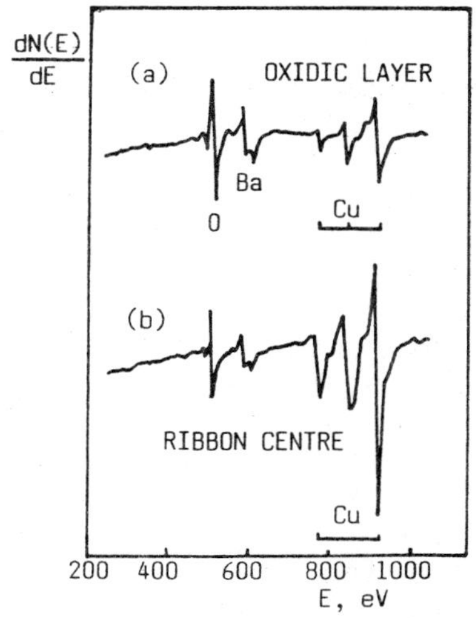

Fig.4. Auger microprobe analysis of sample B

The additional peaks at the high energy in our C-Auger analysis were not detected in the contrary with the results of Parmigiani 1987. The presence of carbon at the grain boundaries in a bulk HTSC $YBa_2Cu_3O_{7-x}$ was also confirmed by authors (Nakahara 1987 and Chisholm 1989), mostly in the chemically bounded state as a $BaCO_3$ compound of 1-4 monolayer thickness. However, Verhoeven (1988) detected the carbon on the grain boundaries in large grained $YBaCu_3O_{7-x}$ with thickness of about 1 monolayer in the graphite-like state. All these authors assumed that the presence of carbon was caused by the C-contamination during the sample preparation.

Contrary to our sample A, for sample B (annealed in oxygen) the carbon was not present at all (Figure 4). Comparing the spectra Figure 3 (sample A), the decrease of Cu peaks with increasing of Ba and O peaks can be seen. It means that on the grain boundary of sample A the "envelopes" rich of Ba oxide are present, with the higher content of graphite-like carbon.

The intensity of $I_{Cu}(x)$ was determined before and that of $I_{Cu}(0)$ after surface sputtering with ion Ar^+. The ratio of $I_{Cu}(x)/I_{Cu}(0)$ at the different grains gave value in the range from 0,24 to 0,49 for Cu (920 eV). We find x=7,2-14,5 Å , which is the thickness of the surface carbon layers (from 4 to 8 monolayer).

We conclude, that the presence of graphite-like carbon in grain boundaries is one of the reasons responsible for low value of the critical current in oxidic layers in microcrystalline Cu-based RQ ribbons.

Cava R J et al 1987 Phys. Rev.Lett. 58 pp 1676
Chisholm et al 1989 Intern.Interfaces in HTCS - preprint
Hajko V Jr et al 1989 Proc.ECer9, Elsevier Appl.Sci.(London) 2 pp 471
Hajko V Jr et al 1988 FNT 14 pp 215
Nakahara S et al 1987 J.Cryst.Growth 85 pp 639
Parmigiani et al 1987 Phys.Rev. 3 pp 7148
Verhoever K et al 1988 Appl. Phys. Lett 52 pp 745
Verkin B I et al 1988 FNT 14 pp 675
Xi X X et al 1988 Physics C 153-155 pp 794

Supercond. Sci. Technol. **4** (1991) S244–246. Printed in the UK

Critical current density and magnetization of Ag-sheathed tapes of Bi(Pb)-Sr-Ca-Cu-O

A.Oota, A.Yata, K.Hayashi and K.Ohba*

Department of Electrical and Electronic Engineering, Toyohashi University of Technology, Toyohashi, Aichi 441, Japan.
*Technical R&D Laboratories, TOPY Industries Ltd., Toyohashi, Aichi 441, Japan.

Abstract. The repetition process of cold-working and sintering yields a maximum transport J_{ct} of 8500A/cm^2 (77K, H=0) in Ag-sheathed tapes of Bi(Pb)-Sr-Ca-Cu-O. The tapes contain the pure high-T_c phase and **c**-axis oriented microstructures. The initial magnetization M(H) and magnetic relaxation M(t) are studied on high-J_c tapes. The J_c and H_{c1} are evaluated from M(H) using Bean's critical-state model and the results are discussed. The creep rate and activation energy are derived from M(t) using the flux creep model and the pinning nature is discussed.

1. Introduction

Since the discovery of high-T_c superconductor (Bednorz and Müller 1986), much effort has been conducted to fabricate superconducting Ag-clad tapes or wires with high current-capacity. The repetition process of cold-working and sintering was found to yield high transport J_{ct} value in Ag-sheathed tapes of Bi(Pb)-Sr-Ca-Cu-O (BPSCCO) by producing a highly textured microstructure (Mimura et al 1989).

We applied the above procedure to fabricate Ag-sheathed tapes of BPSCCO and obtained a maximum J_{ct}=8500A/cm^2 (77K,H=0). We present here the studies on initial magnetization M(H) and magnetic relaxation M(t), together with characterization, for the tapes with J_{ct}=6400, 5400 and 2400A/cm^2 (77K, H=0). They are refered to #6400, #5400 and #2400, respectively.

2. Fabrication and characterization

The calcined powder of the composition $Bi_{1.6}Pb_{0.4}Sr_{2.0}Ca_{2.2}Cu_{3.2}$ was packed into with Ag tubes and such composites were cold-worked into tapes 0.15∿0.2mm thick. They were sintered at 832°C for 100h, cold-worked again and finally re-sintered at 832°C for various periods. The tapes were characterized by XRD, SEM and susceptibility measurements. The XRD and susceptibility studies, together with the SEM observations (Figure 1), reveal that the tape has a single high-T_c (110K) phase and highly textured microstructures with the **c**-axis alignment perpendicular to the rolling surface. The field-dependence of J_{ct} ,determined

0953-2048/91/01S244 + 03 $03.50

from I-V characteristics using criterion $E=1\mu V/cm$ at 77K, is shown for both #6400 and #2400 in Figure 2.

The M(H) was measured at low field (H <100 Oe) after zero-field cooling (ZFC) by a SQUID. The M(t) was measured for both the in-field magnetization (IN-FIELD) and remanent one (REMANENT) by a SQUID after ZFC under H=500 Oe. The measurements were made under a longitudinal configuration of $H \parallel$ tape (i.e., $H \perp c$).

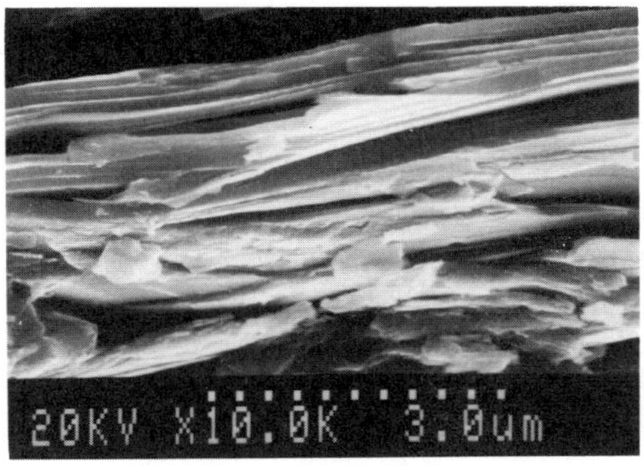

Figure 1. SEM image.

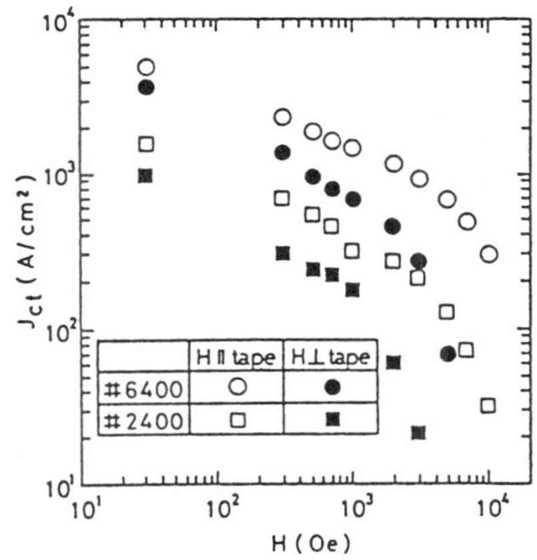

Figure 2. J_{ct} vs H.

3. Results and discussion

The J_c values for both #5400 and #2400 are evaluated from M(H) curves under $H \perp c$ using Bean's critical-state model (Bean 1964) and are compared with the transport J_{ct} in Figure 3. As can be seen, the result shows a fairly good agreement between J_c and J_{ct} at 77K. The J_c increases significantly as temperature decreases and reaches $7x10^4$ and $4x10^4$ A/cm^2 at 5K for #5400 and #2400, respectively.

The lower critical field values, H_{c1}, for the 110K phase deduced from M(H) are shown for #5400 and #2400 in Figure 4. The H_{c1} increases with a slope of -0.2 Oe/K as temperature decreases from T_c and approaches the saturated value of ∿15 Oe at T<20K. The magnitude is much lower than those for La-Sr-Cu-O (Naito et al 1990) and Y-Ba-Cu-O (Krushin-Elbaum et al 1989).

The result of M(t) for both IN-FIELD and REMANEN is analyzed based on the flux creep model (Anderson 1962). The creep rate

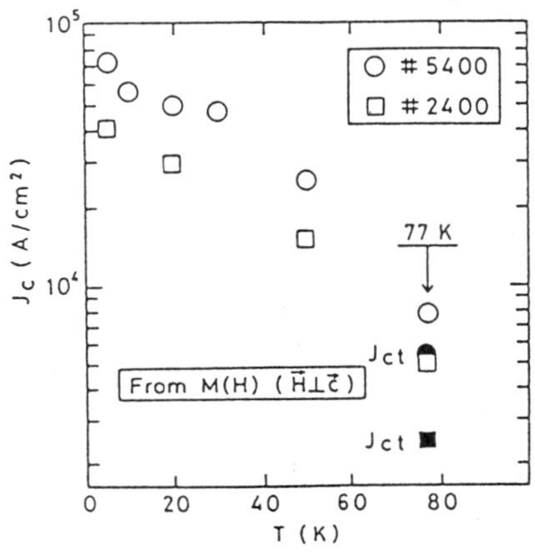

Figure 3. J_c and J_{ct}.

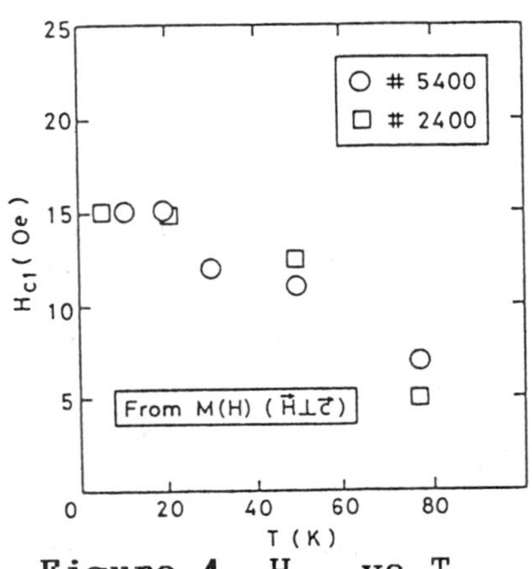

Figure 4. H_{c1} vs T.

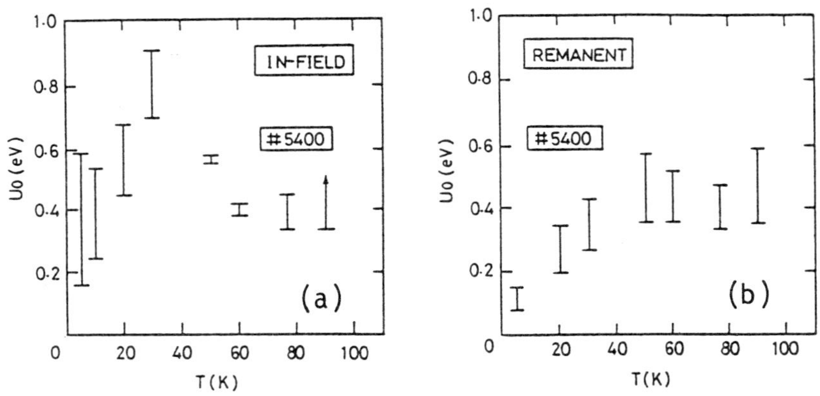

Figure 5. Activation energy.

$R=d(4\pi M)/d\ln t$ is positive (i.e., flux entry) for IN-FIELD, while being negative (i.e., flux expulsion) for REMANENT. The activation energy U_0 is evaluated from R in a way similar to that by Hagen and Griessen (1989). The result is shown for IN-FIELD and REMANENT in Figures 5(a) and (b), respectively. As can be seen, U_0 depends on the measuring process for M(t). It is of the order of 0.4eV at 5K for IN-FIELD and 0.1eV for REMANENT. The U_0 for IN-FIELD shows a peak of 0.8eV around 30K and then decreases as temperature increases, while a peak in it is less pronounced for REMANENT. Such difference in the behavior of U_0 suggests a distribution of pinning strength in the superconductor (Hagen and Griessen 1989), where the flux pinning near the surface is larger than that in its interior. This may be related to the distribution of the degree of texture in the tapes.

4. Summary

The study on M(H) based on Bean's critical-state model yields a fairly good agreement between J_c and J_{ct}. The H_{c1} for the 110K phase under $H\perp c$ increases with a slope of -0.2 Oe/K as temperature decreases from T_c, and approaches a saturated value of ∿15 Oe at T<20K. The magnitude is much lower than those for La-Sr-Cu-O and Y-Ba-Cu-O.

The activation energy U_0 derived from M(t) depends on whether M(t) is taken for IN-FIELD or REMANENT. This suggests a distribution of pinning strength in the superconductor. The U_0 at 5K for H=500 Oe ($H\perp c$) is 0.4eV for IN-FIELD and 0.1eV for REMANENT.

References

Anderson P W 1962 Phys.Rev.Lett.**9** 309

Bean C P 1964 Rev.Mod.Phys.**36** 31

Bednorz J G and Müller K A 1986 Z.Phys.**B64** 189

Hagen C W and Griessen R 1989 Phys.Rev.Lett.**62** 2857

Krushin-Elbaum L, Malozemoff A P, Yeshurun Y, Cronemeyer D C and Holtzberg F 1989 Phys.Rev.**B39** 2936

Mimura M, Kumakura H, Togano K and Maeda H 1989 Appl.Phy.Lett.**54** 1582

Naito M, Matsuda A, Kitazawa K, Kambe S, Tanaka I and Kojima K 1990 Phys.Rev.**B41** 4823

Supercond. Sci. Technol. **4** (1991) S247–249. Printed in the UK

Present State Of Our Research On Silver Sheathed $REBa_2Cu_3O_7$ Superconductors

L. Chaffron and P. Régnier

CEREM/D.T.M./Section de Recherches de Métallurgie Physique
Centre d'Etudes Nucléaires de Saclay
91191 Gif sur Yvette Cedex France

Abstract . Several high Tc superconducting wires and ribbons were prepared by swaging and cold pressing silver tubes filled with various $REBa_2Cu_3O_{7-\delta}$ ceramics. It is shown that the higher the density of the starting green body the better the superconductor. Detailed microstructural characterization of these conductors and critical current density, J_c, achieved as a function of the wire diameter, or the ribbon thickness, are presented. RE = Y or Er was found to give practically equal J_c values, and RE = $Y_{0.6}Er_{0.4}$ substantially higher ones. The highest J_c measured in the present study are 2000 and 4200 A/cm^2 at 77 and 63 K respectively.

1. Introduction

In the present paper, we report on the preparation of short sections of superconducting ribbons and wires, using the powder in tube technique (Jin (1987), Flükiger (1987)).

2. Experimental Procedure

REBaCuO (RE = Y or Er or $Y_{0.6}$ $Er_{0.4}$) powders, reacted and unreacted, were poured into a latex tube and cold pressed under a pressure of 300 MPa. After removing the latex tube, the rod (typically L = 60 mm and ϕ = 4.5 mm) was adjusted in a silver tube and outgassed. The tube was then sealed and swaged down to prepare a 1 mm in diameter wire of about 2 m in length. Some sections were directly sintered at temperatures ranging from 900 to 940 C for 5 to 50 hours, whereas some others were cold pressed before being sintered. Finally the specimens were very slowly cooled down under flowing oxygen to make them superconducting.

3. Microstructural Characterization

The first problem, with the powder in tube technique, is that the powder shrinks as it sinters whereas the tube does not. Moreover and simultaneously, REBaCuO strongly adheres to the sheath. These two features lead to the formation of cracks and porosities which limit the critical current density, J_c. To minimize this effect, Régnier (1990), we have increased, the density of the green powder in the ribbons up to ~ 95% of the maximum theoretical density. As shown in Figure 1, an increase in green density increases the density of the final sintered ceramic however some cracks still exist, most of which are parallel to the ribbons axis and consequently less efficient for limiting J_c. Examination of both the thinner longitudinal and the transverse cross section under low magnification revealed that the cross over section through which the current was flowing was rather irregular both in area and in shape, which probably

0953-2048/91/01S247 + 03 $03.50

results from an instability in the codeformation of the powder and the tube. Observations at higher magnification show the fine porosity due to the incomplete densification and also some narrow cracks which may also be generated as well by the anisotropic change in the dimensions of the unit cell of the ceramic as a function of the temperature as well as the difference in thermal expansion between silver and ceramic. It is worth noting that after 15 h of sintering in air at 925 C, the grains are rather small (typically 5 μm in size) and already have a platelet shape (Figure 2).

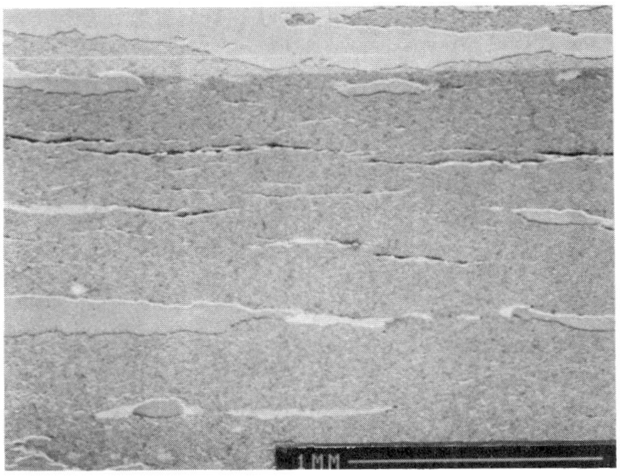

Fig. 1 Cracks and Ag incrustations

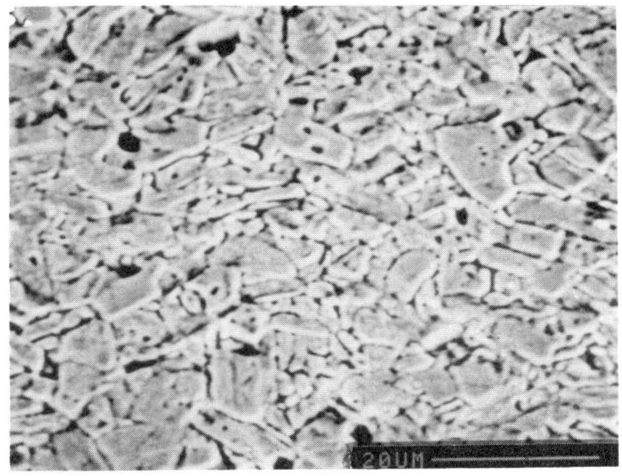

Fig. 2 Grain shape and size

Practically no secondary phase was detected on the X-ray diffraction diagram carried out on the ceramic core of the ribbon but it was noticeable that the thinner specimens were somewhat textured (Figure 3).

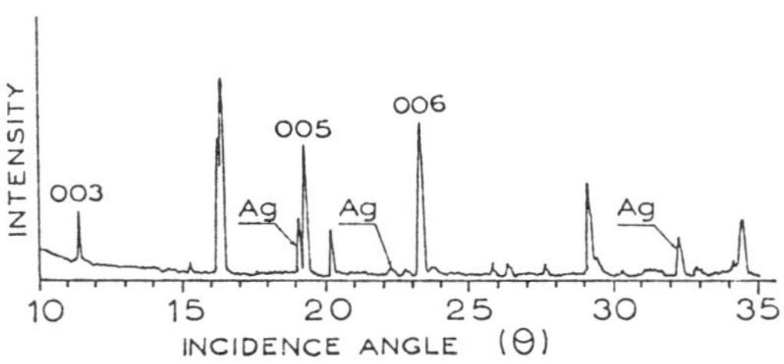

Fig. 3 X-ray diffraction diagram

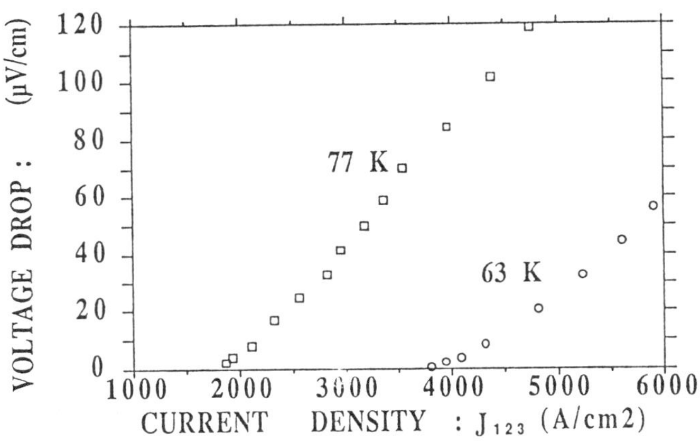

Fig. 4 Voltage drop versus current density

4. Electrical Characterization

Measurements of the resistivity and the critical current density were carried out with direct current using four point probes soft soldered with the silver sheath. Because of this arrangement the measured current, I_m, is divided between the sheath and the ceramic, and the measured resistance is that of the silver clad in parallel on that of the ceramic. It can be easily shown that the current density, J_{123}, which really flows through the ceramic and the intrinsic resistivity of the latter are related to the measured parameters as follow :

$$J_{123} = \left(1 - \frac{\rho_m}{\rho^*}\right) J_m \quad , \quad \rho_{123} = \frac{\rho_m}{1 - \left(\frac{J_o}{J_m}\right)\left(\frac{J_{123}}{J_m}\right)} \quad , \quad \rho_m = \frac{\Delta V_m \, S_{123}}{L \, I_m} \quad \text{and} \quad J_m = \frac{I_m}{S_{123}}$$

where ΔV_m is the measured voltage drop, S_{123} the section of the ceramic, J_o the critical current density corresponding to $\Delta V_m = 0$ and ρ^* the value of ρ_m which should be measured if the ceramic was not superconducting ($\rho^* = \rho_{Ag}\ S_{123} / S_{Ag}$). Consequently, the determination of the genuine parameters of the ceramic required a precise knowledge of S_{123}. To obtain a reliable value, the average area of several polished cross sections was determined. This value was corroborated by a direct measurement of ρ^* after desoxygenation of the conductor. Nevertheless, there are some sections which are smaller than the average through which J_c is higher than that reported here. Figure 4 shows the importance of correcting for the current density which flows through the silver sheath. From this plot J_c was determined on the basis of $\Delta V/L = 1\ \mu V\ cm^{-1}$. Alternatively, as shown in Figure 5, J_c can be determined from the plot giving ρ_{123} versus J_{123}. Finally all the J_c values measured both at 77 and 63 K were plotted as a function of the wire diameter or the ribbon thickness (Figure 6). Such a plot clearly shows that the thinner the ribbon the higher the J_c and that RE = $Y_{0.6}\ Er_{0.4}$ gives significantly higher J_c than RE = Y or Er as expected from the slight increase in T_c observed on this material, Nahm (1990).

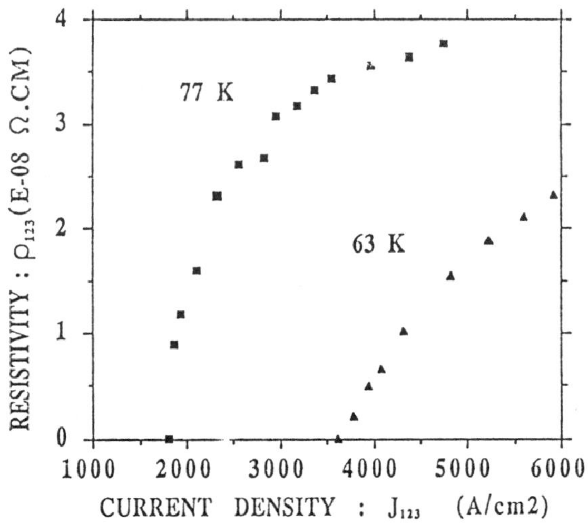

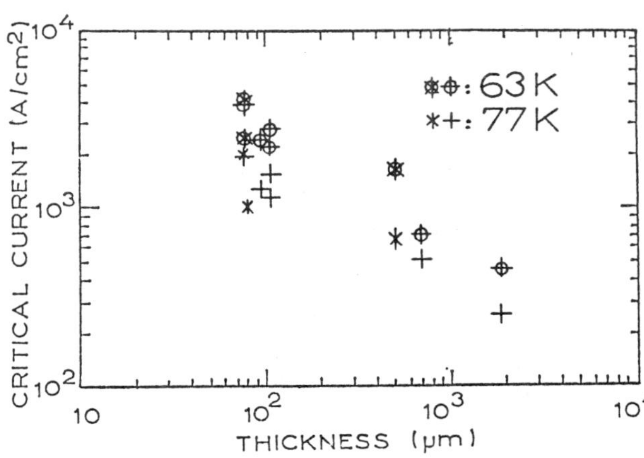

Fig. 5 Resistivity of a 80 μm thick YBaCuO ribbon as a function of the flowing current density

Fig. 6 Critical current density as a function of the ribbon thickness, + are for YBaCuO and * for $Y_{0.6}Er_{0.4}BaCuO$

5. Conclusion

We have succeeded in preparing silver sheathed wires and ribbons of REBaCuO which exhibit critical current densities which are among the best reported up to now, Osamura (1989).

6. Acknowledgments

The Ministère de la Recherche et de la Technologie is sincerely acknowledge for partial funding of the project, and the Comptoir Lyon Alemand Louyot for the co-financement of the CIRCEA scholarship attributed to L. Chaffron. We are also very grateful to P. Régnier junior for preparing the $Y_{0.6}Er_{0.4}BaCuO$ wires.

7. References

Flükiger R, Weiss F, Goldacker W, Seibt E, Wolf T and Muller T 1987 *Proc. 1st Europ. Workshop on High T_c Mat.*, Genova 131
Jin S , Sherwood R C, Van Dover R B, Tiefel T H and Johnson D W 1987 *Jr. Appl. Phys. Lett.* **51** (3) 203-204
Régnier P., Chaffron L., Schmirgeld L. *Submitted for publication in the Proceeding of ICMAS-90, 15-17 Octobre 1990, Grenoble, France*
Nahm K, Kim J S, Lee K B, Shong D W, Cha B Y, Kim C K 1990 *J. of Mat. Science* **25** 1762-1764
Osamura K, Takayama T and Ochiai S 1989 *Supercond. Sci. Technol.* **2** 111-114

Supercond. Sci. Technol. **4** (1991) S250–252. Printed in the UK

Self-similarity in the I-V characteristics of granular $YBa_2Cu_3O_7$

S S Bungre[a], S M Cassidy[a], A D Caplin[a], N McN Alford[b] and T W Button[b]

a Centre for High Temperature Superconductivity, Blackett Laboratory, Imperial College, London SW7 2BZ, UK

b ICI Advanced Materials, Runcorn, Cheshire WA7 4QD, UK

Abstract. The onset of dissipation in the I-V characteristics of granular $YBa_2Cu_3O_7$ shows clear evidence for a finite critical current. At currents above critical, dissipation in the weak links is only slightly field- and temperature-dependent.

1. Introduction

There is strong evidence for a true non-dissipative regime in granular superconductors. For, although these systems are random 3D networks of weak links with an exponentially broad range of strengths, phase coherence through the network plays a vital role.

In granular low temperature superconductors (LTS), as for example the Al-Pb composites examined by England et al. (1987), there is both a persistent remanent magnetization and a sharply defined transport critical current I_c. In granular high temperature superconductors (HTS), remanent magnetization measurements (Laverty et al. (1989)) demonstrate the existence of persistent **inter**granular currents (decays of less than 1% per hour). However, the resistive behaviour of HTS composites is less clear-cut than that of LTS, in part because high resolution SQUID voltmeters are not readily usable with HTS.

2. Parameterization

Near the onset of dissipation, it has been common practice to parameterize the I-V characteristics of HTS materials in the simple power law form $V = kI^{\alpha}$ (Evetts and Glowacki (1988), Kostadinov et al. (1988) and Onogi et al. (1989)), thereby implicitly excluding a finite value of I_c. The exponent α is of course found to vary with field and temperature. However, the striking feature of the measured I-V characteristics in granular HTS is how little the shape of the onset changes (Figure 1). That is, we can write $V = f(I - I_c(H,T))$, with f being independent of field and temperature to a first approximation.

Irrespective of the form of the function f, this observation implies that there is a current $I_c(H,T)$ which, if not exceeded, induces no dissipation. Secondly, it follows that simple power law fits to two identically shaped onsets of different I_c's necessarily produce different values of α. A large I_c gives a large value of α, and depression of I_c with the application of a field or by increasing temperature leads automatically to a smaller α value.

It is still convenient to have a simple parameterization for the I-V characteristics, and a suitable form is

$$V = k(I-I_c(H,T))^n \qquad (1)$$

0953-2048/91/01S250 + 03 $03.50

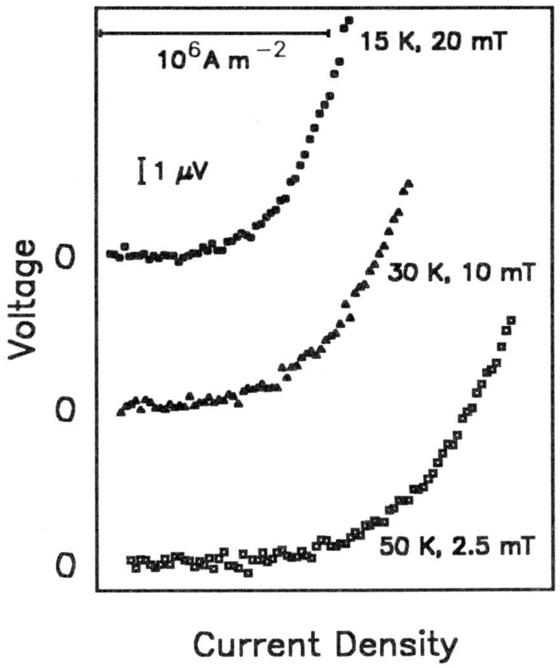

Figure 1 I-V characteristics near the onset, with vertical and horizontal axes displaced. Values of J_c are: 3.9, 7.8 and 8.9×10^6 A m^{-2} (top to bottom).

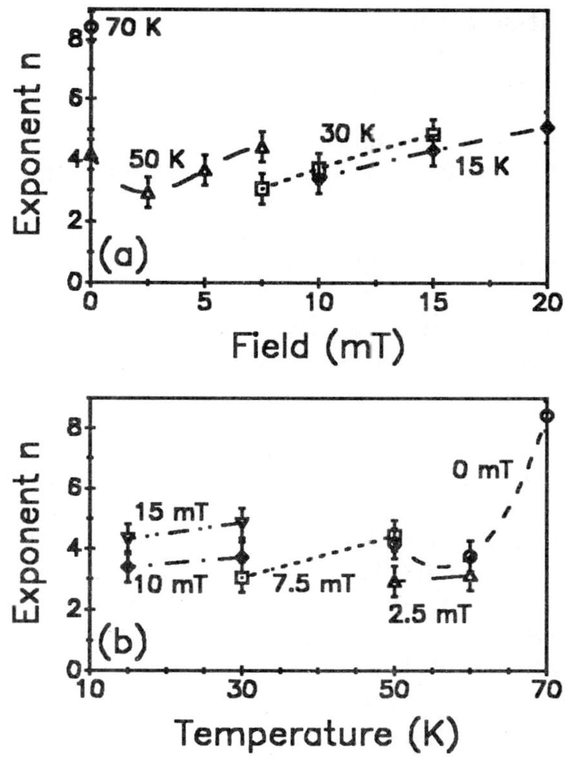

Figure 2 (a) Field dependence of exponent n at fixed temperatures, (b) the temperature dependence at fixed fields. Lines drawn as an aid to the eye.

as used by Meisels et al. (1988), Horng et al. (1989), and others. Furthermore, analysis of the superconducting network using the spin glass analogy suggests that this is an appropriate form (Sergeenkov (1990)).

3. Experimental

The field and temperature dependence of the I-V characteristics of 0.8 mm diameter $YBa_2Cu_3O_7$ rods were measured using a pulsed current technique, as described by Meisels et al. (1988). This avoids Joule heating and allows the characteristics to be studied over a wide temperature range.

By fitting the onset of the I-V characteristics (for $V < 10\,\mu V$) to equation 1, the exponent n was extracted for each temperature and field. It should be emphasized that because there are 3 parameters in the fit (n, k and I_c), and the range of data to be fitted has to be selected, there is considerable flexibility in the fit and the uncertainties in the parameters are large.

Figure 2 shows that over most of the field-temperature plane, the exponent is in the range of 2 to 5. At low fields and low temperatures (i.e. for large critical currents), there are self-field and current re-distribution effects (Bungre et al. 1989). At a given temperature, n tends to increases with field, and at a given field, n increases slightly with temperature. These trends are not artefacts of the parameterization, but reflect a slight steepening that can be seen in the I-V characteristics themselves.

4. Discussion

Our results are qualitatively similar to those of Horng et al. for polycrystalline Bi-Ca-Sr-Cu-O films, where from 20 to 74 K the data fit equation 1 with n in the range 2 to 3. The self-similarity of the I-V onsets demonstrates that, once the critical current of the random granular network has been exceeded, the dissipation is primarily a function of the excess current, and depends only weakly on temperature and magnetic field. That implies that the dissipative behaviour of the weak links, which are now generally accepted to be the grain boundaries, is only weakly field- and temperature-dependent.

We note also that description by any simple power law $V = kI^{\alpha}$, in which the critical current has been set to zero, is bound to produce unphysically large values of the exponent α (values of 20 or 50 are often reported) that vary strongly with field and temperature, and so to distract attention from the almost invariant shape of the I-V characteristics.

In the spin glass model of the superconductor network, the exponent depends upon the magnitude of the order parameter. Sergeenkov's recent analysis suggests that n = 2.0 at T_c, and then increases slowly with either increasing field or decreasing temperature. This value of n is not too far removed from our measurements, but is certainly too low to fit the data well. The predicted field and temperature dependences arise from those of the intergranular coupling, and the disagreement with trends shown in Figure 2 suggests that there is another contribution of the opposite sign.

5. Conclusions

In granular HTS materials, the shape of the onset of dissipation in the I-V characteristics varies little with temperature or field. This self-similarity of the characteristics implies strongly that there is a true critical current, despite the random network nature of these materials. Also, it indicates that the dissipative behaviour of the grain boundary weak links is almost field and temperature independent.

This work has been supported by the Science and Engineering Research Council.

Bungre S S, Meisels R, Shen Z X and Caplin A D 1989 *Nature* **341** 725
England P, Goldie F and Caplin A D 1987 *J. Phys. F* **17** 447
Evetts J E and Glowacki B A 1988 *Cryogenics* **28** 641
Horng H E, Jao J C, Chen H C, Yang H C, Sung H H and Chen F C 1989 *Phys. Rev. B* **39** 9628
Kostadinov I, Mateev M, Petrov I V, Vassilev P and Tihov J 1988 *Physica C* **153-155** 320
Laverty J R, Caplin A D and Male S E 1989 *Physica C* **162-164** 1171
Meisels R, Bungre S and Caplin A D 1988 *J. Less Common Metals* **151** 83
Onogi T, Ichiguchi T and Aida T 1989 *Solid State Commun.* **69** 991
Sergeenkov S A 1990 *Physica C* **167** 339

Superconducting Transport Properties of $YBa_2Cu_3O_{7-\delta}$ Grain Boundary Junctions

R. Gross

Lehrstuhl Experimentalphysik II, Universität Tübingen, D-7400 Tübingen, FRG

P. Chaudhari, M. Kawasaki, and A. Gupta

IBM Thomas J. Watson Research Center, P. O. Box 218, Yorktown Heights, NY 10598

Abstract: We have studied the weak link properties of individual grain boundaries in thin film bicrystals of $YBa_2Cu_3O_{7-\delta}$. The grain boundary junctions (GBJs) show current voltage characteristics that can be closely modeled by the resistively shunted junction (RSJ) model. The magnetic field dependence of their critical current is Fraunhofer pattern like. Their $J_c\rho_N$ products, where J_c is the critical current density of the GBJs and ρ_N their normal resistance times unit area, range between 0.2 and 8mV and scale proportional to about $J_c^{0.6}$.

Grain boundaries in high temperature superconductors have been found to act as weak links limiting the critical current density of these materials. In order to obtain detailed information on their superconducting transport properties and to understand the mechanisms responsible for their weak link behavior, single grain boundaries in $YBa_2Cu_3O_{7-\delta}$ films have been studied intensivly.[1–4] It was found that the grain boundary critical current density J_c is quite sensitive to the misorientation angle, even if the CuO-planes of the adjacent grains are essentially parallel. This behavior may be a general feature of all cuprate superconductors as indicated by recent experiments with BiSrCaCuO-bicrystals[5], which show similar behavior. Here we report on the transport properties of $YBa_2Cu_3O_{7-\delta}$ GBJs fabricated by laser ablation deposition.[6] We have found that all our data on these grain boundaries can be described by a master curve, which relates the $J_c\rho_N$ product of the GBJs to their critical current density. This general scaling behavior, although not completely understood at present, suggests that the superconducting properties of grain boundaries in high temperature superconductors are controlled by an intrinsic mechanism.

The GBJs have been fabricated by patterning narrow lines straddling a grain boundary in thin film bicrystals of $YBa_2Cu_3O_{7-\delta}$. Details on the fabrication of $YBa_2Cu_3O_{7-\delta}$ GBJs using oriented bicrystals of $SrTiO_3$ as substrates are given in Ref.1. In the following we only discuss the transport properties of GBJs with misorientation angles larger than about 15° and width W comparable to or smaller than their Josephson penetration depth λ_J. Low angle or wide ($W \gg \lambda_J$) GBJs show a more complicated behavior not discussed here.

The dependence of the grain boundary critical current density J_c on the misorientation angle for the grain boundaries fabricated by laser ablation deposition was found to be very similar to that of the GBJs fabricated by electron beam coevaporation.[1] However, the ratio J_c/J_c^G for large misorientation angles is about 10 times smaller ($J_c/J_c^G \approx 10^{-3}$) for the GBJs fabricated by laser ablation deposition. This is probably caused by the about 10 times higher grain critical current density J_c^G in the laser ablated films. This observation shows that the magnitude of the grain boundary critical current density is determined by the grain boundary parameters, such as the misorientation angle, and not by the critical current density in the adjacent grains, which is determined by the flux pinning behavior of the grain material.

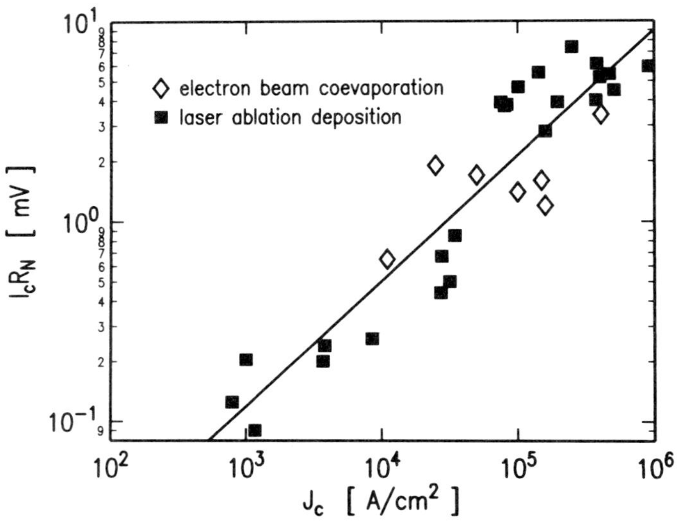

Fig.1: Variation of the I_cR_N product of different GBJs with their critical current density at T=4.2K. The least squares fit to the data (solid line) yields a slope of 0.6.

The current voltage characteristics (IVCs) of the GBJs can be accurately modeled by the resistively shunted junction (RSJ) model,[3,7] in which a sinusoidal current phase relation and a constant normal resistance R_N is assumed. For low angle or wide GBJs deviations from the RSJ behavior, probably caused by the occurrence of one-dimensional flux flow along the grain boundary, are observed. Close to the critical temperature, a finite thermal noise current causing thermally activated phase slippage (TAPS) must be included in the RSJ model to fit the experimental data.[3] TAPS can result in a considerable broadening of the resistive transition of the GBJs and in rounded IVCs as discussed in detail in Ref.3.
The normal resistance R_N of the GBJs was found to be independent of temperature between 4.2K and T_c.[4] The temperature dependence of I_c close to T_c was found to be proportional to $(1 - T/T_c)^2$.[3] For lower temperatures $I_c(T)$ becomes almost linear. It has been shown recently[8] that the observed $I_c(T)$ dependence can be described well by weak link models based on SNINS type junctions (S=superconductor, N=normal metal, I=insulator). The magnetic field dependence of the critical current $I_c(B)$ is similar to that of Josephson junctions with a spatially inhomogeneous current density distribution. Recently, the presence of an inhomogeneous critical current density along grain boundaries has been imaged directly by Low Temperature Scanning Electron Microscopy.[2] Again, low angle or wide GBJs have more complicated $I_c(B)$ curves showing hysteresis for magnetic field sweeps in opposite directions.[1]

The measured IVCs, $I_c(B)$, and $R_N(T)$ dependences, the observation of microwave induced steps in the IVCs,[9] and the experiments, in which dc-SQUIDs were fabricated using GBJs,[10] show that GBJs behave like Josephson junctions. However, in contrast to "classical" SIS-type Josephson junctions, the I_cR_N ($J_c\rho_N$) products of the GBJs, which range between about 0.2 and 8mV, are small compared to the gap energy of

$YBa_2Cu_3O_{7-\delta}$ ($\Delta \sim 15 - 30meV$). Furthermore, the I_cR_N products of the GBJs are not independent of J_c. In Fig.1 the I_cR_N products of a large number of GBJs is plotted versus their critical current density. Although there is some spread, the experimental data show that $J_c\rho_N \propto (J_c)^n$ with $n \simeq 0.6$ or, equivalently, $J_c\rho_N \propto (1/\rho_N)^{2.5}$. Since this behavior is observed for a large variety of different samples with critical current densities ranging between 7×10^2 to $8 \times 10^5 A/cm^2$, we conclude that the behavior is not random but related to a systematic variation of the properties of the grain boundary. A possible reason for this scaling behavior may be the presence of a strong pair breaking process causing a reduced energy gap Δ_i at the grain boundary. Then, ρ_N may be dominated by pair breaking so that $\Delta_i \propto 1/\rho_N$. Since the $J_c\rho_N$ product is a function of Δ_i, then it also will be a function of $1/\rho_N$. The presence of a temperature independent pair breaking process at the grain boundary can account both for the observed temperature dependence and the reduced value of the $J_c\rho_N$ products of the GBJs.[8] There is some evidence that the presence of such pair breaking process is related to variations of the oxygen concentration in the vincinity of grain boundaries. Evidence in support of this comes from the observation that the grain boundary properties shift along the $J_c\rho_N \propto (J_c)^{0.6}$ dependence by postdeposition annealing.

References:

1. P. Chaudhari, J. Mannhart, D. Dimos, C.C. Tsuei, C.C. Chi, M.M. Oprysko, M. Scheuermann, Phys. Rev. Lett. 60, 1653 (1988); J. Mannhart, P. Chaudhari, D. Dimos, C.C. Tsuei, T.R. McGuire, Phys. Rev. Lett. 61, 2476 (1988); D. Dimos, P. Chaudhari, J. Mannhart, Phys. Rev. B41, 4038 (1990); D. Dimos, P. Chaudhari, J. Mannhart, F.K. LeGoues, Phys. Rev. Lett. 61, 219 (1988)
2. J. Mannhart, R. Gross, K. Hipler, R.P. Huebener, C.C. Tsuei, D. Dimos, P. Chaudhari, Science 245, 839 (1989)
3. R. Gross, P. Chaudhari, D. Dimos, A. Gupta, G. Koren, Phys.Rev. Lett. 64, 228 (1990)
4. R. Gross, P. Chaudhari, M. Kawasaki, A. Gupta, to be published
5. N. Tomita, Y. Takahashi, Y. Ishida, Jap. J. Appl. Phys. 29, L30 (1990)
6. G. Koren, A. Gupta, A. Giess, E.A. Segmüller, T.R. McGuire, Appl. Phys.Lett. 54, 1055 (1989)
7. W.C. Stewart, Appl. Phys. Lett. 12, 277 (1968); D.E. McCumber, J. Appl. Phys. 39, 3113 (1968)
8. A.A. Golubov, M.Yu. Kupriyanov, Proc. LT-19, to be published in Physica B
9. S.B. Kaplan. C.C. Chi, R. Gross, D. Dimos, unpublished
10. R. Gross, P. Chaudhari, M. Kawasaki, M. B. Ketchen, A. Gupta, Appl. Phys. Lett. to be published , and Physica C, to be published

Supercond. Sci. Technol. **4** (1991) S256–258. Printed in the UK

Reduction of Measured Critical Currents in HTC Weak Links due to Thermal Activation

M. N. Keene

School of Physics and Space Research, University of Birmingham, Edgbaston, Birmingham B15 2TT.

Abstract

A simple model has been used to determine the effects of thermal activation on the measured critical current of various superconducting junctions. Thermal activation causes a lowering of the measured critical current below the intrinsic critical current of the junction. For HTC junctions, these effects are significant where low temperature intrinsic critical currents are <3mA. This model has been used to calculate the lowering of the critical current for two junction types. These calculations are shown to be consistent with a more rigorous theory for thermal activation.

1. Introduction

Many of the electronic applications of high–Tc (HTC) superconductors will be based upon the Josephson junction. When operated at liquid nitrogen temperatures the HTC junction is subject to thermal fluctuations in energy of approximately 20 times that of conventional helium cooled junctions. This will place powerful constraints on the use of HTC junctions in applications. It is therefore important that the effects of thermal fluctuations are considered when designing and characterizing HTC junctions. In this paper we describe how such effects may be calculated in a way which allows an easy comparison with experimental data without involving complicated numerical models.

The effects of thermal fluctuations on current biased Josephson junctions have been described by Ivanchenko and Zil'berman, (IZ), (1968) and more rigorously by Ambegaokar and Halperin, (AH), (1969). The latter analysis predicts the current/voltage characteristics for a standard resistivity shunted junction (RSJ) model. In this paper we predict the influence of thermal fluctuations on the observed critical current, $I_c^*(T)$ for a junction as we have described previously (Keene 1990). We then compare our results with a more rigorous treatment based on the AH theory and demonstrate consistency between these models.

2. The Model

The Josephson junction biased with a current I which is less than the critical current $I_c(T)$ is a system which exists in a metastable energy minimum with a given phase state S which is separated from adjacent states at $S+/-2\pi$ by potential barriers ΔU_f and ΔU_b. Here the subscripts f and b refer to the barriers inhibiting forwards and backwards motion of the phase. We shall now discuss the time evolution of S at finite temperature as we have described previously (Keene 1990) which is based on the early analysis by IZ. The effect of temperature is to activate the phase state usually over the lower potential barrier ΔU_f. By defining the observed critical current of the

0953-2048/91/01S256 + 03 $03.50

junction as the current at which the first departure from the V=0 state is observed, the value of ΔU_f at $I = I_c^*(T)$ may be expressed as

$$\Delta U_f = kT\ln(\omega_o \Phi_o / V_{min}) \qquad 1$$

where V_{min} is the minimum resolvable voltage in a measurement, ω_o is an exponent prefactor related to the characteristic attempt frequency of the phase in the metastable energy minimum, T is the absolute temperature and k and Φ_o are constants. ΔU_f is related to the junction current, I, and the intrinsic critical current, $I_c(T)$, to a good approximation by

$$\Delta U_f = U_o(T)(1-I/I_c(T))^{3/2} \qquad 2$$

with $U_o(T) = 0.3\Phi_o I_c(T)$ as shown by Fulton and Dunkleburger (1974). Equating equations 1 and 2 gives an expression for the experimentally observed critical current in terms of the intrinsic critical current.

$$I_c^*(T) = I_c(T)[1-(kT\ln(\omega_o \Phi_o/V_{min})/U_o(T))^{2/3} \qquad 3$$

This analysis assumes the junction damping is sufficiently large to restrict the transitions to neighbouring metastable states, and that the transitions occur exclusively over the lower ΔU_f barrier. This latter assumption is good for junctions with $I_c(T)/T \gtrsim$ 16nA/K.

3. Results

We have calculated $I_c^*(T)$ as a function of T for two junction types and for various values of $I_c(0)$. The calculations are based on the Ambegaokar–Baratoff, (AB), (1963) form for $I_c(T)$ which corresponds to ideal BCS tunnel junctions. (Our analysis is not restricted to this form for $I_c(T)$). Figure 1 shows the results of our model for YBCO/Nb and YBCO/YBCO junctions of various strengths, normalized to unity at zero temperature. A comparison with the corresponding curves calculated from the AH

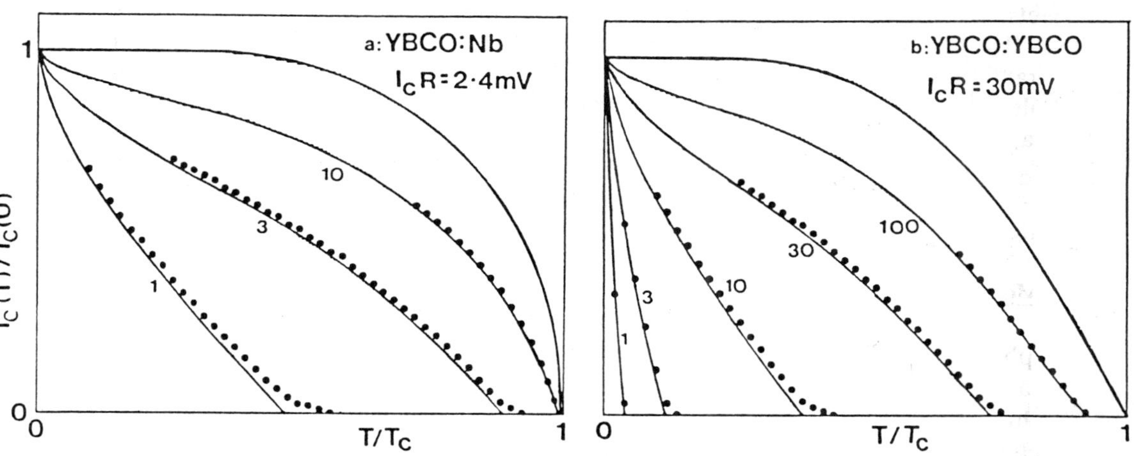

Figure 1, $I_c^*(T)/I_c(T)$ against T (solid lines) for a) YBCO/Nb with T_c = 9.2K, and b) YBCO/YBCO with T_c = 92K for various values of $I_c(0)/\mu A$ as marked. A value of $0.1\mu V$ for V_{min} has been chosen. An exact calculation based on the theory of AH has been superimposed (dots). The upper curves represent the intrinsic $I_c(T)$.

theory are superimposed (dots). (The curvature at low currents is a consequence of backwards jumping which we have chosen to ignore in our model.) This comparison was achieved by calculating the current, using the AH theory, at which the junction voltage equalled our chosen value for V_{min}. The AH theory is based on treating the system as a "Brownian" particle in the same potential as the IZ theory acting in a high damping limit. However, as pointed out by AH, the attempt frequency in the IZ treatment is introduced as an 'ad hoc' assumption, whereas in the AH treatment this quantity is given by Kramers (1940) as

$$\omega_o = RC\omega_j/2\pi = I_cR/\Phi_o \tag{4}$$

where

$$\omega_j = (2\pi I_c/\Phi_o C)^{1/2}$$

for ideal RSJ junctions in the high damping limit where R and C are the junction normal state resistance and capacitance. Knowledge of the I_cR product is required to interpret the AH results in terms of absolute current and voltage so substitution of equation 4 into equation 3 allows a direct comparison between these methods. It is therefore this expression for ω_o which was used to calculate figure 1. It should be noted that the I_cR product at zero temperature is fixed at $\pi\Delta(0)/2e$ for ideal tunnel junctions, where $\Delta(0)$ is the energy gap (AB).

4. Discussion

The expression given in equation 3 gives fair agreement with an exact calculation based on the theory of AH. However, this expression is not limited to the case of well behaved RSJ junctions, but applies to any Josephson junction with a sinusoidal current–phase relationship which is acting in a moderate to high damping limit, provided ω_o is known. One further advantage of this model is that it may be easily reworked to obtain values for $I_c(T)$ from $I_c^*(T)$. Thus from measurements of thermally suppressed critical currents, the behaviour of the intrinsic critical current may be obtained.

Conclusion

We have used a simple model to predict the form of the temperature dependence of the measured critical current of a Josephson junction, and have demonstrated consistency between this model and an equivalent model based on a rigorous theory of thermal activation. Our analysis will therefore serve as a qualitative treatment which is not limited to RSJ models.

Acknowledgements

I would like to thank Timothy Jackson, Dr. Christopher Muirhead and Dr. Colin Gough for helpful discussions and comments.

References

Ambegaokar V. Baratoff A. 1963 Phys. Rev. Lett. **11** 104.
Ambegaokar V. Halperin B.I. 1969 Phys. Rev. Lett. **22** 1364.
Fulton T.A. Dunkleburger L.N. 1974 Phys. Rev. **B9** 40760.
Ivanchenko Y.M. Zil'berman L.A. 1968 JETP Lett. **8** 189.
Keene M.N. 1990 Supercond. Sci. Technol. **3** 312.
Kramers H.A. 1940 Physica **7** 284.

Supercond. Sci. Technol. **4** (1991) S259–261. Printed in the UK

FLUX PINNING ENHANCEMENT BY TWIN BOUNDARIES

W. Kritscha, F.M. Sauerzopf, H.W. Weber
Atominstitut der österreichischen Universitäten
A-1020 Wien Austria

G.W. Crabtree
Materials Science Division, Argonne National Labrotory,
Argonne, IL 60439, U.S.A

J.Z. Liu
Department of Physics, University of California, Davis,
CA 95616, U.S.A

Abstract. We have measured the magnetization of single crystal $YBa_2Cu_3O_{7-\delta}$ with all the twin planes orientated in only one direction. By rotating the magnetic field in the (ab)-plane, we find significant enhancements of the magnetization (up to a factor of 4.6), when the field is directed parallel to the twin boundaries. No angular dependence of the magnetization in the (ab)-plane could be detected at temperatures exceeding ≈40 K.

Since the discovery of high-T_c superconductors, considerable efforts have been made to assess the mechanisms of flux line pinning. One possible defect structure, which could work as a pinning center in $YBa_2Cu_3O_{7-\delta}$, is represented by the twin boundaries (TB's). These were examined more closely in several recent publications (Kwok *et al* 1990). Decoration experiments (Dolan *et al* 1989) showed that vortices were pinned by TB's, but also that there was pinning without them. Magnetization measurements on a single crystal before and after detwinning (Welp *et al* 1990) demonstrated, that the TB's acted as pinning centers. The most pronounced effect in this experiment was found at low temperatures and when the field was perpendicular to the c-axis. Torque measurements (Gyorgy *et al* 1990) at 76 K and with the field perpendicular to the c-axis yielded sharp peaks when the field was parallel to the TB's, whereas at 27 K these peaks were absent. Critical current measurements on crystals with different twin plane densities (Dinger *et al* 1989) suggested, that TB's may play only a minor role. Other magnetization measurements on a single crystal of $YBa_2Cu_3O_{7-\delta}$, which displayed predominantly one TB-orientation (Swartzendruber *et al* 1990), showed only weak pinning effects, when the field was rotated in the (ab)-plane. All of these results demonstrate, that TB's can indeed act as pinning centers. However, some of these papers suffer from an insufficient sample quality or from oversimplified models used to explain the data. Thus a quantitative assessment of the contribution of TB's to flux-pinning has still to be made. In this paper, we report on magnetization measurements made on a high quality $YBa_2Cu_3O_{7-\delta}$ single crystal, which displays TB's only in one single direction, and show that simple models, such as those used previously, cannot be reconciled with the present experimental results.

The platelike single crystal (0.65 x 0.4 x 0.0065 mm^3, weight: 10.8 μg) shows only one twin boundary direction as demonstrated by the polarized light micrograph in Figure 1. The crystal was grown using a self flux method as described by Kaiser (1987). All magnetization measurements were made using an 8 T SQUID-magnetometer. Figure 2 shows the zero field cooled and field cooled diamagnetic transition in a field of 1.6 mT applied perpendicular to the c-axis. This measurement yields a T_c (onset) of 91.5K, and a Meissner-fraction of about 82%, which - in combination with the narrow transition width - bears evidence for the excellent quality of the sample. Full magnetization loops were measured in external fields perpendicular to the c-axis at angles of 0, 22, 45, and 90 degrees with respect to the TB's. The measurements were carried out at temperatures from 5 to 40 K. At higher temperatures (and fields) the signal could not be resolved due to the small sample dimensions. At high fields the output curves of the SQUID - sensor showed heavy distortions due to the inhomogeneity of the magnet, reflecting changes of the magnetization as the sample moved between the pick-up coils. This can be explained as follows. In the present experimental procedure, the sample is kept in the center of the magnet, i.e. in a relative minimum of the external field, during field changes. In the course of the measurement, the sample is then moved to places of higher field, i.e. the locations of the pick-up coils. This leads to only small errors in the magnetization for measurements in increasing field (M^+), but to completely wrong results in decreasing field (M^-). Therefore, only data of M^+ have been used for further evaluation.

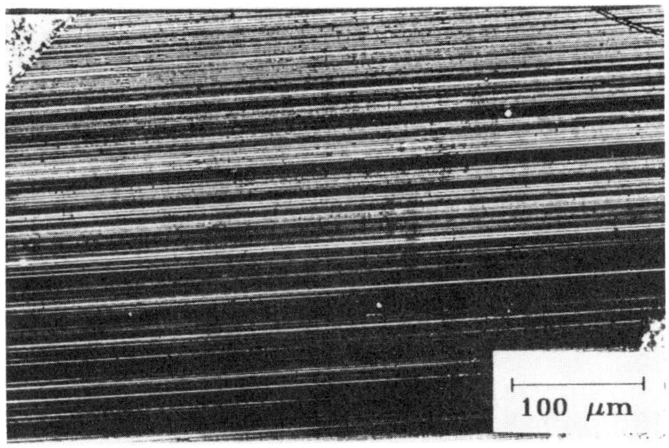

Figure 1: Polarized light micrograph of the single crystal.

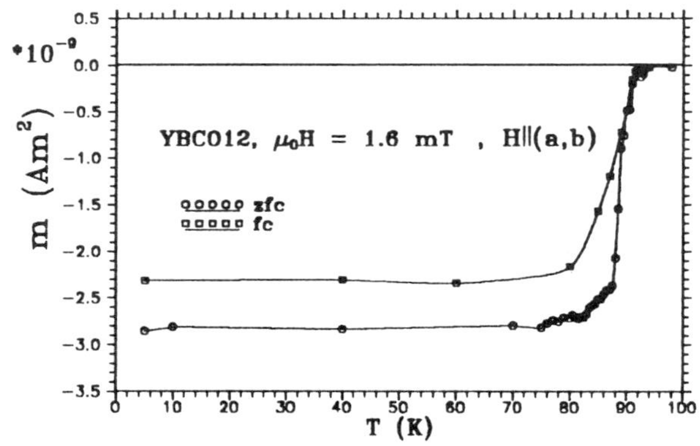

Figure 2: DC transition curves in a field of 1.6 mT.

Figure 3 shows the experimental results (magnetization versus the angle between the applied field and the TB direction in the (a,b)-plane for various temperatures. It can be seen clearly, that a distinct angular dependence of the magnetization

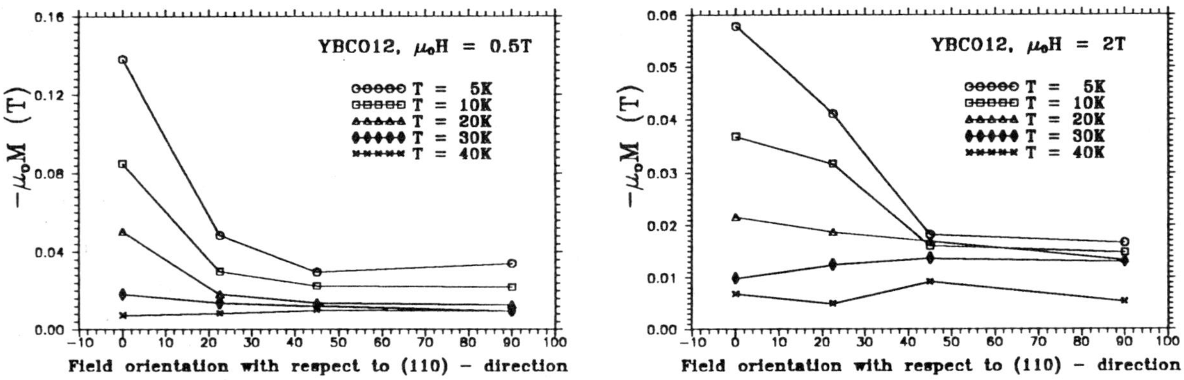

Figure 3: Magnetization at fixed magnetic field versus rotation angle ϑ in the (ab)-plane.

prevails at lower temperatures, whereas the effect disappears at higher temperatures. A pronounced maximum occurs for $\vartheta=0$ (H$\|$TB), with a width of 20 degrees in both directions at low fields, and even broader angular distributions at higher fields. This behaviour cannot be reproduced by the simple model used by Swartzendruber (1990) and by Gyorgy (1990), where a picture of undistorted flux lines rotated with respect to the TB's has been used. In this model, pinning is assumed to be caused by a core pinning mechanism and the change of M with rotation angle ϑ is related to a change in the intersecting volumes of TB's and flux lines. For the present density of TB's, this model would yields no angular variation $M(\vartheta)$ for $\vartheta \geq 0.2°$. Evaluations of a new model including flux line lattice distortions are currently being made.

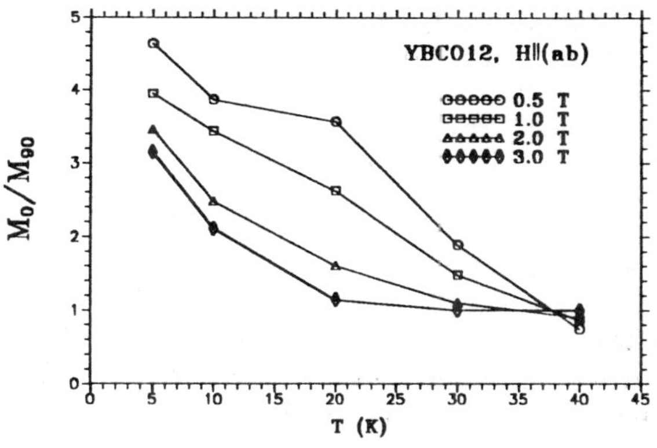

Figure 4: Ratio of the magnetizations, measured parallel and perpendicular to the TB's, respectively, as a function of temperature for different external fields.

As shown in Fig.4, the enhancement of magnetization M_0/M_{90} (field parallel and perpendicular to the twin boundaries) decreases to 1 with increasing temperature. We propose that flux creep prevents us from detecting variations at higher temperatures, since the measuring procedure of our magnetometer involves a waiting time of 11 minutes after any field change, thus leading to an almost complete decay of the low energy deformation of the FLL.

In summary, we have presented measurements of the magnetization as a function of field direction with respect to TB-orientation at various temperatures. A maximum of the magnetization was found at an angle, where the flux lines are oriented parallel to the twin planes. This behaviour cannot be explained by currently used models.

References

Dinger T R, Dolan G J, Keane D, McGuire T R, Worthington T K, Yandrofski R M and Yeshurun Y 1989 *High Temperature Superconducting Compounds: Processing and Related Properties*, edited by S.F.Whang and A.DasGupta (The Metallurgical Society, Warrendale, PA), p.23

Dolan G J, Chandrashekar G V, Dinger T R, Feild C and Holtzberg F 1989, *Phys. Rev. Lett.* **62**, 827

Gyorgy E M, van Dover R B, Schneemeyer L F, White A E, O'Brien H M, Felder R J, Waszczak J V, Rhodes W W and Hellman F 1989, *Appl. Phys. Lett.* **56**, 2465

Kaiser P L, Holzberg F, Chisholm M F and Worthington T K 1987, *J.Cryst. Growth* **85**, 593

Kwok W K, Welp U, Crabtree G W, Vandervoort K G, Hulscher R and Liu J Z 1989, *Phys. Rev. Lett.* **64**, 966

Swartzendruber L J, Roitburd A, Kaiser D L, Gayle F W and Bennett L H 1990, *Phys. Rev. Lett.* **64**, 483

Welp U, Kwok W K, Crabtree G W, Vandervoort K G and Liu J Z 1990, *Appl. Phys. Lett.* **57**, 84

This work was supported in part by Fonds zur Förderung der Wissenschaftlichen Forschung, Wien, under contract # 6873, and by the U.S. Department of Energy, Basic Energy Sciences - Material Sciences, under contract # W-31-109-ENG-38.

Supercond. Sci. Technol. **4** (1991) S262–264. Printed in the UK

Sequential deposition of YBaCuO thin film on sapphire substrates.

N. Hass, U. Dai and G. Deutscher,

Raymond and Beverly Sackler Faculty of Exact Science
School of Physics and Astronomy, Tel Aviv University
69978 Ramat Aviv, Israel.

Abstract. Using a sequential deposition method, thin superconducting YBaCuO films were prepared on sapphire substrates. Our technique is based on the oxidation of a pre-deposited Cu film at 400°C in air, followed by the deposition of a Y_2O_3 layer and a BaF_2 layer. After post annealed at 880°C, the films are superconducting with an onset temperature of Tc(onset)=93K and zero resistance at Tc_0=89K. The films are easily patterned, by applying conventional photolithographic method to the Cu film.

1. Introduction

A variety of methodes has been developed in the past 3 years for the preparation of high temperature superconducting (HTC) thin films. Most of the methods are based on sputtering or laser ablation tachniques.
In this work, the sequential deposition technique previously reported by Hess et al (1988) is used and improved. We demonstrate the possibility of preparing very reproducibly, superconducting thin YBaCuO films on Sapphire substrates, with zero resistance at Tc_0=89K. This result can be very important for microwave electronics applications, since Sapphire is a material widely used in that field.

2. Films preparation

The films are prepared by first depositing a very thin (300A) buffer layer of ZrO_2 on top of a R-plane (1102) sapphire substrate, by a slow evaporation (0.5 A sec) of metallic Zr in 510⁻5 Torr of O_2. The substrate is held at temperature of 350°C during this stage, resulting in a transparent insulating film. The substrate is cooled down to room temperature, and a metallic copper is deposited using E-Beam gun. The film is then taken out of the vacuum and the copper is oxidized for 1 hour at 400°C in air. We believe that this oxidation stage is crucial to the success of this technique. The films are then put back into the vacuum system chamber for the deposition of the Y_2O_3 and BaF_2 layers, using a E-beam gun and a resistive boat respectively. The Y_2O_3 is prepared by slowly evaporating (0.5 A sec) mettalic Y, resulting in a transparent layer, that is Yttrium oxide. A powder of BaF_2 is evaporated, and due to the high stability of this compound, we get a trasparent layer of BaF_2. Silver pads are also evaporated through a mask, in order to provide good electrical contacts to the films. This method of preparing contacts to the films is an improvement to the method suggested by Tessler et al (1988), and has the benefit of maximum simplicity. The post annealing is done at 880°C for 30 minutes in flowing O_2; in the first 15 minutes we use wet-O_2 by passing the O_2 through heated water to help the removal of the F_2 from the films. The high temperature annealing is then followed by a slow cooling to room temperature.

3. Lithography

The films are easily patterned by applying conventional photolithographic method to the Copper film. This stage takes place before the Cu layer is oxidized. After the patterning is finished, the preparation of the film is continued as described above (Sec. 2). The result is a patterned superconducting film, with trasparent material in the areas where the Cu was

0953-2048/91/01S262 + 03 $03.50

removed by the lithography process. This method is simpler than other methods where one patterns the YBaCuO film.

4. Measurements and results

X-ray diffraction spectra indicates that the films are polycrystalline and partially oriented with the c-axis normal to the substrates plane. Other phases than the "123" one, were not detected. The polycrystalline nature of the films was also observed in the Scanning-Electron-Microscope (SEM) pictures, where grains of the order of several microns were seen.

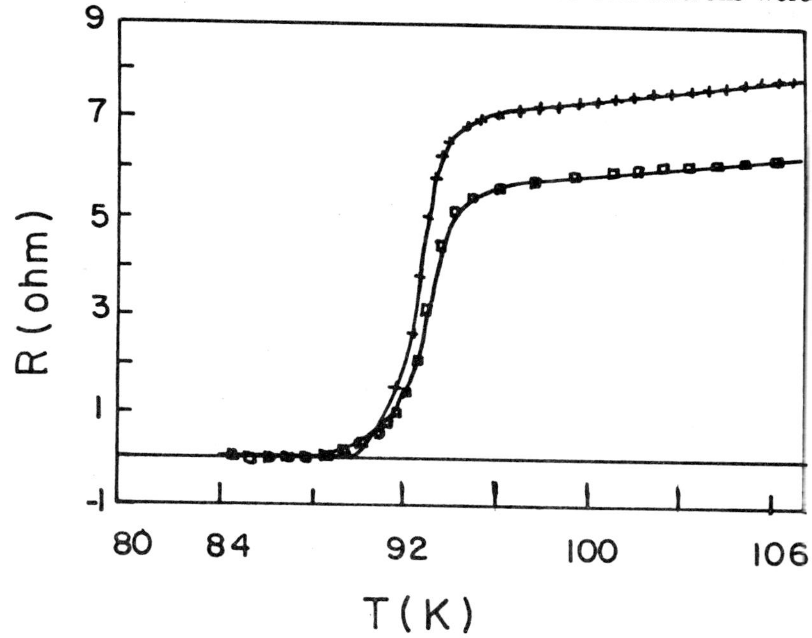

Fig 1: Resistance vs. Temperature
▫ A: Sapphire Substrate with 300A ZrO_2 buffer + B: YSZ Substrate

The resistance as a function of temperature of a 1 μm typical film, can be seen in Fig 1. The onset of superconductivity is observed at 93K while zero resistance is at Tc_0=89K, which is, to the best of our knowledge, the highest Tc reported on sapphire substrates. The ratio between the room temperature resistance and the one at 100K is about 1.6. This result is highly reproducible and is not sensitive to changes of few percents in the composition of the as-deposited films. The critical currents of the films were measured by a four probe DC method using the 1 μV cm criterion. The critical values were of the order of several thousands A cm^2 at 77K, which are typical values for polycrystalline films. The second plot in Fig. 1 is the temperature dependence of the resistance of a film prepared exactly in the method described above, but on yttrium stabilized zirconia (YSZ) substrate. It is noticed that both samples have similar temperature dependences R(T). The critical current densities as well as the morphology of the film on the YSZ substrate are quite similar to those of the films prepared on Sapphire.

5. Discussion

Our results show that it is possible to prepare thin YBaCuO films with non-reduced Tc_0, on sapphire substrates. This high Tc_0 is achieved despite of the fact that we use a post-annealing stage, at a temperature as high as 880°c. To the best of our knowledge, this Tc_0 (89K) is the highest reported on Sapphire substrates. We conjecture that this higher Tc_0 may be due to a reduced interdiffusion related to the sequential deposition that we have used.

We would like to emphasize that the other groups that succeeded in preparing YBaCuO films on sapphire, having Tc_0 > 80K, for example Char et al (1990) and Chromik et al (1990), were using In-situ methods, which benefit from the fact that the highest temperature used in the process (500°C-700°C) is much lower than the 880°c used in the preparation of post-annealed samples.

6. Summary

We prepared thin YBaCuO superconducting films on sapphire substrates with Tc(onset)=93K and Tc_0=89K, using the sequential deposition technique. Our method offers an easy way to pattern the films using conventional photolithography to the Cu film, before evaporating the other constituents of the 123 material (the Y and Ba).

References

Char K, Fork D K, Geballe T H, Laderman S S, Taber R C, Jacowitz R D, Bridges F, Connell G A N and Boyce J B 1990 Appl. Phys. Lett. 56 8

Chromik S, Hanic F, Adam R, Jergel M, Liday J, and Benacka S 1990 Appl. Phys. Lett. 56 22

Hess N, Tessler L R, Dai U, and Deutscher G 1988 Appl. Phys. Lett. 53 8

Tessler L R, Dai U, Hess N and Deutscher G 1988 J. Phys. D: Appl. Phys. 21 1652–1653

Supercond. Sci. Technol. **4** (1991) S265–267. Printed in the UK

As-Grown Superconductivity of Bi(Pb)SrCaCuO Thin Films Formed by Consecutively Sputtering Three Pb-Doped Targets of Slightly Different Compositions

Ken'ichi Kuroda, Kazuyoshi Kojima, Masami Tanioku, Kazuo Yokoyama* and Koichi Hamanaka

Central Research Laboratory, *Materials and Electronic Devices Laboratory, Mitsubishi Electric Corporation, Tsukaguchi, Amagasaki, Hyogo 661, Japan

Abstract. As-grown Bi(Pb)SrCaCuO thin films formed on MgO(100) substrates at various temperatures by RF magnetron sputtering from three Pb-doped targets: $Bi_3Pb_3Sr_2Ca_2Cu_3O_x$, $Bi_2Pb_2Sr_3Ca_3Cu_3O_x$ and $Bi_2Pb_2Sr_2Ca_2Cu_{4.5}O_x$. As-grown films formed at 660°C typically showed a T_c(onset) of 110K and a T_c(zero) of 78K. The T_c(zero) value was higher than that of the film formed from nondoped targets. However, using the Pb-doped targets did not enhance the formation of the high-T_c phase.

§1. Introduction

It is essential to obtain superconductivity in an as-grown state for application in microelectronics. As-grown BiSrCaCuO thin films have been obtained(Fukutomi *et al* 1988, Asano *et al* 1988), but the T_c values of the films were rather lower than those of bulk samples and annealed films. In our previous papers(Kuroda *et al* 1989), we also reported the results of superconducting properties for as-grown BiSrCaCuO thin films formed by RF magnetron sputtering. The films had onset temperatures above 100K, but zero-resistivity temperature was as low as 75K. The main reason for the low T_c(zero) value lies in the difficulty in forming a high-T_c phase (2223 phase) of 110K with good crystallinity in an as-grown state. As is well known, the addition of Pb enhances the formation of the 2223 phase. Therefore, we attempted to form as-grown Bi(Pb)SrCaCuO thin films using targets lightly doped with Pb (Pb/Bi=0.25)(Kuroda *et al* 1990). However, no tace of Pb was detected in the film obtained at a substrate temperature above 600°C. The films formed below 600°C contained Pb elements, but did not show zero resistivity. Thus, we tried to form as-grown films using heavily Pb-doped targets(Pb/Bi=1.0) at various substrate temperatures. In this paper, we describe comparison of T_c values of the as-grown films formed from heavily Pb-doped targets and nondoped targets.

§2. Experimental

The Bi(Pb)SrCaCuO thin films were grown on MgO(100) substrates by RF magnetron sputtering from three targets: $Bi_3Pb_3Sr_2Ca_2Cu_3O_x$, $Bi_2Pb_2Sr_3Ca_3Cu_3O_x$ and $Bi_2Pb_2Sr_2Ca_2Cu_{4.5}O_x$. Each target had a

0953-2048/91/01S265 + 03 $03.50

50% excess of Bi, SrCa and Cu compared to the 2223 phase, respectively, and had a Pb/Bi value of 1.0. The three targets were simultaneously discharged and the substrate was rotated alternately over each target. The substrate was kept directly over the BiPb-rich target for 50 sec, the SrCa-rich target for 20 sec and Cu-rich target for 30 sec per a rotation. The substrates were heated at 540-660°C during the growth. The sputtering gas was neat O_2, and the sputtering pressure was 300mTorr. Discharge power was 100W, and the gap between the target and the substrate was 50mm. The sputtering time was 120min, in which approximately 100nm-thick films were deposited. After deposition, the films were quickly cooled down to room temperature by introducing air into the chamber.

The crystal structure was analyzed by X-ray diffraction(XRD) measurement, and the temperature dependence of resistivity was measured by AC four-probe method using a lock-in amplifier. The chemical composition of the metallic elements in the film was determined by inductively coupled plasma emission spectroscopy(ICP).

§3. Results and Discussion

Figure 1 shows the chemical compositions of the films at various substrate temperatures(T_{sub}). The value of 2(Sr+Ca)/Bi showed a tendency to decrease with a decreasing T_{sub}. Pb was detected in the film when the T_{sub} was lower than 590°C. XRD measurement showed that the films formed at 600-660°C had a 2223/2212 superlattice phase with *c*-axis length of 34Å, as shown in Fig.2. On the other hand, the films formed below 590°C, contained a 2201 phase with *c*-axis length of 24Å and did not contain a 2223 phase with *c*-axis length of 37Å. That is, the films containing Pb elements had no 2223 phase.

Figure 3 shows the temperature dependence of resistivity of the film formed at 660°C, and Fig.4 shows T_c values at various T_{sub}. In Fig.4, the film formed at 660°C from the Pb-doped targets showed zero resistivity at 78K which is higher by 3K than the T_c(zero) of the film formed from Pb-nondoped targets

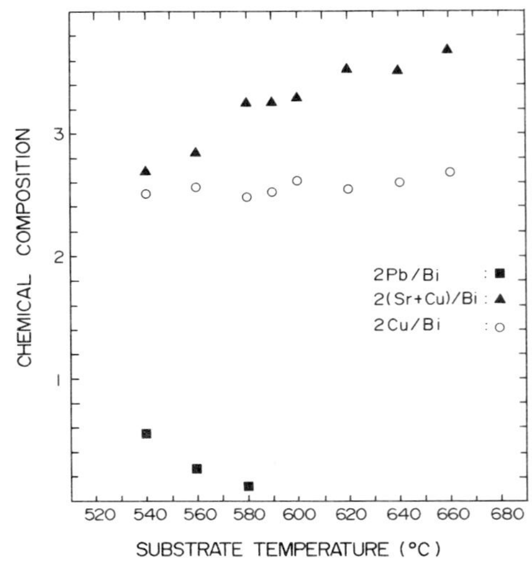

Fig.1 Relation between composition and substrate temperature.

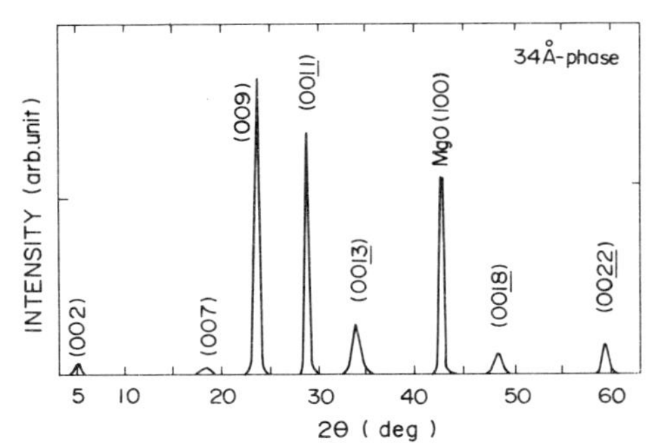

Fig.2 XRD patterns of as-grown film formed at 660°C.

under the same sputtering conditions; the film compositions of the films formed from the Pb-doped and nondoped targets (Bi:Sr:Ca:Cu=3:2:2:3, 2:3:3:3 and 2:2:2:4.5 targets) were Bi:Sr+Ca:Cu=2.0:3.69:2.69 and 2.0:3.56:2.70, respectively. As shown in Fig.4, T_c(zero) values of the films became lower with a decrease in T_{sub}. The film formed at 580°C had a T_c(zero) of 30K; the film composition was Bi:Pb:Sr+Ca:Cu=2.0:0.12:3.25:2.48. On the other hand, the film formed at the same sputtering conditions from nondoped targets showed zero resistivity at 29K; the film had about the same composition except for Pb elements (Bi:Sr+Ca:Cu=2.0:3.24:2.37). From these results, it is supposed that the Pb elements in the films do not influence the T_c value.

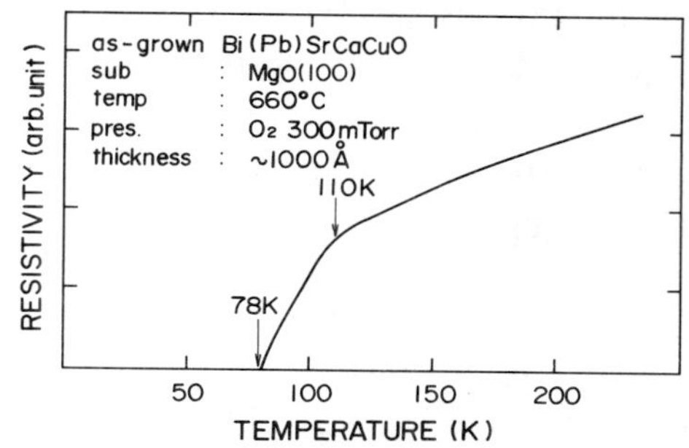

Fig.3 Resistivity vs temperature curve of as-grown film formed at 660°C.

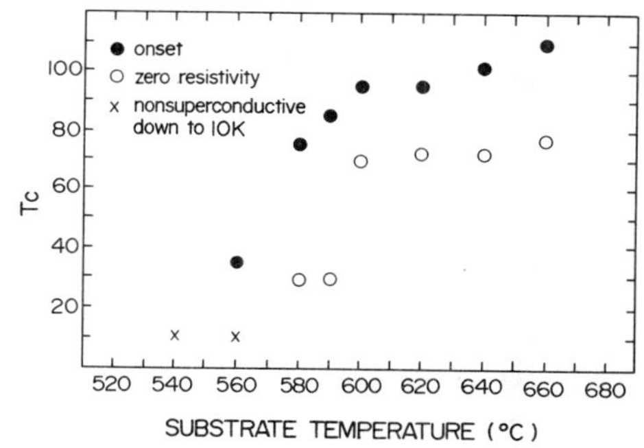

Fig.4 Relation between T_c and substrate temperature of as-grown films.

§4. Conclusions

The film formed at 660°C from Pb-doped targets showed higher T_c(zero) by 3K than that of the film formed under the same sputtering conditions from nondoped targets. However, the films containing Pb elements had no high-T_c phase and did not show discernibly higher T_c(zero) than the film formed from nondoped targets.

References

Asano H, Asahi M, Katoh K and Michikami O 1988 *Jpn.J.Appl.Phys.* **27** L1487

Fukutomi N, Machida J, Tanaka Y, Asano T, Yamamoto T and Maeda H 1988 *Jpn.J.Appl.Phys.* **27** L1484

Kuroda K, Kojima K, Tanioku M, Yokoyama K and Hamanaka K 1989 *Jpn.J.Appl.Phys.* **28** 1586

Kuroda K, Kojima K, Tanioku M, Yokoyama K and Hamanaka K 1990 *Jpn.J.Appl.Phys.* **29** L291

Supercond. Sci. Technol. 4 (1991) S268–270. Printed in the UK

Laser Ablated YBCO Films on Buffer Layers of YSZ and $SrTiO_3$ Deposited on Sapphire and MgO Substrates

M. Schieber*o, Y. Ariel **o, M. Schwartz **, M. Levinsky **, Sh. Shukrun M. Maharizy **, B. L. Zhou *, and S. C. Han *

* Xsirius Superconductivity Inc., 7590 E. Gray Road, Scottsdale, Az. 85260, USA

** Xsirius Superconductivity Materials, Israel. Har Hahotzvim, Kiryat Mada 4, P.O.B. 23566, Jerusalem 91230, Israel.

o School of Applied Science, Hebrew University, Jerusalum, Israel.

Abstract. Thin films of $YBa_2Cu_3O_{7-x}$ (YBCO) were deposited by laser ablation on substrates of sapphire and MgO coated with buffer layers of YSZ (Y_2O_3 stabilized ZrO_2) and $SrTiO_3$ deposited by both OMCVD, organometallic chemical vapor deposition, and laser ablation. The Tc (R=0) of YBCO on YSZ was 91K and on $SrTiO_3$ was 93K. The Jc at 77K were $5x10^3$ and $5x10^6$ A/cm^2 for YSZ and $SrTiO_3$ buffer layers respectively.

1. Introduction

Although recent results [1,2] indicate that under certain conditions YBCO can be deposited directly on sapphire with a Tc of 87°K [2], most published data show much poorer results. Mg0 is also a good substrate for high Tc superconducting (HTSC) thin films. However, the large lattice mismatch reduces both Tc and Jc. The highest Tc on Mg0 reported so far was 88 K by laser ablation [3] and 83 K by OMCVD [4]. YSZ (ZrO_2 with =8%Y_2O_3) has been successfully used as a substrate for YBCO producing films with Tc values which were higher than for bare sapphire. For example, T_c (R=O) values have been reported which are as high as 85-88K for YBCO films deposited on (100) YSZ by laser ablation [5, 6], 87K by spray pyrolysis [7], and 50-80K by OMCVD [8, 9].

ZrO_2 and YSZ have been deposited as buffer layers for YBCO on Si and sapphire with a marked improvement in the Tc [9]. In addition, layers of ZrO_2 have been deposited by OMCVD as the insulating oxide on Si in MIS structures [10,11]. However, this yielded polycrystalline, monoclinic ZrO_2. Some preliminary results for YBCO deposited on a buffer layer will be published elsewhere [12]. $SrTiO_3$ is a substrate which so far has produced YBCO films which have given the highest values of Tc and Jc by both laser ablation [1] and OMCVD [1]. However, it is less available in large sizes than MgO and it is also more expensive.

The present paper will describe Tc and some of the Jc values in YBCO produced on the YSZ and $SrTiO_3$ buffer layers. Most of the YBCO films were produced by laser ablation and some by OMCVD.

2. Experimental.

2.1 OMCVD of YSZ, $SrTiO_3$ and YBCO

A quartz, hot wall type reactor with separate furnaces and gas mass flow controllers for each source was used for the OMCVD of YSZ on sapphire [1]. Zr and Y complexes of 2,2,6,6 - tetramethyl - 3,5 - heptanedione (THD) were used as the source materials and were prepared according to the literature [1,10]. The nominal composition was 50 atomic percent Y. The substrate was (1102) sapphire. The $Zr(THD)_4$ source was held at a temperature of 160°C and the $Y(THD)_3$ source was at 175°C. The Ar flow for the $Zr(THD)_4$ was 300 ml/min and was 500 ml/min for the $Y(THD)_3$ source. An oxygen flow of 500 ml/min was added downstream from the individual source entrances. The nominal pressure was a few Torr. Deposition was done for one hour. After deposition, the samples were given an air anneal at

0953-2048/91/01S268 + 03 $03.50

1200°C for one hour. The same hot wall reactor used for YSZ was used also for $SrTiO_3$. The precursor was a Sr complex of 2, 2, 6, 6 - tetramethyl - 3, 5 - heptandeione, $Sr(THD)_2$. It was prepared and purified in our own laboratory. Tetraisopropyl orthotitanate (Ti $(TIP)_4$) was used as the Ti source and was obtained from Fluka Chemicals. The $Sr(THD)_2$ and the $Ti(TIP)_4$ sources were held at 230°C and 52°C respectively; the Ar flows were 500 ml/min for each source. An oxygen flow of 500 ml/min was added down stream from the individual source entrances and the nominal pressure was a few Torr. The deposition time was one hour at a substrate temperature of 1080°C. A similar hot wall reactor was used for YBCO deposition. The precursors were Y(THD)3, Ba(THD), and Cu(THD) kept at 175, 220, and 110K respectively. The Ar flow was 500, 400 and 300 ml/min for the different precursors respectively. The oxygen flow of 500 ml/min was added downstream and the nominal pressure was a few Torr. The YBCO films were then slowly cooled to room temperature.

2.2 Laser ablation of YBCO and of $SrTiO_3$ on MgO

YBCO was deposited by pulsed laser ablation [5] in an experimental set-up described elsewhere [1]. A 248 nm excimer laser with pulses of -4 J/cm^2 fluence on the target at 4.5 Hz laser repetition rate was used for the deposition. The substrate was glued with silver paint to the heated holder and had a surface temperature of about 750°C. Deposition was carried out in 200 mTorr at a rate of 0.6 A per pulse. The film thickness was 0.30 - 0.35 micron. In-situ oxygen annealing was done by adding 1 atm of 0_2 gas at the end of the deposition process and cooling down slowly with a maintenance time of 30 min at 400°C. The substrates used were (1102) sapphire covered with a buffer layer of YSZ, (100) MgO covered with a buffer layer of $SrTiO_3$ and bare (100) MgO. The thickness of the YBCO films was between 0.3 and 0.8 um. A similar procedure as described for YBCO has been used for laser ablation on (100) MgO substrates. The target was a sintered ceramic stoichiometric $SrTiO_3$ pellet, 20 mm diameter and 2 mm thick.

3. Results and Discussions

The buffer layers of YSZ on sapphire and of $SrTiO_3$ on MgO yielded transparent films. The nominal composition of the YSZ was $Y_{0.5}\,Zr_{0.5}\,0_2$ and for $SrTiO_3$ was the stoichiometric one. The x-ray diffraction, XRD patterns showed only the (100) peaks of YSZ with a small amount of Y_20_3. The lattice constant calculated from the (100) reflections is 5.24 A. which is slightly larger than the lattice parameter of 5.14 A found for $Y_{0.15}\,Zr_{0.55}\,0_{1.93}$. (JCPDS card 30-1468). This is consistent with the larger amounts of Y in our samples [12]. The XRD patterns of the YBCO/YSZ sapphire are published elsewhere [12]. The OMCVD films of $SrTi0_3$ showed only the (100) reflections of $SrTiO_3$ and those of MgO. Similar XRD results were obtained also with the laser ablated films of $SrTiO_3$ on MgO. It is interesting to note that after deposition of YBCO by either laser ablation or OMCVD on the $SrTiO_3$ buffer layer, only the (001) reflections of YBCO and (100) reflections of MgO are still visible as shown in Fig. 1. The complete coverage of the $SrTiO_3$ reflections by the YBCO proves the excellent crystallographic lattice match between these two compounds.

TABLE 1

Values of Tc and Jc of several YBCO films

Substrate	Buffer Layer	Deposition Method Buffer	Deposition Method YBCO	Tc (R=0)	Jc (A/cm^2) at 77K
(1102) Sapphire	YSZ	OMCVD	Laser	91K	5×10^3
(100) MgO		---	OMCVD	84K	10^4
(100) MgO	SrTiO3	OMCVD	OMCVD	88K	5×10^4
(100) MgO		---	Laser	89K	10^6
(100) MgO	SrTiO3	Laser	Laser	93K	5×10^6

The values of Tc (R=o) and Jc (A/cm^2) at 77K of some YBCO films are shown in Table 1. One can see that the Tc (R=O) of laser ablated YBCO on OMCVD deposited buffer layers of YSZ are among the highest ever reported. Also, laser ablated YBCO on lasser ablated $SrTiO_3$ has a very high Tc. The other combinations shown in Table 1 have Tc below 90 K. The R-T curves for YBCO on YSZ and $SrTiO_3$ buffer layers are shown in Fig. 2.

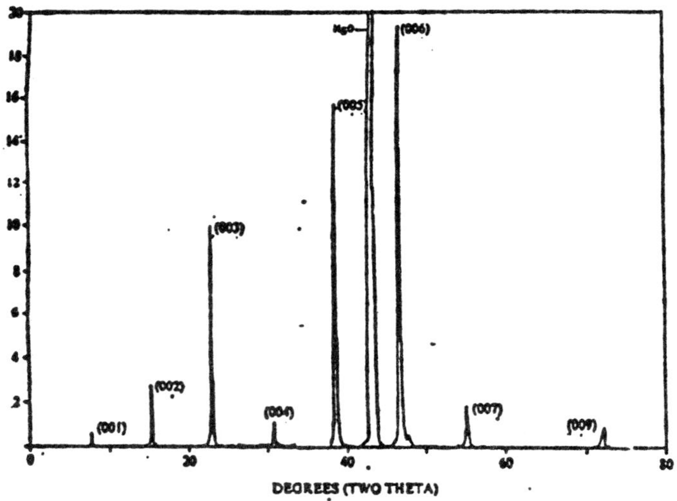

Fig 1. XRD of a laser ablated film of YBCO deposited on a buffer layer of laser ablated $SrTiO_3$on (100) MgO. The $SrTiO_3$ reflections are not seen being covered by YBCO.

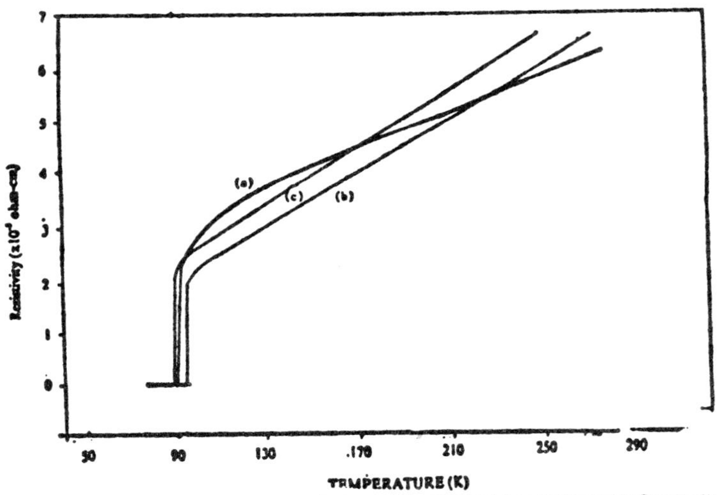

Fig. 2 Resistivity vs. temperature for laser ablated YBCO deposited on a OMCVD buffer layer of YSZ on (1102) sapphire (b) laser ablated $SrTiO_3$ buffer layer on (100) MgO and (c) bare (100) MgO substrate.

The Jc values at 77K of YBCO on YSZ are low, but these can be explained by the unwanted layer of $BaZrO_3$ which is formed by the reaction between YBCO and ZrO_2. The other Jc values are among the higher reported.

<u>REFERENCES</u>

1. See an extensive review on these films of HTSC by physical and chemical methods by Schieber, M., 1991, Proc. 8th American Conference on Crystal Growth, Vail, Co., July 1990, J. of Crystal Growth, to be published.
2. Char, K., Fork, D. K., Geballe, T. H., Laderman, S. S., Taber, R. C., Jacowitz, R. D., Bridges, F., Connel, G. A. N., Boyce, J. B., 1990, Appl. Phys. Lett. 56. 785
3. Kingston, J. J., Wellstood, F. C., Lerch, P., Miklich, A. N., Clarke, J., 1990, Appl. Phys. Lett., 56, 683
4. Lackey, W. J., Carter, W. B., Hanigofsky, J. A., Hill, D. M., Barefield, E. K., Shapiro, M. J., Thompson, J. r., Green, A. J., Moss, T. S., Jake, R. A., Efferson, K. R., 1990, Appl. Phys. Lett. 56, 1175
5. Koren, G., Polturak, E., Fisher, B., Cohen, D., Kimel, G., 1988 Appl. Phys. Lett. 53. 2330
6. Singh, A. K., Ganapathi, L., Tiwari, P., Narayan, J., 1989, Appl. Phys, Lett. 55. 2351
7. Gupta., A., Koren, G., Giess, E. A., Moore, N. R., O'Sullivan, E. J. M., Cooper, E. I., 1988, Appl. Phys. Lett. 52. 163
8. Yamane, H., Kurosawa, H., Iwasake, H., Masumoto, H., Hirai, T., Kobaysashi, N., Nuto, Y., 1988, Japan J. Appl. Phys. 27. 1275
9. Migliuolo, M., Stamper, A. K., Greve, D. W., Schlesinger, T. E., 1989, Appl. Phys. Lett. 54. 859
10. Balog, M., Schieber, M., Michman, M., Patai, S. 1977, Thin Solid Films 41; 247
11. Balog, M., Schieber, M., Michman, M., Patai, S., 1979, J. Electrochem. Soc. 126. 1203
12. Schieber, M., Schwartz, M., Koren, G., Aharoni, E., Appl. Phys. Lett., submitted

Supercond. Sci. Technol. 4 (1991) S271–273. Printed in the UK

HIGH TRANSPORT AND MAGNETIC CRITICAL CURRENT OF BULK YBaCuO *

Shiying Ding, Zheng Yu, Jialie Yan, Yitong Chang, Lou Mo, Xin Jing, Hongtao Ren[+], Qing He[+], Ling Xia[+], Rukun Wang[+]

Physics Department and Solid Microstructure Lab., Nanjing University, Nanjing 210008, PR China
[+]General Research Institute for Non-Ferrous Metals, Beijing 100088, PR China

Abstract. High transport and magnetic Jc of 3×10^4 A/cm^2 have been detected at 77K and various applied fields for $YBa_2Cu_3O_y$ (YBCO) synthesized by an improved melt-textured growth(MTG) process. In order to investigate the mechanism responsible for the enhanced high Jc, the hysteresis at different field-sample orientations have also been measured. It is argued that the high Jc of MTG YBCO results mainly from the reduction of weak links in the sample.

1. Introduction

One of the most important parameters for potential technical application of high temperature superconductor(HTS) is bulk transport critical current density at applied field Jc(H) at 77K. It has been recognized that the weak links in ceramic HTS is the most harmful for getting high Jc of HTS materials. However, significant improvement of Jc of bulk YBCO sample has been made (Jin et al. 1987 1988a 1988b, Salama 1989, Murakami 1989). Very recently, Ren et al.(Ren 1990) has obtained a Jc higher than 2.3×10^4 A/cm^2 in 20 kOe and 77K by the MTG process. It is said that 211 phase is responsible for the high Jc of bulk YBCO (Murakami et al. 1989). In this contribution, we report the transport and magnetic Jc in low and middle field and the influences of field-sample orientation on the magnetic Jc. The mechanism responsible for the enhanced Jc is discussed too.

2. Experiment

The sample prepared by the MTG process which has been described elsewhere (Ren 1990) is a flat plate of $3.75 \times 1 \times 0.5$ mm^3 for magnetic measurement. The other specimen cut from the same bulk is rectangular with cross section smaller than 0.1 mm^2 for transport current measurement. X-ray diffraction pattern shows only the {001} lines. SEM photographs, which was taken from the fracture surface of the sample, show the sample has dense layer-structure with regular alignment of all a-b plane parallel each other and the crystals are plate like with the length of 1 - 2.5 cm. The details of microstructure will be reported elsewhere.

---

* This work is supported by the National Center for Research and Development on Superconductivity.

0953-2048/91/01S271 + 03 $03.50

Transport Jc, which is defined by a criteria of 1 μV/cm electric field strength, was measured by the standard four terminal method in applied magnetic fields whose strength is between 0 - 2 kOe. The magnetic Jc is estimated according to the hysteresis loops which were detected by a vibrating sample magnetometer with maximum of field of 20 kOe. All measurements were carried out in LN_2 (77K).

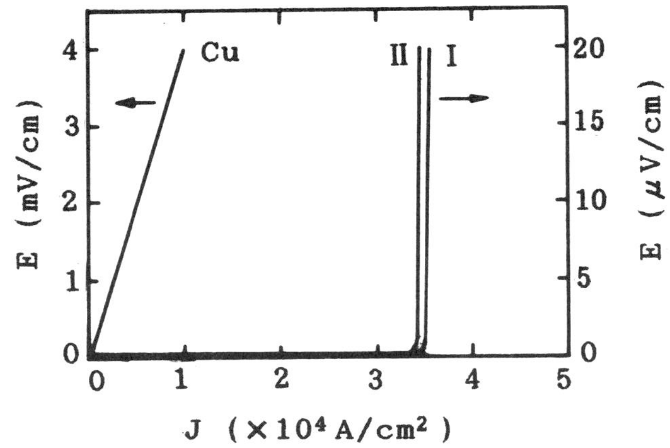

Fig.1 E-J curves of bulk samples at 77K. H // a-b plane.
Cu: a copper bar;
I : the MTG YBCO, H=0;
II: the MTG YBCO, H=800 Oe,

3. Result and Discussion

Fig.1 illustrates three bulk DC E-J curves respectively for a copper bar (Cu curve), the MTG YBCO sample at zero field (curve I) and 800 Oe (curve II). The figure shows a very high Jc// of 3.5×10^4 A/cm^2 at zero field and 77K, and Jc// of 3.4×10^4 A/cm^2 at H of 800 Oe normal to the current, where Jc// represents that the Jc is in a-b plane. These data show a quite different Jc(H) from those of sintered YBCO bulk sample whose Jc is low and sensitive to a small fields, say 50 Oe. Therefore, these data are evidences that the Jc of the MTG sample is not governed by the weak links.

Fig.2 displays the hysteresis loops of the MTG sample at three field-sample orientations shown by the insert where a_1=3.75 mm, a_2=1 mm and a_3=0.5 mm are sample's length, width and thickness respectively. The face of a_1-a_2 is parallel to the crystal a-b plane and a_3 is along the c-axis. While H is parallel to the c-axis the loop I was obtained in fig.2. At assumption without weak links and on the Bean model, the Jc// are estimated by (Yu et al.)

$$\Delta M_1 = 0.5\ Jc_{//}\ a_2(1-a_2/2a_1) \qquad (1)$$

where $\Delta M_1 = M^+ - M^-$, M^+ and M^- are magnetizations at increasing and decreasing applied fields respectively. According the ΔM data of curve I in fig.2 and equ.(1), Jc// = 3.2×10^4 A/cm^2 at H=0 and Jc// = 1.8×10^4 A/cm^2 at H=10 kOe.

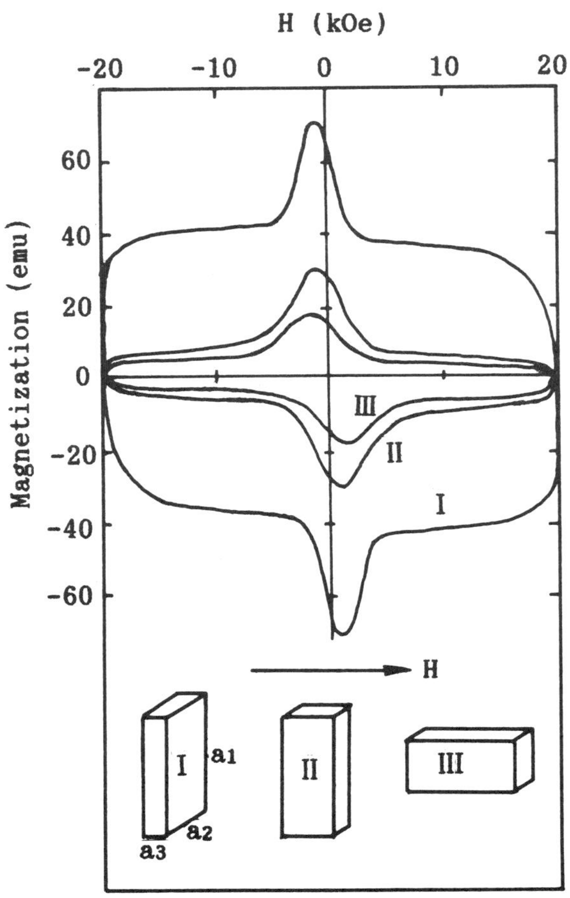

Fig.2 Hysteresis of MTG sample for three field-sample orientations at 77K.

The fact that the transport $Jc_{//}$ is almost same as the magnetic $Jc_{//}$ at H=0 is a strong confirmation that the weak links are almost eliminated for our MTG sample at very low fields. For curve II in fig.2 the magnetic Jc has two values: $Jc_{//}$ and $Jc_{\perp}$ which is normal to the a-b plane. In this case the ΔM is (Yu et al.)

$$\Delta M_2=0.5Jc_{//}\,a_3 - Jc_{//}{}^2\,a_3{}^2/(4Jc_{\perp}\,a_2) \qquad (2)$$

for the same assumption of without weak links. For the curve III in fig.2, in the same way we use (Yu et al.)

$$\Delta M_3=0.5Jc_{//}\,a_3 - Jc_{//}{}^2\,a_3{}^2/(4Jc_{\perp}\,a_1) \qquad (3)$$

We are able to estimate $Jc_{\perp}$ from equ.(2) (or equ.(3)) using $Jc_{//}$ and ΔM_2 (or ΔM_3). This is not done for the limited space. Because $a_1>a_2$, the $\Delta M_3< \Delta M_2$ according to equ.(2) and equ(3). This is just the facts shown in fig.2. This agreement is another evidence that our sample consisted mostly of strong-linked grains at least $H \leqslant 20$ kOe. It is worth remembering that ΔM of any sintered powder YBCO sample with great number of weak links is rather small and independent of field-sample orientation and sample's shape.

Because transport Jc of textured YBCO thin film is well above 10^6 A/cm^2 at 77K and no 211 phase, which is said to be the flux pinning center responsible for the high Jc of MTG sample (MuraKami 1989), has been reported, the role of 211 phase in MTG YBCO material are not clear. In the other hand, we speculate the 211 phase in the high bulk Jc of MTG YBCO sample plays a very important role in significant reduction of harmful weak links.

In summary, we have measured the transport $Jc_{//}$ by DC E-J curves and magnetic $Jc_{//}$ and $Jc_{\perp}$ by hysteresis loops at different field-sample orientations. All these measurements show that the MTG samples have high Jc which may be explained by the significant reduction of weak links in the sample.

References

Murakami M, Morita M and Koyama N 1989 Jpn.J.Appl.Phys.Lett. **28** L1125

Jin S, Tiefel T H, Sherwood R C, Kammlott C W, and Zahurak S M, 1987 Appl. Phys. Lett. **51** 21

Jin S, Tiefel T H, Sherwood R C, Davis M E, Van Dover R B, Kammlott C W, Fastnacht R A and Keith H D 1988a Appl. Phys. Lett. **52** No. 24

Jin S, Tiefel T H, Sherwood R C, Van Dover R B, Davis M E, Kammlott C W, and Fastnacht R A 1988b Phys. Rev. B **37** No.13

Ren H, He Q, Xiao L, Wang L, Yu D, Cui C and Li S 1990 to be published in Cryogenics

Salama K, Selvmanicham V, Gao L and Sun K 1989 Appl. Phys. Lett. **54** 5

Yu Z, Ding S Y, to be published

Supercond. Sci. Technol. 4 (1991) S274–276. Printed in the UK

An Investigation of Inter– and Intragranular Effects in Superconducting YBCO Specimens Using AC Inductive Methods

R A Doyle and D S McLachlan
Department of Physics, University of the Witwatersrand, P O Wits, South Africa, 2050.

Abstract. The complex AC susceptibility [$\chi(T,b_o,B)$] for two cylindrical $YBa_2Cu_3O_{7-x}$ samples has been measured. The critical current densities for the inter– and intragranular components are extracted from these results and the volume fractions of the components are calculated using various models. The solvent used during milling of the starting powders seems to have a marked effect on both of these parameters.

1. Experimental Method

YBCO powder was prepared using the conventional solid state reaction method. The powder was separated into eight batches. Four of these were milled in ethanol for 1,2,4 and 8 hours producing powders with mean grain sizes of 6,3, 1.75 and 1.5 μm. The process was repeated for the other four batches but hexane was used as the milling agent. The ethanol– and hexane milled batches were then cold isostatically pressed into pellets and sintered together. The two samples studied in detail in this paper are ME2W2 (starting material 3μm powder milled in ethanol) and MH4W2 (starting material 1.5 μm powder milled in hexane). The final Archimedes densities are 88.4% of theoretical for ME2W2 and 97% for MH4W2. The microstructures were investigated using optical and scanning electron microscopy. Grain sizes of 5.7 μm and 2.1 μm were measured for ME2W2 and MH4W2 respectively.

The AC susceptibility of each samples was measured at 76 K using series opposed secondary coils, a ratiotransformer bridge and two lock–in amplifiers, one for χ' and χ'' and the other to track the phase of the field. The field (and temperature) sweeps and the data capture were all computer controlled.

2. Experimental Results

Figures 1 (a)–(d) show the AC field (b_o) dependence of both components of susceptibility for samples ME2W2 and MH4W2 at 76 K with DC bias fields (B) of 0,2,10 and 100 mT. These are similar to those previously observed by Gomory et al (1989).

The zero field sweeps clearly show the 'weak link' intergranular material breaking down at b_o's between 0.68 and 1.31 mT and 0.46 and 1.54 mT for samples ME2W2 and MH4W2 respectively. As can be seen from the figures the susceptibility at the shoulders of the 0 T and 10 mT curves are very nearly the same. This suggests that this is the susceptibility characterizing the distributed "good" granular material. The anomalous curve at 2 mT in Figures 1 (a) and (b) is probably due to B and b_o having approximately the same value on the shoulder. In a B of 10 mT one may think that only the intergranular material is in the critical state at 76 K, as B_{c1} (0 K) = 45 mT for the intragranular material (Renker et al 1987). At 100 mT both the inter– and intra grain components are in the critical state.

0953-2048/91/01S274 + 03 $03.50

Supercond. Sci. Technol. **4** (1991) 000–000. Printed in the UK

The analysis due to Campbell (1969) has been applied to the real part of the results at 2,10 and 100 mT. Plots of b_0 vs the normalised "effective" penetration depth, using the formula $x/R = 1 - [S/(b_0k)]^{1/2}$ are shown in Figure 2. (S is the measured signal and k a calibration constant.)

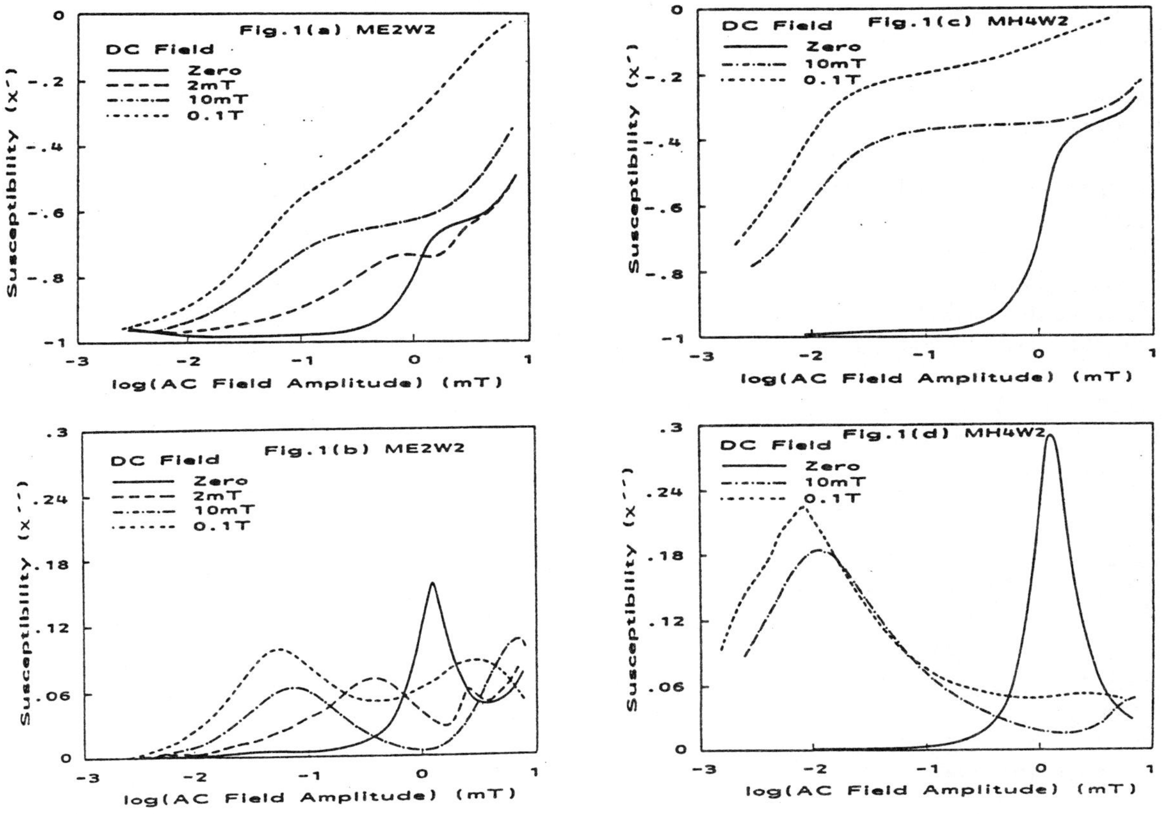

Figures 1(a)–(d): AC Field dependence of the real and imaginary parts of the complex susceptibility at 76 K.

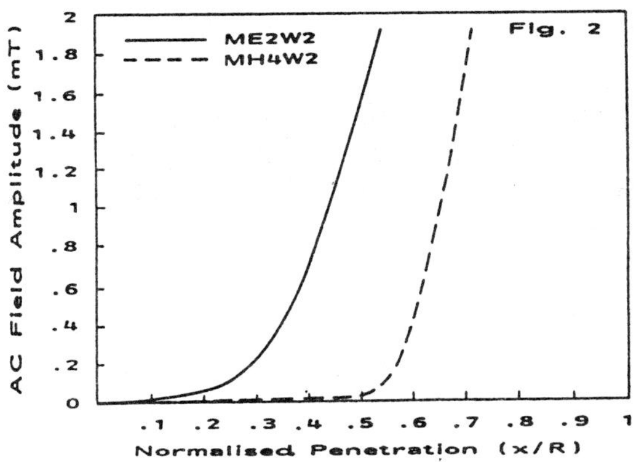

Figure 2: Effective flux profile for the two samples in the critical state at 76 K with a DC bias field of 0.1 T (100 mT)

Table I Sample	J_{cj} (A/cm²)	J_{cg} (A/cm²)	V_g(NP)	V_g(IC)	V_g(BS)	V_g(BAS)
ME2W2	14.9	2.9x10⁵	0.55	0.67	0.45	0.52
MH4W2	1.6	1.4x10⁶	0.20	0.37	0.25	0.27

3. Discussion

The inter– and intragranular critical currents at 100 mT are given in columns 1 and 2 of Table I respectively. These are obtained from the slopes of the near linear regions in the low and high field extremes in Figure 2 divided by the radius of the samples

($\simeq$ 1.5 mm) and the grain sizes respectively (ME2W2 20 μm and MH4W2 3 μm). As the high field (b_o) slopes in Figure **2** are reasonably similar, a major reason for the very different intergranular J_c's for samples ME2W2 and MH4W2 is the normalization with respect to the different grain sizes. The authors feel, as do Kupfer et al (1989) that the correct normalization length may be a multiple of the twin spacing. This could give more similar J_c values. The fact that the positions and magnitudes of the corresponding χ'' are somewhat similar (these have been used to estimate J_c (Matsushita and Ni 1989)) also supports our contention that the intragranular J_c's should be closer than given in Table I.

Column 3 of Table 1 shows the volume fraction of the good granular material (V_g) calculated from the normalised penetration depth (NP) (Kupfer et al 1988). Using the values of χ' from the shoulders at 0 and 10 mT, the V_g's for the cylindrical model, where $V_g = |\chi'|$ (IC), the Bruggeman symmetric (BS) and asymmetric (BAS) effective media equations (McLachlan 1989) are given in columns 4,5 and 6 respectively. As the magnetic properties of the cylinders are reasonably isotropic, a packing of an infinite range of superconducting and non–superconducting spherical grains is assumed in the symmetric case. In the asymmetric case the assumption of spherical superconducting grains completely coated by non–superconducting intergranular material is made.

Finally, the significant difference in the effect of AC and DC fields on the samples should be noted. Note for instance that for $B = 0$ and $b_o = 10$ mT, χ' is –0.5 and –0.3 for ME2W2 and MH4W2 respectively. For $b_o \approx 0$ and $B = 10$ mT the corresponding values are –0.97 and –0.80, showing far less flux penetration. The drop in $-\chi'$ above the shoulders in both samples at about 1.5 mT clearly shows that the AC flux penetrates the intragranular material for b_o above 2 mT and B below 10 mT.

4. Conclusions

The volume fraction of the grains calculated from the measurements depends on the model used. It is suggested that the application of the Campbell technique in these materials can be misleading since the correct size parameter is not obvious. The results of preliminary measurements on other samples support the suggestion here that the milling agent has an effect on the value and field dependence of the intergranular J_c and on the volume fraction of the intergranular material.

5. Acknowledgements

The authors would like to thank P Clayton, M Thomas and W Marais for preparing the samples, and P Clayton, T B Doyle, F R N Nabarro and M Thomas for helpful advice and discussions.

6. References

Campbell A M, 1969, *J Phys C (Solid State Phys)* Ser.2, **Vol.2** pp1492–1501.

Gomory F, Takacs S, Lobotka P, Frohlich K and Plechacek V, 1989, *Physica C*, **160**, pp1–7

Kupfer H, Apfelstedt L, Flukiger R, Keller C, Meier–Hirmer R, Runtsh B, Turowski A, Weich U and Wolf T, 1989, *Cryogenics*, **Vol.29** pp268–280.

Matsushita T and Ni B, 1989, *Jap J of Applied Physics* **Vol.28**, No 3 ppL419–L421.

McLachlan D S , 1989, *Solid State Comm.* **Vol 69**, No.9 pp.925–929.

Renker B, Apfelstedt I, Kupfer H, Politis C, Rietschel H, Schauer W and Wuhl H, 1987, *Proc. 18th Int. Conf. on Low Temperature Physics*, Kyoto pp.1169–1170.

Supercond. Sci. Technol. **4** (1991) S277–279. Printed in the UK

Evidence for Matthiessen's Rule in the Normal State Resistivity of Zinc Doped Yttrium Barium Copper Oxide.

J.R. Cooper, S.D. Obertelli, P.A. Freeman, D.N.Zheng, J.W. Loram and W.Y. Liang
IRC in Superconductivity, University of Cambridge, Madingley Road,
Cambridge CB3 OHE.

Abstract. We have measured the electrical resistivity $\rho(T)$ for a number of sintered samples of $Y_1Ba_2(Cu_{1-x}Zn_x)_3O_7$ with x in the range 0.01- 0.10. Above T_c, $d\rho/dT$ falls smoothly with increasing bulk density and is independent of x as confirmed by measurements on alloys of equal density. Taking $d\rho/dT$ to be independent of x, we find that the extrapolated residual resistivity (ρ_{res}) at $T = 0$ increases linearly with x in the range $0 \leq x \leq 0.07$. These facts suggest that the linear term in ρ arises from an intrinsic scattering mechanism which is independent of x, and that the substituting zinc atoms act as independent scattering centres, in accordance with Matthiessen's rule for normal metals.

1. Introduction

From neutron diffraction studies (Gang Xiao *et al* 1988, Mehbod *et al* 1988) and thermogravimetric analysis (Tarascon *et al* 1988), it is known that zinc substitutes at the Cu(2) 'plane' sites of yttrium barium copper oxide(YBCO). Doping YBCO with zinc causes a large depression of T_c, to the extent that for x = 0.1, T_c of $Y_1Ba_2(Cu_{1-x}Zn_x)O_7$ is around 10K (Fig.1). Although T_c is reduced drastically by doping with Zn, the linear term in the resistivity (ρ) is still present at x=0.07 (see for example Fig.3(b)). We were interested to see whether the magnitude of this term, or equivalently the slope $d\rho/dT$, changed on doping with Zn, and whether we could observe systematic changes in the extrapolated residual resistivity ρ_{res}.

2. Experimental

We have measured the resistivity of Zn doped samples from several preparation batches. They were all prepared by solid state reaction at gradually increasing temperatures with several intermediate grindings. Two sintering stages were used, each one in flowing oxygen for a period of 12 hours. After the first sinter at 935°C, the samples were pressed into a disc at a pressure of 3 - 4 tons cm^{-2} before the final sinter at 940°C. Two batches (1 and 3 in Fig.2) were checked by powder X-ray diffraction and were found to be single phase. Despite nominally identical preparation conditions, the samples had varying bulk densities (D) from 4.4g.cm^{-3} to within 5% of the maximum X-ray density of 6.4 g.cm^{-3}. Resistivity measurements were carried out by a standard four-probe Lock-in technique using currents in the range 10^{-4}-10^{-3}A.

3. Results and Discussion

Specific heat measurements on a series of Zn-doped YBCO samples have recently been reported (Loram *et al* 1990a , 1990b). The corresponding resistivity transtions for the same series are shown in Fig.1, from where it can be seen that the transitions remain sharp for doping levels up to x=0.07. We observe a resistive foot for zinc concentrations x=0 and x=0.01, which is present for all samples studied. This may well be an intrinsic property connected with the double anomaly in the specific heat (Loram *et al* 1990a, 1990b).

The interpretation of resistivity measurements on polycrystalline sintered samples is somewhat complicated. Because the current flows along paths of least resistance, the

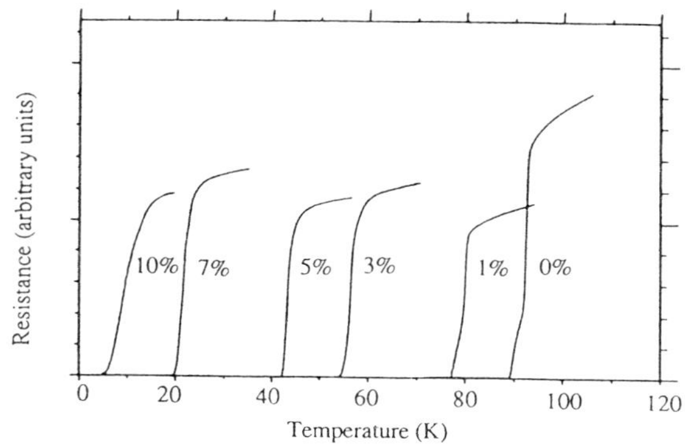

Fig.1 Resistive transitions in the alloy series $YBa_2(Cu_{1-x}Zn_x)_3O_7$ for zinc concentrations in the range x = 0 (0%) to x = 0.10 (10%).

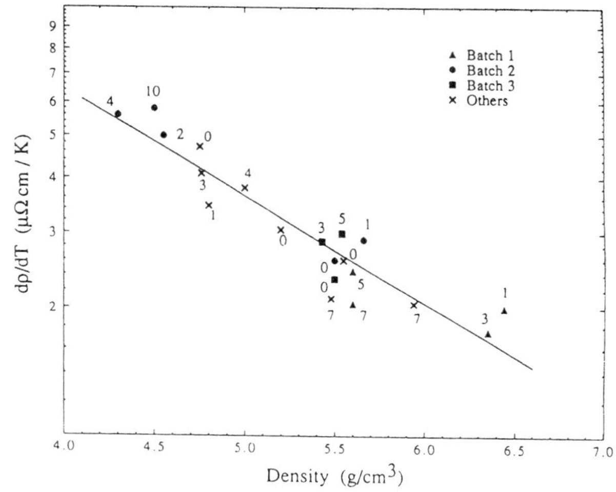

Fig.2 Semi-log plot of dρ/dT versus density for $YBa_2(Cu_{1-x}Zn_x)_3O_7$ samples studied. Label to plotting symbol denotes % concentration of zinc dopant.

resistivity of a sintered pellet is dictated by the in-plane resistivity ρ_{ab}. But for pure YBCO the magnitude of ρ has been found to depend on the bulk density or porosity of the sintered samples (Babić *et al* 1987). Another possible complication is that the process of pressing the powder into a disc and then sintering may lead to preferential grain orientation or texturing, with the c axis tending to lie perpendicular to the plane of the disc (Bonn *et al* 1988).

In Fig.2 we have made a semi-log plot of the values of dρ/dT against D for all samples studied, and find points to be scattered about a single straight line. This demonstrates that within the scatter of our data, which typically had errors of ±5% in ρ and D, dρ/dT is independent of x up to x=0.07. We have checked this directly by preparing another series of samples with x = 0, 0.03, 0.05, together under precisely the same conditions. These had identical densities to within 1% and exhibit closely parallel ρ(T) curves(see Fig.3(a)).

Accepting that dρ/dT was independent of concentration, we were then able to determine the relative values of ρ_{res} for all the alloys in Fig.2 by normalisation to a common value of dρ/dT. We chose to normalise to the value dρ/dT = 1.8 x 10^{-3} mΩ.cm.K^{-1}, obtained by linear extrapolation in Fig.2, which corresponds to the maximum X-ray density of 6.4 g.cm^{-3}. This is approximately three times the value of $d\rho_{ab}/dT$ found for single crystals (Ong *et al* 1988). Thus values of ρ^{ab}_{res} for single crystals may be expected to be three times smaller than those plotted in Fig.4. In Fig.4 one can see that ρ_{res} increases linearly with x up to x=0.07,

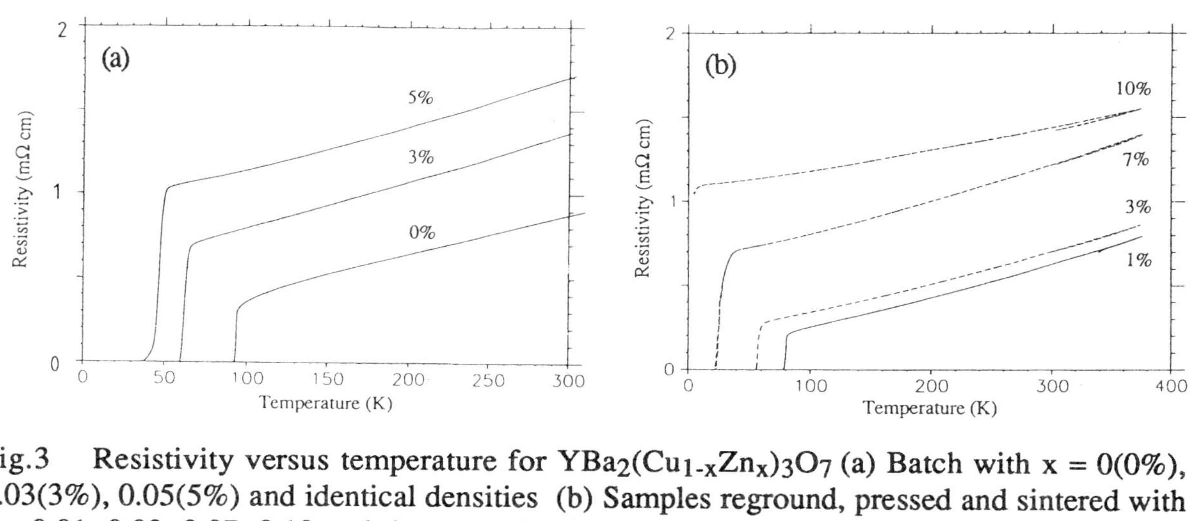

Fig.3 Resistivity versus temperature for $YBa_2(Cu_{1-x}Zn_x)_3O_7$ (a) Batch with x = 0(0%), 0.03(3%), 0.05(5%) and identical densities (b) Samples reground, pressed and sintered with x = 0.01, 0.03, 0.07, 0.10 and the same density to within 5%.

Fig.4 Variation of residual resistivity with zinc concentration for samples of $YBa_2(Cu_{1-x}Zn_x)_3O_7$.

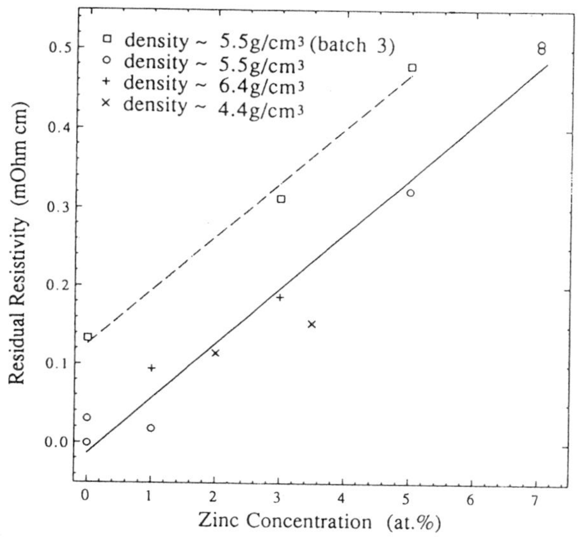

whereupon a more rapid increase sets in (not shown). This increase is independent of D, consistent with our preceding analysis.

As a further check on these results, we have re-ground and re-sintered some of the samples whose resistivities were shown in Fig.1, in an effort to more closely match their densities. Although there is still some spread in D values, the $\rho(T)$ curves, shown in Fig.3(b), are essentially parallel up to x=0.07. However, repeated sintering causes $d\rho/dT$ to deviate below the line in Fig.2, indicating the onset of significant texturing.

In summary, we have given experimental evidence that Matthiessen's rule is obeyed up to x=0.07 for zinc doping, while T_c is correspondingly reduced by a factor of 4.5 to 20K. This is consistent with the assumption made by Halbritter *et al* (1988) for various pure YBCO samples, but it is in disagreement with the work for Zn alloys of Affronte *et al* (1989). We note that the available microwave measurements of ρ_{ab} (Porch) for x=0.05 agree with our conclusions, and that Matthiessen's rule has also been found to hold for nickel-doped samples (Clayhold 1989). These results indicate that the linear resistivity in high T_c oxides arises from a temperature dependent carrier scattering mechanism. To first order this is not affected by Zn substitution and the Zn atoms act as extra independent scattering centres, giving a residual resistivity which is proportional to their concentration.

References

Affronte M, Pavuna D, Martin O, Licci F, Besagni T and Cattani S 1989 *Solid State Comm.* **70** 951

Babić E, Marohnić Z, Prester M and Brnicević N 1987 *Phil. Mag. Letts.* **56** 91

Bonn D A, O'Reilly A H, Greedan J E, Stager C V, Timusk T, Kamarás K and Tanner D B 1988 *Phys.Rev.* **B37** 1574

Clayhold J 1989 private communication of work carried out by J Birmingham

Gang Xiao, Cieplak M Z, Musser D, Gavrin A, Streitz F H, Chien C L, Rhyne J J and Gotaas J A 1988 *Nature* **332**, 238

Halbritter J, Dietrich M R, Küpfer H, Runtsch B and Wühl H *Z.Phys.B Condensed Matter* 1988 **71** 411

Loram J W and Mirza K 1990a To be published in *Electronic Properties of High T_c Superconductors and Related Compounds* (Berlin: Springer Verlag)

Loram J W, Cooper J R, Mirza K R 1990b These proceedings

Mehbod M, Biberacher W, Jansen A G M, Wyder P, Deltour R and Duvigneaud P H 1988 *Phys.Rev.* **B38** 11813

Ong N P, Wang Z Z, Hagen S, Jing T W, Clayhold J and Horvath J 1988 *Physica C* **153-155** 1072

Porch A M private communication

Tarascon J M, Barboux P, Miceli P F, Greene L H, Hull G W, Eibschutz M and Sunshine S A 1988 *Phys. Rev.* **B37** 7458

Supercond. Sci. Technol. 4 (1991) S280–282. Printed in the UK

Specific Heat Evidence for Strong Coupling in $YBa_2Cu_3O_7$

J E Gordon*, R A Fisher, S Kamin*, and N E Phillips

MCSD, Lawrence Berkeley Laboratory, Berkeley, CA 94720, USA

Specific heat data for $YBa_2Cu_3O_7$ are consistent with $\gamma = 15\pm3$ mJ/mole·K^2 and $2\Delta_0/kT_c = 6.8\pm0.6$. These results indicate that strong-coupling effects are present but that the coupling is unlikely to be predominantly a conventional electron-phonon interaction.

In this note an interpretation of specific heat, C, measurements on $YBa_2Cu_3O_7$ is applied to an analysis of data in the vicinity of T_c to obtain information about the strength of the coupling responsible for the superconductivity. Correlations among several sample-dependent parameters have been interpreted (Phillips et al 1990) as showing that typical samples are only 30-90% superconducting; that the volume fraction of superconductivity, f_s, can be determined from the discontinuity, $\Delta C(T_c)$, at the critical temperature, T_c; and that the Sommerfeld constant for the fully normal state, γ, is ~16 mJ/mole·K^2, a value close to that calculated (Massidda et al 1987; Krakauer et al 1988) for the bare density of states, γ_{bs}, indicating that electron-phonon enhancement effects are not significant. Thus, if electron-phonon coupling is responsible for the superconductivity, $YBa_2Cu_3O_7$ should be a weakly-coupled BCS superconductor with $\beta \equiv \Delta C(T_c)/\gamma T_c = 1.43$. However, such a conclusion is clearly at odds with $\gamma \sim 16$ mJ/mole·K^2 and the measured values of $\Delta C(T_c)$.

Figure 1 shows C/T vs T for a $YBa_2Cu_3O_7$ sample made by the citrate pyrolysis technique. An entropy-conserving construction gives $\Delta C(T_c)/T_c \sim 64$ mJ/mole·K^2, ~83% of the value expected for a fully superconducting sample (Phillips et al 1990; see also Junod et al 1990), i.e., $f_s = 0.83$. With $\gamma = 16$ mJ/mole·K^2 and $\beta = \Delta C(T_c)/f_s\gamma T_c = 5.5$ -- almost 4 times the BCS weak-coupling value and clearly indicative of strong-coupling. The dotted and solid curves in Fig. 1 represent, respectively, approximations to C_n and C_s, the normal- and the 83%-superconducting-state values of C. $C_n = C_l + \gamma T$, where C_l, the lattice contribution, is assumed to consist of dilatation and harmonic terms. C_n was obtained by interpolation between the data above 96K and the data in the region 62-65K, the region expected to include the temperature, T_x, at which $C_s = C_n$. Over this interpolation interval the

0953-2048/91/01S280 + 03 $03.50

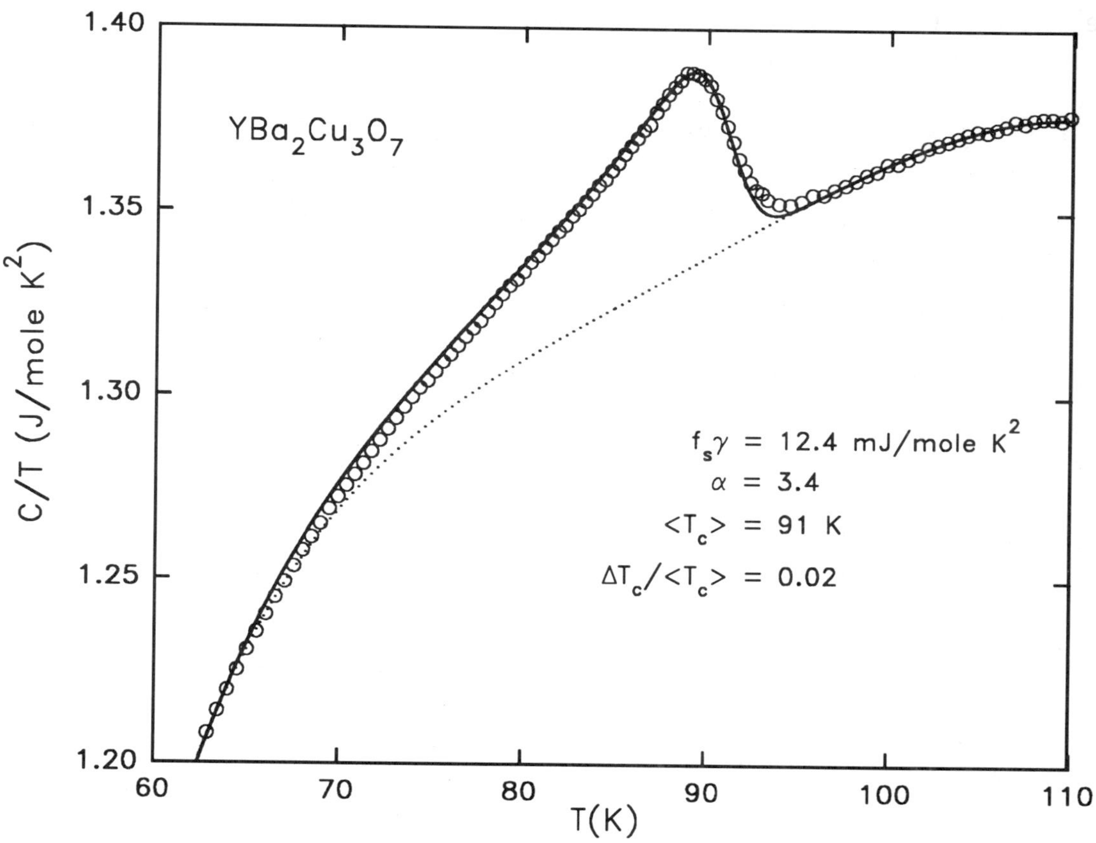

Figure 1. C/T vs T. See text for discussion.

harmonic lattice term can be adequately represented by a polynomial in T, by an expansion of the form described by Gordon et al (1989), or by a superposition of Einstein terms. The solid curve is given by $C_s = C_l + f_s C_{es} + (1-f_s)\gamma T$, where C_{es} is the fully-superconducting-state electronic C, and has been calculated by assuming a Gaussian distribution of transition temperatures centered on $\langle T_c\rangle$ with a width ΔT_c. It is also assumed that the superconducting state can be described by the "α model" (Padamsee et al 1973) in which the gap in the density of states is related to the BCS gap by the factor $\alpha/1.764$. Application of this model involves two adjustable parameters: α which determines the shape of the anomaly, and $f_s\gamma$ which determines the amplitude. The solid curve corresponds to $\langle T_c\rangle$ = 91K, $\Delta T_c/\langle T_c\rangle = 0.02$, $\alpha = 3.4$, and $f_s\gamma$ = 12.4 mJ/mole•K^2, and results in $\Delta C(T_c)/\gamma T_c = 5.3$ and T_x = 63K. The inclusion of a small fluctuation contribution that lowers $f_s\gamma$ to 12 mJ/mole•K^2 would slightly improve the fit in the region near 93K, but is not essential. As required by thermodynamics, the area between the data and the dotted curve in the temperature region $63 \le T \le 96$K is equal to the entropy difference between the normal and superconducting states at T_x. (For this comparison, the normal-state value was taken as γT_x and the superconducting-state value was calculated using the α model.) The results of the fitting procedure are not unique. Reasonably good fits to the data can be obtained with other choices for $f_s\gamma$ and α (but $<T_c>$ and ΔT_c remain essentially the same). We conclude that the data are consistent with the values $f_s\gamma$ = 12.4±3 mJ/mole•K^2 and α = 3.4±0.3. With f_s = 0.83, this range in $f_s\gamma$ corresponds to γ = 15±3 mJ/mole•K^2, in agreement with the value 16 mJ/mole•K^2 obtained on the basis of other considerations (Phillips et al 1990). The values of α consistent with the data are twice the weak-coupled BCS value. They correspond to a gap ratio $2\Delta_0/kT_c$ = 6.8±0.6, in agreement with other measurements (see, eg., Batlogg 1990) and, when combined with the result $\gamma \sim \gamma_{bs}$, imply that the conventional electron-phonon interaction cannot be solely responsible for the superconductivity.

The principal uncertainty in the conclusions derives from uncertainty in C_n, the "background" C on which the analysis is based. Loram and Mirza (1988) have also used the α model to obtain similar values for γ and α. They made differential C measurements with quenched non-superconducting $YBa_2Cu_3O_7$ as the reference. In that case there are uncertainties in the correction to C_l (comparable to those in C_n in this work), and there are also uncertainties in the interpretation of the several linear terms that occur in the data.

This work was supported by the Director, Office of Energy Research, Office of Basic Energy Sciences, Material Sciences Division of the U.S. Department of Energy under Contract DE-AC03-076SF00098. Additional support for J.E.G. and S.K. was provided by a grant from the Research Corporation.

*Permanent Address: Physics Dept. Amherst College, Amherst MA 01002 USA.

References

Batlogg B 1990 Proc. Los Alamos Symp. on HTSC -1989 eds K S Bedell et al (Redwood City: Addison-Wesley) pp 37-93
Gordon J E et al 1989 Solid State Commun 69 625 and erratum Solid State Commun, to be published
Junod A et al 1989 Physica C162 1401
Krakauer H et al 1988 J. Superconductivity 1 111
Loram J W and K A Mirza 1988 Physica C153-155 1020
Massidda A et al 1987 Phys. Lett. A122 198
Padamsee H et al 1973 J. Low Temp. Phys. 12 387
Phillips N E et al 1990 Phys. Rev. Lett. 65 357

Supercond. Sci. Technol. **4** (1991) S283–285. Printed in the UK

INVERSE AC JOSEPHSON EFFECT ON MULTIPHASED HIGH TEMPERATURE OXIDE SUPERCONDUCTORS

W.Hiller, M.Buchgeister, F.Busse, K.Kopitzki, G.Mertler and R.Nebel
Institut für Strahlen- und Kernphysik der Universität Bonn,
Nussallee 14-16, D-5300 Bonn 1, F.R.G.

Abstract. The measurement of the microwave-induced voltage across a granular sample due to the inverse ac Josephson effect provides a technique to detect a superconducting phase in a multiphase system. For the systematic investigation of this technique we prepared powder samples from two Y-Ba-Cu-O compounds of different oxygen content and therefore different critical temperature. We measured the microwave induced dc voltage as a function of temperature for varying concentration of the higher T_c compound. External magnetic field, power and frequency of incident microwaves were also varied.

1. Introduction

Studies of the inverse ac Josephson effect in superconducting granular samples in the form of thin films and bulk material were carried out by many authors (Saxena et al 1974, Chen et al 1987, Gupta et al 1987, Chang et al 1989, Hiller et al 1990). The rf radiation induced dc voltage across the unbiased sample was investigated in dependence of temperature, external magnetic field, frequency and power. The observed dc voltage is attributed to a large number of Josephson junctions in the granular material according to the commonly used model of a weak link network. Since the discovery of high-T_c superconductors the inverse ac Josephson effect was utilized to determine superconductivity transitions in multiphase systems of high temperature superconductors (Chen et al 1987, Gupta et al 1987). This method is more suitable than conventional resistance measurement to establish clearly the transition of a superconducting phase which coexists with a second phase with lower T_c in the same sample. Moreover the measurement of the resistivity of granular material depends on the biasing current because zero resistance can be observed only if the bias current is less than the smallest critical current of all the weak links in the system. In order to investigate the applicability of this technique we prepared powder samples from two Y-Ba-Cu-O-compounds of different oxygen content and therefore different critical temperature. Our unbiased powder samples were irradiated by microwaves in dependence of temperature, microwave frequency and power as well as external weak magnetic field.

2. Experimental

Sintered pellets of $YBa_2Cu_3O_{6.58}$ and $YBa_2Cu_3O_{6.41}$ were prepared applying the usual procedure which starts with the appropriate mixture of Y_2O_3, $BaCO_3$ and CuO. The mixture was calcinated in air, ground and then annealed in flowing oxygen at 930°C for 12 h, pressed to pellets and annealed again. Finally oxygen reduction was performed to the choosen values of oxygen content. The resistance of the pellets was measured by a standard four-probe arrangement. The superconducting transition of the 'high T_c- compound' $YBa_2Cu_3O_{6.58}$ started at around 60 K (onset) and ended at about 54 K (zeroset). The onset of the 'low T_c-compound' $YBa_2Cu_3O_{6.41}$ was about 32 K and the zeroset 23 K. The pellets of the two compounds were pulverized in a ball mill. By mixing the powders of the 'low T_c'-and the 'high T_c-compound' we were able to produce two-phase systems with varying concentration of the 'high T_c-compound'. Our powder samples were prepared by compacting the Y-Ba-Cu-O-grains in a hollow 10 mm long sapphire cylinder of 5 mm inside diameter. The resistance of the granular system was adjusted by applying moderate pressure. The powder samples were irradiated with microwaves of X-band klystrons having a relative frequency stability of about $5 \cdot 10^{-8}$ per hour. The experimental set-up is described in detail by Hiller et al (1990).

3. Results and discussion

We investigated powder samples with values of the relative concentration of the 'high T_c-compound' of 100%, 20%, 2%, 0.2%, and 0%. In all cases, the resistivities of the samples behave like that of semiconductors over the wide temperature range from room temperature down to 2 K. Fig.1 shows as example for the resistivity versus temperature data results for a sample with a concentration of 20%. Discussing similar results Fujita et al (1988) considered granular superconductors as a three dimensionally multiconnected network of two different resistances. Those are the bulk

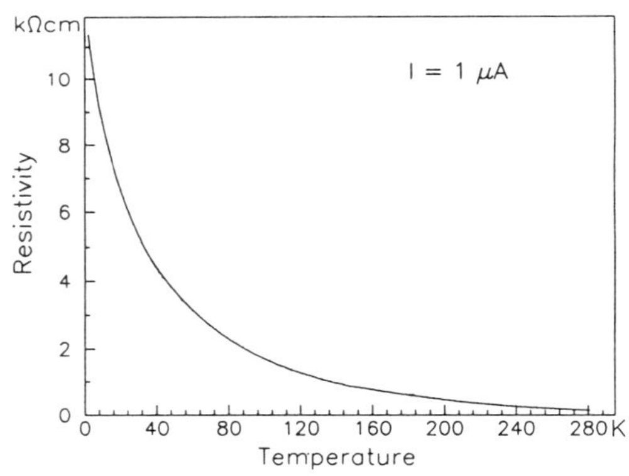

Figure 1: Resistivity versus temperature for the sample with 20% of a 'high T_c-compound'

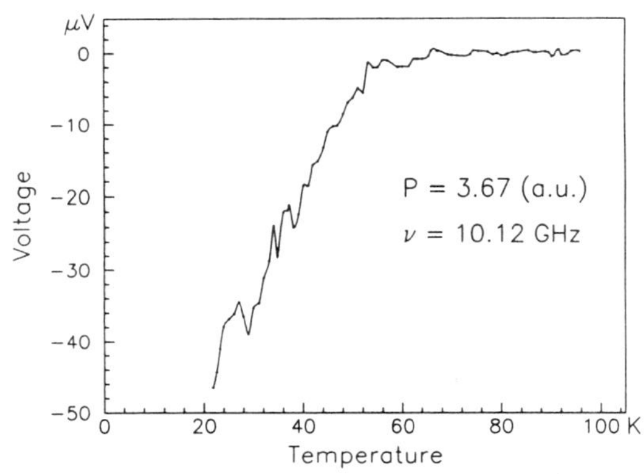

Figure 2: Microwave induced voltage versus temperature for the sample with 20% of the 'high T_c-compound'

resistance of each grain and the contact resistance at the grain boundary. In sintered pellets the resistivity is dominated by the bulk resistance of the grains whereas in our powder samples the resistivities are determined mainly by the contact resistances at the grain boundaries. According to this nonvanishing contact resistance we did not observe zero resistance even at 2 K. Therefore by applying conventional resistance measurement it was not possible to establish either the 'high T_c'-phase or the 'low T_c'- phase. Utilizing the inverse ac Josephson effect the unbiased samples were irradiated by microwaves. We observed below the superconducting transition of either the 'low T_c'-phase or the 'high T_c'-phase a microwave induced dc voltage. Fig.2 shows the typical behaviour of the induced dc voltage versus temperature for the sample of 20% of the 'high T_c-compound' at constant frequency, power and magnetic field. The dc voltage appeared below around 60 K - the critical temperature of the 'high T_c-compound'. For all investigated systems the induced dc voltage is influenced strongly by microwave power, frequency and weak magnetic field. In fig.3 an example of the dependence of the induced dc voltage on microwave power for two different values of the external magnetic field is shown. The sensitivity of the method is demonstrated

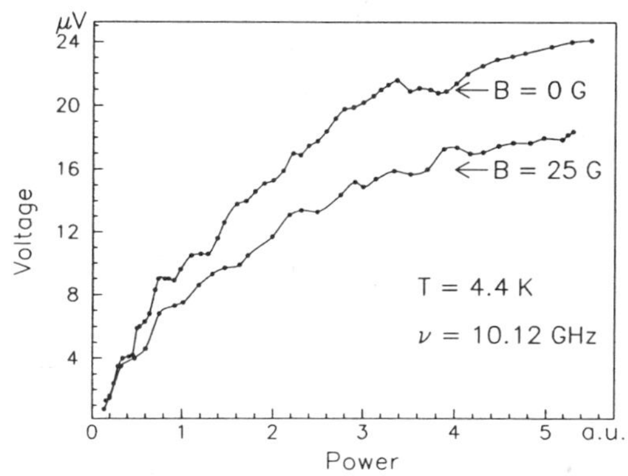

Figure 3: Microwave induced voltage for the sample with 100 % of the 'high T_c-compound' versus power

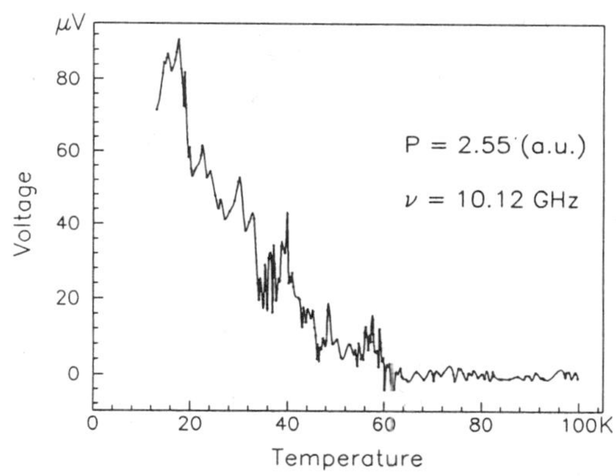

Figure 4: Microwave induced voltage for the sample with 0.2% of the 'high T_c-compound' versus temperature

in fig.4. Even a content as low as 0.2% of the 'high T_c-compound' is readily detected by its microwave response.

In summary the inverse ac Josephson effect can be utilized to establish a superconducting phase of a very low concentration in a multiphase system.

4. Acknowledgements

W. Hiller would like to thank the Friedrich-Naumann-Stiftung and M. Buchgeister the Cusanuswerk for research grants.

5. References

Chang K, Chen J T, Wenger L E and Logothetis E M 1989 *Physica C* **162-164** 1591

Chen J T, Wenger L E, McEwan C J and Logothetis E M 1987 *Phys.Rev.Lett.* **58** 1972

Gupta A K, Jayaram B, Agarwal S K, Gupta A and Narlikar A V 1987 *Phase Transitions (Gordon and Breach, Science Publishers, Inc.)* **10** 29

Fujita H, Ohmichi T and Oda Y 1988 *Jpn.J.Appl.Phys.* **27** L2094

Hiller W, Buchgeister M, Busse F, Kopitzki K, Mertler G and Nebel R 1990 *Proc. Int.Conf.Transp.Prop.of Superconductors, April 29 - May 4 Rio de Janeiro*

Saxena A M, Crow J E and Strongin M 1974 *Sol. State Commu.* **14** 799

Supercond. Sci. Technol. **4** (1991) S286–288. Printed in the UK

Specific Heat Anomalies in the 108K Superconducting System $Y_{1-x}Ca_xSr_2Cu_2Tl_{.5}Pb_{.5}O_7$.

J.W.Loram, K.A.Mirza, R.S.Liu.

IRC in Superconductivity, Madingley Road, Cambridge CB3 OHE, UK.

Abstract. We show the progression with x of the superconducting specific heat anomalies in $Y_{1-x}Ca_xSr_2Cu_2Tl_{.5}Pb_{.5}O_7$. The system is superconducting above x=.5 and T_c shows a broad maximum at x=.8. The size of the specific heat anomaly peaks very much more sharply than T_c, reflecting a correspondingly rapid variation with x of the superconducting pair density, similar to the behaviour previously observed in $La_{2-x}Sr_xCuO_4$. At low temperatures we observe a "magnetic" term which anticorrelates with the size of the anomaly at T_c but no additional electronic term.

1. Introduction

The investigation of a new superconducting system is frequently confined to a determination by resistivity or magnetic measurements of the transition temperature T_c. We have observed in studies of several systems that the superconducting pair density (order parameter) as determined by the magnitude of the specific heat anomaly varies as a function of composition very much more rapidly than does T_c and this information is essential to properly characterise the system. Note that magnetic susceptibility measurements can at best only reveal a superconducting volume fraction and do not reflect the pair density. In this report we describe specific heat measurements on the new 108K system $Y_{1-x}Ca_xSr_2Cu_2Tl_{.5}Pb_{.5}O_7$ (subsequently referred to as the Tl system).

This system is structurally similar to $YBa_2Cu_3O_7$ (Liang et al,1990) with Sr replacing Ba and $Tl_{.5}Pb_{.5}$ replacing the chain Cu(1) in YBCO. It is however tetragonal and therefore has no chain structure. Substituting Ca^{2+} for Y^{3+} adds holes, and the system progresses directly from a semiconducting phase to a superconducting phase at around x=.5 with T_c(max) ~108K at x=.8. The magnetic behaviour of the Tl system has not been established but antiferromagnetic order has been observed in the related compound $YBa_2Cu_2TlO_7$. It is therefore likely that the magnetic phase diagram of the Tl system is similar to that of other oxide superconductors with antiferromagnetic correlations playing an important role in the semiconducting region x<.5.

The system shows increasing metallic conductivity up to x=1. The carrier density estimated by iodometric titration similarly increases to x=1. Close to the metal insulator transition x_{MI} ~ .5 the superconducting resistive transition is very broad but has T_c(onset) comparable with T_c(max). As x approaches the optimum composition x_0=.8 the transition width decreases rapidly to a few K but T_c(onset) changes very little. For x>.8 T_c(onset) drops sharply but the width stays constant.

2. Experimental

To observe the small specific heat anomalies we use a differential technique (Loram 1983) with resolution ~1:10^4 above 30K, in which we measure the difference in specific heat between each sample and a reference Tl sample with x=.2, thus eliminating most of the large phonon term. Sample masses were ~ 1.8g.

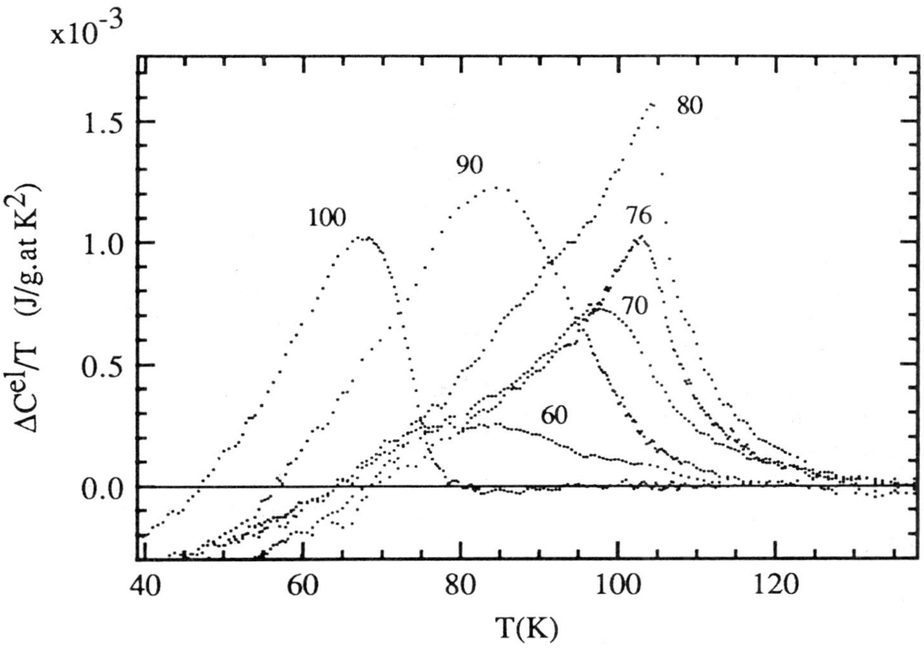

Fig 1. $\Delta C^{el}/T$ for $Y_{1-x}Ca_xSr_2Cu_2Tl_{.5}Pb_{.5}O_7$ where $\Delta C^{el}=C^{el}-C_n^{el}$. Labels show 100x.

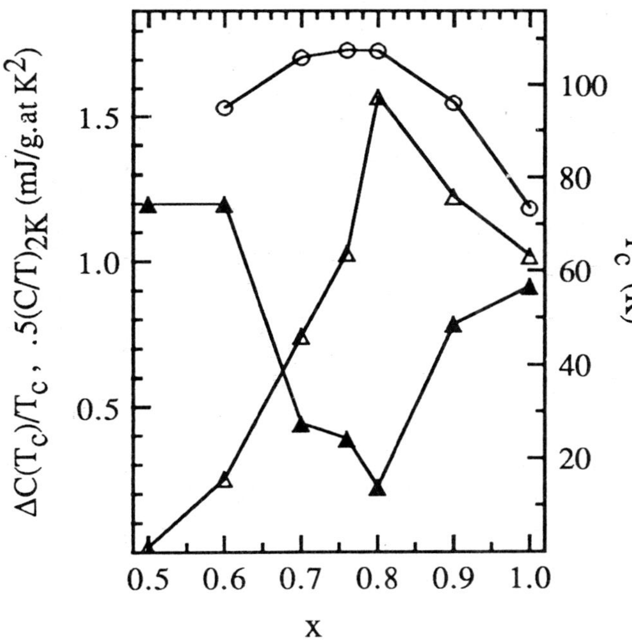

Fig 2. T_c (o) defined as mid-point of the transition.
Step height $\Delta C(T_c)/T_c$ (Δ); Low temperature anomaly $(C/T)_{2K}$ (Δ)

3. Results.

The specific heat results for the region around T_c after correction for a residual phonon background term are shown in Fig 1, and it is clear that the behaviour is quite different for x below and above x_0=.8.

We first observe a very broad anomaly with low entropy (pair density) at x=.6. The anomalies increase rapidly in area and sharpness as x increases to x_0=.8, reflecting respectively a corresponding increase in superconducting pair density and range of phase coherence. In spite of this rapid increase in the pair density, T_c increases only a little, passing through a broad maximum of 108K at x=.8, and the onset temperature is almost independent of x at around 120K. Above x_0, T_c, T_c (onset) and the anomaly size $\Delta C(T_c)/T_c$ all fall steadily with increase in x and metallic conductivity.

At low temperatures (1.5<T<10K) specific heat anomalies typical of spin glass order are observed for x<.5 . In the superconducting region x>.5, $C-C^{ph}$ is approximately **independent of temperature** with no clear evidence for a linear T term . There is, however, a striking anti-correlation of this term with the size of the anomaly at T_c (see Fig 2) which strongly suggests that in this composition range **the low temperature anomaly is electronic in origin.**

4. Discussion

The progression with x of the specific heat anomalies at T_c and at low temperatures is strikingly similar in the present system $Y_{1-x}Ca_xSr_2Cu_2Tl_{.5}Pb_{.5}O_7$ to that observed previously in $La_{2-x}Sr_xCuO_4$ (Loram et al 1989). On first emerging from the semiconducting region the pair density is low **but T_c is already high.** The picture that emerges is one of superconducting fluctuations persisting up to $T_c \sim T_{c,max}$ growing in amplitude with x and finally forming a fully coherent superconducting state over a narrow composition range around x_0. **This contrasts with the conventional view that T_c increases monotonocally with carrier density.** We suggest that this behaviour should not to be viewed as spatially separate superconducting and normal regions but as dynamic fluctuations into and out of the superconducting state due perhaps to pair breaking by static or slowly varying magnetic correlations which persist into the metallic region. These rapidly weaken as the carrier density increases allowing the superconducting order parameter to increase in amplitude and phase coherence.

References.

Liang J.M.,Liu,R.S,Huang Y,Wu.S,Wu P.T,Chen L.J.1990 Physica**C165**, 347.
Loram J.W. 1983 J.Phys **E16**, 367
Loram J.W, Mirza K.A, Osborne J. 1989 Physica **C162-4**,498.

Supercond. Sci. Technol. **4** (1991) S289–291. Printed in the UK

Failure to observe the superconducting fluctuation in $Tl_2Ba_2CaCu_2O_8$ single crystal

Lin Shu-yuan, Lu Li, Wang Xue-mei, and Zhang Dian-lin
Institute of Physics, Chinese Academy of Sciences, Beijing, China.
Duan Hong-min, and A.M. Hermann
Department of Physics, University of Colorado, Boulder, CO 80309, U.S.A.

Abstract The thermopower of $Tl_2Ba_2CaCu_2O_8$ single crystals has been measured from 80K up to ~ 400K using ac method. The giant anomaly just above superconducting transition observed for $YBa_2Cu_3O_7$ sintered samples by Howson et al has not been confirmed in the present sample. The results of present a.c. method are roughly consistent with our previous d.c. measurement.

The new oxide superconductors are fascinating for the study of superconducting fluctuation because the shortness of the coherence length, the low carrier density, and the high transition temperature of the materials promise a large fluctuation which should be visible in experimentally well accessible temperature range by simple estimate (Salamon, M.B., et al., 1989). Recently, Howson et al (Howson, M.A., et al., 1990) observed sharp peaks in the thermopower of $YBa_2Cu_3O_7$ crystals in the vicinity of T_c. The authors argued that the peaks may be due to fluctuation effect which did not observed in previous measurements for the limited temperature resolution of the usual d.c.method.

Here we report the anisotropic thermopower of a $Tl_2Ba_2CaCu_2O_8$ single crystal measured by an improved a.c. method (Zhang Dian-lin, et al., 1990). We failed to find any anomaly at T_c. The overall behavior of the thermopower is basically consistant with our earlier results (Lin Shu-yuan, et al., 1990) obtained by d.c. method.

The crystals were grown by a method described elsewhere (Duan, H.M., Hermann, A.M.,). Powders of BaO, Ca, Cuo,and Tl_2O_3 mixed in appropriate ratio were prebaked at 650 C for half an hour and then heated to 960°C for thorough reaction. After slowly cooled to room temperature, crystals of generally $0.2\times0.2\times0.04$ mm^3 in size were obtained. Analysis by four-circle diffractometer showed that the crystals were of good quality with lattice parameters a=b=5.44 Å and c=29.37 Å. Resistivity measurement showed metallic behavior in both ab plane and c direction with room temperature values 10^{-4} Ωcm and 10^{-2} Ωcm, respectively.

The method of thermopower measurement is schematically shown in Fig.1 (Zhang Dian-lin et al., 1990). A chopped light is introduced to a silver foil 2 of about 0.01 mm in thickness by a quartz bar 1 which shrinks at the low temperature end to make the heating more effective and less inertial. The sample 3 is pressed by a minispring 4 to the Ag foil 2 in a place about 0.5 mm apart from the heat source and where a thin constantan wire 5 is welded at the opposite face of the foil. Since the modulation frequency of the light is 23 Hz which yields for Ag a characteristic

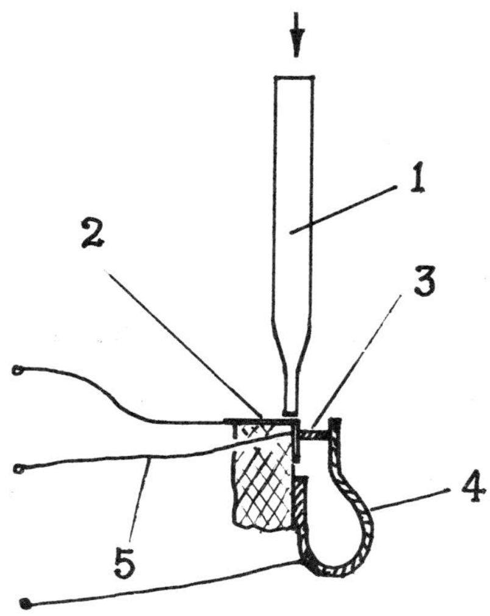

Fig. 1 Schematic of the setup for thermopower measurement. 1)quartz bar, 2)silver foil, 3)sample, 4) spring, 5)thin constantan wire.

damping length of the thermal wave much larger than the thickness of the foil, the sample and the constantan wire should feel the same temperature modulation. The Ag foil serves as a common arm of the thermopower measurement. Over the whole temperature range the temperature differential in the crystal was kept below 10 mK. The thermopower of the sample and the thermocouple was measured separately by two lock-in amplifiers.

The anisotropic thermopower of $Tl_2Ba_2CaCu_2O_8$ crystal was measured from 80K up to ~ 400K. The result is shown in Fig.2. Below ~ 360K, the thermopower in ab-plane, S_{ab}, increases linearly with the decrease of temperature down to ~ 180K, then is rounding up and drops to zero at T_c. The general behavior is consistent with our d.c. measurement (Lin Shu-yuan et al., 1990). But the present measurement gives both a higher thermopower value ($S_{ab(300K)}$ ~ 17μV/K) and its temperature dependence (dS_{ab}/dT ~ 0.076 μV/K) than that obtained in (Lin Shu-yuan et al., 1990). At present we can not tell if it is a systematic error from the different methods or it is a sample-dependent result. However, we notice that the present crystal shows a marked transition tail which extends to ~ 80K (Fig.3), while the samples in our previous measurement finish their transition around ~ 90K (Lin Shu-yuan et al., 1990). The thermopower along c-axis, S_c is higher than S_{ab}, also similar to that in (Lin Shu-yuan et al., 1990), but S_c in present case has about the same , though weaker temperature dependence as S_{ab}, in contract to previous work where S_c is very much temperature independent. So if the long transition

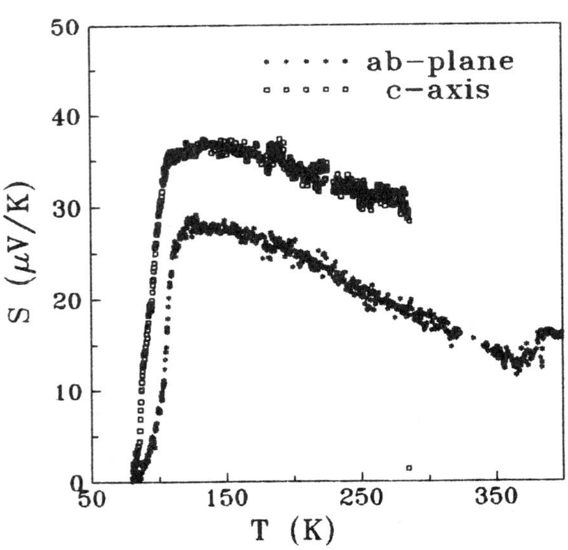

Fig.2 The anisotropic thermopower of $Tl_2Ba_2CaCu_2O_8$ single crystal.

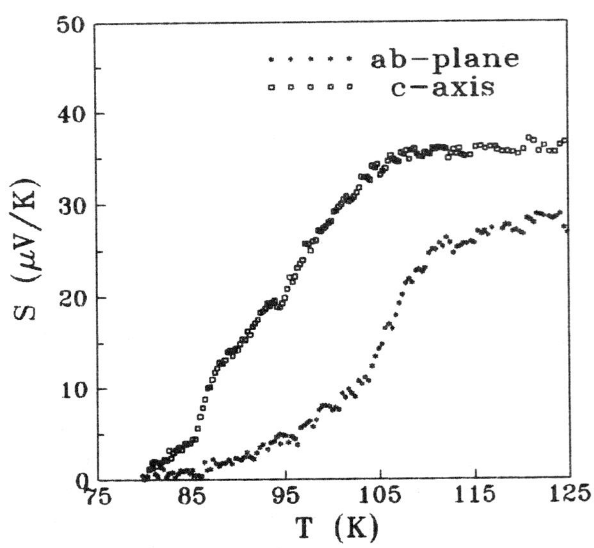

Fig.3 The superconducting transition of the crystal shown by thermopower.

tail in S_{ab} reflects that the quality in the present crystal is not as good as that used in our previous measurement, then it would not be impossible that S_{ab} has made some contribution to S_c. It is interesting to note that above ~ 360K, the slope of S_{ab} changes sign as shown in Fig.2. We have no idea at present what this change over of slope does mean.

The present ac measurement again confirms our previous observation (Lin Shu-yuan et al., 1990) that the transition temperature along c-axis is lower than that in ab-plane (Fig.3). We have used the anisotropy of transition temperature to explain the pressure dependence of the two step transition in $YBa_2Cu_3O_7$ single crystals (Zhang Dian-lin et al.,1989). Although as a thermodynamic quantity T_c should have a unique value for a three dimensional single phase system, it seems not so clear to us that for a two- or quasi-two dimensional material T_c should be isotropic. The anisotropy of superconducting transition has also reported in (Kambe, S., et al., 1989) and (Bucher, B., 1990).

A remarkable result of our present work is the failure to find a fluctuation anomaly in the vicinity of T_c for $Tl_2Ba_2CaCu_2O_8$ single crystal in both ab-plane and c direction. Since the 2212 phase of TlBaCaCuO is generally obtained without oxygen deficiency, we do not expect that the absence of thermopower peak should be related to the the oxygen content as observed for $YBa_2CU_3O_7$ (Howson, M.A., et al., 1990). Furthermore, it is not so clear to us why the oxygen treatment should influence the peaks so strongly if it comes from superconducting fluctuation in origin. We also failed to find any thermopower peak in $GdBa_2Cu_3O_7$ epitaxial films with c-axis perpendicular to the film plane by ac method with high temperature resolution (Zhang Dian-lin, et al., 1990). More work is needed to clarify the matter.

This work is partly supported by the Chinece National Center for Research and Development on Superconductivity.

References

Bucher,B., Karpinski,J., Kaldis,E., and Wachter,P., 1990, Physica C 167, 324

Duan Hong-min and Hermann, A.M., unpublished.

Howson, M.A., Salamon, M.B., Friedmaun, T.A., Rice, J.P., and Ginsberg, D., 1990 Phys.Rev. B41, 300.

Kambe, S., Kitayawa, K., Naito, M., Fukuoka, A., Tanaka, I., and Kojima, H., 1989, Physica C 160, 35.

Lin Shu-yuan Lu li Zhang Dian-lin, Europhys. Lett. (in press)

Salamon, M.B., in "Physical properties of high temperature superconductors 1 " edited by D. Ginsberg (Would Scientific 1989).

Zhang Dian-lin, Lu Li, Huang Jing, Hong-min Duan, Lin Shu-yuan, Ma Bei-hai, Cao Shao-chun, and Jin, B.J., 1989, Phys. Rev. B41, 6692.

Zhang Dian-lin, Wang Xue-min, Lu Li, Lin Shu-yuan, Li, H.C., WAng, R.L., Yi, H.R., and Liu, W., June 25-29, 1990, International Workshop on "Effects of strong disordering in HTSC", Sverdlovsk, USSR.

Supercond. Sci. Technol. 4 (1991) S292–294. Printed in the UK

Thermoelectric Power in Lead-Doped Polycrystalline BiSrCaCuO Superconductors

A.J. López , J. Maza, Y.P. Yadava, F. Vidal, F. García-Alvarado*, E. Morán*, M.A. Señarís-Rodríguez*

Laboratorio de Física de Materiales, Facultad de Física, Universidad de Santiago de Compostela, 15706 Santiago, Spain.
* Dpto. de Química Inorgánica, Facultad de Química, Univ. Complutense, Madrid

Abstract. Measurements of thermoelectric power and electrical resistivity of granular $Bi_{1.5}Pb_{0.5}Sr_2Ca_2Cu_3O_y$ are reported. Rounding of thermoelectric power near and above the critical temperature is analysed in terms of thermodynamic fluctuations of the superconducting order parameter.

1. Introduction

Thermoelectric power (TEP), S , in high-temperature copper-oxide superconductors (HTSC), either ceramic or single crystals, has been measured by several groups (see, e.g., Pekala et al 1988, Ouseph et al 1990, Rao et al 1990, Lu et al 1988, Chen et al 1989, Forro et al 1990). Progress on the critical behavior of TEP in HTSC lags behind that of normal state, and only a few works known to us have addressed this issue (Laurent et al 1988,1989, Dey et al 1990, Howson et al 1990). Indeed, the smallness of the correlation length of the superconducting order parameter in HTSC should entail the existence of large thermodynamic fluctuations. The effect of critical fluctuations on electrical resistivity, ρ, has received considerable effort and advances in the study of its critical behavior (see, e.g., Veira and Vidal 1989 and references therein) may be a valuable reference for the analysis of TEP near the normal-superconducting transition. In this regard, we would like to summarise some distinctive features of both properties from the experimental viewpoint : i) Unlike ρ, S does not depend on the geometrical parameters of the sample. ii) The anisotropy of S is much less than that of ρ; whereas ρ_{ab} (CuO_2 ab planes) is several orders of magnitude lower than ρ_c (c direction) in either the YBCO or BCSCO families, S_{ab} is of the same order of magnitude as S_c (Lu et al 1988, Chen et al 1989). Any measurement of S in a granular sample should thus yield some mixture of S_{ab} and S_c with comparable weights (see later). iii) Unlike ρ, values of S for granular samples are comparable to those obtained for single crystals and so S for granular samples has a more intrinsic character. iv) Near the normal-superconducting transition, the use of finite temperature difference along the sample for measuring S adds a non-intrinsic rounding effect to the intrinsic one due to fluctuations.

Here we present results on S and ρ of a granular sample of nominal composition $Bi_{1.5}Pb_{0.5}Sr_2Ca_2Cu_3O_y$. As first reported by our group (Vidal et al 1988) and later by other authors (Chen et al 1989, Balestrino et al 1989), fluctuation-induced paraconductivity in Bi-based compounds reveals a two dimensional (2D) character in contrast to the 3D character of the YBCO family. It is the aim of this paper to analyse the rounding of TEP searching for a signature of 2D behavior.

2. Experimental Details

Samples of $Bi_{2-x}Pb_xSr_2Ca_2Cu_3O_y$ system, with x = 0.5, were prepared by normal solid state synthesis. TEP was measured by employing the standard d.c. differential method. Temperature gradient across the sample was monitored by two copper-constantan thermocouples and thermo e.m.f. was measured by two copper leads (dia 0.2 mm) and temperature gradient was kept between 1-3K/cm. Absolute value of TEP of samples was calculated after correction for the absolute TEP of the copper (Blatt et al 1976). Electrical resistivity of the samples was measured by employing conventional four probe method.

3. Results and Discussion

Measurements of S and ρ are displayed in Figure 1. TEP is closely linear with temperature and its profile is very similar to previous results (see, eg, Pekala et al 1988, Laurent et al 1989) for the same kind of granular HTSC. Though this linear non-metallic dependence of TEP is still pending explanation, we think that it is just a nearly equal-weight mixture of S_{ab} and S_c. This is consistent with measurements in single crystals (Chen et al 1988) in that the temperature slope of S_{ab} is negative and larger in magnitude than it is the positive temperature slope of S_c.

Figure 2 shows a detailed view of the rounding of S and ρ near the superconducting transition. When analysing critical behavior in terms of mean field theories, one of the central parameters in the mean field critical temperature T_{co}. Being experimently inaccesible, one of the sensible choices is the inflexion-point temperature. Dotted curves correspond to the temperature derivatives of S and ρ, their maxima taking place, respectively, at 107K and 106.4K. The close proximity of both temperatures is a strong indication of its intrinsic character (Veira and Vidal 1989). Correspondingly, we shall take T_{co} = 107K. It is worth pointing out that the inflexion-point temperature is insensitive to the averaging involved in the measurement in S, as long as the derivative curve is symmetrical over the averaging temperature difference.

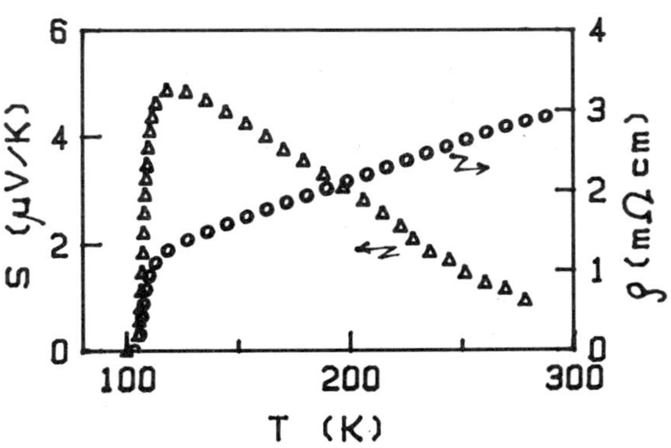

Fig.1 Temperature dependence of TEP (triangles) and resistivity (circles) for the Bi-based compound.

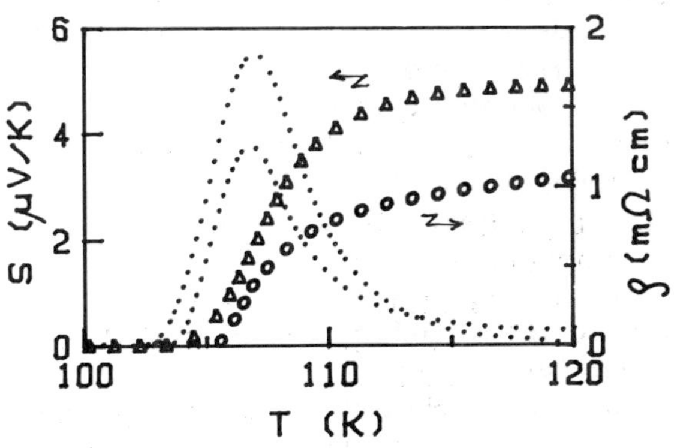

Fig.2 Rounding of TEP (triangles) and resistivity (circles), and their corresponding temperature derivatives.

With regards to the critical behavior of both S and σ, we have sujected them to the same methodological process described elsewhere (see, eg, Veira and Vidal 1989). The resulting excess quantities for both S and σ vs reduced temperature, ε, in a log-log plot are displayed in Figure 3. This Figure encloses the main contribution of this Paper. In sum: i) The excess TEP and the excess conductivity follow virtually the same ε dependence. Writing their relationship as $S = L/\sigma$, where L is an Onsager coefficient, that feature means that in

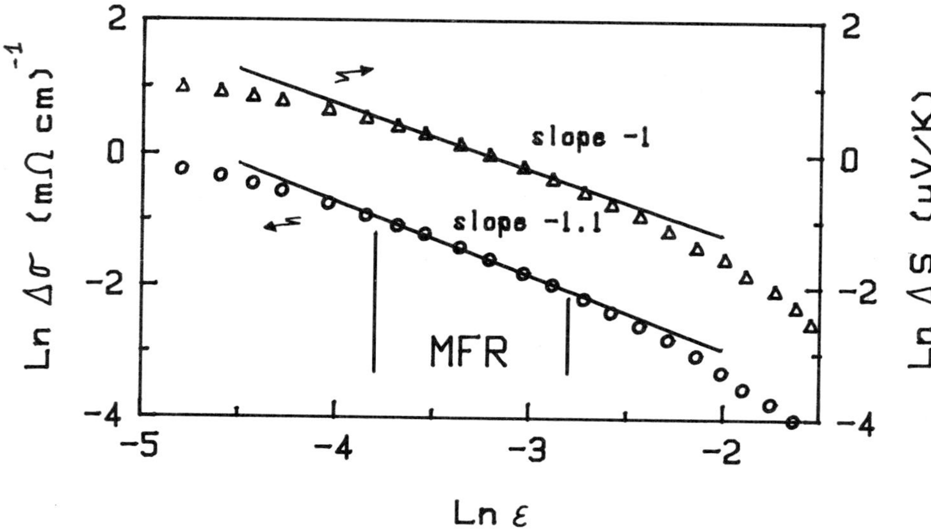

Fig.3 Log-log plot of the excess TEP and excess conductivity for the Bi-based compound.

the mean field region (MFR) the critical behavior of S is dominated by that of σ. Our data do not confirm thus the critical behavior for L predicted by Maki (1974) for superconductors in the dirty limit. ii) Both S and σ show a 2D behavior (characteristic exponent -1) in the MFR. TEP's critical analysis reenforces therefore the first finding by Vidal et al (1988) of the marked 2D character of Bi-based HTSC.

Acknowledgements

First authors from Santiago acknowledge financial support by the CICYT (MAT 88-0769) and by the Programa MIDAS (89-3800), Spain.

References

Balestrino G, Nigro A, Vaglio R, and Marinelli M (1989) *Phys. Rev. B* **39** 12264

Blatt, F.J., Schroeder P.A., Foiles C.L., and Greig D. (1976) *Thermoelectric Power of Metals* (New York: Plenum Press) p.5

Chen G H, Wang J H, Zheng D N, Yan Y F, Jia S L, Yang Q S , Ni Y M and Zhao Z X (1989) *Mod. Phys. Lett. B* **3** 295

Chen G H, Yang G, Yan Y F, Jia S L, Ni Y M, Zheng D N, Yang Q S and Zhao Z X (1989) *Mod. Phys. Lett. B* **3** 1045

Dey T K, Radha K, Barik H K, Bhattacharya D and Chopra K L (1990) *Sol. Stat. Commun.* **74** 1315

Forro L, Lukatela J and Keszei B (1990) *Sol. Stat. Commun.* **73** 501

Howson M A, Salamon M B, Friedmann T A, Rice J P and Ginsberg D (1990) *Phys. Rev. B* **41** 300

Laurent Ch, Patapis S K, Lagnesse M, Vandershueren H W, Rulmont A, Tarte P and Ausloos M (1988) *Sol. Stat. Commun.* **66** 445

Laurent Ch, Patapis S K, Green S M, Luo H.L., Politis C, Durczewski K and M. Ausloos (1989) *Mod. Phys. Lett. B* **3** 241

Lu L, Bei-Hai M, Shu-Yuan L, Hong-Min D and Zhang D L (1988) *Euro. Phys. Lett.* **7** 555

Maki K (1974) *J. Low Temp. Phys.* **14**, 419

Ouseph P J, Ray O'Bryan M (1990) *Phys. Rev. B* **41** 4123

Pekala M and Pajaczkowska A (1988) *Physica C* **156** 497

Rao C N R, Ramakrishnam T V and Kumar N (1990) *Physica C* **165** 183

Veira J A and Vidal F (1989) *Physica C* **159** 468

Vidal F, Veira J A, Maza J, Ponte J J, García-Alvarado F, Morán E, Amador J, Cascales C, Castro A, Casais M T, and Rasines I (1988) *Physica C* **156** 807

Supercond. Sci. Technol. 4 (1991) S295–297. Printed in the UK

ESR study on a ferromagnetic copper oxide : $La_4Ba_2Cu_2O_{10}$

Hiromu Masuda, Fumio Mizuno, Yasuji Yamada, Kazushi Sugawara,
Izumi Hirabayashi, Yuh Shiohara, and Shoji Tanaka
Superconductivity Research Laboratory
International Superconductivity Technology Center
2-4-1 Mutsuno, Atsuta-ku, Nagoya 456, Japan

Abstract. The electron-spin resonance (ESR) of an insulating copper oxide: $La_4Ba_2Cu_2O_{10}$ which shows a ferromagnetic transition at around 5 K has been studied. The ESR spectra show a narrowing with decreasing temperature which is characteristic of the ferromagnetic exchange interaction. The g tensor is uniaxial and the principal values are $g_{\perp} = 2.123$ and $g_{//} = 2.037$. The exchange coupling between dissimilar Cu^{2+} ions in the unit cell explains the features of the ESR spectra of this material and the Cu^{2+} d-holes are in the $d_{x^2-y^2}$-like orbits in the ground state.

1. Introduction

We have recently discovered that a copper oxide compound $La_4Ba_2Cu_2O_{10}$ (the 422-phase) is an tetragonal insulator which shows ferromagnetic transition at around 5 K (Mizuno *et al* 1990). From the effective Bohr magneton ($P_{eff} = 1.66\ \mu_B$/Cu) and saturation magnetic moment ($\mu_s = 0.95\ \mu_B$/Cu), this compound is a localized-type ferromagnet in which Cu^{2+} ions show a ferromagnetic ordering. In this paper, we report the ESR investigation to clarify the relationship between ferromagnetism and the special crystal structure of the $La_4Ba_2Cu_2O_{10}$ compound.

2. Experimental

The single crystals of the 422 phase are grown from flux (BaO-CuO) under pressure of 5 Kg/cm^2. The ESR measurement was carried out with an X-band (9.097 GHz) spectrometer in the temperature range of 4.2-300 K. The ESR sample was sealed to avoid humidity and freezing oxygen.

3. Results and discussion

The angular dependence of the g value at room temperature is shown in Fig. 1. The g tensor is uniaxial and its principal values $g_{//}$ and $g_{\perp}$ are 2.037 and 2.123, respectively. The c-axis corresponds with the principal axis of the g tensor. The temperature dependence of the $g_{\perp}$ and $g_{//}$ values and the width of resonance peak ΔH_{pp} are shown in Fig. 2 (a) and (b), respectively. The difference of $g_{\perp}$ and $g_{//}$ decreases gradually as decreasing temperature between 50 K and 290 K, and more remarkably below 50 K. The widths of the resonance peak: $\Delta H\ pp$'s, in both cases of the magnetic field H perpendicular and parallel to c-axis, decrease gradually from 290 K to around 50 K. Below 50 K, they show the sharp decrease with decreasing temperature. This narrowing is a characteristic of the ferromagnetic exchange interaction. The line shape also changes with the temperature. The analysis plots of the ESR line shape are shown in Fig.3. If the exchange coupling is strong, the ESR line shape takes Lorentzian type, on the other hand, in the case of the

0953-2048/91/01S295 + 03 $03.50

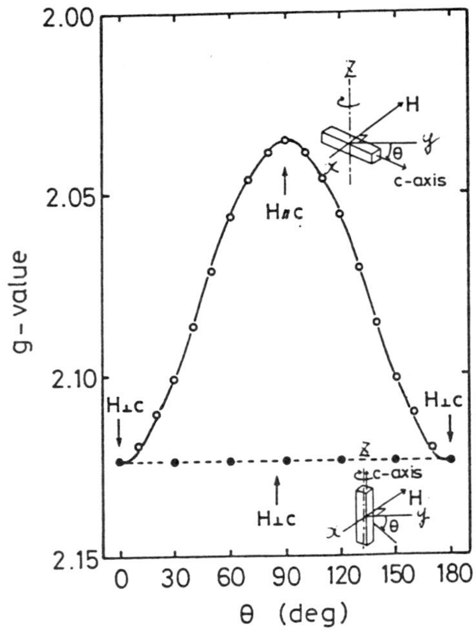

Figure 1. The angular dependence of the g-value for $La_4Ba_2Cu_2O_{10}$ at room temperature.

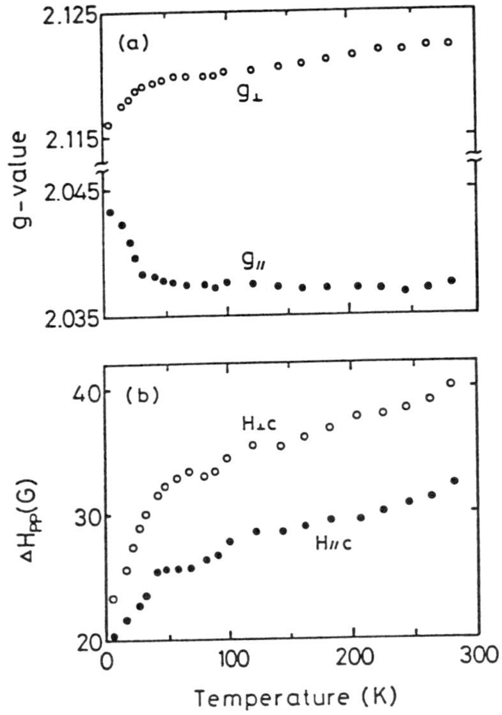

Figure 2. The temperature dependences of (a) the $g_\perp$ and g values and (b) the width of resonance peak *ΔH pp* for $La_4Ba_2Cu_2O_{10}$.

magnetic dilution system, it takes Gaussian type. In the 422-phase, the ESR line shape shows the intermediate shape between the Lorentzian and the Gaussian. It suggests that the exchange coupling does not act perfectly because of the low dimensionality of the spin system in the 422-phase. The ESR line shape changes from the Gaussian type to the Lorentzian type with decreasing temperature.

From the observed ESR signal in Fig. 1, the $g_\perp$ is larger than the $g_{//}$. According to the ordinary ligand field theory, it means the d-hole orbital of the Cu^{2+} ion in the 422-phase is described by the $d_{3z^2-r^2}$ orbital. As shown in Fig. 4, the 422-phase have the square planar CuO_4 groups which are so isolated that one square planar CuO_4 is orthogonal to the nearest neighbor CuO_4 from our neutron diffraction study (Ogawa *et al* 1990). From the expected local field of the CuO_4 plane, the orbital of d-hole of the Cu^{2+} ion should be $d_{x^2-y^2}$. It is not consistent with the ESR results. This inconsistency can be resolved if the exchange coupling is taken into account (Abragam and Bleanley 1970).

The g values of perpendicular and parallel to the local principal axis are denoted by $g_\perp'$ and $g_{//}'$, respectively. When the field is applied parallel to the c axis, the resonance occures at the field corresponding to $H=h\omega/(2g_\perp'\mu_B)$, because both spins at the inequivalent Cu sites have the same $g_\perp'$ value. On the other hand, when the external field is applied perpendicular to the c-axis, and the resonance frequencies are different for the inequivalent Cu spins, the resonance occurs at different external fields corresponding to both $g_\perp'$ and $g_{//}'$. When the exchange interaction acts between a pair of inequivalent ions and its magnitude exceeds the difference of Zeeman energy, two resonance lines amalgamate each other with increasing exchange interaction. At last they are combined at the center of the two resonance lines. Therefore, it is straightforward to deduce the formula of the g tensor

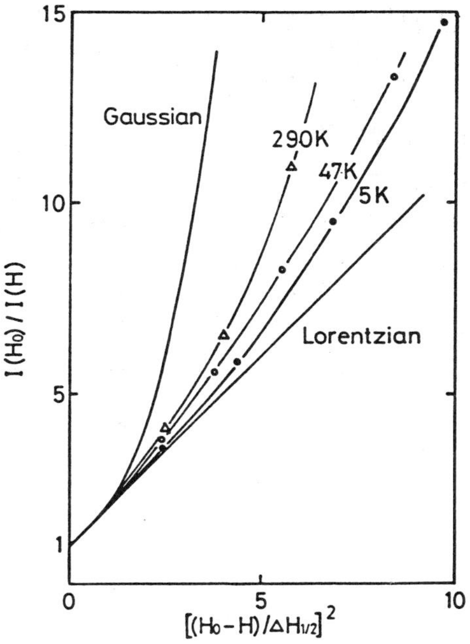

Figure 3. The analysis plots of the ESR line shape of the single crystal $La_4Ba_2Cu_2O_{10}$.

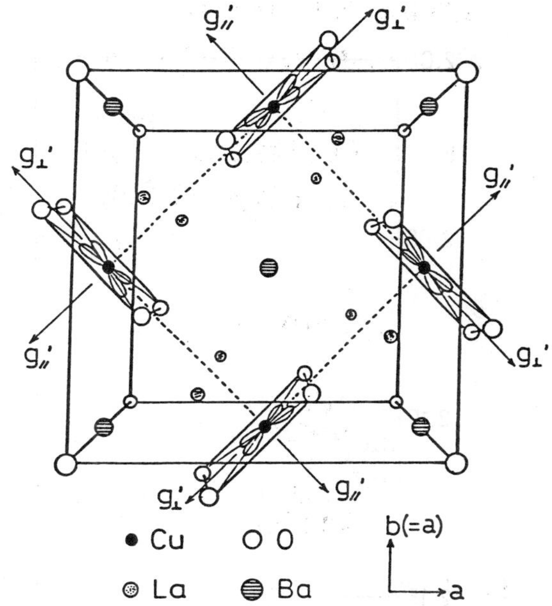

Figure 4. Top view of the crystalline structure of $La_4Ba_2Cu_2O_{10}$. The principal axes of the local g tensor are illustrated.

resonance lines. Therefore, it is straightforward to deduce the formula of the g tensor as $g_\perp = (g_\perp'+g_{//}')/2$. For the experimental values $g_\perp = 2.123$ and $g_{//} = 2.037$ at room temperature, the local principal values of the g tensor can be calculated as $g_\perp' = g_{//} = 2.037$, $g_{//}' = (2g_\perp - g_{//}) = 2.209$. From the relation of the $g_\perp'$ and the $g_{//}'$ values, the ground state of the Cu^{2+} d-hole in the 422-phase should be described by the wave function $d_{x^2-y^2}$ and it agrees with the results deduced from the structure. When the d-hole orbits of the ground state are orthogonal to the nearest neighbor copper one as shown in Fig. 4, only direct exchange interaction acts between the Cu ions. There is no strong antiferromagnetic kinetic exchange due to nonexistence of oxygen between Cu ions. Therefore, it is reasonable to expect that the whole exchange interaction which acts between spins of the Cu^{2+} ion is ferromagnetic.

REFERENCES

Mizuno F, Masuda H, Hirabayashi I, Tanaka S, Hasegawa M and Mizutani U 1990 Nature **345** 788

Ogawa N, Mizuno F, Masuda H, Hirabayashi I, Tanaka S, Mochiku T, Asano H and Izumi F 1990 Physica B **156&166** 1687

Abragam N and Bleanley B 1970 in *Electron Paramagnetic Resonance of Transition Ions* (Oxford Press)

Supercond. Sci. Technol. 4 (1991) S298–300. Printed in the UK

ANOMALOUS ANISOTROPIC CURRENT-VOLTAGE CHARACTERISTICS AND MAGNETIC RESISTANCE OF SINGLE CRYSTALS $YBa_2Cu_3O_{7-\delta}$

M.A. Obolenskii, A.V. Bondarenko, A.M. Kadygrobov , O.A. Mironov[+], S.V. Chistyakov[+], I.Yu.Skrylev[+] and Yu.T. Petrusenko[++]

Kharkov State University, Kharkov 310077, USSR

[+]Institute of Radiophysics & Electronics,UkrSSR Academy of Sciences, Kharkov 310085, USSR

[++]Institute of Physics and Technology,UkrSSR Academy of Sciences, Kharkov, USSR

Abstract. The anomalous anisotropic current-voltage characteristics (CVC) and magnetic resistance in the temperature range ΔT=1-1,5 K above T_c=91,5 K in a single crystal $YBa_2Cu_2O_{7-\delta}$ were studied. The observed S-shape CVC and the resistivity increasing can be explained by repeated Andreev reflection at S-N boundary and reflection at twin contact.

We have measured resistivity anisotropy in ab-plane of single crystals $YBa_2Cu_3O_{7-\delta}$. Typical dimensions of the crystals were 1,2×1,2×0,03 mm^3.Electrical contacts were made from silver paste and were placed in the corners of crystals as it is shown in insert of Fig.1. Resistivity was calculated according to the method described by Montgomery (1971) for thin crystals. Resistivity in both directions usually was $(2\div3)\times10^{-4}$ Ohm·cm and its anisotropy did not exceed 1,2±0.2. But for one of the crystals resistivity anisotropy was $\rho_1/\rho_2\simeq$ 7,5 with $\rho_1\simeq6\times10^{-4}$ Ohm·cm (ρ_1-resistivity along direction parallel to the line which connects contacts 1 and 2). We have seen in optical microscope a line crossing the crystal as it is shown on insert. We think that this line is twin contact (TC) and that this TC has a large resistance which implies large resistivity anisotropy of the crystal. In the temperature range $\Delta T\simeq$1,5 K above T_c=91,5 K the voltage between the contacts 3 and 4 was doubled, though current between the opposite two contacts was maintaned constant. Similar increasing of resistivity above T_c have been observed by Obolenskii (1989) and Makarenko (1988), where measurements have been made by usual four-probe method.

Figures 1a and 2a show current-voltage characteristics (CVC) and magnetic resistance (MR) for the crystal when contacts 1 and 2 were used as current ones,and contacts 3 and 4-as voltage ones. The same characteristics are shown in Fig. 1b and 2b for the case when contacts 2 and 4 were used as current ones and contact 1 and 3 as voltage ones. In the first case we have S-shape CVC, and in the second case - τ-shape one. Fig. 1b, 2b show that both CVC are almost suppressed by magnetic field $H \simeq 10$ kOe. It make us to conclude that a local layer with critical temperature T_c^l larger than one in the rest volume of the crystal (T_c=91,5 K) exists near TC.

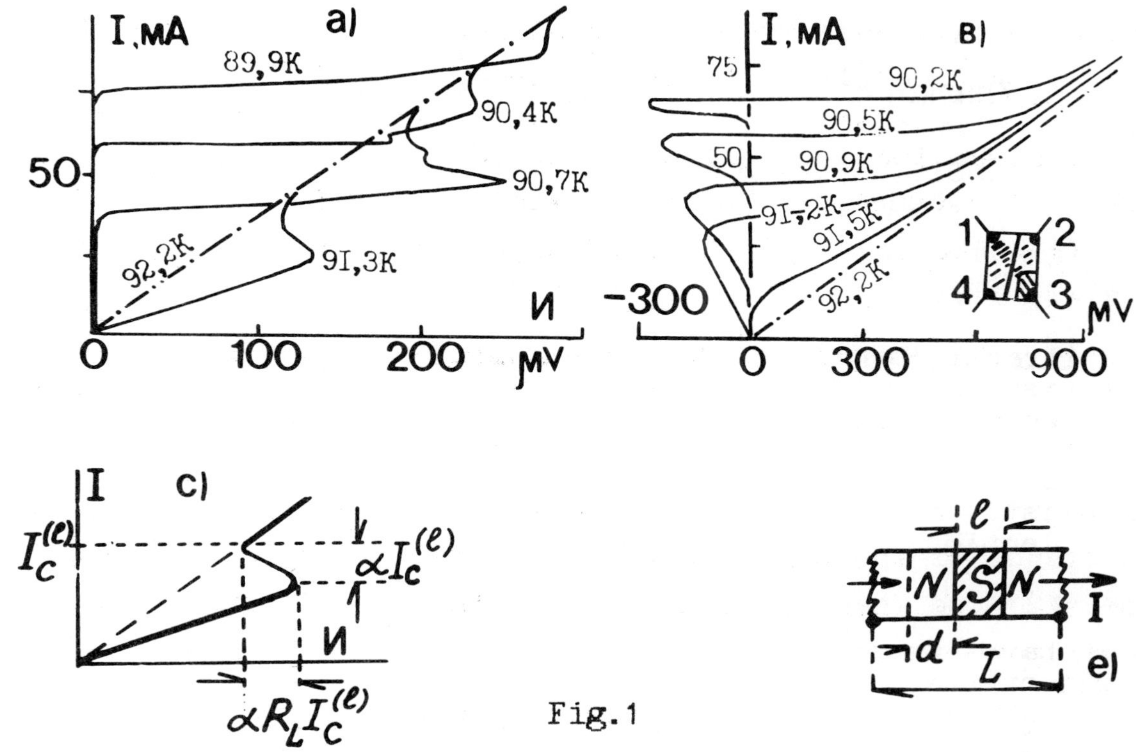

Fig.1

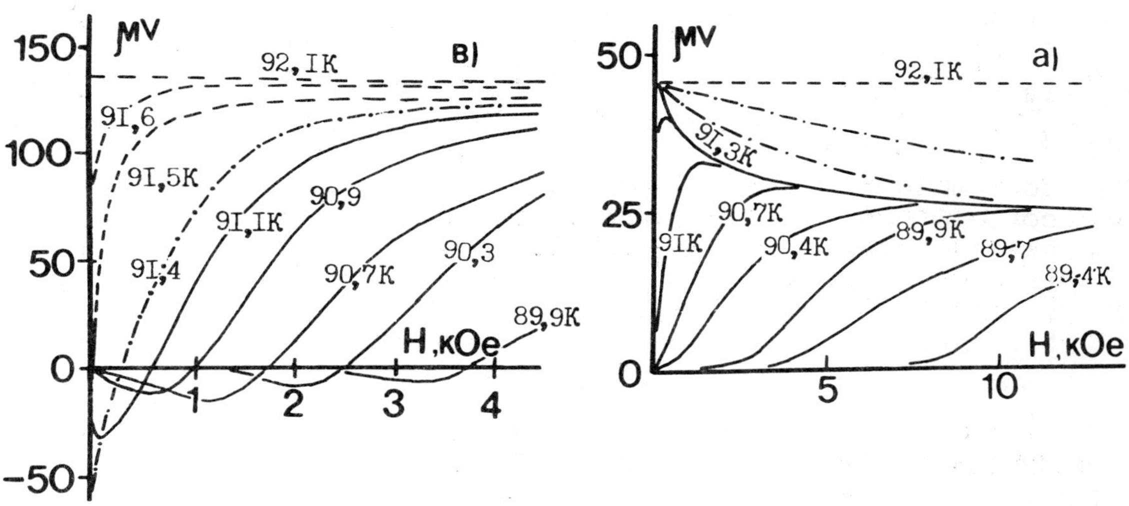

Fig.2

The observed S-shape CVC and the resistivity increasing observed in works by Obolenskii and Makarenko can be explained by repeated Andreev (1967) reflection at S-N boundary and reflection at TC. This mechanism can be realized if the distance between TC and S-N boundary (d) is much less than electron free path (l_o) and when $\varepsilon<\Delta^l$ (Δ^l-superconductive gap of local layer near TC, $\varepsilon_F+\varepsilon$ - electron energy). Total probability of electron reflection at these two boundaries is:

$$|A|^2=2\rho^2(1+\cos\varphi)/(1+2\rho^2\cos\varphi+\rho^4) \qquad (1)$$

where ρ^2 - probability of electron reflection at TC, $\varphi=2\pi dp/\hbar$, p - electron impulse projection on TC-plane. Impulse averaged probability is $|A|^2=2\rho^2/(1+\rho^2)$, and we have $W \equiv |A|^2-\rho^2=(1-\rho^2)\times \times\rho^2/(1+\rho^2)$. Total resistance of the sample shown in Fig.2e (for $\Delta^l \ll kT$) is:

$$R(T)=\begin{cases}(1+\rho^2 l_o/L+W\Delta^l l_o/k_B T_c L)R_L & T_c<T<T_c^l \\ (1+\rho^2 l_o/L)R_L+\delta R^l & \end{cases} \qquad (2)$$

where δR^l - resistance of the local layer in normal state, R_L - resistance of the sample between TC and voltage contacts ($L\gg l_o$). From (2) we can obtain the resistance increase in the temperature range $T_c<T<T_c^l$:

$$\Delta R=\alpha R_L-\delta R^l, \qquad \alpha \equiv W\Delta l_o/k_B T_c L \qquad (3)$$

For the case of $\delta R^p\ll\alpha R_L$ from (2) it is easily shown that

$$U(I)=R_L I\cdot[1+\tau^2 l_o/L+\alpha_o f(T,I,H)] \qquad (4)$$

where $f(T,I,H)\equiv\Delta^l(T,I,H)/\Delta^l(T,I=0,H=0)$, $\alpha_o\equiv \alpha\{\Delta^l(T,I=0,H=0\}$, and $f\sim 1, I\propto f^2(1-f^2)^{1/2}$ (De Gennes,1966). For the case of sample shown in Fig.2b, from (4) and work by De Gennes,1966 it is easily shown that magnetic resistance should be negative and that CVC should be S-shape as it is shown in Fig.2c.

It should be noted that additional contribution into (3) and (4) can be made by "gliding" electron reflection at N-S boundary (Kadygrobov,1988).

Referenses

Montgomery H.C. 1971 J.of Appl.Phys. 47 2971.

Obolenskii M.A. 1989 Sov.Low Temp.Phys. I5 1152.

Makarenko I.N. 1988 JETP Lett. 47 52.

Andreev A.F. 1967 JETP Lett. 53 680.

De Gennes P. 1966 Superconductivity of metals and alloys (W.A.Benjamin,inc. New York - Amsterdam 1966) pp.184.

Kadygrobov A.M. 1988 Sov.Low Temp.Phys. I4 427.

Supercond. Sci. Technol. **4** (1991) S301–303. Printed in the UK

Low-Temperature Specific Heat and Ferromagnetism in $La_4Ba_2Cu_2O_{10}$ and Related Compounds

F. Mizuno, H. Masuda, I. Hirabayashi, S. Tanaka, *M. Hasegawa, *U. Mizutani, #T. Mochiku, #H. Asano, and $F. Izumi
Superconductivity Research Laboratory Nagoya Division, ISTEC, 2-4-1, Mutsuno, Atsuta-ku, Nagoya 456, Japan
* Crystalline Materials Science, Nagoya University, Furo-cho, Chikusa-ku, Nagoya 464, Japan
# Institute of Materials Science, University of Tsukuba, Tennodai, Tsukuba, Ibaraki 305, Japan
$ National Institute for Research in Inorganic Materials, 1-1, Namiki, Tsukuba, Ibaraki 305, Japan

Abstract. We report the low-temperature specific heat and the magnetization measurements of the Cu-O ferromagnet $La_4Ba_2Cu_2O_{10}$ and the related compounds. The overall feature of the specific heat in $La_4Ba_2Cu_2O_{10}$ agrees to that of the mean field approximation with 1/2 spin system and is more 3-dimensional than that of 2-dimensional Cu ferromagnet K_2CuF_4. The measured magnetizations on the single crystals and the estimated anisotropy suggest the Heisenberg system with Ising-like anisotropy due to the comparably large interlayer coupling. In the system of Nd substitution for La, the ferromagnetic transition is not observed because Nd ions having the f-spins may disturb the exchange interaction between interlayer Cu^{2+} ions.

1. Introduction

We have found that $La_4Ba_2Cu_2O_{10}$ (the 422 phase) was a insulating ferromagnet with S=1/2, J=2.6 K, and T_c=5.2 K. [1] The 422 phase has been identified for the first time by Michel *et al.* during their studies on the La_2CuO_4+BaO system. [2] This phase is an insulator and dark brown in colour. The structure is tetragonal with space group P4/mbm and lattice parameters are a=6.8447(9) and c=5.8637(8) Å. [3] The structure selectively contains the edge- and face sharing LaO_8 and BaO_{10} polyhedra, and the square planar CuO_4 is orthogonal to the nearest neighbour CuO_4. All Cu-sites of the 422 phase exists in the same plane and a nearest neighbour distance between Cu ions is 4.84 Å. In this report, first of all, we discuss the characteristics of the spin system from both the results of the magnetic anisotropy in single crystal and low-temperature specific heat. Next, the substitution effects such as Nd for La, Cu for Zn, and nonstoichiometric system denoted for $La_{4-2x}Ba_{2+2x}Cu_{2-x}O_{10-2x}$ are investigated focusing on their magnetism.

2. Experimental

$La_4Ba_2Cu_2O_{10}$, $Nd_4Ba_2Cu_2O_{10}$, $La_4Ba_2(Cu_xZn_{1-x})_2O_{10}$, and $La_{4-2x}Ba_{2+2x}Cu_{2-x}O_{10-2x}$ samples were prepared by solid state reaction method. [1] Single crystals of $La_4Ba_2Cu_2O_{10}$ were grown from BaO-CuO flux, based on the phase diagram for the La_2O_3-CuO-BaO system. [4] Mixtures in a concentration ratio of 1/2 La_2O_3:BaO:CuO=1-1.25:1.8-2.3:2.8-3.3 (molar ratio) were heated in N_2 atmosphere of 5 kg/cm^2 up to 1050 °C, held at this temperature for 5 hrs, cooled to 700 °C at a rate of 3°C/hr, and then furnace-cooled to room temperature.

3. Results and Discussion

3.1 Low-temperature specific heat of $La_4Ba_2Cu_2O_{10}$ ferromagnet

0953-2048/91/01S301 + 03 $03.50

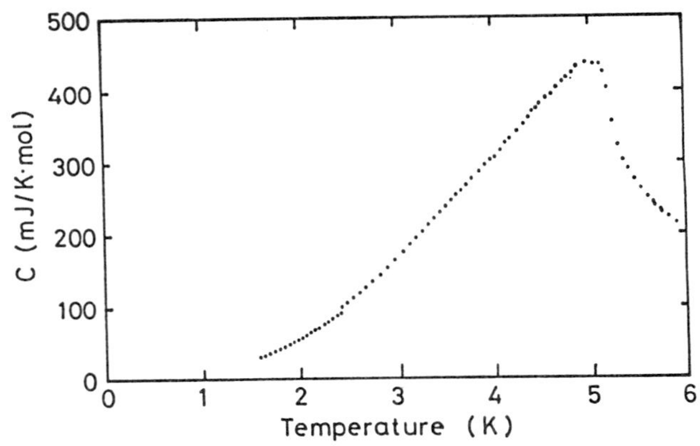

Fig. 1 The low-temperature specific heat for $La_4Ba_2Cu_2O_{10}$ in the range of 1.5-6 K.

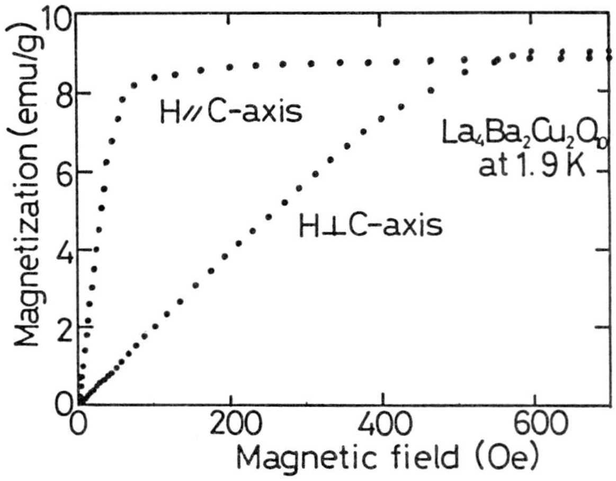

Fig. 2 Magnetization curves parallel and perpendicular to the c-axis at 1.9 K.

The low-temperature specific heat was measured with an ordinary adiabatic DC method using a cylindrical sample in the temperature range of 1.5-6 K. The specific heat of the ferromagnetic sample $La_4Ba_2Cu_2O_{10}$ is shown in Fig. 1. A very large magnetic specific heat is obtained in the low temperature region. The maximum value of the specific heat has a peak at 5.2 K which corresponds to the ferromagnetic transition. The characteristic feature of the temperature dependence below Tc is consistent with that expected on the basis of the mean field approximation for a spin 1/2 3-dimensional ferromagnet. $T^{-d/2}$ dependence due to the spin wave is not observed in this temperature region. A roughly estimation of the magnetic entropy below Tc is 3.6 J/K mol, which corresponds to 63 % of the theoretical total magnetic entropy. Contrary to the larger magnetic entropy below Tc in $La_4Ba_2Cu_2O_{10}$, the large Schottky specific heat which relates to the short-range order above Tc is not observed. This result is because the spin system is more 3-dimensional-like, different from that of typical 2-dimensional ferromagnet K_2CuF_4. It suggests that the interlayer interaction is comparably larger than that of K_2CuF_4 because both the structural difference is reflected on the dimensionality of the spin system. In the case of K_2CuF_4, Cu ion exists not at just above the Cu position in the under layer but at the shifted position with 1/2 lattice periodicity toward each a- and b-direction. On the other hand, in $La_4Ba_2Cu_2O_{10}$, a Cu ion and another one in Cu layer just below belong to the same XY-coordinates. Therefore, Cu distances between interlayers is smaller and the exchange interaction becomes larger.

From the results of the magnetization measurements for the direction parallel and perpendicular to the c-axis on the single crystal of $La_4Ba_2Cu_2O_{10}$, the direction of easy axis is parallel to the c-axis as shown in Fig. 2. This result is consistent with that of the low-temperature specific heat which suggests the Heisenberg system with Ising-like anisotropy due to the comparably large interlayer coupling. On the other hand, in the case of Cu ferromagnet K_2CuF_4, a dipole-dipole interaction is larger than a higher order anisotropic exchange interaction and a-b plane is an easy plane of the magnetization. Therefore, XY-model holds at around T_c and Kosterlitz-Thouless transition has been observed in this system. [5]

3.2 Nd substitution for La ($Nd_4Ba_2Cu_2O_{10}$)

The crystal structure in Ln_2BaCuO_5 system is the same structure as the green phase except for Ln=La and Nd. This structure includes 4 units of Ln_2BaCuO_5 per unit cell and oxygen ions around Cu ion coordinate as a distorted pyramid. In this system, an antiferromagnetic transition occurs at around 10 K. [6] It is interesting to investigate the magnetic properties of $Nd_4Ba_2Cu_2O_{10}$ phase because $Nd_4Ba_2Cu_2O_{10}$ has the same crystal structure as that of

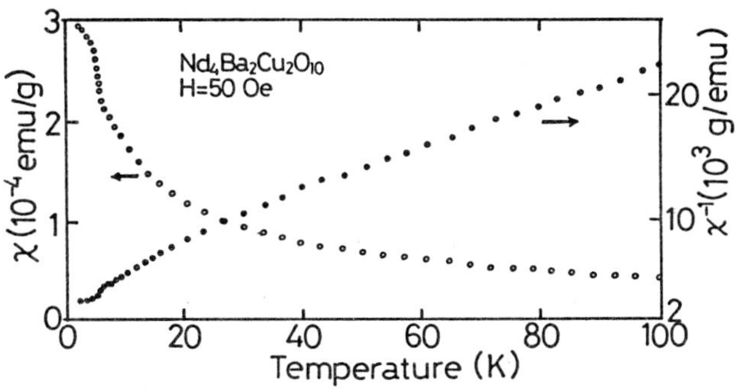

Fig. 3 Temperature dependent susceptibility and its inverse for $Nd_4Ba_2Cu_2O_{10}$ cooled in a field 50 Oe.

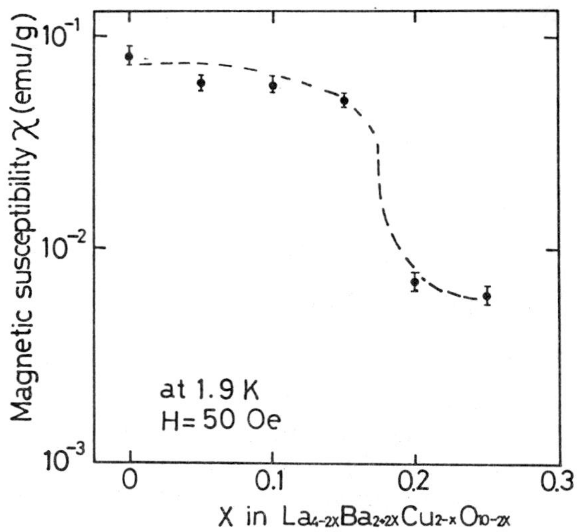

Fig. 4 Magnetic susceptibility of $La_{4-2x}Ba_{2+2x}Cu_{2-x}O_{10-2x}$ in a field of 50 Oe.

$La_4Ba_2Cu_2O_{10}$. The temperature dependence of the DC magnetic susceptibility and its inverse with an external magnetic field 50 Oe for $Nd_4Ba_2Cu_2O_{10}$ is shown in Fig. 3. The ferromagnetic transition is not observed until 1.9 K and a Curie-Weiss plot fitted to the data yields antiferromagnetic interaction below at 4 K. It suggests that interlayer Nd ions which have a magnetic f-spin may disturb the exchange interaction between Cu ions and therefore, ferromagnetism disappears.

3.3 Zn substitution for Cu ($La_4Ba_2(Cu_xZn_{1-x})_2O_{10}$)

We have also tried to realize magnetic dilution by Cu substitution for Zn. We could not achieve the dilution effect because phase separation between $La_4Ba_2Cu_2O_{10}$ and $La_4Ba_2Zn_2O_{10}$ occurs in all composition ranges. The crystal structure of $La_4Ba_2Zn_2O_{10}$ is different from that of $La_4Ba_2Cu_2O_{10}$. In the case of $La_4Ba_2Zn_2O_{10}$, oxygen ions around Zn ion is not quadrilateral but tetrahedral coordination.

3.4 Nonstoichiometric samples for $La_{4-2x}Ba_{2+2x}Cu_{2-x}O_{10-2x}$

In the nonstoichiometric samples for $La_{4-2x}Ba_{2+2x}Cu_{2-x}O_{10-2x}$ denoted by Michel *et al.*, the ferromagnetic signals decrease to 1/10 at x=0.2 as shown in Fig. 4. According to the analysis based on X-ray powder diffraction by Michel *et al.*, they showed that the oxygen deficient sites belonged to CuO_4 planes in their nonstoichiometric sample with x=0.2. Contrary to the Michel's analysis, the square planar CuO_4 groups are complete and the oxygen deficient site belongs to the La-O layers in our preliminary neutron diffraction analysis. The uncompleteness of the La-O interlayers leads to the decay of the ferromagnetism.

References

[1] Mizuno F, Masuda H, Hirabayashi I, Tanaka S, Hasegawa M, and Mizutani U 1990 Nature 345 788

[2] Michel C, Er-rkho L, and Raveau B 1981 J. Solid State Chem. 39 161

[3] Ogawa N, Mizuno F, Masuda H, Hirabayashi I, Tanaka S, Mochiku T, Asano H, and Izumi F Proc. 19th Int. Conf. on Low-Temperature Physics ed D S Betts (Elsevier, Amsterdam) pp 1687-1688

[4] Klibanow D, Sujata K, and Mason T O 1988 J. Am. Ceram. Soc. 71 C-267

[5] Hirakawa K 1982 J. Appl. Phys. 53 1893

[6] Chattopadhyay T, Brown P J, Köbler U, and Wilhelm M 1989 Europhys. Lett. 8 685

Supercond. Sci. Technol. 4 (1991) S304–306. Printed in the UK

Magnetic properties of pure and Fe-doped $Y_1Ba_2Cu_3O_{7-\delta}$ superconductors above T_c and at magnetic fields up to 5.5 T

M A-K Mohamed and J Jung

Department of Physics, University of Alberta, Edmonton T6G 2J1, Canada

Abstract. We report on studies of high magnetic field properties of $Y_1Ba_2Cu_{3-x}Fe_xO_{7-\delta}$ ceramics (x = 0-0.1) around T_c. The total magnetic moment below T_c is a superposition of the diamagnetic and paramagnetic moments. The paramagnetic moment is a linear function of the concentration of Fe and is not affected by possible antiferromagnetic coupling between Fe moments and induced Cu moments.

We report on studies of the magnetic properties of pure $Y_1Ba_2Cu_3O_{7-\delta}$ and Fe-doped $Y_1Ba_2Cu_{3-x}Fe_xO_{7-\delta}$ (x = 0.03, 0.05, 0.1) ceramics over an applied magnetic field range of 0-5.5 T and a temperature range of 80-100 K. Lütgemeier *et al* (1989) and Arai *et al* (1989a) reported that iron atoms substituted for copper in $Y_1Ba_2Cu_3O_{7-\delta}$ carry localized magnetic moments and are in the paramagnetic state. The temperature dependence of magnetic susceptibility in Fe-doped $Y_1Ba_2Cu_3O_{7-\delta}$ ceramics above T_c was studied by Arai *et al* (1989b). They found that the effective magnetic moment per Fe atom calculated from the Curie constant decreases from 5.5 μ_B at $x \leq 0.03$ down to 3.6 μ_B at x = 0.3, suggesting that the reduction of the effective moment is caused by the antiferromagnetic coupling of iron moments with copper moments induced in the direction opposite to the iron moments. On the other hand if a sample of $Y_1Ba_2Cu_3O_{7-\delta}$ is cooled down quickly from high temperature (900°C to 20°C) some oxygen atoms are removed from the chains, causing partial formation of the tetragonal phase. This leads to a partial destruction of superconductivity, and copper atoms appear to have localized magnetic moments (Takabatake *et al* 1987; Kawagoe *et al* 1988; Kadowaki *et al* 1988). Since it has not been clear which process is responsible for the observed magnetic properties, we decided to investigate changes in both the diamagnetic and paramagnetic moments of the pure and Fe-doped $Y_1Ba_2Cu_3O_{7-\delta}$ below and above T_c.

$Y_1Ba_2Cu_{3-x}Fe_xO_{7-\delta}$ ceramics (x = 0-0.1) were prepared by standard solid state reaction method. T_c (dc magnetization) was determined to be between 90-91 K for all samples. The measurement of lattice constants revealed fully oxygenated state of the samples $7-\delta$ = 7.0-6.96. The structure of the samples for x = 0.03, 0.05 is orthorhombic, and for x = 0.1 it is tetragonal.

The measurement of high magnetic field (5.5 T) hysteresis loops below T_c showed the presence of two components of the magnetic moment: diamagnetic and paramagnetic ones (Figs. 1 and 2). In the pure $Y_1Ba_2Cu_3O_{7-\delta}$ both moments are present up to 92 K (Fig. 2) which corresponds to the midpoint of the resistive transition. At this temperature the diamagnetic moment (observed at fields between -500 G and +500 G) is about two-three orders of magnitude smaller than that measured at 90 K. At 95 K only the paramagnetic moment can be seen. The magnetic moment of Fe-doped $Y_1Ba_2Cu_3O_{7-\delta}$, masured below T_c, consists also of the diamagnetic and paramagnetic components. The increase of the concentration of Fe in $Y_1Ba_2Cu_3O_{7-\delta}$ causes the reduction of the diamagnetic moment and the increase of the paramagnetic one (Fig. 1a,b). The total paramagnetic moment measured at 100 K (Fig. 1c) is a linear function of the concentration of Fe in the samples (Fig. 1d).

0953-2048/91/01S304 + 03 $03.50

This linear dependence indicates that the effective magnetic moment per Fe atom does not depend on the concentration of Fe and is not being reduced by possible antiferromagnetic coupling of iron moments with induced copper moments as suggested by Arai *et al* (1989b). It seems also unlikely that the localized magnetic moments of copper, induced by the change of the crystal structure from orthorhombic (x = 0.03, 0.05) to tetragonal (x = 0.1), are able to reduce the effective magnetic moment per Fe atom.

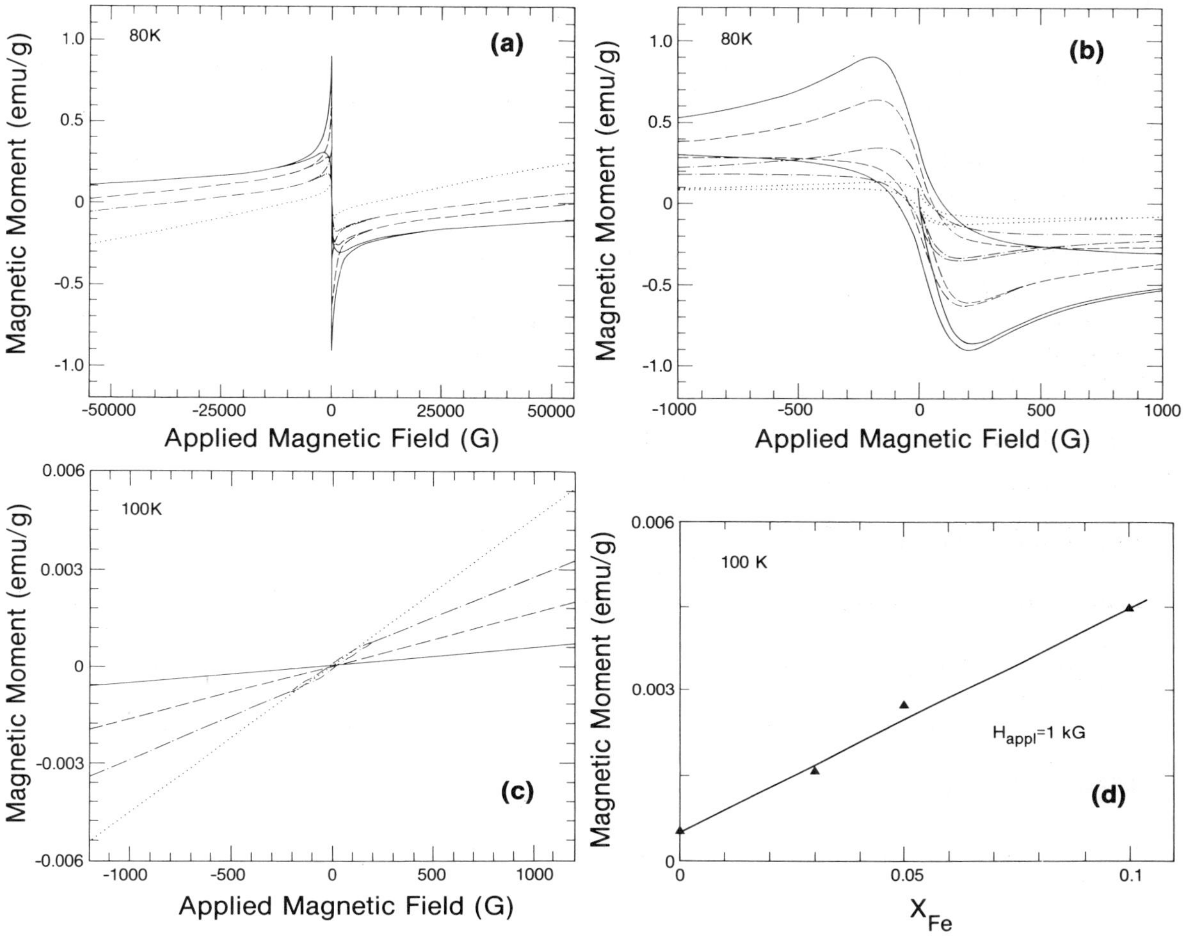

Fig. 1. The high magnetic field (5.5 T) magnetic moments of the pure and Fe-doped $Y_1Ba_2Cu_{3-x}Fe_xO_{7-\delta}$ (x = 0-0.1) ceramics. Solid line: x = 0; dashed line: x = 0.03; dash-dotted line: x = 0.05; dotted line: x = 0.1. (a) The superposition of the diamagnetic hysteresis loops and paramagnetic moments measured at T = 80 K up to 5.5 T; (b) the same for a range of applied fields between -1 kG and +1 kG; (c) the paramagnetic moments at T=100 K; (d) the dependence of the paramagnetic moment at 1 kG and T=100 K on the concentration of Fe.

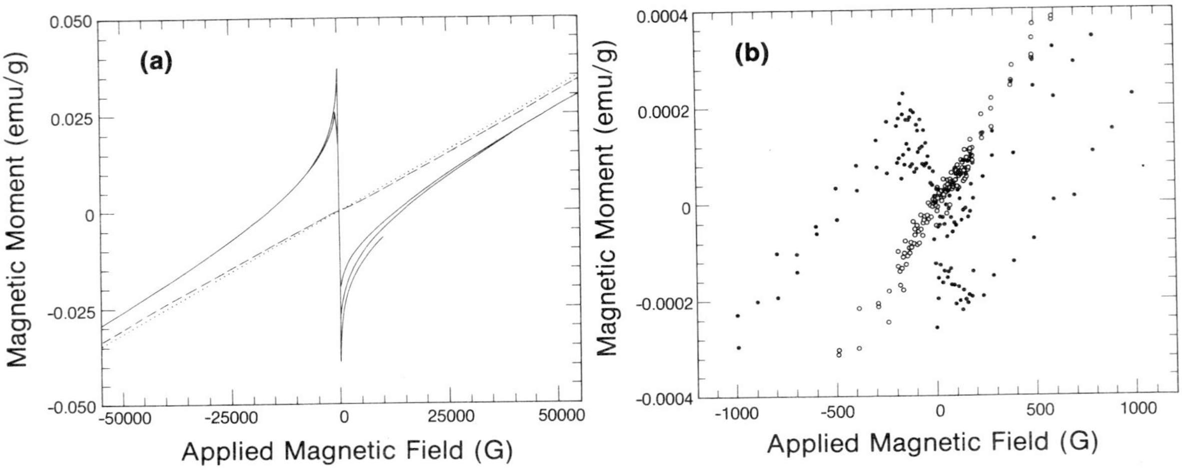

Fig. 2. The change of the high field (5.5 T) diamagnetic hysteresis loop for the pure $Y_1Ba_2Cu_3O_{7-\delta}$ around T_c. (a) Solid line: T = 90 K, dashed line: T = 92 K, dotted line: T = 95 K; (b) Solid symbols: the superposition of the diamagnetic and paramagnetic signals at T = 92 K, open symbols: the paramagnetic signal at T = 95 K.

References:

Arai J and Shimizu H 1989a *Physica* **C 162-164** 1295

Arai J, Shimizu H and Mita M 1989b *J. Phys. Soc. Jpn.* **58** 673

Kadowaki H, Nishi M, Yamada Y, Takeya H, Takei H, Shapiro S M and Shirane G 1988 *Phys. Rev.* **B 37** 7932

Kawagoe T, Mizoguchi T, Kanoda K, Takahashi T, Hasumi M and Kagoshima S 1988 *J. Phys. Soc. Jpn.* **57** 2272

Lütgemeier H, Brand R A, Sauer Ch, Rupp B, Meuffels P M and Zinn W 1989 *Physica* **C 162-164**, 2395

Takabatake T, Ishikawa M and Sugano T 1987 *Jpn. J. Appl. Phys.* **26** L1859

Supercond. Sci. Technol. **4** (1991) S307–309. Printed in the UK

Oxygen Content and Superconductivity of Ca–to–Y Substitution in $YBa_2Cu_3O_y$

N.L.Wang[1),3)], M.C.Tan[1)], J.S.Wang[1)], S.Y.Zhang[2)], H.B.Jiang[1)], J.Sha[3)], and Q.R.Zhang[1),3)]

1) Department of Physics, Zhejiang University, Hangzhou 310027, P.R.China
2) Department of Chemistry, Zhejiang University, Hangzhou 310027, P.R.China
3) Department of Physics, University of Science and Technology of China, Hefei 230026, P.R.China

Abstract. The oxygen content and superconductivity of Ca–to–Y substitution in $YBa_2Cu_3O_y$ was studied. By comparison with that of Pr substituted $YBa_2Cu_3O_y$ system, it is found that the oxygen content as well as the formal oxidation state of Cu is not a factor directly predominating Tc. The depression of Tc is connected with the reducing of the mobile holes in CuO_2 sheets.

1. Introduction

Variable oxygen content and formal oxidation state of Cu is known to be important to the high temperature superconductivity. Upon decreasing the oxygen concentration, the $YBa_2Cu_3O_y$ compound, for example, transforms from a 90K superconductor, to a 60K superconductor, and finally to an insulator. Correspondingly the formal oxidation state of Cu decreases from 2.33 at y=7 to 1.67 at y=6, and the lattice structure undergoes a orthogonal to tetragonal transition due to the oxygen ordering in Cu(1)-O chains. We recently investigated the oxygen contents and formal oxidation states of Cu of Ca substituted $YBa_2Cu_3O_y$ superconductors. By comparison with that of Pr substituted $YBa_2Cu_3O_y$ system, we find that the oxygen content as well as the formal oxidation state of Cu is not a factor directly predominating Tc. We explain the results by using the idea of Y.Tokura et al.[1] of dividing the holes into the localized chain holes and mobile sheet holes. The depression of Tc is connected with the reducing of the mobile holes in CuO_2 sheets.

2. Experiments

The samples with nominal compositions of $Y_{1-x}Ca_xBa_2Cu_3O_y$ (x=0,0.05, 0.1, 0.15, 0.20, 0.30) were prepared by solid state reaction of mixtures of approprite analytically pure Y_2O_3, $CaCO_3$, $BaCO_3$ and CuO powders. They were preheated at 930°C in air for 24 hours, sintered at 940°C in air for 24 hours, then cooled down to room temperature with the furnace. The transition temperature Tc was determined by both resistivity and ac susceptibility measurements. The analysis of the crystal structure was carried out on a RIGACU D-Max-rA^2 rotating target X-ray diffractometer. The oxygen content was determined by idometric titration techniques.

3. Results and Discussions

For $Y_{1-x}Ca_xBa_2Cu_3O_y$ system, the superconducting transition temperature Tc decreased with x increasing X-ray diffraction (XRD) indicated that the samples were the orthogonal single phase within the range of $0<x\leq0.15$. In table I, we presented the results of Tc, lattice parameters, oxygen contents, and formal oxidation state Cu^{+2+p} (i.e.

Table I. Data for $Y_{1-x}Ca_xBa_2Cu_3O_y$ system

	Tc(K)	a(Å)	b(Å)	c(Å)	b-a(Å)	y	p in Cu^{+2+p}
$YBa_2Cu_3O_y$	90	3.810	3.876	11.65	0.066	7.04	0.360
$Y_{0.9}Ca_{0.1}Ba_2Cu_3O_y$	83	3.812	3.876	11.64	0.064	6.87	0.283
$Y_{0.85}Ca_{0.15}Ba_2Cu_3O_y$	80	3.816	3.876	11.66	0.060	6.68	0.168

Table II. Data for $Y_{1-x}Pr_xBa_2Cu_3O_y$ system

	Tc(K)	a(Å)	b(Å)	c(Å)	b-a(Å)	y	p in Cu^{+2+p}
$Y_{0.9}Pr_{0.1}Ba_2Cu_3O_y$	78	3.818	3.872	11.62	0.054	6.95	0.267
$Y_{0.7}Pr_{0.3}Ba_2Cu_3O_y$	49	3.826	3.874	11.64	0.048	7.05	0.267
$Y_{0.5}Pr_{0.5}Ba_2Cu_3O_y$	21.5	3.828	3.876	11.65	0.048	7.10	0.233
$Y_{0.3}Pr_{0.7}Ba_2Cu_3O_y$	—	3.828	3.876	11.63	0.048	7.13	0.186

p, the holes per Cu site) for different x. For comparison, the data for $Y_{1-x}Pr_xBa_2Cu_3O_y$ system were shown in table II, the oxygen contents were measured previously by TGA[2], the p values determined by charge balance by assuming that the effective valence state of Pr is +4[3]. So Pr^{+4} may act as doners and fill holes in CuO_2 sseets. On the other hand, Ca has a valence of +2, it may dope the CuO_2 sheets with holes as the effect of Ca^{+2} substitution for Y^{+3}. However one must be careful about it, for the oxygen content can be changed, resulting in a variable p values. It can be seen that the oxygen content decreases in $Y_{1-x}Ca_xBa_2Cu_3O_y$ system and increases in $Y_{1-x}Pr_xBa_2Cu_3O_y$ system with x increasing. As Y site is far from Cu(1)-O chain, and powder XRD indicates that the lattice parameters, and therefore the ordering of Cu(1)-O chains measured by the differences of b-a, change little, the change of oxygen that comes from the chains is questioned. We suggest that it mainly comes from the oxygen vacancies in CuO_2 sheets (i.e. the occupation rate of the oxygens changing). When Ca^{+2} substitutes for Y^{+3}, Ca^{+2}_2 exerts a weaker attraction on O^{-2} than Y^{+3} does, and more vacancies may be produced. However Pr^{+4} just does the opposite. Because Tc decreases in both systems, we conclude that the oxygen content does not predominate Tc directly. Besides, the formal oxidation states of Cu in $Y_{0.9}Pr_{0.1}Ba_2Cu_3O_{6.95}$ and $Y_{0.7}Pr_{0.3}Ba_2Cu_3O_{7.05}$ are the same but the Tc has a difference of about 30K. This illustrates that the formal oxidation state of Cu can not be a factor directly predominating Tc either.

One important problem is the relationship between the holes determined by the method of chemistry or TGA with the holes in CuO_2 sheets, usually considered as the conducting carriers being responsible for superconductivity. Y.Tokura et al. [1] suggest that the number of holes, as determined chemically, may be divided between localized chain holes and mobile sheet holes. By plotting the oxygen content vs hole concentration p for various substitution in $YBa_2Cu_3O_y$, they could clearly seperate the insulators from the metals by a straight line, $p_c=(y-6.5)/3$. All the samples with $p<p_c$ at certain y are insulating, and the holes must be localized, presumably on the chains. M.W.Shafer et al.[4] has shown that the trends of the hole number on the sheets agrees well with that determined by Hall measurements. Since the measured charge, p, is an average over the charge in

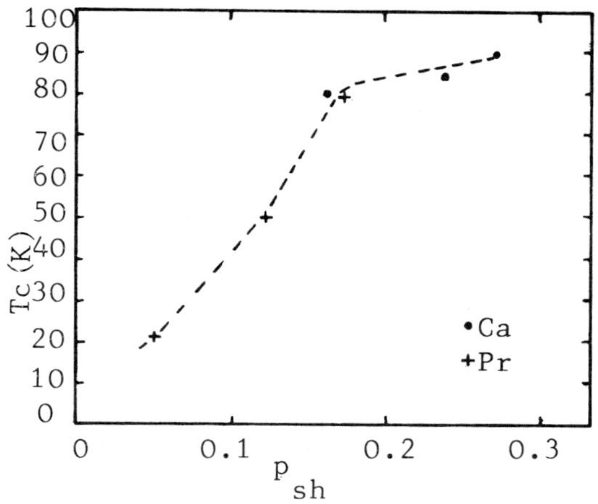

Figure 1. Tc vs holes in CuO_2 Sheets for Ca and Pr substituted $YBa_2Cu_3O_y$

the sheets and chain, or $p=(2p_{sh}+p_{ch})/3$, at the threshold, $p_{sh}=0$, then $p_c=p_{ch}/3$, the mobile sheet holes per Cu site is thus given by $p_{sh}=3(p-p_c)/2$. Based on the above viewpoint, the Tc vs the mobile sheet holes per Cu site for $Y_{1-x}Ca_xBa_2Cu_3O_y$ and $Y_{1-x}Pr_xBa_2Cu_3O_y$ systems is shown in figure 1. It can be seen that the mobile sheet holes decrease in both Ca and Pr substituted $YBa_2Cu_3O_y$ systems as x increases. The $Y_{0.3}Pr_{0.7}Ba_2Cu_3O_{7.13}$ is in the insulating region, which agrees with the experiment, though its average charge p is greater than that of $Y_{0.85}Ca_{0.15}Ba_2Cu_3O_{6.68}$. Therefore the depression of Tc is correlated with the reduction of the mobile sheet holes in both systems. When the mobile sheet holes per Cu site are in the range of 0.16-0.27, the change of Tc is not greater; when less than 0.16, the depression of Tc is rather significant with the reducing of p_{sh}.

In conclusion, we have investigated the oxygen contents, formal oxidation state of Cu of Ca and Pr substituted $YBa_2Cu_3O_y$ systems, and found that neither of them can be a factor directly predominating Tc. The depression of Tc is connected with the reducing of conducting holes in CuO_2 sheets.

Acknowledgement

This work was supported by the Natural Science Foundation of China.

References

[1] Y.Tokura, J.B.Torrance, T.C. Huang, and A.I.Nazzal Phys. Rev. B 38 (1988) 7156

[2] H.Zhang, X.Y.Zhou, Y.Zhao, S. H.Liu, and Q.R.Zhang Solid State Commun. 72 (1989) 75

[3] J.J.Neumeier, T.Bjornholm, M.B. Maple, and Ivan K. Schiller Phys.Rev.Lett. 63 (1989) 2516

[4] M.W.Shafer, T,Penney, B.L. Olson,R.I.Greene, and R.H.koch Phys.Rev. B 39 (1989) 2914

Supercond. Sci. Technol. **4** (1991) S310–312. Printed in the UK

An investigation of the new high-T_c Tl-V-Sr-Ca-Cu-O and Tl-V-Sr-Ca-Y-Cu-O superconductors

R. S. Liu, W. Zhou, R. Janes and P. P. Edwards
IRC in superconductivity, University of Cambridge, Madingley Road, Cambridge CB3 0HE, U.K.

Abstract. Superconductivity up to 105K has been achieved in the Tl-V-Sr-Ca-Cu-O and Tl-V-Sr-Ca-Y-Cu-O systems. The partial substitution of vanadium for thallium in the crystal lattice stablizes the $TlSr_2CaCu_2O_7$ phase and enhances superconductivity up to 80K. Importantly, 20% of Y^{3+} replacement of Ca^{2+} sites in the $(Tl,V)Sr_2CaCu_2O_y$ compound reduces the excess hole concentration and increases superconductivity by about 20K.

1. Introduction

The discovery of a low volume fraction (<1%) superconductivity in the multiphase Tl-Sr-Ca-Cu-O (Nagashima *et al* 1988; Sheng *et al* 1988) system, with T_c ranging from 20K to 80K, has led to widespread research aimed both at identifying the superconducting phase, and stablizing the superconducting transition temperature. The substitution of Pb^{4+} cations (Subramanian *et al* 1988) or $Bi^{3+\sim5+}$ cations (Halder *et al* 1988) into the trivalent Tl sites in $TlSr_2CaCu_2O_7$ compound results in the stabilization of the crystal structure and bulk superconductivity with T_c up to 80K. Similarly, introduction of trivalent rare earth ions (R) into (divalent) Ca sites leads to superconductivity around 80K in the $TlSr_2(Ca,R)Cu_2O_{7-\delta}$ compounds (Sheng *et al* 1989). Liu *et al* (1990) have also found that the combined substitution of Pb and rare earth (e.g. Y) ions into the Tl and Ca sites, respectively, results in a monophasic, septenary compound $(Tl_{0.5}Pb_{0.5})Sr_2(Ca_{0.8}Y_{0.2})Cu_2O_{7-\delta}$ with a T_c up to 110K. All of these studies, therefore, lead one to conclude that inherent "over-doping" of the parent $TlSr_2CaCu_2O_7$ compound can be diminished by high valent element substitution into lower valent metal lattice sites. Vanadium is one such candidate; having a range of oxidation states, including V^{3+}, V^{4+} and V^{5+}, as found, for example, in the oxides V_2O_3, VO_2 and V_2O_5. We report here the synthesis and investigation of a new high-T_c material containing vanadium with a superconducting transition temperature up to 105K.

2. Experimental

Samples with nominal compositions of $(Tl_{1-x}V_x)Sr_2(Ca_{1-y}Y_y)Cu_2O_{7-\delta}$ ($0\leq x \leq 1$ and $0\leq y \leq 1$) were prepared by solid state reaction from stoichiometric mixtures of high purity Tl_2O_3 and/ or V_2O_5 , SrO_2, CaO and/ or Y_2O_3 and CuO powders. The powders were weighed, mixed and pressed into pellets with dimensions of 10mm diameter and 2mm thick using a pressure of 5 ton/cm^2. The pellets were then wrapped with gold foil to alleviate loss of thallium during the heat treatment. The sample was then sintered at 950°C for 3 hours in flowing oxygen, followed by cooling to room temperature at a rate of 5°C/min.

A standard four point probe method was used for the electrical resistivity measurement. Magnetisation data was obtained using a superconducting quantum interference device (SQUID) magnetometer (Quantum Design). X-ray diffraction (XRD) studies were carried out with CuKα radiation using a Spectrolab CPS-120 diffractometer. From the XRD measurements of the compounds with nominal compositions $(Tl_xV_{1-x})Sr_2(Ca_yY_{1-y})Cu_2O_{7-\delta}$ ($0\leq x, y \leq 1$), the majority of the peaks for each compound can be indexed on the basis of the

0953-2048/91/01S310 + 03 $03.50

tetragonal unit cell (P4/mmm) and lattice constants a~3.8Å and c~12Å except for small amounts of $SrVO_3$-type impurity phase. Varying chemical compositions were examined by energy dispersive X-ray spectrometry (EDS) in a Jeol JEM-200CX electron microscope. Electron spin resonance (ESR) spectra were recorded between ca.77K and room-temperature using a Bruker ER200D X-band spectrometer.

3. Results and discussion

Figure 1 shows the temperature dependence of the (normalised) electrical resistivity for samples with nominal compositions of $TlSr_2CaCu_2O_y$ (a), $(Tl_{0.5}V_{0.5})Sr_2CaCu_2O_y$ (b) and $(Tl_{0.5}V_{0.5})Sr_2(Ca_{0.8}Y_{0.2})Cu_2O_y$ (c). The $TlSr_2CaCu_2O_y$ sample clearly exhibits a metallic behaviour [room temperature resistivity (ρ_{300K}) of 7.15 mΩcm], down to 4K except for a very small (but reproducible) resistivity drop around 84K, as shown in Fig. 1(a). This anomaly at 84K may arise from a small amount of the superconducting phase of the $TlSr_2CaCu_2O_y$ present in this sample. The $(Tl_{0.5}V_{0.5})Sr_2CaCu_2O_y$ {the actual composition as determined by EDS is $[Tl_{0.72(4)}V_{0.28(4)}]Sr_{2.1(1)}Ca_{1.04(8)}Cu_2O_y$}sample [Fig. 1(b)] displayed metallic behaviour in the normal state (ρ_{300K}=3.2 mΩcm) and a two-step resistivity drop at 110K and 80K, and $T_{c(zero)}$=57K. Interestingly, the $(Tl_{0.5}V_{0.5})Sr_2(Ca_{0.8}Y_{0.2})Cu_2O_y$ {the actual composition as determined by EDS is $[Tl_{0.71(6)}V_{0.27(4)}]Sr_{1.95(9)}[Ca_{0.86(7)}Y_{0.14(3)}]Cu_2O_y$}sample possessed the largest room temperature resistivity (ρ_{300K}=17.4 mΩcm) but also has the highest superconducting transition temperature, with $T_{c(onset)}$=110K, $T_{c(mid\text{-}point)}$=100K and $T_{c(zero)}$=87K. All the transition temperstures were also observed by the dc magnetisation measurements for these samples.

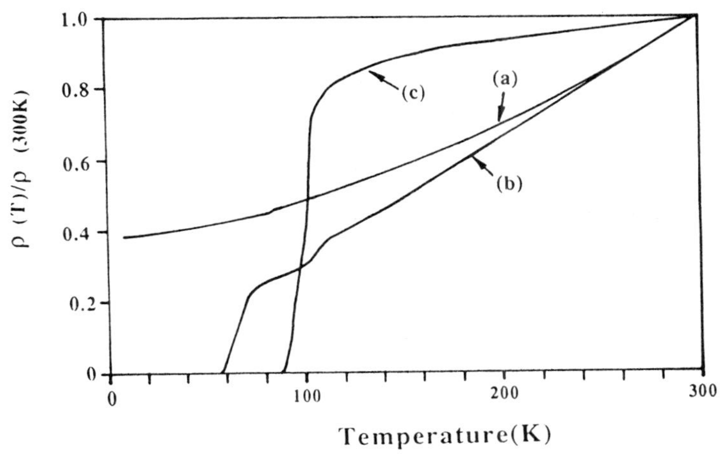

Fig. 1 Temperature dependence of the normalised resistivity for samples with nominal compositions of $TlSr_2CaCu_2O_y$ (a), $(Tl_{0.5}V_{0.5})Sr_2CaCu_2O_y$ (b) and $(Tl_{0.5}V_{0.5})Sr_2(Ca_{0.8}Y_{0.2})Cu_2O_y$ (c).

ESR spectra of the two vanadium containing superconducting phases with nominal compositions of $(Tl_{0.5}V_{0.5})Sr_2CaCu_2O_y$ and $(Tl_{0.5}V_{0.5})Sr_2(Ca_{0.8}Y_{0.2})Cu_2O_y$ were measured between room-temperature and 77K, with a view to probing the valence state of the vanadium ions. The synthetic conditions (see earlier), would almost certainly decompose V_2O_5 to produce V_2O_3 and/ or VO_2. The ESR of localised V^{4+} ($3d^1$), V^{3+}($3d^2$) and V^{2+}($3d^3$) have all been extensively investigated (Abragam *et al* 1986). However, neither superconducting phase gave an ESR signal. The V^{3+} ion has been previously detected as a dopant in Al_2O_3 (Abragam *et al* 1986). It is characterised by a very large spin-lattice relaxation rate (a consequence of strong spin-orbit coupling and accessible, low-lying excited states) meaning the spectrum could only be observed at liquid helium temperatures. Clearly, low temperatures are required to probe the valence state of vanadium via ESR, and in this regard, we were hampered by the appearance of a intense low-field microwave absorption that appeared on cooling the sample through T_c. This response is characteristic of the superconducting state and has been widely studied (Sastry *et al* 1989). Figure 2 shows the low-field microwave absorption from both samples at 77K, the signal from $(Tl_{0.5}V_{0.5})Sr_2(Ca_{0.8}Y_{0.2})Cu_2O_y$ (b) being significantly more intense and broader than that from $(Tl_{0.5}V_{0.5})Sr_2CaCu_2O_y$ (a), equivalent quantities of material being used in each case. The variation in the signal intensity of $(Tl_{0.5}V_{0.5})Sr_2CaCu_2O_y$ as a function of microwave power is also shown as inset in Fig. 2; the signal exhibits an initial enhancement followed by saturation as the microwave power is raised, the latter being accompanied by a significant detuning of the microwave bridge. This response is common to all cuprate superconductors (Janes *et al* 1990).

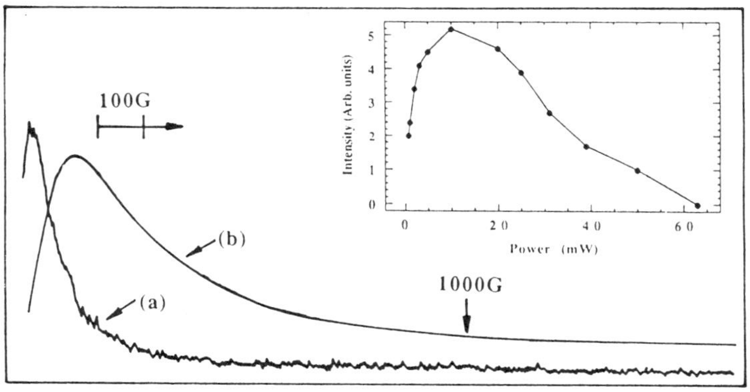

Fig. 2 Modulated low-field microwave absorption measured at 77K for $(Tl_{0.5}V_{0.5})Sr_2CaCu_2O_y$ (a) and $(Tl_{0.5}V_{0.5})Sr_2(Ca_{0.8}Y_{0.2})Cu_2O_y$ (b). Inset shows the variation of the signal intensity as a function of incident microwave power for $(Tl_{0.5}V_{0.5})Sr_2CaCu_2O_y$.

Hence, we conclude that the partial replacement of vanadium for thallium in the crystal lattice stabilizes the $(Tl,V)Sr_2CaCu_2O_y$ phase and enhances superconductivity up to 80K, which is comparable to the corresponding $(Tl,Pb)Sr_2CaCu_2O_y$ phase, with T_c around 85K. Further substitution of 20% Y^{3+} in the Ca^{2+} sites, corresponding to $(Tl,V)Sr_2(Ca_{0.8}Y_{0.2})Cu_2O_y$, effectively reduces the hole concentration and increases T_c to 105K. This is a similar effect to that found in the $(Tl,Pb)Sr_2(Ca_{0.8}Y_{0.2})Cu_2O_y$ system (Liu *et al* 1990). Pb, Bi and V, therefore, can stabilize the $TlSr_2CaCu_2O_y$ structure even though their electronic configurations are quite different. Other 3d elements with accessible 3+ and or 4+ valence states may also be candidates for stabilizing the $TlSr_2CaCu_2Oy$ phase.

We thank the SERC and British Petroleum for support. One of us (R. S. L.) thanks the Materials Research Laboratories, Industrial Technology Research Institute, Taiwan for the finacial support to study in Cambridge.

References

Nagashima T *et al* 1988 *Jpn. J. Appl. Phys.* **27** L1077
Sheng Z Z *et al* 1988 *Phys. Rev.* **B 38** 7074
Subramanian M A *et al* 1988 *Science* **242** 249
Haldar P *et al* 1988 *J. Supercond.* **1** 211
Sheng Z Z *et al* 1989 *Phys. Rev.* **B 39** 2918
Liu R S *et al* 1990 *J. Solid State Chem.* **86** 334
Abragam A and Bleaney B 1986 *Electron Paramagnetic Resonance of Transition Ions* (Dover Publications, New York)
Sastry M D 1989 *Studies of High Temperature Superconductors* (Nova Science Publishers, New York) Chap. 10
Janes R *et al* 1990 *Physica* **C 167** 520

Supercond. Sci. Technol. **4** (1991) S313–315. Printed in the UK

ON THE EFFECT OF Ni SUBSTITUTION IN SUPERCONDUCTING $YBa_2Cu_3O_{7-\delta}$

Nianhua Peng, Jun Yuan, Dongning Zheng and Wei Yao Liang
IRC in Superconductivity, University of Cambridge, Cambridge CB3 0HE, U.K

Abstract. Electrical resistivity, AC magnetic susceptibility, X-ray powder diffraction, EDAX and EELS near edge absorption, have been used to study a series of Ni substituted dense (mass density ~ 6.0 g/cm^3) superconducting $YBa_2Cu_{3-x}Ni_xO_{7-\delta}$ with a wide range of x values. T_c was found to decrease linearly with increasing Ni concentration. We have also found phase separation occuring at high Ni concentration due to the formation of Y_2BaCuO_5, Y_2BaNiO_5 and NiO/BaO which seriously modifies the Ni concentration from its nominal values. For this reason many earlier measurements on the effects of Ni substitution on the physical properties of $YBa_2Cu_3O_{7-\delta}$ may need to be re-examined and this paper attempts to correlate the above physical properties with actual Ni concentrations in the samples.

1. Introduction

Ni substitution for Cu in $YBa_2Cu_3O_{7-\delta}$ has been known (Ginsberg 1989) to depress T_c and broaden the superconducting transition as well as enhance the thermodynamic fluctuations near T_c. However, various authors reported different results concerning these effects. For example, Qian et al (1989) found that nearly all the Ni existed in NiO precipates and concluded that no Ni substitution took place. On the other hand, while Adrian et al (1988) and Le Dang et al (1989) found that T_c reduced linearly with increasing Ni doping, Zhao et al (1988) found a non-linear behaviour over a wide range of Ni concentration. In agreement with Bridges et al (1990) and Le Dang et al (1989), we also found smaller Ni concentrations in $YBa_2Cu_{3-x}Ni_xO_{7-\delta}$ than their nominal values, and this paper presents results of further studies of Ni substitution in $YBa_2Cu_3O_{7-\delta}$.

2. Results and Discussions

$YBa_2Cu_{3-x}Ni_xO_{7-\delta}$ samples of nominal composition x = 0.0025, 0.005, 0.01, 0.02, 0.03, 0.04, 0.05, 0.1, 0.2, 0.25, 0.5 were prepared by solid state reactions starting with stoichiometric amounts of Y_2O_3, CuO, $BaCO_3$ and NiO. The mixed oxides were sintered at 940 °C for 16 hours, cooled, ground and pelletised. The pellets were then annealed in an oxygen atmosphere also at 940 °C for 6 hours, followed by slow cooling to 400 °C; held for 16 hours before slow cooling to room temperature. The density of all the samples was in the range 5.8-6.2 g/cm^3. XRD showed that within the experimental uncertainty the lattice parameters were unchanged with respect to dopant concentration. It should be noted that although XRD found no detectable evidence of impurity phases, careful examinations by means of microprobe EDAX did reveal the presence of small concentrations of Y_2BaCuO_5, Y_2BaNiO_5, $BaCuO_2$, $BaNiO_2$ and NiO. The same experiments also showed that there was a uniform distribution of Ni with a constant spatial variation independent of the Ni concentration. As illustrated in Fig. 1, we found in general, the actual Ni concentration was only about half of the nominal concentration. It appeared that Y_2BaNiO_5 and $BaNiO_2$ could be easily grown under the same condition as for $YBa_2Cu_3O_{7-\delta}$. This suggests that the deviation from nominal concentration can not simply be attributed to the presence of NiO alone in the final product, as argued by Qian et al (1989) and

Bridges et al (1990). Although a linear fit was given to the data presented in Fig. 1 for the range $0.1 \leq x \leq 0.5$, this should not be extrapolated towards low x values of Ni.

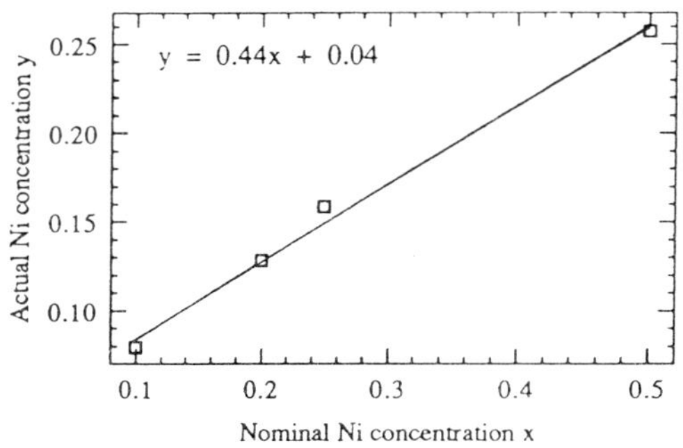

Fig. 1. Comparison between actual and norminal Ni concentrations.

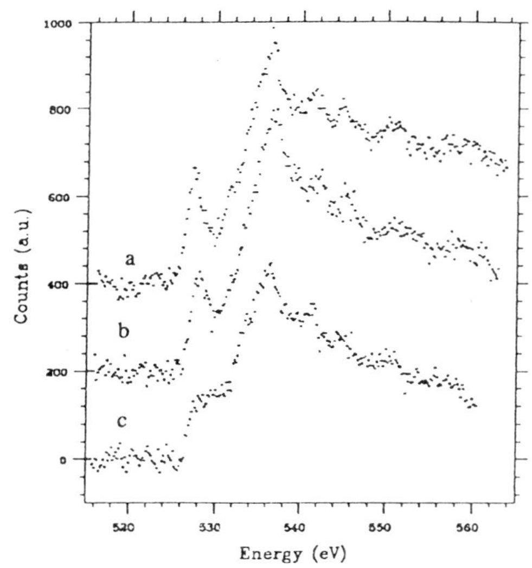

Fig.2. EEL spectra of O 1s pre-edge structure, (a) pristine sample of $YBa_2Cu_3O_{7-\delta}$, (b) $YBa_2Cu_{3-x}Ni_xO_{7-\delta}$ with norminal x = 0.5, (c) $YBa_2Cu_3O_{7-\delta}$ after 30 min. irradiation in high flux electron beam.

Although Ni substitution has been observed to depress T_c of $YBa_2Cu_{3-x}Ni_xO_{7-\delta}$, the degree of the influence by Ni has been found to vary considerably (Westerholt et al 1989, Le Dang et al 1989 and Zhao et al 1988). We believe this is due to the different amounts of impurity phases present, and to a lack of appropriate characterisation of the samples. At the same time the presence of impurity phases also makes accurate determination of oxygen stoichiometry in the bulk specimens impossible in practice. Any reduction in oxygen would further depress T_c in $YBa_2Cu_{3-x}Ni_xO_{7-\delta}$. Usual microprobe analysis techniques such as EDAX are only good for measuring the stoichiometry of heavy elements but not for oxygen. We resort to electron energy loss (EELS) experiments to examine the oxygen content of our samples. Here the intensity of the pre-edge peak (~ 528 eV) due to oxygen 1s core transition has often been related with oxygen concentration. The experiments examine individual grains of approximately 10 μm diameter and usually of a single phase which can easily be determined by the relative intensities of absorption peaks of Ba, Cu and Ni. In this way, grains composed of $YBa_2Cu_{3-x}Ni_xO_{7-\delta}$ or other impurity phases can be easily distinguished. In Fig.2, curve (a) was taken from a fresh specimen of $YBa_2Cu_3O_{7-\delta}$. The strong absorption peak at 528 eV is indicative of small δ with which other spectra are compared. Curve (b) shows the absorption spectrum of $YBa_2Cu_{3-x}Ni_xO_{7-\delta}$, where x is nominally 0.5, while curve (c) shows the spectrum of $YBa_2Cu_3O_{7-\delta}$ taken after the specimen has been irradiated by a high flux of 100 keV electron for 30 minutes. The effects of irradiation on loss of oxygen can be clearly seen in curve (c). By comparison, curve (b) shows little depreciation of oxygen, and this reduction is even smaller in samples of lower Ni concentrations (not shown).

Fig. 3 shows that after correction to the actual concentration of Ni a linear depression of T_c is found. This depression has been observed in other impurity 3d ion substitutions of Cu (Ginsberg 1989). In the case of Ni substitution, the spin 1/2 Cu ion is replaced by a spin 1 Ni ion. This has arguably the same effect on magnetic scattering of carriers and on disruption of the Cu lattice as, for example Zn substitution, since the latter presents an equal but opposite sign spin difference to the Cu ion. This is thought to be the dominant effect although charge transfer will also modify the density of states at E_f. The linear behaviour is consistent with the theory of Abrikosov and Gor'kov (1960) in the low concentration range of magnetic impurity. It would be interesting to try to increase the Ni concentration further (x > 0.5) in order to explore the region where the Abrikosov-Gor'kov theory departs from linearity.

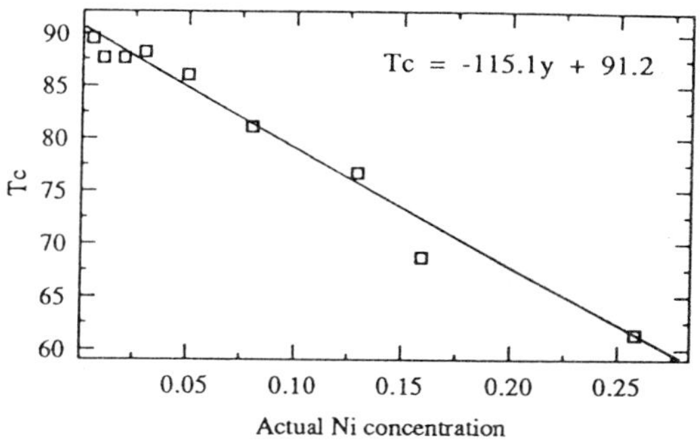

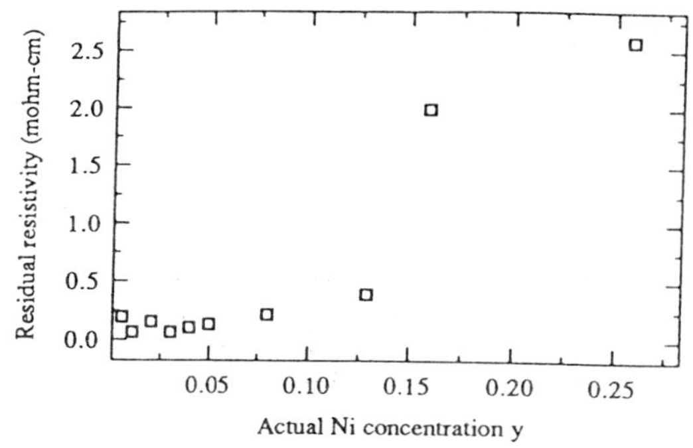

Fig. 3 Linear decrease of T_c with Ni doping.   Fig. 4 Variation of residual resistiity.

While T_c varies linearly with Ni concentration, Fig. 4 shows that the residual resistivity ρ_n (the intercept of the linear resistivity curve at 0 K) varies in a highly non-linear manner, being little effected up to x ~ 0.125, then increasing rapidly, by about an order of magnitude, for higher values of x. Recently Mott (1990) pointed out that disorder would have a small influence on the normal transport properties of oxide superconductors. Our findings of the behaviour of ρ_n are consistent with the interpretation that in this range of Ni substitution, the impurity centres serve only to increase disorder. It is interesting that the magnetic penetration depth λ also follows a similar sudden increase only after x > ~0.125 (Westerholt et al, 1989). This has been further comfirmed by our susceptibility measurements and suggests that $YBa_2Cu_{3-x}Ni_xO_{7-\delta}$ may be appropriately described as a dirty limit type II superconductor.

3. Conclusions

We have carried out measurements of transport, XRD, EDAX and EELS of a series of Ni substituted $YBa_2Cu_{3-x}Ni_xO_{7-\delta}$ compounds. The actual Ni concentration was found to be about half of the nominal values, and there was only a slight indication of oxygen loss in samples of high Ni concentrations. We have also found that while T_c decreased linearly with concentration the residual resistivity had a highly non-linear variation. We believe that the Abrikosov-Gor'kov theory of dirty superconductor could be applied to account for the above behaviour, at least for small dopant concentration.

4. References

A. A. Abrikosov and L.P. Gor'kov, 1960, Sov. Phys.-JETP **12**, 1243

K. Adrian et al, 1987, Physica, **148B**,529

F. Bridges et al, 1990, Phys. Rev. B **42,** 2137

D. M. Ginsberg, 1989, Physical Properties of High T_c Superconductors I, World Scientific

K. Le Dang et al, 1989, Solid State Communications, **72,** 89

N. F. Mott, 1990, Adv. in Phys., **39,** 55

M. Qian et al, 1989, Phys. Rev. B, **39,** 9192

K. Westerholt et al, 1989, Phys. Rev. B, **39,** 11680

Y. Zhao et al, 1988, Solid State Communications, **71,** 31

Supercond. Sci. Technol. **4** (1991) S316–318. Printed in the UK

INTER- AND INTRA-GRAIN MICROWAVE ABSORPTION IN $YBa_2Cu_3O_7$ POWDER

E Buluggiu and A Vera

Istituto di Scienze Fisiche, Università, Via M.D'Azeglio 85, Parma (Italy)

Abstract. Microwave absorption measurements in the transition region of $YBa_2Cu_3O_7$ powder, performed by a conventional X-band ESR spectrometer, give evidence of two distinct dissipative regimes. Besides the well known low field derivative absorption due to viscous flux motion in the inter-grain Josephson junctions network, a weaker peak is observed near Tc more extended in field and essentially related to the intra-grain superconductive transition.

Sample was standard sintered material grinded to a ∿20 μm mean grain dimension and carefully mixed with vacuum silicon grease to avoid non-random orientational effects due to magnetic field.
Derivative signals were recorded after zero-field-cooling with field linearly increasing in the range 0–10 kOe with 2 mW microwave power and 5 Oe and 100 kHz modulation amplitude and frequency, respectively. The low field portion of some typical spectra is shown in Figure 1. More complete spectra are given in the log-log plot in Figure 2.
The null signal in zero field reflects the well known minimum of absorption in the superconducting phase at H=0. As one can see, the signal is due to superposition of two distinct contributions involving markedly different regions of the (H,T) plane. One, present in a large temperature range, is preponderant at very low fields (0–10 Oe). The other one, confined in a temperature range close to Tc, involves a considerably larger field range. This double absorption appears typical of all ceramic superconductors (Kim et al 1988).
The intense low field peak may be with some confidence ascribed to dissipative fluxon motion driven by the microwave supercurrents in the inter-grain boundary regions. The principal rea-

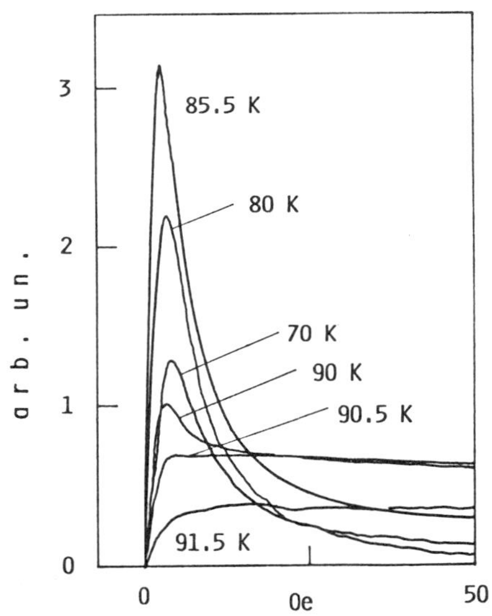

Fig. 1 Low field derivative absorption (X-band) of $YBa_2Cu_3O_7$ powder

0953-2048/91/01S316 + 03 $03.50

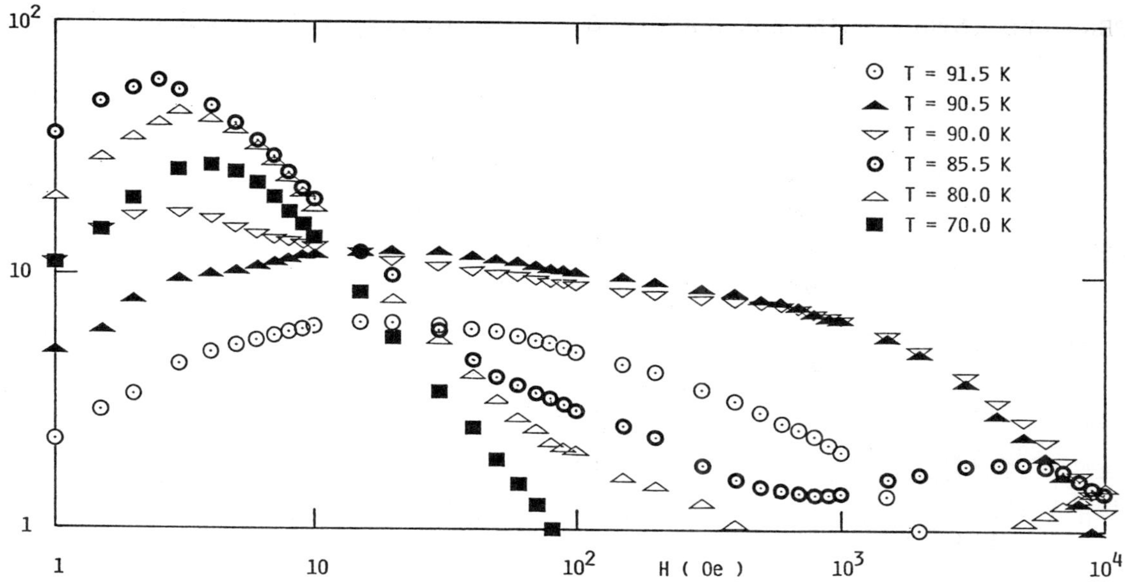

Fig. 2 Log-log plot of X-band derivative absorption of $YBa_2Cu_3O_7$ powder

sons for this are:

a) strong signal suppression is observed in a few Oe field,

b) values and temperature dependence of the field of maximum absorption recall closely the first penetration field $H^*_{c1} \sim H_{c1j}$ (Blazey et al 1987),

c) this spectral portion tends to disappear both in very finely powdered samples (Bohandy et al 1989) and in oriented films (Blazey et al 1989).

Owing to the granularity typical of these compounds, the Josephson coupling may be assumed as the preponderant mechanism to give macroscopical flux exclusion. Thus, the absorption signal may be related (Dulčić et al 1989) to a Fraunhofer-like diffraction pattern of the screening supercurrents through an array of many differently sized Josephson junctions. From the field of the peak an effective junction dimension $d \sim \phi_o/2H\lambda \sim 5$ μm may be deduced. In an alternative superconducting glass vision (Blazey et al 1987), this field may be related to an effective supercurrent loop diameter $d \sim (\phi_o/2H)^{1/2} \sim 2$ μm. In any case we obtain values reminescent of the superconducting grain dimension, coherently with the inter-grain picture. As regards the high field peak near Tc, in general it demonstrates scarcely dependent on the annealing and powdering treatments on the sample. Its salient aspects, as resulting from Figure 2, are:

a) the onset temperature is ~1 K higher than that of the low field signal,

b) the signal suppression induced by field is surprisingly low, with appreciable residual derivative absorption at field as high as 10^3-10^4 Oe for temperatures up to 0.5 K below Tc,

c) the field of maximum intensity (sensibly higher than the previous one) is largely undefined in the range 20-2000 Oe because of the flatness of the involved profiles due also to consistent anisotropy effects on the powder spectrum arising from the pronounced 2D character of these compounds. If we adopt somewhat empirically the previous "Fraunhofer" picture to explain this signal, sub-grain junction dimension of the order of ∿0.1 μm or less would be deduced.

All these features suggest for the peak near Tc a more intrinsic nature possibly connected with intra-grain dissipative processes. This point of view is also confirmed by the observation of an analogous signal in single crystals (Glarum et al 1988) and films (Blazey et al 1989).

Acknowledgements

The Authors are indebted with Profs. D.C. Giori and G. Amoretti for valuable help and discussions. Thanks are also due to Istituto MASPEC (CNR) of Parma for kindly furnishing the sintered material.

This work was partially supported by Consorzio INFM and by CNR-Progetto Finalizzato: Tecnologie Superconduttive e Criogeniche.

References

Blazey K W and Höler A 1989 Solid State Commun. **72** 1199

Blazey K W, Müller K A, Bednorz J G, Berlinger W, Amoretti G, Buluggiu E, Vera A and Matacotta F C 1987 Phys. Rev. B **36** 7241

Bohandy J, Kim B F, Adrian F J and Moorjani K 1989 Phys. Rev. B **39** 2733

Dulčić A, Rakvin B and Požek M 1989 Europhys. Lett. **10** 593

Glarum S H, Marshall J H and Schneemeyer L F 1988 Phys. Rev. B **37** 7491

Kim B F, Bohandy J, Moorjani K and Adrian F J 1988, J.Appl.Phys. **63** 2029

Supercond. Sci. Technol. **4** (1991) S319–321. Printed in the UK

Microwave surface impedance in a coaxial cavity as a material characterisation technique

J C Gallop, W J Radcliffe, T W Button+ & N McN Alford+

National Physical Laboratory, Teddington TW11 OLW, UK & + ICI Advanced Materials, Runcorn WA7 4QE, UK

Abstract

Measurements of surface resistance R_S of a number modes of a coaxial cavity have been made, for which the centre conductor is BiSrCaCuO (BSCCO) rod. Results for $R_S(T)$ have allowed a quantitative estimate to be made of the proportions of two different phases present in this material.

1. Introduction

We report measurements on a microwave coaxial cavity resonator in which the centre conductor is made from a rod of sintered BSCCO. The determination of microwave surface impedance of high temperature superconductors provides a powerful technique for analysing the electromagnetic properties of the material within a distance of around one penetration depth of the surface. The penetration depth is approximately independent of frequency f for $f << 2\Delta/h$, the superconducting energy gap frequency, which condition is easily satisfied for the experiments reported here. Unlike low frequency resistance measurements, which yield no further information once a temperature is reached at which there is a percolating superconducting path from one electrode to the other, the microwave technique samples the entire exposed surface of the specimen and is thus able to detect the presence of any material which has a finite electromagnetic loss. In addition the method is convenient since it is non-contacting and therefore non-destructive.

2. Experimental Details:

Thin (0.75 mm diameter) rods of BiSrCaCuO were prepared as follows: Bi_2O_3, PbO, $Sr(NO_3)_2$, CaO and CuO were mixed in the appropriate stoichiometric proportions and calcined for 5 h at 845°C in air. The calcined powders were processed into rods initially approximately 1 mm in diameter as reported previously [1]. This involves thorough mixing with polymer and non-aqueous solvents to produce a plastic mass which is then extruded through a die. The rods were sintered in air at 850°C for 50 h before being furnace cooled and the

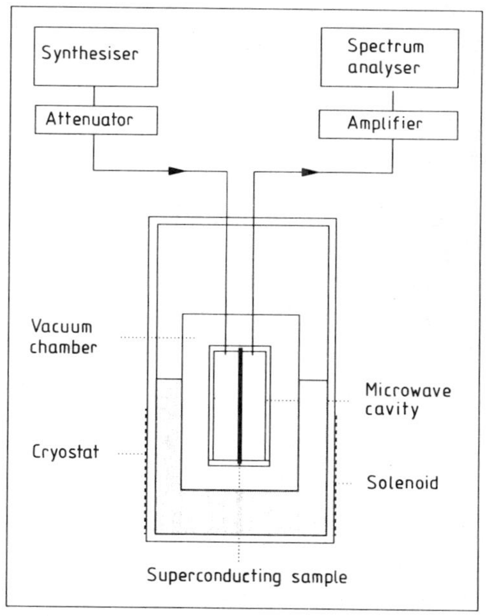

Figure 1. Schematic diagram of the microwave system for measurements on the coaxial cavity.

0953-2048/91/01S319 + 03 $03.50

resulting material had a grain size of about 5 μm. The microwave method used here relies on a cavity perturbation technique to produce quantitative measures of the superconductor's surface impedance Z_s. We have reported a similar set of measurements on a YBCO rod [2] and the experimental technique is described there. A schematic of the cavity construction is shown in fig.1.

3. Estimate of phase proportions

We estimate the fraction of the lower temperature phase using the following argument. The linewidth W_k of the kth resonance is the sum of contributions from various loss mechanisms associated with the cavity:

$$W_k = \frac{\omega_k [R_c(T) + R_s(T)]}{2\pi\Gamma_k} + W_{ke}$$

where ω_k is the angular frequency and Γ_k the geometry factor of the mode, $R_c(T)$ is the temperature dependent surface resistance of the normal metal part of the cavity, $R_s(T)$ is the surface resistance of the superconductor and W_{ke} is a temperature independent width which arises from external coupling losses etc. The outer conductor of the cavity is made from brass which has the advantage that, being an alloy, the temperature dependence of its conductivity is small. It is simple to show that in the local limit when a complex conductivity $\sigma = \sigma_1 + j\sigma_2$ may be used to describe the superconductor

$$R_s = \frac{[\omega\mu_o(\sigma_r-\sigma_2)]^{1/2}}{2^{1/2}\sigma_r} \quad \text{....(1)}$$

where $\sigma_r^2 = \sigma_1^2 + \sigma_2^2$. Furthermore for the BSCCO sample considered here $R_s(T)$ comprises two additive components arising from the two different superconducting phases, whose conductivity in the normal state are similar and which are mainly distinguished by different T_c. We assume that the temperature dependence of each component is the same when scaled by its transition temperature. Then if the frequency and temperature dependent complex conductivity of the two phases are σ_l and σ_h respectively, the proportions of each component being p_l and p_h, the total effective surface resistance becomes

$$R_s = \mu_o\omega\{[p_l^2(\sigma_{rl}-\sigma_{2l})/2\sigma_{rl}^2]^{1/2} + [p_h^2(\sigma_{rh}-\sigma_{2h})/2\sigma_{rh}^2)]^{1/2}\} \quad \text{....(2)}$$

where subscripts l and h refer to the low and high T_c phases respectively. This assumes that the field penetration depth is small compared with the mean grain size, which is believed to be an excellent approximation, except very close to T_c. For a number of HTS samples, both in bulk and thin film forms, we have found that the two-fluid model of superconductivity fits the temperature dependence of both real and imaginary parts of the surface impedance rather closely in the temperature region just below T_c, where most of the change in these parameters is concentrated. Then

$$\sigma_1 = \sigma_n[T/T_c]^4 \quad \text{...(3)}$$

$$\sigma_2 = \sigma_n[1-T/T_c]^4$$

where σ_n is the normal state conductivity of the superconductor at $T = T_c$.

Treating the quantities T_{cl}, T_{ch}, p_l and p_h as unknowns equations (1) and (2) can be fitted to the experimental data yielding the values T_{cl} = 70K, T_{ch} = 110K, and p_l/p_h = 0.03. The T_c's are in line with those found for the two expected BSCCO phases (2212 and 2223 respectively). It is interesting that the proportion of the lower temperature phase is estimated as only 3% (see Fig.2). XRD investigations of the bulk rod indicate that the proportions of the phases averaged over the bulk is much closer to 50% of each

[1]. This discrepancy may be explained by the fact that the microwave technique samples the material only within approximately a penetration depth of the surface. It is quite possible that the proportion of phases is not homogeneous throughout the material and perhaps due to more effective oxygen take-up at the surface the proportion of the desired high T_c phase is much greater here. This picture is supported by observations of sections of the rod in a SEM which indicated a high concentration of the copper rich phase near the surface of the rod, although quantitative estimates of the phase proportions could not be made by this method. If such preferential surface phase segregation is usual it may have significant implications for microwave applications of this material.

Fig.2 Linewidth of TM010 mode vs. T for BSCCO coaxial cavity

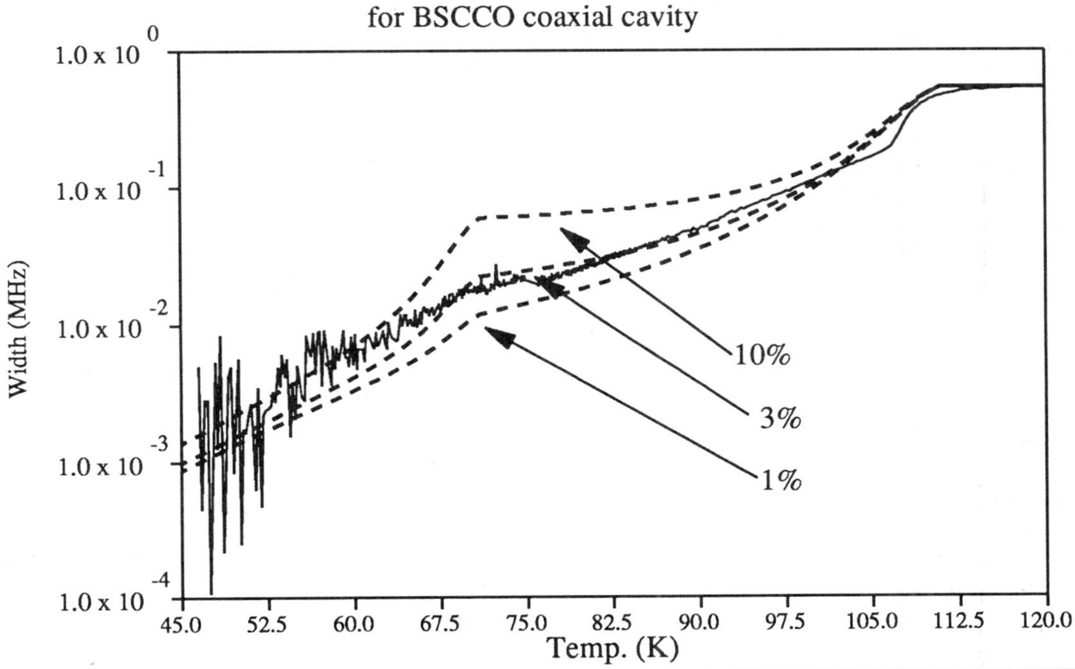

The conductivity of the BSCCO rod at 110K has been measured as 1.65×10^5 S, in reasonable agreement with the value we have found for BSCCO thin films. Substituting this into equation (1) predicts that $R_s = 0.5\ \Omega$, in agreement to within 10% with the experimental values just above T_c.

4. Conclusion

We have demonstrated that the measurement of microwave surface impedance for high temperature superconducting specimens may be used to make quantitative estimates of the proportions present of various discrete superconducting phases, provided each has a distinguishably different T_c. The apparent concentration at the surface of the high T_c phase, believed to be $Bi_2Sr_2Ca_2Cu_3O_{10}$, has not been previously reported and may have significance for applications of these materials. Further work is planned to investigate the d.c magnetic field dependence of the surface impedance of this material, since this may provide useful information on the origin of inter-grain losses in polycrystalline HTS specimens.

References

1. Button T W, Alford N McN, Birchall J D, Wellhofer F, Gough C E & O'Connor D A (1989), Supercond. Sci. Technol. 2 224-9
2. Radcliffe W J, Gallop J C, Langham C D, Alford N McN & Button T W (1990), Supercond. Sci. Technol. 3 151-4

Supercond. Sci. Technol. 4 (1991) S322–324. Printed in the UK

The Effect of Magnetic Field Modulation on the Microwave Responce in High-Tc Superconductors

Y.Hayashi[1], M.Fukui[2], T.Fujita[3], H.Shibayama[3], K.Iwahashi[3] and S.Sako[4]

1.Materials Science,University of Osaka Prefecture,4-804 Mozuume-machi, Sakai,Osaka 591 Japan
2.Nakanihon Automotive College,1301 Sakahogi-cho,Kamo-gun,Gifu 505 Japan
3.Department of Physics,Nagoya University,Chikusaku,Nagoya 464 Japan
4.Department of Physics,Mie University,Tsu, Mie 514 Japan

Abstract. The origin of the strong enhancements of microwave response at the onset of the superconductivity has been investigated. Since the effect is only detectable when the measurements are made by making use of a ESR spectrometer, the effect of modulation is investigated. The origin of the enhancement is considered to be due to strong dependence of the microwave dissipation on the magnetic field.

1.Introduction

Experimental studies of the transmission and the reflection of the microwave power in various compounds of high-Tc superconductors have been made by many investigators. The microwave response under the magnetic field has been measured by ESR spectrometer. The most important result so far obtaind is the apparent increase of the signal corresponding to the microwave dissipation at the onset of the superconductivity(Durney et al 1987,Sastry et al 1988,Shaltiel et al 1989).

Another type of experiment, namely, the resistivity measurement at the microwave frequency without field modulation has been made(Rigby and Rzchowski 1989,Miranda et.al.1990). The resistivity starts to decrease at Tc in accordance with the DC measurements. The result seems to contradict to the sudden increase of the microwave loss at Tc observed by ESR spectrometer. To use a field modulation or not is the main difference in experimental methods.

On the other hand, Pakulis and Osada(1988) measured the microwave loss without field modulation and found that the loss was minimum at zero field and increased sharply as the increase of the magnetic field. Hence, the peak to peak hight of the derivative curve should be strong near zero magnetic field. Although the minimum loss at zero magnetic field has been confirmed, the loss mechanism coupled with modulation effects is still a problem of extensive research(Bodi 1990,Golosovsky 1990).

In a previous paper(Hayashi et al 1989c), we have reported results in the Tl compound and suggested that the critical current in Josephson coupled superconductor was important for the microwave response in the magnetic field. Here, a possible origin of the apparent increase of the microwave loss near Tc is discussed in relation to the modulation.

2.Experimental results and discussion

The sample used in the present experiment is Bi(Pb)-Sr-Ca-Cu-O system.

The method of preparation and the characteristics of the material is the same as that described previously(Hayashi et al 1989a). A conventional ESR spectrometer with 100 kHz magnetic field modulation was employed.

Figure 1 shows typical examples of derivative curves observed at the temperature below Tc. We have obtained similar results in Tl compound at the temperature range very near to Tc, where hysteresis was not observable indicating that there was no trapped flux. The integrated curve was shown to resemble to the static magnetization(Hayashi et al 1989c,1990). In the present case, measurements are made at lower temperature.In order to avoid the hysteresis effect due to trapped magnetic flux, the magnetic field is always sweeped increasing direction from zero to 8000 Oe.

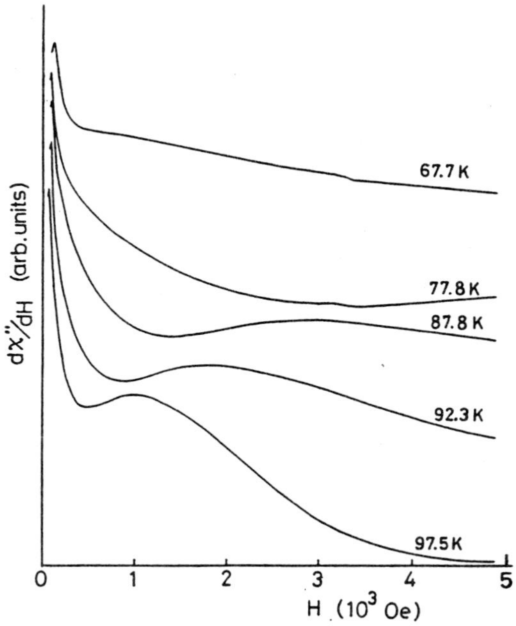

Figure 1. Examples of derivative curves of the microwave absorption in Bi compound(Tc=110 K). At higher temperature above 87.3 K, a minimum appears in each curve.

Broad minima in derivative curves are considerd to represent the reduction of the Josephson critical current by an application of the magnetic field. The temperature dependence of the position of the minimum of the derivative curve was confirmed to follow the same relation as that of the critical current of the Josephson effect near Tc(Hayashi et al 1989c). The minimum in each curve of Figure 1 is considered also to have the same origin. The temperature dependence of the position of the minimum is different from that observed previously because measurements are made at the temperature well below Tc. The minimum diminishes at lower temperature.

Figure 2 dipicts a derivative curve at 26 K. The curve near zero magnetic field region is not shown because the magnetic field is not able to sweep precicely at very low field region* The broad minimum disappears and the ESR signal of Cu ions is clearly seen near 3300 Oe(Hayashi et al 1989b). The derivative curve near zero field is 180 degrees out of phase to the ESR signal which shows that microwave dissipation is minimum at zero field. Nishida and Horai(1990) have made a measurement in the same material as that used in the present experiment without field

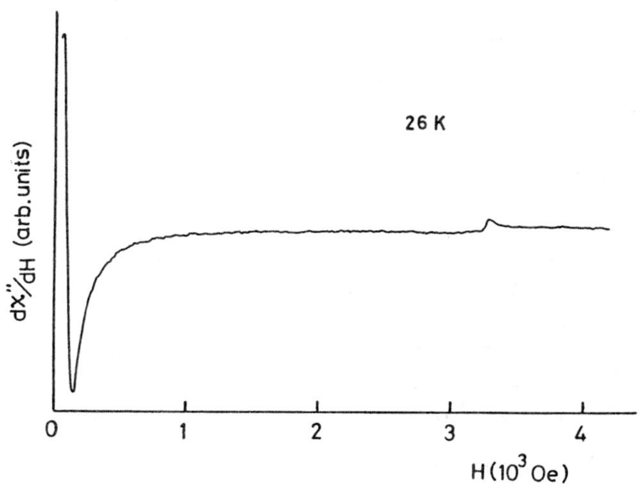

Figure 2. A derivative curve of the microwave dissipation at 26K.

* A very low field part of the previously reported curves where H_{c1} has been defined(Hayashi et al 1989c) is incorrect.

modulation at low temperatures. They have made a careful control of the static magnetic field and confirmed absorption minimum at zero magnetic field. A very steep rise of the microwave absorption by an application of the weak magnetic field has also been confirmed. The origin is considered to be the flux penetration into grain boundaries of the Josephson coupled superconductor. We have found another minimum in the derivative curve and attributed the origin to the reduction of the Josephson critical current flowing through grain boudaries. However, this broad minimum shoud be considered due to another kind of the Josephson effct, namely, caused by the internal Josephson effect. The derivative curve can be recognized as the superposition of two curves both centered at zero field.

Although many of the experimental results are consistent with this result, some of the properties still seem to be unable to understand by the resistive loss alone. Since the flux lines are known to enter easily into the superconductor at very weak field, an application of the modulation field is not a small perturbation in general. Therefore, the appearance of new phenomena can be expected. The detection of the second harmonic of the modulation field(Golosovsky et al 1989, Rakvin et al 1989) and the appearance of a series of equally spaced absorption lines (Blazey et al 1988,Vichery et al1989) have been reported. A very large noise usually encountered in ESR measurements below Tc may also be related to these effects.

Acknowledgements

The authors express their thanks to Prof.K.Horai, Dr.A.Nishida, and Dr.H.Sasakura for their discussion.

References

Blazey K W,Portis A M,Muller K A,Bednorz J G and Holtzberg F 1988 Physica C153-155 56

Bodi A C 1990 Phys.Rev.B41 2127

Durney R,Hautala J,Ducharme S,Lee B,Taylor P C,Zheng D J and Xu J A 1987 B36 2361

Golosovsky M,Davidov D,Rettori C and Stern A 1989 Phys.Rev.B40 9299

Hayashi Y,Fukui M,Sasakura H,Minamigawa S,Fujita T and Nakahigashi K 1989 a Jpn.J.Appl.Phys. 28 L759

Hayashi Y,Sako S,Fukui M,Fujita T,Sasakura H,Minamigawa S and Nakahigashi K 1989b Jpn.J.Appl.Phys. 28 L1531

Hayashi Y,Fukui M,Fujita T,Shibayama H,Iwahashi K and Adachi K 1989c Jpn. J.Appl.Phys.28 L1746

Hayashi Y,Adachi K,Iwahashi K,Shibayama H,Fujita T,Fukui M and Sako S 1990 Advances in Superconductivity II Eds.Ishiguro T and Kajimura K (Tokyo Springer-Verlag)pp555-558

Miranda F A,Bhasin K B,Heinen V O,Kwor R and Kalkur T S 1990 Physica C168 91

Nishida A and Horai K 1990 Solid State Commun.74 947

Pakulis E J and Osada T Phys.Rev.B37 5940

Rakvin B,Pozek M,Paljevic M and Brnicevic N 1989 Solid State Commun.70 729

Rigby Q Li K W and Rzchowski M S 1989 Phys.Rev.B39 6607

Sastry M D,Kadam R M,Babu Y,Dalvi A G I,Gopalakrishnan I K,Sastry P V P S S,Phatak G M and Iyer R M 1988 j.Phys.C 21 L607

Shaltiel D,Bill H,Decroux M,Hagemann H,Junod A,Peter M,Ravisekhar, Triscone Y,Triscone G,Walker E,Yan Y.F and Zhao Z X Physca C 1989 157 240

Vichery H and Beuneu F 1989 Physica C159 823

Supercond. Sci. Technol. **4** (1991) S325–327. Printed in the UK

AC susceptibility of granular superconductors

K.-H. Müller, S.J. Collocott, R. Driver and N. Savvides

CSIRO Division of Applied Physics, Lindfield, Australia 2070.

Abstract. A finite temperature critical state model for a granular superconductor is used to describe in detail the AC susceptibility, $\chi' + i\chi''$, of high-temperature ceramic superconductors. The model shows that the superconducting volume fraction, the average grain size and the inter- and intra-granular pinning force densities determine the temperature and AC magnetic field dependence of χ' and χ''. The reversible behaviour close to T_c and the frequency dependence of the AC susceptibility are shown to be caused by thermally activated flux creep.

1. Introduction

The AC susceptibility of high-temperature superconductors is frequently measured in order to characterize samples. AC susceptibility data strongly depend on the AC magnetic field amplitude and slightly on the AC frequency. At large AC field amplitudes (typically greater than 5 Oe) often two loss peaks are found; a broad peak at low temperature and a narrower peak near T_c. Susceptibility data for ceramic Bi(Pb)SrCaCuO samples often show a clear frequency dependence and a region of reverisible magnetic flux behaviour close to T_c. In this paper we discuss the AC susceptibility using a finite temperature critical state model for granular superconductors.

2. Finite temperature critical state model for granular superconductors

When a magnetic field greater than the Josephson critical field, H_{c1J}, is applied to a granular superconductor, where the grains are coupled by weak-links, flux penetrates the weak-link network in the form of Josephson vortices (Clem 1988), giving rise to an inter-granular magnetic field. At fields greater than H_{c1g}, Abrikosov vortices form and flux now penetrates beyond the London penetration depth into each grain, producing an intra-granular field. The inter-granular Josephson vortices are assumed to be pinned along grain boundaries or at grain boundary discontinuities, while Abrikosov vortices are pinned at defects inside grains. The inter- and intra-granular magnetic field profiles, H_J and H_g, can be obtained by solving the corresponding critical state equation with the appropriate boundary conditions. In the case of a slab sample of thickness $2L$ and space coordinate x, where grains are approximated by long cylinders of average radius R_g with radial coordinate r, aligned parallel to the slab and the applied field, the critical state equations are

$$\frac{dH_J(x)}{dx} = \pm \frac{F_{PJ}}{|\, B_J(x)\, |} \quad \text{and} \quad \frac{dH_g(r,x)}{dr} = \pm J_{cg} \quad , \tag{1}$$

where F_{PJ} is the inter-granular pinning force density and J_{cg} is the critical current density

0953-2048/91/01S325 + 03 $03.50

inside the grains, which is assumed to be field independent for $H_{c1g} < H_J << H_{c2g}$ (Bean critical state). The inter-granular pinning force density, F_{PJ}, is derived from the Anderson flux creep model (Anderson 1962), assuming that the pinning sites for Josephson vortices are on average $2R_g$ (average grain size) apart and the average inter-granular pinning potential U_o is $2R_g$ wide (Müller 1990). One obtains

$$F_{PJ} = \frac{kT \mid B \mid}{2R_g \Phi_o} \text{ arcsinh}[\frac{\nu_h}{2\nu_o} \exp(\frac{U_o(0)(1-T/T_c)^2}{kT} \frac{H_o}{\mid H_J \mid + H_o})] \quad , \tag{2}$$

where ν_o is the attempt frequency for Josephson vortex hopping. The temperature dependence of $U_o(T)$ in Eq.(2) corresponds to SNS grain boundary Josephson junctions. The field H_o is defined as

$$H_o = \frac{\Phi_o}{4\mu_o \lambda_L R_g} \quad , \tag{3}$$

where λ_L is the London penetration depth for grains. In the case of an applied AC magnetic field, the minimal observable hopping rate, ν_h, in Eq.(2) depends on the AC field amplitude H_a^m and the AC frequency f and one finds

$$\nu_h \simeq \frac{8\Phi_o H_a^m R_g^2}{LkT} f \quad . \tag{4}$$

The solutions of Eqs.(1) can be used to determine the spatially averaged flux densities $< B_J >$ and $< B_g >$ where

$$< B_J > = \mu_{eff} \frac{\mu_o}{2L} \int_{-L}^{L} H_J(x) \, \mathrm{d}x \tag{5}$$

and

$$< B_g > = (1-\mu_{eff}) \frac{\mu_o}{R_g^2 L} \int_{-L}^{L} \int_0^{R_g} H_g(r,x) \, r \, \mathrm{d}r \, \mathrm{d}x \quad . \tag{6}$$

Here, μ_{eff} is the effective permeability originating from the superconducting grains (Clem 1988) which depends on the volume fraction f_s of superconducting grains. Finally, the AC susceptibility χ' and χ'' are calculated from the time-dependent spatially averaged total flux density, $< B > = < B_J > + < B_g >$, as

$$\chi' = \frac{2f}{\mu_o H_a^m} \int_0^{1/f} < B > (t) \cos(2\pi f t) \, \mathrm{d}t \quad - \quad 1 \tag{7}$$

$$\chi'' = \frac{2f}{\mu_o H_a^m} \int_0^{1/f} < B > (t) \sin(2\pi f t) \, \mathrm{d}t \quad . \tag{8}$$

3. Results

Figure 1 shows the calculated AC susceptibility for a ceramic $YBa_2Cu_3O_{7-\delta}$ sample, using the parameters f_s=0.85, R_g=10 μm, λ_L=0.4 μm, $\nu_0 = 10^7$ Hz, $J_{cg}(80\text{ K}) = 10^7$ Acm^{-2} and $U_o(80\text{ K}) = 5$ eV. Our calculation agrees well with the experimental data of Goldfarb *et al.* (1987). In contrast to one of our earlier calculations (Müller 1989), here the intra-granular loss peaks appear at a higher temperature than the inter-granular peak for 0.1 Oe, in accordance with experiment. This is because the present model includes the effect of thermally activated inter-granular flux creep which results in a significant

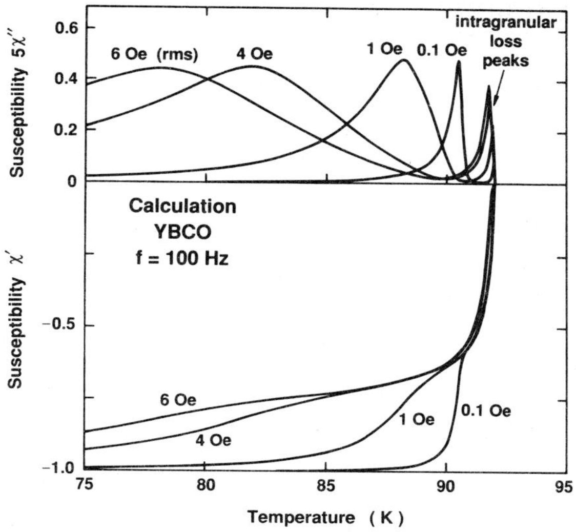

Fig. 1 AC susceptibility versus temperature at different AC field amplitudes for a YBCO ceramic sample.

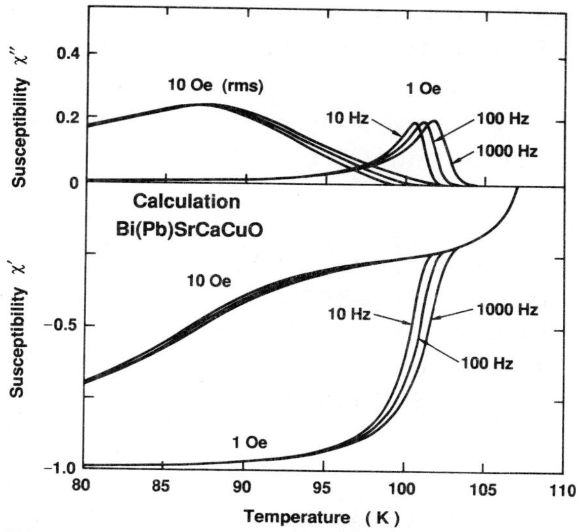

Fig. 2 AC susceptibility versus temperature at different AC field amplitudes and frequencies for a BSCCO ceramic sample.

decrease of the inter-granular pinning force density close to T_c. Though the inter-granular pinning potential U_o is very large, the pinning force density is quite small because the inter-granular potential extends over a large distance equal to the average grain size. This is different for conventional intra-granular Abrikosov vortex pinning, where the potential depth and width are much smaller.

Figure 2 shows the calculated AC susceptibility for a Bi(Pb)SrCaCuO ceramic sample, using the parameters f_s=0.4, R_g=4 μm, λ_L=0.4 μm, $\nu_o = 10^7$ Hz and U_o(80 K)=2.5 eV. Here, the intra-granular part of the susceptibility is neglected. Because of a relatively small inter-granular potential and a small grain size, our calculation shows the AC susceptibility to be strongly frequency dependent which is confirmed by the experimental data of M. Nikolo (to be published elsewhere).

4. Conclusion

A finite-temperature critical state model for a granular superconductor describes in detail AC susceptibility data of high-temperature superconductors. During each AC cycle inter-granular Josephson vortices sweep in and out of the grain boundary weak-link network. At stronger magnetic fields Abrikosov vortex dynamics inside grains contribute to the AC susceptibility. The reversible flux behaviour near T_c and the frequency dependence of the susceptibility are caused by thermally activated flux creep, which depends on the depth of the pinning potential.

References

Anderson P W 1962 *Phys. Rev. Lett.* **9** 309
Clem J R 1988 *Physica C* **153-155** 50
Goldfarb R B, Clark A F, Braginski A I and Panson A J 1987 *Cryogenics* **27** 475
Müller K-H 1989 *Physica C* **159** 717
Müller K-H 1990 *Physica C* **168** 585

Supercond. Sci. Technol. 4 (1991) S328–330. Printed in the UK

ANGULAR DEPENDENCE OF CRITICAL CURRENT DENSITY OF $YBa_2Cu_3O_{7-\delta}$ FILMS IN MAGNETIC FIELDS.

N. Kobayashi, H. Kawabe*, K. Watanabe, S. Awaji, H. Yamane, H. Kurosawa**, T. Hirai and Y. Muto

Institute for Materials Research, Tohoku University, Sendai 980, Japan. *TATSUTA Electric Wire & Cable Co LTD, Higashi-Osaka 578, Japan. **RIKEN Co., Kumagaya 360, Japan.

Abstract. The anisotropic behaviour of the transport critical current density, J_c, of $YBa_2Cu_3O_{7-\delta}$ films prepared by CVD has been investigated in high magnetic fields up to 27 T. The strong angular dependence of J_c is discussed based on the intrinsic pinning model.

1. Introduction

High-T_c superconducting oxides exhibit strong anisotropic properties because of the quasi two dimensional crystal structure. Recently, the anisotropic behaviour of a critical current density, J_c, was both experimentally and theoretically discussed by Roas et al. (1990) and by Tachiki and Takahashi (1989), respectively. Roas et al. (1990) observed the strong enhancement of J_c when the magnetic field was aligned along the CuO layers and Tachiki and Takahashi (1989) attributed such a behaviour to the intrinsic pinning effect arising from an alternate stacking of strongly and weakly superconducting layers. It is of importance to further explore the anisotropic behaviour of J_c, because this behaviour may bring serious problems on applications of the high-T_c superconducting oxides. In this paper, we report the anisotropic properties of the electrical resistance and transport critical current density of $YBa_2Cu_3O_{7-\delta}$ films.

2. Experimental

The $YBa_2Cu_3O_7$(YBCO) films were epitaxially grown on $SrTiO_3$(100) substrates by chemical vapor deposition (Yamane et al. 1989), where β-diketonate complexes of $Y(DPM)_3$, $Ba(DPM)_2$ and $Cu(DPM)_2$ were used as source materials. The films were deposited in an atmosphere of Ar and O_2 (Ar/O_2=3/1) of 10 Torr. After the deposition, an in-situ heat-treatment in an O_2 atmosphere of 1 atm was performed for an hour. The films used in this investigation have a strongly preferred c-axis orientation perpendicular to the substrates. The samples were patterned into dimensions of typically 0.3mm in width, 1mm in length and 1μm in thickness. The electrical resistance and the critical current density

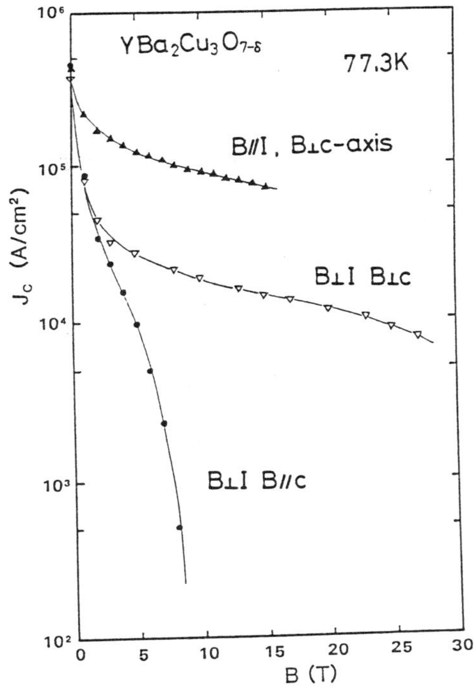

Fig. 1 Field dependence of J_c for both B⊥c and B//c. (J_c criterion is 1 μV/cm)

were measured by usual four probe method with gold electrodes deposited onto the sample surface.

3. Results and Discussion

The T_c values defined by zero resistance is about 90-92 K for almost of our samples. The resistive transition becomes quite broad in magnetic fields, in particular parallel to the c-axis. The B_{c2} value also defined by zero resistance exhibits the anisotropic behaviour. The slope of B_{c2} near T_c, $|dB_{c2}/dT|T_c$, is given to be about -5 T/K and -1 T/K for the magnetic field perpendicular (B⊥c) and parallel (B//c) to the c-axis, respectively. This anisotropy of B_{c2} strongly affects the angular dependence of J_c.

Figure 1 shows the field dependence of J_c for both B⊥c and B//c. The highly oriented samples exhibit high-J_c of more than 5×10^5 A/cm² at 77.3 K in the absence of magnetic field. The J_c value also shows very strong anisotropy in the magnetic field. At 1 T, the J_c value for B//c is slightly larger than that for B⊥c. When the magnetic field exceeds 2 T, the J_c value for B//c is strongly reduced, while for B⊥c J_c still remains above 10^4 A/cm² even at 20 T. Thus, the crossover in the anisotropy of J_c occurs near 2 T.

In Fig. 2, the angular dependence of J_c in the temperature range from 65 K to 80 K is shown as a function of angle, θ, between the magnetic field and the c-axis. The J_c value has a maximum near B⊥c. When the temperature decreases, the anisotropy, $J_c(90°)/J_c(0°)$, is reduced and the peak near B⊥c becomes broader.

While we do not find any other maximum in the angular dependence of J_c, Roas et al (1990) reported that the broad maximum in the angular dependence was observed near B⊥c (θ=0) below 3 T. This broad peak seems to be related with the crossover in the anisotropy observed near 2 T in the field dependence of J_c. These behaviours suggest that the different pinning centers become responsible for J_c depending on the field direction and strength.

The angular dependence of J_c is compared with the theoretical result based on the intrinsic pinning model (Tachiki and Takahashi 1989). Insert shows the theoretical value (solid line) calculated using the $J_c(0°)/J_c(90°)=0.039$ together with the experimental result at 80.4 K and 3 T. Our experimental result shows the broader angular dependence compared with the theoretical one.

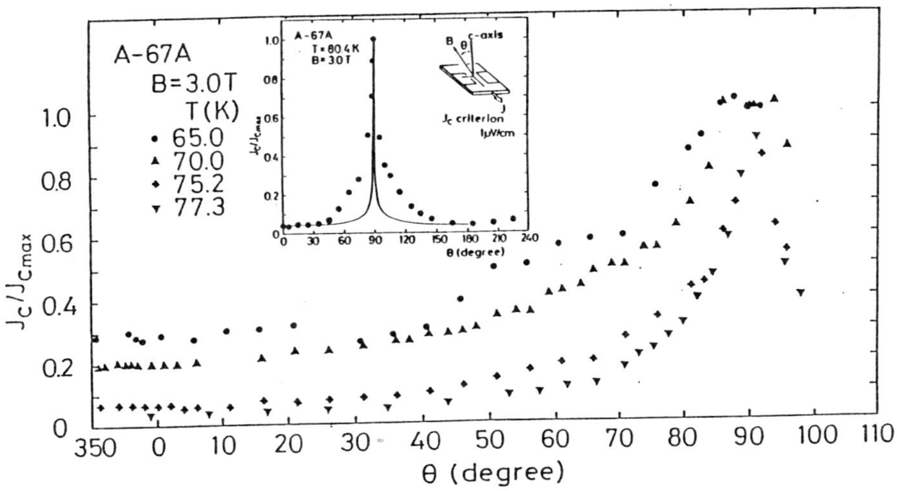

Fig. 2 Angular dependence of J_c. Solid curve in the insert shows the theoretical value calculated from the intrinsic pinning model.

This discrepancy between the experimental result and the theoretical one may be caused from a tilting of the c-axis of small grains. However, the X-ray diffraction pattern of our samples shows very sharp (00ℓ) peak. Therefore, it seems to be difficult to explain the broader angular dependence of J_c by such a tilting.

As alternative explanations, we point out two possibilities for the angular dependence of J_c. One is an anisotropy of B_{c2}. The flux pinning force may be expressed by a well-known scaling law $F_p=KB_{c2}{}^m(B/B_{c2})^p(1-B/B_{c2})^q$ even in the high-T_c oxide superconductors. In the case of $YBa_2Cu_3O_{7-\delta}$, the anisotropy of B_{c2} is not so large and therefore can be described by an effective mass model. Then we could expect a gradual angular dependence of J_c. The second is the existence of another pinning mechanism. We (Watanabe et al. 1989) reported the existence of small disc shaped CuO precipitates embedded in the ab-plane, which were regarded as strong pinning centers. Because these pinning centers have anisotropic structure, we could also expect the strong anisotropy in J_c.

We would like to thank Prof. M. Tachiki, Prof. T. Matsushita and Dr. S. Takahashi for helpful discussions.

References

Roas B, Schultz L and Saemann-Ischenko G 1990 Phys. Rev. Lett 64 479

Tachiki M and Takahashi S 1989 Solid State Commun 72 1083

Watanabe K, Matsushita T, Kobayashi N, Kawabe H, Aoyagi E, Hiraga K, Yamane H, Kurosawa H, Hirai T and Muto Y 1990 Appl. Phys. Lett 56 1490

Yamane H, Kurosawa H, Hirai T, Watanabe K, Iwasaki H, Kobayashi N and Muto Y 1989 Supercond. Sci. Technol. 2 115

Supercond. Sci. Technol. 4 (1991) S331–333. Printed in the UK

Local Hall Probe Magnetometry: a New Technique for Investigation of Magnetic Flux Penetration, Exclusion and Trapping in HTSC.

M. Konczykowski, Laboratoire des Solides Irradiés Ecole Polytechnique, 91128 Palaiseau, France
F. Holtzberg, IBM T.J. Watson Research Centre, Yorktown Heights, NY 10598, USA
P. Lejay, CRTBT - CNRS, B.P. 166X, 38042 Grenoble Cedex, France

Abstract. We propose a novel, local Hall-probe magnetometry technique for the investigation of magnetic phenomena in superconductors. A miniaturized (80 x 100 μm^2 size) Hall sensor allows us to measure magnetic features with a precision approaching that of a single vortex line. We demonstrate the efficiency of this technique by exhibiting magnetic data for $YBa_2Cu_3O_7$ single crystals. These include magnetization curves, and determination of H*, the first field for full penetration .

Since the discovery of high-temperature-superconductors (HTSC) by Müller and Bednorz (1986), a reliable determination of the magnetic characteristics of these materials has been a challenge for the experimentalist. (For a review, see Malozemoff 1989). Most magnetic experiments have been performed on commercial or home-made magnetometers (e.g. vibrating sample magnetometers - VSM) and susceptometers (e.g. superconducting quantum interference devices - SQUIDs). Apart from the cost of the equipment itself there are several drawbacks in the use of such equipment. For example the use of SQUIDs with liquid helium to investigate phenomena near T_c (i.e. above nitrogen temperature) is inconvenient in many circumstances. Moreover, the displacement of an irreversible anisotropic sample in an inhomogenous magnetic field, in VSM and in SQUIDs, can lead to noncontrollable effects on the measured features. Hall-probe magnetometry introduces a simple way to overcome such problems. In the past, the field profiles of superconductors were investigated by this technique (Frankel 1979 , Jung et al 1989, Reich and Nabutovsky 1990). Here we demonstrate the advantages of the use of Hall sensor for the investigation of magnetic properties of HTSC.

The key new technical point of our method is the use of **a miniaturized** Hall sensor with 80 x 100 μm^2 active area, of highly linear slope (typically 50 $m\Omega/G$) which is almost temperature independent below 120K. Such a probe can be made of a narrow-gap semiconductor (Konczykowski 1990) and put directly on the surface of the investigated sample. It should be noted that: (1) The current injected to the probe (typically 500 μA) produces negligible self field. (2) Even a common dc voltmeter allows a resolution of 4 mGs; this corresponds to a single flux line for the active area of the sensor. Improvement by a factor 10 of this resolution can be achieved by use of more sophisticated dc voltmeters or ac bridges.

The calibration of the probe is done as a function of the external field H_a either in the absence of the sample at any temperature or with the sample at temperature T well above the transition temperature T_c. Once the characteristics of the probe, namely $R_H(H_a)$ are established, the probe is ready for measurements of magnetization as a function of temperature, field or time. An external field H_a is applied and the field H_m measured by the probe is detected. As a result of perturbation, by the sample, of the homogenous applied field, we find $\Delta H=H_a-H_m$. This perturbation ΔH is related to the magnetization of the sample, henceforth we refer to it as the magnetization field.

In figure 1 we demonstrate the technique by presenting a virgin magnetization curve recorded at 91K of an YBCO single crystal. The initial linear slope is identified as a Meissner state. The actual slope of ΔH vs H_a in this regime depends on the position of Hall probe. Ideally, it should be -1 (perfect shielding). In reality it is close to -1 only if the probe is put at the centre of a sample much bigger than the probe. Typical values for available crystals range between -0.3 to -0.7. Fig. 1 demonstrates the accuracy and the resolution of the technique. The data is taken at 91 K, 1.3 K below T_c, where the behaviour of the ascending-field branch closely resembles the prediction for ideal type-II superconductors; in particular, above the kink $\Delta H(H_a)$ exhibits the characteristic logarithmic divergence. The sharp kink in the $\Delta H(H_a)$ curve of the figure represents the lower critical field H_{c1} though its actual value is modified by demagnetizing effect. The irreversibiliy which is apparent in the descending-field data represents only a small portion of the total magnetization.

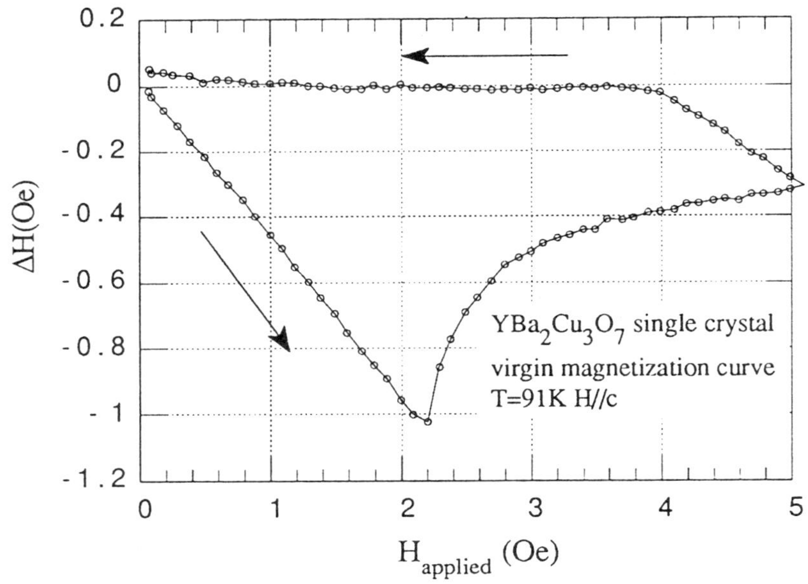

Figure 1. First magnetization curve of $YBa_2Cu_3O_7$ single crystal measured vs applied field. $\Delta H = H_m - H_a$.

The applied magnetic field in this experiment is generated by a low inductance copper coil. Changes in the field follow one of two procedures: (i)The field is increased in steps up to the maximum and then decreased in steps back to zero. We refer to this procedure as 'non-monotonic loop' . (ii) In the 'monotonic loop' the field is turned off after each step. These two procedures are described schematically in Fig. 2.

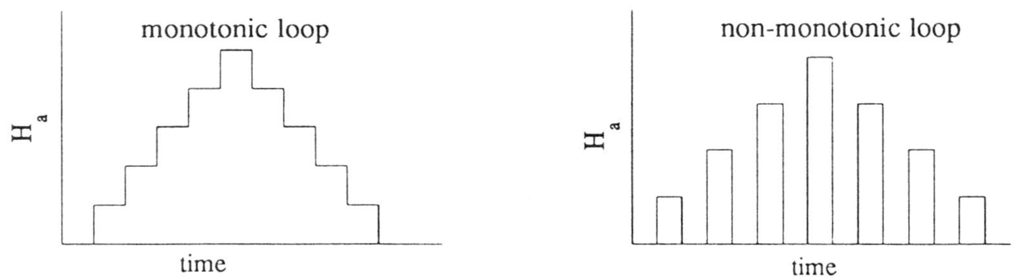

Figure 2. Monotonic and non-monotonic magnetic field time loops.

To further demonstrate the ability of the measurement technique we present in Fig. 3 a typical record of the magnetization, taken during a non-monotonic loop of the applied field.

The circles in the figure are ΔH values measured with the field on, whereas the squares represent data taken with the field off. Note that in the latter case ΔH is just the measured field originating from flux trapped in the sample. The figure demonstrates that the onset of trapped field corresponds to the deviation from linear slope in magnetization vs. applied field curve. This feature confirms our identification of first penetration field.

Finally we describe in Fig. 4 a simple technique for determining H*, the field for complete flux penetration. From the Bean model (1964) it is apparent that the two procedures of Fig. 2 should give the same magnetization values on the ascending branch of magnetization loop, in contrast the behaviours should be essentially different in the descending branch. As can be seen in figure 4 the magnetization field becomes 'reversible' above a relatively low field which we identify as H* . It is obvious that the effect of turning off the field and applying a new field leads to a critical state which depends only on the new value and not on the history, provided that new field is greater than H*.

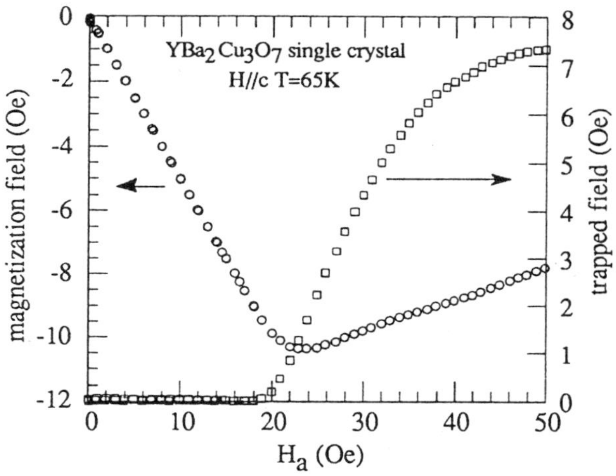

Figure 3. Typical record of virgin magnetization field and trapped field vs. applied field

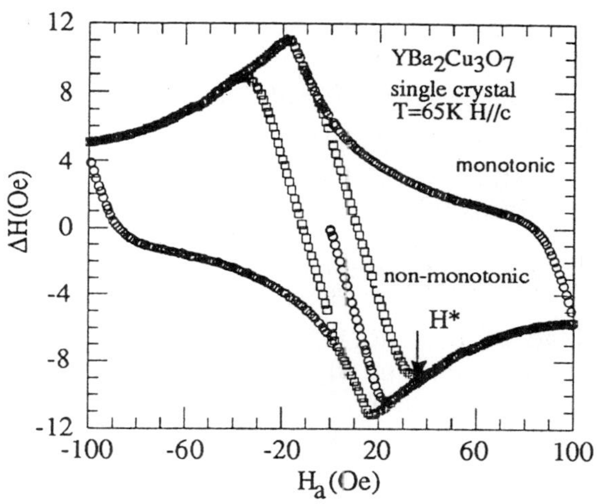

Figure 4. Typical monotonic and non-monotonic magnetic hysteresis loops.

We conclude that local Hall probe magnetometry presents a powerful technique for investigation of magnetic properties.of HTSC. Detailed studies,by this technique, of $H_{c1}(T)$, relaxation of magntezation and other related phenomena will be the object of forthcoming publications.

References.

Bean C P 1964 Rev. Mod. Phys. **36** 31
Frankel D J 1979 J. Appl. Phys. **50** 5402
Jung J, Mohamed M A-K and Franck J.P. 1989 Physica **C162-164** 1183
Konczykowski M 1990 to be published
Malozemoff A P 1989 in Physical Properties of HTSC, ed. Ginsberg D p.71
Müller A and Bednorz J G 1986 Z Physik B 64, 189
Reich S and Nabutovsky 1990 J Appl Phys., to be published

Supercond. Sci. Technol. 4 (1991) S334–336. Printed in the UK

Critical dynamic field and quality of high temperature superconducting ceramics by ac susceptibility measurements in the low field limit.

A. Mehdaoui, D. Bolmont and B. Loegel,
Laboratoire de Physique et de Spectroscopie Electronique, Faculté des Sciences et Techniques, 4, rue des frères Lumière, 68093 Mulhouse, France

Abstract. We present a detailed investigation of the ac field amplitude dependence of the initial complex susceptibility of YBaCuO compounds with various morphologies (bulk,thin film, single crystal) and BiSrCaCuO bulk compounds. We discuss our results together with published data in the critical dynamic field (h_c*) model. h_c* is a measure of the magnetic field above which vortex are unpinned giving rise to energy losses. The highest h_c* are observed in YBaCuO single crystals .A comparison between YBaCuO and BiSrCaCuO polycrystals shows a better quality for the former.The magnetic transitions in YBaCuO compounds show also high sensitivity to grain size .

1. Introduction

One of the most suitable technique for characterizing a superconducting transition in inhomogeneous high temperature superconductors (HTSC) is initial susceptibility which is very sensitive to all phases present, to all the parameters governing samples fabrication and to the thermal and chemical treatments. The imaginary part χ'' of the susceptibility which presents a peak at temperature T_M is strongly influenced by these parameters and the numerous published data show various behaviours, both for the shape of the peak and for the ac field dependence of T_M, making comparisons difficult .

Nevertheless, the critical dynamic field criterion allows such a comparison by modelizing the field dependence of T_M in a phenomenological model (Mehdaoui et al 1989) using Clem's formulation (1988).

2. Results and discussion

The intergrain flux-pinning plays an essential role in the power losses and leads to characteristic dependence of T_M upon applied ac field h_{ac}. The vortex are pinned in the low fields, and the intergranular field varies roughly as $h_{ac}\exp(-x/\lambda_J)$, where λ_J is the intergranular penetration depth(Clem 1988) leading to an exponential dependence between T_M and h_{ac}. Losses take place when h_{ac} reaches the critical value h_c*, resulting in a linear decrease of T_M vs h_{ac} as observed experimentally (Goldfarb et al 1987, Küpfer et al 1988, Gömöry and Lobotka 1988) and as expected from theoretical model when the coupling energy decreases like B^{-1} (Clem 1988, Küpfer et al 1988). We obtain then a phenomenological law for T_M vs h_{ac} containing a low field ($h_{ac}<h_c$*) dominant logarithmic decrease of T_M vs h_{ac} (curve A on figure 1a) and a high field ($h_{ac}>h_c$*) dominant linear dependence of T_M upon h_{ac} (curve B on figure 1a). In order to obtain a practical matter of comparison between various experimental results, we plot $(T_c - T_M)/T_c$ vs a reduced variable h_{ac}/h_c* for each set of experimental results (Figure 1b) and fit them with the theoretical curve of figure 1a. The fitting parameter is h_c*, the critical dynamic field which is a measure of the quality of the

Table 1. Critical dynamic fields h_c* (in Oersteds) for sample S3 and deduced from some selected data .p: polycrystal. single: single crystal. s: sintered. z m: zone melted. p.m.: partially melted.

ref	compound	h_c*	ref	compound	h_c*
Emmen et al 1989	BiSrCaCuO p s	.01	Costa et al 1988	YBaCuO p s	5.0
Emmen et al 1989	BiSrCaCuO z m	.35	Babic et al 1989	YBaCuO p s	8.0
Neumann et al 1989	YBaCuO thin film	.50	Gömöry & Lobotka	YBaCuO p sol-gel	8.5
Gomöry & Lobotka	YBaCuO p s	.65	Mazaki et al 1987	YBaCuO p s	16
Goldfarb et al 1987	YBaCuO p s air	.80	Laborde et al 1989	YBaCuO single	35
Dubots & Cave 1988	$PbMo_6S_8$ p s	1.7	Loegel et al 1990	YBaCuOsingle	50/85
Müller 1989	calculated intergrain	2.1	Müller 1989 calculated intragrain		$10^2/10^3$
Goldfarb et al 1988	YBaCuO p s O_2	2.6	Costa et al 1988	YBaCuO poly p.m.	200
Maeda et al 1988	LaBaCuO p s	3.2	this work	YBaCuO single S3	50

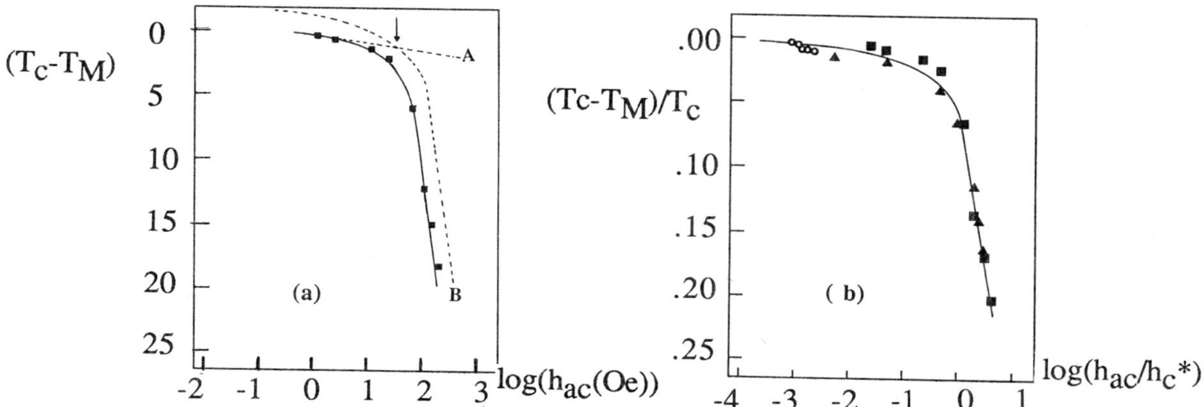

Figure 1 a) Variation of T_M vs applied field h_{ac} ($h_{ac} \perp c$) for single crystal S3. The straight line is a fit corresponding to the sum of a logarithmic term (A) and a linear term (B). h_c* is taken at the intercept of curves (A) and (B). b) same results than for a) in reduced ordinates : ■ single crystal S3; ▲ intergrain and O intragrain contributions deduced from Müller et al (1989)

samples(Mehdaoui et al 1989). The critical state model of Müller et al (1989a, 1989b, Müller 1989) for granular superconductors is also in good agreement with our phenomenological model (Figure 1b).

Table 1 gives the results of our model applied to published data together with our measurements on single crystals and the following main points can be briefly outlined.

i) h_c* and thus quality depends strongly on fabrication methods: sol-gel (Gömöry and Lobotka 1988) or partially molten (Costa et al 1988) YBaCuO alloys have always higher h_c* than the sintered one, and O_2 (Goldfarb et al 1987) treated samples show better h_c* than air treated one. h_c* is even sensitive to the details of the thermal treatment: a set of polycrystalline samples was submitted to the same routine with the exception of the last grinding after which the powder was sieved for fabricating the O_2 treated final pellets (Mehdaoui et al 1990). Although scanning electron microscopy shows no significant differences in grain sizes among the set after the final thermal treatment , we obtain increasing T_c with increasing grain sizes and h_c* values indicate also improved qualities for the highest T_c's (Figure 2).

ii) Bi compounds show always low h_c* values, whatever is the fabrication method. As h_c* is also a measure of the strongness of the flux pinning (Loegel et al 1989, 1990), this confirms the fact that pinning is weaker in Bi compounds than in YBaCuO compounds.

iii) Single crystals present two peculiar features: fine structures and field orientation dependence for the χ" vs T curves. Fine stucture reveals "second phase" and/or defects and

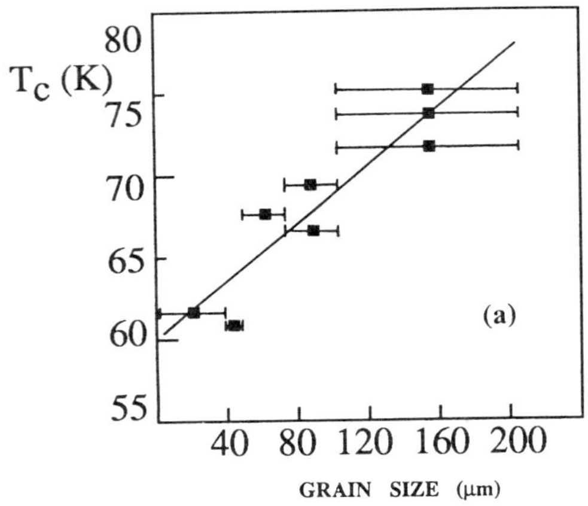

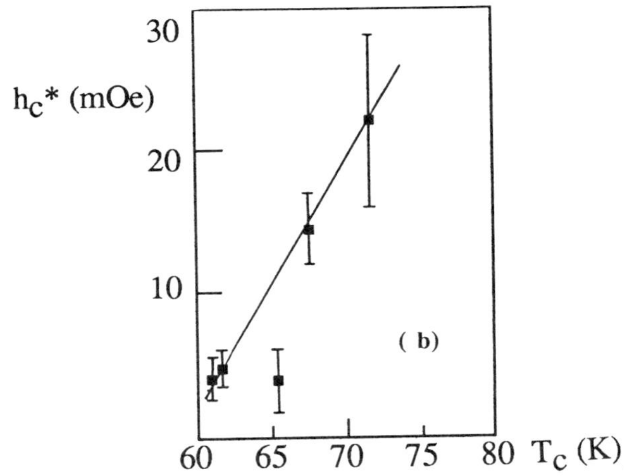

Figure 2 a) Critical temperature vs grain size and b) Critical dynamic field vs critical temperature, for a set of YBaCuO polycrystals (powder sieved before last annealing). Solid line is only a guide for the eye.

field orientation can be explained either by staircase superconducting transition or flux pinning at the Cu-O planes (for a detailed discussion see Loegel et al 1990 and references therein). But in all cases single YBaCuO crystals exhibit the highest observed h_c* which are of the same order of magnitude than the intragrain contribution (in polycrystals) deduced from the critical state model of Mûller (1989).

Acknowledgments.- We are greatly indebted to Dr. G. Collin from the Laboratoire de Physique des Solides at Orsay (France) for providing us YBaCuO single crystals.

References

Babic E,Drobac D, Horvat J, Marohnic Z and Prester M 1989 J. Less Common Metals 151,89
Clem J R Physica C 1988 153-155 50
Costa G A, Ferretti M and Olcese G L, Calzona V, Cimberle M R, Ferdeghini C, Putti M and Siri A S 1988 Sol. State Commun. 68, 923
Dubots P and Cave J 1988 Cryogenics 28, 661
Emmen J H P M, Brabers V A M, de Jonge W J M, van der Steen C, Dalderap J H J, Geppaart P M A and Kopinga K 1989 J. Less Common Metals 151 63
Goldfarb R B, Clark A F, Braginski A I and Panson A J 1987 Cryogenics 27 495
Gömöry F and P Lobotka P 1988 Solid State Commun. 66 654
Küpfer H, Apfelstedt I, Flükiger R, Keller C, Meier-Hirmer R, Runtsch B, Turowski A, Wiech U and Wolf T 1988 Cryogenics 28 650.
Laborde O, Tholence J L, Monceau P, Noel H, Gougeon P, Padiou J, Levet J C and Potel M 1989 Sol. State Commun. 69, 513
Loegel B, Mehdaoui A, Bolmont D, Eckhardt A and Muller F 1990 Cryogenics 30, 623
Loegel B, Bolmont D and Mehdaoui A 1989 Physica C 159, 816
Maeda A, Noda T, Matsumoto H, Wada T, Izumi M, Yabe T, Uchinokuta K and Tanaka S 1988 J. Appl. Phys. 64, 4095
Mazaki H, Takano M, Ikeda Y, Bando Y, Kanno R, Takeda Y and Yamamoto O 1987 Jpn. J. Appl. Phys. 26, L1749
Mehdaoui A, Loegel B, Bolmont D, Collin G, Löhle J and Wachter P 1990 E-MRS Int. Conference on High Superconductor Materials Strasbourg (France) may 29th- june 1st
Mehdaoui A, Loegel B and Bolmont D 1989 J. Appl. Phys. 66 1497
Müller K H, Mac Farlane J C and R Driver R 1989a Physica C 158 69
Müller K H, Mac Farlane J C and R Driver R 1989b Physica C 158 366
Müller K H 1989 Physica C 159 717
Neumann Ch, Ziemann P, Geerk L and Li H C 1989 J. Less Common Metals 151 363

Supercond. Sci. Technol. **4** (1991) S337–339. Printed in the UK

Broadening of the resistive superconducting transition under the magnetic field investigated by the conduction noise measurement

A. Maeda, H. Watanabe, I. Tsukada, I. Terasaki, K. Uchinokura, H. Obara,* and S. Kosaka*

Department of Applied Physics, The University of Tokyo, 7-3-1, Hongo, Bunkyo-ku, Tokyo 113, JAPAN

* Electro-Technical Laboratory, Namiki, Tsukuba, Ibaraki 305, JAPAN

Abstract. The conduction noise of thin films of oxide-superconductors was investigated below and near the superconducting transition temperature, in order to investigate the broadening of the resistive transition under the magnetic field. The result is discussed in terms of the flux-creep model and the fluctuation model.

1. Introduction

High-T_c superconducting oxides provides various interesting topics in superconducting states, one of which is the broadening of the resistive transition of superconductivity under the presence of finite magnetic field (Iye *et al* 1987). This may provide a basic question on the phenomenological description of the superconducting state of high-T_c oxides, where the coherence length is very short.

Two kinds of interpretations on this phenomenon are proposed, that is the interpretation in terms of the giant flux creep phenomenon (Tinkham 1988) and that in terms of the giant fluctuation of superconductivity due to strong two dimensionality (Ikeda *et al* 1989). It remains controversial which mechanism is the origin of the broadening. In this paper, in order to separate one contribution to the resistive broadening from another one, the conduction noise was measured in thin films of $YBa_2Cu_3O_y$ (YBC) and in $Bi_2(Sr,Ca)_3Cu_2O_y$ (80 K phase; BSC-2212) under the presence of the broadening.

2. Experiment

For detailed study of the noise in the metallic system, thin films are indispensable owing to its high resistance. High-quality thin c-axis oriented films which show superconductivity were obtained through the *in-situ* MBE method. Fabrication of the films are described elsewhere (Nakayama *et al* 1989, Obara *et al* 1990). For comparison, films obtained by a simpler method (Terasaki *et al* 1988) were also used in BSC-2212. The voltage noise was measured by using an ac voltmeter after narrow-band amplification under the constant-current condition. The bulk nature of the noise was confirmed by the size-effect study (Maeda *et al* 1989).

3. Experimental Results and Discussions

0953-2048/91/01S337 + 03 $03.50

In normal state, the power spectrum S(f) of the noise was known to show the 1/f spectrum with an empirical formula $S(f)=V^2(a/n)(1/SL)(1/f)$ where V is voltage, S and L is the cross section and the length of the sample, n is the carrier density and a is an empirical constant. Figure 1 shows the histogram of a/n obtained in various films. Although a/n is small in high-quality films, they still remain huge when compared with conventional metals, which is in agreement with our previous work (Maeda *et al* 1989).

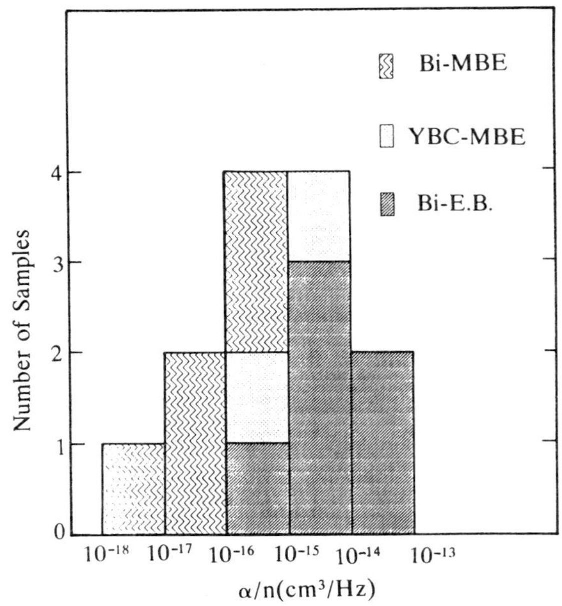

Figure 1 A histogram of the a/n for various films investigated in this study.

Figure 2 shows the temperature dependence of the noise power spectral density S and that of resistivity of an MBE-fabricated YBC film under various magnetic fields. Both shows the broadening under the magnetic field. Similar behavior was observed in an MBE-fabricated BSC-2212 film

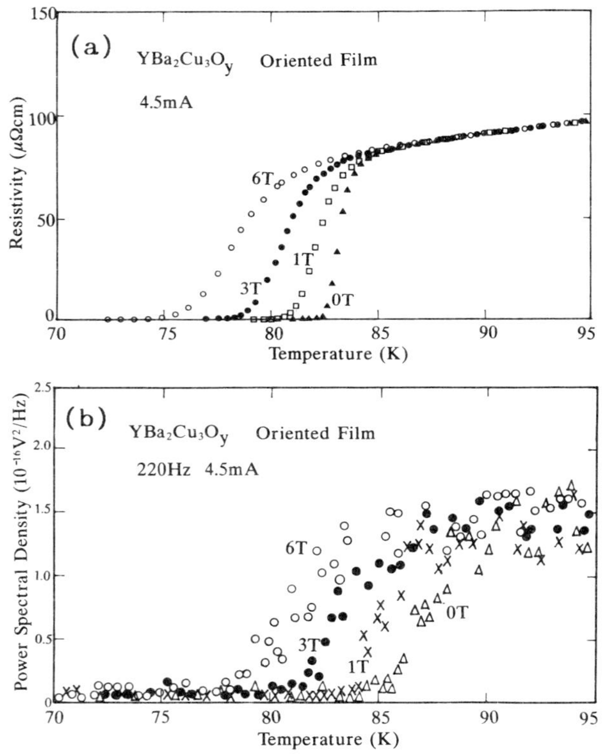

Figure 2 The temperature dependence of (a) the resistivity and (b) the noise power spectral density at 220 Hz of an MBE-fabricated YBC film under various magnetic field perpendicular to the CuO_2 plane.

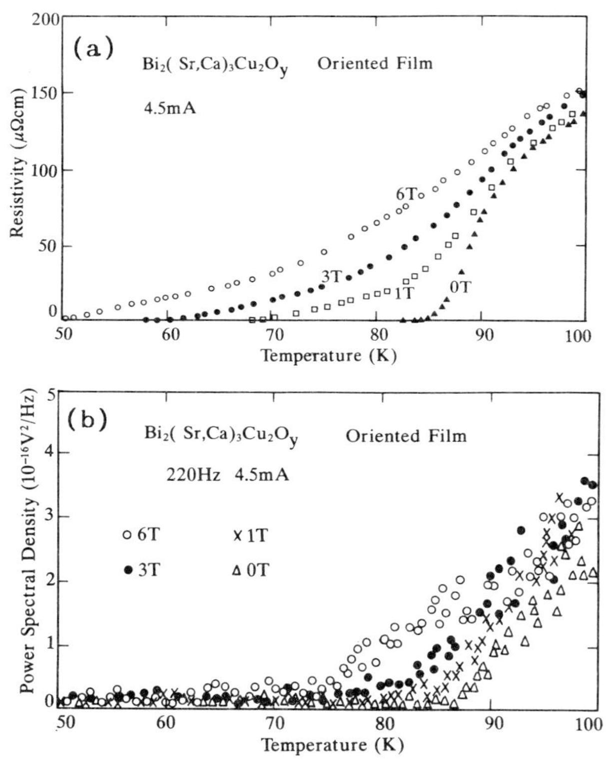

Figure 3 The temperature dependence of (a) the resistivity and (b) the noise power spectral density at 220 Hz of an MBE-fabricated BSC-2212 film under various magnetic field perpendicular to the CuO_2 plane.

(Fig. 3). To be more quantitative, the temperature dependence of the noise in the broadening region was found to scale the square of the temperature dependence of the dc resistivity well, as shown in Fig. 4 (Below we will call this noise "R^2" noise). By the investigation of the frequency dependence the noise in this region was found to have the so-called 1/f spectrum, which is similar to that observed in normal state. Thus, when seen by the conduction noise, the sample behaves as if it were in normal state. This strongly suggests that the appearance of the finite resistance under the magnetic field is mainly due to the large fluctuation effect.

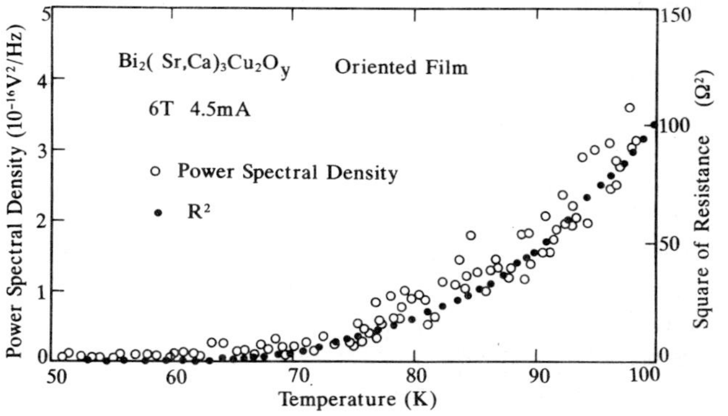

Figure 4 The temperature dependence of the noise power spectral density at 6 T, together with the temperature dependence of R^2.

In films with high resistivity, two other contributions were observed under the magnetic field. One is the extra noise other than the R^2 noise with the Lorentz spectrum, which we think is mainly due to the flux motion between superconducting grains. The second excess noise was observed as a very sharp peak at low temperatures far below the broadening region. Although the quantitative analysis of this peak strongly suggests that this structure is closely related to the flux- line-lattice melting, the absence of the clear peak in high-quality films suggests that the intergrain effect is also related with this phenomenon.

4. Conclusion

The conduction noise of high-T_c thin films was investigated in the so-called "resistive broadening region" under the finite magnetic field. The noise was found to be proportional to R^2 and the power spectrum is 1/f type. This suggests that the resistive broadening of high-T_c superconductors originates in the fluctuation effect due to strong two dimensionality.

References

Gammel P L, Schneemeyer L F, Waszczak J V and Bishop D J 1988 *Phys. Rev. Lett.* 61 1666

Ikeda R, Ohmi T and Tsuneto T 1989 *J. Phys. Soc. Jpn.* 58 3499

Iye Y, Tamegai T, Takeya H and Takei H 1987 *Jpn. J. Appl. Phys.* 26 L1057

Maeda A, Nakayama Y, Takebayashi S, Uchinokura K 1989 *Physica* C160 443

Nakayama Y, Ochimizu H, Maeda A, Kawazu A, Uchinokura K, Tanaka S 1989 *Jpn. J. Appl. Phys.* 28 L1217

Obara H, Kosaka S, Umeda M and Kimura Y 1990 *Physica* C to be published

Terasaki I, Nakayama Y, Uchinokura K, Maeda A, Hasegawa T, Tanaka S 1988 *Jpn. J. Appl. Phys.* 27 L1480

Tinkham M 1988 *Phys. Rev. Lett.* 61 1658

Supercond. Sci. Technol. **4** (1991) S340–342. Printed in the UK

Range of applicability for the critical state model and the interaction between the Abrikosov and Josephson vortices

E V Blinov, E B Sonin, A K Tagantsev and K B Traito

A F Ioffe Physico-Technical Institute of the USSR Academy of Sciences
Leningrad 194021, USSR

Abstract. The minimal a.c. magnetic field which is able to penetrate into the hollow ceramic Y-Ba-Cu-O cylinder was studied experimentally and theoretically as a function of the d.c. magnetic field. The critical state model predicts reversible dependence on the d.c. field in disagreement with observed irreversibility which is explained by interaction between Abrikosov vortices in grains and Josephson vortices in weak links.

1. Introduction

A lot of hysteresis magnetic phenomena observed in the type II superconductors are interpreted in terms of the critical state models which are entirely determined by the shape of the critical current *versus* magnetic field dependence $J_c(H)$. However, it is possible to detect experimentally some quantities for which those models predict no hysteresis. One of those quantities is the value H_{pen} of an a.c. magnetic field $H_\sim$ at which the field begin to penetrate into the hole of a hollow cylinder. In this paper we investigate the process of an a.c. magnetic field penetration into a hollow cylinder made of the high-T_c ceramic superconductor and the effect of the d.c. magnetic field $H_=$ on this phenomenon ($H_=$ and $H_\sim$ are assumed to be parallel to the cylinder axis).

2. Experiment

The sample was made of sintered Y-Ba-Cu-O ceramics and has the shape of the hollow cylinder (the length = 30 mm, the outer radius R = 8 mm and the wall width d = 4 mm). The a.c. magnetic field frequency is 50 Hz. The a.c. magnetic field in the hole is detected by means of a pick-up coil. Zero field cooling is made down to 77 K. Then the d.c. magnetic field is increased up to a value H^*, decreased down to zero, and then increased again. Curves 1, 2, and 3 in Figure 1 correspond to the first increasing of the d.c. field from 0 to 200 Oe, the decreasing from 200 Oe to 0, and the new increasing of the magnetic field, respectively.

The experiment shows that the H_{pen} is not a single valued function of $H_=$ but depends also on the maximum value $H_=$, H^*, reached during the cycle. The value of H_{pen}, H^o_{pen} measured after the cycle: $H_==0 \rightarrow H_==H^* \rightarrow H_==0$, is depicted as a function of H^* in the insert to Figure 1.

3. Discussion

At $H_\sim=H_{pen}$ the critical state magnetic field for the maximal and minimal fields outside of the cylinder, $H_= \pm H_\sim$, is determined by the critical state equation $dH/dr=(4\pi/c)J_c(H)$ (r is the distance from the cylinder axis) with the boundary conditions $H=H_= \pm H_\sim$ at r=R and $H=H_o$ at r=R-d. The field H_o inside of the cylinder and the field H_{pen} are determined from the condition that two solutions for the maximal and minimal field give the same field H_o at r=R-d. If the dependence $J_c(H)$ is single-valued as assumed in the critical state model. that procedure yields a single-valued dependence $H_{pen}(H_=)$ in disagreement with the experiment (Blinov *et al* 1989). The dependence $H_{pen}(H_=)$ is irreversible and as a result of it the field H^o_{pen} depends on H^* at H^*<10 Oe and H^*>100 Oe as it is shown in the insert in Figure 1. It is expected that reversible magnetic properties in the interval of fields 10 Oe <H^*<100 Oe are determined mostly by the intergranular Josephson vortices. Deviation from the critical state model at low fields < 10 Oe where large intergranular vortices expanding over many weak links and grains (hypervortices) may exist (Sonin and Tagantsev 1989) will be analyzed elsewhere. Here we shall discuss high fields H^* >> 100 G large enough for penetration of Abrikosov vortices into the grains.

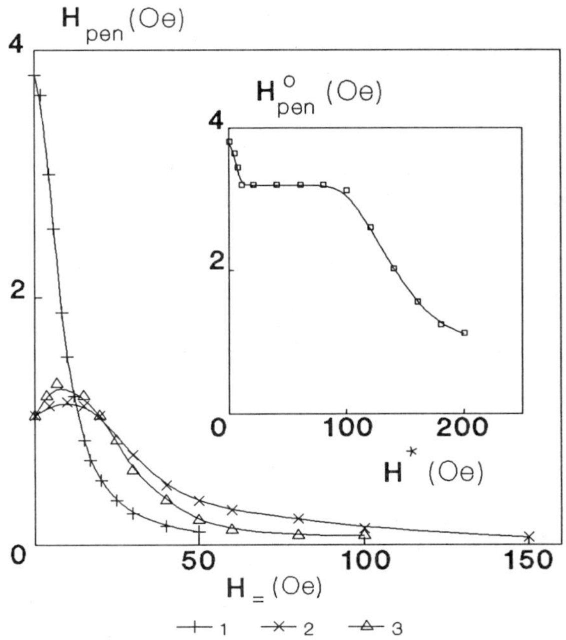

Figure 1 The experimental dependence of H_{pen} on the constant magnetic field $H_=$: 1 - the first increasing of $H_=$ from 0 to H^*=200 Oe, 2 - the decreasing of $H_=$ from 200 to 0 Oe, 3 - the second increasing of $H_=$ from 0 to 200 Oe. Insert: the dependence of H^o_{pen} on the maximal field H^* during the cycle.

The hysteresis phenomena observed in the experiment (difference between curves 1 and 2 in Figure 1) may be interpreted as follows. In the case of the first increasing of the d.c. field from the zero-field-cooling state (curve 1), the sample contains only the Josephson vortices in the weak links between the grains, while in the case of decreasing of the d.c. field (curve 2), in addition to the Josephson vortices in the weak links the trapped Abrikosov vortices in the grains are present. According to Evetts and Glowacki(1988) the trapped Abrikosov vortices induce the strayed magnetic fields in weak links that influence the weak link critical currents and lead to hysteresis of $J_c(H)$. However, the vortices trapped in grains influence the critical currents not only through the intergranular strayed field (their magnitude depends on the distribution of the trapped vortices over the grains and the shape of the sample) but also through the field of the trapped vortices inside of the grains.

In order to show it let us compare the average critical current J_c=<j_c sin kx> for the Josephson junction (the weak link) in the cases when the grains forming the junction have or have not trapped the magnetic flux. An electrodynamical analysis shows that the phase difference gradient $k=\partial\varphi/\partial x$ is proportional to $j_x \propto \partial H/\partial y$. Here axes x and y are directed along and normally to the junction correspondingly. On the other hand, $\partial H/\partial y \propto (H - H_g)$, where H_g is the average field inside of grains. If without the trapped flux ($H_g \cong 0$), $J_c \propto 1/H$ (the envelope of the Fraunhoveer curve for the Josephson junction in the magnetic field), then $J_c \propto 1/(H - H_g)$ for grains with trapped flux. So the field H_g of the trapped vortices shift the dependence $J_c(H)$ and explains the maximum at nonzero magnetic field on curve 2 (Figure 1). The analysis supposed that the field of the trapped Abrikosov vortices inside of the grains was homogeneous. In fact it is inhomogeneous and the Abrikosov vortices near the weak links may be considered as pinning centers for intergranular Josephson vortices. In the limit of the low density of the Abrikosov vortices inside of the grains one may use the theory of Fistul(1989) who supposed that positions of the Abrikosov vortices are random and uncorrelated. If $n\lambda^2 \ll 1$ and $nL\lambda \gg 1$ Fistul's theory gives

$$J_c \propto \sqrt{\frac{n}{1.37n^2 + (n - 4H_=/\Phi_o)^2}} \tag{1}$$

Here Φ_o is the flux quantum, L is the length of the weak link and n is the density of trapped Abrikosov vortices. The experimental curve 2 obtained for the decreasing field fits well to the dependence Eq.(1) with $n=2.1\ 10^8\ cm^{-2}$. So irreversibility and the shift of the dependence $J_c(H)$ may be interpreted as due to interaction of the trapped Abrikosov vortices and intergranular Josephson vortices. The theory of Fistul was used also for interpretation of the experiments by Fisher *et al* (1990).

In conclusion we consider the effect of strayed fields in weak links connected with the shape of the sample. A simple magnetostatics calculation gives that the field in the weak link,

$$H_i = \frac{H - H_g m}{1 - m(1-\mu)}, \tag{2}$$

differs from the external magnetic field H if the demagnetization factor m differs from zero. Here $\mu=\lambda/L$ is the effective magnetic permeability of the granular material (Sonin and Tagantsev 1989) and λ is the London penetration length for grains. For our experimental geometry the demagnetization corrections are not considerable.

Acknowledgements

The authors thank Dr. B A Glowacki for interesting discussion and Drs. M A Borik and V T Udovenchik for preparing of the sample.

References

Blinov E V, Stepanov Yu P and Fleisher V G 1989 *Reports to II All-Union conference on High T_c Superconductivity (in Russian)* v2 pp 92-3
Evetts J E and Glowacki B A 1988 *Cryogenics* **28** 641
Fisher L M, Il'in N V, Podlevskikh N A and Zakharchenko S I 1990 *Sol. St. Comm.* **73** 687
Fistul M V 1989 *Pis'ma Zh. Eksp. Teor. Fiz.* **49** 95
Sonin E B and Tagantsev A K 1989 *Phys.Lett.* **A140** 127

Supercond. Sci. Technol. 4 (1991) S343–345. Printed in the UK

XAFS Determines Mini-Clusters of Ni-O in Ni Doped $YBa_2Cu_3O_{7-\delta}$

J.B.Boyce[1], F.Bridges[2], T.Claeson[3], T.H.Geballe[4], and J.M.Tarascon[5]

[1]Xerox Palo Alto Research Center, 3333 Coyote Hill Rd, Palo Alto, CA 94304, USA
[2]Department of Physics, University of California, Santa Cruz, CA 95064, USA
[3]Department of Physics, Chalmers University of Technology, S-41296 Göteborg, Sweden
[4]Department of Applied Physics, Stanford University, Stanford, CA 94305, USA
[5]Bell Communications Research Laboratory, Red Bank, NJ 07701, USA

Abstract. X-ray Absorption Fine Structure (XAFS) measurements probe the very local environment of atoms. A sizeable amount of the Ni atoms in Ni substituted YBCO was shown to form Ni-O clusters with bond lengths characteristic of NiO even when x-ray diffraction indicated single phase samples. Subtracting the amount of Ni in NiO precipitates, T_c decreases linearly with Ni concentration and extrapolates to zero at 8.5% Ni.

1. Introduction

The superconducting transition temperature, T_c, of $YBa_2Cu_3O_{7-\delta}$ (YBCO) decreases as Cu is replaced with other elements. Ni is one of the most effective dopants. T_c decreases rapidly at low concentrations of Ni, the decrease becomes less dramatic at higher Ni content. X-ray diffraction indicates single phase samples (Tarascon et al 1988). The a and c lattice parameters of the orthorombic (or tetragonal, at high Ni concentration) structure vary smoothly, at least up to 10% substitution. NiO and/or other phases have been reported to occur at higher Ni concentrations(Qian et al 1989). A problem is that x-ray measurements generally are not sensitive to small amounts of Ni rich phases like small particles, precipitates, or mini-clusters.

In this study we intend to show that it is possible to use X-ray Absorption Fine Structure (XAFS) to estimate the amount of Ni that is replacing Cu substitutionally at random. Using the corrected amount of Ni in solution, we will plot T_c as a function of Ni concentration and find a linear variation.

XAFS is a local probe where the interference structure caused by back-scattered electrons from surrounding atoms determines the very local environment of absorbing atoms. This structure can be analyzed to give the distances to surrounding shells of neighbor atoms, number and kind of atoms in the shells, and the spread in neighbor separations. We have previously used the method to analyze a number of high T_c superconducting systems, e.g. samples with varying oxygen concentration (Boyce et al 1988) and with Fe or Co dopants (Bridges et al 1989). Much of the analysis of this work is based upon the methods worked out there.

2. Experiments and Data Analysis

The samples were prepared by solid state reaction. X-ray diffraction indicated that they were single phase. Variations in lattice parameters, T_c, Meissner effect, and other details are discussed by Tarascon et al (1988). Fine powder was brushed onto tapes which were stacked together to give approximately two x-ray absorption lengths at the Cu K edge. The transmission and fluorescence x-ray data were collected at the Ni K (8337 eV) and Cu K (8982 eV) edges. A Si(400) monochromator with an exit slit height of 1mm gave an energy resolution of about 1 eV.

The energy space data were converted to k-space and Fourier transformed to r-space following a normal procedure. In order to carry out a detailed analysis of the Ni XAFS we first developed a set of single peak standards for Ni-O, Ni-Y, Ni-Ba, Ni-Cu, and Ni-Ni. To fit the first neighbor environment of oxygen around nickel in YBCO, i.e. the Ni-O peak, we used both the first peak of the compound NiO and the Cu-O peak from Cu_2O. The latter standard is cleaner because of the larger separation between the metal-oxygen and metal-metal peaks in Cu_2O. A spherical wave correction (McKale et al, 1988) was used to convert the Cu-O standard to Ni-O. For the multi-Gaussian fit to the second neighbor region (Ni-Y, Ni-Ba, Ni-Cu, Ni-Ni) at 3.2-3.9 Å, we used Cu-Y and Cu-Ba standards that were developed for our study of oxygen depleted YBCO (Bridges et al 1988) and applied the spherical wave correction for the change of the core atom from Cu to Ni. Furthermore, Ni-Ni was extracted from the first peak of the XAFS data for Ni foil; again it could be used to generate a Ni-Cu standard by a theoretical spherical wave correction.

3. Results

The Fourier transformed XAFS for pure YBCO, YBCO with 3 and 10% Ni, as well as NiO are shown in figure 1. Note that the r-space XAFS for the low Ni concentration sample shows similarities with the pure YBCO, the reduced height of the second neighbor multi-peak, we beleive, is due to an increased destructive interference between Ni-Y, Ni(1)-Ba, Ni(2)-Ba, and Ni-Cu peaks, most likely produced by relative displacements of the Ba and Ni atoms. The higher doped sample has structure that we think is a result of NiO contamination. We will concentrate on the latter aspect.

Low Ni concentration (x=0.033). The XAFS of both the first and second neighbor regions could be fit to standards giving numbers and distances of Ni-O, Ni-Y, Ni-Ba, and Ni-Cu neighbors similar to those of corresponding Cu-X in undoped YBCO (Boyce et al 1988). The best fits were obtained for a uniform distribution of Ni on the (chain) Cu(1) and (plane) Cu(2) sites. The Ni(2) site was slightly distorted while the Ni(1) site was essentially undistorted. The results will be discussed elsewhere.

High Ni concentration (x=0.067, 0.1). The r-space XAFS for the strongly doped YBCO differed considerably from both the undoped and the lightly doped (x=0.033) YBCO. The first neighbor (Ni-O) peak was reduced as if there were a destructive interference from another distribution. The x=0.1 Ni K edge XAFS could be fit using the data for the x=0.033 sample to represent YBCO(Ni_x) and adding Ni-O (r≈2.08 Å) and Ni-Ni (r≈2.95 Å) to represent NiO precipitates (or mini-clusters). The fit of the first (Ni-O) peak region indicated that roughly 50% of the Ni occurs as NiO in the 10% sample.

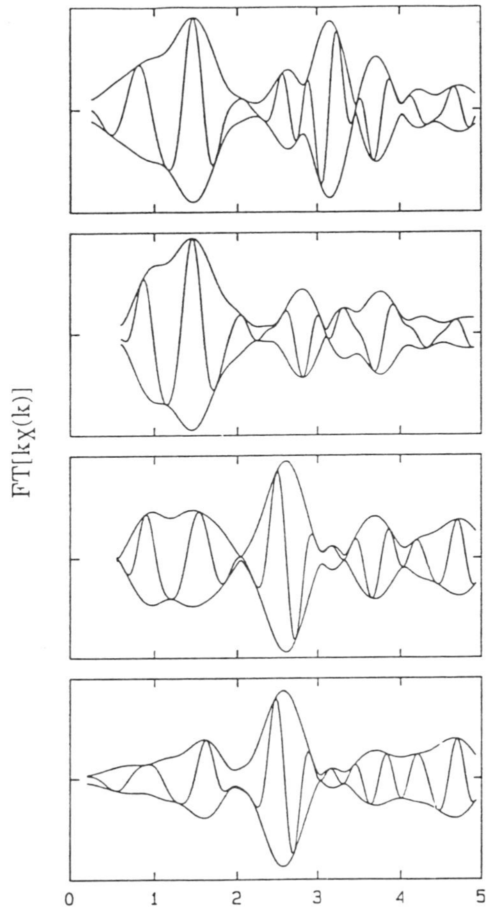

Figure 1 A comparison of the r-space FT(kχ(k)) for the Cu K edge for YBCO (top), the Ni K edge for YBCO($Ni_{0.033}$) (2nd fr top), YBCO($Ni_{0.1}$) (3rd fr top), and NiO (bottom). The transform range is a window from 3.1 to 11.5 $Å^{-1}$, Gaussian broadened by 0.3 $Å^{-1}$. The vertical scales for the top three spectra are the same, while the scale of the bottom curve is enlarged a factor of two.

Also the second neighbor region is strongly affected. A strong peak at about 2.6 Å in the high concentration samples shows strong similarities with the Ni-Ni peak in NiO, see figure 1. It occurs in a region where the YBCO(Ni_x) data has relatively small amplitude. Fitting the x=0.1 data over the range 2.35 to 2.8 Å to a sum of YBCO($Ni_{0.033}$) and Ni-Ni we obtained a good fit with a Ni-Ni separation of 2.97±0.03 Å. This is close to the value 2.95 Å for Ni-Ni in NiO. About 45% of the Ni was in an environment as in NiO. This agrees with the 50% fraction obtained from the first neighbor fit. Similar results were obtained for the x=0.067 sample.

4. Discussion

The amount of Ni that is in solution in YBCO can be determined once we have a good estimate of the fraction of Ni in the form of an NiO like environment. If we replot the T_c (defined as the midpoint of the resistance vs temperature curve as given by Tarascon et al (1988)) as a function of x using the corrected solubility concentration, we find that T_c decreases linearly with x rather than flattening out. T_c would go to zero at a Ni concentration of 8.5%. This is close to the value for Zn doped YBCO.

5. Conclusion

The XAFS data allowed us to estimate the amount of Ni that resides in a NiO like environment. Subtracting this part, we find that (i) Ni substitutes nearly uniformly with roughly 67% on the Cu(2) site and 33% on the Cu(1) site or possibly with a slight preference for the Cu(2) site, and (ii) T_c decreases linearly with Ni concentration.

Acknowledgment

The experiments were performed at SSRL, which is funded by Department of Energy under contract DE-AC03-82ER-13000, Office of Basic Energy Sciences, Division of Chemical Sciences, and the National Institutes of Health, Biotechnology Resource Program, Division of Research Resources. Partial support was given by AFOSR F49620-89-C-0017 and the Swedish Natural Science Research Council.

References

Boyce J B, Bridges F, Claeson T and Nygren M 1988 Phys. Rev.B **39** 6555

Bridges F, Boyce J B, Claeson T, Geballe T H and Tarascon J M 1989 Phys. Rev.B **39** 11603

McKale A G, Veal B W, Paulikas A P, Chan S-K and Knapp G S 1988 J. Amer. Chem. Soc. **110** 3763

Qian M, Stern E A, Ma Y, Ingalls R, Sarikaya M, Thiel B, Kurosky R, Han C, Hutter L and Aksay I 1989 Phys. Rev.B **39** 9192

Tarascon J M, Barboux P, Miceli P F, Greene L H, Hull G W, Eibschutz M and Sunshine S A 1988 Phys. Rev.B **37** 7458

Supercond. Sci. Technol. **4** (1991) S346–348. Printed in the UK

Quantification of Spatial Fluctuations of Oxygen Stoichiometry in $YBa_2Cu_3O_{7-\delta}$

N D Browning, J Yuan§, L M Brown

Cavendish Laboratory, Madingley Road, Cambridge, CB3 OHE, UK .

§ IRC in Superconductivity, Madingley Road, Cambridge, CB3 OHE, UK

Abstract . High quality electron energy loss spectra (EELS) with high spatial resolution have been acquired using a parallel recording system attached to a scanning transmission electron microscope (STEM). We have found that the fine structure of the Oxygen K-edge shows features which change as oxygen is removed. We have developed a simple method to quantify this change and applied it to estimate the oxygen stoichiometry in a series of spectra obtained from a bulk sintered sample of $YBa_2Cu_3O_{7-\delta}$.

1. Introduction

The electrical properties of $YBa_2Cu_3O_{7-\delta}$ (YBCO) are a strong function of the oxygen stoichiometry in the range $0<\delta<1$. As YBCO can vary from an anti-ferromagnetic insulator at $\delta=1$ to a superconductor when $\delta=0$, it is particularly important to be able to characterise the oxygen stoichiometry in bulk sintered material where fluctuations in oxygen can occur at grain boundaries. The electronic structure near the fermi level is very sensitive to oxygen stoichiometry and as a measure of the unoccupied states near the fermi level the pre-edge of the oxygen 1s absorption edge reflects this sensitivity in the range of $0<\delta<1$. As YBCO is unstable, noticeable damage occurs after a short exposure to the fast electron beam. This poses certain limitations on the acquisition time and hence, the signal to noise ratio of the core loss spectra collected. By using a CCD camera as a parallel recording device (McMullan et al 1989) this problem can be largely overcome. On the other hand, the effect of the electron irradiation is to reduce the oxygen content of the sample from 7 to 6. This allows one to observe the variation of the oxygen absorption edge as a function of the oxygen stoichiometries. Coupled with the nanometer probe available on a dedicated STEM, we think that this provides the best method of detecting small oxygen variations on a microscopic scale.

2. Experimental

The superconducting pellets are prepared using the standard methods for full oxygen stoichiometry and are then crushed for study in the microscope. This involves grinding the pellet between two microscope slides and dusting the powder onto a holey carbon film covered aluminium microscope grid before quickly transferring it to the microscope. The electron energy loss spectroscopy was carried out on a VG HB501 dedicated STEM operating at 100 kV. The microscope was operated in the standard image configuration with a probe of 6 mrad convergence semi-angle. The collection semi-angle of the spectrometer is 24 mrad.

3. Results

The spectra in figure 1 show the variation in the pre-edge features of the Oxygen K-edge as a function of electron irradiation. Each spectrum was taken with a 5 second exposure in spot

0953-2048/91/01S346 + 03 $03.50

mode giving a total exposure time of less than 2 minutes. As a comparison, the spectrum at the bottom of the fig. 1 is from the sample $YBa_2Cu_3O_{6.1}$ specially prepared by a quenching method with the oxygen stoichiometry measured by TGA. As one can see, in 2 minutes time the oxygen content of the sample has changed from originally near 7 to 6 per unit cell. To demonstrate the microanalytic capability of this technique, we obtained several spectra (Fig.3) from points 100nm apart (Fig.2) on a thin flake from a different batch of samples prepared under a different condition. By referencing to fig. 1, we may deduce that oxygen deficiency δ can change from ~ 0 to ~1 in positions about 1000nm apart (The method used to determine the oxygen stoichiometry is outlined in the next section). In bulk sintered materials, therefore, vast differences in oxygen stoichiometry can exist between neighbouring grains.

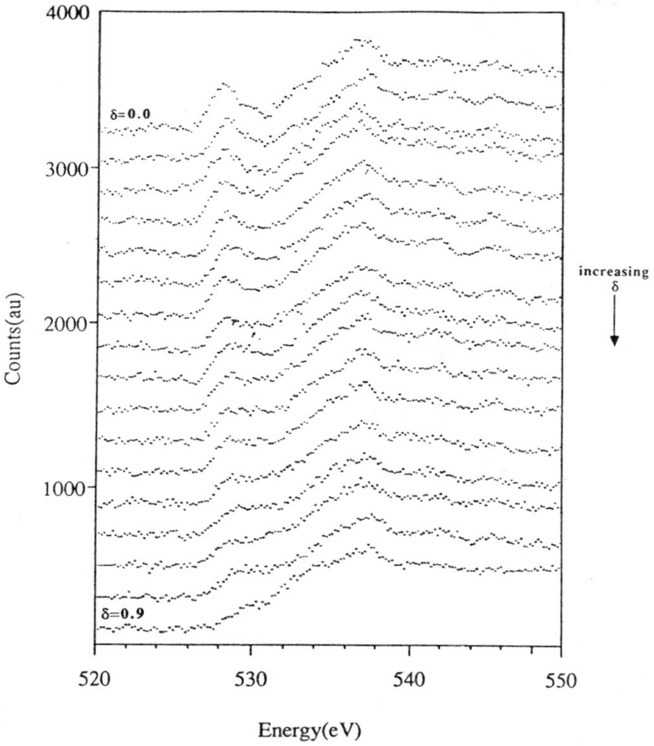

Fig 1. Oxygen k-edge spectra for YBCO showing the decrease in the pre-edge feature from full oxygen stoichiometry at the top to oxygen 6.1 at the bottom, with increasing electron beam irradiation.

4.Quantification of spectra

It has been shown by Nücker *et al* (1988) that the intensity of the pre-edge peak of the Oxygen K-edge can be related to the oxygen stoichiometry in YBCO. The pre-edge is known to consist of two main oscillator strengths (Nucker *et al* 1989), therefore to simulate the spectra three Gaussian functions were used, with the third one modelling the main peak. The choice of Gaussian lineshape was made, not as a means of reproducing the spectra accurately, but to provide a relatively simple way of estimating the area of each peak.

For each spectrum in the damage sequence shown in Fig.1, the areas of the three Gaussians were calculated. For all spectra the positions of the Gaussians were kept constant with the only changes being the height and width. The Gaussians were situated at 527.8eV, 529.6eV and 536.5ev. The results showed that the ratio of the first peak to the third peak decreased with exposure time. The ratio of the second peak to the third peak remained fairly constant

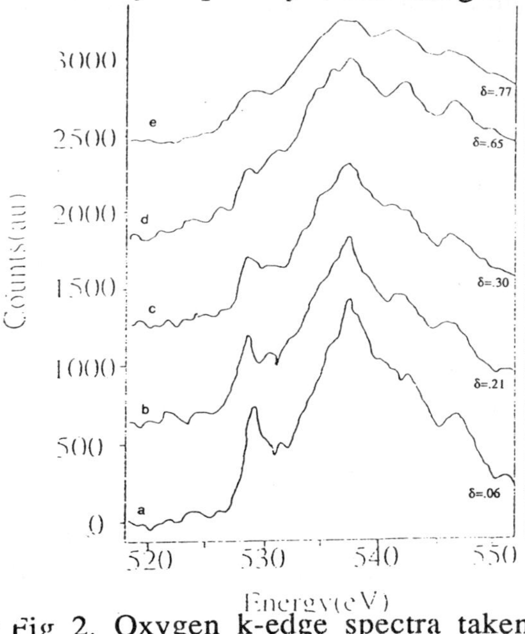

Fig 2. Oxygen k-edge spectra taken 100nm apart at points a,b,c,d,e in Fig 2.

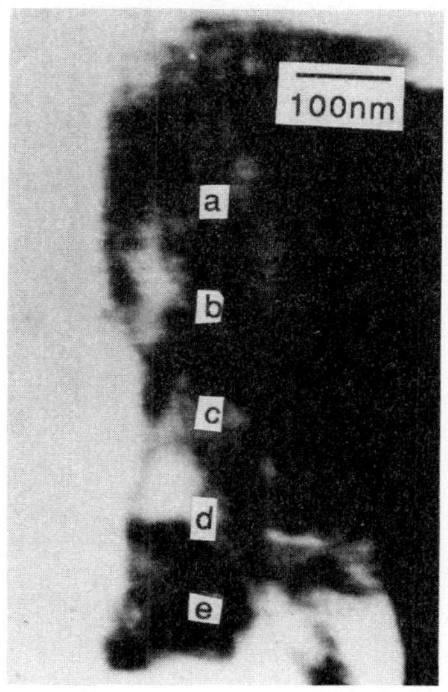

Fig 3. Bright field image of a thin flake of YBCO

for all spectra. The oxygen stoichiometry was estimated from theratio of the first peak at 527.8 eV to the main peak at 536.5 eV (Fig.4.). The graph shows an uneven rate of decrease in the relative intensity of the first pre-edge peak with exposure time and may be related to the ordering of oxygen vacancies ordering. Cava *et al* (1987) has shown that such ordering has a direct effect on the superconducting transition temperature of the sample. Using this method to calculate the oxygen stoichiometry of the $YBa_2Cu_3O_{6.1}$ sample gave a value for the oxygen stoichiometry of 6.09. This agreement gives confidence in the method.

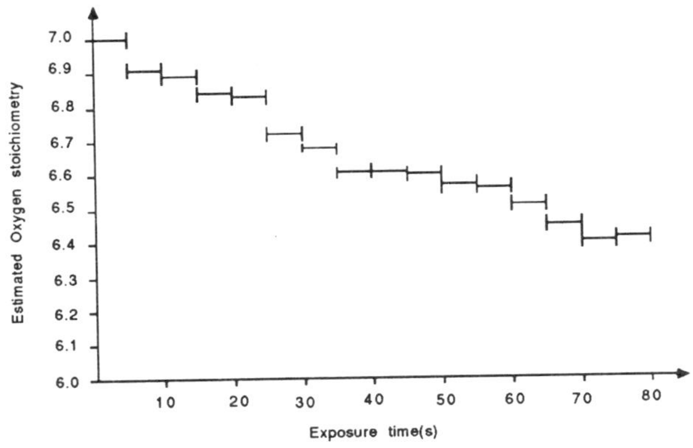

Fig 4. Oxygen loss as a function of beam exposure for the series of spectra shown in Fig 1. The horizontal lines represent a 5 second exposure time.

5. Discussion and Conclusions

Using the special sensitivity of the pre-edge structure of the oxygen K-edge to the oxygen stoichoimetry of YBCO, we have shown that it is possible to quantitatively estimate the local oxygen content from 7 to 6 with an accuracy about 0.1. Because of the possible uncertainty about the oxygen content of the original sample used in fig. 1, the calibration may be a slightly overestimated. However, we believe the error is unimportant with the accuracy of measuring the relative area from experimental spectra. Another potential error may arise from the beam damage during the acquisition. This could be particularly serious in microanalysis with a high concentration of the electron beam over a very small area. Our initial results suggest that, given a right balance between the requirement for good signal to noise ratio and the need for minimum acquisition time, fluctuations in oxygen stoichiometry can be measured in microscopic detail and large variations are found. Because of the large collection angle used, we believe that the anisotropy of the pre-edge feature is not important. We want to pursue this technique to correlate the possible oxygen nonstoichiometry with microstructure. At the moment, the measurements have been limited to samples produced by mechanical crushing. It has the advantage of the introducing minimum chemical alteration to the sample, but as the sample is most likely to break at the grain boundary or other weak points, interesting microstructural features can not be preserved for study. On the other hand, large thin areas containing multiple grain boundaries can be produced by ion beam milling . Unfortunately the ion beam also causes serious damage to the near surface layer. Unless such layer is removed or somehow minimised, EEL spectroscopy can not produce results representative of the undamaged materials. We are working on ways to overcome such problems.

Acknowledgements

We thank R.S.Liu and N.Peng of the IRC in superconductivity for the supply of YBCO samples. The SERC for a studentship (NB).

References

Cava R J, Batlogg B, Chen C H, Rietman E A, Zahurak S M, Werder D, 1987 *Phys Rev B* **36** 5719

McMullan D, Rodenburg J M, Murooka Y, McGibbon A J, 1989 *Inst.Phys.Conf.Ser.* **98** 55

Nücker N, Fink J, Fuggle J C, Durham P J, Temmerman W M, 1988 *Phys Rev B* **37** 5158

Nücker N, Romberg H, Xi X X, Fink J, Gegenheimer B, Zhao Z X, 1989 *Phys Rev B* **39** 6619

Supercond. Sci. Technol. 4 (1991) S349–351. Printed in the UK

Intergranular bondings in $YBa_2Cu_3O_{7-x}$ ceramics and their character versus sample density

G V Golubnichaja A Ya Kirichenko N T Cherpak

Institute of Radiophysics and Electronics Academy of Sciences of the Ukrainian SSR Kharkov 310085 USSR

V T Zagoskin Yu G Litvinenko

Institute of Single Crystals Kharkov 310141 USSR

Abstract. The dependence of the radiofrequency energy absorption Q^{-1} (f=2-5 MHz) upon temperature, density and static magnetic field H has been studied for samples of hot-pressed YBCO ceramics. The experimental data suggests that in the ceramics with γ > 5 g/cm most intergranular bondings are in the form of Josephson bridges. Lowering of the ceramics density results firstly in a narrowing of the bridges' widths followed by a break down and gradual transition to SIS type bondings and from them to the percolation threshold.

1. Introduction

Electromagnetic energy losses in high-T_c superconductors must depend on their structure. For the radiofrequency (f=2 5 MHz) losses' study in $YBa_2Cu_3O_{7-x}$ samples the inductive technique has been used. Ceramic samples have been produced using the hot pressure technology and they had the following characteristic of superconducting (SC) transition: $T_c \sim 92K$, $\Delta T = 1-1.9K$ when decreasing the density γ from 5.6 to 4.0g/cm^3.

2. Experimental data

Dependences of the losses Q^{-1} both at T=300K (Q^{-1}) and directly before transition into the SC state (Q_{92}) versus γ are shown in Fig.1 by curves 1 and 2 respectively (Q is the quality of the RF resonance. The losses in the absence of the sample have been subtracted) The losses' increase with increasing γ at 300 K corresponds to the increase of sample homogeneity and strengthening of the bondings between separate grains. The dependence $Q_{92}(\gamma)$ is nonmonotonic with a maximum in the region of $\gamma \lesssim 4.6-4.8$ g/cm^3. In the SC state at T=77K the losses Q_{77} decrease smoothly with increasing γ (curve3 in Fig.1).

Correlation between hysteretic curves of $\chi(H)$ and $I_c(H)$ (Dolgin and Smirnov 1989) can also be extended to the hysteretic behaviour of Q(H)(Aronov et al. 1989). The dependence of $Q^{-1}(H)$ (Fig.1) was observed when the H field was switched on after cooling the sample to 77K. A decrease of γ is accompanied by an increase of the field effect on the sample losses (Q^{-1} $(H=170)/Q^{-1}$). Simultaneously both the value of Q_R^{-1}/Q_0^{-1} in the residual magnetic field and the area bounded by the hysteretic loop grow.

These dependences confirm that in the main weak bondings between grains in a polycrystal contribute to losses in a low field. The bondings are weakened as the density decreases and the sample approaches percolation threshold (PT).

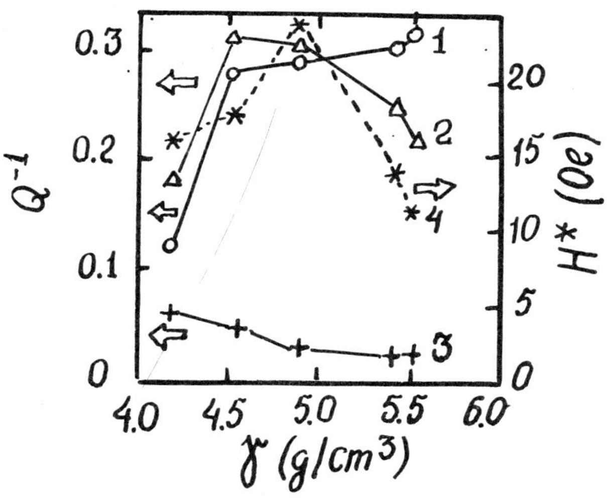

Fig.1

If we also take into account the conclusion reached by Aomine and Isuru(1987) that there is a connection between the location of the extremum on the curve of $I_c(H)$ in a falling field and the microbridge(MB) width of the weak link which bonds the grains we can try to decide what is the primary type of Josephson contact (JC). The dependence of $H^*(\gamma)$ is shown in Fig.1, curve 4. H^* is a field when an extremum of $Q^{-1}(H)$ in the decreasing field takes place.

3. Interpretation

The growth of H^* observed with decreasing γ at large γ can mean a narrowing of the MB. So, we can speak about bondings in the form of JC MB in the samples of th the higher γ . Any narrowing of MB with decreasing γ must result in growing losses including growth of Q_R^{-1} . The increase of γ intensifies the bondings between the grains, widening the MB and decreasing the number of MB, that is accompanied by a lowering of the hysteretic loop area of Q Q^{-1} (H). In the normal state the derivative $dQ^{-1}/dT > 0$ must be observed (Fig.1). The γ-dependence of H for $\gamma < 4.7$ g/cm^3 (Fig.

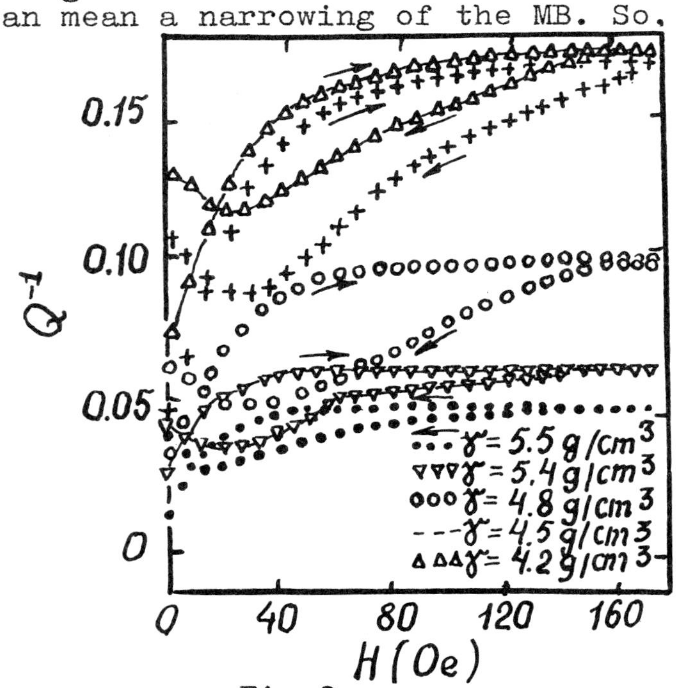

Fig.2

1) changes its character that can be connected with a change of the bonding character, in particular with breaks of the MB and the appearance of SIS contacts having a tunnel character; that explains a negative derivative $dQ^{-1}/dt < 0$ in the normal state for the samples of the lower γ. The weakening of links in the sample with $\gamma = 4.19$ g/cm^3 results in a weakening of the influence of H on $Q^{-1}(H=170)/Q_0^{-1}$ and Q_R^{-1}/Q_0^{-1} and in a decrease of the hysteretic loop area of $Q^{-1}(H)$. Degradation processes in samples in contact with water are accompanied by the approach to the PT and by a possible way out of the limits of PT. All of these effects lessen the losses and the loop area of $Q^{-1}(H)$. Superconductivity is however well preserved in the simple grains themselves, as is indicated by magnetic measurements when the SC transition is absent in resistive measurements. So, decreasing in the metaloxides γ results at first in a narrowing of the MB, followed by breaking of MB and a gradual transition from SIS type bondings to the PT.

Aomine T and Tsuru K 1987 Jap.J.Appl.Phys. 26 2013

Aronov I Ye Kirichenko A Ya Krokhin A A and Cherpak N T 1989 Problems of High-T_c Superconductivity (Kiev: IMF AN Ukr SSR) pp 95-9

Dolgin A M and Smirnov S N 1989 Superconductivity: physics, chemistry, technique 2 104

Supercond. Sci. Technol. **4** (1991) S352–354. Printed in the UK

Use of YBCO Wires to Extend the Temperature Range of a SQUID Voltmeter

H.W.Lean
IRC in Superconductivity, Madingley Rd, Cambridge CB3 OHE, UK.

N.McN.Alford
ICI Advanced Materials, The Heath, Runcorn, Cheshire, UK.

Abstract. We describe modifications to a conventional SQUID voltmeter involving the use of YBCO wires in the superconducting potentiometer circuit. This modification allows measurements to be made of voltages across samples at temperatures well above the 8K limit set by the transition temperature of the original superconducting wires.

1. Introduction

For many years SQUID voltmeters have been used in investigations which have necessitated the measurement of very small voltages (about 10^{-13}V) at low temperatures (typically up to 7K). The equipment described in the present paper has been used in work on the thermopower (Battersby and Waldram 1984) and resistance (Lean and Waldram 1989) of SNS structures fabricated with conventional superconductors. More recent work has focussed on measurements of voltages across Ag/YBCO interfaces (Lean et al 1990) for which it was realised that it would be useful to be able to extend measurements to temperatures above the 8K limit imposed by the apparatus. This paper describes preliminary work on a simple modification which, in principle, should allow data to be taken on samples at temperatures up to 90K.

Figure 1 shows the electrical arrangement of a SQUID voltmeter. The circuit is essentially a superconducting potentiometer with the SQUID used as the null detector. An SHE rf SQUID was used in the present system. The voltage across the sample is compared to that across a $10^{-6}\Omega$ standard resistor; the null condition is achieved by applying a current to the standard resistor either manually or fedback from the SQUID electronics. Because the SQUID itself is essentially a current meter (giving FSD for 0.5μA into the superconducting input coil) an important parameter of the system is the total resistance of the potentiometer loop, R_{loop}; the voltage sensitivity will be inversely proportional to this resistance. By making the circuit of superconducing wires it is possible to reduce R_{loop} to the sum of the resistance of the sample and of the standard resistor.

Temperature control is achieved by mounting the sample on a copper block suspended inside a vacuum can on a weak thermal link. The rest of the circuit is held at the temperature of the helium bath surrounding the can (usually 1.2K). The SQUID sensor itself is immersed in the helium bath and the standard resistor is mounted on a copper heat sink attached to the inside of the can. It is the requirements of heat flow which dictate that the potentiometer circuit be made of superconducting wire; the loop resistance could be reduced sufficiently by the use of thick copper wire but the heat leak between the sample and helium bath would then be too large to allow high sample temperatures to be achieved. It was

0953-2048/91/01S352 + 03 $03.50

realised that the 8K limit on the use of the apparatus imposed by the transition temperature of the Pb/Sn coated wires used could be increased if the sections of the circuit close to the sample were remade using YBCO wire. This would appear to be an ideal application for currently available wires because they will be able to be held rigid and will only need to pass a small current. In principle the use of YBCO wires should allow measurements to be made on samples at temperatures up to their transition temperature of about 90K. In this paper we report work which shows that the technique is feasible; measurements have already been successfully made at temperatures well over 8K.

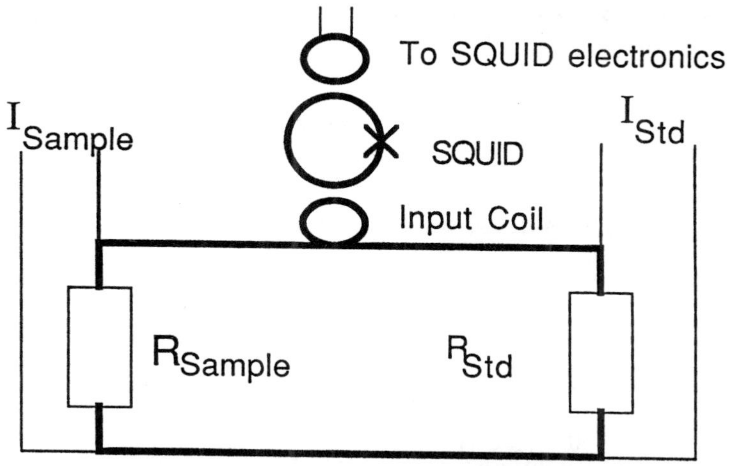

Fig. 1. Electronic Arrangement of SQUID Voltmeter. The thick lines represent superconducting wires.

2. Modifications to Apparatus

The YBCO wire used in the present work was prepared by viscous processing techniques described in detail by Alford et. al. (1988). The wire has a diameter 0.7mm and has no silver cladding. It was clear from the outset that the crucial question was going to be how to make contacts to the wire. The contacts should (preferably) give superconducting joints to the Pb/Sn wire at the 1.2K end and as low as possible contact resistance to the sample at the high temperature end. Previous work (Lean et al 1990) had shown that low resistance contacts to sintered YBCO could be produced by the fairly standard recipe of evaporating Ag contacts, reannealling in flowing O_2 at 800°C for 1 hour and then soldering to the Ag with indium. A preliminary test with a small piece of the wire showed that, as with the standard sintered material, this sort of contact had a resistance of a few $\mu\Omega$ at 4.2K and about 3 times this at 60K. Below T_c of the indium the junction is superconducting with critical current of probably tens of mA at 1.2K. This type of contact was used in the present work. At the 1.2K end of the wires the Pb/Sn wire was embedded in the indium melted onto the Ag contact. At the sample end the indium is used to solder the silver contact to the sample (which in these tests was simply a piece of copper or indium). There is a potential problem with this arrangement since at higher temperatures the higher resistivity of the In may start to make a significant contribution to R_{loop}.

3. Results and Discussion

The system was usually operated with the helium bath at about 1.2K. This was because previous work had shown that noise levels increased with the helium above the lambda point associated with vibrations caused by bubbles in the He, especially when the sample was heated. It was found that with the sample at the temperature of the helium bath the performance of the system was essentially the same as with the usual Pb/Sn wires; ie R_{loop} about $10^{-6}\Omega$ and a noise level of about 8×10^{-15} $VHz^{-1/2}$. The noise in the system is then dominated by Johnson noise from the standard resistor (or the sample if that has a greater resistance), and it is not practical to reduce the standard resistor further since the loop time constant (given by L_{loop}/R_{loop}) would become impractically long.

In the trials which have been carried out so far the sample has been heated to a maximum temperature of about 46K. As the temperature is increased above the indium transition R_{loop} increases to about $10^{-5}\Omega$; this corresponds to the In/Ag/YBCO joints at the sample end of the wires no longer being superconducting. As the sample is further warmed R_{loop} is then constant until some temperature in the range 8-12K where it increases again by about another order of magnitude and then stays constant until about 35K. Above 35K R_{loop} starts to increase again reaching about $10^{-3}\Omega$ at 46K. It is thought that the rather large resistances seen above 15K can only be caused by the Pb/Sn coated wires at the nominally 1.2K end of the YBCO wires being driven normal by heat conduction along the wires from the sample. The current heat sinking arangement involves glueing the wires at the 1.2K end to the copper block with Ge varnish. The problem with this arrangement is that the YBCO has a thermal conductivity about 10 times greater than the varnish although there should only be a very thin layer of varnish between the copper and the YBCO. Clearly this arrangement needs to be improved upon; a recent idea has been the use of 22 guage Cu wires embedded in the In joints and heat sunk to the copper block. With the current arrangement the noise level increases to about $3x10^{-13}$ $VHz^{-1/2}$ at 30K and $8x10^{-13}VHz^{-1/2}$ at 46K. These rather high noise levels seem to be consistent with Johnson noise from the loop resistance and should therefore be reduced when the heat sinking is improved allowing R_{loop} remain low at higher sample temperatures.

A problem which has been found is that the contacts to the YBCO wires appear to degrade on thermal cycling. In one case R_{loop} was found to increase by a factor of 500 at low temperatures after cycling between 300K and low temperatures twice. This is clearly a potential disadvantage with the system since the need to replace the wires between runs would significantly increase the turnaround time and work required.

The next stage in the development of the system (once the heat sinking has been improved as mentioned above) will be to take measurements to higher temperatures and to try measurements on real samples (YBCO/Ag junction structures). It seems likely that the ability to carry SQUID voltage measurements to higher temperatures will be of benefit to our research program in the near future. This approach is clearly potentially useful in any application where small voltages across a low resistance sample need to be measured at temperatures below about 90K. It may find use in measurements of thermopower since the superconducting potentiometer arrangement allows an absolute measurement of the thermopower of the sample and in practice the voltages to be measured are often limited by the amount of heat which can be passed down the sample.

3. Conclusions

We have described a method involving YBCO wires to extend measurements with a standard SQUID voltmeter to higher temperatures. The trials of the system carried out so far have revealed several problems but have also demonstrated the feasibility of the method.

Aknowledgements

We would like to thank J.R.Waldram and E.J.Tarte for helpful discussions.

References

Alford N McN et al J Mat. Sci. **23** (1988) 761

Battersby S.J. and Waldram J.R. J. Phys. F **14** (1984) L109

Lean H.W. and Waldram J.R. J. Phys. Cond. Matt. **1** (1989) 1285

Lean H.W. et al Proceedings of LT19, XIX International Conference on Low Temperature Physics, Brighton, August 1990.

Supercond. Sci. Technol. **4** (1991) S355–357. Printed in the UK

Thermal Conductivity of High Tc YBaCuO Bulk Prepared by Melt Growth Technique

H Yamamoto, M Sano, S Ozawa and M Tanaka
Nihon University, College of Sci. & Technol.,
Narashinodai, Funabashi-shi, Chiba 274, Japan
M Kaneko, Y Matsubara and T Ogasawara
Nihon University, Atomic Energy Research Institute,
Kanda Surugadai, Chiyoda-ku, Tokyo 101, Japan

Abstract A $YBa_2Cu_3O_7$ sample with Tc of about 80 K and Jc of about 3000 Acm^{-2} at 50 K was prepared by a melt powdering melt growth technique. Thermal conductivity λ of the sample was about 10^{-2} $Wcm^{-1}K^{-1}$ at 15 K. Addition of about 10wt.%Ag increased a value of λ about two times as large. Temperature dependence of λ suggested the sample includes many defects. The sample is a hopeful material for superconducting current leads.

1. Introduction

The discovery of high Tc oxide superconductors enables application systems which are operated at a comparatively high temperature region. Recently Murakami et al(1989) developed a new class of a high Jc $YBa_2Cu_3O_7$(YBCO) superconductor prepared by a Melt Powdering Melt Growth(MPMG) process. The feasibility of the MPMG bulk has been studied from various points of view. We remark that the thermal conductivity is expected to be low because the high Jc MPMG bulk includes many fine Y_2BaCuO_5 particles as pinning centers.

In this work, crystal structure, superconductivity and thermal conductivity were studied for the MPMG YBCO sample. From the fundamental properties obtained the samples are evaluated as a material for superconducting current leads which will be used in a superconducting magnet system operated by a solid conduction cooling(Furuyama et al 1990).

2. Experimental Procedure

The sample was prepared by the MPMG process as following. Mixed powders of Y_2O_3, $BaCO_3$ and CuO with the mole ratio of 1:4:6 were calcined at 900 °C for 24 h. The calcined powder was heated and melted in a Pt crucible at 1400 °C for 15 min. The melt sample was cooled down to room temperature on a copper plate. Then the solid sample was crushed into powders and pressed into a disk with diameter of about 2 cm and thickness of about 1 cm. In this stage 10 wt.% Ag was added to several samples. After the disk was again heated up to 1100°C and held for 20 min, it was cooled down rapidly to 1000°C with a rate of 2°$Cmin^{-1}$ and slowly to 500°C with a rate of 0.16°$Cmin^{-1}$. Finally thermal anneal at 450°C for 24 h was done in the atmosphere of oxygen.

Thermal conductivity λ was measured by a steady heat flow method. A schematic configuration of a measurement system is shown in Fig. 1. The sample was cut into a rectangular of which the size was typically about 6x5x4 mm^3. The surface of the sample was coated by evaporated indium

0953-2048/91/01S355 + 03 $03.50

film after polish by a sand paper. Figure 2 shows the sample surface before and after the metalization. Two samples were strongly sandwiched by three copper blocks on which carbon-glass(CGR) thermometers were embedded. Heat flux flows through the two samples from the heater 2. System temperature was controlled by the heater 1. In this measurement system the observed λ is the value including the term from the thin indium plates used for thermal contacts.

3. Result and Discussion

The x-ray diffraction indicated that the sample included a small amount of Y_2BaCuO_5 or Y_2O_3 in $YBa_2Cu_3O_7$ matrix. The YBCO and the Ag-added sample showed Tc of about 80 K as shown in Fig. 3. Resistivity ρ slightly decreased by the addition of Ag. The resistivity in the vicinity of Tc was about 4 mΩcm.

A temperature dependence of Jc of the sample was measured with transport current induced by pulse magnetic field(Yamamoto et al 1990). The result is shown in Fig. 4. The value of Jc was about 3000 Acm^{2} at 50 K and increased as decreasing temperature.

Typical results of $\lambda(T)$ are shown in Fig. 5. In the range of temperature from 10 K to about 50 K λ decreased as decreasing temperature as a function of T^n. The value of n of the YBCO sample was smaller than that of Ag-YBCO sample. The low value of n suggests that the sample includes many deffects. The value of λ was an order of 10^{-2} $Wcm^{-1}K^{-1}$ in the range of measurement temperature within the maximum error of 10 %. A value of λ of a single crystal YBCO(Kirk et al 1989) is much larger than that of a sintered one(Noto et al 1989). Then the variation of λ is attributed to differences in crystal structure and/or porosity. The value of λ of the MPMG sample is an intermediate value between that of the single

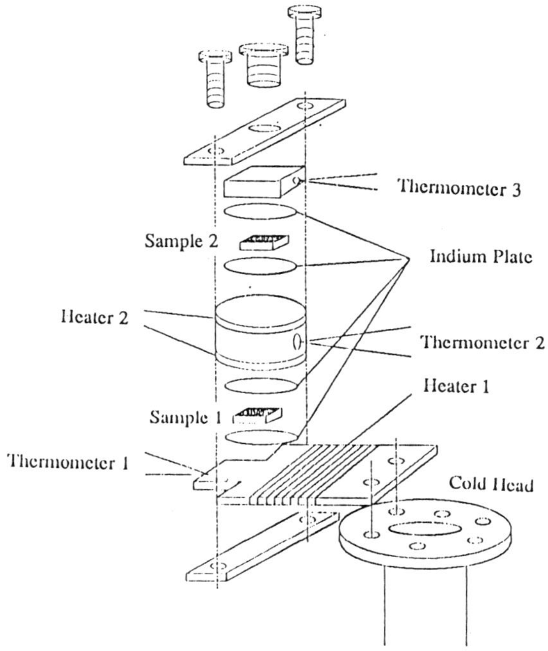

Fig.1 Measurement system of thermal conductivity.

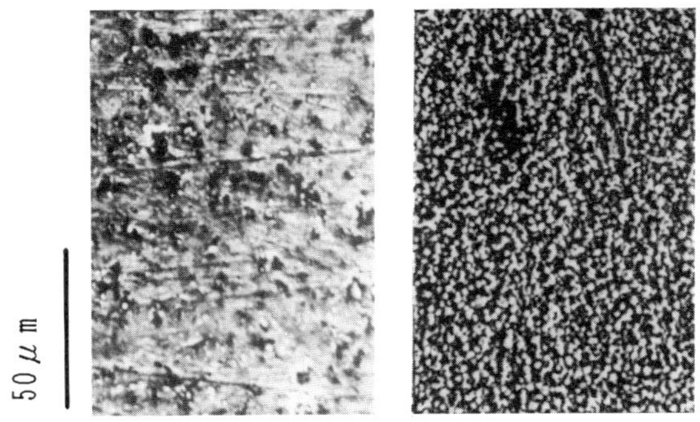

Fig.2 SEM photographs of surface (a) of the polished sample, (b) of the In coated sample.

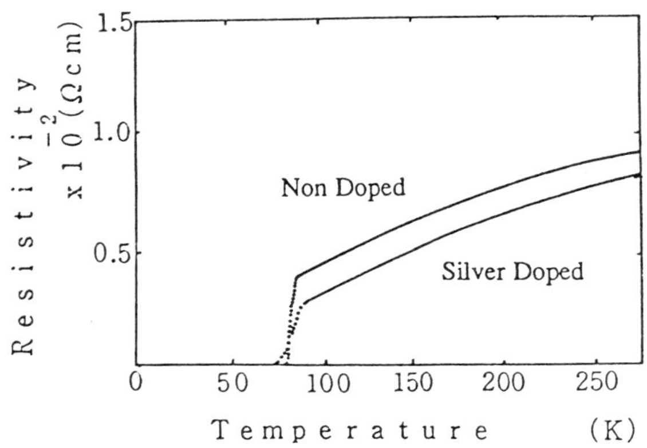

Fig.3 Resistivity vs. temperature of MPMG YBCO samples.

crystal and the sintered one.

Hull(1989) discussed about high temperature superconducting current leads and concluded that a heat flow into a heat sink is characterized by the quantity $\rho\lambda$. The value of $\rho\lambda$ in the vicinity of T_c is estimated about 4×10^{-5} WΩK^{-1} of which the value is nearly an order of magnitude larger than that of a pure metal. The result reveals a high feasibility of the sample for an application of superconducting current leads.

4. Summary

Thermal conductivity of the high Jc MPMG $YBCO_2$ bulk is fairly small, about 10^{-2} Wcm^{-1}K^{-1} at 15 K. The temperature dependence of λ revealed that the sample includes many defects expectedly. The addition of about 10wt.%Ag increases λ about two times as large. Obtained results reveal that the MPMG bulk is a very hopeful material for superconducting current leads.

Acknowledgments

The authors would like to thank Kawashima A and Seki M for their assistances in the preparation of samples.

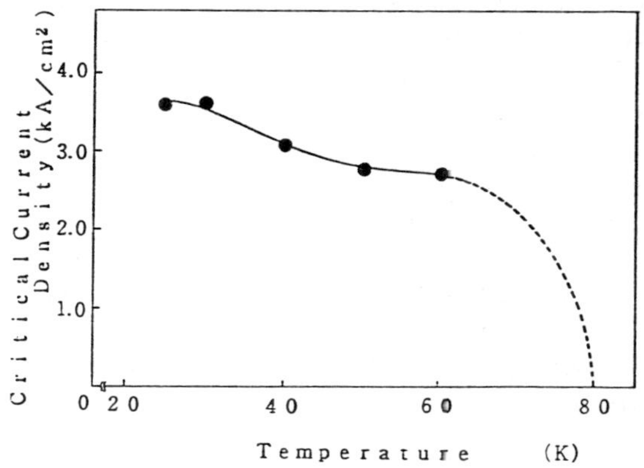

Fig.4 Critical current density vs. temperature of a MPMG YBCO sample.

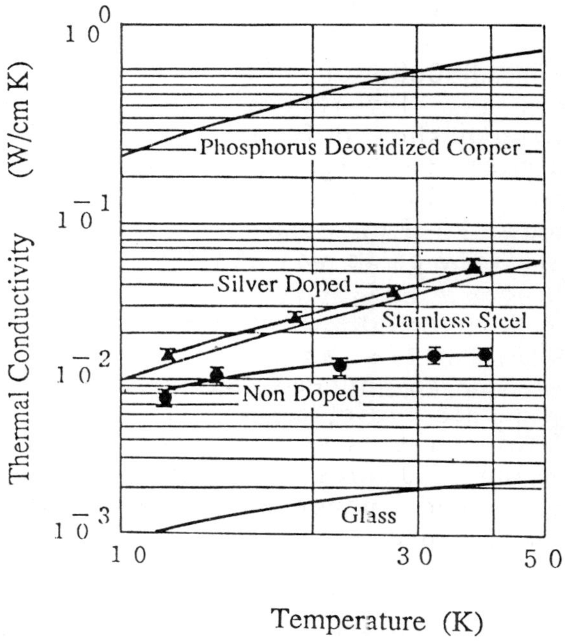

Fig.5 Thermal conductivity vs. temperature of MPMG YBCO samples and several materials.

References

Furuyama M, Yamamoto H, Tanaka M, Kaneko M, Matsubara Y and Ogasawara T 1990 Advances in Cryogenic Engineering 35 625

Hull J R 1989 Cryogenics 29 1116

Kirk W P, Kobiela P S, Tsumura R N and Pandey R K 1989 Ferroelectric 92 151

Murakami M 1990 Mod.Phys.Lett. 4 163

Noto K, Watanabe K, Morita H, Mori K, Sasakawa M, Isikawa Y, Sato K, Isihara T, Inukai E, Fujimori H and Muto Y 1989 Cryogenics 29 648

Yamamoto H, Ozawa S, Sano M, Kouma N, Tanaka M, Takizawa T, Kanbara K and Katoh K 1990 Proc. 65th Topical Symposium of the Magnetics Society of Japan MSJ65-12 73(in Japanese)

Supercond. Sci. Technol. **4** (1991) S358–360. Printed in the UK

Formation of $YBa_2Cu_3O_{7-x}$ Thick Films by a Diffusion Process Between Y_2BaCuO_5 and ($BaCuO_2$, CuO)

P.A. Freeman, N. Özkan, P.D. Honneyball and B.A. Glowacki

IRC in Superconductivity, University of Cambridge, Madingley Road, Cambridge, CB3 OHE, UK.

Abstract. The preparation of thick films of $YBa_2Cu_3O_{7-x}$ by diffusion between Y_2BaCuO_5 substrates and a surface layer of a barium cuprate mixture ($BaCuO_2$, CuO) was investigated. The influence of the substrate microstructure and density on the morphology and properties of the diffusion layer were determined by scanning microscopy and transport current measurements.

1. Introduction

The application of high-T_c oxide superconductors in the future is dependent upon the development of suitable fabrication processes and techniques, particularly those for the manufacture of wires and tapes capable of carrying high current density. Sadakata et al (1989) produced dense layers of $YBa_2Cu_3O_{7-x}$(123) with critical currents in excess of $10^3A/cm^2$, at temperatures well below 1015°C (peritectic decomposition temperature of 123 in air) by a diffusion reaction at the interface between two non-superconducting materials. The diffusion reaction proceeds by diffusion of elements to a high melting point, more stable compound which acts as a substrate from a low melting point compound. The advantage of this technique, apart from the lower processing temperatures required, is that the contamination of 123 by the substrate is negligible. Freeman et al (1990) have previously studied in detail the diffusion process between Y_2BaCuO_5 (211) substrates sintered at 1000°C and a low melting point barium cuprate mixture containing Ba:Cu in the nominal atomic ratio 3:5 (subsequently referred to as 358). The diffusion layer obtained by a solid-liquid diffusion process contained three separate regions; a thick dense surface layer of 123 (Region I) containing large elongated grains with twin domains, an intermediate region (Region II) consisting of finer grained 123 with CuO inclusions and a third region (Region III) adjacent to the 211 substrate which could not be identified as a single phase by XRD but whose chemical composition was similar to that of Region II. The thickness of the diffusion layer was not uniform and the critical current of the material was low due to the presence of microcracks and impurity phases at grain boundaries. In this paper we have investigated the influence of pressing conditions, sintering temperature and time on the density and morphology of the 211 substrate. The relationship between the density and morphology of the 211 substrate on the properties of the 123 diffusion layer formed from the above barium cuprate mixture were then investigated.

2. Experimental

The diffusion materials, Y_2BaCuO_5 and the barium cuprate mixture (Ba:Cu=3:5), were prepared by conventional solid state reactions in air starting from Y_2O_3, $BaCO_3$ and CuO as described by Freeman et al (1990). The 211 substrates were prepared by pressing 211 powder into 13mm pellets to loads in the range of 2-8 tons followed by sintering to temperatures between 1000°C-1200°C for 18-72 hours. Samples for the investigation of the diffusion process were prepared by placing a pressed pellet of 358 onto the surface of a sintered 211 pellet. The diffusion reaction between 211 and 358 was studied by heating the multicomponent systems to temperatures of 930-955°C in flowing oxygen for 6 hours followed by slow cooling (60°C/hour) to 450°C, an anneal for 6-12 hours at 450°C and slow cooling (180°C/hour) to room temperature. The composition and structure of the starting materials and the final products was studied by X-ray diffraction, electron probe microanalysis , scanning electron microscopy and optical microscopy. The transport critical current of the specimens was measured in liquid nitrogen both with and without an applied magnetic field (H_e=15mT) for two orientations of field, parallel and perpendicular to the surface (S) of the diffusion layer. In all cases I_c was perpendicular to H_e. The transport critical current was also determined under a remanent magnetic field, H_r, which is defined as the field trapped within the sample after the removal of the applied magnetic field, H_e. The anisotropy parameter K_A has been defined as the ratio of I_c for $H_e \parallel S$ to I_c for $H_e \perp S$. The criterion of electric field for the determination of J_c from I-V characteristics was $E=1\mu Vcm^{-1}$.

3. Results

The density and microstructure of the 211 substrates was found to be dependent upon pressing load, sintering time and sintering temperature. Figure 1 illustrates the morphology of a 211 substrate obtained by extrusion prior to sintering. The material consists of agglomerates and fine particles. Figure 2 illustrates a 211 pellet which was pressed to 10 tons and sintered to 1100°C for 12 hours in air. The average grain size has increased, with rod shape and angular grains being formed by liquid phase sintering. The density of the substrates were found to increase significantly with an increase in sintering temperatures up to 1150°C. For sintering temperatures above 1100°C the surface of the pellets contained regions which appeared to have melted. The chemical composition of such areas differed from the 211 stoichiometry thereby giving a lower melting temperature for this material compared with the decomposition temperature of 211. Differential thermal analysis in air at a heating rate of 10°C/min of a 211 substrate sintered to 1100°C for 24 hours showed that the peritectic decomposition of the 211 material took place at approximately 1275°C. Partial melting of impurity phases was observed at lower temperatures (1048°C and 1103°C). The substrate density for sintering times greater than 24 hours did not increase significantly. For a given sintering temperature and time, the pressure to which the powder was pressed had a significant effect upon the densification of the substrate. Low pressures (2 tons) gave a lower density substrate with reduced grain growth due to the poor connectivity between particles. In all cases the theoretical density for 211 was not achieved .

Figure 1
Surface of an extruded 211 tape before sintering

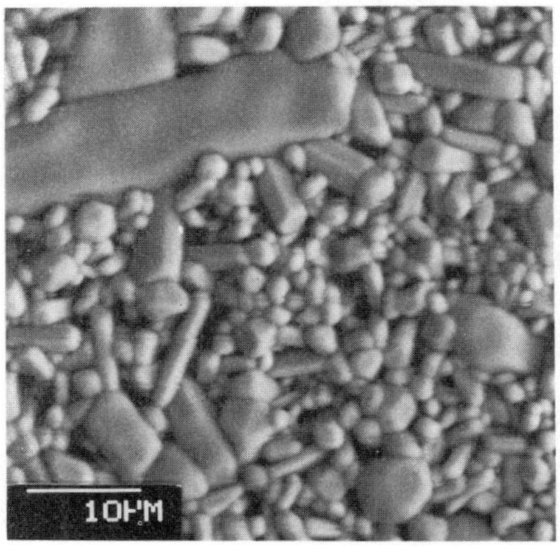

Figure 2
Surface of 211 pellet pressed to 10 tons and sintered at 1100°C for 12 hours

The relationship between the density of the 211 substrate and the thickness of the superconducting diffusion layer was very apparent. The thickness of the diffusion layer decreased with an increase of substrate density. From this observation it was concluded that

Figure 3
Surface of the remaining porous barium enriched phase at the interface with the diffusion layer

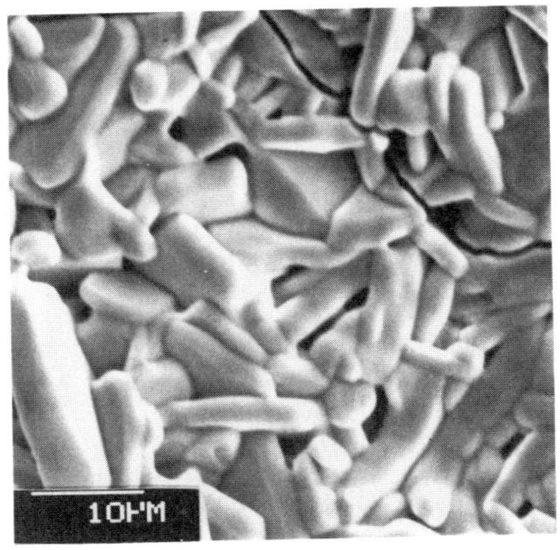

Figure 4
Surface of a diffusion layer at the Barium-Copper Oxide interface

the penetration of the liquid phase, formed by the partial melting of the 358 material, into the substrate is significantly reduced with increasing substrate density. Rather than being rapidly infiltrated into the bulk of the substrate, the liquid phase was found to spread evenly across the surface of the substrate forming a uniformly thick diffusion layer. The diffusion layer obtained on the higher density substrates (sintered above 1000°C) contained only two regions, I and II, whose composition and structure was as described previously in the introduction. Region I however was not easily distinguishable and the major fraction of the diffusion layer formed on the higher density substrates in most instances was Region II. In all cases, a porous surface layer of barium cuprate remained above the diffusion layer, Figure 3. The overall composition of the porous layer was previously estimated to contain Ba:Cu = 3.5:4.16 (Freeman et al 1990).The diffusion layer obtained on the more dense substrates contained a significant number of microcracks, Figure 4. The measured critical current values obtained for the diffusion layers (containing mostly Region II) formed on the higher density substrates were a factor of 5-10 times lower than the average value of 500 A/cm^2 obtained for diffusion layers (containing Regions I and II) formed on the lower density substrates (sintered at 1000°C, 18hrs). The microcracks observed were probably responsible for the reduction of the critical current and may be a consequence of the difference in the lattice parameters for the 211 and 123 phases. The difference in the parameters appears to be very significant for diffusion layers formed on the higher density 211 substrates. The anisotropy parameter, K_A, increased slightly with an increase in substrate density. The reason for this increase however is not completely clear since Region II is characterised by randomly orientated grains of 123 with oval shaped CuO inclusions.

4. Conclusions

From these investigations it was concluded that an increase in the density of the substrate did not result in an improvement in the transport critical current density of the 123 diffusion layer. Since it is not clear, at this stage, how the current is distributed throughout the different regions of the diffusion layer and why, after densification of the 211 substrate, the major phase formed consisted of a mixture of 123 and CuO, further investigations of the relationship between the microstructure of the diffusion layer and the critical current are necessary.

References

Freeman P A and B A Glowacki 1990 Presented at the 7th Cimtec World Ceramics Conference,Trieste, Italy

Sadakata N, Sugimoto M et al 1989 IEEE Trans. on Magnetics **25** 2180

Supercond. Sci. Technol. **4** (1991) S361–363. Printed in the UK

Effect of Cold Rolling on Microstructure, Grain Alignment, and Transport Critical Current Anisotropy of Silver Clad $(Bi, Pb)_2Sr_2Ca_2Cu_3O_x$ Superconducting tapes.

Wai Lo and B.A. Glowacki
Interdisciplinary Research Centre in Superconductivity, University of Cambridge, Madingley Road, Cambridge CB3 OEH, UK.

Abstract. Processes of manufacture of the superconducting $Ag\text{-}Bi_2Sr_2Ca_2Cu_3O_x$ tape form conductors by consolidation of powders in the silver tubes, followed by cold swaging and rolling were presented and discussed. The morphology and microstructure of the tape form conductors were investigated by SEM and X-ray microanalysis. Influence of the processing conditions and microstructure in respect to critical current value, and critical current anisotropy as a function of magnetic field, were discussed in detail.

Introduction

Both the Y-based "123" and Bi-based "2223" phase can be well aligned mechanically. The $(Bi, Pb)_2Sr_2Ca_2Cu_3O_x$ phase ("2223") however is much more promising for future applications because of its high critical temperature value (T_c=110K) as well as its ease of alignment during mechanical cold deformation, a common technique in materials processing. In addition, the highest Jc value for polycrystalline silver clad ceramic wires, 17000 A cm^2 at 77K, has been achieved for this phase (Hikata et al. 1989). Although the preparation of tape form silver clad wires is a rather simple task, the processing parameters for the manufacture of a superconducting tape capable of carrying high current at 77K appears to be complicated. To achieve high values of J_c in silver clad wires it is necessary to anneal the high T_c phase at a temperature around 840-850°C (Lo et al. 1990) under an atmosphere of composition N_2:O_2=13 by volume (Dou et al. 1989). In our paper the influence of the rolling sequence on the critical current anisotropy absolute value as well as current anisotropy are presented. Critical current anisotropy measurements can provide data about alignment of the local current paths as well as the morphology of the pinning centres in tape form metallic and ceramic superconductors (Glowacki 1985, Evetts et al. 1988, Glowacki et al. 1987).

Experimental Details

Precursor of $(Bi, Pb)_2Sr_2Ca_2Cu_3O_x$ was prepared by mixing Bi_2O_3, PbO, $SrCO_3$, $CaCO_3$ and CuO in the appropriate ratios followed by calcination at 845°C for 90h in air with intermediate grinding. Details of the preparation procedures for the above phases have been described elsewhere (Chen et al. 1990). The final product, "2223" phase (characterised by T_c=110K) were reground to powder forms. The powder, as ground, was packed mechanically into silver tubes of OD 6mm and ID 5mm. These composites, containing powder packed to a density of up to 40% by volume, were swagged to a diameter of 1.56mm and then cold rolled to various thicknesses. The tape form conductors obtained were sintered at 840°C for 20 hours under reduced oxygen atmosphere, N_2:O_2=13.

Transport measurements were performed in liquid nitrogen for the current perpendicular to the applied magnetic field, H_e=15mT, for two orientations of field, parallel and perpendicular to the compaction axis (P). The anisotropy parameter K_A has been defined (Evetts et al. 1988) as the ratio of I_c for H_e P to I_c for H_e P. The criterion of electric field for the determination of J_c from I-V characteristics was E=1μVcm^{-1}.

Results

Significant differences in the morphology between the surface of the ceramic core and its internal part after the first rolling can be observed in Fig. 1. The surface labelled part A was in immediate contact with the silver sheath, while part B is a fracture surface which is almost parallel to the rolling direction. The difference in morphology is related to the preferential cleavage which takes place under shear between the silver metal and ceramic core along the c-plane, i.e. along the $(Bi_2O_2)^{2+}$ layers (which are assumed to be the cleavage planes) (Kugimiya et al. 1988). Slippage of crystals during cold deformation takes place along the c-plane, and as a consequence leads to texturing. Structural examination of the ceramic ribbon (extracted from the silver sheath) by XRD indicated that in comparison to a randomly oriented powder the ribbon surface exhibited some texture along the c-direction. Co-deformation and flow of ceramic particles during rolling did not result in a uniform ceramic core dimension and as a consequence, the surface is locally distorted from the theoretical rolling plane by up to 30 degrees. Microcracks mostly perpendicular to the rolling direction were also observed on the surface.

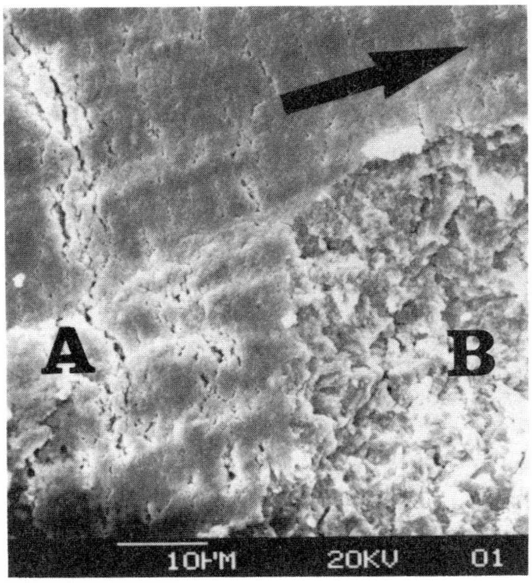

Fig. 1 Scanning electron micrograph of the Ag-$(Bi, Pb)_2Sr_2Ca_2Cu_3O_x$ interface after first rolling (silver was peeled off before SEM; arrow represents rolling direction).

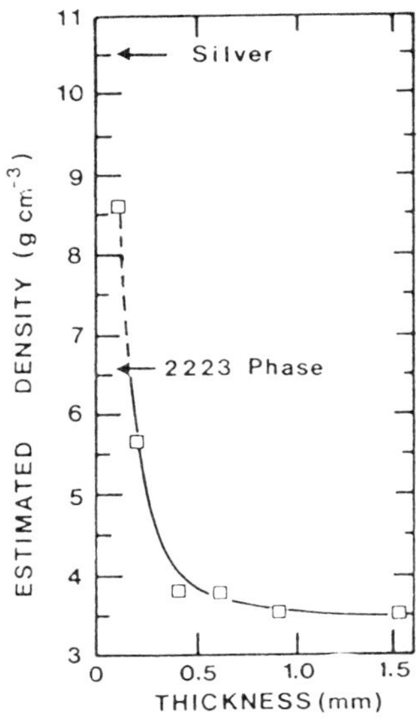

Fig. 2 Estimated density of superconductor core vs gross thickness of tape.

After sintering, the surface of ceramic became much flatter and smoother. The grains are all about well aligned with the c-axis approximately perpendicular to the surface of the tape although some steps and therefore some misalignments still exist. The grains are also closely packed, but the orientations among them are irregular.

Results of systematic studies of the effects of cold rolling on the estimated density, critical current and anisotropy parameter are presented in Figs. 2,3 and 4 respectively. The critical current density increases slowly with decreasing thickness at first (see Fig. 3), but suddenly doubles its value as the thickness is reduced from 0.4mm to 0.2mm, followed by an even bigger jump to achieve a value of $1266 A/cm^2$ when the thickness is reduced to 0.1mm. Similar behaviour is also observed in the variation of density with thickness, Fig 2. However an unexpected increase of density to $8.57 g/cm^3$ for a thickness of 0.1mm was observed. Because this density is higher than the theoretical density of the material (about $6.55 g/cm^3$) silver (density=$10.5 g/cm^3$)may play essential role in this suspicious behaviour.

In estimating the densities of the superconductor core, it was assumed that the cross-sectional areas of the core was uniform. Over-estimation of the density certainly invalidates such as assumption, and the implication is a non-uniform cross-sectional areas of the core along the tape. The cracks in Fig. 1 (taken from a tape with a thickness of about 0.2mm) further suggests such a possibility. Such cracks would be expected to develop under successive rolling. If silver characterised by density of 10.5g/cm^3 fills the cracks, it would be calculated as part of the core, which would then result in an over-estimation of superconductor density since the density of silver is higher than $(Bi, Pb)_2Sr_2Ca_2Cu_3O_x$. In view of this possible cause for density over-estimation, the actual value of critical current density for the 0.1mm thick tape should be about 2000A/cm^2.

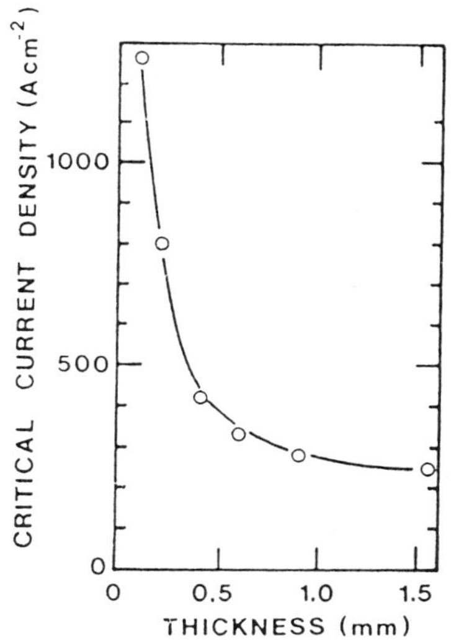

Fig. 3 Critical current density vs gross thickness of tape.

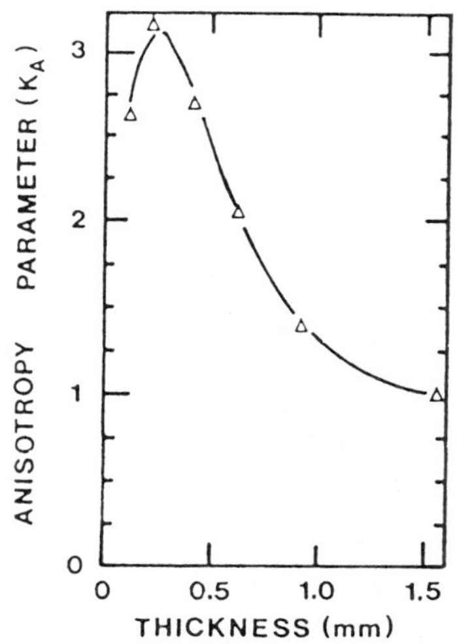

Fig. 4 Transport critical current anisotropy parameter vs gross thickness of the tape.

The drop in the anisotropy value can also be explained similarly. Enlarged cracks are likely to cause non-uniformity along the tape was well as across the tape, and as a result, the global alignment of grains is reduced although local alignment is retained. The overall effect is that a magnetic field applied perpendicular to the tape surface may not be perpendicular to a-b plane of the crystal lattice, and a magnetic field parallel to tape surface does not exclude flux penetration of the a-b plane. The anisotropy parameter is thus lowered. However, a more detailed explanation of the reduction of anisotropy parameter value below 0.2mm requires further investigation of the relationship between texture, intergrain connectivity and critical current anisotropy with respect to the degree of deformation.

References

Chen YL, Lo W, Zhen YS and Stevens R 1990 *J. Mater. Sci.*

Dou SX, Song KS, Liu HK, Sorrell CC, Apperley MH, Gouch AJ, Savvides N and Hensley DW 1989 *Physica C* **160** 533.

Evetts JE and Glowacki BA 1988 *Cryogenics* **28** 641.

Glowacki BA 1985 *J. Mater. Sci. Lett.* **4** 389.

Glowacki BA and Evetts JE 1987 *Materials Research Society, Symposium Proceedings* **99** 419.

Hikata T, Nishikawa T, Mukai H, Sato K and Hitotsuyanagi H 1989 *Jpn. J. Appl. Phys.* **28** L1204.

Kugimiya K, Kawashima S, Inoue O and Adachi S 1988 *Materials Lett.* **6** 277.

Lo W and Glowacki BA May 1990 *ICMC '90 Topical Conference, High-Temperature Superconductors, Materials Aspects, Garmisch-Partenkirchen (FRG).*

Supercond. Sci. Technol. 4 (1991) S364–366. Printed in the UK

SPATIAL NONUNIFORMITY OF INTERGRAIN JOSEPHSON JUNCTIONS AND MAGNETIC FIELD DEPENDENCE OF HTSC-CERAMICS CRITICAL CURRENT

E.Z.MEILIKHOV

Kurchatov Institute of Atomic Energy, 123182 Moscow, USSR

Abstract. The double-step behavior of critical current vs magnetic field in HTSC- ceramics of various compositions is assumed to be the consequence of the similar behavior of the critical current for the single intergrain Josephson weak links with the significant structure fluctuations. Parameters of the critical current spatial modulation in these junctions are determined.

The weak links between the single-crystal fragments of the material are known to be the origin of the low critical current density in HTSC-ceramics. As a rule these links are the contacts between the individual ceramics grains or the twin planes inside the grains. The peculiarity of the bulk HTSC-ceramics distinguished them from the ordinary superconductors consists in that after fast and significant drop of the critical current in weak magnetic fields it slightly changes at the further enhancement of the field B. The detailed investigations show that in the higher fields the secondary drop of the critical current $I_c(B)$ occurs. This phenomenon is universal and does not depend on the ceramics composition (Ekin et al 1989).

Our interpretation of the $I_c(B)$ double-step behavior in HTSC-ceramics relates to the nonuniformity of the individual Josephson intergranular contacts. Corresponding spatial nonuniformity of the critical current density $j_c(x)$ may be of two kinds (Barone and Paterno 1982): the "edge" one where j_c increases at the contact edges and the stochastic one where j_c is randomly distributed within the contact area. It is possible to distinguish these two situations on the basis of the behavior of the individual contact critical current $i_c(B)$ versus magnetic field. Such dependencies are known (Mannhart et al 1988):although $i_c(B)$ drops with the magnetic field but it never comes to zero. That corresponds to random structure fluctuations (nonuniformity of the intergrain layer) leading to the random fluctuations $j_c(x)$ within the contact area (Barone and Paterno 1982).

The problem dealing with influence of j_c-fluctuations on the field dependence of individual Josephson junction critical current was formulated and solved by Janson (1970) (see also Barone and Paterno 1982). Corresponding magnetic field dependence $I_c(B)$ keeps an "embryo" of all mentioned above peculiarities of the HTSC-ceramics critical current $I_c(B)$.

0953-2048/91/01S364 + 03 $03.50

The procedure of $I_c(B)$ calculation in the model of Josephson medium (the network consisted of the superconducting grains connected by the independent Josephson junctions) is described in our recent publication (Meilikhov and Gershanov 1989). $I_c(B)$ is found by maximization of the function $I_c(B) = \mathrm{Const}\,(P-P_c)^t\, i^*(P)$ where $P \equiv P(i^*,B)=\int_{i^*}^{\infty} f_B(i_c)di_c$ is the fraction of the weak links with the critical currents $i_c>i^*$, $f_B(i_c)$ is the distribution function for the link critical currents, t and P_c are percolation parameters (for cubic lattice of links $t\cong1.5$, $P_c= 0.25$) and $i_c^*(P)$ is the function reciprocal to $P(i^*)$. Also in (5) the way is considered for the calculation of the function $f_B(i_c)$ if one knows the functions $i_c(B)$ and $f_0(i_c)$ (for B=0). It was shown that weak links in YBCO-ceramics are well described by the "cut" power distribution function $f_0(i_c)\sim i_c^n$ with $n\simeq2$.

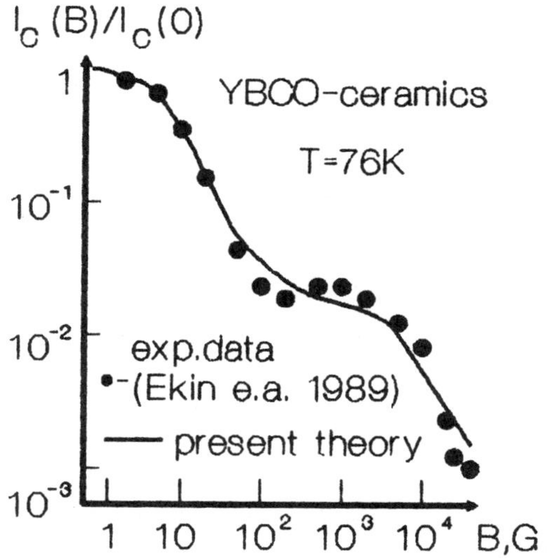

Fig.1 Critical current vs field B

The calculated dependencies $I_c(B)$ are defined by three parameters: 1)effective contact dimension a (defining the relation between B and dimensionless magnetic field $b=2\pi\lambda_g aB/\Phi_0$, λ_g is the grain penetration depth, $\Phi_0=hc/2e$); 2)correlation length r of the random critical current density $j_c(x)$; 3)mean coefficient K of $j_c(x)$ modulation. The fitting of the calculated functions $I_c(B)$ to the experimental data (Ekin et al, 1989) for different ceramics gives the values of all these parameters. The example of the fitting for YBCO-ceramics is shown on figure 1.

The parameters found are listed in the table. The great value of modulation coefficient K is an indication of the large nonuniformity of

PARAMETERS OF HTSC-CERAMICS

Composition	$YBa_2Cu_3O_x$	$Bi_{0.7}Pb_{0.3}SrCaCu_{1.8}O_x$	$Tl_2Ba_2Ca_2Cu_3O_x$
$T_c(R=0)$, K	92*	111*	113*
$j_c(B=0)$, A/cm	180*	150*	500*
$\lambda_g(T=77K)$, μ	0.4	0.4	0.2
a, μ	2.5	0.8	0.8
r, nm	1.2	1.2	1.3
K	0.85	5.0	2.5
Δ/r	1.8	15	5

* Data from 1]; $j_c(B=0)$ at T=76K.

$j_c(x)$ distribution and defined values of the correlation lengths r are of order of the lattice constant (along axis c). Both these facts permit one to suggest that the weak link supercurrent flows through the "contacts" of "superconductive" CuO_2-planes. In this case the question rises whether these planes belong to the neighbour grains or to the neighbour fragments of the single grain (for instance, two domains divided by the twin plane).

In the first case the regular alternation of CuO_2-planes is violated in the vicinity of the contact and the distances between ones randomly change. In the second case the regular arrangement of the CuO_2-planes remains and the spatial distribution of the critical current density along the contact area is described by the periodic (in the range $0<x<a$) function $j_c(x)$. Function $I_c(B)$ is presented by the spatial spectrum of $j_c(x)$ (2) and the envelope of the spectrum of any physically "reasonable" (that is smooth enough) periodic function falls relatively quickly (at least as 1/B) with the magnetic field increase. Therefore appearance of the plateau in this spectrum (and corresponding plateau in the magnetic field dependence of critical current $I_c(B)$) is hardly possible. Hence one should choose the first case from two ones mentioned, that is the case corresponding to the random $j_c(x)$ distribution within intergrain contacts. It is confirmed by cited in (Chaudhary et al, 1988) magnetic field characteristic for one of YBCO- film intergrain contacts that resembles "white noise" one.

Taking into account the great value of the modulation coefficient K (see table) one would imagine the intergrain contact as a set of relatively rear weak links between CuO_2-planes belonged to the neighbour grains. The mean distance between these links $\Delta \gg r$ (see table). It is schematically shown on fig.2. The value Δ=2.2 nm for YBCO-ceramics listed in the table surprisingly coincides with the distance between dislocations coming out the boundary between misaligned single-crystal grains (misorientation angle in basal plane $\simeq 10^\circ$) (Dimos et al, 1988). It permits to suggest that Josephson critical current density increases just in the points where the dislocations come out the intergrain boundary.

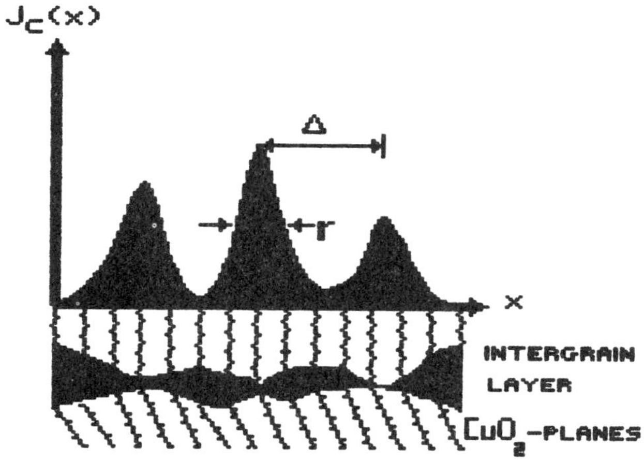

Fig.2 Local current density

The model considered permits to came to a number of the practically important conclusions concerning the dependence of $I_c(B)$ behavior in HTSC-ceramics on the the various technology factors.

Barone A. and Paterno G. 1982 Physics and Applications of the Josephson Effect (John Wiley & Sons)

Chaudhary P. et al 1988 Phys.Rev.Lett. **60** 1653

Dimos D. et al 1988 Phys.Rev.Lett.**61** 219

Ekin W. et al 1989 Physica **C160** 489

Mannhart J. et al 1988 Phys.Rev.Lett.**61** 2476

Meilikhov E. and Gershanov Yu. 1989 Physica **C160** 431

Yanson I. 1970 JETP **58** 1511

Supercond. Sci. Technol. **4** (1991) S367–369. Printed in the UK

Current Conduction and Macroscopic Quantum Interference in BiSrCaCuO Screen Printed Films

S. Mészáros, K. Vad, G. Halász, N. Hegman and G. Katona*

Institute of Nuclear Research, H-4001 Debrecen, P.O.B. 51.,Hungary,
*Microelectronic Co. "MEV", H-1325 Budapest, P.O.B. 21.,Hungary

Abstract Electric current conduction properties of screen printed superconducting BiSrCaCuO films on alumina substrates were investigated. The film contains the low T_c 2212 phase. The temperature dependence of H_{c2} and of critical current were determined from the magnetic field dependence of resistance-temperature curves and from current-voltage characteristics. The film cross section dependence of critical current density and change in the shape of current-voltage curves due to the cross section reduction of the film were observed. Quantum interference inherent in the film was demonstrated.

1. Introduction

The screen-printed-film technique is a promising preparation method of high T_c superconducting materials. This can be performed on different substrates including cheap polycrystalline materials. It enables us to use high T_c superconductors to connect electronic elements and to prepare two dimensional layouts of complicated shape, as detector loops, gradiometers, etc. The eletric current conduction properties of this type of films were studied to elucidate the factors determining critical current, electric field generation and magnetic flux motion. As it is well known the electric field generated in superconductors depends on current density, temperature and magnetic field. The latter may significantly differ from the externally applied field due to diamagnetism and magnetic flux trapping. The usual experimental procedure is to measure the current, temperature and external magnetic field dependence of the longitudinal voltage. The voltage is generated mainly at intergrain junctions in high T_c polycrystalline materials and the character of these junctions can be of SIS, SNS or bridge type. BiSrCaCuO screen-printed films were prepared on alumina substrate and their current conduction properties were studied experimentally. The cross section dependence of the critical current density of these films was also studied in order to get more direct information on the character of intergrain junctions.

2. Sample preparation

The samples were prepared on polycrystalline alumina substrates of 96 and 99 % purity and thickness between 20 and 40 μm by the method described by Besenyei et al (1989). The films are equipped with Ag-Pt current and voltage contacts prepared by the screen-printing technique with less then one Ohm resistance to the film. The distance between the potential contacts is 6 mm, the width of the film is 1 mm. The film contains the low T_c 2212 phase which was checked by X-ray diffraction.

3. Measurements

The experimental arrangement and measurement method was the same as used by Mészáros et al (1990) in the study of similar properties of $YBa_2Cu_3O_{7-x}$ ceramics. Resistance-temperature curves were measured by pulsed current method in magnetic fields parallel to the direction of the current. The sample temperature was set by a temperature controlled helium gas stream and magnetic field was produced by a Nb-Ti superconducting solenoid. Calibrated carbon glass thermometer was put into the same gas stream near to the samples to monitor the temperature. The combination of the pulsed current method and the convective cooling ensured the reliable sample temperature measurement and control. Current-voltage characteristics were measured in zero magnetic field by the pulsed current method up to electric field of 80 mV/cm. The dependence of the critical current and current-voltage characteristics on the width of the film was studied at 4.2 K by immersing the sample into liquid helium. In this case the characteristics taken by the pulsed and continuous current method were identical. The modulation of the voltage by weak external magnetic field was studied at 4.2 K by passing continuous current through the sample and generating weak magnetic field perpendicular to the film by a small coil driven by a triangular wave current. To study the cross section dependence of current-voltage characteristics the width of the film was systematically reduced by scraping a part of the film between potential contacts. The number of intergrain junctions in this region was reduced and their individual characteristics were not averaged out. At a small enough cross section current pulses were able to cause permanent change in the film including total destruction as well. This effect was used to reduce the number of effective junctions by controlled current pulse treatment of the film immersed in liquid He similarly to the method used for YBaCuO thin films by Robbes et al (1990). Destruction was avoided by shunting the potential contacts by a room temperature resistor.

4. Results and discussion

The resistance-temperature curves were measured at current densities 30 mA/cm^2 to 30 A/cm^2. The magnetic field dependence was measured in longitudinal magnetic fields from the remanent field of the cryostat i. e. a few hundred μT up to 5 T. The apparent feature of the R-T characteristics measured on the films (see Figure 1) is the lack of separation of two regions on the curves as it was observed for YBaCuO

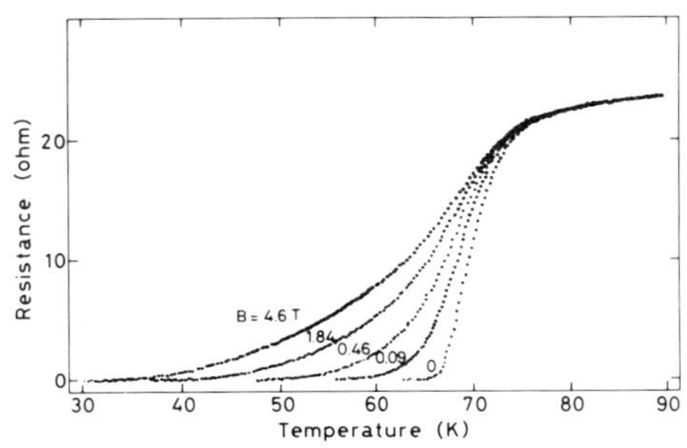

Figure 1
Resistance-temperature curves of a typical sample in magnetic fields

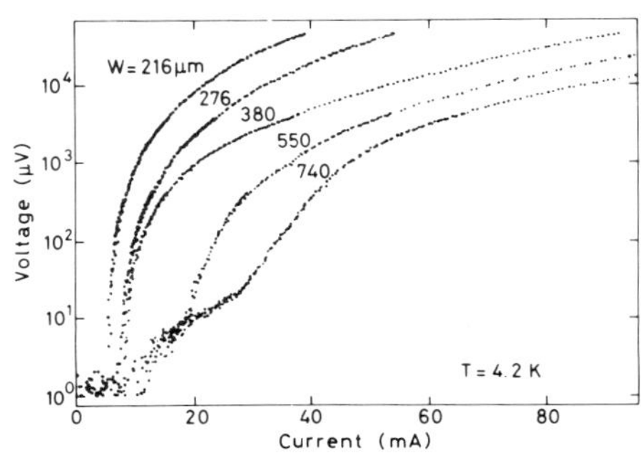

Figure 2
Current-voltage characteristics for different film widths at 4.2 K

ceramics by Mészáros et al (1990). The separation of resistance contributions of grains and grain boundaries was not possible in this case. H_{c2} values were calculated from the shift of zero resistance temperature due to magnetic field. The $H_{c2}(0)$ value was calculated using WHH theory and found to be 12 T. The characteristic feature of I-V curves is that neither simple power-law nor exponential dependence of voltage on current was found. The size dependence of current-voltage characteristics is shown in Figure 2. At larger widths voltage appears at a current independent of the width and all curves follow the same line up to a turning point where a steep increase of voltage takes place. The position of the turning point depends on the width. At smaller widths the steep increase starts directly at zero voltage state. Quantum interference is clearly seen for all widths (see Figure 4) and for voltages above the turning point but is invisible below it. Critical currents were determined from the current-voltage characteristics using 3 μV/cm electric field criterion and the position of the turning point. These values are displayed in Figure 3. The temperature dependence of critical current shows steep increase as the temperature decreases.

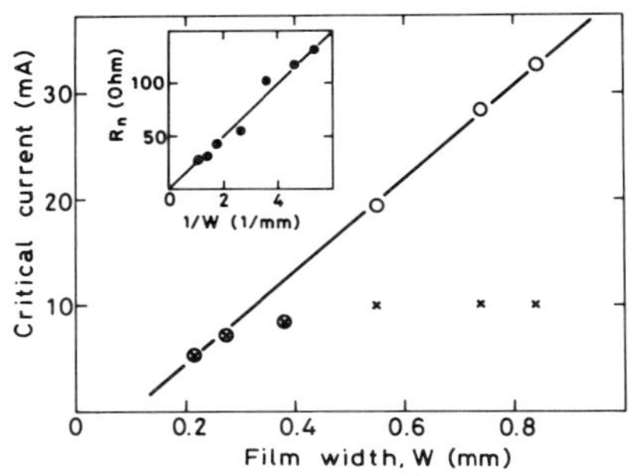

Figure 3
Film width dependence of critical current determined from
x 3 μV/cm criterion
o turning point position
Inset shows the width dependence of the normal state resistance

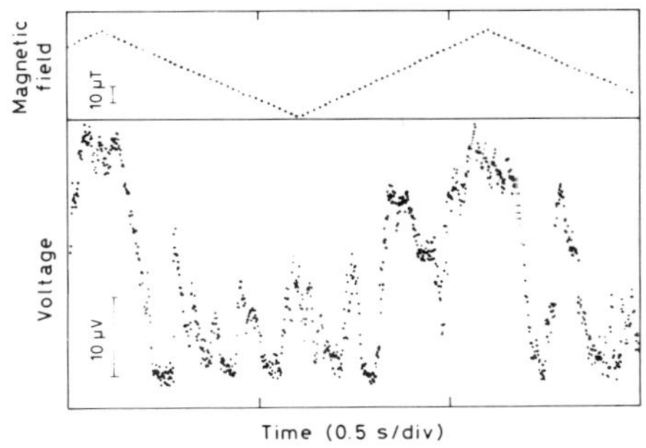

Figure 4
Quantum interference for a film of 186 μm width
-upper curve: magnetic field of the coil
-lower curve: voltage at fixed current

These observations can be interpreted as follows: the voltage is generated at the intergrain Josephson junctions and there is a special voltage generation at low voltages for films of larger widths. In the latter case the voltage may be produced by the flux creep of Josephson vortices and flux lines trapped in superconducting rings. At higher voltages the film operates as a multijunction DC SQUID biased above the critical current (De Bruyn Ouboter and De Waele 1970).

References

Besenyei E, Katona G, Arató P and Kele A 1989 Supercond. Sci. Techn. **2** 220

Mészáros S, Vad K, Halász G, Kedves F J and Balanyi Sz 1990 Physica C **167** 139

De Bruyn Oubuter and De Waele 1970 Progr. in Low Temp. Phys. Vol. 6. ed C J Gorter (Amsterdam-London: North Holland) pp 243-288

Robbes D, Miklich A H, Kingston J J, Lerch Ph, Wellstood F C and Clarke J 1990 Appl. Phys. Lett. **56** 2240

Supercond. Sci. Technol. **4** (1991) S370–372. Printed in the UK

Magnetization of 2212-Melt Material before and after Irradiation with 400 MeV Oxygen Ions

H.-W. Neumüller*, G. Ries*, W. Schmidt*, W. Gerhäuser*†, S. Klaumünzer‡

*Siemens AG, Zentrale Forschung und Entwicklung, Postfach 3220, D-8520 Erlangen, FRG
†Physikalisches Institut, Universität Erlangen-Nürnberg, Erwin-Rommel-Str. 1, D-8520 Erlangen, FRG
‡Hahn-Meitner-Institut, Glienicker Str. 1, D-1000 Berlin 39, FRG

Abstract: Dense melt processed $Bi_2Sr_2CaCu_2O_8$ superconductor was irradiated below T_c with 400 MeV oxygen ions up to a dose of $3 \times 10^{15}\,cm^{-2}$. After a dose of $1.5 \times 10^{15}\,cm^{-2}$ the irreversible magnetization showed a marked increase at low temperatures, while the adverse effect of T_c reduction seemed to prevail for $T > 60\,K$; at $3 \times 10^{15}\,cm^{-2}$ no further increase or even a decrease was observed. On the other hand the logarithmic time decay $1/M\ \mathrm{d}M/\mathrm{d}\ln t$ remained nearly unchanged. Irradiation initially shifts the irreversibility line to higher fields followed by a subsequent drop after a dose of $3 \times 10^{15}\,ions/cm^2$.

1. Introduction

Many studies of radiation effects on the current carrying properties have already been reported on YBaCuO-based HTSC in literature (Roas *et al* 1989). Fewer experimental data exist with respect to the Bi-Sr-Ca-Cu-O system (BiSCCO) where most of literature is concerned with the dose dependence of critical temperature and specific resistivity (Iwase *et al* 1989). Recently substantial enhancement of magnetization was achieved with neutron irradiation in single crystals below 40 K (Kritscha *et al* 1990). An overview of published work on particle irradiation of BiSCCO is given by Neumüller *et al* (1990b). Our irradiation experiments were performed on polycrystalline melt processed 2212-BiSCCO samples of high density. This material has been thoroughly investigated by resistive and magnetization measurements in the nonirradiated condition (Neumüller *et al* 1990a, Neumüller and Ries 1989). Particle irradiation was considered to be the first step to improve the observed weak flux line pinning in this BiSCCO compound by introducing additional lattice defects.

2. Preparation and experimental details

Details of the sample preparation and experimental techniques were described in a previous paper (Neumüller *et al* 1990b). The irradiation was performed with 400 MeV oxygen ions at the low temperature facility of VICKSY at the Hahn-Meitner-Institut in Berlin. The projected range of this projectile is 0.5 mm. This is nearly twice as large as the target thickness and hence no implantation occurs. During the irradiation the temperature of the flow cryostat was held constant at 70 K. After the irradiation the samples were warmed up to room temperature. The magnetization measurements were performed using a commercial SQUID magnetometer.

3. Critical temperature

The phase transition measured both by the resistive and inductive method shifted nearly parallel during irradiation. No broadening of the transition occured, which indicates a homogeneous damage distribution and shows that the depression of T_c is due to intragranular damage. T_c(midpoint) dropped from $T_{c0} = 85.7\,K$ to $T_c = 81\,K$ and $T_c = 76.4\,K$ for $1.5 \times 10^{15}\,ions/cm^2$ and $3 \times 10^{15}\,ions/cm^2$ respectively measured inductively after warm up to room temperature. A more detailed analysis is given by Neumüller *et al* (1990b).

0953-2048/91/01S370 + 03 $03.50

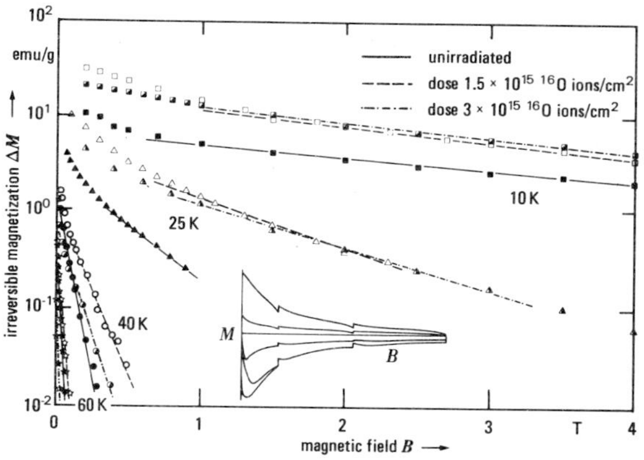

Figure 1: Irreversible magnetization of melt processed $Bi_2Sr_2CaCu_2O_8$ superconductor before and after irradiation with oxygen ions. Insert: magnetization loop at 10 K

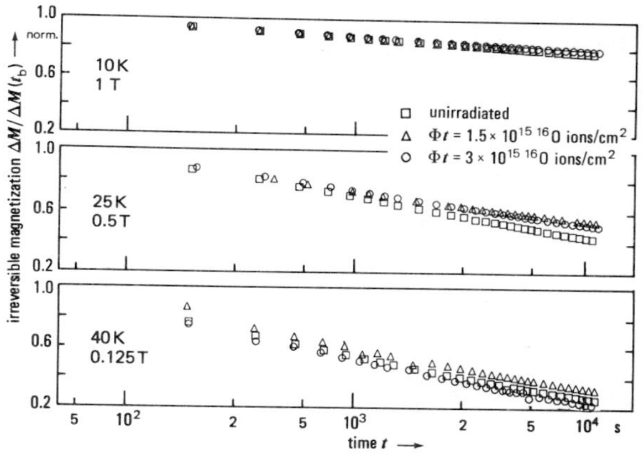

Figure 2: Time and dose dependence of irreversible magnetization normalized to the first data point taken at $t_b = 50\,\mathrm{s}$

4. Magnetization

In figure 1 the irreversible magnetization ΔM both prior to and after irradiation with two doses 1.5×10^{15} and 3×10^{15} ions/cm^2 is plotted versus field. ΔM is given by the difference between the magnetization branches for increasing and decreasing field; all values were taken 110 s after settling of the field. The data clearly show that the magnetization currents are enhanced below 60 K by irradiation with the lower dose, in particular their drop with increasing field is reduced. At 10 K the magnetization of the BiSCCO material shows an irreversible behaviour up to the highest measured field of 5.5 T and the irradiation increases the magnetization at most by a factor of three. A further increase of the number of irradiation defects by doubling the dose gives no further enhancement but a saturation behaviour at 10 K and 25 K at $B \gtrsim 1\,\mathrm{T}$ and even a reduction in the low field regime was observed for all temperatures. At 70 K magnetization for the two doses is found to be even smaller than before, which may be attributed to the decrease of the critical temperature (Kritscha *et al* 1990).

In order to investigate the nature of intrinsic and irradiation induced pinning centers, we measured the time decay of the irreversible magnetization $\Delta M(t)$. The values normalized to the first data point are plotted in figure 2 versus the logarithm of time for different applied fields between 10 K and 40 K. The irradiation induces no significant change in the flux creep rate. Our interpretation is that the irradiation with 400 MeV oxygen ions obviously increased the number of effective pinning centers but the irradiation induced defects have pinning energies U_0 similar to those of the intrinsic defects present in the samples before irradiation. A detailed evaluation of U_0 is given by Neumüller *et al* (1990b).

Figure 3 shows the dose dependence of the irreversibility lines (IL) (Gupta *et al* 1989) defined by the onset of detectable irreversibility in the isothermal DC-magnetization. The samples have pronounced reversibility of the hysteresis curve above 20 K, which is characteristic for the BiSCCO compounds (van den Berg 1989). The difference of the IL in the nonirradiated condition reflects the different intrinsic defect densities present in single crystals and polycrystalline material.

We find a shift to higher field for low dose and saturation or even reversal above 40 K for the high dose of 3×10^{15} O ions/cm^2. This behaviour is also correlated to the dose dependence of irreversible DC magnetization in this work (see Figure 2) and in literature whithin comparable defect concentrations after neutron irradiation in YBCO (Wisniewski *et al* 1988) and BiSCCO (Przyslupski *et al* 1988) compounds respectively. For the low dose we calculated 7.4×10^{-4} dpa using the TRIM computer code (Biersack 1980). The corresponding IL is very close to that of a 2212-BiSCCO single crystal irradiated up 6×10^{17} n/cm^2 (Kritscha *et al* 1990) for which we

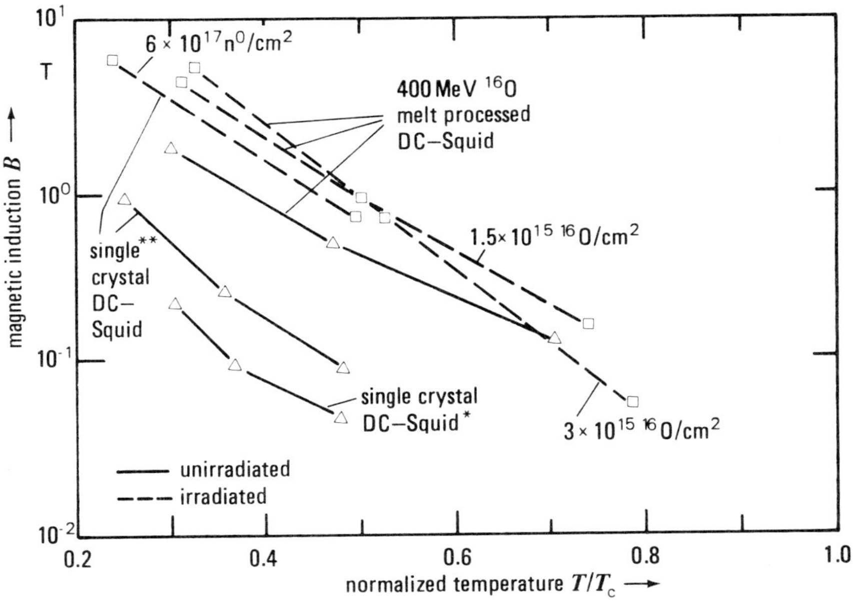

Figure 3: Dose dependence of the irreversibility line of 2212-BiSCCO after neutron and fast ion irradiation (*Gerhäuser *et al* 1990, **Kritscha *et al* 1990)

estimated about half the defect concentration. There is evidence that the IL seems projectile independent at least for neutrons and fast ions.

In summary the irradiation induced defects can obviously increase the intragranular currents in BiSCCO not as effectively as in YBCO material (Wisniewski *et al* 1988). The origin of saturation of irreversible magnetization at relatively low defect concentration is not yet understood.

Acknowledgement

The authors would like to thank Mrs. W. Ruppert and Mr. M. Hollfelder who did most of the measuring work. This work was partly supported by the EC Brite-Euram program.

References

Biersack J P and Haggmark L G 1980 *Nucl. Instrum. Methods* **174** 257

Gerhäuser W, Neumüller H W, Schindler W, Saemann-Ischenko G 1990 to be published

Gupta A, Esquinazi P, Braun H F, Gerhäuser W ,Neumüller H-W, Heine K, Tenbrink J 1989 *Europhys. Lett.* **10** (7) 663

Iwase A, Watanabe M, Iwata T and Nihira T 1989 *Jap. J. Appl. Phys.* **28** L1939

Kritscha W, Sauerzopf F M, Weber H W, Crabtree G W, Chang Y C and Jiang P C 1990 *Europhys. Lett.* to be published

Neumüller H-W and Ries G 1989 *Physica C* **160** 471

Neumüller H-W, Ries G, Bock J and Preisler E 1990a *Cryogenics* **30** 639

Neumüller H-W, Ries G, Schmidt W, Gerhäuser W and Klaumünzer S 1990b Proc. of the E-MRS, May 29 – June 1, Strasbourg, to be published

Przyslupski P, Wisniewski A, Kolesnik S, Dobrowolski W, Pajacskowska A, Pytel K and Pytel B 1988 *Physica C* **153–155** 345

Roas B, Hensel B, Saemann-Ischenko G and Schultz L 1989 *Appl. Phys. Lett.* **54**(11) 1051; *Jap. J. Appl. Phys.* **28** L1521

van den Berg J, van der Beek C J, Kes P, Mydosh J A, Menken M J V and Menowski A A 1989 *Supercond. Sci. Technol.* **1** 249

Wisniewski A, Baran M, Przyslupski P, Szymczak H, Pajaczkowska A, Dobrzynski L, Waliszewski J, Biernacka M, Pytel K and Pytel B 1988 in: *Proc. Int. Discussion Meeting on HTSC* (Feb. 7–11 1988, Mauterndorf Austria) ed. Weber H W (New York: Plenum) p 279

Supercond. Sci. Technol. **4** (1991) S373–375. Printed in the UK

Behaviours of transport critical currents in low magnetic fields for $YBa_2Cu_3O_{7-\delta}$ single crystalline films

T Aomine†, T Nishizaki‡, I Fujii§, T Terashima||, K Iijima¶, K Hirata¶, K Hayashi¶, K Yamamoto¶ and Y Bando||

†Department of Physics, Kyushu University, Fukuoka, Japan
‡Department of Information Systems, Kyushu University, Fukuoka, Japan
§Ube Laboratory, Ube Industries Ltd., Ube, Japan
||Institute for Chemical Research, Kyoto University, Uji, Japan
¶Research Institute for Production Development, Kyoto, Japan

Abstract. Behaviours of the transport critical currents I_c in increase and decrease of low magnetic field H are reported for $YBa_2Cu_3O_{7-\delta}$ single-crystal films. The I_c-H curves for "good" films with the small values of electrical resistivity ρ showed no hysteresis, but for "no good" films with the large values of ρ showed the hysteresis. From a dependence of I_c-H curves on angles between the film surface and H, bridge structures in the "no good" films were suggested.

1. Introduction

High T_c superconducting ceramics prepared by the solid state reaction have low critical current density and their critical currents I_c show hysteresis in increase and decrease of low magnetic field H (Noto et al 1987). The hysteresis in superconducting ceramics is similar to that in superconducting bridges made artificially. The hysteresis has been interpreted as effect of bridges at grain boundaries of superconducting ceramics (Aomine and Tsuru 1987). Here we refer to the constricted part and the bank connecting it together as bridge. It is not difficult to recognize the existence of bridge in the porous ceramics with various sizes of grains. Considering from the origin of the hysteresis, it was expected that the I_c of single crystalline films in low values of H showed no hysteresis. Aomine et al (1990) found that the I_c for "good" $YBa_2Cu_3O_{7-\delta}$ single crystalline films showed no hysteresis and for "no good" films showed the hysteresis. Whether the hysteresis existed or not was related to the electrical resistance R. In this work we present experimental results of the I_c-H and of the R for the other $YBa_2Cu_3O_{7-\delta}$ single crystalline films and suggest the bridge structures in "no good" films.

2. Origin of the Hysteresis in Bridges

The hysteresis in bridges made artificially was explained by Mizuno and Aomine (1983) in the following: In bridges the bank traps the flux for the magnetic field sweep, so a screening current in increase of |H| is larger than that in decrease of |H|. When the sum of a transport current and the screening current equals a pair-breaking current, the transport current corresponds to the I_c in bridges made artificially. The above explanation leads to that the I_c shows the hysteresis; the I_c in increase of |H| is smaller than that in decrease of |H|.

3. Experimental Procedure

Samples have been prepared by the activated reactive evaporation (Terashima et al 1988). Critical current and electrical resistance measurements were carried out by the same methods as in the previous work (Aomine et al 1990). A magnetic field was supplied by Helmholtz coils and applied to the sample surface at angles θ and ϕ as shown in Figure 1.

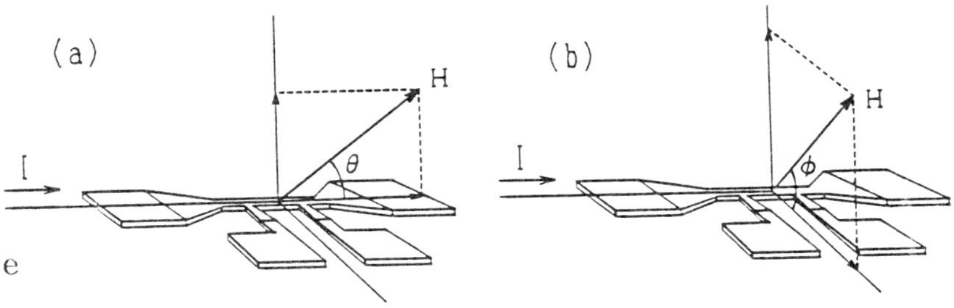

Fig. 1 A schematic diagram of the samples carrying a transport current I in a magnetic field H applied (a) at θ and (b) at ϕ.

4. Experimental Results and Discussions

Figure 2 shows experimental results of transport critical current density J_c as a function of H for the three samples. Here we increased H from zero to a maximum value, then decreased again to zero, next increased |H| to a maximum value in the reverse direction, and lastly decreased |H| to zero. This is one cycle. The I_c-H curves with increase of H from zero at a virgin state were dissimilar to those from the second cycle on which were the same. The I_c-H curves in Figure 2 were not those at the virgin state. Sample Y348A had no hysteresis, but the other samples had the hysteresis. The hysteresis disappeared near the critical temperature. The J_c-H curves for sample Y279A at ϕ = 30° and θ = 90° and for sample Y279B at ϕ = 0 and θ = 90° have not been familiar. The J_c-H curves did not cross during the above cycle of H. Those features of J_c-H curves have been observed for the following bridges made artificially; the constricted part is one-sided for the centre of banks (Aomine and Yamasaki 1990). Next we present the experimental results about the electrical resistivity ρ to search the origin of hysteresis of J_c-H. The temperature T dependence of ρ near the transition from the normal state to the superconducting state is shown in Figure 3. Samples Y348A and Y325A had "good" electrical properties; the value of ρ at 290 K was small, e.g. 242 $\mu\Omega\cdot$cm, the transition curve was sharp and smooth and

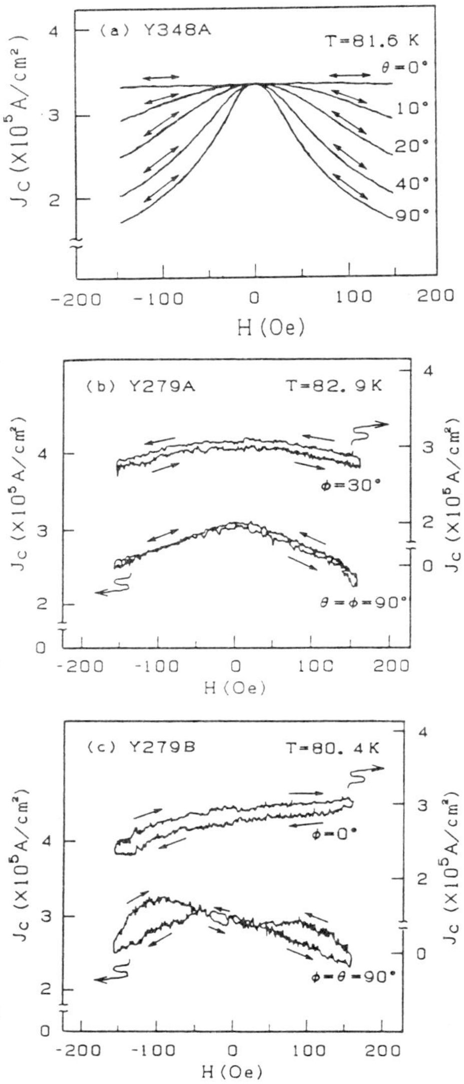

Fig. 2 Transport critical current density J_c vs. applied field H for the three samples. Arrows express the field-sweep direction.

the straight line of ρ-T curve at the normal state passed through the origin. On the other hand, samples Y279A and Y279B had "no good" electrical properties; the value of ρ at 290 K was large, e.g. 505 μΩ·cm, the transition curve was broad and had a shoulder, and the straight line of ρ-T curves at the normal state did not pass through the origin. Although for sample 325A the J_c-H curves are not given in the figure, we note that the hysteresis of J_c-H was not observed. Thus we noticed that samples with the "good" electrical properties and with "no good" electrical properties showed no hysteresis and the hysteresis, respectively. The hysteresis in superconducting ceramics was tried to be explained by Evetts and Glowacki(1988), Matsushita et al (1988), and Chen and Qian (1989) in addition to us. Here they considered that samples included Josephson junctions. We do not think that "no good" films include Josephson junctions. This is supported by the facts that the J_c for "no good" films is only about one third of the that for "good" films and does not show a H dependence of the Josephson junction even if we admit the existence of only one Josephson junction. The bridge structures which are the origin of the hysteresis are suggested from the results in Figure 2 (b),(c), as shown in Figure 4. Those structures are models satisfying the experimental results of I_c-H.

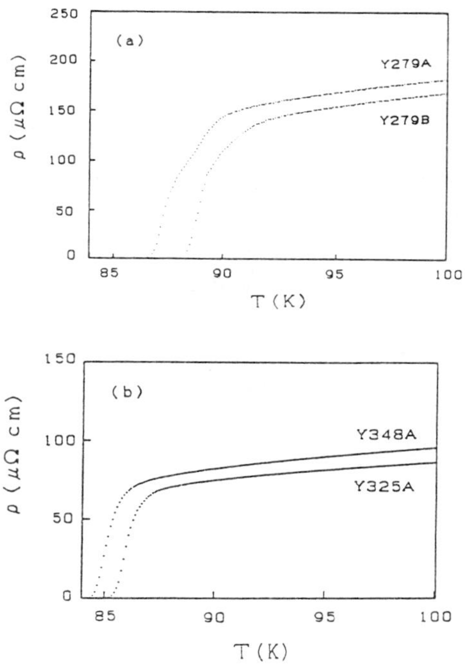

Fig. 3 Electrical resistivity ρ vs. temperature T.

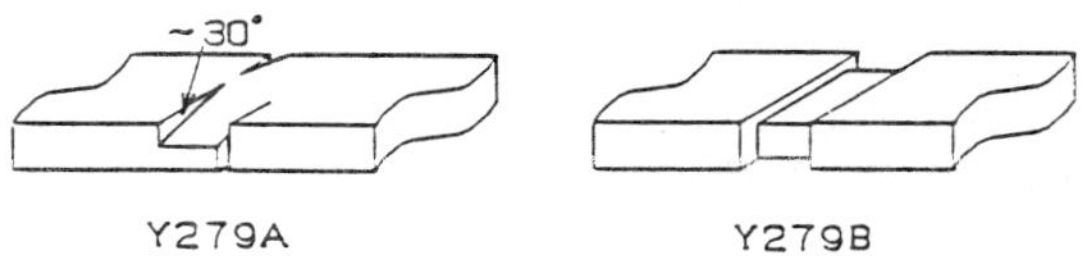

Fig. 4 Suggested bridge structures.

References

Aomine T and Tsuru K 1987 *Jpn. J. Appl. Phys.* **26** L2013

Aomine T, Nishizaki T, Fujii I, Terashima T, Iijima K, Hirata K, Hayashi K, Yamamoto K, and Bando Y 1990 *Jpn. J. Appl. Phys.* submitted for publication

Aomine T and Yamasaki S 1990 private communication

Chen K Y and Qian Y J 1989 *Physica C* **159** 131

Evetts J E and Glowacki 1988 *Cryogenics* **28** 641

Matsushita T, Ni B, Yamafuji K, Watanabe K, Noto K, Morita H, Fujimori and Muto Y 1988 *Advances in Superconductivity* ed K Kitazawa T Ishiguro (Tokyo Springer-Verlag)pp393

Mizuno K and Aomine T 1983 *J. Phys. Soc Jpn.* **52** 4311

Noto K, Morita H, Watanabe K, Murakami T, Koyanagi T, Yoshii I, Sato I, Sugawara H, Kobayashi N, Fujimori H and Muto Y 1987 *Physica* **148B** 239

Terashima Y, Iijima K, Yamamoto K, Bando Y and Mazaki H 1988 *Jpn. J. Appl. Phys.* **27** L91

Supercond. Sci. Technol. 4 (1991) S376–378. Printed in the UK

YBaCuO Superconducting Films with T_c=90K Grown Off-1:2:3 Composition in Low Oxygen Pressure

V. Matijasevic, R.H. Hammond, P. Rosenthal, K. Shinohara, A.F. Marshall, and M.R. Beasley

Department of Applied Physics, Stanford University, Stanford, CA 94305 USA

Abstract. YBaCuO films were grown in-situ by reactive coevaporation at low oxygen pressures. Films made at lower pressures (1-10mTorr) and in non-equilibrium conditions have a c-axis lattice constant (c_o) which is expanded compared to the films made at higher pressures (>100mTorr). We have also found that at lower oxygen pressure off-1:2:3-composition and Ba-deficient films have higher T_c's (88-90K) and lower c_o than do films on-1:2:3 stoichiometry. We suggest that metal-atom point-like defects are quenched into the films at lower oxygen growth pressures.

1. Introduction

In-situ growth of YBaCuO thin films by various physical vapor deposition techniques has now been well established. Correlations have been found between the pressure and temperature conditions for in-situ growth and the thermodynamic stability phase diagram for $YBa_2Cu_3O_y$, as shown by Hammond and Bormann (1989). Fig. 1 shows the stability line for the tetragonal $YBa_2Cu_3O_6$ compound as determined by Bormann and Nölting (1989). In general, successful film growth has been reported above this line and best superconducting properties are for growth processes close to the line. During growth the tetragonal phase is formed which then transforms to the orthorhombic phase in cooldown.

Empirical results also indicate that there is a minimum molecular oxygen pressure (i.e. without use of activated oxygen species) of about 1mTorr necessary for growth of the $YBa_2Cu_3O_y$ structure. We have investigated growth in and near non-equilibrium conditions at "minimal" oxygen pressures (1-10 mTorr of O_2), shown by the shaded region in Fig. 1. Films in this region have been reported to have slightly depressed transition temperatures, T_c ~ 80-85K, and expanded c-axis lattice constants, c_o, but high critical currents and low surface resistances; see for example Eom et al (1990) and Laderman et al (1990). However, films made under higher oxygen pressures, conditions corresponding to the small circle in Fig. 1, have bulk values for c_o. We have examined the compositional, pressure and temperature dependencies of the films made in the low pressure region.

2. Experimental Procedure and Results

Films were made by reactive electron-beam coevaporation of three metal sources as described by Missert et al (1989) and Matijasevic et al (1990). <100>-oriented MgO was used as the substrate. Deposition temperatures were 550-780°C and the rate of deposition used was 4Å/sec. A high differential pressure was maintained around the substrates with oxygen introduced about 1cm away. Oxygen pressure at the substrate was measured with a manometer gauge. T_c's depended on the precise conditions of growth, and in the best cases were as high as 90K. T_c's were measured by dc transport and also by an ac two-coil mutual inductance probe. The latter showed that the films can carry currents above 10^3 A/cm^2 just

0953-2048/91/01S376 + 03 $03.50

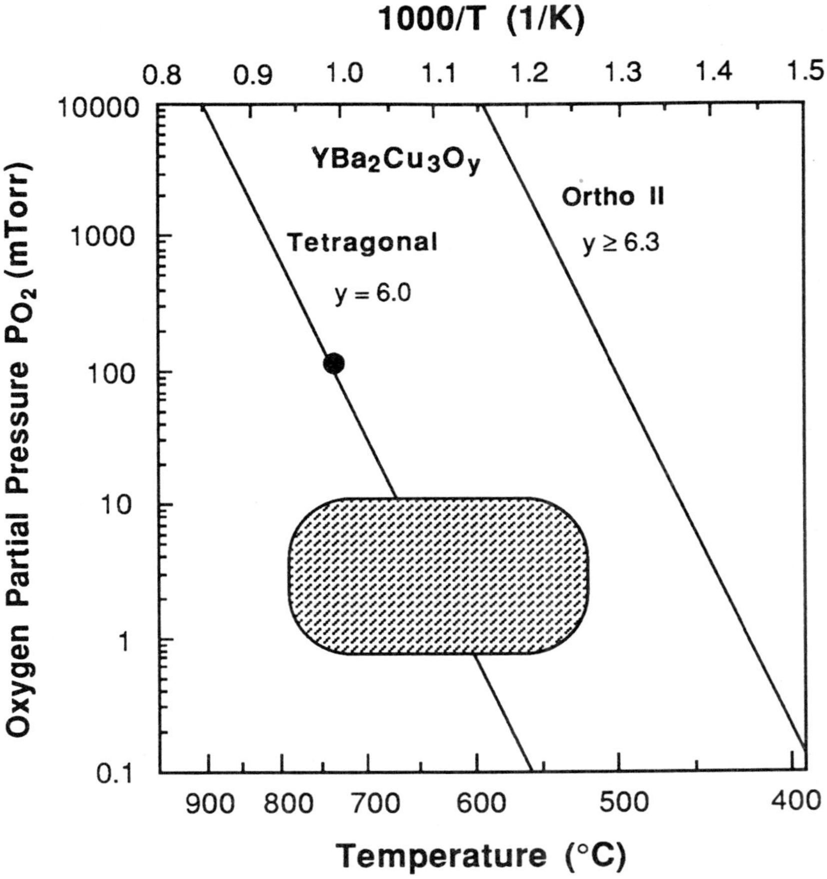

Fig. 1 The p-T diagram with the thermodynamic stability line for $YBa_2Cu_3O_y$ drawn in. The shaded region shows the region where films were studied as a function of composition. The small circle represents where bulk values of c_o were obtained.

below the transition. Also, critical currents at 4.2K are typically above 10^7 A/cm^2 as measured by magnetization. This demonstrates that the observed T_c's correspond to bulk film values. X-ray diffraction with the scattering vector perpendicular to the film was used to determine the phases present and in particular the c-axis lattice constant of the $YBa_2Cu_3O_y$ structure.

Superconducting films were made across the whole region depicted in Fig. 1. No clear boundary is detected as the temperature is increased even 200°C to the left of the thermodynamic stability line, i.e. the $YBa_2Cu_3O_y$ structure was made on both sides of the line. However, the c-axis lattice constant for the as-made films is longer than the bulk value of 11.68Å and ranges from 11.70 to 11.85Å. In fact this c-axis expansion is a general feature of in-situ films. It appears more prominent as the pressure is reduced. Fig. 2 shows c_o as a function of the growth pressure of our films. This figure also includes data points from 100mTorr of ozone growth and from 0.2mTorr growth which was ion beam assisted.

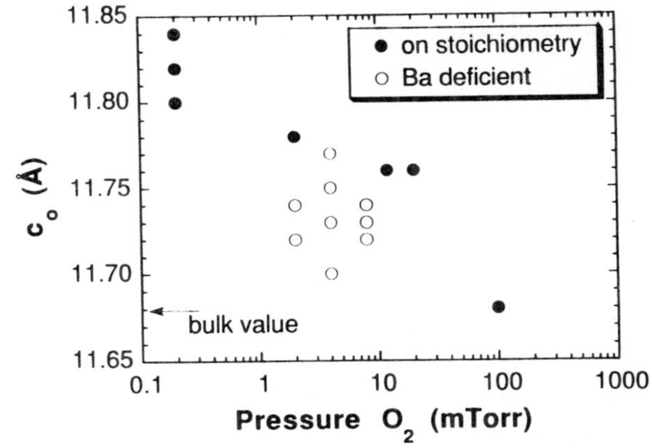

Fig. 2 c-axis lattice constant as a function of the oxygen partial pressure during growth

We have investigated variations of T_c and c_o as a function of the metal deposition ratios during growth. The deposition ratio is determined by the overall average composition of the film from a post-deposition microprobe analysis. For films grown at 1-10mTorr we find that in spite of second phase inclusions, the films have higher superconducting transitions and less expanded c_o's when the films are made Ba-deficient. Fig. 3 shows T_c and c_o as a function of the Ba/Y ratio. At a Ba/Y ~1.4, T_c and c_o reach their extreme values.

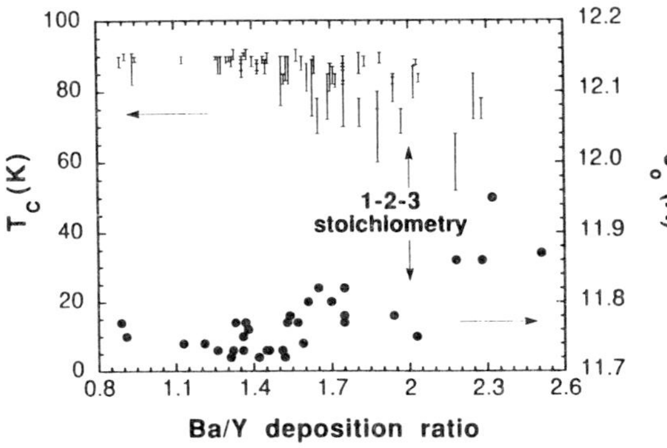

Fig. 3 Superconducting transitions (bars represent onset and zero resistance) and c-axis lattice constants as a function of Ba/Y deposition ratio

3. Discussion

The c-axis expansion of in-situ films could be due to several factors. Insufficient oxygen in the structure is known from experiments on bulk materials to expand c_o. We have checked our samples for this by doing extensive low-temperature anneals. In general c_o cannot be reduced by a low-temperature (<800°C) anneal in oxygen. However, c_o can be expanded even more by an anneal in argon, depleting oxygen from the chain layers of $YBa_2Cu_3O_y$. This indicates that the c_o expansion is not just due to lack of oxygen in cooldown. Another possible explanation for c_o expansion is substrate epitaxy induced strain in the film. This can be ruled out on the basis of the facts that expansion occurs on various substrates with different signs of mismatches as well as that c_o expansion occurs also on grains that have the a-axis perpendicular to the substrate.

We suggest, however, that cation disorder may be increasingly present at reduced oxygen pressures, as proposed by Matijasevic et al (1990). In particular, Ba for Y substitution seems likely at lower pressures. This could be caused by the fact that growth conditions are such as to not fully oxidize the Cu atoms to the Cu^{II} state in the $YBa_2Cu_3O_y$. The lack of oxygen in the plane layers during growth would favor Ba^{II} over Y^{III} between the plane layers, and thus preferentially substitute Ba for Y. This would expand c_o due to the larger size of Ba ions and also due to less oxygen in the structure. T_c may only be slightly depressed. By growing films in Ba-deficient conditions we would then dope the structure with extra Y and compensate for the Ba self-doping effect, bringing the lattice constant down to its bulk value. Such a Ba/Y cation-disordered structure may be a metastable phase, explaining the growth in non-equilibrium conditions.

References

Bormann R and Nölting J 1989 *Appl. Phys. Lett.* **54** 2150

Eom C B, Sun J Z, Lairson B, Streiffer S K, Marshall A, Yamamoto K, Anlage S, Bravman J, Geballe T H, Laderman S, Taber R, and Jacowitz R 1990 submitted to *Physica C*

Hammond R H and Bormann R 1990 *Physica C* **162-164** 703

Laderman S S, Taber R C, Jacowitz R D, Moll J L, Eom C B, Hylton T L, Marshall A F, Geballe T H, and Beasley M R 1990 submitted to *Phys. Rev. B*

Matijasevic V, Rosenthal P, Shinohara K, Marshall A F, Hammond R H, and Beasley M R 1990 submitted to *J. Mater. Res.*

Missert N, Hammond R H, Mooij J E, Matijasevic V, Rosenthal P, Geballe T H, Kapitulnik A, Beasley M R, Laderman S S, Lu C, Garwin E L, Barton R 1989 *IEEE Trans. Magn.* MAG-**25** 2418

Supercond. Sci. Technol. **4** (1991) S379–381. Printed in the UK

Pressure dependence of the resputtering of Y-Ba-Cu-O thin films grown by DC magnetron sputtering

T. I. Selinder*, G. Larsson*, U. Helmersson* and S. Rudner**

* Thin Film Group, Linköping University, S-581 83 Linköping, Sweden
**National Defence Research Establishment, Box 1165, S-581 11 Linköping, Sweden

Abstract. $YBa_2Cu_3O_{6+x}$ films were grown using DC magnetron sputtering from fully presputtered stoichiometric compound targets. A planar geometry was used with the substrates opposing the target. For deposition at high substrate temperatures the film composition was strongly pressure dependent. Films were stoichiometric when grown at pressures above 9.0 Pa, and at 600°C. At lower pressures resputtering by energetic oxygen occurred. The generation of these energetic particles is discussed and a method to avoid deterious bombardment is presented.

1. Introduction

To reproducibly grow high quality films of the high temperature superconducting materials, accurate control of the film growth process is required. We have chosen the magnetic field enhanced diode sputtering technique (magnetron sputtering) in a planar geometry. An extensive understanding of the process is essential to avoid complications. Stoichiometry and reproducibility problems have been solved in a number of ways by others; spherical symmetry of the target (Wehner 1988), very high sputtering gas pressure (Poppe 1988), or off-axis geometry and substrates perpendicular to the target (Eom 1989). The objective of this paper is to present results on the Y-Ba-Cu-O thin film sputtering process and to discuss variations in the deposition rate and the stoichiometry as a function of some controllable parameters. We have studied the effect of energetic bombardment on the stoichiometry of the as-deposited material. It was concluded that energetic oxygen is responsible for the adverse effects observed on the film composition. The deterious bombardment can be avoided by thermalisation using high enough sputtering pressure, or by other means decreasing the energy of the particles incident on the substrate.

2. Experimental

A high vacuum DC-magnetron sputtering system was used, with a base pressure of 3×10^{-5} Pa. During deposition the pump was baffled to a pumping speed of 50 Pa l/s. Pure argon sputtering gas at pressures ranging from 0.4 to 11 Pa was used. The magnetron source was operated by a constant current, 750 mA. The targets were $YBa_2Cu_3O_x$ ceramic discs, 5 cm in diameter. As prepared, the onset critical temperature of the targets was around 92 K.

The depositions were made on substrates placed parallell to the target surface and 6.0 cm directly below the target center. Single crystal (100) MgO substrates were used in most depositions. When films were deposited with a negative bias voltage applied to the substrate Ta metal discs (ø=0.3 cm) were used as substrates to ensure good electrical contact to the film. In some experiments a biased tungsten wire mesh was placed in front of the magnetron, in order

0953-2048/91/01S379 + 03 $03.50

to alter the energy of energetic charged particles. The substrate temperature was monitored by a Chromel–Alumel thermocouple mounted in a hole drilled in a MgO dummy substrate.

The compositions of the as-deposited films were determined by Energy Dispersive X-ray analysis (EDX). The uncertainty in units of atomic percent in the compositions is estimated to less than ±2 for the metals. Films that were found to be stoichiometric by EDX were post-annealed in oxygen, and further characterised by XRD and resistivity measurements to verify the superconducting phase. Typically these films were strongly oriented with the c-axis parallel to the [100] substrate normal. The resistive transition occurred in the 80-90 K range.

3. Results and Discussion

The film stoichiometry was affected by changes in the total pressure, particularly at higher substrate temperatures. This can be seen in Figure 1, where the metal composition is plotted as a function of the sputtering gas pressure for a substrate temperature of 600 °C. The films had correct composition at argon pressures above approximately 9 Pa. At this pressure stoichiometric films could be produced at substrate temperatures up to at least 760 °C.

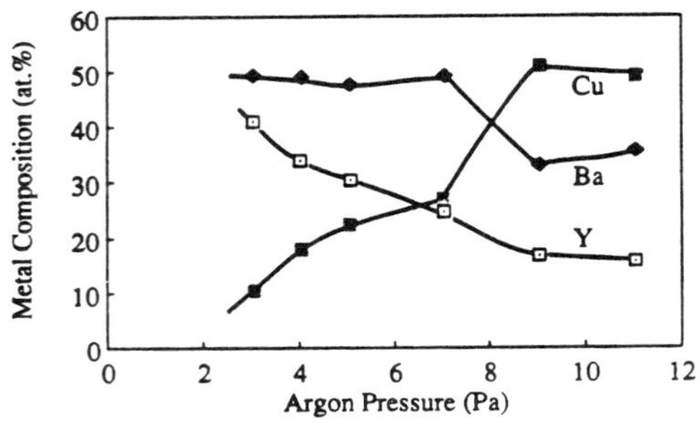

Figure 1. The film composition as a function of sputtering gas pressure. Pure argon was used as sputtering gas, and the substrate temperature was 600 °C.

To verify that the deviation in stoichiometry was due to energetic particle bombardment, bias sputtering was done. Negative voltage was applied to the substrate in order to draw argon ions from the plasma. The film composition as a function of bias potential (argon ion energy) is shown in Figure 2, for an argon pressure of 3.0 Pa. The copper deficiency was more pronounced for bias values of approximatly 100 V and above. In this experiment, for all bias values there was an overlapping effect of resputtering by oxygen, causing films to be nonstoichiometric even at floating potential.

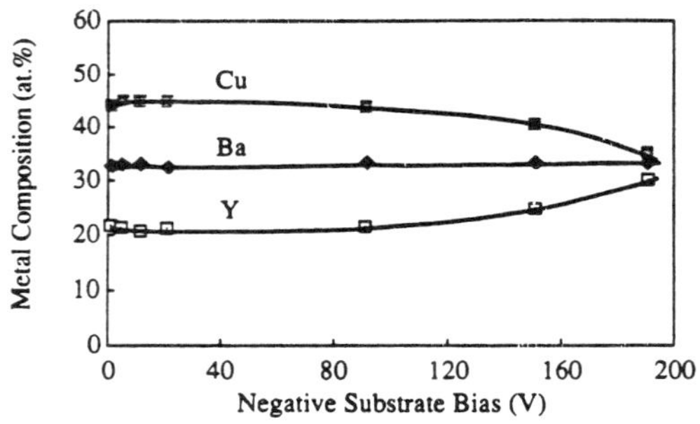

Figure 2. The film composition as a function of substrate bias. The substrate temperature was 200 °C and the sputtering gas pressure was 3.0 Pa (pure argon). The bias current amounted to approximately 0.4 mA/cm^2.

Energetic particles bombarding the substrates, when growing oxides, are formed because oxygen on the target surface will very easily become ionised. The negative ions are then accelerated across the dark space sheath. During magnetron sputtering at pressures in the Pa range this sheath is only of the order of 0.1 cm thick. This is less than the mean free path for both charge and momentum transfer interactions between the oxygen ions and atoms/ions in the plasma (Maniv 1989). In our case this resulted in oxygen ions with an energy of 100 to 180 eV, according to the actual potentials. Once out in the plasma charge transfer interactions between the oxygen ions and the argon in the discharge are likely to occur for pressures above approximately 1 Pa (Tominaga 1982), resulting in a beam of neutral and energetic oxygen atoms, as has been shown in time of flight experiments (Tominaga 1984).

In pure argon sputtering the partial pressure of oxygen can be substantial due to the steady state flow of oxygen from the target and the limited pumping speed of the vacuum system. It is not trivial to reduce the number of energetic negative ions or their energy by increasing the argon gas pressure, since the cathode sheath thickness decreases with increasing pressure (Rossnagel 1989). A collisionless acceleration is still probable, and the beam of energetic oxygen ions entering the plasma persists even at very high pressures. The cross section for momentum transfer (collisions) is very low for energetic as compared to thermal particles (Robinson 1978), allowing a fraction of the energetic particles to reach the substrate with negligable energy loss. The number of neutrals that arrive at the substrate after a collisionless transport decreases approximately as the inverse pressure. The film growth rate is roughly proportional to the inverse pressure as well, according to our measurements. Thus the ratio between the number of energetic oxygen and the number of depositing atoms arriving at the substrate does not decrease with pressure as fast as is expected intuitively. At very high pressures charge transfer interactions during acceleration in the dark space will significantly reduce the oxygen bombardment: the ions will not accelerate across the whole dark space and do consequently not aquire full energy, before they are neutralised. When this happens both the energy and number of energetic oxygen atoms incident on the substrate decreases rapidly on further increase of pressure. In our experiment a pressure of 9.0 Pa is sufficient to thermalise the majority of the energetic oxygen particles.

A possible way to get around the use of high sputtering pressures is to use a retarding potential, in order to decrease the ion energy. We have initial results that show an increase in the Cu/Y ratio from 1.7 to 2.6 by using a -70 V bias on a grid between the magnetron and the substrate, during sputtering at 0.4 Pa. A low pressure during deposition would normally result in off-stoichiometric films, as described above, since the bombarding oxygen would arrive at the substrate with full energy. In our experiment the pressure was low enough to keep charge transfer interactions on a low level and the oxygen ions could be decelerated by the negative potential on the grid. The use of lower sputtering gas pressure also gave a significantly higher deposition rate.

4. Conclusions

A strong pressure dependence of composition was observed for Y-Ba-Cu-O films grown by DC magnetron sputtering. At elevated substrate temperatures films were stoichiometric if grown at pressures above 9.0 Pa. For lower pressures there was a substantial drop in the copper content of films. This was due to resputtering by energetic oxygen neutral bombardment during growth. This conclusion was substantiated by results from films intentionally subjected to argon ion bombardment during deposition. A second possible way to remedy the off-stoichiometry at lower pressures was to decrease the initial energy of the bombarding species. The oxygen ions could be slowed down by a retarding potential on a wire grid in front of the magnetron, provided that the pressure was low enough and consequently the oxygen ions were not neutralised. The low pressure also enhanced the deposition rate.

5. References

C.B. Eom, J.K. Sun, K. Yamamoto, A.F. Marshall, K.E. Luther and T.H. Geballe 1989 *Appl. Phys. Lett.* **55**, 595

S. Maniv and W.D. Westwood 1980 *J. Vac. Sci. Technol.* **17**, 743

U. Poppe, J. Schubert, R.R. Arons, W. Evers, C.H. Freiburg, W. Reichert, K. Schmidt, W. Sybertz and K. Urban 1988 *Solid State Comm.* **66**, 661

R.S. Robinson 1978 *J. Vac. Sci. Technol.* **16**, 185

S.M. Rossnagel 1989 *Thin Solid Films* **171**, 125

K. Tominaga, S. Iwamura, Y. Shintani and O. Tada 1982 *Japn. J. Appl. Phys.* **21**, 688

K. Tominaga, M. Kume, T. Yuasa and O. Tada 1984 *Japn. J. Appl. Phys.* **24**, 35

G.K. Wehner, Y.H. Kim, and A.M. Goldman 1988 *Appl. Phys. Lett.* **52**, 1187

Supercond. Sci. Technol. **4** (1991) S382–384. Printed in the UK

Role of atomic oxygen produced by ECR plasma in the oxidation of $YBa_2Cu_3O_{7-X}$ thin films studied by in-situ resistivity measurement.

K. Yamamoto*[a], B.M. Lairson[b], C.B. Eom[b], R.H. Hammond[a], J.C. Bravman[b] and T.H. Geballe[a], Stanford University, Stanford, CA 94305 USA

Abstract In-situ resistivity measurements have been performed to monitor the effect of atomic oxygen on films subjected to an ECR oxygen plasma. The resistivity is shown to be determined by the flux of atomic oxygen. The steady state concentration in the film is found to depend on the kinetics of oxygen dissociation at the film surface. We determined the activation energy for the decomposition of O_2 at the surface of films with mixed a+c axis orientation and c-axis orientation to be ~1.3eV and ~ 2.1eV, respectively.

1. Introduction

For the synthesis of in-situ grown $YBa_2Cu_3O_{7-X}$ superconductors, several deposition techniques including off-axis magnetron sputtering[1], laser ablation and reactive evaporation have been used. Among them, atomic oxygen has played an essential role in achieving in-situ $YBa_2Cu_3O_{7-X}$ growth, especially for film growth at low oxygen pressures[2]. In this paper we investigate the role of atomic oxygen in the oxidation process by making resistivity measurements of $YBa_2Cu_3O_{7-X}$ thin films in the presence of molecular oxygen at a variety of pressures and in a flux of atomic oxygen.

2. Experiments

In-situ $YBa_2Cu_3O_{7-X}$ films were deposited on several single crystal substrates (MgO, $LaAlO_3$) by 90° off-axis magnetron sputtering[1]. The first is a mixed a(50%)+c oriented film (Tc~84K) on (100) $LaAlO_3$; the other is a c-axis oriented film (Tc~85K, Jc~3×10^7 A/cm^2 at 4.2K) on MgO(100). The resistivity of these films was measured using 4-point DC transport at constant temperature. The contacts were tested for ohmic behavior before the in-situ measurements were made. The current-voltage measurement was controlled by a 9816 HP computer.

To produce the atomic oxygen, a small compact permanent magnet ECR source was used[3]. The ion current and its energy were measured by Langmuir probe and energy analyzer to be ~0.1mA/cm^2 and ~23eV, respectively. The atomic oxygen flux was measured by the oxidation rate of Ag and the etching rate of standard photoresist on a Quartz crystal monitor and found to be ~2×10^{15} atom/cm^2·sec.

3. Results and Discussion

In Fig.1 the resistivity of an a+c mixed $YBa_2Cu_3O_{7-X}$ film on $LaAlO_3$ in the presence of the ECR plasma is shown as a function of time, at a constant temperature of 470°C. The original resistivity is the equilibrium value at the applied oxygen pressure of 0.2mTorr. After the ECR plasma was established, the resistivity drops rapidly and approaches a steady-state value. When the ECR plasma was turned off, the resistivity gradually increases and reaches the original value.

As has been previously shown[4], we find that the equilibrium resistivity in the presence of molecular oxygen is strongly dependent on the oxygen pressure, at a given temperature. On the right side of Fig.1, the measured equilibrium resistance at several values of ambient molecular oxygen (with no plasma) are labeled for a temperature of 470°C. It can be seen that the steady state resistivity of the film at P_{o2} = 0.2mTorr in a flux of $2x10^{15}$ atom/cm^2·sec corresponds to the resistivity of the film in 200mTorr of molecular oxygen, with no plasma. We define this molecular oxygen pressure at the equivalent oxygen pressure.

The temperature dependence of this equivalent oxygen pressure is shown in Fig.2. There are several important features in the data presented in Fig. 2. First it is noted that the activity of atomic oxygen increases with decreasing temperature (the equivalent oxygen pressure increases with decreasing temperature). We attribute this to the existence an activation barrier at the film surface for the dissociation of O_2 into atomic oxygen, which is then free to diffuse into the bulk of the film. As the temperature increases, this barrier is more easily overcome, so that a smaller pressure of O_2 is required to yield an equivalent amount of atomic oxygen at the film surface.

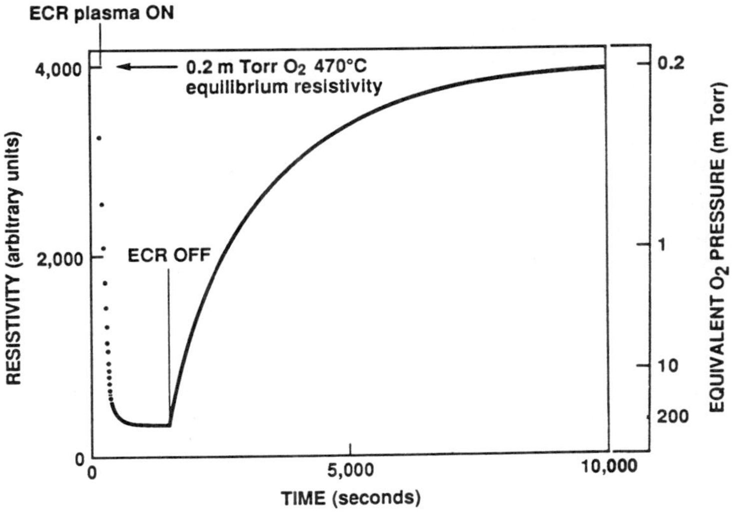

Fig.1 Time dependence of resistance for an a+c mixed $YBa_2Cu_3O_{7-X}$ film in ECR plasma at 470°C. On the right-hand axis, the equilibrium resistance at several oxygen pressures is shown.

A second feature of the data in Fig.2 is the orientation dependence of the equivalent pressure. We believe this to be caused by differences in the decomposition process on the two different film surfaces, with the mixed a+c oriented film having many more dissociation sites per unit area and a lower activation energy to produce atomic oxygen on the surface. The slope shows the activation energy of the following reaction for species adsorbed on the surface.

$$O_2 \longleftrightarrow O \qquad (1)$$

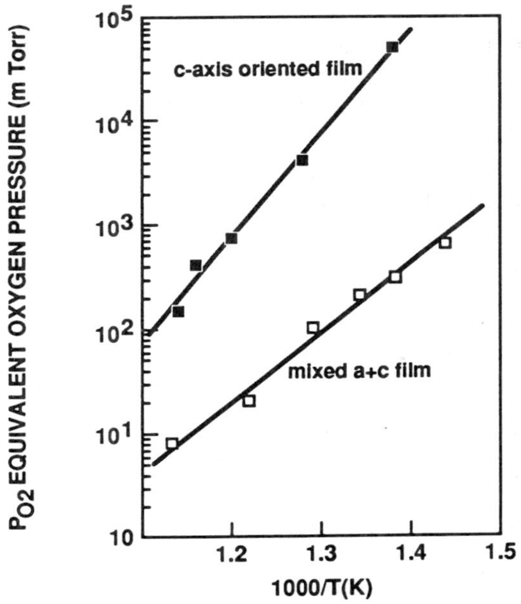

Fig.2 The temperature dependence of the equivalent oxygen pressure. The filled squares and the open squares represent the equivalent pressure for c-axis and in a+c mixed oriented films, respectively.

From the slope in Fig.2 we determined an activation energy for this reaction to be ~1.3eV and ~2.1eV for a+c axis and c-axis orientation, respectively.

This data can be applied to making the following observation concerning in-situ deposition. If deposition is performed at quite high oxygen pressure (100mTorr) and at high temperature (700°C), the $2x10^{15}$ atom/cm^2·sec atomic oxygen flux has very little effect on the oxidation process, since this flux corresponds to ~15mTorr and ~1.5mTorr of molecular oxygen for c-axis and a+c mixed films, respectively.

The time dependence of conductivity (inverse of resistivity) of c-axis oriented $YBa_2Cu_3O_{7-X}$ on MgO is shown in Fig.3 for a substrate temperature of 580°C as a function of atomic oxygen flux. This data shows that with increasing flux, conductivity increases. This is consistent with our expectation that the equivalent oxygen pressure increases with atomic oxygen flux.

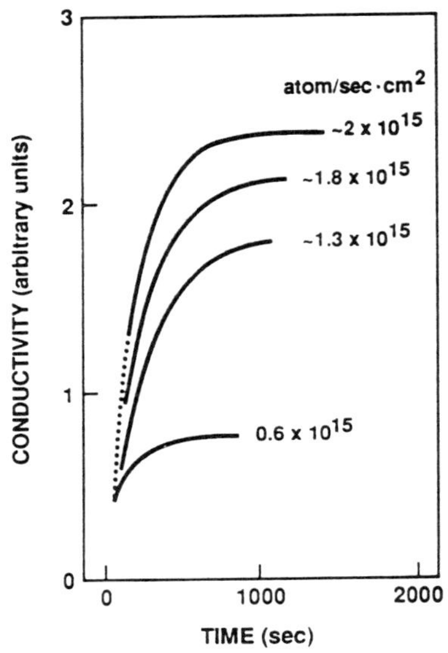

Fig.3 The time dependence of conductivity for a c-axis oriented $YBa_2Cu_3O_{7-X}$ film on MgO at 580°C at several values of atomic oxygen flux.

4.Conclusion

The high activity of atomic oxygen has been discussed. We have demonstrated that atomic oxygen increases the oxygen content in $YBa_2Cu_3O_{7-X}$ thin films. The magnitude of the increase depends on the orientation of the film surface, which we attribute to the differences in the kinetics of the oxygen dissociation process. This implies that the phase diagram of oxygen content of $YBa_2Cu_3O_{7-X}$ might be changed as a function of atomic oxygen flux. This change would increase with increasing atomic oxygen flux and with decreasing temperature. More details of this study and further research such as kinetics of diffusion will be published elsewhere[5,6].

5.Acknowlegement

The authors would like to thank D. L. Keith and W. Holmes for their help in developing the in-situ resistivity measurement system and acknowledge encouragement and support from other members of the Stanford superconductivity group(KGB). Discussion with Dr. J.Z. Sun was very useful. This work has been supported in part by the Air Force Office for Scientific Research, by the Stanford Center for Materials Research under the NSF-MRL program, by EPRI contact RP8009-11, and by the Kanegafuchi Chemical Industry Co., LTD.

References

* On leave from Central Research Lab., Kanegafuchi Chemical Industry Co., LTD., 1-Chome, Yoshida -Cho. Hyogo-Ku, Kobe 652, JAPAN

[a] Department of Applied Physics

[b] Department of Materials Science and Engineering

(1) C.B.Eom, J.Z. Sun, K.Yamamoto, A.F.Marshall, K.E. Luther, S.S. Laderman and T.H. Geballe, Appl. Phys. Lett. 55,(1989)595

(2) N.Missert, R. H. Hammond, J.E. Mooij, V. Matijasevic, P. Rosental, T.H. Geballe, A. Kapitulnik, M.R. Beasley, S. Laderman, C. Lu, E. Garwin and R. Barton, Ieee Trans. Magn. MAG 25(1989)2418

(3) K.Yamamoto and R. H. Hammond, submitted to J. Vac. Sci. Tech.

(4) K.N. Tu, N.C. Yeh, S.I. Park and C.C. Tsuei, Phys. Rev. B 39(1989)304

(5) K. Yamamoto et al, submitted to Apl. Phys. Lett.

(6) K. Yamamoto et al, to be submitted to J. Appl. Phys.

Supercond. Sci. Technol. **4** (1991) S385–387. Printed in the UK

Formation and Superconducting Properties Of As Grown Bi-Ca-Sr-Cu-O Films By AC Sputtering

H.C.Yang+, H.C.Chen+, H.E.Horng* and J.C.Yu*

+ Department of Physics, National Taiwan University, Taipei, Taiwan, R.O.C.
* Department of Physics, National Taiwan Normal University, Taipei, Taiwan, R.O.C.

Abstract. We have in situ grown Bi-Ca-Sr-Cu-O films by ac sputtering. The substrates were heated to 630 - 670°C during ac sputtering. The sputtering was first carried out in a mixture of Ar(70%) and O_2(30%) at a pressure of 0.2 mbar. Then the mixture of Ar(70%) and O_2(30%) was replaced by pure O_2 and the sputtering was kept for another one hour. Superconducting $Bi_2CaSr_2Cu_2O_y$(2122) films with zero resistance at 70-75 K can be reproduced. The critical fields of the as grown films parallel to c-axis and ab-plane has been measured. The estimated zero temperature coherence lengths are ξ_c = 3.3 Å and ξ_{ab} = 36.1 Å respectively for the (2122) phase.

Thin films of high-T_c superconductors in the system Bi-Ca-Sr-Cu-O have been prepared by different methods, for example, magnetron sputtering (Fu et al 1989), laser ablation (Fork et al 1988), and electron beam evaporation (Kuroda et al 1988) In this work we reports the in situ growth and superconducting properties of Bi-Ca-Sr-Cu-O films by ac sputtering.

Bi-Ca-Sr-Cu-O films were prepared by ac sputtering. Two targets with the same composition of $Bi_{1.5}Pb_{0.5}Ca_{1.5}Sr_{1.75}Cu_2O_y$ were mounted in the ac sputtering system face to face with a distance of about 1.5 cm. The substrate was MgO(100) which was heated to about 650° during sputtering. The sputtering was first carried out in a mixture of Ar(70%) and O_2(30%) at a pressure of 0.2 mbar for about 8 hours. After that process the sputtering gas was replaced by pure O_2 and the sputtering was kept for another one hour. Finally the sample was cooled down to room temperature in about

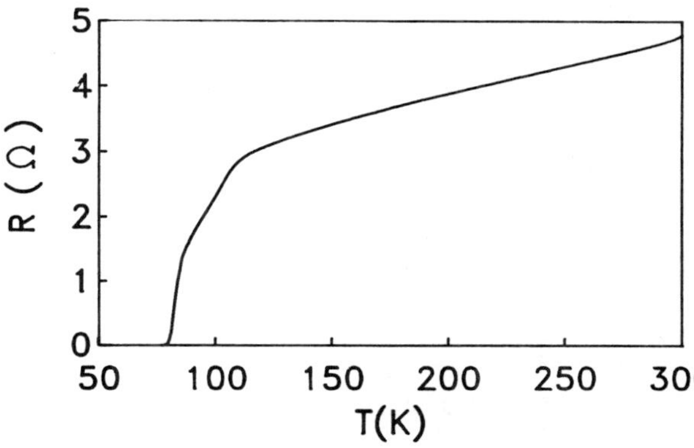

Figure 1. Temperature dependence of resistance of the as grown film.

one hour. The extra sputtering process in pure O_2 was found to be successful in preparing Bi-Ca-Sr-Cu-O films. Films grown this way are shiny and smooth. Films with zero resistance at 70 - 78 K can be reproduced. Typical thickness of the films was about 0.5-1.5 μm.

Figure 1 shows typically the temperature dependence of resistance of the as grown film. The film was in situ grown at 655 °C. The film shows an onset at 110 K and zero resistance at 78 K. The X-ray diffraction shown in Figure 2 indicates that the film is highly oriented with c-axis normal to (100) plane of the MgO substrate. The diffraction peaks can be identified with low T_c $Bi_2CaSr_2Cu_2O_y$(2122) phase with c = 3.05 nm indicating that the film is mainly the (2-1-2-2) phase.

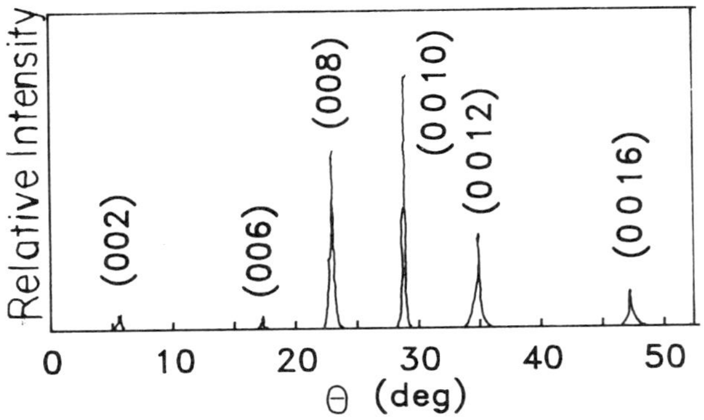

Figure 2. Powder X-ray diffraction pattern of the as grown film.

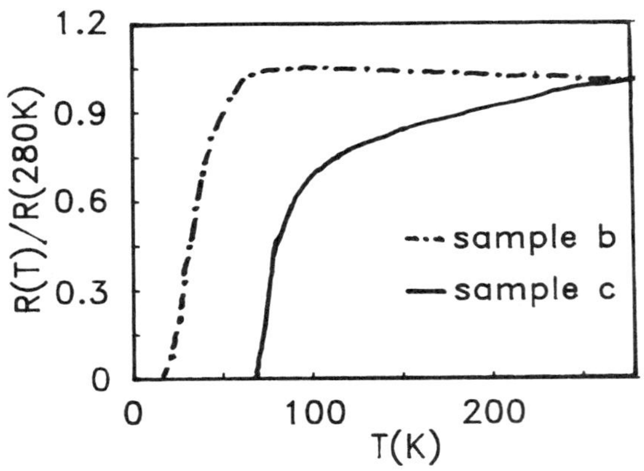

Figure 3. Temperature dependence of resistance for samples b and c. Sample c was subjected to an extra sputtering process in pure O_2 while sample b did not.

In preparing high T_c Bi-Ca-Sr-Cu-O films we have employed an extra pure O_2 sputtering process and found that this extra process is very useful in preparing the superconducting film. This has been demonstrated in Figure 3 for samples b and c. Samples b and c were in situ grown at 630 °C. However, sample b was not subjected to an extra sputtering process in pure O_2 whereas sample c did. One can see that sample c shows a better superconducting property compared with that of sample b. The data indicate that this extra sputtering process in pure O_2 plasma does help the incorporation of O_2 into the sample, therefore it results in improving the superconductivity of sample c.

We have measured the resistance as a function of temperature in magnetic fields parallel to c-axis and ab-plane respectively. The current to the sample was 1 mA. The resistance was measured by a four probe method. The critical fields estimated from 50% resistive transition show a linear temperature dependence as shown in Figure 4. From the linear temperature dependence of the critical fields we obtain $dH_{c2,ab}/dT = -6.7$ T/K and $dH_{c2,c}/dT = -0.61$ T/K in fields parallel to ab-plane and c-axis respectively. The anisotropic factor, $(dH_{c2,ab}/dT)/(dH_{c2,c}/dT)$, is 10.9 for the (2122) phase. We note that $dH_{c2,c}/dT$ is close to single crystal data (Iye Y et al 1988) of $YBa_2Cu_3O_{7-y}$. However, the value of $dH_{c2,ab}/dT$ obtained in this work is 1.5 - 2.0 times larger than that of $YBa_2Cu_3O_{7-y}$ single crystal data which are in the range of -3.0 T/K to -4.0 T/K. The zero temperature coherence lengths are ξ_{ab} = 36.0 Å and ξ_c = 3.3 Å. In calculating zero temperature coherence lengths , we have used the relations $\xi_c(0)/\xi_{ab}(0) = H_{c2,c}/H_{c2,ab}$ and $\xi_{ab}^2 = \phi_o/2\pi H_{c2,c}$, where ϕ_o is the flux quantum. The zero temperature critical fields were obtained from (Werthamer et al 1966) $H_{c2}(0) = 0.69T_c(dH_{c2}/dT)_{Tc}$.

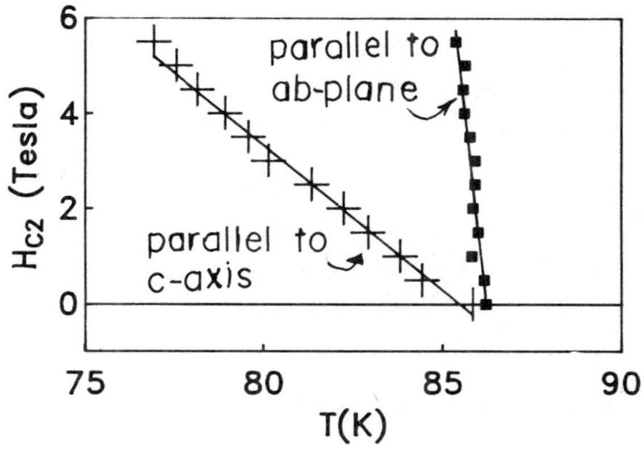

Figure 4. Critical fields as a function of temperature obtained from 50% resistive transition for fields parallel to ab-plane and c-axis respectively.

In conclusion we have in-situ grown Bi-Ca-Sr-Cu-O films by ac sputtering with an extra sputtering process in pure O_2. This extra sputtering process in O_2 helps the corporation of oxygen into the film. Superconducting films with zero resistance at 70-78 K can be reproduced for the (2122) phase. The estimated zero temperature coherence lengths are ξ_c = 3.3 Å and ξ_{ab} = 36.0 Å for the (2122) phase.

This work was supported by National Science Council of R.O.C. under Contract No. NSC79-0208-M002-38.

References

Fu S.M. et al. 1989 Physica **C162-164**, 619

Fork D K, Boyce J B, Ponce F A, Johnson R I, Anderson G B, Connell G A N, Eom B and Geballe T H 1988 Appl. Phys. Lett. **53** 337

Kuroda K, Mukaida M, Yamamoto M and Miyazawa S 1988 Jpn. J. Appl. Phys. **27** L625

Iye Y et al. 1988 In superconducting materials ed Nakajima S and Fukujama H (JJAP Series 1) PP46-8

Werthamer N. R. et al. 1966 Phys. Rev. **147** 295

Supercond. Sci. Technol. **4** (1991) S388–390. Printed in the UK

Anomalies in Elastic and Thermal Properties Associated with the Superconducting Transition of Single Crystal $Bi_2Sr_2CaCu_2O_y$

O.–M.Nes, M.Castro[1], M.Slaski, T.Laegreid, K.Fossheim, N.Motohira[2] and K.Kitazawa[2]

SINTEF Applied Physics, and Division of Physics, The Norwegian Institute of Technology, N–7034 Trondheim, Norway
[1] Aragon Institute of Material Science, University of Zaragoza, Spain
[2] Department of Industrial Chemistry, University of Tokyo, Japan

We have measured the in–plane Young's modulus sound velocity $v_e(T)$ at kHz frequencies and the specific heat $C_p(T)$ of $Bi_2Sr_2Ca_1Cu_2O_y$ single crystals with T_c=93.2K. In the region of T_c a softening is observed in v_e, corresponding to a discontinuity in $d\ln v_e/dT$ of 6.2ppm/K and an indication of a change in $\Delta v_e/v_e$ of the order of 10ppm. From $\Delta C_p/T_c \simeq 16mJ/(molK^2)$ we find $|\partial T_c/\partial\sigma_i| \approx 0.1K/kbar$ and $-\partial^2 T_c/\partial\sigma_i^2 \approx 10^{-3} - 10^{-2}$ K/kbar^2 (i=1,2).

1 Introduction

Studies of elastic and thermal properties are valuable ways of searching for conventional and unconventional bulk behaviour in the new high–T_c materials. It is of particular interest to perform experiments on single crystals, since these may reflect more of the intrinsic properties of the superconductors as compared to ceramics.

Here we present measurements of Youngs modulus E and internal friction Q^{-1} of $Bi_2Sr_2CaCu_2O_y$ (Bisco) single crystals using a vibrating reed (VR) technique at kHz frequencies and combine the result with specific heat (C_p) measurements performed on the same material to obtain the stress derivatives of T_c.

2. Sample and Experimental

Single crystals with nominal composition $Bi_2Sr_2CaCu_2O_y$ (2212) were grown by a floating zone technique (Motohira et al, to be published), yielding samples that were easily cleaved into crystalline platelets along the basal plane. X–ray powder diffraction patterns confirmed a 2212 phase with less than 2% of other phases present. AC–susceptibility measurements showed onset of the superconducting transition at 93.2K.

A small, thin bar–shaped crystal was extracted from the as–grown material for VR measurements. The sample was mounted as a cantilevered reed by clamping one end between two copper blocks. The free part was l=3.33mm long, w=0.62mm wide and d=0.04mm thick, even though its shape did not completely match that of a perfect bar. Flexural vibrations were driven and detected with electrostatically coupled electrodes at the free end (Raychaudhuri and Hunklinger 1984). The fundamental mode of the resonance frequency is given as $f_0=1.615\cdot 10^{-1}\cdot d\cdot l^{-2}\cdot v_e$, where v_e is the Youngs modulus sound velocity along the reed, $v_e=(E/\rho)^{1/2}$, ρ being the density. A recently constructed, automated set–up monitored f_0 and Q^{-1} (which is given by the half–width of the resonance curve) as a function of temperature with mK stability at each measured point. The spread of datapoints in $\Delta v_e/v_e$ at T≈95K was typically less than 0.2ppm. According to Xiang et al (1989) and Wang et al (1990) one might expect considerable elastic anisotropy in the ab–plane, but since we did not determine the orientation of the a– and b–axis the measurement may yield $E_1=S_{11}^{-1}$,

0953-2048/91/01S388 + 03 $03.50

$E_2=S_{22}^{-1}$ or some combination of the elastic compliances S_{ij}. Specific heat measurements were performed using an AC–calorimeter (Stokka and Fossheim 1982).

3 Results and Discussion

The temperature dependence of f_0 and Q^{-1} between 80 and 300 K for the 2212–crystal during cooling and heating is shown in Figure 1 with $|dT/dt|\simeq 0.1$ K/min. f_0 increases monotonically with decreasing temperature, implying a hardening $\Delta E/E \simeq 4.7\%$ in this temperature range, consistent with Xiang et al (1989). Assuming the sample has the shape of a perfect beam, $v_e=3\cdot10^3$m/s and $E=7\cdot10^{10}$Pa at 300K with $\rho=6.70$g/cm^3. This is not unreasonable compared to E=38.8 GPa reported by Sihan et al (1989) for 2212–ceramics. We observe no appreciable hysteresis in f_0 during cooling and heating. The three absorption peaks in Q^{-1} at T=145, 225 and 285K coincide with the three main softenings in f_0, as can be clarified by considering df_0/dT (O.–M.Nes et al, to be published).

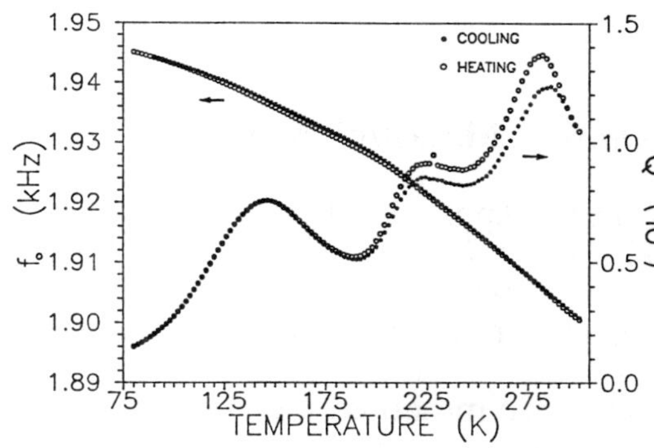

Fig. 1 Temperature dependence of resonance frequency f_0 and internal friction Q^{-1} of $Bi_2Sr_2CaCu_2O_y$ single crystal

Here we focus on the anomalous behaviour of v_e in the region of the superconducting transition, and we therefore performed heating and cooling measurements at a rate $|dT/dt|\simeq$ 0.02K/min from 80 to 105K. Figure 2 shows dv_e/dT for these data.There is a pronounced jump and an altered slope between 92 and 95K, and we associate it with the superconducting transition. In order to extract the value of the discontinuity, we simply fitted the two regions above 95 and below 88K in Figure 2 to a linear expression and obtained the jump from the difference in these fits. The "softening" during the superconducting transition then corresponds to a smeared discontinuity $\Delta(d\ln v_e/dT))=(6.2\pm0.5)$ppm/K and $\Delta(v_e^{-1}\cdot d^2v_e/dT^2)=(0.3\pm0.1)$ppm/K^2 at 93.2K. We further estimate the softening in $\Delta v_e/v_e$ to be of the order of 10ppm at T=93.2K by subtracting a 2nd order polynomial from $\Delta v_e/v_e$. Regarding Q^{-1}, it is difficult to decide whether we observe an anomaly or not near T_c at this stage.

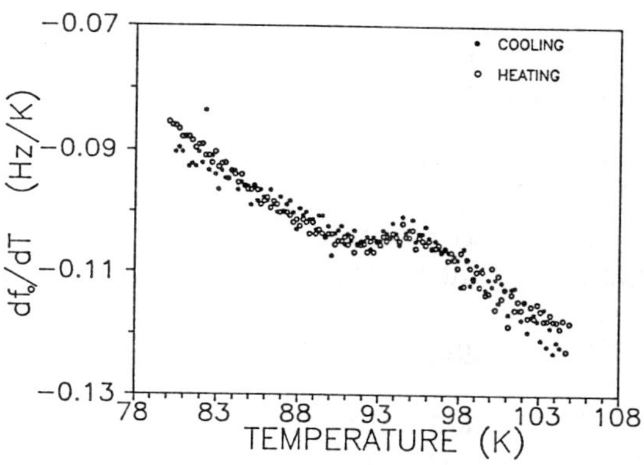

Fig. 2 Temperature derivative of the resonance frequency f_0 of single crystal $Bi_2Sr_2CaCu_2O_y$ close to T_c =93.2K

Due to high reproducibility we can probably rule out errors created by thermal gradients. Changes in f_0 due to influence of thermal expansion on the dimensions l and d of our reed can not explain the result (when using values from Schnelle et al, (1989) valid for a polycrystalline 2223 sample). Thus our observations are roughly consistent with Xiang et al (1989), who found that $\Delta E/E<5\cdot10^{-5}$ and $\Delta(d\ln E/dT)<4\cdot10^{-6}$K^{-1}, even though we find $12\cdot10^{-6}$K^{-1} for the latter one.

Specific heat measurement of a 2212 crystal from the same batch as the VR sample is shown in Figure 3, plotted as C_p/T. For calibration we used C_p=259J/(molK) at 180K (Okazaki et al 1990). The C_p–anomaly is centered at approximately 95K, corresponding to

T_c, and we have indicated how we found $\Delta C_p / T_c \simeq 16$ mJ/(molK^2) and $T_c \simeq 94.5$K from a simple extrapolation method taking into account preservation of entropy. Simi– lar values of $\Delta C_p/T_c$ were reported by others (Okazaki et al 1990, Wohlleben at al 1990) for ceramic 2223– and 2212–phases).

Thermodynamics for a second order phase transition yield at T_c (Testardi 1975, Millis and Rabe 1988) $\Delta E_i / E_i = - E_i \cdot (\Delta C_p/T_c) \cdot (\partial T_c / \partial \sigma_i)^2$ and $\Delta(d\ln E_i / dT) \simeq -(E_i \cdot \Delta C_p / T_c) \cdot (\partial^2 T_c / \partial \sigma_i^2 + (\partial T_c / \partial \sigma_i)^2 \cdot 2z / T_c)$, where σ is the stress, z is of order unity and i=1,2 in our case. Using our measured values, we estimate from the first equation that $|\partial T_c / \partial \sigma_i| \approx 0.1$K/kbar, quite comparable to $dT_c/dp = +0.2$K/kbar (Beille et al, 1988) mea– sured on a 2212–crystal. The second equa– tion then gives $-\partial^2 T_c / \partial \sigma_i^2$ of the order of 10^{-3} to 10^{-2}.

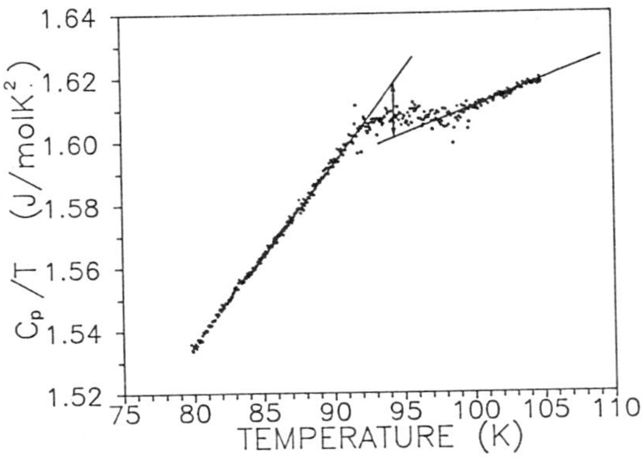

Fig. 3 Specific heat plotted as C_p/T vs T of single crystal $Bi_2Sr_2CaCu_2O_y$ near T_c. The lines show how we found $\Delta C_p / T_c \simeq 16$mJ/(mol$K^2$)

4. Conclusions

We measured Young's modulus sound velocity v_e of a $Bi_2Sr_2CaCu_2O_y$ crystal and found a softening in the region of T_c=93.2K corresponding to a smeared discontinuity of 6.2ppm/K in $d\ln v_e/dT$ and 0.3ppm/K^2 in $\Delta(v_e^{-1} \cdot d^2 v_e/dT^2)$. The anomaly in $\Delta v_e/v_e$ was estimated to be of the order of 10ppm. Specific heat measurements gave $\Delta C_p / T_c \simeq 16$ mJ/(molK^2) and combining our results we found $|\partial T_c/\partial \sigma_i| \approx 0.1$K/kbar and $-\partial^2 T_c / \partial \sigma_i^2$ of the order 10^{-3} to 10^{-2} K/kbar^2. Our experimental observations provide no evidence for an unconventional behaviour in the elastic properties.

Acknowledgements The authors would like to thank L.T.Sagdahl for providing AC-susceptibility data, E.D.Tuset for assisting during C_p measurements and both for help in preparing this manuscript. This work was supported by NAVF, NORSK HYDRO A/S and the NTH foundation.

References

Beille J, Dupendant H, Laborde O, Lefur Y, Perroux M, Yen Y F, Wang J H, Zheng D N and Zhao Z 1988 Physica C **156** 448

Millis A J and Rabe K M 1988 Phys.Rev.B **38** 8908

Okazaki N, Hasegawa T, Kishio K, Kitazawa K, Kishi A, Ikeda Y, Takano M, Oda K, Kitaguchi H, Takada J and Miura Y 1990 Phys.Rev.B **41** 4296

Raychaudhuri A K and Hunklinger S 1984 Z.Phys.B–Condensed Matter **57** 113

Schnelle W, Braun E, Broicher H, Weiss H, Geus H, Ruppel S, Galffy M, Braunisch W, Waldorf A, Seidler F and Wohlleben D 1989 Physica C **161** 123

Stokka S and Fossheim K 1982 J.Phys.E **15** 123

Testardi L R 1975 Phys.Rev.B **12** 3849

Wang Y–N, Wu J, Shen H–M, Zhu J–S, Chen X–H, Yan Y–F and Zhao Z–X 1990 Phys.Rev.B **41** 8981

Wohlleben D, Braun E and Schnelle W 1990 Proc. Int. Conf. on Superconductivity (World Scientific Publishing Corp)

Xiang X–D, Chung M, Brill J W, Hoen S, Pinsukanjana P and Zettl A 1989 Solid State Comm. **69** 883

Supercond. Sci. Technol. **4** (1991) S391–393. Printed in the UK

A Study of the Split Transition in $YBa_2Cu_3O_{7-\delta}$ by Specific Heat and Resistivity Measurements.

J.W. Loram, J.R. Cooper and K.A. Mirza.

IRC in Superconductivity, Madingley Road, Cambridge CB3 OHE, UK.

Abstract Fully oxygenated ceramic $YBa_2Cu_3O_{7-\delta}$ ($\delta \sim 0$) exhibits a double transition with a sharp specific heat peak at 88K (T_{c1}) and a mean field like step at 92.5K (T_c) indicative of a multicomponent order parameter. With small increase in δ, T_c - T_{c1} decreases systematically (and reversibly) and is zero at $\delta = 0.05$. Several interpretations are considered and we conclude that the evidence favours an unconventional pair interaction. From a preliminary analysis of the fluctuation component of the specific heat and resistivity we find a fourfold decrease in coherence volume as δ increases from 0 to 0.1.

1. Introduction

Using a high precision differential technique we are currently investigating the superconducting specific heat anomaly in $YBa_2Cu_3O_{7-\delta}$ as a function of δ. We present preliminary results for the "90K plateau" region $0 \le \delta \le 0.175$ which show a splitting of the transition which varies systematically with δ.

Of the many published measurements of the superconducting specific heat anomaly in $YBa_2Cu_3O_{7-\delta}$ very few show evidence of a split transition (reviewed by Junod 1990 and Choy et al 1990) with only one recorded example in a single crystal sample (Inderhees 1988). In view of this it is perhaps surprising that we have observed structure in **all** our fully oxygenated undoped YBCO samples (Fig 1a) taking the form of a peak at T_{c1}~88K and a step at T_c=92.5K These temperatures agree well with similar features in other published results (Choy et al 1990) but are around 1.5K higher than the sharper features observed by Butera et al (1988).

2. Experimental

The same ceramic YBCO sample (mass~4g) is used throughout, the oxygen content 7-δ being varied by reannealing the sample in flowing oxygen and quenching into liquid N_2 from predetermined temperatures T_Q = 420, 470, 520 and 570 °C for δ=.025, .05, .01 and .175 respectively. Changes in δ are determined from changes in weight before and after heat treatment and have an uncertainty of ±0.01. The true value of δ for the nominally fully oxygenated "δ=0" sample (furnace cooled in oxygen to 150 °C) cannot be determined until we have completed the series, but we would expect δ~.03 for this sample. Other quoted values of δ represent differences from this starting value.

To resolve the small specific heat anomalies we use a differential technique (Loram 1983) to measure directly the difference in specific heat between each sample and a $YBa_2(Cu_{.93}Zn_{.07})_3O_7$ reference (Loram et al 1990a) thus eliminating most of the large phonon term. A correction has been made for the small residual phonon specific heat, and the curves in Figs 1b,c show $\Delta C^{el}=C^{el}(T,\delta)-C^{el}(T,\delta)_{normal}$. Resistivity measurements are made on a slice cut from the specific heat sample and subjected to the same heat treatments and AC susceptibility measurements have been made on a similarly treated powder sample.

2. Results

The specific heat results for $0 \le \delta \le .175$ are shown in Fig 1b. The anomalies are sharp up to δ=.05 then broaden and decrease rapidly in height as found by Junod (1989). We attribute this behaviour to the suppression of superconducting order on the CuO chains by the removal of chain oxygen. This not only reduces the total pair density (anomaly height) and phase coherence (anomaly sharpness) but also makes the order parameter more 2-dimensional with a larger fluctuation contribution above T_c. A similar effect is found on substituting Co^{3+} for chain Cu(1) (Loram et al 1990b). Note that a small **increase** in T_c to around 94K is observed in both cases.

For δ=0 we see clear evidence of structure with a peak at T_{c1}=88K and a step with maximum negative slope at T_c=92.5K (splitting ΔT_c=4.5K). We next increased δ to .05 and found a single sharp anomaly with T_c=92.8K but no splitting. On reducing δ to .025 a reduced splitting reappeared with ΔT_c=2.2K. In each case the features at T_{c1} and T_c are sharp to a few tenths of a K (Fig 1c). We consider that this systematic behaviour with δ on a single specimen and our regular and consistent observations of identical splitting on earlier fully oxygenated samples are strong evidence that the effect is both reproducible and reversible.

We have made a preliminary analysis of the fluctuation contribution to the specific heat above T_c. Above around 96K this term varies as $(T-T_c^*)^{-1}$ (where $T_c^* \sim$ 90K for $\delta \leq .05$)) consistent with 2-dimensional Gaussian fluctuations. The magnitude of this fluctuation term increases with δ and corresponds to a quasi-2D coherence volume decreasing from ~1600 $Å^3$ for δ=0 to ~ 400 $Å^3$ for δ=.1.

The resistivity of samples with $\delta \leq .1$ is shown in Fig 1d. In each case the resistivity has maximum slope at a temperature T_ρ which correlates well with the upper specific heat transition T_c (Fig 1e), but then shows a "foot" to zero resistance at $T_{\rho 0}$ approximately mid-way between the specific heat transitions T_c and T_{c1}. We have observed similar behaviour in most fully oxygenated YBCO samples though the relative heights of the foot and normal state resistance vary somewhat.

The AC susceptibility (not shown) correlates well with the resistivity. With falling temperature the diamagnetism increases slowly below T_ρ then increases rapidly with maximum slope at the zero resistance point $T_{\rho 0}$ for all samples with or without splitting. **No change in slope of χ_{AC}** could be detected at the lower specific heat transition T_{c1} for δ=0 and .025.

3. Discussion

Structure in the specific heat anomaly (reflecting **discontinuous** changes in curvature or slope of the free energy) is **definitive** evidence for a multicomponent order parameter (Joynt 1987, Choy etal 1990). We now consider the evidence that this is a bulk intrinsic effect.

* The comparable magnitude of the specific heat anomalies at T_{c1} and T_c show that each transition is associated with a **bulk** phenomenon involving all (or at least half) of the sample. This eliminates nucleation of superconductivity on grain or twin boundaries at T_c, or a minority phase with T_c different from the bulk T_c as possible explanations.
* The absence of a susceptibility anomaly (change in slope) at T_{c1} precludes the possibility of two coexisting majority superconducting phases for δ=0 and .025 with different T_c.
* The anomalies at T_{c1} and T_c are **sharp** and cannot be explained as a spread of T_c due, for example, to an inhomogeneous oxygen distribution. This conclusion is reinforced by the non-sequential order δ=0, .05, .025 in which the samples were prepared and measured.
* The resistivity and susceptibility data show conclusively that long range superconducting order commences at the upper transition T_c and is fully established at $T_{\rho 0}$, well above the lower transition T_{c1}. Thus although granularity may account for the "foot" in the resistance it cannot explain the lower transition in the specific heat at T_{c1}. We conclude that the structure in the resistivity is not **directly** related to the splitting of the specific heat transition but that the decreasing breadth of the resistive transition with δ results from the increasingly rapid increase in superconducting entropy (average order parameter) below T_c. In support of this we find by integrating the specific heat curves in Fig 1b that the entropy change in the temperature interval $T_\rho(90\%)$ - $T_{\rho 0}$ (ie 90% to zero resistance) is approximately **constant** (~3% of the normal state entropy at T_c) for $0 \leq \delta \leq .05$ reflecting the non-zero magnitude for the bulk superconducting order parameter required before tunneling through weak links is complete.

The evidence discussed above strongly supports a bulk intrinsic origin for the split transition in YBCO and is consistent with an anisotropic (and thus multicomponent) superconducting order parameter as discussed by Joynt (1987),Choy et al (1990) and Annett (1990). For non s-wave pairing, coupling between the superconducting order parameter and an environment of different symmetry may yield a direction dependent T_c. The sensitivity of T_{c1} and the insensitivity of T_c to chain oxygen deficiency δ suggest that these may represent the different transition temperatures (for very low δ) for c-axis and a-b plane superconductivity (Loram etal 1990b).

The extreme sensitivity of the splitting T_c-T_{c1} to the oxygen deficiency δ and the high resolution required to observe the effect may explain the infrequent observation of a double transition in the specific heat of YBCO. Its very rare appearance in measurements on single crystals probably reflects the difficulty in achieving a fully oxygenated single crystal sample.

References

Annett J.F. 1990, Advances in Physics **39**, 83-126.
Butera R.A., 1988, Phys.Rev **B37**, 5909
Choy T.C, Das M.P,Hongxing He., 1990. Phase transitions **20**, 1-25.
Joynt R. 1987 Physica Scripta **36**, 175
Junod A, Eckert D, Graf T, Triscone G, Muller J.1989 Physica **C162-164**, 482
Junod A., 1990, "Physical Properties of High Temperature Superconductors 2" Ed.D.M.Ginsberg, World Scientific
Inderhees S.E. etal 1988 Phys.Rev.Letts. **60**, 1178
Loram J.W., 1983, J.Phys **E16**, 367
Loram J.W, Mirza K.A, Freeman P.F., 1990a., Submitted to Physica **C**
Loram J.W, Mirza K.A. Freeman P.F, Tallon J.L.,1990b, These proceedings, **PS** 1.24

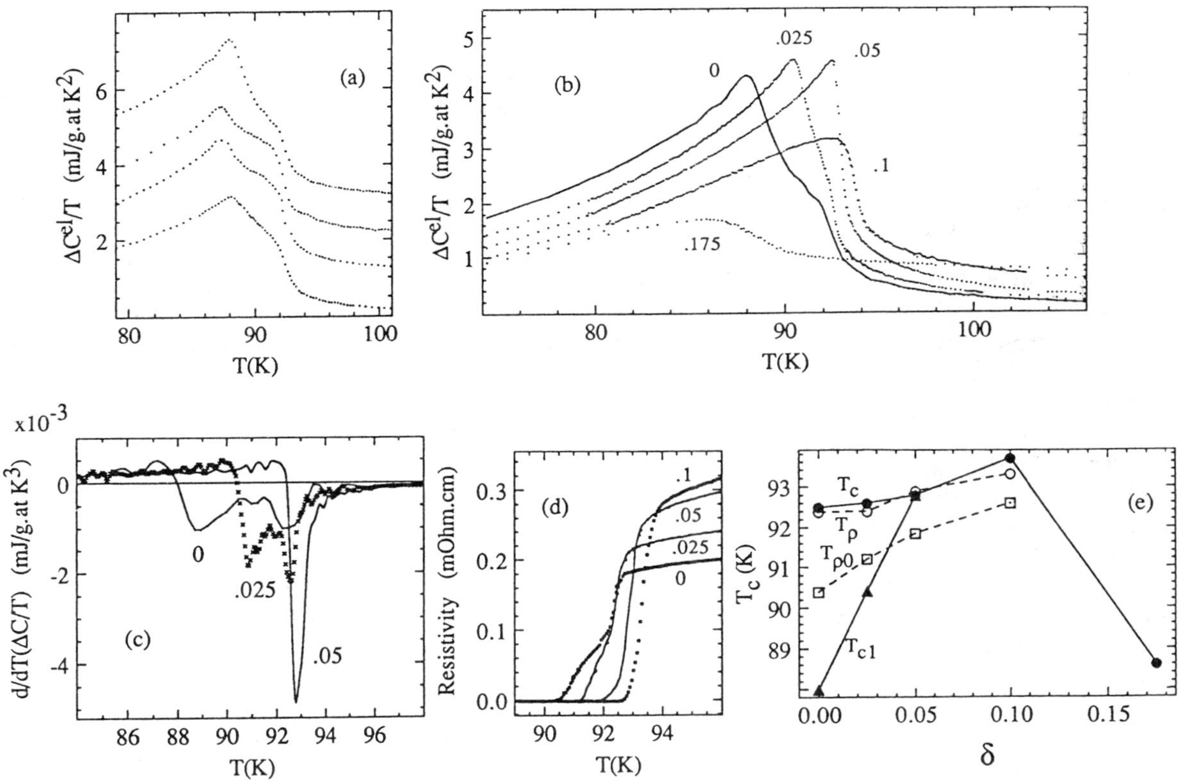

Fig1(a) The split transition in four $YBa_2Cu_3O_{7-\delta}$ samples with δ~0 from three laboratories.
(b) Specific Heat anomalies in $YBa_2Cu_3O_{7-\delta}$ as a function of δ. Labels show δ.
(c) Temperature derivative of the Specific Heat anomalies in $YBa_2Cu_3O_{7-\delta}$
(d) Resistivity of $YBa_2Cu_3O_{7-\delta}$. Labels show δ.
(e) T_c vs δ for $YBa_2Cu_3O_{7-\delta}$. T_c and T_ρ are the temperatures for maximum slope of specific heat and resistivity respectively. T_{c1} is the temperature of the specific heat peak. $T_{\rho 0}$ is the temperature for zero resistance.

Supercond. Sci. Technol. **4** (1991) S394–396. Printed in the UK

Carrier Density Dependence of Transport Properties of Metallic Cuprate Systems

T. Tamegai, K. Koga and Y. Iye

The Institute for Solid State Physics, The University of Tokyo, Roppongi, Minato-ku, Tokyo 106, Japan

Abstract. Metallic cuprate systems, $La_{2-x}Sr_{1+x}Cu_2O_{6+y}$, $La_{8-x}Sr_xCu_8O_{20-y}$ and $La_4BaCu_5O_{13+y}$ were investigated through transport, magnetic and structural measurements. $La_{2-x}Sr_{1+x}Cu_2O_{6+y}$ does not show superconductivity in a wide range of carrier density in spite of the presence of the antiferromagnetic ordering in the carrier depleted sample. In $La_{8-x}Sr_xCu_8O_{20-y}$, an anomaly in the magnetic susceptibility accompanied by a structural change was found in a narrow range of the strontium content. A large magnetic field dependence of the resistivity and the Hall coefficient in $La_4BaCu_5O_{13+y}$ indicate the presence of a band with a small effective mass.

1. Introduction

It has been established that the presence of the two-dimensional Cu-O planes and the carrier density in an appropriate range are the necessary conditions for the occurrence of the high-T_c superconductivity. There are a few cuprate systems which do not show superconductivity in spite of similarity with the high-T_c compounds. In this paper, we report changes due to the carrier density variation in some metallic cuprates. Three representative metallic cuprate systems, $La_{2-x}Sr_{1+x}Cu_2O_{6+y}$ with Cu-O plane, $La_{8-x}Sr_xCu_8O_{20-y}$ and $La_4BaCu_5O_{13+y}$ with three-dimensional Cu-O networks are treated in this paper.

2. Experiments and Results

Samples used in this study were prepared by a standard solid-state reaction of La_2O_3, $SrCO_3(BaCO_3)$ and CuO with repeated firing in oxygen atmosphere. Oxygen contents of the samples were controlled by quenching them from high temperature (T_q) into liquid nitrogen. Electrical resistivity (ρ) and magnetoresistance (MR) were measured by a standard four-probe method. Hall coefficient (R_H) measurements were done by use of a split-type superconducting magnet which generates a magnetic field of 5 Tesla as described in our previous paper (Tamegai and Iye. 1989). Magnetic susceptibility was measured by a pendulum magnetometer based on the Faraday method under the field of 1.18 Tesla. μSR measurements were performed at Meson Science Laboratory at KEK, Tsukuba.

2.1 $La_{1.9}Sr_{1.1}Cu_2O_{6+y}$

Temperature dependence of ρ for $La_{1.9}Sr_{1.1}Cu_2O_{6+y}$ with different T_q are shown in Fig. 1. With increasing T_q, ρ at room temperature smoothly increases. Low temperature increase of ρ is also pronounced for higher T_q samples and no trace of superconductivity was observed. Hole number per copper calculated from R_H at room temperature (p_H) and determined by iodometric analyses (p) as a function of T_q are shown in Fig. 2. As T_q increases, p changes from 0.3 to 0.05 due to the decrease in oxygen content. The values of p_H for samples with higher T_q agree well with p.

0953-2048/91/01S394 + 03 \$03.50

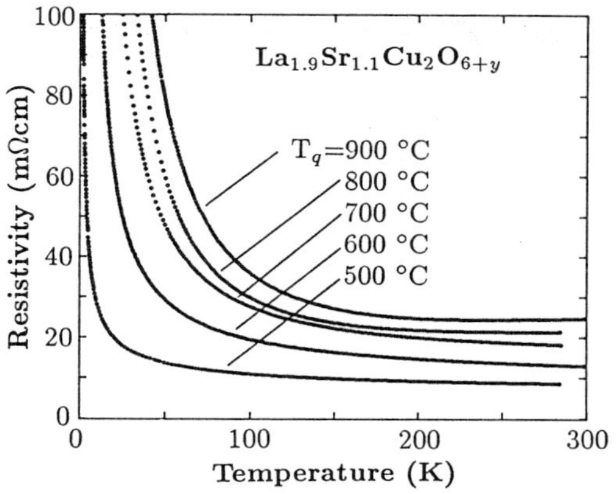

Fig. 1 Temperature dependence of the resistivity in $La_{1.9}Sr_{1.1}Cu_2O_{6+y}$ with different T_q.

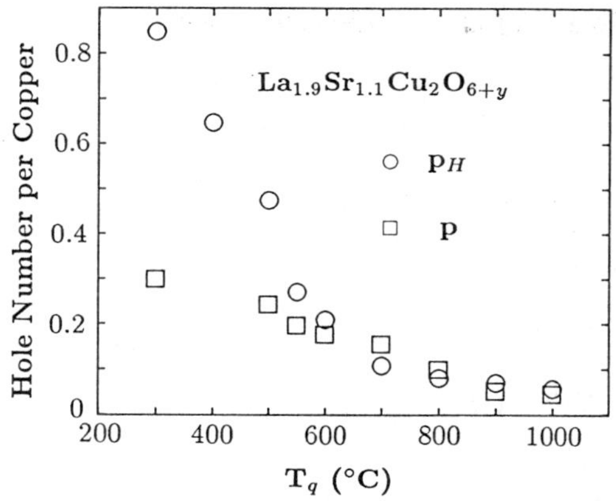

Fig. 2 Hole number per copper determined from the Hall coefficient at room temperature p_H (circle) and from iodometry p (square) as a function of T_q.

Tokura et al. (1988) established that the key parameter for high-T_c superconductivity is the carrier density in the Cu-O plane. Kondo et al. (1988) gave a theoretical support to this through the calculation of the Madelung potential in $YBa_2Cu_3O_7$. Though we have successfully changed the carrier density from p=0.3 to p=0.05, superconductivity was not realized at any carrier density. According to the structural analysis on $La_{2-x}Sr_{1+x}Cu_2O_{6+y}$ by Nguyen et al. (1984), the site between two neighboring Cu-O planes are randomly occupied by La and Sr. Such a random occupation induces an excess oxygen in the same plane. The random potential exerted to carriers in the Cu-O plane by the above structural feature must cause a localization in this system as probed in the transport measurement by Tamegai and Iye. (1989), which results in the destruction of the superconductivity. Details of the transport properties in $La_{2-x}Sr_{1+x}Cu_2O_{6+y}$ including the influence of Sr content variation will be reported elsewhere (Tamegai et al. 1990).

For all high-T_c cuprate systems known to date, there exists an antiferromagnetic (AF) insulator phase next to the high-T_c phase. In order to see if the $La_{2-x}Sr_{1+x}Cu_2O_{6+y}$ system follows the same pattern, we carried out μSR measurements. To make the carrier density as low as possible, the sample was heated at 900 °C in nitrogen atmosphere for 10 hours followed by a slow cooling to room temperature. Figure 3 shows the temperature dependence of the paramagnetic fraction of $La_{1.9}Sr_{1.1}Cu_2O_{6-y}$ determined μSR measurement under zero magnetic field. A gradual decrease of the paramagnetic fraction below 40K suggests the presence of a magnetic ordering below this temperature. A detailed analysis clarified the presence of a short range AF ordering below 20K. The nature of the AF ordering is considered to be similar to that observed in the spin-glass phase in $La_{2-x}Sr_xCuO_4$ with x ≈ 0.05. Details of the μSR experiment will be reported separately (Shiratake et al. 1990). The reason for such a low T_N is most probably due to the presence of the small but finite amount of carriers supplied by Sr.

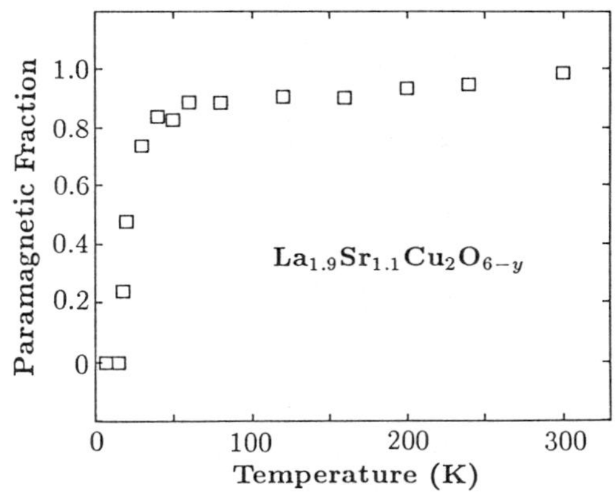

Fig. 3 Temperature dependence of the paramagnetic fraction in $La_{1.9}Sr_{1.1}Cu_2O_{6-y}$ determined by zero field μSR measurement.

Cava et al. (1990) have very recently discovered superconductivity in a related compound $La_{1.6}Sr_{0.4}CaCu_2O_6$. Their finding supports our discussion on the absence of the high-T_c superconductivity in $La_{2-x}Sr_{1+x}Cu_2O_{6+y}$ due to randomness. They started from $La_2CaCu_2O_6$, in which La and Ca ordered almost completely and no excess oxygen was present, and made the effect of random potential minimum. Carriers were supplied by substituting Sr for La.

2.2 $La_{8-x}Sr_xCu_8O_{20-y}$

Anomalies in the temperature dependences of ρ and R_H in $La_{8-x}Sr_xCu_8O_{20-y}$ with $x \leq 1.6$ were reported by Tamegai and Iye (1989). Magnetic susceptibility (χ) measurement revealed a new anomaly around 150 K for a narrow range of the strontium concentration around x=1.6 as shown in Fig. 4. As the anomaly in χ occurs only around x=1.6, it is considered to have a different origin as that in the transport properties. The shape of the temperature dependence of χ reminds us of the weak ferromagnetism observed in $La_{2-x}Sr_xCuO_4$ which is closely related with a structural change from tetragonal to orthorhombic. Low temperature X-ray diffraction measurement in $La_{6.4}Sr_{1.6}Cu_8O_{20-y}$ revealed the presence of an anomaly in the lattice constants around 150 K. A more detailed study of the structural change using single crystal is desired.

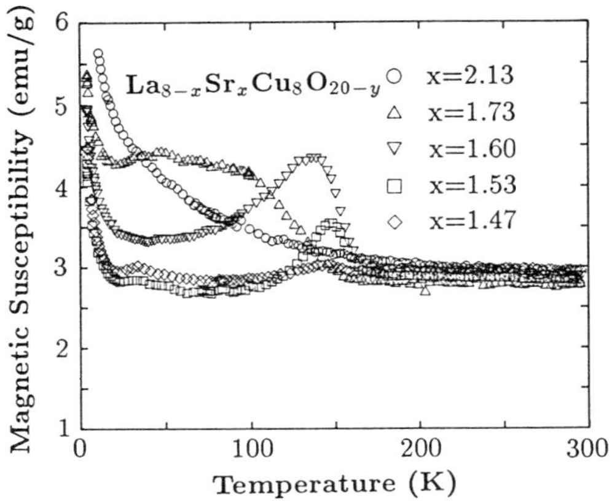

Fig. 4 Temperature dependence of the magnetic susceptibility in $La_{8-x}Sr_xCu_8O_{20-y}$.

2.3 $La_4BaCu_5O_{13+y}$

A large MR in $La_4BaCu_5O_{13+y}$ was reported by Tamegai et al (1989). We extended the measurement including the magnetic field dependence of R_H. As shown in the Fig. 5, $La_4BaCu_5O_{13+y}$ shows an large field dependence of the MR which is proportional to H^2 at high temperatures and shows a tendency of saturation at low temperature. R_H is also strongly magnetic field dependent. Such a large field dependence of MR and R_H indicate the presence of a band with a small effective mass, which seems consistent with the large plasma frequency reported by Kaplan et al (1988). In spite of the structural similarity, MR in $La_{8-x}Sr_xCu_8O_{20+y}$ is much less compared with that in $La_4BaCu_5O_{13+y}$, which reflects a difference in the band structure due to different Cu-O networks.

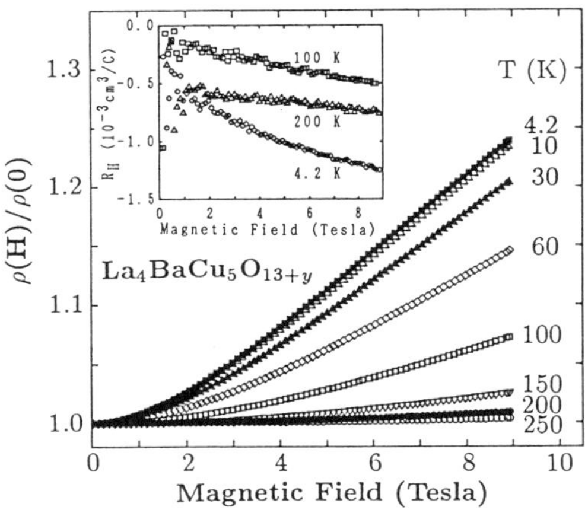

Fig. 5 Magnetic field dependence of the resistivity and the Hall coefficient in $La_4BaCu_5O_{13+y}$.

Acknowledgment

We acknowledge the collaboration with S. Okuma, S. Shiratake and N. Nishida in the μSR measurements and with A. Hayashi and Y. Ueda in the low temperature X-ray diffraction measurements. This work is partly supported by Grant-in-Aid for Priority Areas of Scientific Research "Mechanism of Superconductivity" from the Ministry of Education, Science and Culture, Japan.

References

Cava R J et al 1990 Nature **345** 602

Kaplan S G, Noh T W, Sulewski P E, Xia H, Sievers A J, Wang J and Raj R 1988 Phys. Rev **B38** 5006

Kondo J, Asai Y and Nagai S 1988 J. Phys. Soc. Jpn **57** 4334.

Nguyen N, Er-Rakho L, Michel C, Choisnet J and Raveau B (1980) Mat. Res. Bull. **15** 891

Shiratake S et al. to be published

Tamegai T and Iye Y 1989 Physica C **159** 181

Tamegai T, Okada C, Iye Y 1990 Physica C to be published

Tokura Y, Torrance J B, Huang T C and Nazzal A I (1988) Phys. Rev. **B38** 7156

Supercond. Sci. Technol. **4** (1991) S397–399. Printed in the UK

The Dielectric Functions of High-T_c Cuprates Obtained from the Optical and Electron Energy Loss Measurements

I. Terasaki, T. Nakahashi, S. Takebayashi, A. Maeda and K. Uchinokura

Department of Applied Physics, The University of Tokyo,
7-3-1 Hongo, Bunkyo-ku, Tokyo 113, JAPAN

Abstract Various dielectric functions of high-T_c cuprates, $(La,Sr)_2CuO_4$ and $Bi_2Sr_2CaCu_2O_{8+\delta}$, and their related compounds were obtained by two kinds of measurements —optical reflectivity and electron-energy-loss spectroscopy. Comparative study of the obtained spectra has revealed that the excitations associated with CuO_2 plane in high-T_c cuprates are concentrated in lower-energy region and are well isolated from the excitations in the insulating blocks.

1. Introduction

Oxide superconductors have been synthesized vigorously, and formed a large family during the last few years. Most of them have a characteristic CuO_2 plane responsible for high-temperature superconductivity, and a number of scientists have studied the physical properties of the CuO_2 plane. As one more characteristic feature of Cu-oxide superconductors, we can point out their layered structure, where the conducting CuO_2 and other insulating blocks are alternately stacked along the c axis. For example, the combination of the $(La,Sr)_2O_2$ layer with the CuO_2 plane makes $(La,Sr)_2CuO_4$. The insulating blocks which isolate CuO_2 planes have been systematically classified by Tokura (1990). He has shown that a superconductor can be designed by combining these insulating blocks with the CuO_2 plane, and actually discovered a new superconductor (Tokura *et al* 1989).

Tokura's work suggests that a new superconductor will be easily obtained if a new insulating block is discovered. It is then quite important for basic study and application to clarify a common property of the insulating blocks, because it can be a good guide to search for a new insulating block and a new superconductor. From this motivation, we studied the dielectric functions of various kinds of $3d$-transition metal oxides including high-T_c cuprates. We adopted two different techniques, optical reflectivity and electron-energy-loss spectroscopy (EELS). These two kinds of measurements are complementary, since reflectivity gives information on the excitation for $\boldsymbol{k}=0$ and EELS does for $\boldsymbol{k}\neq 0$. The dielectric functions were obtained up to 35 eV, which is high enough to cover most of the excitations of valence electrons.

2. Sample Preparation and Experimental Results

All the samples except for $La_{0.7}Sr_{0.3}CoO_3$ were prepared as single crystals by the usual CuO(or Bi_2O_3)-flux method. More detailed information and the characterization of the crystals are given elsewhere (Terasaki *et al* 1990, Tarascon *et al* 1989). Polycrystalline $La_{0.7}Sr_{0.3}CoO_3$ was synthesized by the traditional solid-reaction method (Jonker 1969).

0953-2048/91/01S397 + 03 $03.50

Optical reflectivity spectra at room temperature were measured by two systems. The Seya-Namioka monochromator with synchrotron radiation as a light source was employed for high-energy region (from 2 to 40 eV), whereas the rapid-scanning Michelson interferometer was used for low-energy region (from 0.03 to 3 eV). The electron-energy-loss spectra were measured by the pulse-count technique in reflection mode. The energy of incident electrons and the energy resolution are 1 keV and 0.5 eV, respectively.

The measured reflectivity and EELS spectra were transformed into the dielectric functions through the Kramers-Kronig relation. Figures 1 and 2 show the loss functions $\mathrm{Im}[-1/\varepsilon(\omega)]$ of various materials obtained from reflectivity (solid curves) and EELS (dotted curves). Note that Sr content of $(La,Sr)_2CuO_4$ is about 10 %. The consistency between the optical and EELS spectra is satisfactory, although the EELS spectra are somewhat smeared due to the energy resolution of 0.5 eV. One would notice that some common characteristic structures are seen in all spectra in Fig. 1, and also in all spectra in Fig. 2. Looking more quantitatively, all the spectra in Fig. 1 have two distinct peaks near 13–14 and 28–29 eV, which correspond to the characteristic excitations in the LaO layer. As for the characteristic features in the Bi_2O_2 layer, a broad maximum near 18–20 eV and a pair of sharp peaks around 27.5 and 29 eV, are seen in Fig. 2.

3. Discussion

As was seen in the preceding section, all the spectra with a common insulating block resemble one another. It means that excitations in the insulating layers are mainly located in the high-energy region, and they hardly affect the low-lying excitations which dominate transport and magnetic properties. Such low-lying excitations are associated with the electrons in the layer of 3d-transition metal oxide, e.g., CuO_2 or CoO_2 layer.

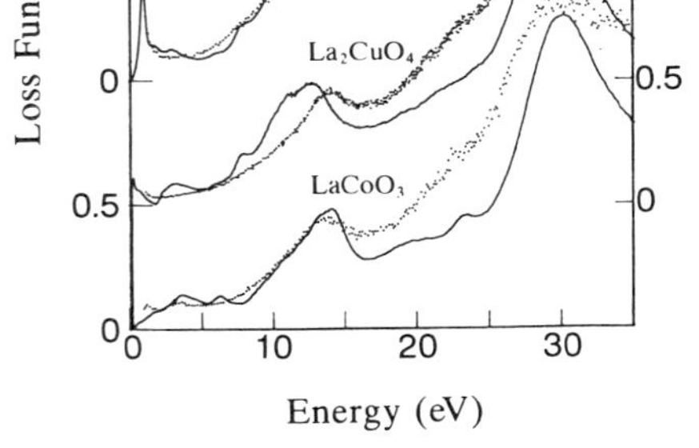

Fig. 1 The loss functions of materials with LaO layer. Solid curves are obtained from reflectivity and dotted curves from EELS.

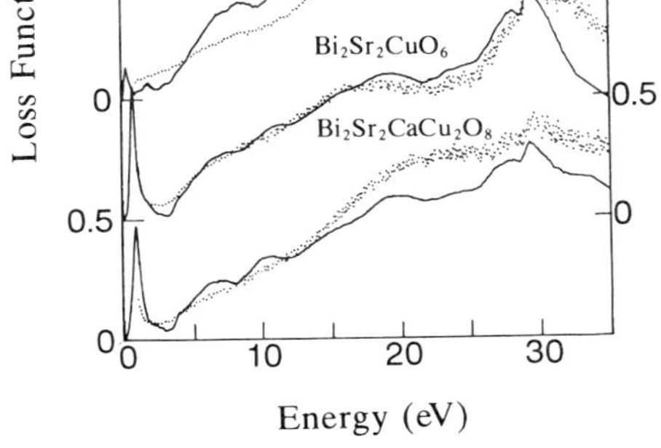

Fig. 2 The loss functions of materials with Bi_2O_2 layer. Solid curves are obtained from reflectivity and dotted curves from EELS.

We can estimate the energy range of low-lying excitations assuming that they strongly depend on the carrier density. The doping effect on the loss function in Cu and Co systems is shown in Fig. 3, where the loss functions obtained from reflectivity are shown. Figure 3 shows that the spectra for $(La,Sr)_2CuO_4$ exhibit the carrier-density dependence below 3 eV. A similar doping effect is seen in the spectra for the Bi-based cuprates (Terasaki *et al* 1990). On the other hand, the spectra for $(La,Sr)CoO_3$ are varied with doping above 4 eV. This suggests that the low-lying excitations in the Cu oxide are more concentrated in low-energy region than those in the Co oxide. We then point out as a characteristic feature in Cu oxides that the excitations in the CuO_2 layer are well isolated from higher-energy excitations in the insulating layers such as LaO or Bi_2O_2 layer.

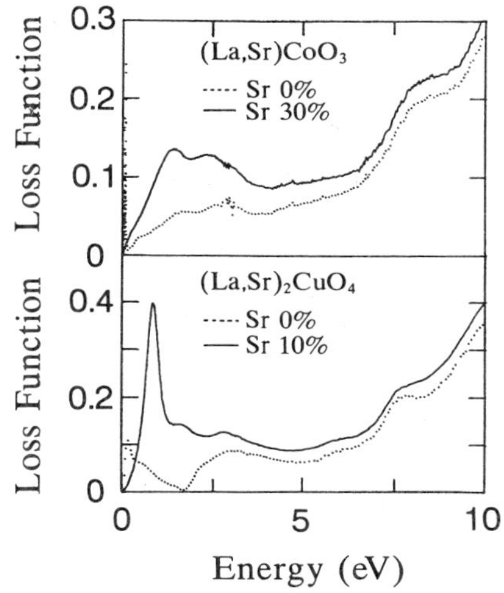

Fig. 3 The loss functions (obtained from reflectivity) of Cu and Co oxides in low-energy region.

4. Summary

Dielectric functions of high-T_c cuprates and their related compounds are studied by reflectivity and electron energy loss spectroscopy. The excitations in the insulating layer of high-T_c cuprates, LaO and Bi_2O_2 layers, are obtained. It has been revealed that the excitations in the insulating layers of cuprates are similar to their related compounds, and that the dielectric function is well understood by dividing them into each excitation in the insulating block, just as the crystal structure can be understood by Tokura's classification. The doping-dependent excitation is well isolated from the higher one in cuprates, whereas the former overlaps with the latter in Co oxides.

Authors would like to thank S. Uchida, S. Tajima, H. Eisaki, T. Ido, and T. Arima for their collaboration. They also appreciate S. Suga and M. Seki for technical assistance of optical measurement in Synchrotron Radiation Laboratory, Institute for Solid State Physics, The University of Tokyo. This work is supported partially by the Grant-in-Aid for Scientific Research from the Ministry of Education, Science and Culture.

References

Jonker G H 1969 Philips Res. Rep. **24** 1

Tarascon J M, Ramesh R, Barboux P, Hedge M S, Hull G W, Greene L H, Giroud M, LePage Y, McKinnon W R, Waszczak J V, and Schneemeyer L F 1989 Solid State Commun. **71** 663

Terasaki I, Nakahashi T, Takebayashi S, Maeda A and Uchinokura K 1990 Physica **C165** 152

Tokura Y, Arima T, Takagi H, Uchida S, Ishigaki T, Asano H, Beyers R, Nazzal A I, Lacorre P and Torrance J B 1989 Nature **342** 890

Tokura Y 1990 Parity **5** No.2 34 (in Japanese)

Supercond. Sci. Technol. **4** (1991) S400–402. Printed in the UK

Seebeck Effect of Weak Links in a High-T_c Superconductor

A V Ustinov†, M Hartmann, R P Huebener, G Yu Logvenov†, and V V Ryazanov†

Lehrstuhl Experimentalphysik II, University of Tuebingen, D-7400 Tuebingen, FRG
† Institute of Solid State Physics, USSR Academy of Sciences, Chernogolovka, 142432 Moscow district, USSR

Abstract. We have observed experimentally the analog of the superfluid helium fountain effect in a polycrystalline $YBaCuO$ film. In a superconductor a temperature gradient induces the counterflow of superconducting electrons and quasiparticles. The supercurrent creates a phase difference at the intra- and intergrain weak links. In the presence of a temperature gradient parallel to an electric current we have detected an asymmetry of the current-voltage characteristic. Reversing the direction of the temperature gradient also reverses the sign of the current asymmetry. We estimate the thermoelectric coefficient for the quasiparticles in the $YBaCuO$ sample.

1. Introduction

The response of the quasiparticles to an applied temperature gradient in superconductors is effectively shorted out by the superconducting electrons (Ginzburg 1944). As a result the thermopower S (Seebeck effect) vanishes in the superconducting state. The normal current density j_n is locally cancelled by a counterflow of supercurrent with density j_s . A physical separation of these two currents is possible only in inhomogeneous or anisotropic superconductors (for a review see Ginzburg and Zharkov 1978, Van Harlingen 1981). High-T_c superconducting materials combine both of these properties, and a study of the thermoelectric effects in these materials is of fundamental interest.

In a superconductor a phenomenon analogous to the superfluid helium fountain effect can be used for detecting the counterflow of both currents I_n and I_s in the presence of a temperature gradient. (Clarke and Freak 1972). In a Josephson junction or weak link the superconducting phase difference φ between both electrodes depends only on the supercurrent according to the Josephson relation $j_s = j_c \sin\varphi$, where j_c is the critical current density. The temperature gradient ∇T applied to a weak link with the resistivity ρ_n generates a supercurrent with density

$$j_s = -j_n = -\frac{S\nabla T}{\rho_n}, \tag{1}$$

i.e.,$\varphi \neq 0$. If an external supercurrent of density j_e also passes through the weak link, a non-zero voltage will appear when the total current density exceeds j_c , i.e. for

$$j_e = j_c \pm \frac{S\nabla T}{\rho_n}. \tag{2}$$

Hence, flow of the supercurrent due to ∇T may be detected from the asymmetry of the critical current and of the current-voltage characteristic (IVC) with respect to the origin of the current axis. For conventional superconductors such a *dc* thermoelectric Josephson effect has been observed by Kartsovnik *et al* (1981) in low-resistive SNS junctions using a SQUID-voltmeter. Without the external current through the junction, a non-zero voltage was detected at the temperature gradient $\nabla T > \nabla T_c = \rho_n j_c/S$. For $j_e = 0$ the *ac* thermoelectric Josephson effect has been also detected in SNS junctions (Panaitov *et al* 1984).

Many of the properties of the high-T_c superconductors are strongly influenced by the presence of intra- and inter-granular weak links. We suppose, that the application of a temperature gradient thermoelectrically induces a supercurrent I_s , which causes a phase difference at the weak links inside the bulk superconducting material. If the external current I_e is also applied to the sample, I_e and I_s superimpose and the apparent critical value of I_c is changed. In this paper we show experimentally, that the weak link detection of the thermoelectric current is easily performed in a high-T_c superconductor.

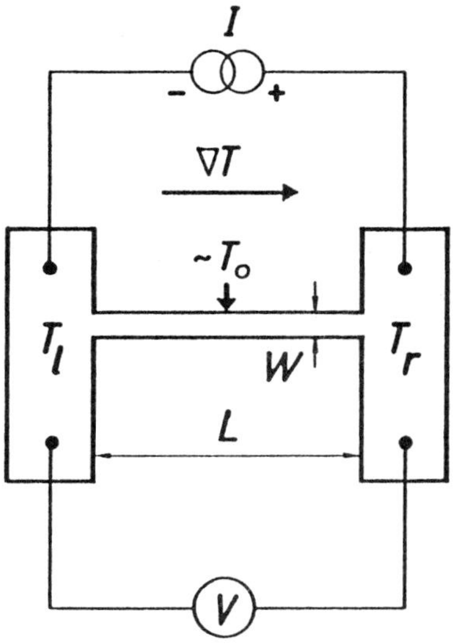

Fig.1 Configuration of the experiment.

2. Experimental

The experimental setup is shown in Figure 1. We have studied a $1.8\mu m$ thick polycrystalline $YBaCuO$ film prepared on an Y-stabilized ZrO substrate by multilayer electron beam evaporation. The film had a resistivity of about $7m\Omega cm$ in the normal state at $T = 100K$ and a zero-resistance temperature of about $70K$. Taking a material not with the best superconducting parameters was important for having a sufficiently large number of weak-links with a low value of the critical current. The film geometry was defined by mechanically scribing on the substrate in form of the wide bridge with the dimensions $W \approx 0.1mm$ and $L \approx 1.0mm$. The details of the experiment may be found elsewhere (Ustinov *et al* 1990).

3. Results and discussion

The influence of ∇T on the IVC is shown in Figure 2. In the two cases for both directions of the temperature gradient the middle of the bridge was kept at approximately the same temperature $T_0 \sim 61K$, while $\nabla T \approx 20K/cm$ was applied. The solid line shows the IVC in each case. The dashed lines are the same curves rotated by 180° around the origin of coordinates. Comparing the shift between the dashed and the solid curve, it is easy to see the magnitude and the direction (sign) of the asymmetry of the IVCs. The direction of the asymmetry was well reproducible. In a small magnetic field of a few gauss the critical current was drastically decreased and the asymmetry was also vanishing.

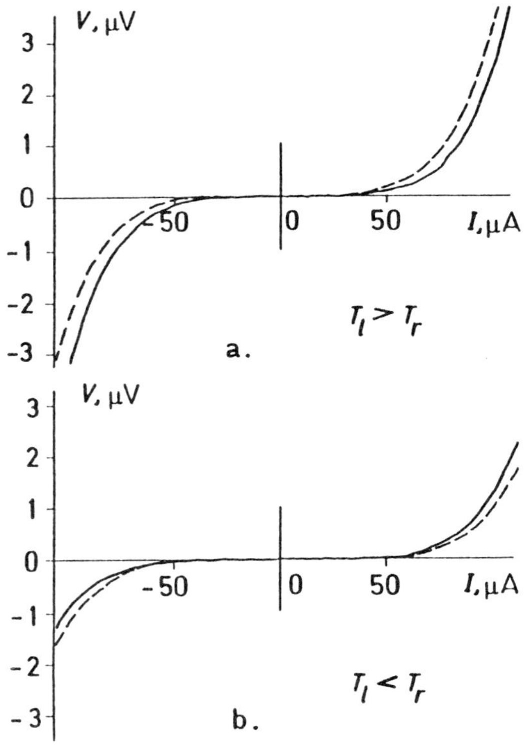

Fig.2 IVCs in the presence of the temperature gradient.

Our observations lead to the conclusion, that we deal with *the interaction of the thermoelectrically induced flow of the supercurrent I_s with the intergranular weak links in the polycrystalline film.* From the value of the current asymmetry using (2) we estimate the thermopower generating the normal current I_n, yielding $S \approx 0.2\mu V/K$. From the mutual direction of I_e and ∇T we also conclude, that the sign of S is positive. It has been shown experimentally (Ouseph and O'Bryan 1990), that above T_c the thermopower of $YBa_2Cu_3O_{7-x}$ is very sensitive to the oxygen content, and that it varies from about $-6\mu V/K$ to $10\mu V/K$ for x ranging between 0.00 and 0.19. The sign reversal of the thermopower occurs around $x \sim 0.08$. From the parameters of our polycrystalline film we expect the oxidation not to be complete. So it is not surprising to have a positive and small estimated value for the thermopower below but not too far from T_c. The spatial distribution of I_e is strongly influenced by inhomogeneities in the sample, while I_s is following ∇T and is therefore more homogeneous. This means that the correct value of S is likely larger than that estimated above.

4. Acknowledgements

This work was supported by the Bundesminister für Forschung und Technology (project 13N5482). A V U is pleased to acknowledge support by the Alexander von Humboldt Stiftung.

5. References

Clarke J and Freake S M 1972, *Phys. Rev. Lett.* **29** 588.
Ginzburg V L 1944, *J. Phys. USSR* **8** 148.
Ginzburg V L and Zharkov G F 1978, *Sov. Phys. Usp.* **21** 318.
Kartsovnik M V, Ryazanov V V, and Schmidt V V 1981, *JETP Lett.* **33** 356.
Ouseph P J and O'Bryan M R 1990, *Phys. Rev.* **B41** 4123.
Panaitov G I, Ryazanov V V, Ustinov A V, Schmidt V V 1984, *Phys. Lett.* **A100** 301.
Ustinov A V, Hartmann M, and Huebener R P 1990, *Europhys. Lett.* to be published.
Van Harlingen D J 1982, *Physica* **B109-110** 1710.

Supercond. Sci. Technol. **4** (1991) S403–405. Printed in the UK

London Penetration Depth in Bi-based High-T_c Compounds

M. Weber[1], A. Amato[1], P. Birrer[1], V.N. Duginov[3], V.G. Grebinnik[3], F.N. Gygax[1], S. Kapusta[3], B.F. Kirillov[4], E. Lippelt[1], H. Maletta[2], V.G. Olshevsky[3], A.V. Pirogov[4], V.Yu. Pomjakushin[3], A.N. Ponomarev[4], A. Schenck[1], V.G. Storchak[4] and V.A. Zhukov[3]

[1]Institut für Mittelenergiephysik der ETH Zürich, CH-5232 Villigen PSI
[2]Institut für Festkörperforschung, Forschungszentrum Jülich, D-5170 Jülich 1
[3]Joint Institute for Nuclear Research, P.O. Box 79, Dubna, 10100 Moscow, USSR
[4]I.V. Kurchatov Institute of Atomic Energy, Kurchatov sq.1, 123182 Moscow, USSR

Abstract. The muon spin relaxation was measured in the high-T_c superconductors $Bi_2(Sr_{1.3}Ca_{0.7})CuO_{6.2}$, $Bi_2Sr_2CaCu_2O_{8.16}$, $(Bi_{1.85}Pb_{0.15})Sr_2CaCu_2O_{8.18}$ and $(Bi_{1.83}Pb_{0.25})Sr_{1.97}Ca_{1.97}Cu_{3.07}O_x$. The data were analysed in order to determine the London penetration depth λ_{ab} parallel to the basal plane. In all investigated samples λ_{ab} revealed a temperature dependence which can be explained with the framework of conventional BCS theory in the weak coupling limit.

1. Introduction

The London penetration depth λ is one of the most important parameters characterizing a superconductor, since it is determined by the effective mass m^* and the concentration of carriers n_s involved in the pair formation, via the clean limit formula

$$\lambda = \sqrt{m^* c^2 / 4\pi n_s e^2} \quad . \tag{1.1}$$

Its temperature dependence is determined initially by the pairing mechanism and actually also by pinning centers for magnetic flux lines due to impurities (e.g. Gross 1986).

μSR spectroscopy allows to determine the magnetic field distribution $F(B)$ within a sample. In the mixed state of an isotropic type II superconductor, $F(B)$ is a function of the penetration depth λ. In case of a highly anisotropic superconductor ($\lambda_c/\lambda_{ab} \gtrsim 5$, where λ_c and λ_{ab} are respectively the values of the penetration depth $\|$ and $\perp$ to the single crystal's c-axis), the second moment of the polycrystalline field distribution due to a regular triangular vortex lattice is given by (Barford 1988)

$$< \Delta B^2 > = 0.00371 \; \Phi_0^2 / (1.23 \cdot \lambda_{ab})^4 \quad . \tag{1.2}$$

Under the assumption of a Gaussian distribution $F(B)$, the μSR spin depolarization rate σ is equal to $\gamma_\mu \cdot < \Delta B^2 >^{1/2}$, where γ_μ is the gyromagnetic ratio of the muon. Hence, using the generalization of formula (1.1) to the anisotropic case (Barford 1988) and formula (1.2), σ can be related to λ_{ab} and n_s/m^*_{ab} (m^*_{ab} denotes a component of the anisotropic mass tensor) as

$$\sigma \propto 1/\lambda_{ab}^2 \propto n_s/m^*_{ab} \quad . \tag{1.3}$$

Uemura *et al* (1989) have found a universal linear relation between T_c and σ for all investigated specimens of high-T_c cuprate superconductors with low carrier doping (Figure 1). The universal validity of this relation would impose severe restrictions on any theory of high-T_c superconductivity.

0953-2048/91/01S403 + 03 \$03.50

2. Experimental Results and Discussion

We have measured the muon spin relaxation in the high-T_c superconductors listed in Table 1 in a transverse magnetic field of 3 kG using the field-cooling procedure. The experiment was perfomed at the Synchrophasatron of JINR in Dubna, USSR. X-ray powder diffraction diagrams show single and pure phases in case of the one- and two-CuO-layered systems and a small admixture of a foreign phase in $Bi_{1.83}Pb_{0.25}Sr_{1.97}Ca_{1.97}Cu_{3.07}O_x$. Resistivity and magnetization measurements revealed a sharp onset of T_c in all specimens. The exact composition of the samples was determined by chemical analysis.

In a first step, our data were analyzed by fitting a Gaussian depolarization function to the μSR signals of the samples in the superconducting state. Since our chosen fit function and our demands for the sample quality were exactly the same as described by Uemura *et al* (1989), the values we obtained for the depolarization rate $\sigma(T \to 0\ \mathrm{K})$ may be compared to his results (Figure 1). It is obvious that the "universal" relation between T_c and $\sigma(0)$ does definitely not hold for $Bi_2(Sr_{1.3}Ca_{0.7})CuO_{6.2}$, while the (σ, T_c) positions of the two- and three-CuO-layered systems follow Uemura's curve within the experimental accuracy.

The strongly deviating results in $Bi_2(Sr_{1.3}Ca_{0.7})CuO_{6.2}$ may be explained as follows (Emery 1990): The average interplane distance between adjacent CuO planes d is about 6 ± 1 Å for each of the specimens considered by Uemura *et al*, but also for the two- and three-CuO-layered systems listed in Table 1. On the other hand for $Bi_2(Sr_{1.3}Ca_{0.7})CuO_{6.2}$ it amounts to about 12 Å. The μSR line width measured in polycrystalline samples is proportional to the **three**-dimensional carrier density per volume divided by m^*_{ab}, according to equation (1.3). Uemura's relation may be explained in terms of a non-interacting **two**-dimensional electron gas when the energy scale of the mediating boson is comparable to or higher than the Fermi energy. For this special case however, T_c is proportional to the **two**-dimensional carrier density divided by m^*_{ab} and should therefore be plotted against $(\sigma \cdot d)$. The thus modified relation holds also for $Bi_2(Sr_{1.3}Ca_{0.7})CuO_{6.2}$.

The values of $\lambda_{ab}(T \to 0\ \mathrm{K})$ listed in Table 1 were determined by a more sophisticated technique which takes into account that the actual distribution $F(B)$ in a superconductor in the vortex state is non Gaussian: We compared the measured μSR spectra with several simulated ones, based on field distributions calculated for systematically varied values of λ_{ab}. In the case of good agreement between measured and calculated spectra, we took the values of λ_{ab} assumed for these specific spectra as the true intrinsic ones of the corresponding samples (Figure 2). The method is described in detail by Weber *et al* (1990). A careful analysis showed that the ratio of the square root of the true second moment of such a field distribution and the value of the depolarization rate σ obtained by a fit of a Gaussian damped fit function to the corresponding μSR spectra is constant. This means that the fitted σ is proportional to the true values of n_s/m^*_{ab} and of λ_{ab}^{-2}. For this reason Uemura's T_c versus σ plot is physically reasonable. But also the temperature dependence of $< \Delta B^2 >^{1/2}$ is correctly reflected by the fitted values of σ.

Figure 3 displays the temperature dependence of $< \Delta B^2 >^{1/2}$ in $Bi_{1.85}Pb_{0.15}Sr_2CaCu_2O_{8.18}$, which is similar to the one observed also in the three other investigated samples. It is perfectly explained by the framework of conventional BCS theory in the weak coupling limit. The description by the two-fluid model, which is valid in the strong coupling limit, is significantly worse. The observation of BCS like behaviour in the weak-coupling limit is rather surprising, since in this case the linear relation between T_c and n_s/m^*_{ab} (Figure 1) cannot be expected (Uemura 1988).

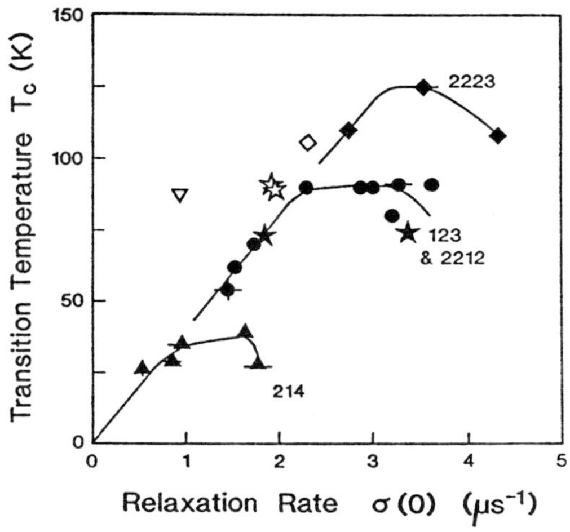

Fig. 1 Compilation of transition temperatures T_c for various members of different high-T_c families plotted versus the corresponding muon spin relaxation rates $\sigma(0)$. Solid points are measured by Uemura *et al* (1989). Open points are obtained at Dubna, USSR. Different symbols refer to different families: • 123, ▲ 214, ⋆ 2212, ◆ 2223, ∇ $Bi_2(Sr_{1.3}Ca_{0.7})CuO_{6.2}$.

Compound	T_c [K]	$\lambda_{ab}(0)$ [Å]
$Bi_2(Sr_{1.3}Ca_{0.7})CuO_{6.2}$	87	2300 (200)
$Bi_2Sr_2CaCu_2O_{8.16}$	91	1400 (100)
$Bi_{1.85}Pb_{0.15}Sr_2CaCu_2O_{8.18}$	88	1350 (100)
$Bi_{1.83}Pb_{0.25}Sr_{1.97}Ca_{1.97}Cu_{3.07}O_x$	107	1200 (100)

Table 1 Compilation of T_c measured with a SQUID susceptometer and of the London penetration depth parallel to the CuO-layers $\lambda_{ab}(0)$ determined by means of μSR spectroscopy for four different Bi-based high-T_c compounds.

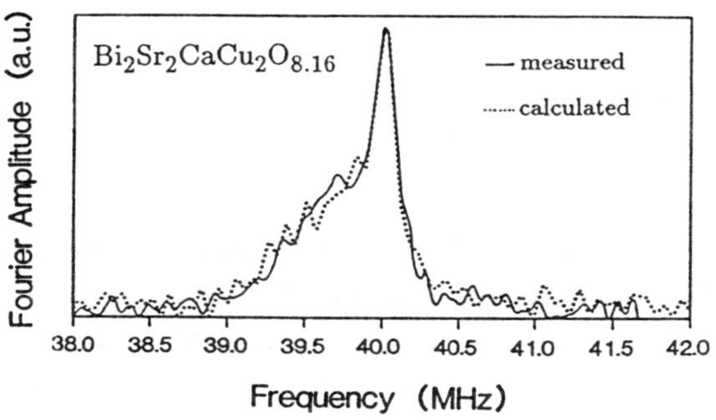

Fig. 2 Comparison of the μSR line shape measured in $Bi_2Sr_2CaCu_2O_{8.16}$ at 11.6 K (solid line) with the corresponding simulated time spectrum based on a theoretical polycrystalline field distribution that refers to $\lambda_{ab} = 1400$ Å (dotted line). The real parts of the complex Fourier transforms are shown.

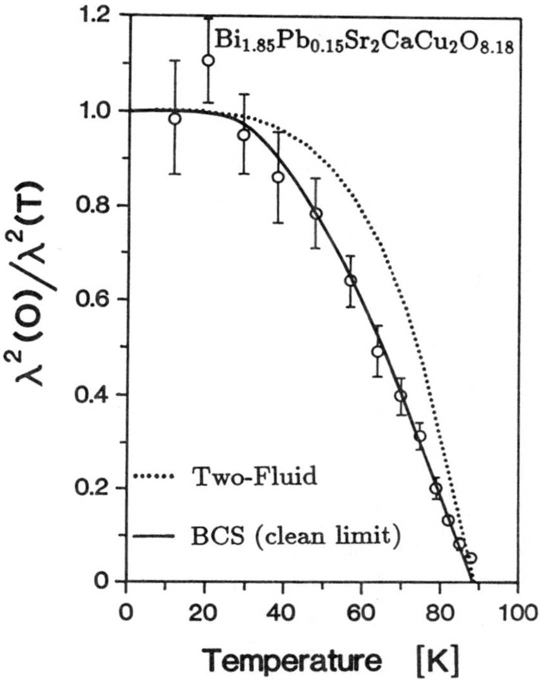

Fig. 3 Temperature dependence of $< \Delta B^2 >^{1/2} \propto 1/\lambda_{ab}^2$. Solid and dotted lines refer respectively to the conventional BCS-theory in the weak coupling limit and to the two-fluid model. $< \Delta B^2 >^{1/2}$ is obtained by two component fits.

References

Barford W and Gunn J M F 1988 *Physica* **C 156** 515
Emery V J 1990 *private communication*
Gross F *et al* 1986 *Z. Physik* **B64** 175
Uemura Y J *et al* 1988 *Phys. Rev.* **B38** 909
Uemura Y J *et al* 1989 *Phys. Rev. Letters* **62** 2317
Weber M *et al* 1990, *Proceedings of the International μSR Conference, Oxford* (to be published in Hyperfine Interactions)

Supercond. Sci. Technol. **4** (1991) S406–408. Printed in the UK

Nuclear Magnetic Resonance in Cobalt-doped $YBa_2Cu_3O_7$

D P Tunstall and W J Webster

Department of Physics and Astronomy, University of St Andrews, North Haugh, St Andrews, Fife KY16 9SS, UK.

ABSTRACT

Four samples of YBCO doped with cobalt have been characterised using several techniques, including measurements of T_c, a.c. susceptibility, resistivity, magnetic susceptibility, and yttrium nuclear magnetic resonance (nmr). The last two techniques were applied over a temperature range of 200 K to 300 K, with the aim of elucidating the local or non-local character of the magnetic changes produced by the doping process.

These changes occur almost exclusively on the CuI sites, and the appearance of a Curie term in the susceptibility is not due to the conventional cause of such a term i.e. a local moment.

INTRODUCTION

Cobalt-doping of yttrium barium copper oxide, to produce $YBa_2(Cu_{1-x}Co_x)_3O_7$, is known[Tarascon et al (1988) and Kistenmacher (1988)] to affect the superconducting transition temperature, T_c, of $YBa_2Cu_3O_7$ (YBCO). Figure 1 summarises the effect of doping on T_c by three different techniques.
Cobalt goes onto the CuI sites in YBCO[Tarascon et al (1988)], and is believed to exist on these sites in a triply ionised state, but pulling in extra oxygen into the structure. The transition at low doping concentrations from orthorhombic to tetragonal structure is correlated with the appearance of this extra oxygen. There is evidence[Terada et al (1990)] that the net effect of the doping process is a reduction in the hole concentration.

There are magnetic measurements[Tarascon et al (1988) and Yamada et al (1989)] that show that the effect of cobalt-doping in YBCO is to introduce extra terms of both Curie-Weiss- and Pauli-like character into the susceptibility. Our magnetic measurements, in the range 200 to 300 K, permit a simple breakdown of the extra susceptibility, $\Delta\chi$, due to cobalt-doping as:

$$\Delta\chi = \chi_o + \frac{C}{T} \qquad (1)$$

where χ_o is temperature-independent. Figure 2 demonstrates our measurements of C, giving an effective magnetic moment per cobalt atom of 4 ± 0.2 Bohr magnetons, rather more consistent with local moments of Co^{2+} character (3.9 β) than with Co^{3+} (4.9 β). Figure 2 also shows the measurements of the temperature independent term χ_o (see equation 1) in our samples.
We turn now, having characterised the electrical and magnetic states of the samples, to address the principal topic of this article; given the extra magnetism described in eqn (1)

0953-2048/91/01S406 + 03 $03.50

what further information about the magnetism can be gleaned from ^{89}Y nmr shift measurements? In particular, we consider the distribution of the extra magnetisation, given that the ^{89}Y nucleus couples[Alloul et al (1988)] primarily to the magnetism around the CuII sites in YBCO.

In this context we should note a related nmr study of MacLaughlin and collaborators where the effect of praesodymium doping of YBCO has been investigated[Reyes et al (1990)]. There the conclusion is that magnetic doping via praesodymium hinders superconductivity mainly through its associated electron doping, deriving from at least the partial presence of Pr^{4+}, rather than through any direct magnetic pair-breaking.

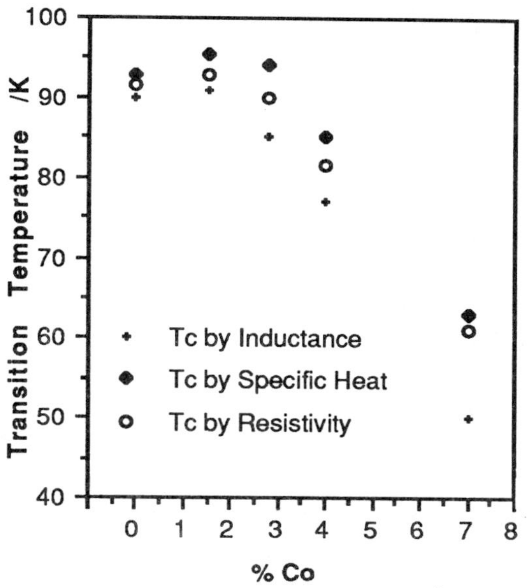

Figure 1. The variation of transition temperature as a function of Co concentration in YBCO-measured using three techniques.

RESULTS

Figure 3 shows our results at room temperature for the Knight shift as a function of cobalt concentration. All ^{89}Y shifts are referenced with respect to that in a 1 M solution of YCl_3. The samples were fine powders of YBCO packed tightly into test-tubes. In general, the nmr lines get broader, and the precision of the Knight shift measurement less, as the cobalt concentration is increased, consistent with the appearance of randomly oriented dipolar fields associated with the cobalt.

The trend of the data in Figure 3 is consistent with an ^{89}Y shift independent of cobalt concentration, or very weakly dependent on it. The susceptibility in the same samples is varying considerably as we have seen in Figure 2. For example, the total susceptibility change introduced by 4.0% doping with cobalt at 300 K is equivalent to seven times the susceptibility of undoped YBCO.

The temperature appears to cause little change in the shifts in our samples. In the same temperature range, Figure 2 demonstrates that the total susceptibility is considerably modified.

DISCUSSION

The results above demonstrate that the ^{89}Y shifts are rather concentration and temperature independent, over the 0 to 4 atomic % cobalt-doping and 200 to 300 K temperature ranges. This is in strong contrast with oxygen deficiency studies[Alloul et el (1988)] in YBCO, where large ^{89}Y shift changes accompany rather small susceptibility changes. The magnetic changes induced by cobalt-doping clearly couple only very loosely, if at all, to the ^{89}Y nuclei.

From this nmr study then cobalt-doping acts very directly to change the susceptibility of the chains, and the thus much enhanced chain susceptibility exists

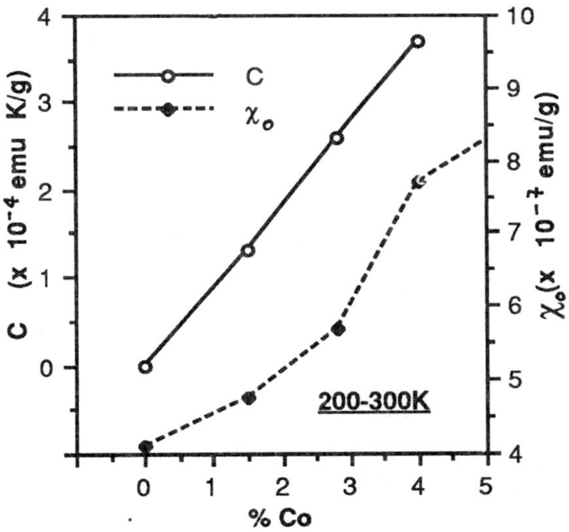

Figure 2. The Curie constant & the temperature independent susceptibility of YBCO as a function of atomic % of Co

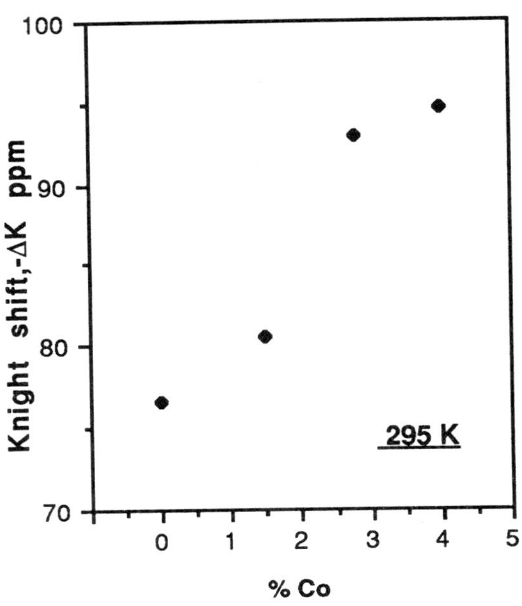

Figure 3. The room temperature Knight shift shown as a function of atomic % of Co in YBCO

side by side with the planes susceptibility, with little or no enhancement of the latter. Furthermore the Curie component on the chains introduced by the cobalt-doping, does not look to be due to localised cobalt moments. The existence of such moments would create very large dipolar fields at the yttrium sites, and the random nature of the doping would then lead to unobservably large linewidths for ^{89}Y nmr. As in the praesodymiun study[Reyes et al (1990)] pair-breaking via local moments does not appear to play a role as the explanation of the reduced T_c in cobalt-doped YBCO. Rather we must look towards the hole- or electron-doping introduced by the added oxygen and cobalt as the reason for the effect on T_c. It may well be that both the small rise in T_c at low cobalt doping and the large drop in T_c at higher cobalt doping have their origin in this variation of hole concentration.

CONCLUSION

Temperature- and concentration-dependent studies of a range of cobalt-doped YBCO samples establish that all magnetic changes introduced by the cobalt occur on the chains. However the Curie term so introduced into the full expression for the magnetic susceptibility must not be thought of as being generated by localised moments. As a consequence, hole- or electron-doping introduced by the cobalt-doping appears to be the mechanism for T_c variation in these systems.

ACKNOWLEDGEMENTS

We would like to acknowledge the strong collaboration with the Cambridge IRC in this work. The samples came from that laboratory and many of the characterisation measurements displayed in this article were performed there. In this context we mention in particular Dr Peter Edwards, Miss Pamela Freeman and Dr John Cooper.

REFERENCES

Alloul H, Mendels P, Collin G and Monod P 1988 *Phys Rev Lett* **61** 746.
Kistenmacher T 1988 *Phys Rev B* **38** 8862.
Reyes A P, MacLaughlin D E, Takigawa M, Hammel P C, Heffner R H, Thompson J D, Crow J E, Kebede A, Mihalisin T and Schwegler J 1990 *J Appl Phys* **67** 5032.
Tarascon J M, Barboux P, Miceli P F, Greene L H, Hull G W Eibschutz M and Sunshine S A 1988 *Phys Rev B* **37** 7458.
Terada S, Kobayashi N, Iwasaki H, Tokiwa A, Kikuchi M, Syono Y and Muto Y 1990 *to be published in the Proc of 19th Low Temperature Physics Conference*, Brighton, August 1990.
Yamada Y, Oda Y, Kawaji N, Yamagata H and Asayama K 1989 *J Phys Soc Japan* **58** 3053.

Supercond. Sci. Technol. 4 (1991) S409–411. Printed in the UK

The a-b Anisotropy of the Normal State Resistivity of Untwinned $YBa_2Cu_3O_{7-\delta}$

U. Welp [1,2], S. Fleshler [2,3], W. K. Kwok [2], J. Downey [2], Y. Fang [2], G. W. Crabtree [2], J. Z. Liu[4]

(1) Science and Technology Center for Superconductivity
(2) Materials Science Division, Argonne National Laboratory, Argonne, IL 60439 USA
(3) Department of Physics, Purdue University, West-Lafayette, IN 47907 USA
(4) Department of Physics, University of California-Davis, Davis, CA 95616 USA

Abstract. We present measurements of the **a-b** anisotropy of the normal state resistivity ρ_a/ρ_b of untwinned $YBa_2Cu_3O_{7-\delta}$. ρ_a/ρ_b increases with increasing temperature and reaches a (sample dependent) value of 1.2 to 1.85 near room emperature.

1. Introduction

In addition to the anisotropy between the basal plane and the c-axis the orthorhombic structure of $YBa_2Cu_3O_{7-\delta}$ shows also anisotropies due to the presence of the Cu-O chains along the b-direction. Band structure calculations (Massida et al., 1987; Yu et al., 1987) predict enhanced conductivity along this direction. Recent measurements of infra-red reflectivity (Schlesinger et al., 1990; Koch et al., 1989) on detwinned $YBa_2Cu_3O_{7-\delta}$ seem to confirm this prediction.

Here we present measurements of the anisotropy ρ_a/ρ_b of the normal state resistivity on six detwinned crystals of various quality. We find the resistivity along **a** to be larger than along **b**. ρ_a/ρ_b reaches values ranging from 1.2 to 1.85 at room temperature. The results are discussed in a model of a parallel circuit of planes and chains.

2. Results and Discussion

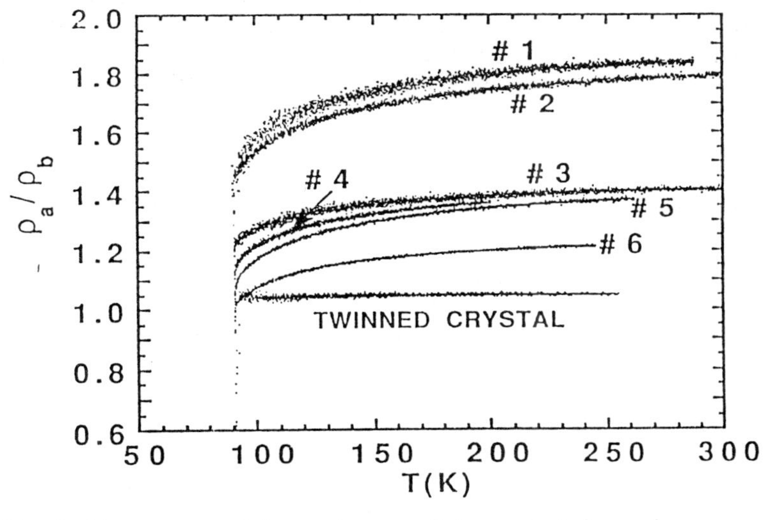

Fig. 1 Temperature dependence of the resistivity anisotropy ρ_a/ρ_a for the six detwinned crystals and the twinned crystal.

0953-2048/91/01S409 + 03 $03.50

The a-b anisotropy of the resistivity was measured on six $YBa_2Cu_3O_{7-\delta}$ crystals grown by a self flux method (Kaiser et al., 1987). The crystals were detwinned by annealing under uniaxial stress (Welp et al., 1989). The Montgomery method (Montgomery, 1971) was used to determine the anisotropy ρ_a/ρ_b. In order to obtain absolute values for the resistivity the contacts on two crystals were extended to the conventional four probe geometry after completing the measurements of ρ_a/ρ_b.

The temperature dependence of ρ_a/ρ_b for the six detwinned crystals is shown in Figure 1. ρ_a/ρ_b increases with increasing temperature and seems to saturate at temperatures higher than 250 K at a sample dependent value of 1.2 to 1.85. In all cases, ρ_a is larger than ρ_b. This result is in agreement with recent measurements of the infra-red reflectivity (Schlesinger et al., 1990; Koch et al., 1989) and bandstructure calculations (Massida et al., 1987; Yu et al., 1987), which indicate additional metallic conductivity along the Cu-O chains. The temperature independent value of ρ_a/ρ_b = 1.05 for the twinned crystal indicates that the Montgomery method gives reliable results.

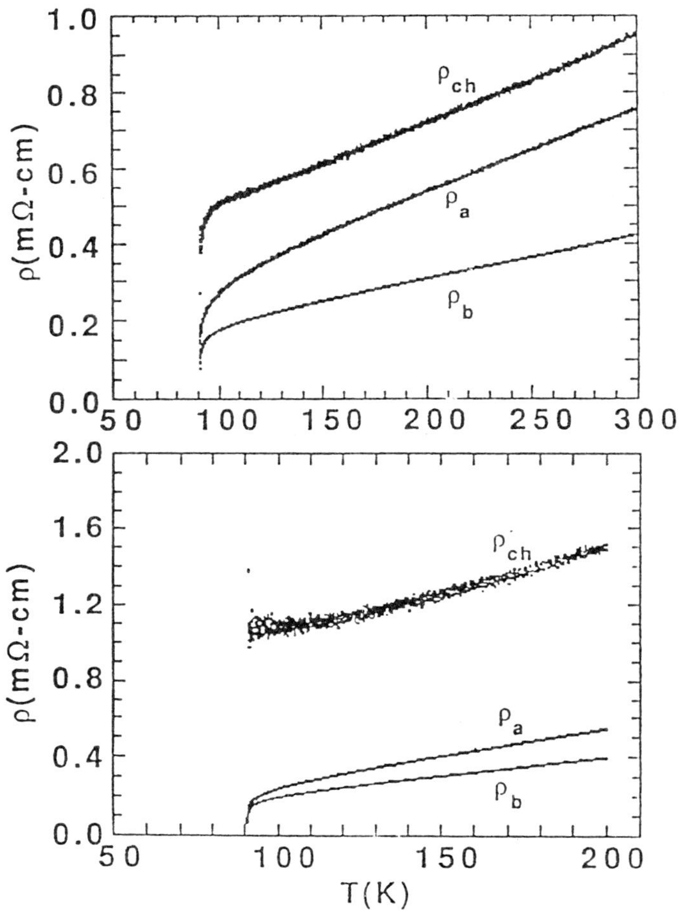

Fig. 2 Temperature dependence of the resistivity along the crystallographic axes and of the chain contribution for the crystal with ρ_a/ρ_b = 1.8 (upper panel) and the crystal with ρ_a/ρ_b = 1.35 (lower panel).

Figure 2 shows the temperature dependence of the resistivity along the two orientations for the crystal with ρ_a/ρ_b = 1.8 (top panel) and with ρ_a/ρ_b = 1.35 (lower panel). ρ_b has been measured after extending the contacts to the conventional geometry, whereas ρ_a has been calculated from the data in Fig. 1. For the following we assume that the chains and planes act as a parallel circuit for a measurement along **b**, whereas along **a** only the planes are seen. Then the chain contribution of the resistivity can be calculated from $\rho_{ch} = \rho_b / (1-\rho_b/\rho_a)$ which is also shown in Fig. 2. We observe that for the crystal with the large value of ρ_a/ρ_b the chain contribution is of similar size as ρ_a and ρ_b and varies linearly with temperature nearly to T_c. In contrast, for the crystal with ρ_a/ρ_b = 1.35 ρ_{ch} is about a factor of five larger than ρ_a, ρ_b and shows the tendency to saturation below 120 K. The absence of a downturn in the chain resistivity suggests that fluctuations occur only in the planes, which also would explain The temperature dependence of the anisotropy. Using the parallel resistor formula and a linear temperature dependence for the chains ($\rho_{ch} = c_0 + c_1$ T) and planes ($\rho_a = a_0 + a_1$ T), we obtain

$\rho_a/\rho_b = 1 + \{1/(c_0 + c_1 T)\}/\{1/(a_0 + a_1 T) + \sigma_{fl}\}$ where σ_{fl} is the fluctuation conductivity. For large T, ρ_a/ρ_b will saturate at $1 + a_1/c_1$, while at T_c ρ_a/ρ_b will drop sharply to 1 since σ_{fl} diverges. This picture of superconductivity confined to the planes with metallic chains acting as reservoir of charge is consistent with the absence of anisotropy in H_{c2} (Welp et al., 1989) and H_{c1} (Umezawa et al., 1990) between **a** and **b** and IR data (Schlesinger et al., 1990) finding no superconducting gap in the chains at 35 K, as well as with the anisotropic normal state resistivity reported here. The large sample dependence of ρ_a/ρ_b is attributed to the different degrees of disorder and interruptions in the chains (Welp et al., 1990). This might account for some differences of ρ_a/ρ_b in a similar recent work (Friedmann et al., 1990)

3. Acknowledgements

Work supported by the U.S. Department of Energy, BES under contract #W-31-109-ENG-38 (SF, WKK, GWC, YF, JD, JZL) and the National Science Foundation under contract # STC-8809854 (UW), Science and Technology Center for Superconductivity. SF acknowledges partial support from the Division of Educational Programs, Argonne National Laboratory.

4. References

Friedmann T A, Rabin M W, Giapintzakis J, Rice J P, Ginsberg D M, 1990 Phys. Rev. **B** (submitted)
Kaiser D L, Holtzberg F, Chisholm M F, Worthington T K, 1987 J. Cryst. Growth **85** 593
Koch B, Geserich H P, Wolf Th, 1989 Solid State Commun. **71** 495
Massidda S, Yu J, Freeman A J, Koelling D D, 1987 Physics Letters A **122** 198
Montgomery H C, 1971 J. Applied Physics **42** 2971
Schlesinger Z, Collins R T, Holtzberg F, Feild C, Blanton S H, Welp U, Crabtree G W, Fang Y, Liu J Z, 1990 Phys. Rev. Letters (August)
Umezawa A, Crabtree G W, Welp U, Kwok W K, Vandervoort K G, Liu J Z, 1990 Phys Rev **B**(submitted)
Welp U, Grimsditch M, You H, Kwok W K, Fang M M, Crabtree G W, Liu J Z, 1989 Physica **C 161** 1
Welp U, Fleshler S, Kwok W K, Downey J, Fang Y, Crabtree G W, Liu J Z, 1990 Phys. Rev **B** (submitted)
Yu J, Massidda S, Freeman A J, Koelling D D, 1987 Physics Letters A **122** 203

Supercond. Sci. Technol. **4** (1991) S412–414. Printed in the UK

Tunneling spectroscopy in $(Pb_xBi_{1-x})_2Sr_2CaCu_2O_8$ crystals

J.J. Wnuk, R.T.M. Smokers, F.W. Nolden, L.W.M. Schreurs, Y.S. Wang and H. van Kempen

Research Institute for Materials, University of Nijmegen, Toernooiveld, NL-6525 ED Nijmegen, The Netherlands

Abstract. Tunneling experiments on single crystals of the $(Pb_xBi_{1-x})_2Sr_2CaCu_2O_8$ compound have been performed with the aid of a low temperature tunneling microscope. Tunneling in the crystal c-axis direction resulted in a parabolic current-voltage dependence for voltages up to 0.2 V, unless the tungsten tip was pushed into the crystal. By changing the configuration to perpendicular to the c-axis, clear energy gap effects were observed. Tunneling between two crystals resulted in conductivity characteristic to a Superconductor - Insulator - Superconductor (SIS) junction.

1. Introduction

The single crystals were grown by the self-flux method as described elsewhere (Wnuk *et al.* 1990). The critical temperature, as determined by the zero resistance criterion, ranged from 80 to 90 K for different crystals. This was also confirmed by the magnetization measurements. The $(Pb_xBi_{1-x})_2Sr_2CaCu_2O_8$ crystallises in the form of thin platelets oriented along the *a-b* plane. The crystals are easily cleavable and clean surfaces are readily produced in this way. High anisotropy of the physical properties of these crystals (more than $4 \cdot 10^4$ in resistivity, over 40 in the upper critical field) indicate almost two-dimensional character of the electronic structure. The coherence lengths in the *a-b* plane and in the *c*-axis direction, as estimated from the upper critical fields, ranged from 18 to 45 Å and from 0.4 to 1.5 Å, respectively, for our crystals. This means that the tunneling experiments are extremely hard to perform for this material and measurements of the energy gap in the *c*-axis direction are most probably impossible without breaking through the crystal surface.

2. Results

The tunneling experiments were performed with the aid of the low temperature tunneling microscope and three different tip-sample configurations were used. Tunneling with a tungsten tip in the crystal *c*-axis direction resulted in a parabolic $I(V)$ dependence for voltages up to 0.2 V unless the tip was pushed into the crystal. The latter led to $I(V)$ characteristics indicating a multiple of interconnected weak links.

In order to perform tunneling measurements in the direction perpendicular to the *c*-axis, we used part of the solidified melt with visible crystal edges rather than extracted single

0953-2048/91/01S412 + 03 \$03.50

crystals. This allowed to keep the crystals in place easily and prevented eventual distortion of the thin platelets while cooling to liquid helium temperature. Both tungsten and platinum-iridium (90% Pt - 10% Ir) tips were used in this case resulting in clear energy gap structure visible in the conductance, $\sigma(V) = dI/dV$, characteristics (Figure 1).

The third configuration consisted of two single crystals with their edges crossed to produce small area tunnel junction with tunneling taking place primarily perpendicular to the *c*-axis. Measurements were performed at two temperatures, namely 4.2 and 77 K (in helium and nitrogen baths). Results shown in Figures 2 and 3 indicate that in both cases well defined SIS junctions were obtained. In particular, the background conductance, always disturbing the results of the SIN (Superconductor-Insulator-Normal metal) junctions measurements, is much less pronounced here.

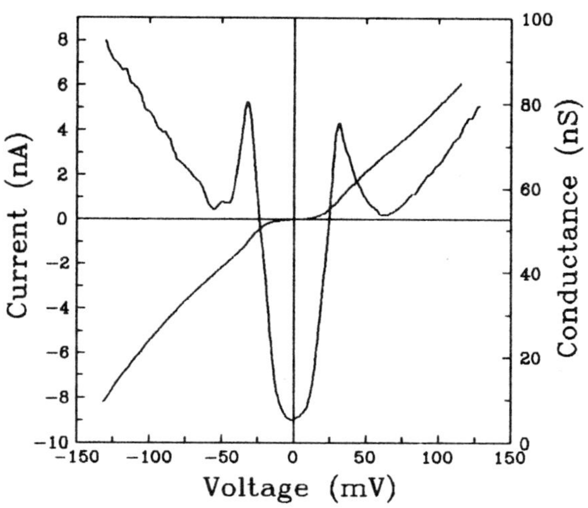

Fig. 1 Plot of $I(V)$ and $\sigma(V)$ of a SIN junction with a tungsten tip measured at 4.2 K.

In all the cases, the tunneling conductance was carefully calibrated to allow for further analysis. Looking at Figures 1 and 2 one may notice low values of the conductance minima at zero voltage. This shows the high quality of the junctions and, if the picture of highly anisotropic energy gap is valid, marks small injection angle of the tunneling particles.

3. Discussion

Because of the lack of space, the detailed analysis of the data will be presented elsewhere. We would like to stress here only the major points. First of all, there is no doubt that all three experiments which results are shown in Figures 1-3 reflect the same phenomenon. From the conductivity of the SIN junction one may, within a few percent, obtain the conductivity of the SIS junction at 4.2 K without any adjustable parameters. Standard formulas are used and the only assumption is the general parabolic shape of the background normal metal tunneling conductivity. In the case of the 77 K experiment, one additional parameter has to be used, as the tunneling conductivity is smeared much stronger than the finite temperature would suggest. We used the standard model of the finite quasiparticle lifetime (Dynes *et al.* 1978) to account for the additional rounding and in this way obtained the conductivity presented in Figure 3 from that shown in Figure 1 with accuracy better than 1%.

That quasiparticle tunneling has been observed and not another effect one may conclude from the exponential dependence of the tunneling current on distance observed in most cases, the accordance between the used experimental configurations and the obtained results, and the applicability of the standard expressions of the tunneling theory for the data analysis.

The size of the gap in the quasiparticle tunneling density of states visible in the tunneling conductance depends on the theoretical model used and none of the methods traditionally used can give an accurate value before such a model is formulated. For comparison with literature, a practical approach has been applied. The value of the energy gap parameter should lay between the point where the reduced conductance reaches unity for the first time and the conductance maximum. This gives $\Delta_0 \approx 22 - 31$meV for the crystal with $T_c = 83$ K and $\Delta_0 \approx 24 - 34$meV for the crystals with $T_c = 90$ K.

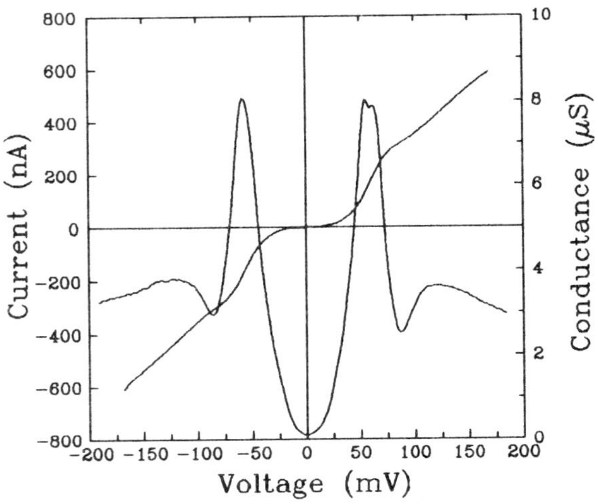

Fig. 2 Tunneling current and conductance versus voltage measured at 4.2 K. The SIS-type junction was obtained between two single crystals.

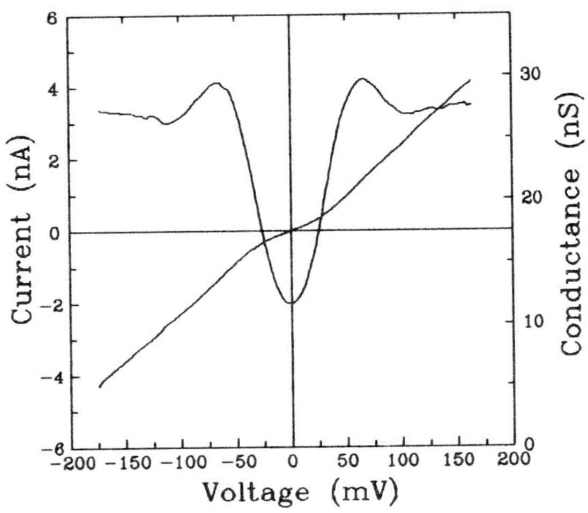

Fig. 3 The $I(V)$ and $\sigma(V)$ dependences obtained at 77 K for a SIS junction.

4. Conclusions

Evidence for the energy gap in the quasiparticle tunneling density of states of the $(Pb_xBi_{1-x})_2Sr_2CaCu_2O_8$ compound is given based on tunneling experiments performed with crystals of this material. The energy gap may not be full, at least not on the whole Fermi surface and is most probably highly anisotropic. Other conclusions coming from the results presented here go beyond the scope of this short report and will be published separately.

Acknowlegments

This work is a part of the research program of the Stichting voor Fundamenteel Onderzoek der Materie (FOM) and was made possible by financial support from the Nederlandse Organisatie voor Wetenschappelijk Onderzoek (NWO).

Literature

Dynes R.C., Narayanamurti V. and Garno P.J. 1978 *Phys. Rev. Lett.* **41** 1509

Wnuk J.J., Schreurs L.W.M., Wang Y.S., Linden van der P.J.E.M. and Eggenkamp P.J.T. 1990 *Mat. Res. Bull.* **25**, to appear in No 9

Supercond. Sci. Technol. 4 (1991) S415–417. Printed in the UK

The Carrier Characteristics of Single Phase Superconducting Oxides $(Bi,Pb)_2Sr_2Ca_2Cu_3O_y$ Under Different Heat Treatments

Zhenhui HE, Yiping SUN*, Jian SHA, Huaqing YING*, Xiangai CHEN, Xianhui CHEN, Jianyi JIANG*, Shunxi WANG, Jiaju DU*, Liezhao CAO, Qirui ZHANG+

Department of Physics, University of Science and Technology of China, Hefei 230026, P.R.China
*Institute of Solid State Physics, Academia Sinica, Hefei 230031, P.R.China
+Department of Physics, Zhejiang University, Hangzhou 310027, P.R.China

Abstract. Single phase samples $(Bi,Pb)_2Sr_2Ca_2Cu_3O_y$ have been prepared under different heat treatments. The midpoint transition temperature (Tc) is determined by both ac susceptibility and resistance measurement. The resistance-temperature coefficient is found changed from positive to negative when the sample is fast cooled. The Hall coefficient measurement shows that Tc increases monotonically with the effective carrier concentration (n_H) and that n_H increases with temperature linearly. We draw the conclusion that the reduction of the carrier concentration is caused not by the reduction of the oxygen content, but by the weak localization of the carriers which is unfavorable to High Tc.

The 110K superconductivity of Bi-Pb-Sr-Ca-Cu-O has been discovered for more than two years[1]. But many properties, such as the relation between Tc and the oxygen content of a certain phase, the role that the carrier concentration plays in this system, have not been fully understood. Thorough experimental analyses were hardly done because of the difficulty of preparing single phase samples in different conditions. Several months ago, Du et al.[2] were successful in preparing single phase samples $(Bi,Pb)_2Sr_2Ca_2Cu_3O_y$ in the same nominal composition but under different annealing conditions.

In this paper, four single phase samples were prepared in the way described in ref.2, i.e., fast cooled in LN_2 after sintered at 850°C for 200 hours. Then, sample A and B were annealed at 480°C for 24 hours and at 400°C for 12 hours, respectively, leaving sample D untreated. The one fast cooled in air

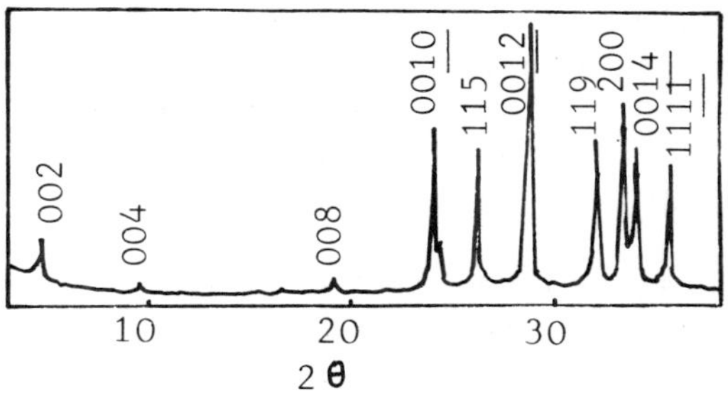

Figure 1. The X-ray diffraction pattern of sample A

was sample C. The X-ray diffraction patterns showed that all of the four samples were in 2223 single phase. (Fig. 1 represents one of them) From these diffraction data, their lattice parameters were indexed and listed in table 1. The orthorhombicity seems to degrade slightly from A to B.

The sharpest drops in the resistance-temperature relation, shown in figure 2, correspond respectively to the onset transitions in the ac susceptibility measurements. The

0953-2048/91/01S415 + 03 $03.50 © 1991 IOP Publishing Ltd

Table I. The lattice parameters of the $(Bi,Pb)_2Sr_2Ca_2Cu_3O_y$

sample	a	b	c	(b-a)/b
A	5.396	5.376	37.074	0.0037
B	5.398	5.388	37.037	0.0019
C	5.398	5.388	37.033	0.0019
D	5.378	5.376	36.887	0.0004

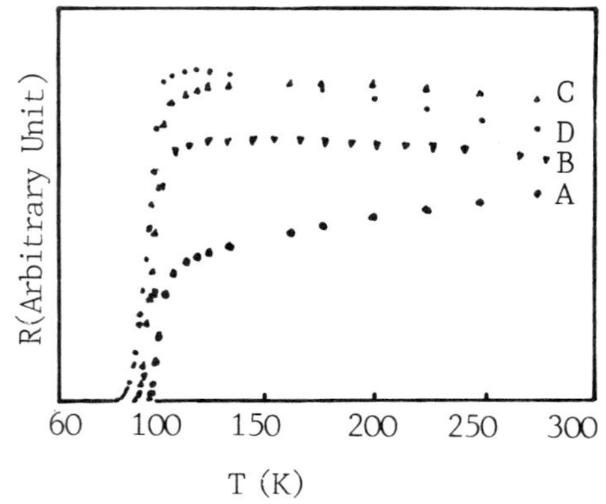

Figure 2. The resistance-temperature curves of the samples A, B, C and D

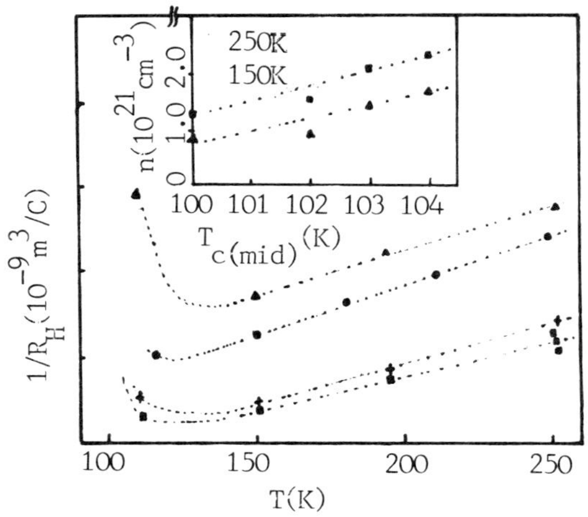

Figure 3. The relation between n_H and temperature for the four samples. Inset: the relation between n_H and Tc at two given temperatures

value of the resistance-temperature coefficient changes from positive to negative, which is characteristic of the weak localization of the carriers.

Measured in the way similar to that reported in ref.3, the Hall effect shows that n_H increases linearly with temperature (see figure 3). This is similar to the result reported by Xia et al.[3] and that observed in 123 phase system. The similar carrier behavior in the Bi and the Y systems implies that the two systems have similar band structure or even a similar pairing mechanism. A two-band model has been proposed to explain such a linear dependence. As can be seen in the inset of fig.3, Tc increases with n_H monotonically, in contrast to the results in ref.2 where the undoped sample has a higher n_H but a lower Tc than does the Pb-doped sample.

In high Tc superconducting compound, n_H is generally related to the position of the Fermi level at its band and its localized level if there is any. If the localization effect can be neglected, there is an optimum position of the Fermi level which gives an optimum n_H for superconductivity. This is supported by Shafer's [4] results. If the localization effect can not be neglected, then, this effect tends to lower n_H and Tc. On the other hand, Uemura et al. [5] found an optimum ratio of n_H/m^* for the highest Tc, (where m* is the effective mass of electron). Later, this was also found to be associated with disorder [6].

In the case of ref.2, though relatively great difference exists in n_H, both Pb-doped and undoped samples have similar metallic behavior. Provided that the band structure is identical for the Pb-doped and undoped samples, the difference in superconducting properties between the two samples may be due to the doping effect only, i.e., the different position of the Fermi level as mentioned above. In our case, the LN_2 cooled sample D, which is expected to maintain the high temperature phase in which atomic disordering is most likely to occur, has the lowest n_H and the lowest Tc as well as the largest value of the negative resistance-temperature coefficient, while sample A is the opposite. Therefore, though there may be different oxygen contents in the identical phase samples, doping effect would not be the dominant factor to influence Tc. Instead, localization effect becomes important. However, the range of both Tc and n_H is too narrow to draw a general conclusion that Tc varies monotonically with n_H for the Bi system.

Acknowledgements

This work was supported in part by the Natural Science Foundation of China, and by the center of superconductivity research of China.

References

[1] e.g., C.Michel et al., Z.Phys. B68 (1987) 421
[2] Du Jiaju et al., (submitted to Z.Phys. B)
[3] Xia Jiangsheng et al., J. Phys. C (1990) (in press)
[4] M.W.Shafer et al., Phys. Rev.B 39 (1989) 2914
[5] Y.J.Uemura et al., Phys. Rev. Lett. 62 (1989) 2 17
[6] D.R.Harshman et al., High Tc Up Date,3 (1989) 2
[7] C.S.Grader et al., Phys. Rev. B 38 (1988) 844

Supercond. Sci. Technol. 4 (1991) S418–420. Printed in the UK

SOME PROPERTIES OF $Tl_{0.4}Pb_{0.5}Sn_{0.2}Sr_2Ca_{0.8}Y_{0.2}Cu_2O_{7-\delta}$ SUPERCONDUCTORS

Hong Ma, Yuandong Dai, Lin Wang, Tong Chen and Shousheng Yan
Department of Physics, Peking University, Beijing 100871, P.R. China

Yufen Chen and Dingyuan Zhou
Department of Chemistry, Peking University, Beijing 100871, P.R. China

The structure of $(Tl_{0.5}Pb_{0.5})Sr_2CaCu_2O_y$ was stabilized by partial substitution of Ca by Y(+3) and (Tl,Pb) by Sn(+4). In a sample with nominal composition $Tl_{0.4}Pb_{0.5}Sn_{0.2}Sr_2Ca_{0.8}Y_{0.2}Cu_2O_{7-\delta}$, the obtained T_c(R=0) is 103K. The behavior of thermopower and magnetic irreversibility is close to those of $Tl_2Ba_2Ca_2Cu_3O_y$. In view of the $YBa_2Cu_3O_7$-like structure of the sample, our results imply that the situation between CuO_2 planes has significant influence on the properties of the oxide superconductors.

The thallium superconductors not only have high transition temperatures, but also make us understand the high Tc superconductivity further by comparing their properties with yttrium system's. Subramanian et al. (1988) reported that the Tc of $(Tl_{0.5}Pb_{0.5})Sr_2CaCu_2O_y$ with partial substitution of Pb for Tl, is about 85K. The crystal structure of this compound is very similar to $YBa_2Cu_3O_{7-\delta}$ (Y-123). They both have two CuO_2 planes which are mainly responsible for superconductivity. The difference is the substitution of (Tl,Pb)O layers for Cu-O chains. Therefore, comparing their properties will be beneficial to reveal the influence of that between CuO_2 layers.

The structure of $(Tl_{0.5}Pb_{0.5})Sr_2CaCu_2O_y$ might not be very stable, because the plus valence is 12.5 while the minus one is 14 when all the oxygen sites are fully occupied and all the positive ions are in normal valence state. Even if there is a little oxygen deficiency like that in Y-123, the whole minus valence is still greater than the plus one. The structure maybe stabilized by doping ions with higher plus valence. Liu et al.(1989) reported that the transition temperature can be raised by partial substitution of Y(+3) for Ca(+2). We further partially substitute Sn(+4) for (Tl,Pb) and obtained samples with Tc(R=0)=103K (Chen et al., 1990). The nominal composition is $(Tl_{0.4}Pb_{0.5}Sn_{0.2})Sr_2(Ca_{0.8}Y_{0.2})Cu_2O_{7-\delta}$. During the sample preparing process, we found that Tl is more stable as expected. The R(T) curve is very sensitive to oxygen content, and the samples show oxygen reversibility effect, just like that in Y-123. The occupation fraction of oxygen in (Tl,Pb,Sn)O layers has certain range beyond which the transition temperature can't be high. This is different from Y-123, where the more the oxygen is absorbed during the cooling process, the higher orthorhomicity the structure is, hence the higher the Tc is. Thus the cooling condition must be carefully controlled in order to obtain the highest Tc.

The irreversibility behavior of this sample was determined by measuring the zero-field-cooled and field-cooled magnetization. Figure 1 shows two typical results. At lower field (H=800 Oe), the irreversibility point T^* is about 89.5K, closing to Tc; at higher field (H=2T), T^* is much lower than

0953-2048/91/01S418 + 03 $03.50

Tc, approximately 34K. However, the T^* of Y-123 is always close to Tc and is not so sensitive to the field. Therefore the irreversibility behavior of this sample is roughly identical with the typical behavior of Tl-system, but greatly different from Y-system's. These results indicate that the situation between CuO_2 layers has greet influence on the magnetic properties. Recently, based on the experimental results in Bi-system, de Rango et al. (1990) suggested that the magnetic irreversibility can be interpreted by the existence of a breakdown field at which a proximity effect induced superconductivity is destroyed. The structure of oxide superconductors can be considered as superlattice composed of superconducting layers and weakly superconducting layers or normal layers. There may be a kind of correlation between the layers due to the proximity effect. At a certain field, the irreversibility point T^* correspond to the breakdown of this proximity effect. The vertices correlation between layers then disappeared and the flux flow phenomenon appeared. According to this theory, the differences result from properties of the block between CuO_2 layers, which is weakly superconductive in Y-123 case and normal in this sample. The normal (Tl,Pb,Sn)O layers greatly reduced the correlation between CuO_2 layers. The correlation is easily destroyed and T^* is much lower than Tc.

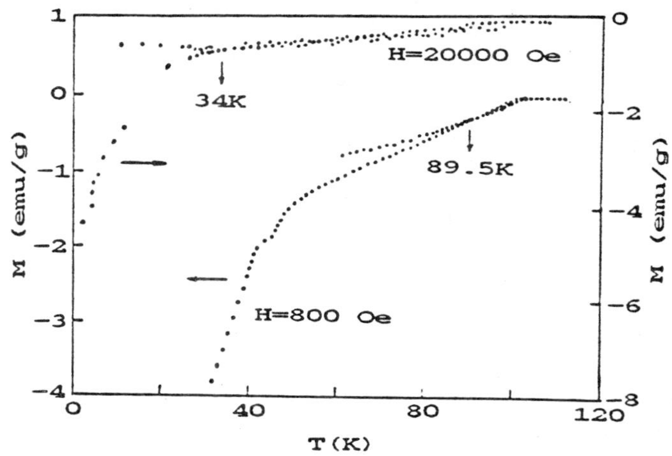

Fig. 1 Irreversible behavior of the sample used in this work at two different applied fields. The value of T^* is 34K and 89.5K respectively.

Figure 2 shows thermopower (S) versus temperature (T) curve of this sample (curve a). For comparison, the thermopower of ab plane in Tl-2223 (curve b, after Zhang et al., 1990) and Y-123 (curve c, after Ma et al., 1989) are also presented here. The thermopower of ceramic sample is usually close to that of corresponding ab plane. It can be seen that the S(T) behavior of this sample is similar to that of Tl-2223, but very different from that of Y-123. It is noticed that the S(T) of different oxide superconductors can be well described using the formula $S = A + BT$. The natural explanation of this is that the constant term A can be considered as the contribution of localized carriers and the linear term BT is the contribution of mobile carriers. Fitting the experimental results, we get $A=8.86\mu V/K$, $B=-1.85 \times 10^{-2}$ $\mu V/K^2$. According to Chaikin and Beni's discussion (1976), when K_BT is much less than the on-site Coulomb interaction, S can be written as $S = (k_B/e)\ln[x/2(1-x)]$, where x is the number of electrons per site. From A's value, the filling factor x is determined as 0.689 electrons per site corresponding to 0.311 holes per Cu-site. For Y-123 sample, x is about 0.332 holes per Cu-site. It is only a little greater. If we suppose that the localized carriers come from Cu^{+2} as it was pointed out by Johnston (1989) in La-Sr-Cu-O, it seems that the change arose from the substitution of (Tl,Pb,Sn)O layer for Cu-O chains is not very great, only that the number of localized carriers decreases a little. As to the linear part of

S(T), it is qualitatively close to that predicated by Allen, Pikett and Krakaner (1988) using the model of $\tau(\mathcal{E})$=const. For Y-123, looking at the S(T) behavior at high temperature region, it changes with temperature linearly, having a very small slope about $-7.77\times10^{-4}\mu V/K^2$. According to nearly free electron model, the greater the S's absolute value is, the lower the carrier density is. Our results seem to indicate that the mobile carrier density for this sample is nearly the same as that of Tl-2223, but much lower than that of Y-123(in order of magnitude). If the mobile carriers come from the vacancy at oxygen sites in CuO_2 layers, it is puzzling that although these two samples both have two CuO_2 planes, the density of mobile carriers has such a great difference. Perhaps the difference cannot be simply analyzed by nearly free electron model. The difference might come from the band structure or the different energy-dependent relations of the relaxation time, which might be affected by the substitution of (Tl,Pb,Sn)O layers for Cu-O chains.

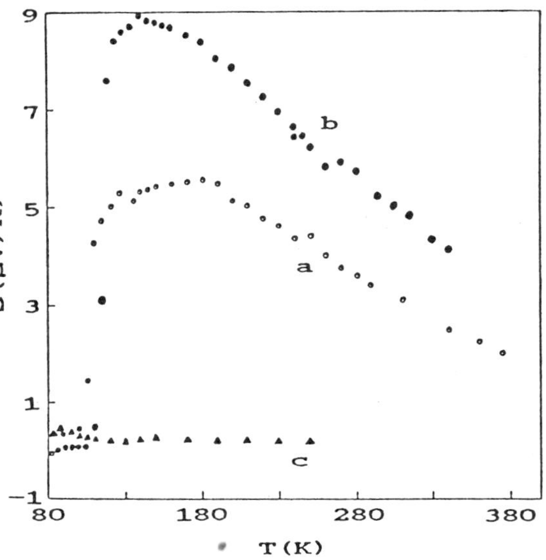

Fig. 2 Thermopower vs. temperature of the sample used in this work (a), and, for comparison, the c-axis-oriented Tl-2223 film (b, after Zhang et al., 1990) as well as the ab plane of Y-123 (c, after Ma et al., 1989).

To sum up, it is found that the irreversibility and thermopower behavior of $(Tl_{0.4}Pb_{0.5}Sn_{0.2})Sn_2(Ca_{0.8}Y_{0.2})Cu_2O_{7-x}$ are very similar to those of Tl-2223 and Bi-system, but very different from those of Y-123. Since both of them have similar structure, only that the (Tl,Pb,Sn)O layers substitute for Cu-O chains. The result seems to indicate that the situation between CuO_2 layers has great influence on properties of the high Tc superconducting materials although their superconductivity is mainly determined by CuO_2 layers.

Acknowledgments - This work is supported by the Chinese National Center for Research and Development on Superconductivity. The authors are grateful to many cooperators for their help and interest.

References

Allen P B, Pickett W E and Krakauer H 1988 *Phys. Rev. B37 7482*

Chaikin P M and Beni G 1976 *Phys. Rev. B13 647*

Chen Y F et al. 1990 *preprint*

de Rango P et al. 1990 *preprint*

Johnston D C 1989 *Phys. Rev. Lett. 62 957*

Liu R S et al. 1989 *Jpn. J. Appl. Phys. **28** 1179*

Ma H et al. 1989 *Phys. Rev. **B40** 9374*

Muller K A and Takashige M et al. 1987 *Phys. Rev. Lett. **58** 1143*

Subramanian M A et al. 1988 *Science **242** 249*

Zhang H et al. 1990 in *Proceedings of 19th Inter. Conf. on Low temp. Phys.*

Supercond. Sci. Technol. **4** (1991) S421–423. Printed in the UK

The Effect of Ln Substitution for Ca in $Bi_2Sr_2CaCu_2O_{8+y}$

R. Yoshizaki[+], J. Fujikami, M. Akamatsu and H. Ikeda[+]

Institute of Applied Physics, [+]also Cryogenics Center,
University of Tsukuba, Tsukuba, Ibaraki 305, Japan

Abstract. The dependence of Tc on the substitution of lanthanide (Ln) for 20% Ca was investigated for the 2212-phase BSCCO with maintaining the maximum Tc condition. The Tc increases with decreasing the ionic radii of the Ln atoms.

1. Introduction

One of the key parameters which determine the transition temperature of the high-Tc superconductors is hole concentration. In order to study substitution effect, it must be desired to keep hole concentration at the maximum Tc condition. One of such experiments was carried out in YBCO; full substitution of Ln for Y (Asano et al. 1987). This substitution gives rise to a distinct but small change of Tc, although the substitution causes the change of lattice parameters and the Cu(2)-O-Cu(2) length in the CuO_2 plane.

In the present study, we substituted Ln or Y for Ca in the 2212-phase BSCCO. The amount of the substitution was fixed to 20%, since it gave the maximum Tc of the system. Systematic changes of Tc and the lattice parameters were observed. The results are compared with those of YBCO because of the similar, double pyramidal structures of YBCO and BSCCO.

2. Experiments

The samples were prepared by employing solid-phase reaction technique. Powders of Bi_2O_3, $SrCO_3$, $CaCO_3$, CuO and Ln_2O_3 were mixed with the stoichiometry of $Bi_2Sr_2Ca_{1-x}Ln_xCu_2O_y$ (Ln=La, Nd, Eu, Gd, Dy, Ho or Yb, or Y) and calcined at 820 °C for 24h. The products were ground and sintered at 880 °C for 24h. They were reground and formed into pellets. The second sintering was done at 890-903 °C for 100h. The sintering temperature was chosen as the lower temperature for the lighter Ln elements.

X-ray powder diffraction measurement was performed for each sample, and the Rietveld analysis of the spectrum ensured it single phase of 2212. Superconducting properties of the samples were obtained from the magnetic measurement by using a SQUID magnetometer. The amount of oxygen , 8+y, was estimated from an iodometric titration method.

3. Results and Discussion

Hole concentrations of the as-sintered BSCCO are so-called overdoped because of including excess oxygens of y=0.3 in the system. The maximum Tc was obtained from the substitution of Y for Ca by 20% (Yoshizaki et al. 1989). In the present systems, we confirmed that the maximum Tc was also obtained by 20% substitution for the small atom Y and the large one La as shown in Figure 1. Thereby, we fixed the Ln content at x=0.2. Since the hole concentration is a function of oxygen content, we confirmed the amount of oxygens for the samples to be constant. The result is represented by triangles in Figure 2. The oxygen contents are almost constant around 8.26 for the entire samples. Accordingly, the hole densities per Cu are in the range of 0.16 as shown by circles in Figure 2. Thus we conclude that the present samples have an appropriate hole concentration satisfying the maximum Tc condition.

The diamagnetic susceptibility at 5 K in the field cooling process was in the range of -5×10^{-3} emu/g, and the bulk superconductivity was attained for the entire samples. The superconducting transition is sharp enough to estimate Tc with less ambiguity. We defined Tc from the temperature corresponding to -1×10^{-4} emu/g and determined Tc within the error of ±0.5 K taking into account of the reproducibility.

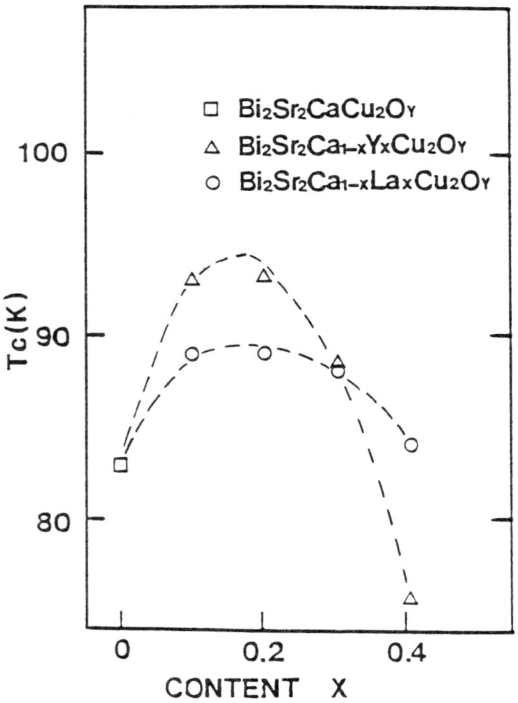

Fig. 1 Tc vs. x for the substitution of Y or La for Ca.

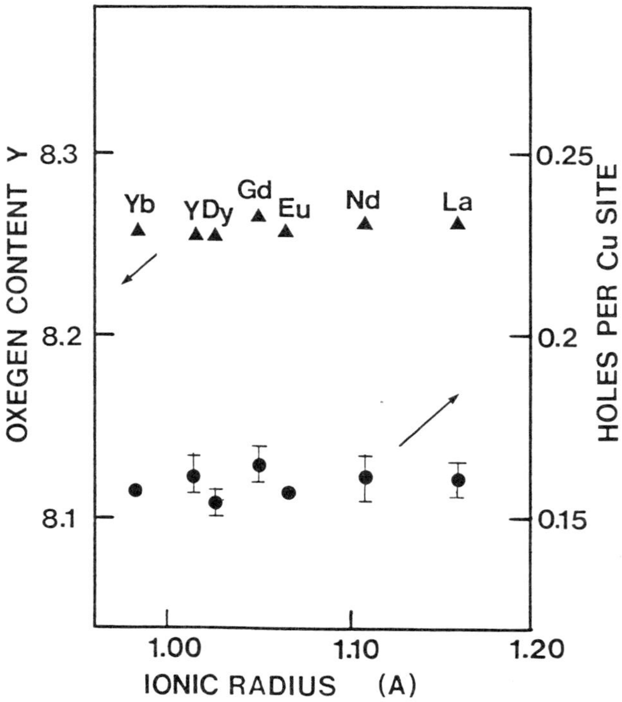

Fig. 2 Oxygen content (triangles) is plotted against ionic radius of Ln. Hole densities are shown by circles.

The obtained Tc is plotted by solid squares as a function of the ionic radius of Ln or Y (Shannon 1976) in Figure 3. Systematic change of Tc is observed; Tc increases with decreasing the ionic radius of Ln or Y. When we consider the origin of this change, it is worthy to compare it with the results of the substituted YBCO system, in which entire Y are replace by lanthanides. The variation of Tc for the YBCO system is represented by squares in Figure 3 (Yoshizaki et al. 1990). In this system, the tendency of Tc against the ionic radius is similar to those for BSCCO for the ions bigger than Gd. From the structural point of view, the Y site is believed to be stable for the cations with the larger ionic radius than 1.025 A, since those ions can fulfill the Y-site space and contact with the neighbouring eight oxygen atoms. In the case of BSCCO, on the other hand, the replacement was performed only 20%, and the mean ionic size of $Ca_{0.8}Ln_{0.2}$ satisfy the stability condition mentioned above for the entire lanthanides.

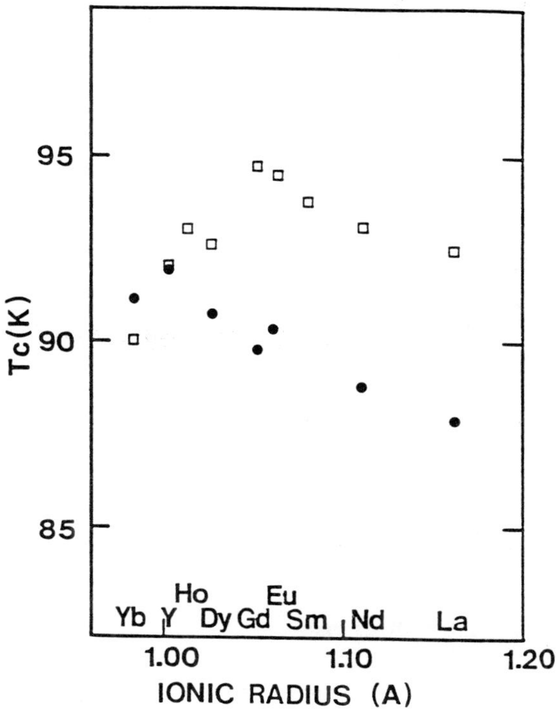

Fig. 3 Tc is plotted as a function of ionic radius of Ln for BSCCO (circles) and YBCO (squares).

Structural analysis of the Ln substituted BSCCO suggests that the Cu-Cu bond length of the CuO_2 plane and the distance of the planes along the c axis are correlated to the Tc; Those lengths decrease with decreasing the ionic radius of Ln. When we compare the results in BSCCO and YBCO, local shrinkage of the CuO_2 planes around the Ln atoms would affect Tc probably.

Acknowledgement

This work is partly supported by the Grant-in-Aid for Science Research on Priority Areas, "Mechanism of Superconductivity" from the Ministry of Education, Science and Culture.

References

Asano H, Takita K, Katoh H, Akinaga H, Ishigaki T, Nishino M, Imai M and Masuda K 1987 Jpn. J. Appl. Phys. **26** L1410

Shannon R D 1976 Acta Cryst. **A32** 751

Yoshizaki R, Kurahashi H, Ishikawa N, Akamatsu N, Fujikami J and Ikeda H 1989 Proceedings of the Tsukuba Seminar on High Tc Superconductivity (Tsukuba, Tsukuba University) pp 21-26

Yoshizaki R, Fujikami J, Ishigaki T and Asano H 1990 submitted to Physica C

Supercond. Sci. Technol. 4 (1991) S424–426. Printed in the UK

Superconducting properties of $Bi_2(Sr,Ca)_3Cu_2O_y$ investigated by electromagnetic field in microwave and radio-frequency regions

A. Maeda, T. Shibauchi and K. Uchinokura

Department of Applied Physics, The University of Tokyo, 7-3-1, Hongo, Bunkyo-ku, Tokyo 113, JAPAN

Abstract. The temperature dependence of the surface impedance at 9 GHz and 24 GHz were investigated in single crystals of $Bi_2(Sr,Ca)_3Cu_2O_y$, together with the temperature dependence of the resonant frequency of an LC resonator including the crystals at around 20 MHz. From these results, the magnitude and the temperature dependence of superconducting energy gap and those of the penetration depth were studied.

1. Introduction

The superconducting properties of the high-temperature superconductors based on copper oxides have been studied extensively, because they are closely related to the novel mechanism of high-temperature superconductivity. For this purpose, various efforts using many techniques have been devoted to the determination of the magnitude and the temperature dependence of the microscopic parameters which describe superconductivity.

Measurement of the surface impedance in the microwave region was one of the powerful tools to study the superconducting energy gap and the magnetic-field-penetration depth, at least for conventional superconductors (Waldram 1964). In high-T_c superconductors, owing to the strong two-dimensionality of the electric conduction, superconducting properties are expected to be fairly different from those in conventional BCS superconductors (Bardeen *et al* 1957).

In this paper, we studied the temperature dependence of the surface impedance Z_s in single crystals of $Bi_2(Sr,Ca)_3Cu_2O_y$ (BSC-2212), which is thought to be the most "two-dimensional" among various high-T_c oxides, and tried to deduce the magnitude and the temperature dependence of the superconducting energy gap Δ and the magnetic-field penetration depth λ. For more precise determination of the temperature dependence of λ, an rf resonance method was also used. The results were compared with several theoretical predictions, including a model which regards high-T_c oxides as a stacking of the two-dimensional superconducting sheets (Ye *et al* 1990).

2. Experiments

The surface impedance at 9 GHz and 24 GHz was measured by the transmission-cavity-resonance method, using HP8340B sweeper and HP8757C network analyzer. A TE113 brass cavity was used, and the surface of the cavity was coated with silver. The sample was placed at the center of one of the cavity walls (Kobayashi *et al* 1987). The surface impedance of the sample Z_s was approximately obtained as the difference between the surface impedance of

0953-2048/91/01S424 + 03 $03.50

the cavity including sample Z and that without sample Z_0. That is, $Z_s = R_s + iX_s = Z - Z_0 = K\{(Q^{-1}-Q_0^{-1})+2i(f-f_0)/f_0\}$ where Q and Q_0 are the quality factors of the cavity at the resonance frequency with and without the sample, respectively, f and f_0 are the resonance frequencies of the cavity with and without the sample, respectively, and K is a geometrical factor which can be calculated easily. As for X_s, f_0 is linear in temperature near the superconducting transition temperature T_c. Thus, we regard the deviation of f(T) from the linear part as the contribution from a superconductor. The change in the penetration depth λ was related to the deviation of f from the linear part as $\delta\lambda = K\pi\mu(\delta f)$, where μ is the permeability. For precise determination of the temperature dependence of λ, the resonance frequency of an LC resonating circuit including a sample was measured using a HP4194 gain-phase analyzer in the radio-frequency range. The circuit was driven by an ultra-stable HP3325B oscillator. With this system, we can measure the resonance frequency with an accuracy of 10^{-7} at around 20 MHz. The estimation of λ was made using the same formula shown above with a different geometrical factor. In this case, although the temperature dependences of f_0 and f are complicated, the reproducibility is excellent. Thus, subtraction was performed easily.

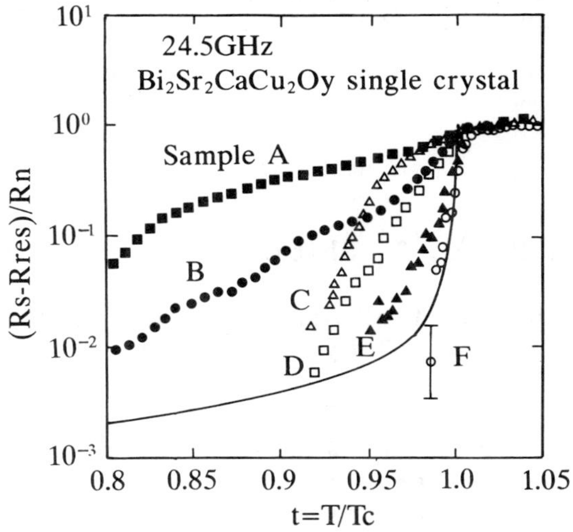

Figure 1 The temperature dependence of the normalized R_S at 24 GHz for 6 BSC-2212 crystals. The solid curve represents the BCS theoretical prediction.

3. Experimental Results and Discussions

Figure 1 shows the temperature dependence of the surface resistance R_s of 6 samples of BSC2212-single crystals. All data are normalized to the R_s at the superconducting transition temperature T_c, and the residual surface resistance R_{res} which is independent of temperature has been subtracted. Because of the strong sample-dependence, we are obviously not in the final stage of determining firmly the temperature dependence of the R_s and deducing that of energy gap of this material. But the temperature dependence of the normalized R_s of the sample with the sharpest transition is similar to that expected by the BCS theory in the local limit. We also measured the temperature dependence of the frequency shift which is proportional to the surface reactance X_s of the sample with the sharpest superconducting transition (Fig. 2), and obtained λ_{ab}(0 K) of about 2000 Å (Fig. 3)(The suffix represents the direction of the

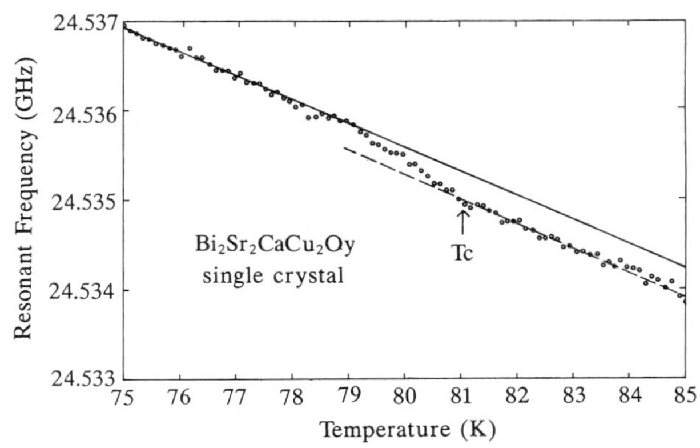

Figure 2 The temperature dependence of the resonance frequency of the cavity including a BSC-2212 crystal. The solid and dashed lines are guides for eye.

shielding current).

Figure 4 shows the temperature dependence of the difference in the normalized resonance frequency shift $\{(\delta f)/f\}$ in the radio-frequency region, which is proportional to the change in λ_{ab} ($\delta \lambda_{ab}$). Using this plot, we can determine the temperature dependence of $\lambda_{ab}(T)/\lambda_{ab}(0\ K)$. The fit to the several theoretical curves shows that clear deviations from the BCS or empirical two-fluid formula (Pippard 1948) exist at low temperatures, whereas the multilayer model (Ye *et al* 1990) gives a good fit down to low temperatures. This is in sharp contrast to the results obtained in $YBa_2Cu_3O_y$ (YBC) where the BCS formula is thought to hold well (Krusin-Elbaum *et al* 1989, Sridhar *et al* 1989). The origin of the difference may be due to the difference in the degree of two dimensionality. We also measured the frequency shift in different directions and tried to obtain $\lambda_c(T)$. In the case of λ_{ab}, the complicated geometrical problem often prevents us from obtaining the magnitude of $\lambda_{ab}(0\ K)$, whereas in the case of $\lambda_c(0\ K)$ we do not face such a problem. At present the temperature dependence measured is, however, far from those expected by the models introduced above, which also prevents us determining $\lambda_c(0\ K)$. Further studies are necessary.

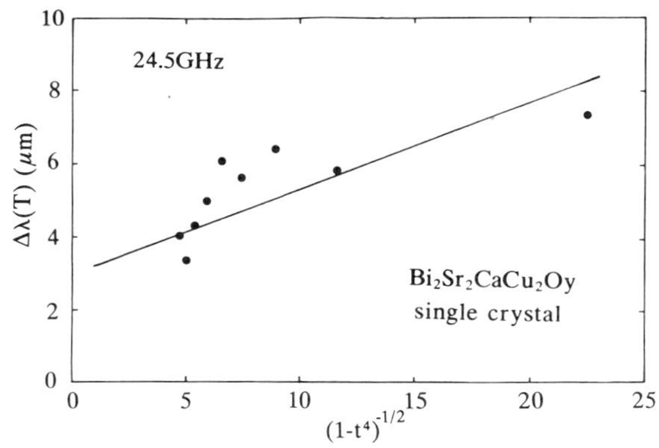

Figure 3 The fit of the data in Fig. 2 near T_c to the two-fluid formula.

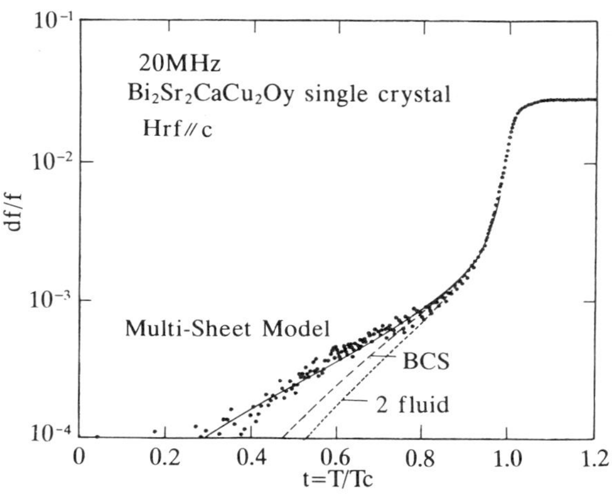

Figure 4 The temperature dependence of the change in the resonating frequency at 21 MHz when the rf field is perpendicular to the CuO_2 plane. The curves represent three formulae. See the text for detail.

4. Conclusion

We investigated the electrodynamic properties of BSC-2212 single crystals. The penetration depth λ_{ab} is well fitted by the multilayered model and the magnitude of $\lambda_{ab}(0\ K)$ was found to be about 2000 Å. This result suggests that the BSC-2212 is strongly two-dimensional. We believe this strong two-dimensional nature is common to all other high-T_c oxides. Thus, we can expect many interesting properties in the superconducting state, which deserves extensive investigation.

References

Bardeen J, Cooper L N and Schrieffer J R 1957 *Phys. Rev.* 108 1175

Kobayashi M, Okabe K, Tanuma S, Sankawa I, Sato M, Konaka T and Ishihara K 1989 *IEEE Trans. Mag.* 25 997

Krusin-Elbaum, Greene R L, Holzberg F, Malozemoff A P and Yeshurun Y 1989 *Phys. Rev. Lett.* 62 217

Pippard A B 1948 *Nature* 162 68

Sridhar S, Wu D H and Kennedy W 1989 *Phys. Rev. Lett.* 63 1873

Waldram J R 1964 *Adv. Phys.* 13 1

Ye Z, Umezawa H and Teshima R 1990 *Solid State Commun.* 74 1327

Supercond. Sci. Technol. **4** (1991) S427–429. Printed in the UK

Microwave Surface Resistance in High-T_c Superconducting Ceramics

M. Sato & T. Konaka
NTT Transmission Systems Laboratories, Tokai, Naka, Ibaraki, 319-11, Japan

Abstract To clarify the origins of high microwave surface resistance R_s in high-T_c superconducting ceramics, the R_s of different sized grain samples (from μm to mm) are measured as a function of temperature and magnetic field using the resonant cavity method. It is demonstrated that the profiles of the R_s-T and the R_s-H curve strongly depend on the grain size. Since T and H dependence becomes greater as the grain size is decreased, it is concluded that the R_s of ceramics is mainly caused by weak-links located at the grain boundaries.

1. Introduction

Based on the BCS theory, high-T_c superconductors are estimated to have a much lower R_s in the microwave frequency range than ordinary low-T_c superconductors because of their high critical temperatures (Kennedy 1988). Therefore high-T_c superconductors are expected to be suitable for microwave applications. However, it has been reported that the R_s of ceramic samples is higher than that of copper above about 10GHz at liquid N_2 temperature (Kennedy 1989). This high R_s in ceramics prevents large size applications, for example, filters, cavity resonators and antennas. In this paper, we report the grain size dependence of R_s-T and R_s-H curves and demonstrate that the high R_s in ceramic samples is caused by weak links located at the grain boundaries.

2. Sample preparation

Three kinds of $YBa_2Cu_3O_y$ ceramic samples with very different sized grains were prepared; the small sized grain samples (~μm in typical grain diameter) were prepared using the coprecipitation method; the middle-sized grain samples (~10μm) were prepared by an ordinary mixed powder method; the large-sized grain samples (~mm) were prepared with a partial melt method (Salama 1989). The sintered ceramics were shaped into an appropriate size of about 3mm in thickness and 26mm in diameter for measurement. One side of each disk was polished with polishing sheets to achieve a flat surface and then washed in alcohol. Before measurement, the samples were reannealed at 950°C in an O_2 flow for 5 hours and cooled slowly at a rate of 0.5°C/min.

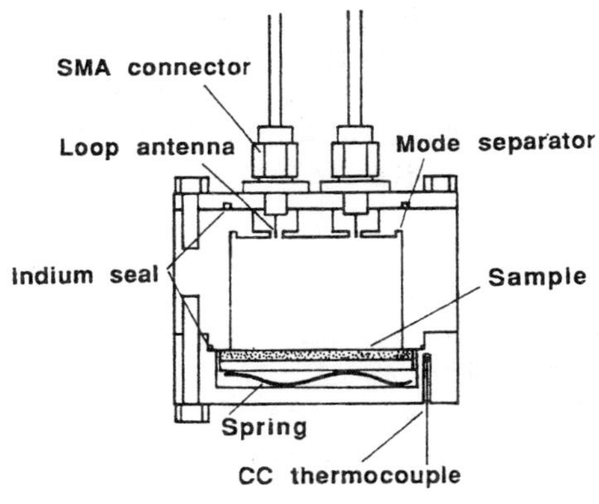

Fig. 1 The cylindrical TE_{011} mode cavity with resonant frequency of 17GHz.

3. Measurement

Surface resistance was measured using the resonant cavity method with a transmission type gold-plated copper cylindrical TE_{011} mode 17GHz cavity. Figure 1 shows the schematic of the cavity with a sample substituted for one end plate. The microwave field is excited and

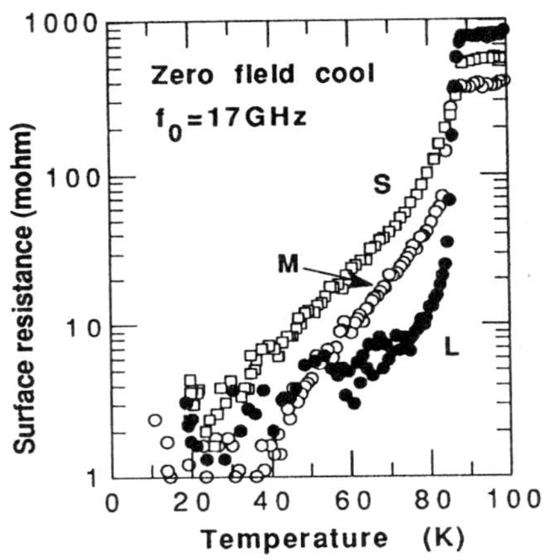

Fig. 2 Temperature dependences of zero-field-cool R_S in the samples; (S) small sized grain sample (~μm), (M) middle sized grain sample (~10μm) and (L) large sized grain sample (~mm).

received by small loop antennas located on the SMA connectors. An external static magnetic field (<225Oe) is applied perpendicularly to the sample surface. The cavity with the sample is cooled (>20K) by inserting it into a liquid He vessel.

4. Results

Figure 2 shows the R_S-T curves of the three different grain sized samples. Above T_c, the R_S of the samples shows the same linear temperature dependence. In the low temperature region below T_c, however, the manner in which the R_S decreases is quite different for each sample. It should be noted that the sharp R_S drop near T_c becomes large when the grain size is large and the slope of R_S-T steepens as the grain size decreases.

The R_S-T curves under a magnetic field are more informative. Figure 3 shows the R_S-T curves observed under a magnetic field and Fig. 4 shows the R_S-H curves for the samples. It is interesting that the degree of sharp R_S drop near T_c is not reduced by the magnetic field and the slope of $\log R_S$-T in the low temperature region becomes small. While the increase of R_S induced by the magnetic field is large in the small-sized grain sample, no noticeable R_S increase is observed in the large-sized grain sample. Furthermore, as shown in Fig. 4, the field induced R_S is mainly brought about by a low magnetic field which is lower than H_{c1} of superconducting grains (Sridhar 1989).

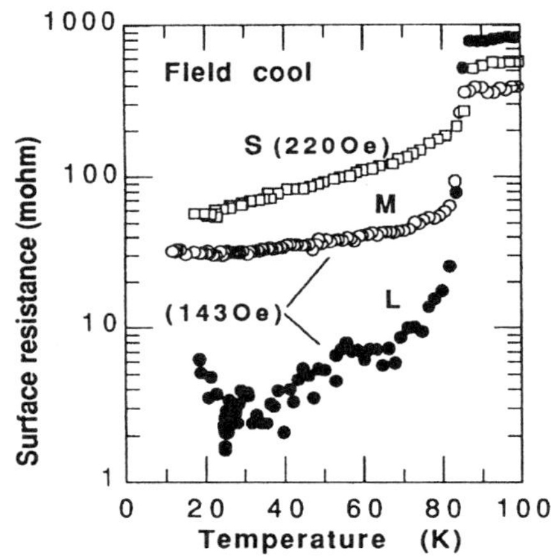

Fig. 3 Temperature dependences of field-cool R_S in the three different grain sized samples.

5. Discussion

Because the field induced R_S becomes high when the grain size is decreased, and because decreasing the grain size makes the boundary density high, this R_S increase is concluded to be a property of the grain boundaries. Furthermore, because the weak links between the grains are broken by a low magnetic field because of their low Jc, the field induced R_S is considered to be brought about by weak link breaking. It is expected that the influence of the weak links is strong in the small sized grain sample, but not in the large sized grain sample. Therefore the steep gradient in R_S-T of the small sized grain sample is understood to represent the temperature dependence of the weak link R_S. The weak temperature dependence in the large sized grain sample is assumed to be caused by other phases because of its metallic temperature dependence and its lack of field induced R_S.

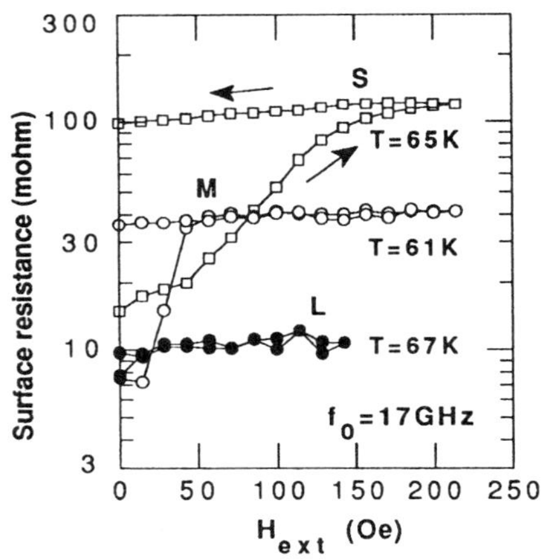

Fig. 4 Magnetic field dependences of R_S under initial field application after zero-field cooling.

On the other hand, because the rapid R_S drop near T_c becomes large when the grain size is increased, and because this drop is not influenced by a magnetic field, it is concluded that the R_S drop is a property of the grain R_S. It is also expected, from the resonant frequency shift of the cavity, that the R_S near T_c comes from the grain properties. Figure 5 shows the resonant frequencies of the cavities with the samples as a function of temperature. The frequency shifts observed near T_c are considered to be caused by the shortening of the penetration depth from the metallic skin depth to the superconducting penetration depth λ. The λs calculated from the frequency shift and R_S in nornal state are 4.2, 1.4 and 4.6μm for the sample S, M and L, respectively. Because the effective λ is the average of λ_{grain} and $\lambda_{weak\text{-}link}$ ($>>\lambda_{grain}$), the short λ of the middle sized grain sample represents the reduced influence of the weak links. The large sized grain sample has the long λ, however, it is thought to be caused by the intragrain other phases or cracks, not by the weak links, since no large R_S increase induced by a field is observed below T_c.

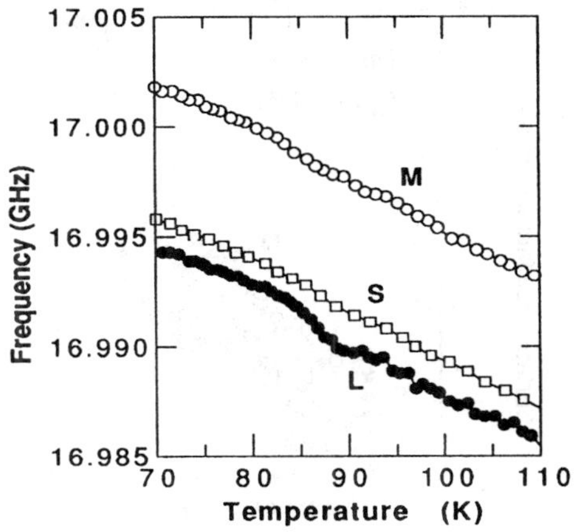

Fig. 5 Temperature dependences of resonant frequency of the cavities with the samples.

As mentioned above, it is understood that the main origin of R_S in ceramic samples is different for samples with different grain sizes and in different temperature regions. We propose that the ceramic R_S is presented as a summation of R_{Sgrain}, $R_{Sweak\text{-}link}$ and $R_{Sothers}$ (other phases and surface imperfections), as shown in Fig. 6. Moreover it is expected that the origin is different in different frequency regions because of their different frequency dependences. While the R_S of a superconductor is proportional to $\omega^{1.5\sim2.0}$, that of metal is proportional to $\omega^{0.5}$. Therefore, it is expected that in high frequency regions the weak link R_S or the grain R_S is dominant, and that in low frequency regions the R_S of other phases (metallic) is dominant.

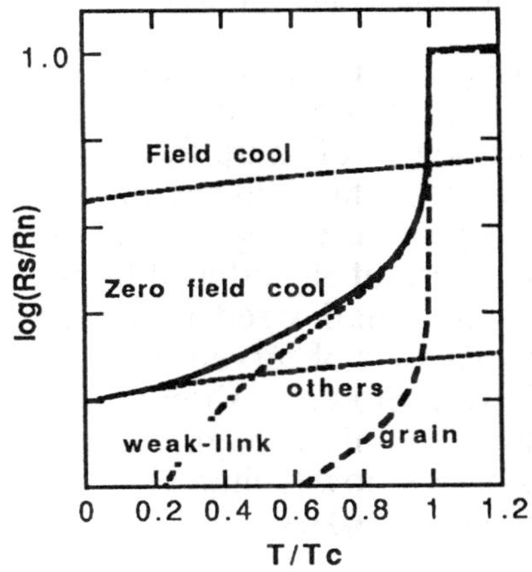

Fig. 6 The R_S model for high-T_c super-conducting ceramics.

Based on our model, to obtain a low R_S at liquid N_2 temperature, it is necessary to increase the grain size to reduce the $R_{Sweak\text{-}link}$. We have obtained an R_S lower than that of Cu at liquid N_2 temperature with large-sized grain samples. However, the problem of reducing $R_{Sweak\text{-}link}$ is replaced by the problem of reducing $R_{Sothers}$ which is caused by cracks in the large-sized grain samples. Furthermore it should be noticed that the R_S-T curves of ceramic samples cannot be used to estimate the energy gap without recognizing each R_S component.

6. Conclusions

We demonstrated that the profiles of the R_S-T curve and the R_S-H curve strongly depend on the sample grain size. It is concluded that the origin of the high R_S in high-T_c superconducting ceramics is the R_S of the weak links located at grain boundaries. We proposed a model in which the R_S of the ceramic samples is composed of R_{Sgrain}, $R_{Sweak\text{-}link}$ and $R_{Sothers}$.

References

Kennedy W L and Sridhar S 1988 Solid State Commun. **68** 71
Kennedy W L, Zahopoulos C and Sridhar S 1989 Solid State Commun. **70** 741
Salama K, Selvamanickam V, Gao L and Sun K 1989 Appl. Phys. Lett. **54** 2352
Sridhar S, Wu D, Kennedy W 1989 Phys. Rev. Lett. **63** 1873

Supercond. Sci. Technol. **4** (1991) S430–432. Printed in the UK

Oscillations in the imaginary part of the susceptibility as a function of field upon excitation of superconductors by microwaves

Keshav N. Shrivastava

School of Physics
P.O. Central University, Hyderabad 500134, India.

Abstract. We find oscillations of frequency hc/2e in the imaginary part of the susceptibility as a function of the magneic field. The superconductor $ErBa_2Cu_3O_{7-\delta}$ upon excitation by microwaves of frequency 9.2 GHz shows well sustained oscillations as a function of magnetic field. The intensity of the spectral lines due to flux quantization varies with temperature showing a minimum at the flux-lattice melting temperature.

We wish to report our finding that the high-temperature superconductors when placed in the electron-spin resonance spectrometer with 100 kHz field modulation at an angle of $\pi/2$ radians from the external d.c. field, H, give oscillations in $d\chi''/dH$ as a function of H. These oscillations depend on size of the sample, modulation amplitude, rate of scanning and the temperature. We find that the oscillations are casued by Josephson currents in the superconductors which give rise to flux-quantized eigen states (Shrivastava 1988a,b). There is a wide variety in the shape, form and the number of oscillations which are observed. The intensity of these oscillations varies with temperature going to zero at T_c (Shrivastava 1987a). The flux motion affects the width and the intensity of oscillations so that when the Abrikosov flux-lattice melts. We find motional narrowing of lines and hence the flux-lattice melting temperature is found. The oscillations are time reversal invariant as is seen by scanning negative fields which makes d/dH also negative. In some spectra at 4.2 K the oscillations appear like noise which can be resolved by reducing the range of magnetic field sweep and increasing the sweep rate. Alternatively, increasing the temperature results into well resolved lines. The broad absorption line, $d\chi''/dH$ shows hysteresis, i.e., the absorption depends on the direction of the field sweep. The hysteresis loop area, $Hd\chi''/dH$, is almost negligible at 1.1 K but it increases with increasing temperature till a peak is reaching just below T_c. For temperatures close to T_c the intensity reduces and hence the area also reduces going to zero at T_c. In some of the real materials more than one chemical formula is present and the samples used for the microwave studies are multigrain bulk powders. With increasing temperature one of the formulas becomes normal while the other formula remains superconducting so that junctions of the superconducting and normal formulas are formed leading to very strong flux-quantized transitions. These lines are time dependent. If the field is switched off suddenly the lines persist for a few seconds with some of the lines surviving for as long as about one hundred seconds. There is a long tail in the number of lines per second as a function of time. The lines depend on the sweep rate of the magnetic field with time keeping the number of lines per second of the sweep time a constant. This time constant is characteristic of flux creep. The lines were first shown in 1987 by Shrivastava (1987b) and do not seem to have been published in the low critical temperature superconductors. In Fig.1 we show the microwave absorption, $d\chi''/dH$ as a function of the magnetic field,

0953-2048/91/01S430 + 03 $03.50

for several samples of $ErBa_2Cu_3O_{7-\delta}$ ($\delta \simeq 0.2$). There is a wide variety of oscillations, (a) a wide line, (b) single line, (c) wiggle like lines, (d) equally spaced lines, (e) hundreds of sharp lines, (f) bunching lines, (g) the time reversed lines, ...etc. In zero-filed cooled samples, the wide line shows hysteresis, $d\chi''/dH$ vs H but the hysteresis area is very small at a temperature of 4.2 K and increases with increasing temperature until a peak is reached beyond which it reduces becoming zero at T_c. When the sweep rate is properly adjusted hundreds of lines are seen as in Fig. 1(e). If only a small field range is scanned lines occur in several groups with oscillating intensities. Due to time reversal invariance, the same spectrum for reversed field should be obtained as for the forward field. When the sign of H is changed, the sign of $d\chi''/dH$ also changes, so that reverse field spectrum should occur in the third quadrant of coordinates. The wide line of Fig.1(a) does just that and the full line is seen by reversing the field. For the case of Fig.1(e) the field reversed lines are shown in Fig.1(g). They are time reversal invariant as they should be. In Fig.2 we show the hysteresis of the broad line with low gain of the spectrometer showing crossing of lines at large modulation amplitude and noncrossing at small modulation. It can be seen that small equally spaced lines are riding over the hysteretic background. There are two methods of seeing the equally spaced lines. One can scan rapidly at 4.2 K or vary the temperature until above 50 K or so, particularly near 77 K. Very sharp lines are seen characteristic of motional narrowing of magnetic resonance lines. The intensity shown in Fig.3 shows a shallow minimum as expected for a flux liquid at the flux-lattice melting temperature, T_{FM}. The depinning of flux lines also occurs below the T_c but the depinning temperature T_D is not resolved from T_{FM}.

The flux-quantized imaginary part

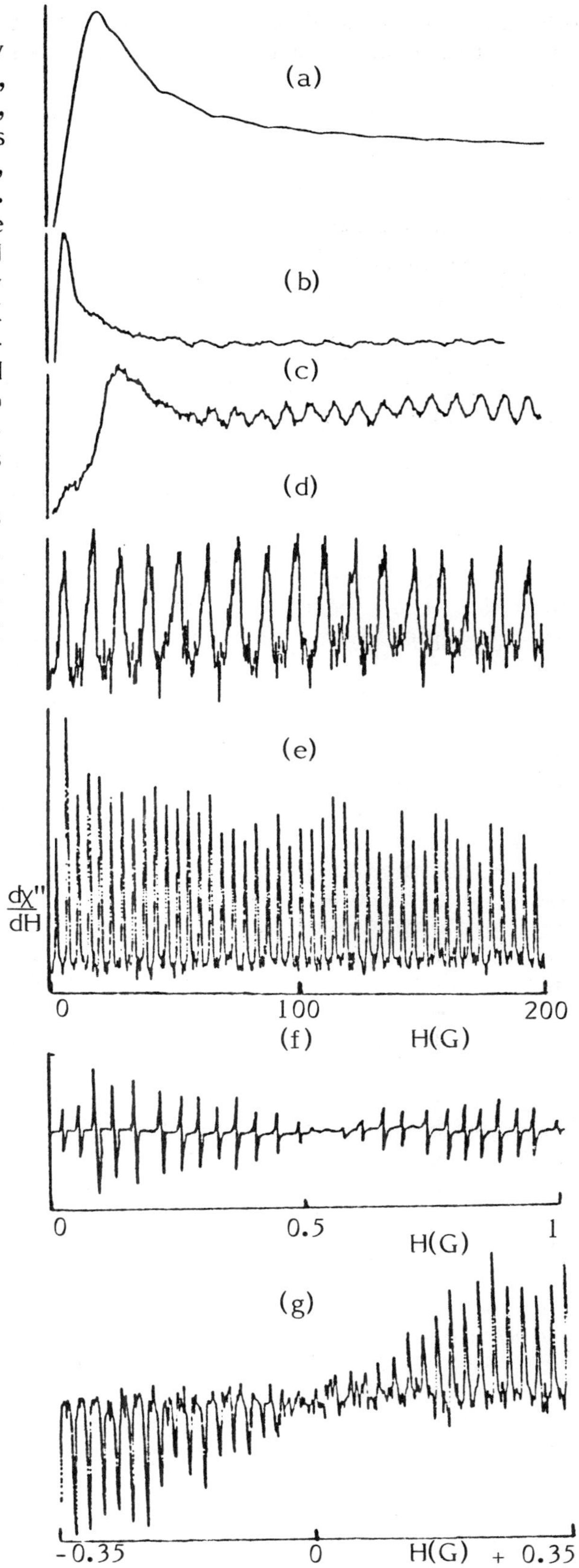

Fig.1. $d\chi''/dH$ of microwaves at 9.4 GHz for seven samples of $ErBa_2Cu_3O_{7-\delta}$ ($\delta \simeq 0.2$).

of the susceptibility is given by Shrivastava (1988a,b) as

$$\chi'' \sim \frac{\delta\Omega}{(\omega-\omega_n)^2-(\delta\Omega)^2} \qquad (1)$$

where n is an integer (n=0,1,2,....). The eigen values due to Josephson type current owing to Lorentz force are given by

$$\hbar\omega_n = E_n = nJS\phi \qquad (2)$$

There are equally spaced eigen values which cause equally spaced lines in $d\chi''/dH$ vs H. When the microwave frequency $h\nu$ equals the current, a large number of equally spaced resonances are excited. For ν =9.4 GHz, the current is estimated to be $J \sim 0.9\times10^3$ A/cm^2. In our samples of area S=0.01 cm^2, the relation

$$h\nu = n\,JS\phi \qquad (3)$$

is well obeyed. Unlike the Bitter pattern technique we do not have a static field nor we have any critical field beacuse of a finite rate of sweep of the magnetic field, dH/dt. Some of the lines in $d\chi''/dH$ show signs of emission similar to those found in the phenomenon of self-induced transparency so that the field depends on distance and time (Atsarkin et al 1990). In a flux liquid (Shrivastava 1990, Chakravarty et al 1990) we expect motional narrowing of the microwave absorption lines as found.

Atsarkin, V.A., et al, 1990, Solid State Commun. 74, 603.
Chakravarty, S , et al., 1990, Phys. Rev. Lett. 64, 3187.
Shrivastava, K N, 1987a, J. Phys. C 20, L 789 ; 1987 (b), Conference on High Temp. Supercond. 14-15 Dec. 1987, University of Bristol, (Inst. of Physics, London) Abs. No.P.30 (Edited by R. Evans, B.L. Gyorffy and J.A. Wilson)
1988a, Solid State Commun. 68, 259
1988b, Solid State Commun. 68, 1019
1990, Phys. Rev. B 41, 11168.

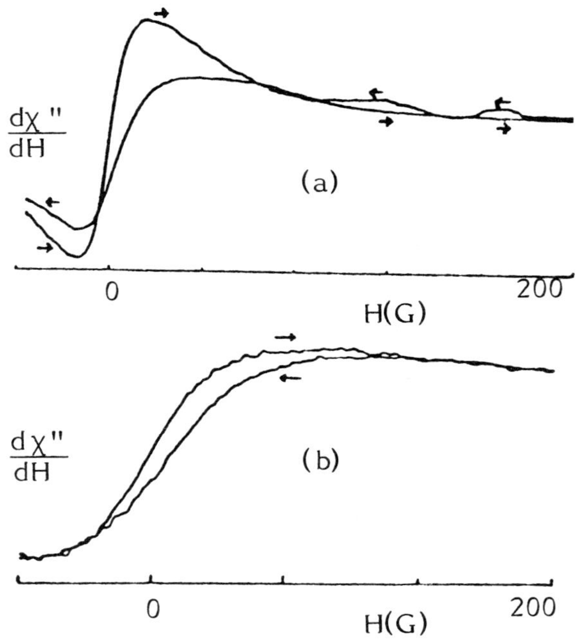

Fig.2. The broad line $d\chi''/dH$ vs H (a) large modulation, 5 G and (b) small modulation, 0.5 G.

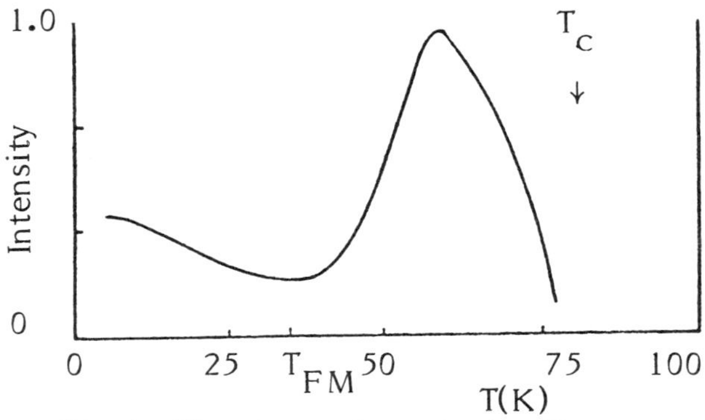

Fig.3. The intensity of the lines as a function of temperature showing the flux-lattice melting temperature T_{FM}. The existence of the flux liquid in the high T_c systems is thus quite clear. The depinning of the flux lines also occurs at T_D not resolved from T_{FM}.

Supercond. Sci. Technol. **4** (1991) S433–435. Printed in the UK

A Study of Dissipation Mechanisms in High–T_c Superconductors

Z. Przelozny, T.R. Finlayson, D.A. Garvie, D.B. Lowe and T.F. Smith

Department of Physics, Monash University, Clayton, Victoria, Australia. 3168

Abstract Magnetic losses and time dependence of magnetization for high–T_c, polycrystalline samples are being investigated. From magnetization data the flux–density–profile and critical–state–model parameters are obtained. The critical–state model has been modified to include the observed time dependence of magnetization which is described in terms of thermally activated flux–flow theory. The field and temperature dependences of the creep rate and the activation energy have been found and are discussed.

1. Introduction

The study of flux line configurations in high–T_c superconductors provides a unique opportunity to investigate the physical properties of these materials. In high–current applications, the most intriguing problems, such as the low critical current and the time decay of magnetization inside the superconducting grains (Salama et al. 1989), have to be solved. The limitation of transport current by the low value of junction tunneling current, the short coherence length (which result in the low activation energy) and high application temperatures, are the origins of these problems.

In recent work (Finlayson et al. 1990) we have explored the low–field critical state in YBaCuO polycrystalline samples using a.c. susceptibility and magnetization data. It is clear from these studies that the temperature scaling of the intergrain critical current is controlled by the coupling between grains and grain size. In this paper the study of the critical state is extended over a broad temperature range using low– and high–field magnetic loss measurements.

2. Pinning Forces

For fields less than the first penetration field of the superconducting grains, H_{c1g}, any magnetic irreversibility is due to weak links between grains. For fields higher than the first penetration field of these links, H_{c1J}, Josephson vortices enter into the junction. Since a Josephson vortex has no normal core, any pinning force, f_p, may be produced by a magnetic interaction of the form $\phi_0 M_{rev}/\lambda_J$, where ϕ_0, M_{rev} and λ_J are the flux quantum, reversible magnetization and Josephson penetration depth, respectively. For fields close to H_{c1J}, $M_{rev} \approx H_{c1J}$ and the pinning force becomes $\phi_0 j_m$. (j_m denotes the maximum intergrain current.) For grains of average radius r_g ($< \lambda_J$)

$$f_p = c j_m^{3/2} \tag{1}$$

where $c = r_g^{3/2}\phi_0^{1/2}/\ln(\lambda_J/r_g)$ (Finlayson et al. 1990). On adding a d.c. field, B,

$$f_p = (c/2) j_m^{3/2} \ln[\phi_0/(4Br_g^2)]. \tag{2}$$

0953-2048/91/01S433 + 03 $03.50

For fields higher than H_{c1g}, the intergranular field is uniform, and the polycrystalline sample behaves almost like a superconducting powder. The large differences in the irreversible behaviours for Bi and Y compounds (Yeshurun et al. 1989) suggest that twin boundaries contribute to the flux pinning with a force $\Delta M_{rev}/\lambda_L$ (Campbell and Evetts 1972), where ΔM_{rev} is the difference in magnetization (or H_{c1}) within the twinned unit cells and λ_L the London penetration depth. The second possible pinning mechanism is the core interaction described by $f_p = H_c^2\xi$, where ξ is the coherence length. For high–κ superconductors the second mechanism dominates. For both the magnetic and the core interactions the temperature dependence of the pinning force is determined by $f_p \sim (1 - T/T_c)^{3/2}$.

3. Magnetic Flux Profiles

A simple critical–state model is given by $j_c\phi_0 = k(T)H^n$. The exponent n specifies the flux profile. For a slab specimen of thickness, D, n can be found directly from the ratio of maximum flux density at fields $H_m \le H^*$, (at H^* the field penetration depth equals D/2), to the flux density trapped after decreasing the field from H_m to zero, from the expression, $B_{max}/B_{rem} = 2^{1/(1-n)}$.

We note that this does not fit the intragranular flux profile for polycrystalline high–T_c superconductors. The high anisotropy of the first penetration field and non–uniform demagnetization field, make the remanent flux less than that for an isotropic super–conductor. To study the flux profile of the intragranular material, the losses, given by $W = \int BdH$, must be determined. For $H_m < H^*$, $4\pi\chi'' = \mu_{eff}f(n)[2H_m/(j_cD)]^{(1-n)}$, where the imaginary component of susceptibility, χ'', equals W/H_m^2. μ_{eff} is the effective permeability and f(n) is a numerical function of the exponent n. The maxima for χ'' versus field correspond to the inter– and intra–granular fields, H^*_l and H^*_h, for low and high fields, respectively. The critical current is related to H^* via $j_c = 2H^*/D$, where D is either the sample thickness (low fields) or the grain size (high fields).

4. Time Dependent Critical State

The effects of thermally assisted flux flow on mixed–state properties depend on the relative magnitudes of the pinning potential well, U, and kT. For U both > and < kT the electric field during thermally activated flux flow is such that, in an experimental time scale,

$$j_c \sim 1 - (kT/U)\ln(t/t_0). \qquad (3)$$

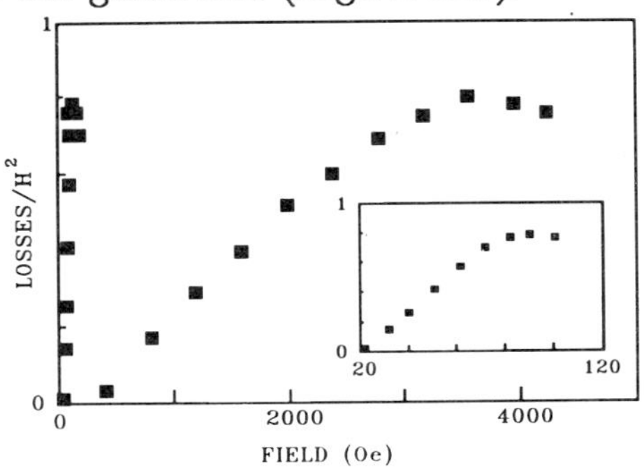

Fig. 1. χ'' (at 5K), calculated from magnetization losses. Low–field data is shown in the inset.

5. Results and Discussion

The data are from magnetization or a.c. susceptibility, both of which have already been described (Finlayson et al. 1990). The materials were plasma–sprayed deposits, heat treated to enhance their superconducting properties (Pawlowski et al. 1990). The data presented here are from material heat treated in flowing oxygen at 875°C for 20 hrs and furnace–cooled.

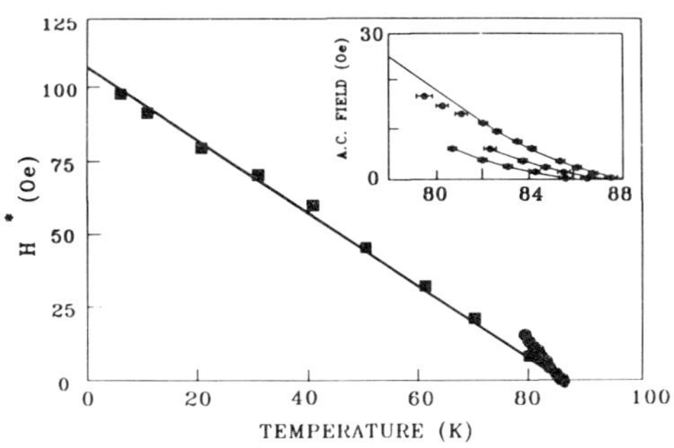

Fig. 2. H^*_l versus temperature from magnetization ( ) and a.c. susceptibility ( ). The inset is explained in the text.

Firstly, the critical state is considered. χ'' versus field at the various temperatures (e.g., Figure 1) exhibits two linear regions and two peaks at H^*_l and H^*_h, respectively. The linear dependences for χ'', indicate field–independent critical states persisting almost to H^* in each case. For the low–field region this was consistent with $B_{rem} = 0.5B_{max}$, as found from trapped flux measurements. $H^*_l(T)$ is shown in Figure 2, obtained from magnetization data for $5K \leq T \leq 80K$ and determined directly from maxima in $\chi''(T)$ via a.c. susceptibility measurements for $80K \leq T \leq 88K$. We note that $j_c = j_c(0)[1 - T/T_c]^p$ where $p = 1$ and 2 for SIS and SNS type of intergrain junctions, respectively. The linear temperature dependence of the field H^*_l from Figure 2 implies the SIS type of Josephson medium. The non–linear temperature dependence close to T_c (the inset of Figure 2) is ascribed to the crossover from one– to two–dimensional Josephson vortices (Finlayson et al. 1990). The lines are fits of (1) and (2) to $j_c \sim (1 - T/T_c)^{3/2}$ for d.c. fields of 0, 10 and 20 Oe, respectively. Using (2), the best fit to the data up to 20 Oe was given by $H^* = 130(1 - T/T_c)^{3/2}(5.4 - \ln H_{dc})$.

Linearity of χ'' $(H < H^*_h)$ leads to the following intergranular critical state equations:

$$-4\pi M = (1 - \mu_{eff})H_m\{1 - H_m[1 + (kT/U)\ln(t/t_0)]/(2r_g j_{c0})\} \quad \text{for } H_m < H^* \quad (4)$$

and $$-4\pi M = (1 - \mu_{eff})(H^*/2)[1 - (kT/U)\ln(t/t_0)] \quad \text{for } H_m \geq H^*. \quad (5)$$

In Figure 3 the creep rate $dM/d\ln t$ is plotted versus field for a temperature 5K. Below H^* the creep rate (from (4)) is proportional to H^2. For $H = H^*$, (5) yields $U \approx 0.02$ eV, the lower limit found for a YBaCuO single crystal (Yeshurun et al. 1989). The constant creep rate for $H \geq H^*$ implies a field–independent activation energy. As $U = j_c\phi_b L_b X_p$, where ϕ_b is a flux bundle of length L_b and X_p is the width of a pinning barrier, then $\phi_b = \phi_0$. Thus j_c and U are field independent and the pinning force varies with temperature as $(1 - T/T_c)^{3/2}$.

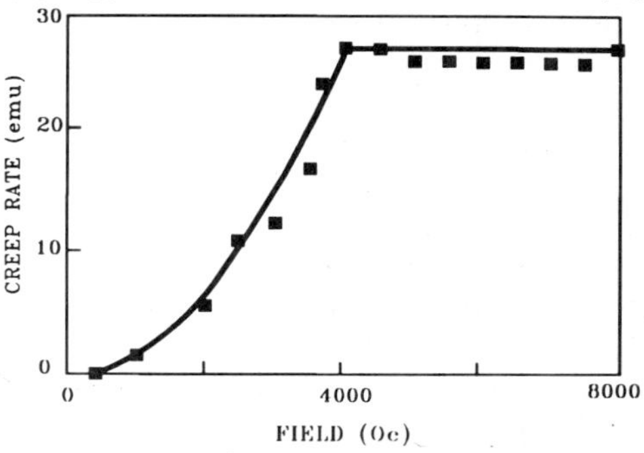

Fig. 3. Flux creep rate versus field at 5K.

The temperature dependence for the zero–field–cooled (ZFC) and field–cooled (FC) magnetizations at 1 kOe are plotted in Figure 4. The solid and broken lines for the ZFC case are fits from the above, with and without flux creep and with $U = j_{c0}\xi^2\phi_0$. Measurements of the creep rate versus temperature at 1 kOe, show a maximum rate in the vicinity of 30K above which it is seen (from Figure 4) that a satisfactory fit to the ZFC magnetization is possible only with the inclusion of flux creep.

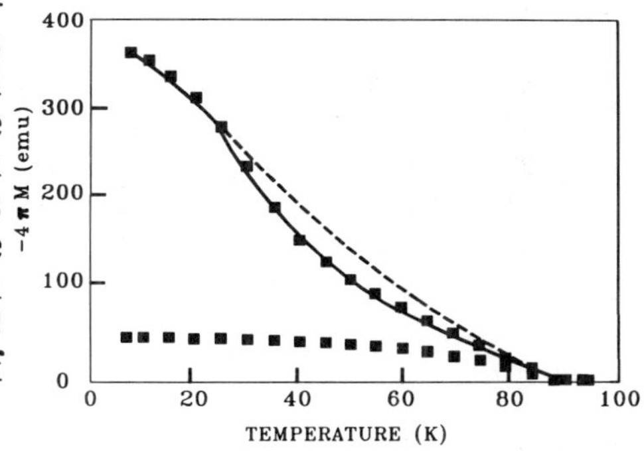

Fig. 4. ZFC and FC magnetization data for a field of 1kOe. The dotted and full lines are with and without flux creep.

Acknowledgements

This work has been supported by an I R & D Board GIRD Grant under Project No.15011.

References

Campbell and Evetts J E 1972 Adv. Phys. **21** 199

Finlayson T R, Garvie D A, Lowe D B, Przelozny Z and Smith T F 1990 Cryogenics **30** 467

Pawlowski L, Hill A, McPherson R, Garvie D, Przelozny Z and Finlayson T 1990 Proc. National Thermal Spray Conference (Long Beach, USA) 21–24 May, 1990 (ASM International 1990) (in press)

Salama K, Sevamanickam Y, Gao L and Sun K 1989 Appl. Phys. Lett. **54** 2352

Yeshurun Y, Malozemoff A P, Worthington T K, Yandrofski R M, Krusin–Elbaum L, Holtzberg F H, Dinger T R and Chandrashekhar G V 1989 Cryogenics **29** 258

Supercond. Sci. Technol. 4 (1991) S436–438. Printed in the UK

Effective Activation Energy in High T_c Superconductors

M.Reissner, R.Ambrosch and W.Steiner

Institut für Angewandte und Technische Physik, T.U. Wien, Austria

Abstract. Measurements of the time dependence of the magnetization are reported for ceramic samples from the different classes of high T_c superconductors. The obtained variation in the activation energies for the different classes points to pinning mechanisms characteristic for each structure type.

Explaining the observed logarithmic time dependence of the magnetization within the model of thermally activated flux creep, it was assumed that, in addition to defects, the small coherence length of the high T_c materials also causes pinning mechanisms which are associated with the structure itself (e.g. Tinkham 1988, Dinger et al. 1989). Therefore we investigated the relaxation behaviour of ceramic samples from different classes of superconductors (YBCO, BSCCO, LSCO, NCCO), prepared in various manner and with different substitutions, to check if typical values for the activation energies of the different classes are present.

The change of magnetization with time was determined by d.c. magnetic measurements using a vibrating sample magnetometer in the field range $0.1mT \leq B_a \leq 6.7T$ and for temperatures $1.8K \leq T \leq T_c$. The shape of the samples was cylindrical to allow corrections of the field values by appropriate demagnetizing factors. After cooling the samples in zero field, the specimens were brought into the critical state by slowly raising the field up to the measuring field and the decay in magnetization was monitored for time intervals up to 2 hours.

In the YBCO-system samples prepared from starting materials with different grain sizes ranging from 0.14μm to 0.17μm (batches A, B and C), pressuring forces between 20 and 150MPa, sintering temperatures (T_s) between 880 and 930°C and sintering times between 4 and 128h were investigated (Reissner et al. 1990a). From the Bi/Tl-compounds Bi-samples with 2212 and 2223 structure as well as one Tl-sample with 2223 structure were measured. In $Bi_{1-x}Pb_xSrCaCu_{1.8}O_y$ the samples with x = 0.1, 0.2 and 0.3 exhibit a relative amount of the 2223 to the 2212 phase of ~0.78, ~1.0 and ~0.85, respectively. The other samples of this series consist of 2212 phase with T_c~70K (Reissner et al. 1990b). Three samples were selected from the La-system with 0.25, 0.50 and 1.00% of Fe substituted for Cu and 7.5%Sr for La. From the T'-phase superconductors three samples (7.5%Ce) were investigated, one of them with 1.00%Fe.

For all compounds the temperature dependence of the normalized creep rate ($S = |1/M_0(dM/d\ln t)|$) exhibits distinct maxima (T_{max}), e.g. BSCCO inset Figure 1b, which are shifted to lower temperatures for higher applied fields. T_{max} measured in fields according to B_{min} (defined by dM/dB = 0), seems to be typical for each class of superconductors (inset Figure 2). The ratios T_{max}/T_c exhibit values of ~0.27, ~0.4, ~0.2 and ~0.4 for YBCO, BSCCO (and the Tl-sample), LSCO and NCCO, respectively. Different preparation conditions as well as substitutions of Pb in BSCCO and Fe for Cu in LSCO and NCCO influence T_{max} much less than the change of the structure type. Both a high value of S at 4.2K and a dependence of S(T), which does not extrapolate to

0953-2048/91/01S436 + 03 $03.50

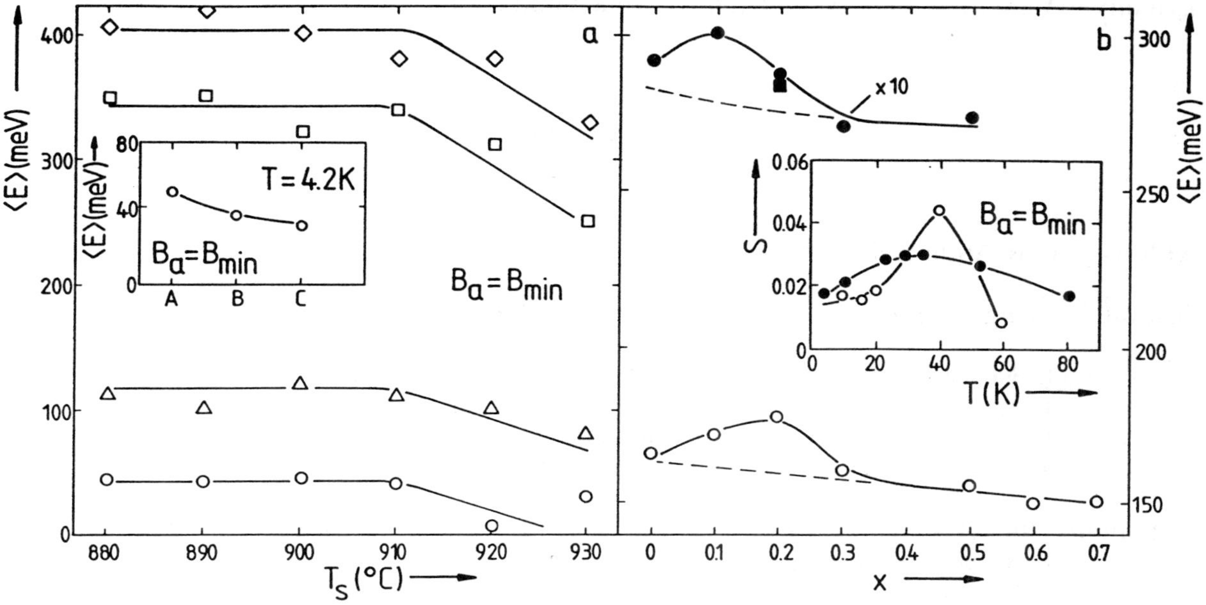

Figure 1: (a) Influence of the sintering temperature (T_s) on the effective activation energy <E> for YBCO at 4.2K (○), 20K (△), 40K (□) and 46K (= $T_c/2$, ◇); Inset: <E> in fields $B_a=B_{min}$ for three differently prepared series. (b) Influence of lead content on <E> for $Bi_{1-x}Pb_xSrCaCu_{1.8}O_y$ at 4.2K (●), 35K (=$T_c/2$ for 2212, ○) and 53K (=$T_c/2$ for 2223, ■). Inset: temperature dependence of the normalized creep rate S for x=0 (○) and 0.2 (●).

$S \to 0$ for $T \to 0$, are sometimes observed. It must be checked by further experiments, if this behaviour is an indication of non thermally activated flux creep.

The effective activation energy <E> was determined within the model of Hagen et al. (1989), where a distribution of activation energies is assumed to explain the decrease of S at high temperatures. In this model the time dependence of the magnetization is given by

$$M(t,T) = M_0(T)\ [1 - \frac{kT}{\langle E(T)\rangle}\ ln(1 + \frac{t}{\tau})]$$

where $M_0(T)$ is the magnetization at $t = 0$ and τ the relaxation time. In the analysis $10^{-12} \leq \tau \leq 10^{-6}$s was assumed (Hagen et al. 1989) and $M_0(T)$ was replaced by $M(t = 60s,T)$ leading to

$$\frac{1}{S} + 17.9 \leq \frac{\langle E(T)\rangle}{kT} \leq \frac{1}{S} + 31.7\ .$$

The dependence of <E> on T_s at B_{min} for series B in YBCO is given in Figure 1a. An increasing significance of the high energy portion in the activation energy distribution with rising temperature can explain the large increase of <E> with increasing temperature for the same T_s. Batches A and C resemble the behaviour of B, but with slightly different values for <E> (inset Figure 1a). The enhancement of <E> in batch A, which is also reflected by a larger width of the transition, is caused by a smaller grain size and a broader grain size distribution of the starting material. <E> decreases with increasing sintering time, whereas no systematic changes with respect to different pressure forces were found. As a further example the influence of lead substitution on <E> for $Bi_{1-x}Pb_xSrCaCu_{1.8}O_y$ is shown in Figure 1b. Within the preparation conditions for this series <E> is nearly constant for the lead containing 2212-samples and is slightly increased for the 2223 containing compounds (Figure 1b). The influence of Fe substitution for both LSCO and NCCO is, if present, very small.

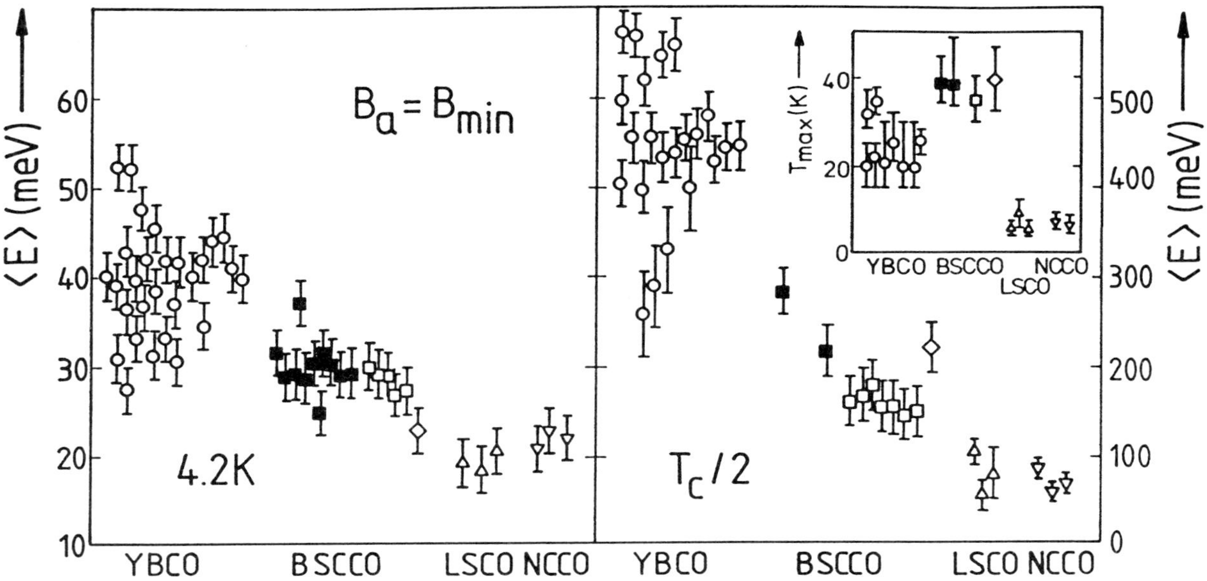

Figure 2: Effective activation energy <E> for YBCO (○), BSCCO (2223 ■, 2212 □)/Tl (◇), LSCO (△) and NCCO (▽). Inset: Temperature of the maximum in the normalized creep rate measured in $B_a = B_{min}$.

Within the chosen preparation conditions <E> varies for YBCO around an average value which is remarkably higher (both at 4.2K and at $T_c/2$) than the typical values found in the Bi/Tl- system and in LSCO as well as in NCCO at comparable temperatures (Figure 2). At 4.2K for BSCCO the observed difference between 2212 and 2223 (Figure 1b) is smeared out, if samples prepared with different conditions and starting compositions are compared. The appearance of larger differences in <E> at $T_c/2$ reflects different temperature dependences of the effective activation energy for each class. Since <E> increases more strongly with temperature if the centre of gravity for the distribution of activation energies is shifted to lower values, we can conclude that the dominant energies are in YBCO at the low and in LSCO as well as in NCCO at the high energy side of the distribution.

To sum up, beside the variation in <E> due to different preparation conditions and different substitutions, typical average values were found for the activation energy of the different classes of high T_c superconductors, pointing to pinning mechanisms which are characteristic for each structure type.

Acknowledgement

We want to thank the groups of N.Brnicevic, O.Eibl, G.Gritzner, H.Gruber, H.Kuzmany P.Rogl and W.Wruss for leaving the samples. The work was supported by the Austrian Fonds zur Förderung der wissenschaftlichen Forschung, grant number P7061P.

References

Dinger T R, Dolan G J, Keane D, McGuire T R, Worthington T K, Yandrofsky R M and Yeshurun Y 1989 *Proc. High Temp. Supercond. Oxides, TMS - AIME,* Las Vegas

Hagen C W, Griessen R P and Salomons E 1989 *Physica C* **157** 199

Reissner M, Ambrosch R, Steiner W, Stroh R, Schmid W and Wruss W 1990a *Proc. ICMC '90* in print

Reissner M, Ambrosch R, Steiner W, Kuzmany H and Schaffarich P 1990b *Physica B* in print

Tinkham M 1988 *Helvetica Physica Acta* **61** 443

Supercond. Sci. Technol. **4** (1991) S439–441. Printed in the UK

Recent High Pressure Studies on the Cu-Oxide Superconductors at I.S.S.P.

Nobuo Mori, Hiroki Takahashi and Chizuko Murayama

Institute for Solid State Physics, University of Tokyo, Roppongi, Minato-ku, Tokyo 106 Japan

Abstract. The pressure derivative of T_c in some hole-doped Cu oxide compounds is shown to change the sign from positive to negative at a certain pressure with a maximum of T_c, which seems to be a general feature in these compounds.

1. Introduction

High pressure experiments on the high T_c superconductors have proved to be of great interest in revealing the characteristic features in these compounds. For example, large and positive pressure effects on T_c were confirmed for almost all the hole-doped Cu oxide compounds(Griessen 1987, Wijngaaden and Griessen 1989). On the contrary, anomalous absence of pressure effects on T_c was revealed in some electron-doped ones(Murayama *et al.* 1989). These differences have been argued from the view point of the differences in their local crystal structures. With these considerations in mind, many groups have attempted to seek a relation between the pressure dependence of T_c and structural changes in these compounds experimentally(Kaldis *et al.*1989, Hewat *et al.*1990, Nelmes *et al.*1990), and especially the role of apical oxygens in the pyramids or octahedra has been paid attention from the theoretical view point(Matsukawa and Fukuyama 1990).

In this paper we present additional information about the pressure effects on T_c of some typical hole-doped Cu oxide compounds having apical oxygens.

2. Experimental

As typical hole-doped Cu oxide compounds we used the polycrystalline samples of $Nd_{1.3}Ce_{0.2}Sr_{0.5}CuO_{4-y}$, $YBa_2Cu_4O_8$, $(YCa)Ba_2Cu_4O_8$, $Tl_2Ba_2CuO_{6+y}$, and $Tl_2Ba_2CaCu_2O_8$ prepared by solid state reactions, having various superconducting temperatures. The pressure dependence of T_c was obtained by measuring the electric resistance with a dc four-lead method. A piston-cylinder device(below 2GPa) and a cubic-anvil device(up to 8GPa)(Mori and Takahashi 1990) were used alternatively to generate and retain nearly hydrostatic pressure in a Teflon cell filled with a fluid pressure medium, a mixture of Fluorinert No. FC70 and No. FC77. The force applied to the piston or anvil was changed only at room temperature and controlled to be kept constant during the measurements upon cooling and warming runs with using a high pressure cryostat. The pressure was determined by using the Bi I-II and III'-V phase transition points at room temperature of 2.55 GPa and 7.7 GPa, respectively. The pressure below room temperatures was estimated to be controlled within 3%. The temperature was measured with a calibrated Pt(Co) resistance thermometer placed near the sample.

0953-2048/91/01S439 + 03 $03.50

3. Results

Figure 1 shows the temperature dependence of relative resistances normalized at 130K for $YBa_2Cu_4O_8$ as an example obtained by using our cubic-anvil apparatus. We defined the superconducting transition temperature T_c as the temperature at which the resistivity drops to 50% of its extraporated normal state value. In Fig.2 the pressure dependence of T_cs for the samples of $Nd_{1.3}Ce_{0.2}Sr_{0.5}CuO_{4-y}$, $YBa_2Cu_4O_8$, $(YCa)Ba_2Cu_4O_8$, $Tl_2Ba_2CuO_{6+y}$, and $Tl_2Ba_2CaCu_2O_8$ are shown as functions of pressure. It is seen that the T_cs of these samples exhibit a wide variety of pressure dependences. The most remarkable findings are as follows. The pressure coefficient of T_c changes from positive to negative value at a certain pressure where the T_c has a maximum value depending on the samples as observed in $(YCa)Ba_2Cu_4O_8$ and $Tl_2Ba_2CaCu_2O_8$. The highly hole-doped $Tl_2Ba_2CuO_{6+y}$ shows a large pressure effect on T_c with negative coefficient. And the T_c of $Nd_{1.3}Ce_{0.2}Sr_{0.5}CuO_{4-y}$ doubles under the pressure of 8GPa compared to the value at ambient pressure.

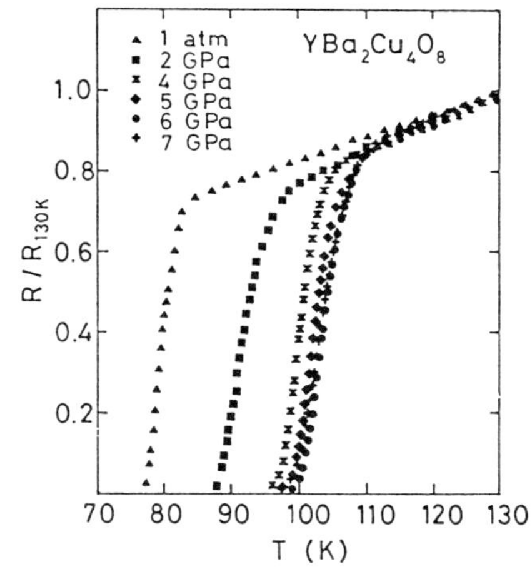

Fig. 1

In Fig.**3 a relation** between the pressure coefficients of the logarithm of T_c, $d\ln T_c/dP$, and T_c are shown. The values of pressure derivative $d\ln T_c/dP$ at each T_c are evaluated from a tangent of the smoothed curves in Fig. 2. It seems that $d\ln T_c/dP$ changes almost symmetrically with respect to T_c, that is, the negative pressure coefficients have almost the same magnitude in absolute value as that of positive ones with the same T_c, suggesting that in a given sample T_c changes with pressure having a general trend as shown schematically in the insert. This is a new additional feature for the hole-doped Cu oxide compounds.

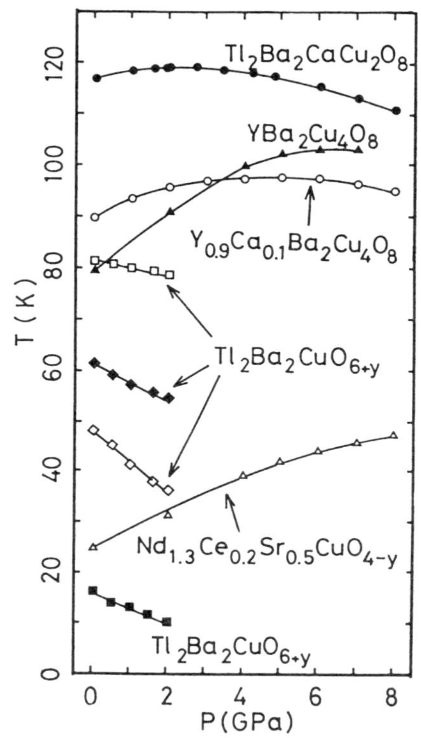

Fig. 2

4. Discussions

One of the possible interpretations of these results is that the concentration of holes in the CuO_2 planes increases with increasing pressure. From this point of view, the appearance of maximum T_c implies there exists the optimum hole concentrations in CuO_2 planes at a certain pressure to bring the best condition to the superconductivity. And the further increase in carriers destroys the superconductivity. This feature resembles well the one of the variation of T_c with hole concentration as observed in some hole-doped compounds at ambient pressure (See a review article of Shafer and Penny 1990). Although the change of carrier concentration with pressure in the CuO_2 planes has not yet been confirmed, there are some indications that it takes place. One is that the distance between copper and apical oxygen decreases with pressure larger than that expected from the change in c-axis direction in case of $YBa_2Cu_4O_8$(Kaldis *et al.* 1989). This result seems to reflect the increase of hole concentration in CuO_2 planes, because the induced holes(positively charged carrier) attract still more the negative apical oxygens. The other is that the possibility of the redistribution of hole carriers was shown also in $YBa_2Cu_4O_8$ from the calculation of Madelung energy(Yamada *et al.*1990). Experimentally, however, the Hall coefficients do not depend so much on pressure for the pollycrystalline $YBa_2Cu_3O_7$(Parker and Friend 1988) and La-Sr-Cu-O compounds(Tanahashi *et al.*1989). To confirm **these explanations,** high pressure experiments on the Hall coefficients of single crystals are highly desirable .

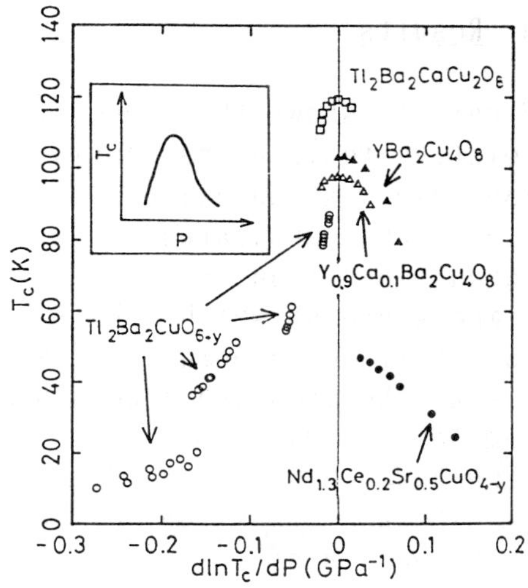

Fig. 3

Finally, until now a comparison between theory and experiment for T_c has been done mainly on the basis of the large positive value of dT_c/dP (Griessen 1987, Wijngaaden and Griessen 1989), though our experimental results indicate more general features which should be explained by theory for the hole-doped Cu oxide superconductors.

Acknowledgments

This paper constitutes a brief report of work involving important contributions of many peoples. The work on $Nd_{1.3}Ce_{0.2}Sr_{0.5}CuO_{4-y}$ was in collaboration with Drs. S.Uchida, H.Takagi and Y.Tokura, the work on $YBa_2Cu_4O_8$ with Drs. Y.Yamada, T.Matsumoto, and Y.Komada, the work on $(YCa)Ba_2Cu_4O_8$ with Drs. T.Miyatake, M.Kosuge, N.Koshizuka and S.Tanaka, and the work on $Tl_2Ba_2CuO_{6+y}$ and $Tl_2Ba_2CaCu_2O_8$ with Drs. Y.Shimakawa, T.Manako, and Y.Kubo. More detailed accounts of the experimental results will be published elsewhere.

References

Griessen R 1987 *Phys. Rev.* **B36** 5284

Hewat A W , Fisher P, Kaldis E, Karpinski J, Rusiecki S and Jilek E 1990 *Physica* **C167** 579

Kaldis E, Fisher P, Hewat A W, Hewat E A, Karpinski J, and Rusiecki S 1989 *Physica* **C159** 668

Matsukawa H and Fukuyama H 1990 *J. Phys. Soc. Jpn.* **59** 1693

Mori N and Takahashi H 1990 *J. High Pressure Institute Japan* **28** (in the press)

Murayama C, Mori N, Yomo S, Takagi H, Uchida S, and Tokura Y 1989 *Nature* **339** 293

Nelmes R J, Wilding N B, Hatton P D , Caignaert V, Raveau B, McMahon M I and Piltz R O 1990 *Physica* **C166** 329

Parker I D and Friend R H 1988 *J. Phys.* **C21** L345

Shafer M W and Penny T 1990 *Eur. J. Solid State Inorg. Chem.* **t.27** pp191-220

Tanahashi N, Iye Y, Tamagai T, Murayama C, Mori N, Yomo S, Okazaki N and Kitazawa K 1989 *Jpn. J. Appl. Phys.* **28** L762

Wijngaaden R J and Griessen R 1989 *Studies of High Temperature Superconductors* ed A V Narlikar (New York: NOVA Science) pp 29-80

Yamada Y, Matsumoto T, Kaieda Y and Mori N 1990 *Jpn. J. Appl. Phys.* **29** L250

Supercond. Sci. Technol. **4** (1991) S442–444. Printed in the UK

Variations in the Field Broadening of the Resistive Transition in Oxygen Deficient Single Crystals of $YBa_2Cu_3O_{7-\delta}$

G. W. Crabtree,[2] W. K. Kwok,[2] U. Welp,[1] H. Claus,[3] K. G. Vandervoort,[1,3] S. Fleshler,[2,4] Y. Fang[2]

(1) Science and Technology Center for Superconductivity, Argonne, IL 60439 USA
(2) Materials Science Division, Argonne National Laboratory, Argonne, IL 60439 USA
(3) Department of Physics, University of Illinois at Chicago, Chicago, IL 60680 USA
(4) Department of Physics, Purdue University, West Lafayette, IN 47907 USA

Abstract. Resistive measurements in an oxygen deficient single crystal show evidence of flux motion resistivity and pinning by twin boundaries. The shape of the transition for H along **c** changes as a function of field. Values of the activation energy are derived.

1. Introduction

The resistive transition in fully oxygenated single crystals of $YBa_2Cu_3O_{7-\delta}$ shows unusual field broadening and dissipation in the superconducting state just below T_c, which has been widely attributed to flux motion effects. In this paper we present resistive data on an oxygen deficient single crystal with T_c=23.5 K. Unlike in the 90 K crystals, the shape of the resistive transition changes with field, becoming more concave downward for high fields along the **c** direction. This behavior is similar to that seen in the new organic superconductor κ-(BEDT-TTF)$_2$Cu[N(CN)$_2$]Br, with T_c=11.6 K. The angular dependence of the resistivity shows a very small Lorentz force component, a large background component, and pinning by twin boundaries, as in the 90 K crystals. DC magnetization shows a diamagnetic onset well above the temperature of zero resistivity. In addition, we see an aging effect in the resistivity which is consistent with ordering of oxygen vacancies in the chains.

2. Results and Discussion

The single crystal was grown by a flux method and had an initial T_c in excess of 91 K with a width of less than 0.5 K in a field of 1 Gauss as determined by dc magnetization. Oxygen was removed by annealing in 1.5 Torr of oxygen for seven days at 530 °C. The crystal then had a diamagnetic onset of 29 K and a transition width of ~5 K in 1 Gauss, with zero resistivity occurring at 23.5 K. Ac resistance data were taken at 17 Hz as a function of field and temperature with a four probe technique, and dc magnetization was measured in a commercial SQUID magnetometer. Figure 1 shows the resistive transition in fields along the **c** axis. At low fields, field broadening occurs as in the 90 K crystals, with the onset nearly constant and a tail appearing at low temperatures. At higher fields, the shape of the transition changes qualitatively, with the resistivity nearly constant at the onset and falling more sharply to zero at lower temperature. This form is remarkably similar to that seen in the new organic superconductor κ-(BEDT-TTF)$_2$Cu[N(CN)$_2$]Br (Kwok et al., 1990a) with T_c=11.6 K and a layered structure consisting of conducting planes separated by ~15 Å.

The origin of the field broadening was further explored with dc magnetization measurements in a field of 1 T and with the dependence of the resistivity on the angle between the current and field. The dc magnetization was measured as described earlier (Welp et al., 1989) and gave the

0953-2048/91/01S442 + 03 $03.50

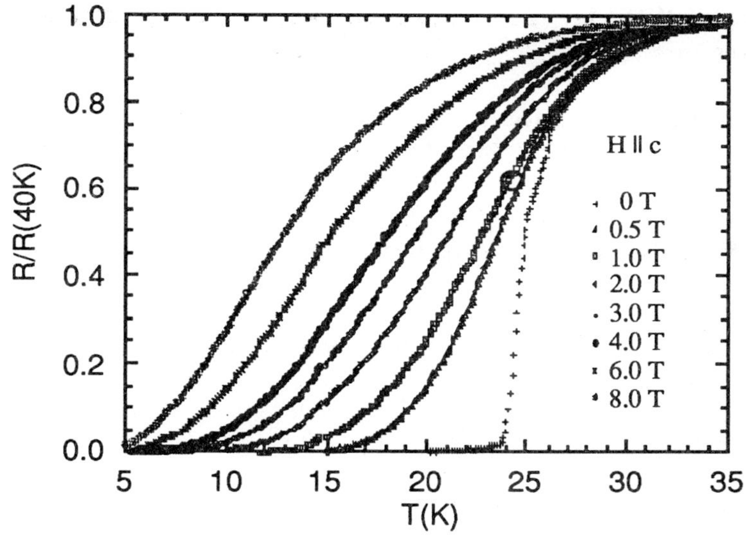

Fig. 1 Resistance of an oxygen deficient crystal of $YBa_2Cu_3O_{7-\delta}$ for H along **c** in various fields. The open circle on the 1T curve indicates the superconducting transition temperature in 1T determined from dc magnetization.

upper critical field point indicated on the 1 T curve in Figure 1. As in the 90 K crystals, the resistivity remains large well below the diamagnetic transition, indicating significant dissipation in the superconducting state. If this dissipation is due to flux motion the resistivity should depend on the angle between the field and the transport current as $\sin^2\theta$ due to Lorentz force effects (Kwok et al. 1990b). Figure 2 shows the angular dependence of the resistivity at a constant field strength of 1.5 T at 23.02 K, as the field direction is varied relative to the current, both of which are in the **ab** plane. There is a small variation following a $\sin^2\theta$ dependence, accounting for about 20 % of the total resistivity, while 80 % of the resistivity does not depend on angle and thus is not due to simple Lorentz force induced flux motion. Figure 2 shows sharp minima separated by 180° implying strong pinning by twin boundaries in the oxygen deficient crystal. These twin boundaries were not visible in polarized light images of the crystal surface. The pattern of minima implies the crystal has twin boundaries in

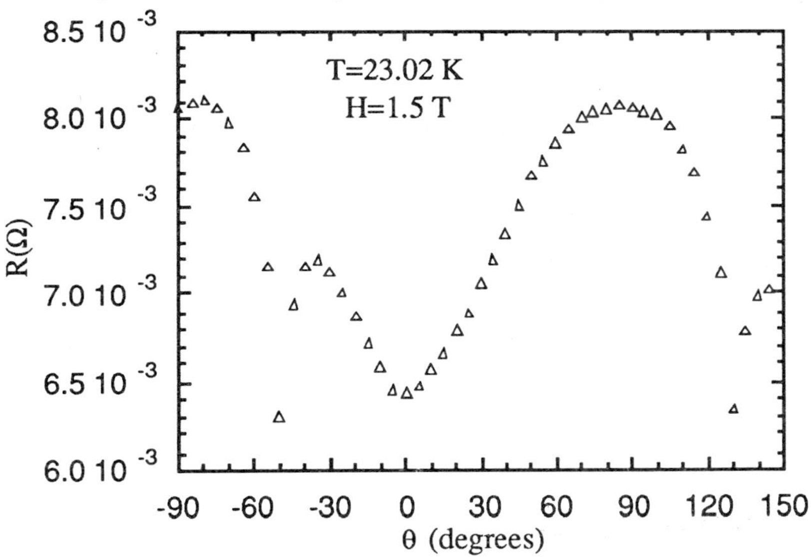

Fig. 2 Angular dependence of the resistance at 23.02 K in a constant field strength of 1.5 T as a function of the angle between field and current. Both field and current are in the **ab** plane.

only one of the two possible (110) type directions, and that the flux motion responsible for the angular dependent resistivity is essentially completely inhibited by twin boundary pinning.

The activation energy of the resistivity in various fields was derived from Arrhenius plots shown in Figure 3. Fits to the approximately linear part give steadily decreasing activation energies as the field increases, with U_{eff}/k ranging from 94 K to 31 K. These activation energies are about a factor of 800 smaller than those seen in similar measurements on 90 K crystals, where we find $U_{eff}/k = 8 \times 10^4$ K at 2.0 T.

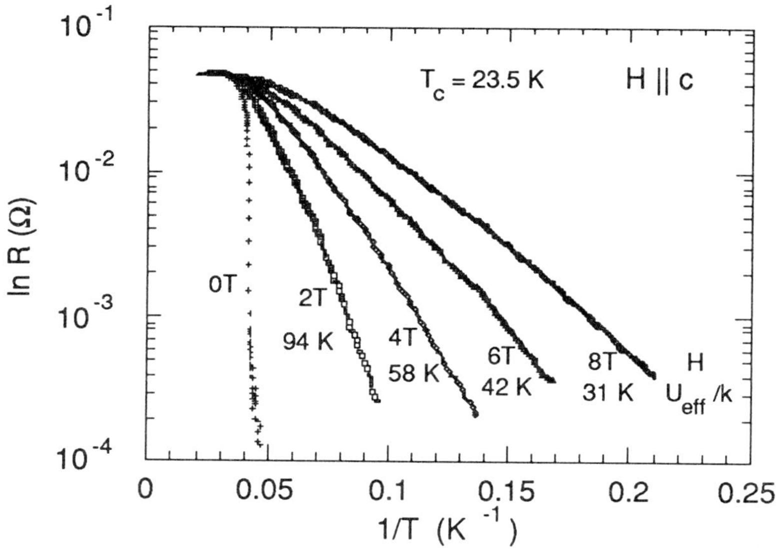

Fig. 3 Arrhenius plots for the resistance data of Fig. 1. The activation energies U_{eff}/k from a fit to the approximately linear parts of the curves are indicated

Finally, we see a new time dependent feature in the the resistivity of the oxygen deficient crystal due to ageing during the course of the measurements while the crystal was cycled repeatedly from 4.2 K to room temperature. The zero of resistivity increased from 23.5 K to 27.0 K, and the normal state resistivity at 40 K decreased by 20 %. This is the first observation of ageing effects in the resistivity. Both changes are consistent with effects previously observed in the magnetization and with the explanation of ageing as an ordering of oxygen vacancies in the chains (Veal et al., 1990).

3. Acknowledgements

This work was supported by the U. S. Department of Energy, Basic Energy Sciences--Materials Science under contract #W-31-109-ENG-38 (GWC, WKK, SF, YF) and the National Science Foundation--Office of Science and Technology Centers under contract #STC-8809854 (UW, KGV). KGV and SF acknowledge partial support from the Division of Educational Programs, Argonne National Laboratory.

4. References

Kwok W K, Welp U, Carlson K D, Crabtree G W, Vandervoort K G, Wang H H, Williams J M, Stupka D L, Montgomery L K, Thompson J E, 1990a, preprint

Kwok W K, Welp U, Crabtree G W, Vandervoort K G, Hulscher R, Liu J Z 1990b Phys. Rev. Letters **64** 966

Veal B W, Paulikas A P, You H, Shi H, Fang Y, Downey J W 1990 Phys. Rev. B in press

Welp U, Kwok W K, Crabtree G W, Vandervoort K G, Liu J Z 1989 Phys. Rev. Letters **62** 1908

Supercond. Sci. Technol. **4** (1991) S445–447. Printed in the UK

Effects of Electron Irradiation on Irreversibility Line in Superconducting $YBa_2Cu_3O_7$

M. Konczykowski, F. Rullier-Albenque, Laboratoire des Solides Irradiés,CEREM, Ecole Polytechnique, 91128 Palaiseau, France

Y. Yeshurun, E.R. Yacoby, A. Shaulov*
Department of Physics, Bar-Ilan University, Ramat-Gan, Israel

J. Gilchrist, CRTBT - CNRS B.P. 166X, 38042 Grenoble Cedex, France

Abstract. We have investigated the effects of 2.5 MeV electron irradiation on the irreversibility line of a sintered $YBa_2Cu_3O_7$ by applying dc and ac measuring techniques. We find that after electron irradiation, both the superconducting transition temperature T_c and the irreversibility temperature T_{irr} , determined by either technique, decrease; the decrease in T_{irr} scales with the decrease in T_c.

Conventional zero-field-cooled/field-cooled (zfc/fc) magnetization measurements led to the discovery (Müller et al , 1987, Yeshurun and Malozemoff, 1988) of the irreversibility line (IRL) in the magnetic phase diagram of high-temperature superconductors (HTSC). The importance of this line to the basic understanding of HTSC and to their potential applications, has motivated intensive studies of the IRL by other techniques. Of these, the most efficient, in terms of time consumption, are techniques which are based on ac magnetometry. There are several variants of these measurements. Thus, e.g. the IRL may be deduced from a study of the field and temperature dependence of the peak in the loss signal (Malozemoff et al, 1988) , from the onset of third harmonics in the ac response (Shaulov and Dorman, 1988), and from screening efficiency to ac fields (Gilchrist and Konczykowski, 1990, Konczykowski and Gilchrist, 1990).

In the present article we deal with the effect of irradiation on the IRL. Irradiation has been used in numerous experimental works to induce defects in a controlled way. The main goal in these works was the enhancement of critical currents J_c by introducing new, more efficient, pinning centres for flux trapping. In most studies, however, J_c was affected only insignificantly. Only recently a dramatic improvement of J_c was achieved by van Dover et al (1989) by neutron irradiation of a single crystal of $YBa_2Cu_3O_7$ (YBCO). Civale et al (1990), however, showed, by proton irradiating their YBCO crystal, that the improvement in J_c is not accompanied by changes in the IRL. The main goal of this paper is to explore the effect of electron irradiation on the IRL.

The sample used in this experiment is a sintered YBCO with transition temperature of T_c=90.5 K and dimensions 4 x 4 x 0.1 mm^3. Low temperature (20 K) irradiation by 2.5 MeV electrons with a Van de Graaff accelerator produced damage consisting of homogeneously distributed isolated Frenkel pairs on all sublattices of YBCO (Vichery et al, 1989). However the agglomeration of small defect clusters is expected when the sample is warmed to room temperature for transfer to the measuring cryostat.

For determining the IRL we used four experimental techniques: (1) Conventional zfc/fc magnetization measurements. (2) Ac screening efficiency measurements under zfc and fc conditions. In these measurements the sample is put between the driving and the detecting coils (see Fig. 1 in Konczykowski and Gilchrist, 1990) and the 'transmittivity' of the ac signal is measured.(3) Measurements of the imaginary component of the transmittivity. The IRL is defined via the peak in this signal. (4) Measurements of the onset of odd harmonic components in the ac

* permanent address: Philips Laboratories, Briarcliff-Manor, NY1050, USA

response (fundamental frequency was 20 kHz). In a forthcoming publication we compare the four techniques.

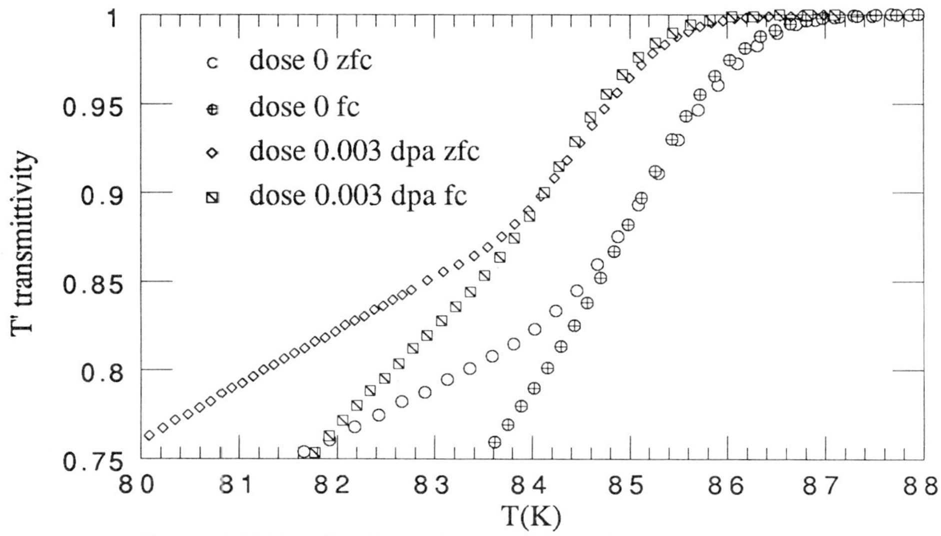

Figure 1. $YBa_2Cu_3O_7$ polycrystal irradiated at 20K by 2.5 MeV electrons. H_{dc}=250 Oe, H_{ac}=180 mOe, f=11.7 kHz.

All these techniques - except number (2) - have been employed in the past for determination of the IRL. Their adequate interpretation enables the IRL to be obtained even for polycrystalline material. To demonstrate the new technique (2) we show typical data in Fig. 1. The split of zfc and fc branches of the ac screening efficiency of polycrystal, corresponds to the split of the zfc/fc curves in conventional measurements of intragranular dc magnetization. Figure 1 also demonstrates the main result of this paper, namely that as a result of the irradiation, T_c and the irreversibility temperature T_{irr} are slightly decreased. However, **the reduction in T_{irr} scales with the reduction in T_c.**

To further demonstrate this important point we summarize in Fig. 2 our results for the irreversibility line - before and after irradiation - measured by the conventional zfc/fc magnetization (squares) and by the onset of the third harmonic signal (circles).The open symbols denote data taken before irradiation, whereas the filled ones refer to the irradiated sample (0.003 displacements per atom, as for Fig. 1). The solid lines are fits to the unirradiated data assuming a power-law dependence of field on the reduced temperature. The difference between the ac and dc data described in Fig. 2 is apparently due to the dependence of the IRL on the measuring frequency, as reported by Malozemoff et al (1988). We find that not only T_{irr} increases with frequency but also the slope (i.e. the exponent in the power-law) depends strongly on frequency. We find a slope of 1.6 for the dc magnetization data, in agreement with previous experiments (e.g. Yeshurun and Malozemoff, 1988); the exponent deduced from the onset of third harmonic is 2.5 and that deduced from transmittivity measurements (not shown here) yields 3. This point will be discussed further elsewhere.

As a result of 0.003 dpa elecron irradiation damage, the critical current J_c, as deduced from measurements of magnetization curves at low temperatures, is increased by a factor of 2. Further indirect evidence for the increase in J_c is deduced from the lowering of the height of the peak in the third harmonic signal measured as a function of temperature.

Figure 2 demonstrates that the IRL, described as a function of the **reduced** temperature, is not affected by irradiation. Our results are in agreement with those of Civale et al (1990). It is important to note, however, that in their experiment protons produced defects of 30 Å in size, somewhat larger than the coherence length, while in our experiment electrons cause damage on a much finer scale, perhaps the finest possible scale (Vichery et al, 1989). The fact that neither experiment shows any significant change in the IRL may indicate that only much larger defects might be effective in shifting it.

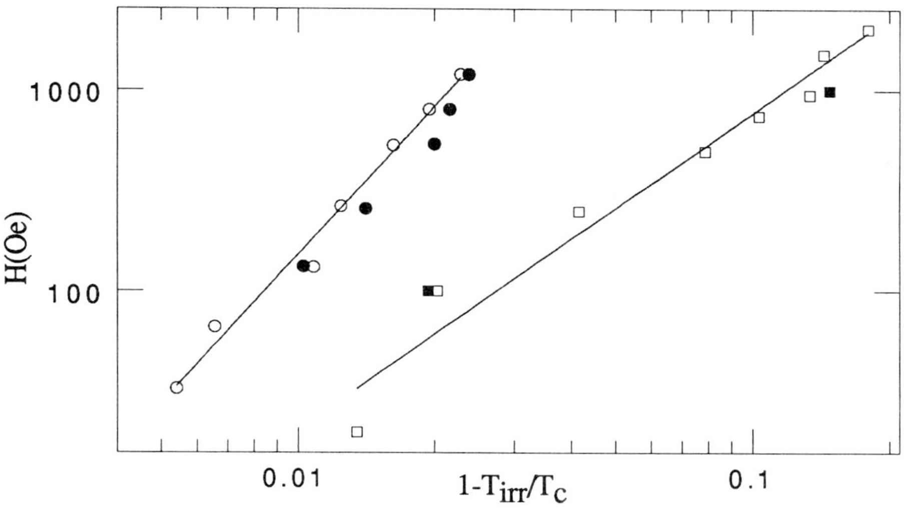

Figure 2. Irreversibility line for $YBa_2Cu_3O_7$ polycrystal from zfc/fc measurements (squares) and from onset of third harmonic signal (circles) before (open symbols) and after (filled) irradiation.

The conclusion that the IRL is not altered by irradiation whereas J_c is increased seems to contradict the simple flux creep interpretation. However, Malozemoff and Fisher (1990) have recently pointed out that normalized relaxation rates S, in numerous experiments, exhibit unexpected universality, being independent of J_c and interpreted the results within vortex glass (Fisher et al, 1990) and collective flux creep (Feigel'man et al, 1989, Nattermann, 1990) models. We believe that their surprising observation may be correlated with the behaviour of the IRL for a given measuring technique. The IRL may be defined as the loci of points in the field-temperature plane for which the relaxation rates are of the same order of magnitude as the measuring frequency. Constancy of S would thus imply a constancy of the IRL.

Acknowledgements.The group at Bar-Ilan acknowledges a partial support from the Ministry of Science and Technology. A partial support from the Maurice Pollack Foundation is acknowledged.

References

Civale L, Marwick A D, McElfresh M W, Worthington T K, Malozemoff A P, Holtzberg F H, Thompson J R and Kirk M A 1990 Phys.Rev.Lett. , to be published

Feigel'man M V, Geshkenbein V B, Larkin A I and Vinokur V M 1989 Phys.Rev.Lett. **63** 2303

Fisher D S, Fisher M P A and Huse D A 1990 Phys Rev B, to be published

Gilchrist J and Konczykowski M 1990 Physica C **168** 123

Konczykowski M and Gilchrist J 1990 Physica C **168** 131

Malozemoff A P, Worthington T K, Yeshurun Y, Holtzberg F and Kes P H 1988 Phys.Rev **B 38** 7203

Malozemoff A P and M P A Fisher, 1990 Phys.Rev.B, to be published

Müller K A, Takashige M and Bednorz J G 1987 Phys.Rev.Lett. **58** 408

Nattermann T 1990 Phys.Rev.Lett. **64** 2454

Shaulov A and Dorman D 1988 Appl. Phys. Lett. **53** 2680

van Dover R B, Gyorgy E M, Schneemeyer L F, Mitchell J W, Rao K V, Puzniak R and Waszczak J V (1989) Nature **342** 55

Vichery H, Rullier-Albenque F, Pascard H, Konczykowski M, Korman R and Favrot D 1989 Physica C **159** 689

Yeshurun Y and Malozemoff A P 1988 Phys.Rev.Lett. **60** 2202

Supercond. Sci. Technol. **4** (1991) S448–450. Printed in the UK

INFLUENCE OF THE SPACING OF $Cu-O_2$ SHEETS ON THE SUPERCONDUCTING TRANSITION TEMPERATURE

K.SHIMIZU , H.NOBUMASA and T.KAWAI*
Composite Materials Lab.,Toray Ind.Inc.,2-1,Sonoyama 3-chome, Otsu, Shiga 520, Japan
*The Institute of Scientific and Industrial Research, Osaka University, Ibaraki, Osaka 567, Japan

Abstract. Relationships between *Tc* and crystal structures of copper-oxide superconductors are examined from the view point of the interaction between $Cu-O_2$ sheets. The *Tc* correlates with the Cu-Cu distance between the different $Cu-O_2$ sheets, and there is a critical length around 4-6Å. For longer distances, the *Tc*'s are less than 50K. For shorter distances, on the other hand, the *Tc*'s are distributed from 50 to 125K. In the latter case, *Tc* seems to linearly decrease with the Cu-Cu distance.

1. Introduction

In the copper based oxide superconductors, the *Tc* certainly becomes higher with the increase of the number of $Cu-O_2$ sheets in a unit, but it shows a tendency of saturation for more than three. This result may indicate the existence of the interaction between $Cu-O_2$ sheets. There has been, however, no report in which the experimental data were classified paying attention to the interactions between $Cu-O_2$ sheets. We classified the data of the distances between $Cu-O_2$ sheets which seem to influence directly on this interaction and Tc (Nobumasa 1989). We found that there is a distinct relationship between *Tc* and the spacing of $Cu-O_2$ sheets, especially, Cu-Cu distance. From this viewpoint the *Tc* difference between $Y_1Ba_2Cu_3O_{7-y}$ and recently discovered $(Ln_pBa_{1-p})_2(Ln_{1-q}Ce_q)_2Cu_3O_y$ (Ln=Nd,Sm,Eu), both of which have three $Cu-O_2$ sheets in one unit, can be explained.

2. Results

In Fig.1, *Tc*'s versus distances between the nearest $Cu-O_2$ sheets (Cu-Cu distance) are plotted for the known copper oxide superconductors. The data adopted from papers (see the ref. in Nobumasa 1989) have been obtained by Rietveld analysis of the X-ray or neutron powder diffraction and by the X-ray four-circle diffraction study of single crystals. In this figure, superconductors are clearly divided into two groups. One is the A-group in which the distances are less than 4Å. The other is the B-group in which the distances are more than 6Å. The *Tc*'s in the B-group are less than 50K, except $Tl_2Ba_2Cu_1O_y$. On the other hand, those in A-group are distributed from 50 to

125K. In the A-group, *Tc* seems almost linearly proportional to the Cu-Cu distance.

3. Discussion

We can consider that the layers separating the Cu-O_2 sheets more than 4-6Å are blocking or intercepting the interaction and maintaining the two dimensionality. In this sense we call these layers blocking layers. These are, for example, Bi_2O_2 layer in Bi-Sr-Ca-Cu-O, Tl_2O_2 or TlO layer in Tl-Ba-Ca-Cu-O, PbO-Cu-PbO layer in Pb-Sr-(Ca,Ln)-Cu-O, rock salt $(LaO)_2$ layer in La_2CuO_4 or rock salt and fluorite layers in $(Ln,Ce,Sr)_2CuO_4$. All the compounds of A and B groups in fig.1 contain such blocking layers. The (Sr,Ca)-Cu-O which has no such a blocking layer but consists of only Cu-O_2 sheets does not become a superconductor. The existence of the blocking layers with the length more than 4-6Å which block the interaction between Cu-O_2 sheet and maintain two dimensionality seems to be necessary for all the copper oxide superconductors.

For the compounds of one Cu-O_2 sheet sandwiched by blocking layers(B-group), the distance between the nearest Cu-O_2 sheets becomes longer than 4-6Å, so that the interaction between Cu-O_2 sheets may become very small and *Tc* is not so high.

For the $(Ln_pBa_{1-p})_2(Ln_{1-q}Ce_q)_2Cu_3O_y$ (Ln=Nd,Sm,Eu) superconductors studied in detail (Nobumasa 1989), oxygen deficient Cu-O_2 sheets at the center can be considered as blocking layers as in the case of Cu-O chains in Y-Ba-Cu-O, and the distance across $(Ln,Ce)_2O_2$ fluorite layers becomes the nearest one to be 6.2Å (see fig.2(a)). The *Tc* of this material is located at point P in fig.1. We see that this superconductor belongs to the B-group because the distance between Cu-O_2 sheets is long enough, though there are three Cu-O_2 sheets in the unit. Accordingly *Tc* does not go up higher than 50K.

Generally, superconductors which have more than two Cu-O_2 sheets between the blocking layers have higher *Tc* than those which have only one Cu-O_2 sheet. This may be because there are some interactions between closely located Cu-O_2 sheets, for example, a magnetic one. In these cases, the atomic layer consisting of cations among Cu-O_2 sheets, such as Ca in Bi or Tl compounds and Y in Y compounds, may mediate the interaction. Accordingly, we can call this thin atomic layer a mediating layer. The relation of *Tc* with the interaction between those Cu-O_2 sheets through this mediating layer have been discussed theoretically by Wheatley(1988), Sawa(1989), Eab(1989) and Appel(1990).

In this study, we would like to emphasize that all the systems of copper-oxide superconductors, La,Y,Bi,Tl and Pb systems, can be classified only by the Cu-Cu distance between Cu-O_2 sheets, which seems to indicate that the interactions between Cu-O_2 sheets are essential.

Finally, from the discussion above, we may be able to explain the *Tc* difference between $Y_1Ba_2Cu_3O_{7-y}$ (90K) and $(Ln_pBa_{1-p})_2(Ln_{1-q}Ce_q)_2Cu_3O_y$ (Ln=Nd,Sm,Eu) (40K), of which the crystal structures resemble each other. The largest difference in the crystal structure is which layer is sandwiched between Cu-O_2 bottom plane of pyramidal structure (fig.2(a),(b)). In $Y_1Ba_2Cu_3O_{7-y}$, the Y atom is sandwhiched so that the distance between Cu-O_2 sheets becomes short and this Y atom forms the mediating layer. Accordingly, *Tc* of this compound is as high as 90 K. On the other hand, in $(Ln_pBa_{1-p})_2(Ln_{1-q}Ce_q)_2Cu_3O_y$, a thick $(Ln,Ce)_2O_2$ layer is sandwiched to form the blocking layer so that the distance between Cu-O_2 sheets becomes longer than 6Å . Therefore, *Tc* of this compound could not become higher than 50 K.

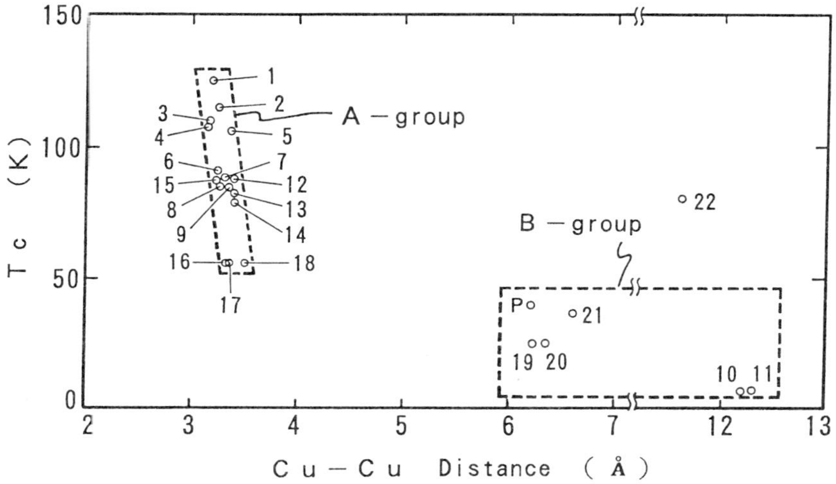

Fig.1 Correlation between *Tc*'s and Cu-Cu distances.

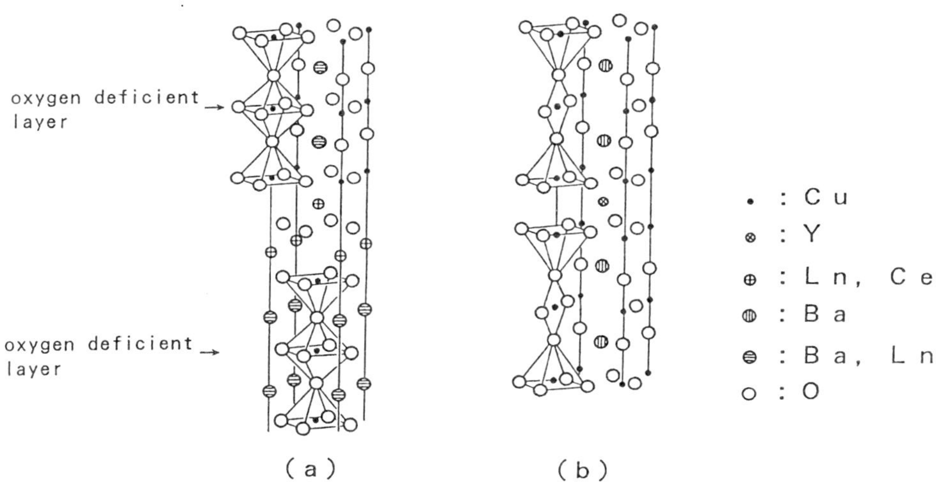

Fig.2 Crystal structures of $(Ln_pBa_{1-p})_2(Ln_{1-q}Ce_q)_2Cu_3O_y$ (a) and $Y_1Ba_2Cu_3O_{7-y}$ (b)

References

Appel J and Fay D 1990 Phys. Rev.B, 41 873
Eab C H and Tang I M 1989 Phys. Rev.B, 40 4427
Nobumasa H, Shimizu K and Kawai T 1990 Physica C 167 515
Nobumasa H, Shimizu K, Kitano Y, Tanaka M, and Kawai T 1989 Jpn.J.Appl.Phys.28 L1948
Sawa Y, Tanaka Y and Tsuda M 1989 Phys. Rev.B, 39 9113
Wheatley J M,Hsu TC and Anderson P W 1988 Nature 333 121

Supercond. Sci. Technol. **4** (1991) S451–453. Printed in the UK

The Effect of Yttrium and Copper Deficiency on the Superconducting Properties of Y-Ba-Cu-O System

M. Coldea, T. Petrisor and A. Giurgiu

Department of Physics, University of Cluj, 3400 Cluj-Napoca, Romania

Abstract. The controlling factor in determining T_c in $Y_{1-x}Ba_2Cu_3O_y$ ($0 \leq x \leq 0.20$) and $YBa_2Cu_{3(1-x)}O_y$ ($0 \leq x \leq 0.20$) superconductors is the formal average copper valency and not any specific oxygen content. The results reveal the importance of the local structure of the constituent atoms on the superconducting properties of Y-Ba-Cu-O system.

1. Introduction

Complementary to the experimental works on the effects of substitution for copper in high-T_c superconductors, the study of the effects of copper deficiency and of the other constituent elements are very useful and important in exposing novel features of the high-T_c superconductivity.

It is now well established that the degree of oxidation, or the formal average copper valency(FACV), is an important parameter for the superconducting properties(Greene and Bagley 1990).

In the superconductor $YBa_2Cu_3O_y$, the transition temperature T_c varies dramatically as oxygen is depleted by annealing the sample in vacuum and shows the well known two-plateau behaviour with two marked drops for y=6.75 and y=6.4, respectively(De Jongh 1988). The oxygen content in the cuprates can also be altered by cation substitutions and by varying the Y and Cu stoichiometry. If Y and Cu atoms are removed from Y-Ba-Cu-O system, the oxygen content will decrease.

Samples of $Y_{1-x}Ba_2Cu_3O_y$ and $YBa_2Cu_{3(1-x)}O_y$ with Y and Cu deficiency were studied with the intent of determining the limit of single phase behaviour and the relation between copper-oxygen content and the transition temperature T_c.

2. Results and discussion

All the samples studied in this work were prepared by a solid-state reaction in air. Powders of Y_2O_3, $BaCO_3$ and CuO, all with a purity of 99.99%, were mixed, pressed into pellets and heated at 950°C for 24 hours. The reacted pellets were reground, pressed again into pellets and sintered at 950°C for 12 hours at oxygen pressure of 1.2 bar. The electrical resistance of all samples was measured by the d.c. four-probe method.

The compound $YBa_2Cu_3O_y$, without Y and Cu deficiency, has a narrow phase

0953-2048/91/01S451 + 03 $03.50

transition($\Delta T_c < 5K$ and T_c=94K) and exhibits metallic behaviour above the transition. These data confirme the good quality of the samples and that they were sufficient oxygenated.

Figure 1 shows the temperature dependence of the resistance for Cu-deficient compounds $YBa_2Cu_{3(1-x)}O_y$ with x=0.05, 0.10, 0.15 and 0.20. The samples with $x \leq 0.15$ exhibit a single phase behaviour, whereas the sample with x=0.20 shows a multiphase behaviour. The resistances in the normal state point out a progressive transition from the metallic behaviour($x \leq 0.05$) with $dR/dT > 0$ to the semiconductor one($x \geq 0.10$) with $dR/dT < 0$.

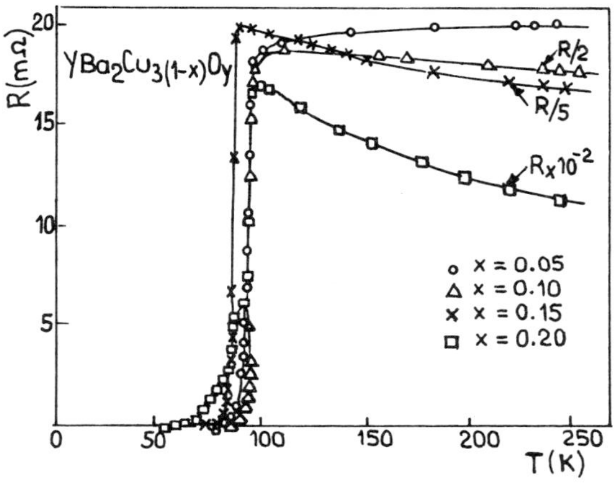

Fig. 1 Temperature dependence of the resistance of $YBa_2Cu_{3(1-x)}O_y$

In these compounds there is a discontinuous drop in T_c at x=0.15(Figure 2), closed to the limit of single phase behaviour and which corresponds to a measured oxygen content of y=6.4(Tao et al 1990). No variation of T_c versus Cu-deficiency between x=0 and 0.10 was detected. The large values of T_c for $x \leq 0.15$ indicate that the copper vacancies in $YBa_2Cu_{3(1-x)}O_y$ are concentrated in the Cu(1) sites.

The compound $YBa_2Cu_{2.40}O_y$(x=0.20) exhibits three superconducting transitions at T_c=94K, 86K and 76K, showing the multiphase character of this sample. One can distinguish the phases for x=0 and x=0.15, the phase with T_c=76K probably has the Cu-deficiency of x=0.20. Due to the high values of the resistance and of the slope dR/dT, the properties in the normal state of this sample are dominated by the phases with high Cu-deficiency.

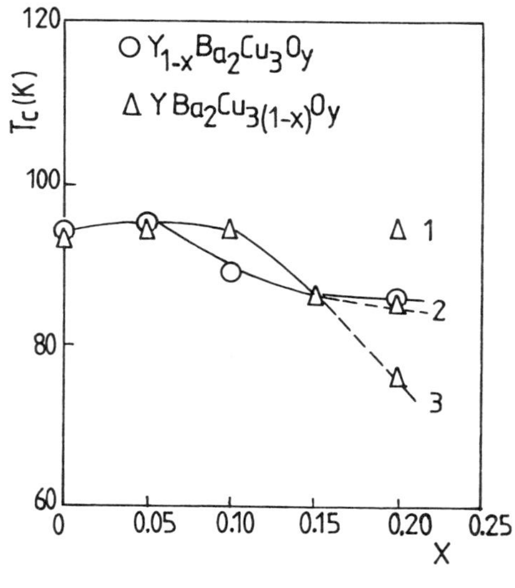

Fig. 2 Variation of T_c as a function of Y and Cu-deficiency in $Y_{1-x}Ba_2Cu_3O_y$ and $YBa_2Cu_{3(1-x)}O_y$. The numbers 1, 2 and 3 represent the three phases in the sample $YBa_2Cu_{2.40}O_y$

Experimental results for Y-deficient compounds $Y_{1-x}Ba_2Cu_3O_y$ with x=0.05, 0.10 and 0.20 are given in Figure 3. The transition temperature exhibits a maximum at x=0.05(T_c=95K), a pronounced decrease at x=0.10 and decreases monotonously with a tendency towards saturation as x increases up to the limit of single phase behaviour which probably one extends up to the value of x=0.25 (Figure 2). For a precise determination of this limit are needed measurements on the samples with higher Y-deficiency.

The most important result of this study is that the transition temperature for $Y_{1-x}Ba_2Cu_3O_y$ and $YBa_2Cu_{3(1-x)}O_y$ does not scale with oxygen content. Sample $YBa_2Cu_{2.55}O_y$ (x=0.15) has T_c=86K, although its oxygen content y is 6.4. At this oxygen content, the transition temperature of the parent compound $YBa_2Cu_3O_y$ approaches OK. We may therefore try to interpret the results in terms of a correlation between T_c and formal average copper valency which is (3x+2y-7)/3 for $Y_{1-x}Ba_2Cu_3O_y$ and (2y-7)/3(1-x) for $YBa_2Cu_{3(1-x)}O_y$. The FACV of $YBa_2Cu_3O_y$ for y=6.92 is 2.28. For Cu-deficient compound $YBa_2Cu_{2.55}O_y$ (with T_c=86K), x=0.15 and y=6.4 which also gives FACV = 2.27. In fact this indicates that the controlling factor in determining T_c in these compounds is the FACV and not any specific oxygen content. A similar conclusion was drawn with regard to the transition temperatures versus FACV in $Ca_xY_{1-x}Ba_2Cu_3O_y$ (Morris et al 1990). Supposing that this is also true for Y-deficient compounds, we can calculate the oxygen content in $Y_{1-x}Ba_2Cu_3O_y$. Taking also FACV=2.28, results for $Y_{0.90}Ba_2Cu_3O_y$ (with T_c=89K), x=0.10, the value y=6.77. Unfortunatelly, we could not verify experimental this value. It should also be noted, that at this oxygen content, in the parent compound $YBa_2Cu_3O_y$, takes place a pronounced decrease in the transition temperature.

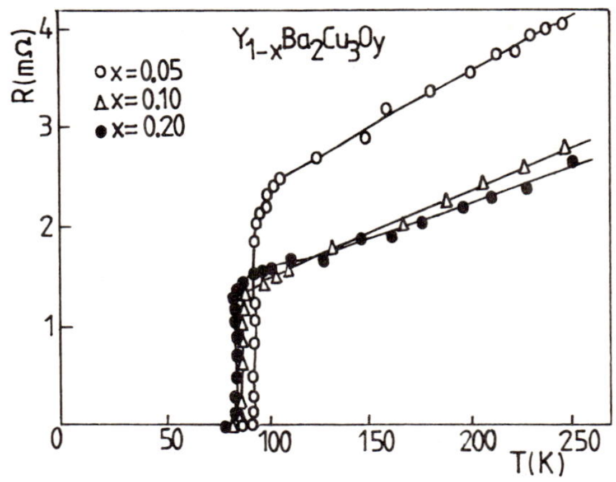

Fig. 3 Temperature dependence of the resistance of $Y_{1-x}Ba_2Cu_3O_y$

The high values of T_c in Cu-deficient compounds could be explained supposing that the Cu vacancies create a local orthorhombic character of the structure. On the other hand, because of a change in bond length Cu1-O4, affecting the charge transfer to the CuO_2 planes, the transition temperature in $Y_{1-x}Ba_2Cu_3O_y$ is modified by the Y-content. So, the local structural order influences indirectly the superconductivity in these compounds.

References

Greene L H and Bagley B G 1990 Physical Properties of High Temperature Superconductors II ed D M Ginsberg (World Scientific Press, Singapore)

De Jongh L J 1988 Physica C 152 171

Morris D E, Narwankar P K and Sinha A P B 1990, to be published

Tao Z, Alburger D E, Jones K W, Yao Y D and Kao Y H 1990, to be published